W9-BON-700

MEASUREMENT ERRORS IN SURVEYS

MEASUREMENT ERRORS IN SURVEYS

Edited by

Paul P. Biemer
Research Triangle Institute

Robert M. Groves
U.S. Bureau of the Census and The University of Michigan

Lars E. Lyberg
Statistics Sweden

Nancy A. Mathiowetz
U.S. Bureau of the Census

Seymour Sudman
University of Illinois at Urbana-Champaign

A Wiley-Interscience Publication
JOHN WILEY & SONS, INC.
New York • Chichester • Brisbane • Toronto • Singapore

Library of Congress Cataloging in Publication Data:

Measurement errors in surveys / editors, Paul Biemer . . . [et al.].
p. om. — (Wiley series in probability and mathematical statistics, ISSN 0271-6356. Applied probability and statistics)
"A Wiley-Interscience publication."
Includes bibliographical references.
1. Error analysis (Mathematics) I. Biemer, Paul. II. Series
QA275.M43 1991
511′.43--dc20 91-20598
ISBN 0-471-53405-6 CIP

Printed in the United States of America

10 9 8 7 6 5 4 3 2 1

Printed and bound by Courier Companies, Inc.

Morris H. Hansen

This book is dedicated to the memory of Morris Hansen, born in 1911 who died on October 9, 1990 in Washington, D.C. In addition to his many important contributions to survey sampling, Morris and his colleagues at the U.S. Bureau of the Census pioneered the development of non-sampling error research in the early 1950s and 1960s. Those accomplishments, as well as his recent ones, continue to influence the development of the field of survey nonsampling error.

CONTRIBUTORS

Frank M. Andrews, Survey Research Center, Institute of Social Research, The University of Michigan, Ann Arbor, Michigan

Barbara Bickart, Department of Marketing, University of Florida, Gainesville, Florida

Paul P. Biemer, Research Triangle Institute, Research Triangle Park, North Carolina

Johnny Blair, Survey Research Center, University of Maryland, College Park, Maryland

Norman M. Bradburn, University of Chicago and NORC, Chicago, Illinois

David Cantor, Westat, Inc., Rockville, Maryland

Rachel Caspar, Research Triangle Institute, Research Triangle Park, North Carolina

Edith D. de Leeuw, Vrije Universiteit, Amsterdam, The Netherlands

Wil Dijkstra, Vrije Universiteit, Amsterdam, The Netherlands

Don A. Dillman, U.S. Bureau of the Census, Washington, D.C. and Washington State University, Pullman, Washington

John E. Donmyer, A.C. Nielsen Co., Northbrook, Illinois

Judith Droitcour, U.S. General Accounting Office, Washington, D.C.

Solomon Dutka, Audits and Surveys, Inc., New York, New York

W. Sherman Edwards, Westat, Inc., Rockville, Maryland

Donna Eisenhower, Westat, Inc., Rockville, Maryland

James L. Esposito, National Center for Health Statistics, Hyattsville, Maryland

Trena M. Ezzati, National Center for Health Statistics, Hyattsville, Maryland

Ron Fecso, National Agricultural Statistical Service, U.S. Department of Agriculture, Washington, D.C.

Gösta Forsman, University of Linköping, Linköping, Sweden

Barbara H. Forsyth, Research Triangle Institute, Research Triangle Park, North Carolina

Floyd J. Fowler, Jr., Center for Survey Research, University of Massachusetts, Boston, Massachusetts

Lester R. Frankel, Audits and Surveys, Inc., New York, New York

Wayne A. Fuller, Department of Statistics, Iowa State University, Ames, Iowa

Robert M. Groves, U.S. Bureau of the Census, Washington, D.C. and Survey Research Center, The University of Michigan, Ann Arbor, Michigan

Daniel H. Hill, Opinion Research Institute, The University of Toledo, Toledo, Ohio

Hans-Jürgen Hippler, Zentrum für Umfragen, Methoden, und Analysen (ZUMA), Mannheim, Germany

David Holt, Department of Social Statistics, University of Southampton, Southampton, United Kingdom

Joop J. Hox, University of Amsterdam, Amsterdam, The Netherlands

Michael L. Hubbard, Research Triangle Institute, Research Triangle Park, North Carolina

Jared B. Jobe, National Center of Health Statistics, Hyattsville, Maryland

Daniel Kasprzyk, National Center for Education Statistics, Washington, D.C.

Ita G. G. Kreft, University of California at Los Angeles, Los Angeles, California

William Kruskal, Department of Statistics, University of Chicago, Chicago, Illinois

Judith T. Lessler, Research Triangle Institute, Research Triangle Park, North Carolina

Lars E. Lyberg, Statistics Sweden, Stockholm, Sweden

Nancy A. Mathiowetz, U.S. Bureau of the Census, Washington, D.C.

Barbara Means, SRI International, Menlo Park, California

Geeta Menon, Department of Marketing, New York University, New York, New York

Nico Molenaar, Vrije Universiteit, Amsterdam, The Netherlands

David Morganstein, Westat, Inc., Rockville, Maryland

Ingrid M. E. Munck, Statistics Sweden, Stockholm, Sweden

Colm O'Muircheartaigh, London School of Economics and Political Science, London, United Kingdom

Jeroen Pannekoek, Central Bureau of Statistics, Voorburg, The Netherlands

Teresa L. Parsley, Research Triangle Institute, Research Triangle Park, North Carolina

Frank W. Piotrowski, A. C. Nielsen Co., Northbrook, Illinois

J. N. K. Rao, Department of Mathematics and Statistics, Carleton University, Ottawa, Ontario, Canada

Willem E. Saris, University of Amsterdam, Amsterdam, The Netherlands

Nora Cate Schaeffer, Department of Sociology, University of Wisconsin, Madison, Wisconsin

Irwin Schreiner, U.S. Bureau of the Census, Washington, D.C.

Norbert Schwarz, Zentrum für Umfragen, Methoden, und Analysen (ZUMA), Mannheim, Germany

Stuart Scott, U.S. Bureau of Labor Statistics, Washington, D.C.

Adriana R. Silberstein, U.S. Bureau of Labor Statistics, Washington, D.C.

Christopher J. Skinner, University of Southampton, Southampton, United Kingdom

Johannes H. Smit, Vrije Universiteit, Amsterdam, The Netherlands

Tom W. Smith, NORC and University of Chicago, Chicago, Illinois

S. Lynne Stokes, Department of Management Sciences and Information Systems, University of Texas, Austin, Texas

Seymour Sudman, Survey Research Laboratory, University of Illinois, Urbana, Illinois

Gary E. Swan, SRI International, Menlo Park, California

John Tarnai, Washington State University, Pullman, Washington

D. Roland Thomas, Department of Mathematics and Statistics, Carleton University, Ottawa, Ontario, Canada

Johannes van der Zouwen, Vrije Universiteit, Amsterdam, The Netherlands

Piet G. W. M. van Dosselaar, Central Bureau of Statistics, Voorburg, The Netherlands

Wendy Visscher, Research Triangle Institute, Research Triangle Park, North Carolina

Kirk M. Wolter, A. C. Nielsen Co., Northbrook, Illinois

CONTENTS

PREFACE

Survey measurement error, as the term is used in this book, refers to error in survey responses arising from the method of data collection, the respondent, or the questionnaire (or other instrument). It includes the error in a survey response as a result of respondent confusion, ignorance, carelessness, or dishonesty; the error attributable to the interviewer, perhaps as a consequence of poor or inadequate training, prior expectations regarding respondents' responses, or deliberate errors; and error attributable to the wording of the questions in the questionnaire, the order or context in which the questions are presented, and the method used to obtain the responses. At the time survey responses are collected, all of these factors may intervene and interact in such a way as to degrade response accuracy. The consequences of measurement errors are survey results that, while ostensibly accurate and credible may in reality be quite inaccurate and misleading. Measurement error does not include the errors of nonobservation (such as nonresponse error), processing errors, or other errors occurring after data collection. Thus, these errors are intentionally not discussed in the present book.

Given the importance of the topic of measurement error in survey work, the Survey Research Methods Section (SRM) of the American Statistical Association (ASA) in 1986 determined that survey measurement error should be the topic of an SRM-sponsored conference and approached Paul Biemer to develop the idea. Two years later, while Biemer was visiting Statistics Sweden, the groundwork was laid for a measurement error conference and an edited monograph of the conference-invited papers. It was decided that the conference should seek the participation of researchers worldwide. Further, the scope of the conference was expanded from one which focused only on the statistical aspects of measurement error to one which also enveloped important research being conducted in sociology, psychometrics, psychology, market research, and other disciplines as well. A preliminary outline for the book, which would also serve as a general framework for the conference sessions, was developed.

By the fall of 1988, an organizing/editing committee was formed consisting of: Paul P. Biemer (as Chair), Robert M. Groves, Lars Lyberg, Nancy A. Mathiowetz, and Seymour Sudman. Gösta Forsman was enlisted to assist in planning and conducting the conference. The committee contacted numerous research organizations for monetary contributions. The committee also reviewed and finalized the monograph outline and began to identify and contact researchers throughout the world as potential authors. Abstracts were requested and 128 abstracts were received from researchers interested in writing for the monograph. From these, the committee selected 32 to become the chapters in this monograph and developed the conference program. Kate Roach at John Wiley & Sons was contacted and Wiley agreed to publish the book.

Four professional organizations were asked to sponsor the conference: the American Association of Public Opinion Research (AAPOR), the American Marketing Association (AMA), the American Statistical Association (ASA), and the International Association of Survey Statisticians (IASS). All four organizations enthusiastically agreed. Two of these (ASA and AAPOR) also contributed funds to support the project. In addition, the following research organizations contributed funds:

Australian Bureau of Statistics
Central Statistical Office of Finland
Istituto Centrale di Statistica, Italy
National Agricultural Statistics Service
National Center for Health Statistics
National Science Foundation
Nielsen Media Research
NORC
Office of Population Censuses and Surveys, United Kingdom
Research Triangle Institute
Statistics Sweden
U.S. Bureau of Labor Statistics
U.S. Bureau of the Census
U.S. General Accounting Office
Westat, Inc.

Without the financial support of these organizations, the conference and edited monograph would not have been possible.

The International Conference on Measurement Errors in Surveys was held on November 11–14, 1990 in Tucson, Arizona. It drew 421 attendees from 15 countries. The program consisted of 60 invited papers (including the 32 chosen for the present book) and 70 contributed papers. Additionally, two short courses dealing with measurement errors were presented. The number of presented papers and the number of attendees surprised and delighted the committee who, in the early planning stages, had anticipated a much smaller program and audience, because of the specialized topic of the conference.

During the conference, the committee was assisted by eight graduate student fellows. Six of these were selected from applicants worldwide to attend the conference and the two others were sponsored by Iowa State University to attend the conference and to assist the committee. These eight fellows were: Tom Belin, Harvard University; Karen Bogen, University of Michigan; Joseph Croos, Iowa State University; Stasja Draisma, Vrije Universiteit, The Netherlands; Daniel Merkel, Northwestern University; Steven Pennell, University of Michigan; Todd Rockwood, Washington State University; and Todd Sanger, Iowa State University. The fellows took notes during discussions between section editors and authors and during floor discussions after the monograph papers were presented. At least one fellow was available at every session to ensure that the session would run smoothly.

In designing the book, the committee did not intend to merely publish a conference proceedings volume. Rather the aim was to collectively write a book dealing with the most important issues in the field of survey measurement error, attempting whenever possible to integrate diverse perspectives. Thus, each chapter has undergone extensive editing, review, and revision. The book is organized into five sections. The section titles and their editors are:

Section A: The Questionnaire (Seymour Sudman)
Section B: Respondents and Responses (Nancy A. Mathiowetz)
Section C: Interviewers and Other Means of Data Collection (Lars Lyberg)
Section D: Measurement Errors in the Interview Process (Robert M. Groves)
Section E: Modeling Measurement Errors and Their Effects on Estimation and Data Analysis (Paul P. Biemer)

The reader will notice that the sections are not mutually exclusive. The difficulty in knowing exactly where a paper should be placed merely reflects the complex nature of survey data collection.

In designing this volume, difficult decisions had to be made about what to exclude. Some topics were excluded because they were only marginally related to the main theme of survey measurement error. Other topics were omitted because we could identify no current research, despite the fact that new knowledge might be needed. Readers will note an emphasis on population and household surveys. We have, however, included three chapters on business and establishment surveys (Chapters 7, 12, and 18) and one on crop yield surveys (Chapter 17).

Each section editor had responsibilities as a secondary editor for at least one other section as well. The authors of the chapters, in addition to their extensive writing and revising activities, were also involved in the review of the other monograph chapters. They were encouraged to seek outside reviews for their chapters on their own. Thus, the monograph reflects the efforts and contributions of scores of writers, editors, and reviewers.

The diversity of orientations of the authors for the monograph made it impossible to impose a unified terminology and set of notation across all chapters. Two chapters (Chapter 1 and Chapter 24) have been provided as an attempt to link the terminological and notational conventions of the various disciplines represented in the book. Further, except for one section (Section E), the statistical level of the monograph is quite accessible by graduate students in sociology, psychology, or marketing research. Section E, however, which deals with the more theoretical statistical side of measurement error research, requires a fairly thorough grounding in survey sampling and mathematical statistics, at the level of *Introduction to Mathematical Statistics*, 3rd ed., by R. V. Hogg and A. Craig, 1970.

Although the present book can serve as a course text, the primary audience is researchers having some prior training in survey research and/or survey methods. Since it contains a number of review articles on measurement error research in several disciplines, it will be useful to researchers actively engaged in measurement error research who want an introduction to the techniques and research issues in the field from a different theoretical perspective. The book will also be useful to the survey methodologist or survey practitioner who wants to learn more about the causes, consequences, and cures of survey error in order to improve the quality of surveys through better design, data collection, and analytical techniques. This book is far from the final word on measurement error. It does, however, reflect current knowledge in 1990, to the best of our editorial judgment. As a group, we hope that its publication will stimulate future research in this exiting field.

Gösta Forsman deserves great appreciation for all the activities he performed so ably for the conference and book project. We are also truly grateful to Lee Decker of ASA, who tirelessly and efficiently handled an enormous number of logistical details associated with the conference.

Sincere thanks go to Cheryl Crawford and Mary Kay Martinez, at New Mexico State University, and to Linda Miller, at Research Triangle Institute, who performed many clerical and secretarial functions associated with the project. Thanks are due to Graham Kalton and Wayne Fuller for their efforts, while chairing the Survey Research Methods Section of the ASA, in promoting the idea of a conference on nonsampling error. We are appreciative of the efforts of Pat Dean, Dan Kasprzyk, and Lynne Stokes, who assisted in the review of a number of manuscripts, and Pat Ellis, who prepared the final copy of the reference list. Our employing organizations also deserve great appreciation for supporting our activities in assembling the book: New Mexico State University and Research Triangle Institute (Biemer); Statistics Sweden (Lyberg); Survey Research Center at the University of Michigan (Groves); National Center for Health Services Research (Mathiowetz); U.S. Bureau of the Census (Groves and Mathiowetz); and University of Illinois at Urbana–Champaign (Sudman).

PAUL BIEMER
ROBERT GROVES
LARS LYBERG
NANCY MATHIOWETZ
SEYMOUR SUDMAN

Research Triangle Park, North Carolina
Washington, D.C.
Stockholm, Sweden
Washington, D.C.
Urbana, Illinois

November 1991

INTRODUCTION

William Kruskal
University of Chicago

This volume summarizes the International Conference on Measurement Errors in Surveys, held in November 1990 in Tucson, Arizona. I agreed to be the opening speaker as an opportunity to present some thoughts to the conference participants: primarily to illustrate how pervasive in society are concepts of measurement error, how important they are to understanding our world, and thus to motivate this volume.

Dare we ask what is truth? Some readers of this volume are surely members of the AARP, the American Association of Retired Persons, which has a close relationship with the NRTA, the National Retired Teachers Association. Members receive a journal called *Modern Maturity*, a somewhat bland, inspirational magazine that elaborates on Robert Browning's "Grow old along with me." In last March's *Modern Maturity* I was delighted to find an excellent article on philosophy by Steven Ross (1990), City University of New York. Ross treats the big, big questions of epistemology and ethics in a down to earth, up-to-date, and lucid way that is relevant to a big question for this conference: is there a true value for what is measured by an observation or a survey?

Ross sketches the traditional positivist position and the attacks on it by Wittgenstein and Quine. We are left with no sharp lines — perhaps even no fuzzy lines — between theory and observation. Many of us feel uncomfortable with that position, for if there is no true value lurking in the platonic underbrush how can we talk sensibly about error? And if we cannot talk sensibly about error, how can we design surveys and do inferences?

There are other philosophical issues for surveys: for example, should we regard interviewers as introducing fixed but unknown systematic effects each, or should we think of their individual effects as randomly chosen? Or should we do both . . . perhaps in separate analyses? Or again

Measurement Errors in Surveys.
Edited by Biemer, Groves, Lyberg, Mathiowetz, and Sudman.

ISBN: 0-471-53405-6

should we be more realistic (aha, note my recourse to realism as a goal) and introduce models with stochastic dependence among interviewers. A similar distinction arises in so-called dual systems analyses . . . of which I will say more later. One of the two samples may be taken idealistically as perfect in terms of measurement error; at another extreme, the two samples might be regarded as replicates, i.e., independent and identically distributed.

How we handle these questions can make major differences in our analyses of error structures and hence yield major differences in conclusions. Our choice may also be relevant to debates among survey organizations, for example, about whether it is better to maintain a relatively fixed, highly trained interviewing staff or to arrange frequent turnover, short-term training, and consequently possible lower esprit de corps?

There is a widespread desire to cling to faith in an objective reality out there; I confess that I feel it strongly on Mondays, Wednesdays, and Fridays. Lionel Trilling (1951, p. 4) paraphrased the historian Vernon L. Parrington (before criticizing him):

> There exists . . . a thing called *reality*; it is one and immutable, it is wholly external, it is irreducible. Men's minds may waver, but reality is always reliable, always the same, always easily to be known.

The last part of that statement — "always easily to be known" — is not part of my thrice a week mind-frame, and I expect that few statisticians would agree that reality is easily known. Indeed we earn our livelihoods because of difficulties in knowing reality, or whatever each of us accepts as a substitute for, or reflection of, reality.

Another American historian, Oscar Handlin, said more recently (1979, p. 405):

> Truth is absolute, it is as absolute as the world is real . . . the historian's vocation depends on this minimal operational article of faith.

The opposing viewpoint, that there are no true values, just different ways of measurement, has had statistical proponents for many years, perhaps most notably W. E. Deming (e.g., Deming, 1950, Ch. 1). Another example is a 1939 (p. 135) quotation by Walter A. Shewhart:

> . . . Consider in contrast the concept of the true value X' of a quality characteristic, for example the length of a line AB, or the velocity of light. I am not able even to conceive of a physical operation of observing or experiencing a true length X'. You may argue that there are ways of measuring the length of a line, by any one of which you may obtain a sequence of observations; you may even argue that the limiting average $\bar{X}'$ is equal to X'. But the physical operation is a

method of obtaining $\bar{X}'$, not X'. Whether $\bar{X}' = X'$ we shall never know. The true length X' is the given, unknowable, unapproachable, ineffable. It is removed from the pale of observation as securely as $\sqrt{-1}$ is removed from the realm of real numbers; there is not even the question of approximating $\sqrt{-1}$ with the rational and irrational numbers.

Attacks on some sort of absolute truth have a long literary history as well. I have a hard time trying to understand continental literary figures like Jacques Derrida, but their frequent scorn for simple truth comes through clearly. One continental literary figure, now turned national leader, Vaclav Havel, is relatively understandable. He was recently quoted in Senator Daniel Patrick Moynihan's newsletter (April 20, 1990). Havel, speaking to a joint meeting of the Congress about events in eastern Europe, said that he had "one great certainty: Consciousness precedes Being, and not the other way around, as the Marxists claim." Moynihan adds that "this is a real issue to intellectuals such as Havel; an issue men and women have died for; it holds that the beliefs create the 'real' world and not vice versa."

I might happily continue along these lines. For example, I might repeat to you Higgins's Law from a John McPhee *New Yorker* article on maritime trade (April 2, 1990, p. 74):

> Use one electronic [location] system and you always know where you are. Use more than one and you're always in doubt.

A great potential text for a statistical sermon. Or I might present a book review of Miles Orvell's recent *The Real Thing* (1989), in part named after Henry James's story about genuine aristocrats who were far less able to pose for an artist illustrating a book about aristocrats than were lower-class models. That summary rides brutally over Jamesian subtleties, but I have no choice. We must go on to other subtopics, keeping in mind the breadth of our interests, and awaiting Bruce Spencer's conference paper on error in true values (Spencer, 1990).

Let me first, however, note a topic that would merit treatment at length if we had the length. There may be true values that lie behind each individual measurement: your true age, your true belief about celestial music, the true income of family i. Then there may be true values for an entire population. For additive qualities like money that is not such a big step. For beliefs and opinions, true value for the entirety can be troublesome.

This is a good time to present a brief outline of topics. I have dealt briefly with the problem of true values. Next I turn to scope and ask how widely or narrowly the concept of survey might be construed. I then develop a fundamental approach to measurement error, with consideration of the apparent human need for a single figure of merit. Also discussed are the bottom-up and top-down versions of total survey error. We end with a few remarks about the study of error structure.

Scope; what are surveys? Let me place aside various claims or restrictions. It is artificial to limit our measurement error discussion to social surveys, to person-to-person question and answers, or to finite populations. We want to be able to cover not only standard questions about age, sex, ethnicity, prices, incomes, etc., not only attitude questions about public figures and policies, but also questions about luminosity of stars, pollen counts in Chicago, how fast Peter Pan flies (for studies of myth), the economic backgrounds of members of Parliament in 1806, medical diagnoses, and limitlessly on. Robert Groves (Chapter 1, this book) writes about many of these vistas. At present I pause mainly to express a caution about overdefensiveness. Social surveys are sometimes given rough handling by so-called hard scientists, cynical newspaper columnists, and congressmen at budget hearings. Surely some of those attacks are justified, but a great many are not, and the attacks often come from glass houses. All of us, no doubt, have inspected glass houses of biology and medicine. As to physical sciences, many of you have doubtless seen the famous graph of thermal conductivity of copper that is poles away from one traditional view of physics as a pure science. It has been popularized by J. S. Hunter and appears in Turner and Martin (1984, p. 15). I do not at all wish to denigrate physical science, but I wonder how much of its literature is open to sharp statistical criticism. Problems arise in connection with estimating fundamental constants of nature, for example, see W. L. Wiese et al. (1966) on atomic transition probabilities. Other problems arise in contexts with political and economic facets: examples are cold fusion, anomalous water, and space probe reliability. There are other attacks on the crystal house of physics, for example, a 1983 book by Nancy Cartwright with the vigorous title *How the Laws of Physics Lie*. A related lively discussion is given by Huber (1990).

In a spirit of continued professional iconoclasm, let me mention some other traits often said to characterize the field of sample surveys, although I am not persuaded about any. Thus I see no crisp lines between surveys and other parts of statistics.

Emphasis on finite populations, to repeat a point, is frequently mentioned, but it does not strike me as of the essence. Further, *all* statistical populations are finite because of rounding resolution of measuring instruments; our ruler only measures to 1/16 inch or whatever. Traditional continuous models are useful approximations, not gospel.

A related trait is the usual exceedingly wide family of probability models common in sample survey theory . . . as opposed to much narrower parametric models in traditional statistics. Yet that boundary strikes me as hazy because, on the one hand, there is a movement toward more nearly parameterized models for sampling studies; on the other hand, nonparametrics uses extremely wide models, similar to those in sampling theory, that depend on broad structures only, like symmetry, independence, and equidistribution.

Also related is the feeling by some that survey sampling deals primarily with totals and averages, leaving more complex quantities to other hands. I think that simply misleading.

Some say that survey sampling deals only with *descriptive* statistics, as opposed to *inferential* statistics. Yet inferential questions arise widely in surveys. In particular, surveys have had central roles in properly experimental studies.

Finally, some colleagues believe that the essence of survey sampling is the fundamental presence of *labels*, identifiers of the sampled units by name, number, etc.. That's a fruitful assertion but I think labels neither necessary nor sufficient for sample surveydom.

That leaves me without sharp boundaries for this subtaxonomy of statistics, a condition of mild discomfort yet honesty.

Approach to measurement error. It is high time to turn to our main theme: an approach to measurement error that seems to me basic and not always acknowledged. To start with, let's suppose that we have a population of people (or cars or stars or trout) and that we have a measurement system for *age*. It might be self-report, lookup in administrative files, counting scales or teeth, whatever. Let's say age in years for simplicity and let's, at least pro tempore, suppose that there is a true age for each individual of interest, i.e., in a specified population. We can think in principle of attaching to each individual a distribution for observed age. For Jim Brown, true age 31, that distribution might be

observed age	21	...	29	30	31	32	33	...	41
probability	0.1		0.075	0.1	0.5	0.075	0.05		0.1

That is, the observation error in measuring Jim Brown's age to the closest year is zero half the time, usually two years or less, and with a nonnegligible chance of a ten-year error exactly.

Our motivations for attacking error of age and other measurements are mainly to gain improved understanding of the process, thus making better decisions based on the measurements and deciding what components should get priority for improvement.

Now one might consider such separate hypothetical distributions, one for each individual, but I suspect that many of us would like to simplify by supposing the observed age distribution to depend only on the true age. In any case, this approach pushes one to consider explicitly what simplifying assumptions are being made.

Starting with such distributions — usually hypothetical — of observed age, the next logical step is introduction of sampling. (I know that this volume is on measurement, not sampling, yet the two broad topics are intertwined.) For any suggested sampling procedures and any set of little distributions like the example, a putting-together provides one with a table having columns (say) for true age in years, rows for observed age, and cell entries for proportions. I think of all entries as summing to one. A proportion 0.017 in column 16 and row 18 would mean that with probability 17 out of 1,000 a probabilistically chosen population member will in fact be 16 and measured as 18 years old.

When true value is a questionable concept, and even when it makes sense, the joint distribution for two or more modes of measurement might naturally be considered. For example, age might be measured (1) by asking a person directly for age, (2) by asking for year of birth, (3) by asking another household member, (4) by looking at a birth certificate, (5) by interviewer guess; there might well be others and the empirical study of their interconnections presents a constant task for survey organizations. Instead of starting with a univariate distribution for Jim Brown's age, then, we might start with a bivariate distribution in which rows and columns correspond to observed age by mode 1 and mode 2 respectively; one cell, for example, would give the probability that measurement mode 1 provides 21 years and mode 2, 29 years. Then as before one might introduce a sampling scheme and obtain a three-dimensional table in which one dimension corresponds to true age and the other two dimensions correspond to the two measurement modes. Or one might omit the true age dimension as meaningless. Decisions such as the latter one might be appropriate for attitude studies or for surveys of beliefs about heaven.

Age is, of course, a metric variable and a relatively objective one . . . unless one gets down to fine detail, as in some evaluations of astrology. One can also go through an analogous discussion for qualitative measurements: ethnicity, state of birth, breed of horse, chromosome on which a gene sits, and so on. The central point is that an approach of the kind sketched pushes one into fundamental thought about measurement error.

Simplification becomes necessary in practice, and in the case of quantitative variables like income, height, length of life, boiling point, etc., it may be natural to look at diagonal differences in the basic table, that is, to look at the observed value *minus* the true value . . . or, when true value is questionable, to look at the difference between observed values for two modes.

Caution: we should by no means automatically jump to differences. For example, if we are interested in age-heaping, round numbers play a special role. Or if we are working with circular observations, like time of day for births or day of year for draft lotteries, then simple differences might be quite misleading. Again, there is no law that we must look at differences; ratios, for example, may be better.

Now back to differences for simplicity. We are led to the familiar canonical decomposition

$$\text{observed value} = \text{true value} + \text{diagonal difference}$$

or its analog for two measurement modes. We are now in a standard position, and we may go on to customary discussions, for example of systematic error and dispersion, usually but not necessarily expressed by the expected value of the diagonal difference and by variance.

A single figure of merit. It is often felt useful to have a single number for the error of an observational system, and a technically comfortable one is *mean squared error*, the expectation of the squared difference from the true value. Almost by definition

$$\text{mean squared error} = \text{variance} + \text{systematic error squared.}$$

I hasten to note, however, that this quadratic additivity may be a weak reason for settling on mean squared error. Dispersion and systematic error often enter a problem in quite different ways.

The idea of a single number reflecting error magnitude of an observational system has a powerful appeal, especially if we are comparing two or more observational systems. Yet the use of a single figure of merit can be misleading, in part because many of us would almost automatically adopt mean squared error without adequate reflection.

The single figure of merit question comes up in areas other than sample surveys, for example in refined measurements of the kind done by the U.S. National Institute of Standards and Technology. I think of relevant publications by Shewhart (1939), by Churchill Eisenhart (e.g., 1963), and others. Another area in which univocal error concepts arise is that of industrial process control, where there may be a designated interval around some target value of a dimension, a voltage, etc. The presence of that interval may indeed permit sensible univocality. A recent relevant article is by A. F. Bissell (1990).

In all these contexts one has, first, the problem of discovering the sources of systematic error, and then of estimating magnitudes. If we knew them exactly enough, we presumably would correct exactly enough, so we must have a mode of stating magnitude roughly, perhaps via an upper bound for absolute error. There are major puzzles here.

For some more on these basic themes, see my segment of the O. D. Duncan Panel report on surveying subjective phenomena (Turner and

Martin, 1984, Sec. 4.1). For a start on axiomatic approaches to measurement error see the masterful trilogy by an eminent quartet (Patrick Suppes, David M. Krantz, R. Duncan Luce, and Amos Tversky; 1971, 1989, 1990). See also work of E. W. Adams (e.g., 1965, 1966). I must cite a wonderful article on nonsampling error by Frederick Mosteller (1978).

Total survey error, bottom-up or top-down. One is often interested in components of error for a survey characteristic. These may be components of bias, components of random error, or other, more complex, entities. For example, consider again so simple a characteristic as age. Aspects of bias might be the desire of teenagers to appear older, the desire of some middle-agers (note my careful androgeny) to appear younger, and the desire of the very old to have made it to that round century. We've already mentioned age-heaping.

Random components might come from simple forgetfulness, interviewer mistakes, coding mistakes, optical reading smudges, etc. Correlations might be introduced via a family informant who is not sure of everyone's age. And so on. There may be many sources of error and we never, never can be sure that we have even thought of all those peaky sources, much less know enough about their effects on the measurement. For example, is use, contrary to instructions, of a ball-point pen instead of a #2 lead pencil a mistake? What does it lead to?

(Another point — I reluctantly skim over it — is that random errors are always in terms of hypothetical repetitions of the survey with some things fixed and others not: interviewers, questions, cities, weather, etc.)

Ideally, we want to have a formula giving the error in a given characteristic as a function of the many components. In fact, we rarely know that much. We often write down a linear or other simplified expression and press ahead heroically. Note that crudeness in our formula, for example nonadditivities, may express themselves as interactions or correlations.

That age example is surely one of the simpler ways in which error sources are expressed. A more complex (but à la mode) example is error in a capture–recapture estimator for the population of a small area . . . a block. I choose that example not only because it is complex, but also because it is under intense current discussion and debate in connection with the 1990 census. Capture–recapture — usually called dual system in census circles as opposed to wildlife circles — is complex and clever, but it has a possibly soft underbelly of assumptions that are dubious. Perhaps the most dubious of these assumptions are about homogeneity and independence: that capture and recapture are stochastically independent events with homogeneous probabilities. Assumed away are the wily, experienced trout who elude all nets, the hard-core libertarians who will have nothing to do with so intrusive an arm of government as the census, the households that live two or three to a single apartment and feel — often correctly — that such crowding if known will get them into trouble with authorities. And so on.

Friends who are enthusiastic about dual system for census modification say that at least the estimate modifications tend to go in the correct directions from the more nearly raw count. I think that not necessarily so if one looks at two or more small areas and asks what can happen in their relative modifications.

Analysts list eight or so sources of error for dual systems methods and some sort of formula for putting those errors together. Next we need estimates of the *magnitudes* of these errors: amounts of bias, size of variances, correlations as appropriate. Those magnitudes may come from other surveys or censuses, from fresh empirical studies (perhaps a special postenumeration survey), and (inescapably but sadly) from expert guesses. Finally it all gets put together so that we have an estimated total error structure from the bottom up.

Note the inherent difficulties: (1) we never know all the sources of error; (2) we know only imperfectly the functional way in which errors operate and combine; (3) we rarely know well the magnitude of the ground level or fundamental errors.

In contrast, consider top-down total survey error. Sometimes — not often — one has separate ways of measuring what the survey is intended to measure, so a thorough calibration is possible; thus top-down. A canonical example is described in the book called *Total Survey Error* by Ronald Andersen, Judith Kasper, and Martin Frankel (1979). The topic is medicine and health; respondents are asked, among other questions, about visits to physicians and hospital stays. Then, with some complications, the physicians and hospitals are asked about the same people. Thus in principle one has a direct calibration or confrontation. Other path-breaking studies of this kind had been carried out by Norman Bradburn and Seymour Sudman on such topics as voting at all, arrest for drunken driving, etc. (see Bradburn, Sudman et al., 1979).

One weakness of this approach is that you can only do it for questions where calibrating information exists. There will generally be questions in a survey for which there are no calibrating questions, so one reasons — weakly — by analogy.

Another weakness is that when the calibration points to problems with a particular question, it may not be clear where the major problems lie: misunderstanding, memory lapses, fear or embarrassment, etc. And of course the calibrating responses will have their own error structures.

In the case of the 1990 census I know of only two possible sources of top-down calibrating information. The first is demographic analysis, a process that the U.S. Bureau of the Census has been doing for years. Its two big problems are, first, it applies only nationally, not for subsections like states, much less blocks as mentioned above; second, migration figures are mighty shaky, especially for undocumented individuals.

The second possible census calibrating information is that obtained from the so-called ethnographic approach, sometimes known as participant observation. Here, specially chosen and trained observers go to small population clumps and get to know families well so that census information can be checked or calibrated in fine detail. I think this is a highly promising approach, but there are at least three difficulties. First the ethnographer might appear to be an intrusive, big-brother agent of the distant inimical government. Second is the ever-present possibility that the selection and training process through which the observers inevitably pass will tend to select for people of particular kinds whose traits will bias the endresults. Third, ambiguities of definition will forcefully come to our attention. For example, to which household does an impoverished teenager belong when he sleeps, eats, leaves clothing, etc. in three separate households? There are, of course, partial answers to these problems.

I note that there is one paper at this conference on the survey vs. the ethnographic approach (Campanelli, Martin, and Salo, 1990). Will the authors persuade their readers that these two approaches are antagonistic rather than — as I hope — cooperative?

Generally speaking it is good to have both bottom-up and top-down studies. Would that I had serious wisdom about allocation of energy and time between these two approaches. I can, however, say generally that far too small a fraction of resources is put into investigation of error structure. Something there is that doesn't love an instrumental study.

There are of course good instrumental studies, for example, the study by Camilla A. Brooks and Barbara A. Bailar (1978) on the nonsampling errors of the U.S. Current Population Survey. As the authors properly point out, however, there are big gaps in our understanding despite the efforts of many brilliant and energetic thinkers. A 1990 U.S. Census Bureau study of error structure (Jabine et al., 1990) covers a great deal of ground. In particular it makes the top-down vs. bottom-up distinction but in a different terminology: micro- and macro-evaluation.

Conclusion. You may feel that the standards I have sketched for error structure study are too stringent and impossible to meet. A fair comment; yet it seems to me that, both as a profession and as individuals, we must keep trying and trying to improve our understanding of error. Even though we never get there. In that respect, understanding accuracy is like Truth, Beauty, and Justice.

I have ranged broadly in this assay; the topic of measurement error deserves nothing less. In doing so, I hope to convince you of the central role that notions of measurement error play in our use of surveys. I also hope that you have been alerted to diverse approaches to measurement error, many of which you will see discussed in this volume. The editors designed the volume to be an interdisciplinary mosaic of studies on measurement error. It is a rare treat to see such alternative perspectives combined in one place.

ACKNOWLEDGMENTS

Among the many colleagues who have made helpful suggestions, I list the following subject to conventional disclaimers:

Barbara A. Bailar, Tom Belin, Norman Bradburn, Churchill Eisenhart, Robert M. Groves, J. Stuart Hunter, Benjamin King, Frederick Mosteller, John W. Pratt, Howard Schuman, Jacob Siegel, Tom Smith, Stephen Stigler, Seymour Sudman, and Ronald Thisted.

MEASUREMENT ERRORS IN SURVEYS

CHAPTER 1

MEASUREMENT ERROR ACROSS THE DISCIPLINES

Robert M. Groves
U.S. Bureau of the Census and University of Michigan

1.1 MEASUREMENT ERROR IN THE CONTEXT OF OTHER SURVEY ERRORS

Among all the diverse outlooks on the survey method there is a common concern about errors inherent in the methodology. In this context *error* refers to deviations of obtained survey results from those that are true reflections of the population. Some choose one error to study; others choose another. Some study ways to eliminate the error; others concentrate on measuring its impact on their work. But all of them share the preoccupation with weaknesses in the survey method.

There appear to be at least two major languages of error that are applied to survey data. They are associated with two different academic disciplines and exemplify the consequences of groups addressing similar problems in isolation from one another. The two disciplines are statistics (especially statistical sampling theory) and psychology (especially psychometric test and measurement theory). Although other disciplines use survey data (e.g., sociology and political science), they appear to employ similar languages to one of those two.

Some attention to terminological differences is necessary to define *measurement error* unambiguously (see Deming, 1944; Kish, 1965; Groves, 1989; Chapter 24, this volume). A common conceptual structure labels the total error of a survey statistic the *mean squared error*; it is the sum of

---

Views expressed are those of the author and do not necessarily reflect those of the U.S. Census Bureau.

Measurement Errors in Surveys.
Edited by Biemer, Groves, Lyberg, Mathiowetz, and Sudman.

ISBN: 0-471-53405-6

all variable errors and all biases (more precisely, the sum of variance and squared bias). *Bias* is the type of error that affects the statistic in all implementations of a survey design; in that sense it is a constant error (e.g., all possible surveys using the same design might overestimate the mean years of education per person in the population). A variable error, measured by the *variance* of a statistic, arises because achieved values differ over the units (e.g., sampled persons, interviewers used, questions asked) that are the sources of the errors. The concept of variable errors inherently requires the possibility of repeating the survey, with changes of units in the replications (e.g., different sample persons, different interviewers).

A *survey design* defines the fixed properties of the data collection over all possible implementations within a fixed measurement environment. Hansen, Hurwitz, and Bershad (1961) refer to these as the "essential survey conditions." For example, *response variance* is used by some to denote the variation in answers to the same question if repeatedly administered to the same person over different trials or replications. The units of variation for response variance are different applications of the survey question to the same person. In general, variable errors are based on a model of replicated implementation of a survey design, whether or not such replication is ever actually conducted.

Errors of nonobservation are those arising because measurements were not taken on part of the population. *Observational errors* are deviations of the answers of respondents from their true values on the measure; for our purposes, these are *measurement errors*. The most familiar observational error is probably that associated with a respondent's inability to provide the correct answer (e.g., failure to report correctly whether he visited a physician in the last six months). The answer provided departs from the true value for that person. If there is a tendency to make such an error throughout the population, the overall survey proportion will depart from the true population proportion, yielding a measurement bias.

The final level of conceptual structure concerns the alternative sources of the particular error. Errors of nonobservation are viewed as arising from three sources — coverage, nonresponse, and sampling (see Groves, 1989, for definitions). *Observational errors* are conveniently categorized into different sources — the interviewer, the respondent, the questionnaire, and the mode of data collection. *Interviewer errors* are associated with effects on respondents' answers stemming from the different ways that interviewers administer the same survey. Examples of these errors include the failure to read the question correctly (leading to response errors by the respondent), delivery of the question with an intonation that influences the respondent's choice of answer, and failure to record the respondent's answer correctly. There are also effects on the quality of the respondents' answers from the wording of the question or flow of the questionnaire, which are labeled *instrument error*. Research has found that small, apparently innocuous, changes in wording of a question can lead to large changes in responses. Thus, *respondent error* is another source of observational error. Different respondents have been found to provide data with different amounts of error, because of different cognitive abilities or differential motivation to answer the questions well. Finally, there are observational errors associated with the mode of data collection. It has been found, for example, that respondents have a tendency to shorten their answers to some questions in telephone interviews compared to answers on personal visit interviews.

Another class of errors arises after the measurement step for sample units — coding and classification of data, deterministic and statistical imputations, and changes to the data in editing and processing. These kinds of errors cause departures from the desired value for variables collected in the survey. They can clearly be conceptualized as having variable and fixed components. Variable components of the error arise from different coders or substantive editors producing codes to be stored in the data records related to measures obtained from the sample unit. These will not be considered measurement errors in this chapter, but *processing errors*.

1.2 NOTIONS OF FIXED AND VARIABLE ERRORS

I have asserted in past work that three questions can be answered to remove misunderstandings about different concepts of measurement error:

a. What is the statistic of interest when errors are being considered?

b. Which features of the data collection are viewed to be variable over replications and which are fixed?

c. What assumptions are being made about the nature of those persons not measured, or about properties of the observational errors?

We will keep these questions in mind while reviewing notions of measurement error in different disciplines.

The answer to the first question will determine whether observational errors of constant magnitude across all persons have any effects on the statistic of interest. For example, if all respondents underestimate their own weight by five pounds, the mean weight of the population will be underestimated, but correlations of reported weight and other variables will not be. The impact of errors at the point of data collection varies over different statistics in the same survey. Another example concerns a measurement procedure for which each person's error is constant over trials, but the sum of the errors over all persons is zero. That yields a case of a biased indicator of the person's true value, but an unbiased estimate of a population mean. (Note that such a case would violate the assumptions of true score theory as defined in Lord and Novick, 1968).

The second question (b) determines whether a problem is viewed as a bias or as a component of variance of the statistic. One of the most common instances of this in the experience of sampling statisticians is the following. A researcher who is the client of the sampling statistician, after having collected all of the data on the probability sample survey (for simplicity, let's assume with a 100 percent response rate), observes that there are "too many" women in the sample. The researcher calls this a "biased" sample; the statistician, viewing this as one of many samples of the same design (and being assured that the *design* is unbiased), views the discrepancy as evidence of sampling variance. In the view of the sampler the sample drawn is one of many that could have been drawn using the design, with varying amounts of error over the different samples. If the design is properly executed, the proportion of women would be correct *in expectation* over all these samples. The statistician claims the sample proportion is an unbiased estimate of the population proportion. This is a conflict of models of the research process. The sampler is committed to the view that the randomization process on the average produces samples with desirable properties; the analyst is more concerned with the ability of this single sample to describe the population.

The third question (c) determines whether some types of errors in the statistic of interest are eliminated by the model assumptions. This is perhaps the most frequent source of disagreements about error in statistics. The simplest example is the use of true score theory to eliminate response biases, as we saw above. Certainly if true score theory were accepted, but statistics were calculated that were affected by expected values of measures (e.g., intercepts in regression models), then the existence of error in the statistics would be dependent on that assumption. Another example concerns the fitting of a regression model with data that a survey researcher would claim are subject to large sampling bias. The analysis regressed a measure of annoyance with airport noise on a measure of the distance from the respondent's home to the airport. The sample included (a) some persons so far from the airport that the noise was just noticeable by equipment that simulated normal hearing capabilities, and (b) some persons in the neighborhood closest to the airport. The survey researcher claims that the omission of persons living in intermediate neighborhoods will bias the regression coefficient. The model builder claims that the omission has nothing to do with the error in the regression coefficient. The difference in viewpoints is that the model builder, following ordinary least squares theory, asserts the unbiasedness of the coefficient, *given that the model is well-specified.* He is specifying a linear model, hence, the persons in intermediate neighborhoods are not needed to obtain an unbiased estimate of the slope coefficient. In contrast, the survey researcher doesn't accept the assumption behind the regression estimator and claims that there may be a bias in the coefficient because people in intermediate neighborhoods may behave differently than may be implied by observation of the closest and farthest away neighborhoods.

1.2.1 Notions of Measurement Error in Survey Statistics

The most elaborated view of survey error held by some survey statisticians comes from those interested in total survey error (e.g., Hansen, Hurwitz, and Pritzker, 1964; Fellegi, 1964; Bailar and Dalenius, 1969; Koch, 1973; Bailey et al., 1978, Lessler et al., 1981). This perspective retains the possibility of fixed errors of coverage and nonresponse. In addition to variability over samples, it acknowledges variability in errors over different trials of the survey. Underlying this is the notion that the survey at hand is only one of an infinite number of possible trials or replications of the survey design. Respondents are assumed to vary in their answers to a survey question over trials, leading to "simple response variance" (Hansen, Hurwitz, and Bershad, 1961). Respondents may also be viewed as varying in their decision to cooperate with the interview request over trials. The same respondent might cooperate on one trial but refuse on the next, without having a well-developed rationale that yields predictable behavior. This would lead to variable nonresponse errors over trials, a nonresponse error variance term.

The interviewer is often treated as a source of error in this perspective, and is most often conceptualized as a source of variable error. That is, each trial of a survey is viewed to consist both of a set of sampled persons (one replication of the sample design) and a set of interviewers (one set selected to do the work, from among those eligible). Both sets are viewed to change over trials, both the sample and the interviewer corps and the assignment of interviewers to sample persons. The variable effects that interviewers have on respondent answers are sometimes labeled *correlated response variance* in this perspective (Bailey, Moore, and Bailar, 1978). This arises from the notion that errors in responses might be correlated among sample persons interviewed by the same person. A generalization of this perspective would permit interviewers to differ in their effects on coverage and nonresponse error also. This would lead to variation over trials in these errors because of different interviewer corps. There are very few examples, however, of this generalization.

In this literature, true to its "describer" roots, the statistic of interest is a sample mean. It defines the "mean squared error" as

$$E_{stia}[\bar{y}_{stia} - \bar{X}]^2 = E_{stia}[(\bar{y}_{stia} - \bar{y}_{....})^2 + (\bar{y}_{....} - \bar{X})^2]$$

where $E_{stia}[\,]$ = expectation over all samples, s, given a sample design; over all trials, t; over all sets of interviewers, i, chosen for the study; and over all assignment patterns, a, of interviewers to sample persons

$\bar{y}_{stia}$ = mean over respondents in the s-th sample, t-th trial, i-th set of interviewers, a-th assignment pattern of interviewers to sample persons, for y, the survey measure of the variable X in the target population

$\bar{y}_{....}$ = expected value of $\bar{y}_{stia}$ over all samples of respondents, all trials, all sets of interviewers, and all assignment patterns

$\bar{X}$ = mean of target population for true values on variable X.

The bias of the mean is $\bar{y}_{....} - \bar{X}$. The variance of the mean is seen to have components arising from sampling, variable measurement errors, as well as covariance terms (see Fellegi, 1964; Chapter 24, this volume).

1.2.2 Terminology of Errors in Psychological Measurement

When moving from survey statistics to psychometrics, the most important change is the notion of an unobservable characteristic that the researcher is attempting to measure with a survey indicator (i.e., a question). Within survey statistics, in contrast, the measurement problem lies in the operationalization of the question (*indicator*, in psychometric terms). That is, the problem is not the impossibility of measuring the characteristic, but the weakness of the measure. The psychometrician, typically dealing with attitudinal states, is more comfortable labeling the underlying characteristic (*construct*, in psychometric terms) as unobservable, something that can only be approximated with an applied measurement. [The existence of an observable true value is labeled an example of Platonic true scores (Bohrnstedt, 1983, p. 71)].

There are two influential measurement models that will be discussed in this section. In the first, *classical true score theory*, all observational errors are viewed as joint characteristics of a particular measure and the person to whom it is administered. Errors in responses are acknowledged. In such measurement however, the expected value (over repeated administrations) of an indicator is the true value it is attempting to measure. That is, there is no measurement bias possible, only variable errors over repeated administrators. That is, each asking of a question is one sample from an infinite population (of trials) of such askings. The *propensity distribution* describes the variability over trials of the error for the particular person. The only concept of error akin to those of the survey statistician is the variance of the error term, the *error variance* (Lord and Novick, 1968). The true score on the construct, X_{gj}, of a person, j, on the indicator, g, is defined as the expected value of the observed score; that is, $X_{gj} = E_t[y_{tgj}]$, where y_{tgj} = the response to indicator g, on the t-th trial for the j-th person, and $E_t[\ ]$ = expectation with respect to the propensity distribution, over trials of the indicator's administration. Thus, the model of measurement is $y_{tgj} = X_{gj} + \varepsilon_{tgj}$ where ε_{tgj} = the error for the g-th indicator committed by the j-th person on the t-th trial. In a population of persons it is assumed that $\text{Cov}(X_g, \varepsilon_g) = 0$ where $\text{Cov}(\ )$ is the covariance over trials and persons in the population.

Although classical true score theory provides the basis for much of the language of errors in psychometrics, it is found to be overly restrictive for most survey applications. The need to acknowledge possible biases in survey measurements is strong. In psychometrics different labels are given to this kind of measures. Most survey measures will be labeled as sets of *congeneric measures* or indicators in a *multiple factor model*, where measurement errors can yield biases in indicators of underlying constructs, and indicators can be influenced by various methods of measurement. A set of congeneric measures is a set of indicators of different underlying characteristics. The different characteristics, however, are all simple linear functions of a single construct.

An additional change when moving to the field of psychometric measurement is the explicit use of models as part of the definition of errors. That is, error terms are defined *assuming* certain characteristics of the measurement apply. In classical true score theory (a model), the most important assumption is that if an indicator were administered to a person repeatedly (and amnesia induced between trials), the mean of the errors in the respondent's answers would be zero. That is, the indicator is an "unbiased" measure of the respondent's characteristic, in the sense used by survey statisticians. (Here the parameter of interest is the single respondent's value on the construct.) This is not as strong an assumption as it may appear to be because psychometricians often view the scale on which their measurements are made as rather arbitrary (e.g., a 0 to 100 scale in their view has all the important properties of a scale from −50 to +50). This fact arises because the statistics of interest to psychometricians are not generally means or totals for persons studied, but rather correlation coefficients, relative sizes of variance components, factor loadings, and standardized regression coefficients. All of these statistics are functions of variance and covariance properties of measures, not of means (expected values).

Recall that with most of the psychometric perspective, only variable errors exist. Two terms, *validity* and *reliability*, are frequently used to label two kinds of variable errors. The notion of *theoretical validity*, sometimes *construct validity*, is used to mean "the correlation between the true score and the respondent's answer over trials." Note well that validity is not to be simply equated with *unbiasedness*, as used by survey statisticians. Under true score theory all the questions produce unbiased estimates of the person's true value on the trait. Instead, a completely valid question is one that has a correlation of 1.0 with the true score. Since validity is based on correlations, it is defined only on a population of persons (who vary on the true values), not on a single person. That is, there is no concept of a valid measure of a single person's attribute.

The other error concept used in psychometrics, when only one construct is under examination is *reliability*, the ratio of the true score variance to the observed variance (Bohrnstedt, 1983, p. 73). *Variance* refers to variability over persons in the population and over trials within a person. With this definition of reliability, it can be noted that the concept is not defined for measurements on a single person, only on a population of persons *and* reliability has a value specific to that population. Each population will produce its own reliability magnitude on a particular measure.

When dealing with populations of persons, true score theory adds another assumption about the errors, that their values are uncorrelated with the true values of the persons on any of the trials. With this assumption a simple mathematical relationship between reliability and theoretical validity follows, namely,

$$\begin{aligned}\text{Validity} &= \text{Cov}(X,x)/(\sigma_X\sigma_x)\\ &= \text{Cov}(X, X+\varepsilon)/(\sigma_X\sigma_x)\end{aligned}$$

$$= \sigma_X^2 / (\sigma_X \sigma_x)$$

$$= \rho_x^{\frac{1}{2}}$$

where Cov(X,x) = covariance of true scores and observed scores over trials and persons in the population and $\rho_x = \sigma_X^2 / \sigma_x^2$ is the reliability index.

The theoretical validity of a measure is merely the square root of its reliability. This relationship shows how different the concepts of reliability and validity, on one hand, are from variance and bias, on the other. Given this definition, the traditional statement that "no measure can be valid without also being reliable, but a reliable measure is not necessarily a valid one," is not true. In contrast, a sample statistic may have an expected value over samples equal to the population parameter (unbiasedness), but have very high variance from a small sample size. Conversely, a sample statistic can have very low sampling variance (from an efficient sample design) but have an expected value very different from the population parameter (high bias).

There are two terms, however, that need clarification. *Random measurement error*, as used by Andrews (1984), refers to "deviations (from true or valid scores) on one measure that are statistically unrelated to deviations in any other measure being analyzed concurrently." In the language of survey statistics this would refer to lack of correlation between two variables in their response deviations. *Correlated measurement error* means "deviations from true scores on one measure that *do* relate to deviations in another measure being concurrently analyzed." Thus, correlated measurement error means something very different than the *correlated response variance* used by survey statisticians. The latter refers to correlations among respondents contacted by the same interviewer (or other administrative units) in deviations obtained on *one* indicator. The correlated measurement errors could arise from the fact that two indicators share the effects of a common method of measurement. Such a viewpoint is central to the multitrait multimethod approach to estimating construct validity. This alternative measurement model retains all the basic concepts of error, but necessarily alters the computational forms of error estimates (see next section).

1.3 MEASUREMENT ERROR MODELS IN SURVEY METHODOLOGY RESEARCH

This section reviews the various techniques used in survey research to investigate measurement errors. For each of them it reviews measurement error models implicit or explicit in the technique.

1.3.1 Measurement Error Research in "Cognitive Laboratories"

Research on survey measurement error is sometimes difficult to perform in the uncontrolled setting of the survey interview. For example, if unusual additional data must be collected to study measurement errors, the very collection of those data may interfere with the survey interview. Although this is not the typical design, some studies use nonsurvey settings to learn about components of error.

The use of laboratory experiments to study survey measurement error properties is currently increasing because of the interest in possible insights from cognitive psychological theories into the causes of respondent errors in surveys. Cognitive theories have by and large been tested and refined in controlled laboratory settings, using student subject pools. The survey response task resembles some of those studied in the laboratory tests and has led to speculation about the direct applicability of the cognitive theories to the survey setting. Tests of these speculations are using both laboratory settings (Lessler and Sirken, 1985) and more traditional surveys. For example, one technique used in the cognitive research laboratories is protocol analysis or "think aloud" techniques (see Royston et al., 1986). This asks the respondent to verbalize his thoughts while he is answering the question. This attempt to observe the cognitions "as they happen" is an effort to learn what tasks the questions present to the respondent and the methods used to perform them. Respondents are sometimes asked to paraphrase the survey question in order to provide insight into their comprehension of the question's intent. Respondents are asked to assign levels of confidence to the answers they provide. The number of seconds passing between the delivery of the question and the response (a "response latency" measure) is recorded as an indicator of the depth of cognitive processing required to obtain an answer.

Most of the laboratory studies do not specify a formal measurement error model. However, the experimental situations often permit the assessment of the true value of measures taken on the subjects. In those cases the actual error in response is determined. Most often the implicit measurement model seems to focus on measurement bias, assuming that the experimental variables being manipulated produce constant effects over replications for all subjects. When the true values of measures are not easily observed, experimental treatment groups are compared. For example, alternative wordings of the questions might be compared to measure the effects of wording on the response task. These comparisons are used to estimate whether differences in statistics are to be expected under the different conditions. Few of the experiments attempt to measure variation in these differences over replications of the measurement. They thus exclude measurement error variability from their designs.

However, there do seem to be cognitive–theoretical assertions that imply impact on what might be called variable errors. For example, let us examine the work that shows that improved cuing improves recall of autobiographical material (Loftus and Marburger, 1983). The theories specify that in the absence of the thorough search of memory involved with dense cuing, access to memory is likely to be driven by the most available memory connected to those cues that are presented. Since this process is unique to each individual, the impact is a decrease in reliability of the measure, and an increase in response variance. In a formal model this might be expressed as $y_{jt} = X_j + \varepsilon_{jt}$, where ε_{jt} = deviation from the true value X_j for the j-th person at time t. In the absence of sufficient cues, the variance of the error term in the above expression is increased. In some recall tasks, the existence of certain cognitive heuristics will affect whether the $E(\varepsilon_{jt}) = 0$; that is, whether, in expectation, the respondent is expected to recall the event or whether there are cognitive shortcuts which will produce consistent underreporting or overreporting. For example, in recall of the date of a remembered event, there is a tendency to report dates closer to the interview (so-called "forward telescoping").

1.3.2 Measures External to the Survey

One way to estimate measurement error directly is to compare the result of a survey measurement to another indicator of the same characteristic, one likely to be more accurate. There are two methods that are often used to do this; one obtains external data on individual persons in the survey, another compares external population parameter values with survey-based estimates of the same quantity.

Perhaps the most common method of assessing measurement error is the use of a "record check study" or "validation study." Such a study generally assumes that information contained in the records is without error, that is, the records contain the true values on the survey variables. Record check studies are usually employed to estimate measurement errors that are conceptualized as biases. They often explicitly compare measurement procedures (e.g., different questioning strategies) to see which will match the record data most closely. That procedure is judged to have the lowest "measurement bias." Generally omitted from this view is the possibility of measurement variance differing between the two approaches. This is clearly not a limitation of record check study design.

Most could be used to estimate variable measurement errors, but most investigators who use record check studies conceptualize the errors as fixed over replications of the survey. One measurement error model that applies to these studies is $y_i = X_i + \varepsilon_i$ or response (for person i) = true value + random error, where measurement bias is defined by $E(y) - E(X)$. In most record check studies, there is little interest in $\mathrm{Var}(\varepsilon)$.

1.3.3 Split Sample Experiments Concerning Measurement Error

There are large literatures in survey methodology based on comparisons of results in random subsamples of a survey. For example, measurement errors associated with question wording are usually investigated this way. Schuman and Presser (1981) compared two forms of many different survey questions intended to measure the same construct by asking one question of one random half sample and another of the complement half sample. Survey estimates based on the two different questions were compared to measure the effect of question wording. Labeling the differences as functions of measurement bias, however, is not used by Schuman and Presser, nor by most other researchers in this tradition. Instead, the result prompts the use of social psychological and psycholinguistic arguments to explain the difference in results. They note different connotations of words. They call attention to the different grammatical structures in the questions. The arguments appear to imply that the questions measure different concepts, which although closely related to one another, are subtly different. This line of research is quite distinctive from the others reviewed in this section. Throughout most of the work the researchers do not explicitly address the question of whether the two forms of questions are viewed as measures of the same underlying concept, subject to different response biases, or whether they are really measures of two different concepts. Instead, the focus of the research is on learning what features of the questions produce the differences. By not explicitly addressing the underlying concept to be measured by the questions, notions of measurement error are not key to the research.

Split sample experiments are also used to study measurement biases associated with interviewer behavior. In a large set of experimental studies Cannell and his colleagues (see Cannell et al., 1981, for a summary) have explored the use of controlled interviewer behavior during the survey interaction. For an "experimental" random half-sample, the designs (1) ask for a commitment of the respondent to give correct answers to the interviewer's questions, (2) provide instructions to the respondent regarding the intent of individual questions, and (3) give standardized feedback to the respondents regarding the quality of their performance. This set of interviewer behaviors was compared to another (called the "control" procedure) which allowed more flexibility, but was the natural result of traditional interviewer training. In terms of the survey error structure we are using in this book, the work could be classified as attempting to estimate measurement bias associated with training and interviewing procedures.

Most of the measurement error literature employing split sample methods (a) do not use explicit measurement models in their work, and (b) do not include variable measurement errors in the approach. If it were to be specified, a formal measurement error model applicable to this work probably resembles the following:

response = true value + form effect + random error, or

$$y_{ij} = X_i + M_{ij} + \varepsilon_{ij}$$

where y_{ij} = response obtained for the i-th person using the j-th method or form

X_i = true value of the characteristic for the i-th person

M_{ij} = effect on the response of the i-th person of using the j-th method

ε_{ij} = deviation for the i-th person from the average effect of the j-th method.

Although there is no formal model employed, several features of the measurement do seem obvious from the literature. First, the method effect is generally not viewed to be constant over all subgroups [M_{ij} probably varies over different persons (i)]. *This is an important distinction between this literature and those we will encounter later.* A favorite correlate is educational attainment, most often used as a proxy indicator for cognitive facility with the question and answering task. Second, partially because of this, across persons the method effect is correlated to the true values of the concept being measured. Thus, in contrast to the models studied below, the covariance of the true values and the method effects are assumed to be nonzero, $\text{Cov}(X_i, M_{ij}) \neq 0$. For example, those who had actually experienced some health event were the only group susceptible to the change in interviewer behavior studied by Cannell, et al. (1981).

Research using split sample methods sometimes acknowledges variable measurement errors, but rarely does it attempt to measure them. (Indeed, the split sample design itself is poorly suited to such estimation). That is, using the model above, for example, there is little empirical work in this field concerning the variance of ε_{ij} over replications for the same person. There is no empirical investigation about whether the reliability (in the psychometric sense of the term) of one form of measurement is higher than another. That is, over repeated administrations of one question wording does the stability of answers differ from that of another wording? Instead, the emphasis is on error that is viewed to be constant over replications of the design.

1.3.4 Interpenetration for Estimation of Variable Measurement Error

Sometimes the investigator is faced with several possible measurement procedures that are viewed as essentially equivalent ways of obtaining the needed data. By "essentially equivalent" is meant that any differences in results arise from an unknown or unmeasured cause. We will discuss two examples of this — interviewers as a source of variable measurement error in surveys and the ordering of questions or response categories as a source of measurement error. In both cases, there will typically be no prior guidance for the choice of best procedure (i.e., the best interviewer or the best ordering).

To estimate the component of variance in survey statistics attributable to some survey design feature, interpenetration of the design is used. For example, random subsets of the sample are given different orderings of the questions or response categories. In one sense, we view this research as related to the split sample methods, when the number of methods is very large (instead of two forms of a question wording, we are interested in hundreds of possible orderings of the questions). In another sense, however, the approach is very different from that using split sample methods. This is because the researcher is not generally interested in measuring bias associated with each of the methods, but rather variation in results due to method. An effort is made to (a) include all possible, *equivalent* methods in each trial of the survey, (b) induce in each survey statistic the variability inherent in the measurement process, and c) measure the variance component in the survey statistics corresponding to the different methods. Thus, the goal is direct estimation of variable measurement error, not measurement bias.

The measurement models for interviewer effects are identical to those applicable to the split sample investigations. Typically, however, additional constraining assumptions are placed on terms in the model. These are generally not introduced because they enhance the realism of the model, but rather because they aid the empirical estimation of terms in the model. The Hansen, Hurwitz, and Bershad (1961) model, sometimes referred to as the U.S. Bureau of the Census response error model, is based on:

response = true value + interviewer effect + random error, or

$$y_{ijt} = X_i + M_{ij} + \varepsilon_{ijt}$$

where y_{ijt} = observed response on a survey item for the i-th person, the j-th interviewer on the t-th trial

X_i = the true value for the variable for the i-th person

M_{ij} = effect on the response of the i-th person of the j-th interviewer

ε_{ijt} = a deviation of the i-th person by the j-th interviewer on the t-th trial from the true value plus the average interviewer effect.

First of all, it should be noted how similar this model is to the one implicitly used by those employing split sample methods. The "interviewer effect" here replaces the form or method effect in the prior discussion. The difference between the two approaches lies in sets of assumptions about the terms.

The model does not focus on possible biasing effects of interviewers (M_{ij} terms) or of the individual errors (ε_{ijt} terms). That is, over all interviewers and persons, the expected value of the interviewer effect terms is zero; $E(M_{ij})=0$. Similarly, the expected value of the ε_{ijt} terms is zero; $E(_{ijt})=0$. These assumptions are not too consequential for the estimation of interviewer variance, but they do affect estimates of total mean squared error. That is, if the entire set of interviewers tend to induce overreports or underreports of the X_i, then the sample mean will be biased regardless of which interviewers are chosen for the study. This possibility is eliminated by the assumption of $E(M_{ij})=0$. Finally, the random error terms, ε_{ijt}, are viewed to be completely nonsystematic disturbances to the obtained response, whose values are unrelated to which interviewer is assigned or the true value for the respondent being examined. That is, $\text{Cov}(X_i,\varepsilon_{ijt})=0$ and $\text{Cov}(M_{ij},\varepsilon_{ijt})=0$. To any researcher who has observed interviewers in action, this assumption usually causes some concern. Many vivid examples of how interviewers adjust their behavior to idiosyncrasies of respondents are available. These cases violate this assumption to the extent that different interviewers adjust their behavior in different ways *and* different measurement errors result from these differences. These assumptions are desirable not because they enhance the realism of the model, but because they aid in the estimation of error terms.

The real focus of the model is describing the effects of interviewers on descriptive sample statistics, like the sample mean. If a simple random sample of persons of size n were selected, partitioned at random into J

subsets of equal size m, so that each of J interviewers would have an equal workload, then the variance of the sample mean would be

$$\begin{aligned}\mathrm{Var}(\bar{y}_t) = {} & [(N-n)/(N-1)]\mathrm{Var}(X_i)/n] \\ & + [\mathrm{Var}(M_j) + \mathrm{Var}(\varepsilon_{ijt})]/n][1 + \rho(m-1)]\end{aligned}$$

where $\rho = \mathrm{Var}(M_j)/[\mathrm{Var}(M_j) + \mathrm{Var}(\varepsilon_{ijt})]$ is an intraclass correlation coefficient associated with interviewers. This correlation measures the extent to which response errors ($y_{ijt} - X_i = M_j + \varepsilon_{ijt}$) made by respondents of the same interviewer are correlated. Do interviewers influence respondents to make response errors of a similar type? To the extent that this correlation is large and positive, they do. This is the origin of the term *correlated response variance.*

The above expression for Var ($\bar{y}_t$) has two components. The first term is the sampling variance of the mean, reflecting the fact that different samples of persons will produce different mean values. The second is a measurement variance or response variance component. The reader will note that the correlation of measurement errors within interviewer workloads acts to inflate the response variance component. It is a multiplier effect on the other term, $[\mathrm{Var}(M_j) + \mathrm{Var}(\varepsilon_{ijt})]/n$, sometimes called the *simple response variance.*

If each interviewer were assigned only one respondent, there would be no correlations among measurement errors across respondents. In that case only the simple response variance would exist. This is the case in which $m = 1$, so the multiplier effect of the interviewer influence does not exist. This observation also leads to the prescription to use more interviewers on a survey, each completing smaller numbers of interviews, in order to reduce the measurement variance associated with interviewers. Some confusion arises among most practitioners when that prescription is given. How can interviewer effects be eliminated by having each interviewer complete only one interview? Two comments are in order. First, there is no claim that interviewer effects are eliminated. The model merely states that the correlated component of the error is eliminated. Further, designs using very small interviewer workloads may experience higher *bias* terms because of reduced positive effects of field experience by the interviewers.

1.3.5 Repeated Measurements of the Same Persons

There is a large body of methodological literature in the social sciences which attempts to learn about measurement error by taking several measurements on the same person. In some designs these measurements are viewed as being fully equivalent to one another, as replicates which have the same expected values. Thus, they are used to investigate variability in response behavior over trials or replications (i.e., "measurement variance," "response variance," "reliability"). Sometimes reinterviews of the same person are conducted, asking the questions a second time. Sometimes multiple measures of the same concept are used in the same survey. Other designs attempt to understand measurement errors in one item by introducing other measures in the survey. These include direct questioning of the respondent about the understanding of a survey question, use of other questions to test comprehension of a question, recording of interviewer observations about the respondent behavior during the questioning, use of supervisory monitoring or taping of interviews, and questioning of other persons about the respondent.

These various techniques differ radically in the kind of information provided to the researcher about measurement error. The designs with repeated measures of the same quantity (or related quantities) on a person generally produce quantitative estimates of measurement error components; the designs with auxiliary measures about the questioning or response process generally provide indicators about the conformity of the measurement to the specified survey design (e.g., training guidelines for interviewer behavior, requisite respondent understanding of the questions). In this section we describe these different procedures and comment on their relative strengths and weaknesses.

Any reinterview design that produces estimates of response variance or reliability is based on a model of the measurement process (sometimes explicitly stated; other times only implicit in the procedures). A key premise is the stability of true values over trials. Although the concept of repeated trials does not require time to pass between trials, practical applications of the design do require repeated measurement over time. Thus, the researcher at some point is required to specify that the true value of the measure for a respondent is the same at the time of the first measurement and the second. That is,

response at trial 1 = true value + random error, or

$$y_{i1} = X_i + \varepsilon_{i1}$$

response at trial 2 = true value + random error, or

$$y_{i2} = X_i + \varepsilon_{i2}$$

where y_{i1} = the obtained response for the i-th person at time j
X_i = the true value of the measure for the i-th person
ε_{ij} = the response deviation from true the value for the i-th person at time j.

Note that relative to the prior models that include a method or interviewer effect on response, these simple models do not incorporate such an effect. The two models assume that $E[\varepsilon_{i1}] = E[\varepsilon_{i2}] = 0$, that is, the expected values of the two measures is the same quantity, X_i, a value that is constant over time. This assumption really has two components: (a) the assumption that the measurement on the second trial is an exact replicate of that on the first trial (e.g., same question, same context to the question, same interviewing procedures, same medium of data collection), and (b) the assumption that the underlying characteristic that is being measured has not changed between the first and second trial for the person being measured. With such a model, the difference between the two answers, $y_{i1} - y_{i2} = (X_i - X_i) + (\varepsilon_{i1} - \varepsilon_{i2}) = \varepsilon_{i1} - \varepsilon_{i2}$, is merely the difference over trials in measurement error. One added assumption permits an estimation of the variance in errors over trials. That assumption is that the errors are uncorrelated over trials, that $\text{Cov}(\varepsilon_{i1}, \varepsilon_{i2}) = 0$. With this additional constraint, the variance in errors over trials can be measured by $E[y_{i1} - y_{i2}]^2 = E[\varepsilon_{i1} - \varepsilon_{i2}]^2$, that is, the squared difference of observations can yield estimates of error variance.

Let us examine the plausibility of the various assumptions. Most can be violated through the respondent's memory of the first trial answer. That memory may prompt the respondent to give exactly the same answer (to appear consistent) or to give a different answer (to appear open-minded, flexible, especially on attitudinal items). Under this possibility, $E(\varepsilon_{i1}) = 0$, but once the ε_{i1} is committed, then ε_{i2} is determined to have a similar or dissimilar value. In this case, there is no guarantee that $E(\varepsilon_{i2}) = 0$. Further, the errors are correlated; $\text{Cov}(\varepsilon_{i1}, \varepsilon_{i2}) \neq 0$. Thus, on the respondent side, memory effects can violate the assumptions. On the interviewer side, knowledge of the prior answer can induce correlation of errors across trials. Note well that the violation of the assumption may yield an overestimate or underestimate of the error variance (the direction depends on the sign of the correlation of errors). For these reasons, reinterview studies to estimate measurement error variance must be used carefully.

Another approach also uses replicated measures to estimate measurement error, but uses multiple measurements of the same characteristic in a single survey. The approach is connected with a radically different set of estimation procedures, firmly based in the psychometric literature. In this approach measurement error associated with a particular method of data collection and/or a particular question can be assessed. *Measurement error* here is defined as a component of variance in the observed values of indicators, not corresponding to variability in the true values of the underlying measures. In the terminology of this book, it corresponds to variable errors of measurement only. *Method* has been used to mean the mode of data collection (personal visit, telephone, or self-administered), the format of the question (5-point scale, 10-point scale, open questions), the respondent rule (self response, proxy response), or various other characteristics of the measurement.

The key difference between this technique and those reviewed above is that a substantive theory is the basis on which the expected values of two survey questions are equated. Typically, there are no empirical data external to the survey to support this premise. Rather, the substantive theory asserts the equivalency, and measurement errors are estimated, *given the theory.*

When several characteristics are measured, each with the same set of different methods, the multitrait multimethod matrix results, originally proposed by Campbell and Fiske (1959). Implicit is a measurement model,

response = population mean + influence of true value + method effect + random error, or

$$y_{ijkm} = \mu_k + \beta_{km}X_{ik} + \alpha_{jm}M_{ij} + \varepsilon_{im}$$

where y_{ijkm} = the observed value of the i-th person using the j-th method to measure the k-th characteristic using the m-th indicator

μ_k = the mean value of the k-th characteristic for the population studied

β_{km} = "validity" coefficient for the m-th indicator of the k-th underlying characteristic

X_{ik} = for the i-th person, the true value of the k-th characteristic

α_{jm} = "method effect" coefficient for the m-th indicator of the j-th method

M_{ij} = for the i-th person, the common effect of using the j-th method

ε_{im} = a random deviation for the i-th person on the m-th indicator.

This is sometimes referred to as the *common factor model* (Alwin, 1974) because multiple indicators, each subject to "specific" error (method effect) and random error, are used for an underlying trait (factor). To the extent that the influence of the true value (X_{ik}) on the response is large, the indicator is said to have high validity. With such a model and appropriate numbers of indicators, traits, and measurement methods, empirical estimation of some components of measurement error is permitted. Specifically, the researcher can estimate what portion of the variance in observed indicators corresponds to variation in true scores (in X_{ik}), what portion to variation in method effects (in M_{ij}), and what portion to random error (in ε_{im}). The method effects concern the error-producing features of the differences among methods. The random errors contain all other sources of deviation from the true value.

With this measurement model, reliability is defined as

$$\mathrm{Var}(\beta_{km}X_{ik} + \alpha_{jm}M_{ij})/\mathrm{Var}(y_{ijkm})$$

and validity is defined as

$$\frac{\mathrm{Var}(\beta_{km}X_{ik})}{\sqrt{\mathrm{Var}(y_{ijkm})\mathrm{Var}(X_{ik})}}.$$

Note that, with the introduction of systematic error, the simple relationship between reliability and validity is broken. Reliability will always be higher than validity, as defined immediately above. Indeed, if β_{km} is zero (the indicator reflects only the method effect), then reliability can be 1.0, but validity would be zero.

There are several assumptions that are necessarily made about the nature of the terms in the model. Whenever survey estimates are based on models, it is important to conceptualize what the assumptions really mean. Sometimes it helps to give examples of violation of the assumptions:

(a) $E(\varepsilon_{im})=0$; that is, except for random variation, the model perfectly determines the value of an item for each person. (This assumption could be violated if some of the indicators also share some other source of influence; for example, for a study of three methods, if two of the methods were subject to more effects of social desirability than the third. Thus the model omits another term reflecting social desirability effects.)

(b) $\mathrm{Cov}(\varepsilon_{im},\varepsilon_{i'm})=0$, for all $i \neq i'$; that is, over persons in the population, deviations from one measurement model are uncorrelated with the deviations from another. (This assumption could be violated with the same case as above, a misspecification of the measurement model to omit another shared influence on two indicators.)

(c) $\mathrm{Cov}(X_{ik},\varepsilon_{im})=0$, for all k, m; that is, over persons in the population, the deviations from the expected values under the model are uncorrelated with the true values of the trait. (This can be violated if the nature of response errors is such that persons with high values on the trait tend to make different kinds of errors than those with low values; for example, if income is the underlying constraint, persons with high incomes may underestimate incomes and persons with low incomes may overestimate, regardless of what measurement method is employed.)

(d) $\mathrm{Cov}(M_{ij},\varepsilon_{im})=0$, for all i, j, m; that is, over persons in the population (i), the deviations from expected values under the model are uncorrelated with the magnitudes of the effects of the measurement method on the person's response to an item. (An example of violation of this assumption concerns the use of different scale labels (e.g., "very good" to "very bad" versus "excellent" to "poor".) The assumption is violated in the case that for some traits being measured, persons who interpret "very good" to mean something more positive than others tend to become subject to more social desirability effects than those who interpret it in the normal manner.)

The model does permit the true values of the traits to be correlated with the method factors over different persons. That is, persons with high values on the underlying traits may be more or less affected by the method of measurement than those with low values on the trait. (This flexibility makes the approach very attractive since it frees the investigator from one of the restrictive characteristics of most of the models in the interpenetrated designs for interviewer variance.)

1.4 CONDITIONING INFERENCE ON ASSUMPTIONS ABOUT ERRORS VERSUS CONDITIONING INFERENCE ON ERROR ESTIMATES

One interesting difference among those concentrating on variable measurement error is the active use of model-based estimates of the errors at the point of inference. This is typified by the use of the covariance structure models of Jöreskog and his colleagues (Jöreskog and Sörbom, 1989) and by certain random effects models in econometrics.

We note that these procedures yield estimates of the instability of analytic statistics (measures of relationships between traits) in a way not dissimilar to the use of probability sampling models to provide estimates of standard errors. The difference here is that the instability arises from measurement errors conceived to vary over administrations of the item. These measurement errors are specified to have a particular form by the analyst of the data. We note that this trend in the use of measurement error models in analysis (among modelers) has a parallel in the combination of response variance estimates with sampling variance estimates in descriptive statistics (see Bailey et al., 1978; Fellegi, 1974).

In contrast to adjustments by describers to inference for variable error, there are fewer measurement bias adjustments that are utilized prior to the inferential step. (This is especially curious in view of the fact of the common adjustment for bias due to errors of coverage or nonresponse in descriptive statistics (see Little and Rubin, 1987).) I have seen some attempts at this. For example, the Michigan Survey of Consumer Attitudes changed respondent rules in the 1970's from a head of household to a random adult. A transition period in which both rules were used provided an estimate of bias difference between the two rules. The estimate is used to estimate the index as if it were measured under the head rule.

There have been proposals for measurement bias adjustment in descriptive statistics. LaVange and Folsom (1985) build regression-based estimates of bias due to forward telescoping in unbounded measurement of victimization experience. They propose to use the adjustments to incorporate an unbounded recall interview used at the beginning of the seven-wave panel survey of the U.S. National Crime Survey program.

Others propose a slightly different approach to incorporating bias terms into estimation processes — turn the biases into variable errors by random assignment of measurement techniques to subsets of respondents. That is, assume there are two question wordings which are different indicators of the same construct. Ask the two questions of two random half-samples and incorporate the differences between the two groups into a variance estimator for the mean (thus reflecting both sampling error and measurement error). Such a design is practical when there is great assurance that the two items measure the same concept.

Finally, there are possible designs which use an expensive measurement technique (less prone to measurement bias) on a subset of the sample and cheaper one (more prone to bias) on the complement. If the cheaper measure is a subset of the more expensive measure (e.g., unaided recall of an event), then the subsample assigned the expensive method can be used to build an adjustment function between the cheap and the expensive method.

1.5 THE TRUE VALUE MORASS

A paper on measurement error across disciplines must contribute to the discussion of the existence of true values. Researchers dealing with attitudinal and other psychological attributes often assert that the notion of true values is not applicable to their work. They are comfortable with notions of validity and reliability, but not with bias, as a deviation from some true value.

Some assert a fundamental difference in measurement properties between, for example, an attitude toward the performance of the president in his job and the unemployment status. Both are concepts often measured in surveys. They note that there is no "objective" reality to the attitude. It is, they note, merely the answer to a question (e.g., "Do you think George Bush is doing a good, fair, or poor job as President these days?"). This may be something never considered by the respondent prior to the time the question is asked. There may be no relatively stable affective valuing of the respondent toward Bush's performance. The answer is judged to be of value only if it is related to other properties of the individual. For example, choice of political party, voting behavior, or consumer behavior might be related to those measures.

Let's examine similar properties of the concept of unemployment. The U.S. Current Population Survey, not unlike other labor force surveys in the world, attempts to measure attributes of a sample of people in the week preceding the survey. They ask, "What was . . . doing most of LAST WEEK — Working, keeping house, going to school, or something else?" Now let us examine whether there is a *fundamental* difference in the two concepts and their measurement problems.

First, whether the respondent was "working for pay" or not may have never been considered by him/her. What does "most" mean — the majority of the hours, the majority of the days? What does "last week" mean — the seven days previous to the interview, the week that began with Monday no more than 14 days previous to the interview? What does "working" mean — on the job, not on vacation from the job; performing tasks related to core goals of the work organization, not doing other activities at the place of employment or elsewhere; performing activities in return for money, not performing activities in return for goods or services. What is included in "something else" — does it include sleeping, traveling, etc.? The question may not have been considered by the person before the interview; the respondent may have not conceptualized his activities as "working" or "not working." The activities may be inherently enjoyable to the person in the manner of "play." There may be no stable property of the self schema of the individual that is related to "working" or "not working." The answer may merely depend on how the question is worded.

Others will acknowledge all of these limitations of the *question* but maintain that it attempts to measure something basically more *real*, that is, something that can be observed. They assert that if honest judges observed the individual over a course of time, they could agree on the labeling of the individual as "working" or "not working." They note that the problem with the measure is a reflection of the fallibility of words, not the uncertainty of the underlying concept.

But, others retort, how can you handle the situation of one performing activities after which another pays him money but the respondent answers he is not working and is active for the fun of it. (Maybe that could be handled with a better definition for that circumstance, some reply.) Can the person be "working" if he is not aware of it? How can judges make a determination of the respondent's working status if they can only observe him?

There is no end to this debate. We could extend the debate to physical properties of inanimate objects. For example, is there a fundamental difference between the "weight of a potato" and the attitude toward how the President is doing his job? Be careful. Keep in mind that measured weight varies by humidity, problems of defining where the potato begins and ends (do you count dirt on the potato and what is dirt?). Keep in mind that gravity varies over parts of the earth.

I see no fundamental measurement difference between many of the attributes we call "facts" and those we call "attitudes." Each gain conceptual clarity as they are defined in relation to other attributes, whether observable or not. The import of this perspective is that notions of fixed and variable errors can be made useful in attitudinal measurement to the extent that the relationships to other variables are well mapped.

1.6 CONCLUSIONS

One conclusion from the diverse concepts of measurement error discussed above is the observation that notions of bias and variable errors take their meaning from uses of the data. To trivialize the point, reported years of formal education has one set of error properties as an indicator of experience with formal schooling, and another set of properties as an indicator of cognitive abilities. Or from another perspective, the identification number for a data set record can be considered to be a measure of income, one with very deep measurement error problems.

Some of this viewpoint is at the heart of conceptualization of measurement errors. True score theory notes that $E[y_{jt}+\varepsilon_{jt}]=X_j$ by assertion. That is, if y is not a unbiased indicator of X, then it is an unbiased indicator of some other concept, say X'. This viewpoint is seen in real data. For example, Schuman and Scott (1987) note that some wordings of attitudinal questions can easily be interpreted as "valid" or "unbiased" indicators of subdimensions of a complex concept, while being less desirable indicators of the whole.

The link of measurement error to uses of the data may be one source of debate among researchers about levels of measurement error. This is one time that the question "What is the statistic of interest when errors are being considered?" is helpful in debates about measurement error (see Section 1.2). The tailoring of measurement error models to the substantive use of the data makes this linkage explicit, and this is a great benefit of the approach.

1.7 ORGANIZATION OF THIS VOLUME

The chapters that follow comprise five main book sections dealing in detail with many of the issues raised in this chapter. The initial chapters in each section (Chapters 2, 8, 13, 19, and 24) are intended to provide more specific overviews of subfields. They are organized about two main dimensions: sources of measurement errors (the questionnaire, respondents, interviewers, and interaction of the three) and statistical models estimating measurement errors. Thus, the volume reflects diverse perspectives used in the design and conduct of surveys, and in the analysis of survey data. The reader will note that the authors bring with them different conceptualizations of measurement error related to the data. These differences are reflected in the research they mount to investigate the quality of survey data, in the models they specify to reflect the errors of interest, and in their incorporation of errors into their analysis. Many of these differences are, however, superficial. For example, in Sections A–D many chapters will address multiple error sources simultaneously, [e.g., interviewer and respondent (Chapter 23), mode of data collection and questionnaire (Chapter 5), interviewer and questionnaire (Chapter 22), and respondent and questionnaire (Chapter 9).] In addition, many of these chapters, although focused on cause of measurement error or efforts to reduce measurement errors, utilize statistical models to estimate the errors. Finally, Section E, focusing on statistical models of measurement error, grapples with how the psychological and sociological approaches toward measurement error reviewed in Sections A–D might be summarized in mathematical functions. This volume contains one of the few presentations of these different disciplinary approaches to survey measurement error. The reader will confront many of the conceptual, linguistic, and statistical differences reviewed above. Because the chapters were jointly presented at a single conference there have been unique attempts to bridge the disciplinary boundaries (see especially Chapters 19, 22, 24, and 25).

The value of this volume's interdisciplinary approach is that it may serve to stimulate a blending of methods and conceptual strategies. This blending could profitably arise from a joint focus on the joint effects of design features on measurement, e.g., joint studies of interviewer effects and question effects (see Chapter 19), as well as a blending of psychometric and statistical survey models of measurement error (see Chapter 24). Such blendings may reduce the conceptual and linguistic problems now facing researchers working with survey data. It may also lead to more successful efforts to reduce those errors and adjust inferences from data subject to those errors. That is a benefit worth pursuing.

SECTION A

THE QUESTIONNAIRE

CHAPTER 2

THE CURRENT STATUS OF QUESTIONNAIRE RESEARCH

Norman M. Bradburn
University of Chicago and NORC

Seymour Sudman
Survey Research Laboratory, University of Illinois at Urbana–Champaign

2.1 A BRIEF HISTORY OF MEASUREMENT ERROR RESEARCH ON QUESTIONNAIRES

Almost from the start of public opinion polling in 1935, there was a recognition that question wording had a significant effect on the answers that people gave. George Gallup was a strong proponent of the use of split forms to measure the effects of questionnaire wording. Cantril (1967) gave an interesting example. In June, 1941, after Germany invaded Russia, Gallup asked half of the respondents the question:

> Some people say that since Germany is now fighting Russia, as well as Britain, it is not necessary for this country to help Britain. Do you agree or disagree with this?

The other half of the respondents were asked:

> Some people say that, since Germany will probably defeat Russia in a few weeks and then turn her full strength against Britain, it is more important than ever that we help Britain. Do you agree or disagree with this?

Support for aid to Britain was 73 percent for the first question (disagree) and 71 percent on the second question (agree). It was quite clear in this case that public opinion was solidly behind Britain and that measure-

Measurement Errors in Surveys.
Edited by Biemer, Groves, Lyberg, Mathiowetz, and Sudman.

ISBN: 0-471-53405-6

ment errors were insignificant. President Roosevelt when shown these results was particularly relieved that his policies were supported, regardless of question wording.

Many other split ballot studies, however, revealed large differences between different question wordings, especially for complicated questions and on new issues where people had little information and had not yet made up their minds. A famous example that is often cited is given by Rugg (1941). When asked "Do you think the United States should forbid public speeches against democracy?," 54 percent of respondents said *yes*; when asked "Do you think the United States should allow public speeches against democracy?," 75 percent of respondents said *no*.

Initially it was thought that questions asking about behavior were not subject to questionnaire effects, but it was soon found that behavior questions were also impacted by question wording and context. These effects were first found when different organizations used different wordings and obtained different results. Ultimately this led to split-form experiments.

The U.S. Bureau of the Census measured the reliability of its questions with extensive reinterview programs (U.S. Bureau of the Census, 1963, 1964). Changes in status between the two interviews clearly indicated measurement unreliability although the causes of this unreliability then needed to be determined by additional research.

The standard method for measuring validity was to check responses against records. This procedure was used for surveys of hospitalization and crime victimization, although it was soon recognized that records had their own problems and the limited availability of such records prevented widespread use of record checks (U.S. National Center for Health Statistics, 1965, 1968, 1969).

In 1972, Sudman and Bradburn developed a model of response effects in surveys and quantified their model by a meta analysis of some 900 studies in the literature (Sudman and Bradburn, 1974). Their model explored causes of response effects that are discussed in other sections of this book, but it also focused on questionnaire variables such as question length, the difficulty of the words used, open versus closed questions, the position of the question in the questionnaire, the salience of the questions and the effect of aided and bounded recall procedures. Although they examined both behavior and attitude questions, Sudman and Bradburn gave more attention to behavioral questions.

The use of secondary analysis methods to study questionnaire effects has continued to the present. In this section, Chapter 6 by Molenaar examines the effects of content, wording and position in the questionnaire on response.

Schuman and his colleagues at the Survey Research Center, University of Michigan began a research program in 1971 to study questionnaire effects on attitude questions and summarized their results in their book, *Questions and Answers in Attitude Surveys* (Schuman and Presser, 1981). The topics they considered included question and response order effects, open and closed questions, the effects of middle alternatives, balance in questions and the tone of wording.

An exciting development in the past decade has been the recognition that both cognitive psychologists and survey researchers have contributions to make to the others' research. Cognitive theories have already led to a fuller understanding of the task of a survey respondent (see Chapters 8 and 10) and how aspects of the questionnaire, especially context effects, affect responses (Hippler et al., 1987; Sudman and Schwarz, 1989).

As we learn and understand more, we realize how enormously complex the survey method really is, as complex as human speech and social interactions. It is unrealistic to expect that most measurement errors will ultimately be eliminated. Optimistically, we should attempt to reduce those effects that can be reduced using the best survey methods and to understand and be able to measure the size of effects that cannot be reduced.

2.2 THE SOCIAL AND COGNITIVE CONTEXTS OF THE QUESTIONNAIRE

The design of questionnaires typically focuses attention on the measurement of the variables of interest in the study. The investigators' concern is naturally on making sure that the questions adequately reflect the issues being studied and measure what they want to measure. In so doing, investigators may forget that the survey interview takes place within a social context which partakes of larger social and cognitive processes that may affect it.

Surveys are a special type of social activity. They involve contact between people who typically are strangers and an interview which is a special type of conversation between individuals who speak more or less the same language. While the selection of respondents, and the way in which they are contacted and induced to participate in the survey, are governed by general rules of survey design and practice, the process also takes place in the context of a larger set of social norms governing relations between strangers, general canons of politeness and ways of treating strangers. Understanding the way in which surveys fit into these broader social norms helps us understand how to increase participation and make the interview more pleasant for respondents.

The survey interview is a special kind of conversation because it has a definite structure and rules of its own that make it different from casual conversations between friends or instrumental conversations such as those that occur at work or in economic transactions. The survey interview consists of a social situation in which one person (the interviewer) asks questions to another (the respondent), and the questions typically have been written by a third party, the researcher. Although it may be a special kind of conversation and have some rules of its own, it is also a conversation and is subject to more general rules that govern voluntary social interactions. (See Chapter 19.)

Everything that goes on in an interview has the potential to influence respondents' answers. The wording of the questions, of course, is of paramount importance, but many other aspects of the situation may also be important — where the interview takes place, the perceived sponsorship of the survey and the purposes to which the information is put, whether the data are perceived to be kept confidential or not, the professionalism of the interviewer — to mention only a few. Many of these aspects of the situation will have different effects depending on the characteristics of the situation and the content of the questionnaire. And, of course, there will be individual differences in the degree to which respondents are influenced by any of these factors. They are, however, factors which we need to understand better if we are to improve our measurement techniques.

Although most of the discussion in this chapter and in the book relates to household and population surveys, we should recognize that surveys of businesses and institutions are also of very great importance and that such surveys share common measurement problems with household and population surveys, but also have unique problems. In this section, Dutka and Frankel, in Chapter 7 discuss these issues. Chapter 12 applies cognitive principles used in models of household survey response to establishment surveys.

Several models for understanding the question/answer interchange within a more general framework of information processing have been developed. These models are discussed by several authors in this book so we do not repeat that discussion here (see Mathiowetz et al., Chapter 8, and Schwarz and Hippler, Chapter 3).

2.3 HOW THE QUESTIONNAIRE AFFECTS RESPONSE

2.3.1 Question Wording

2.3.1.1 Understanding Question Words

We are all aware that language is basically ambiguous and that words can have different meanings to the person who says them and to those who hear them. This is one of the principle sources of humor. In ordinary conversation, people who are well acquainted and speaking about familiar topics will usually understand each other, and if in doubt, will

Table 2.1. Means and Standard Deviations of Responses to: "How Often Were You Bored or Excited?"

Response Categories	Excited	Bored
Means		
"Not too often"	6.65 (327)	4.15 (552)
"Pretty often"	12.95 (495)	13.72 (127)
"Very Often"	17.73 (247)	17.39 (99)
Standard Deviations		
"Not too often"	8.57	5.71
"Pretty often"	12.11	10.64
"Very often"	15.00	13.09

ask clarifying questions. If misunderstandings occur even here, it is evident that many more will occur in the survey situation, where the conversation is between strangers and where the setting discourages questions from the respondent. Given a question with ambiguous words, respondents may not ask for clarification, but may interpret the question as best they can and then answer the question.

Belson (1981, 1986) has been one of the most active researchers in demonstrating that significant respondent misunderstandings occur. This is demonstrated by intensive post-interview evaluations, where respondents are asked what they thought a word or question meant. As examples of the kinds of different interpretations respondents give, Belson (1986, p.13) cites the following (we paraphrase slightly):

> The question was "do you think children suffer any ill effects from watching television programs with violence in them?" "Children" was interpreted from babies to teenagers — even up to the age of 20 years; some respondents took the word to mean your own children or grandchildren, nervous children or children who have been brought up properly.

Bradburn and Miles (1979) asked respondents in a post-interview to define what the words "very often," "pretty often" and "not too often" meant to them in terms of days per month. These words had been used as answer categories in questions dealing with feeling states such as "How often were you bored or excited?" The data are shown in Table 2.1. While the words convey some meaning, it is also clear that the standard deviations are about as large as the means, suggesting great variability in how these words are interpreted.

Note that the examples we have given are not especially difficult words. Payne (1951, Chapter 10) in his classic book, *The Art of Asking Questions*, gives a Rogue's Gallery of Problem Words:

> *any, anybody, anyone, anything*: may mean *every, some* or *only one.*

fair: meanings include *average, pretty good, not bad, favorable, just, honest, according to the rules, plain, open.*

just: may mean *precisely, closely, barely.*

you: may be used as a singular or collective pronoun.

Currently, researchers who use cognitive laboratory methods ask respondents either during or after the interview what the question means. It must be pointed out, however, that it is not always possible to change a question so that it will be completely unambiguous. Sometimes trying to clarify the meaning of words requires significantly longer and more complex questions which create their own difficulties.

2.3.1.2 Words as Cues

One situation where longer questions or a series of questions have been shown to improve the quality of responses is when the words provide cues for respondents that both help improve memory search as well as understanding the question. There are several reasons why this is the case.

Different respondents may store the same information in different ways and even the same respondent may store the information requested in a single question in many different locations. A well-known example is the question that asks for information on total household income. While some people may store this information as a total, many more will store their own wages and salaries separately from the wages and salaries of other household members and will store information on interest from savings accounts, dividends from stocks and other investments all in separate locations.

It is also well known (Krosnick, 1991b) that respondents usually do not make strenuous efforts to search their memories completely. Rather, they do just enough memory search to arrive at some kind of estimate. Ordinarily, this means accessing the most readily available information. The words used in a question, and in earlier questions, provide cues that help to make more information easily accessible and, thus, improve reporting.

As one further illustration of research in this area, Bradburn, Sudman, et al. (1979) demonstrated that there were substantial increases in the level of reporting of drinking of beer and liquor, use of drugs, and sexual activity when longer questions with cues were used.

2.3.1.3 The Emotional Content of Words

We all recognize that some words have strong emotional content and that this content can strongly influence responses to questions. The Rugg

(1941) "Forbid–Allow" example given at the start of the chapter was an early demonstration of this. Indeed when surveys are used for persuasion rather than information, opposing sides on an issue will frequently obtain substantially different results because of the different words used. A well-known finding (Roshco, 1978) was that the word "Communist" had a strong negative content, at least until Glasnost. A proposed action described as intended to "fight Communism" always obtained higher support than the same action with these words omitted.

Some might consider this result as a special case of attribution theory. If an action or opinion is attributed to someone or something that is favorably regarded, such as the President of the United States or the Supreme Court, it will receive higher favorability ratings. Thus, mention of the Supreme Court decision increases agreement with a woman's right to an abortion.

2.3.2 The Question

2.3.2.1 Length of Questions

Common sense suggests that questions that are short, use simple language and avoid too many qualifying clauses or phrases will be easiest for respondents to comprehend. Indeed experienced questionnaire designers would agree that common sense is a good guide in this case. The use of such questions, however, leads many critics of surveys to charge that surveys, particularly public opinion surveys, simplify complex issues, tap opinions superficially, and force respondents to reduce the subtleties of their thoughts to a few simple response categories. There is clearly tension between the desire to simplify language, shorten the questions to promote comprehension, and reduce the mental work needed to interpret questions and the desire to limit the possible interpretation of questions so that respondents are all thinking about the same thing when answering the question.

Research on question length has mostly been concentrated on the effects of question length on the accuracy of behavioral recall. Marquis and Cannell (1971) showed that longer questions actually produced better behavioral reports of symptoms and doctor visits than shorter questions. As noted earlier, similar results were found by Bradburn, Sudman, et al. (1979) for questions on sensitive topics such as alcohol and drug use. Longer questions typically provide more cues to stimulate the memory search and more time for the respondents to search their memories. Such effects are consistent with what would be expected from memory research, which indicates that more cues and longer time for recall produce better memories.

2.3.2.2 Specificity of the Question

Questions sometimes use concepts or images that may have specific meaning for the investigator but, in fact, have a number of diverse meanings for the public at large. For example, Fee (1981) found that the term "big government" evoked four quite different images and meanings for respondents and that it was not clear that respondents were all responding to the same question meaning, even though they had all been asked the same question wording. Smith (1981) found similar disparate meanings for the word "confidence."

The fact that there can be multiple meanings to the same question increases the importance of adequate developmental work for questionnaires. There are several techniques, notably focus groups and cognitive interviewing, that can be used to investigate the way in which people think about topics that are the subject of a questionnaire. Use of such techniques on the one hand helps us develop questions in a language that respondents use to think and talk about the subject and, on the other hand, allows us to test out questions to see that the phraseology used in the questions are understood by respondents in that way we intended. The process is similar to that of translating a questionnaire into another language and then backtranslating it into English.

Concern over possible misunderstanding of questions has led some investigators to write very detailed questions that try to specify the exact referent of the questions. Suchman and Jordan (1990) have recently criticized survey questions such as those in the U.S. Health Interview Survey, that attempt to make a number of specifications within a question so that the respondents will be answering the question with the same definitional framework as that of the investigator. They argue that making the question carry all of the burden of standardizing the meaning of questions produces complex questions that violate the normal rules of conversation, which allow for more give and take between conversational partners in order to clarify the meaning of questions.

A possible solution to this dilemma is to simplify the questions and give the interviewers more freedom to probe and work with the respondents to ensure that they are understanding the questions in the way they were intended. Moving in this direction implies a more active role for the interviewer as a collaborator of the investigator than has been the case. It also requires that the interviewer be well trained on the purpose of the questions and what is being measured.

2.3.3 Response Categories

2.3.3.1 Response Categories as Cues

Response categories in a closed end question have multiple and sometimes unexpected effects not only on the way respondents answer that question but also the way they answer subsequent questions. Chapter 3 (this volume), by Schwarz and Hippler, discusses this topic in detail. Using the cognitive framework, responses may be seen as cues for understanding the question, generating a response, formatting an answer and editing that answer. From this perspective, response categories provide additional information similar to that contained in other questions (see the discussion of context effects below) and in interviewer appearance and behavior which respondents use as cues.

A question may have several dimensions; the response categories tell the respondent in what dimension the researcher is interested. While not all respondents understand the meanings of the response categories, there are fewer misunderstandings with closed than with open questions, if the categories have been carefully developed and tested.

Response categories also provide cues that aid in retrieving stored information. A well-known illustration is that it is far easier for us to remember newspapers and magazines we have read in the recent past if we are given a list, rather than simply using unaided recall. A corollary of this fact is that items not included in the response categories but included in an "all-other" category will be significantly less likely to be recalled (Belson and Duncan, 1962). A significant problem occurs if some respondents are unable to fit their answers into the response categories provided. This can occur if the researcher has not done sufficient thinking and testing of the alternatives, or sometimes because the researcher wishes to force the respondent to choose one of the answers given. When this happens, respondents become frustrated and, especially on mail surveys, refuse to cooperate.

For behavioral questions, respondents assume that the answer categories given reflect the researcher's knowledge of how most people answer (Schwarz and Hippler, 1987). Thus, the middle category is taken to be an average response and the end categories are assumed to be unusual. Especially for sensitive behaviors, response categories may be used by the respondent to edit reported behavior. If the level of actual socially undesirable behavior falls in the top category, respondents will underreport by selecting a lower category (Bradburn, Sudman, et al., 1979). On the other hand, if that level of behavior is in the middle category, more respondents will be willing to report it (Schwarz and Hippler, 1987).

Respondents will also infer their place in the real distribution of behavior by comparing their pre-edit answer to the response categories and thus concluding they are above or below average on a given activity. This conclusion may then have an impact on how they respond to subsequent questions.

2.3.3.2 Scales

Most users of numerical scales believe that the optimum number of points on a scale ranges from 5 to 11. Theoretically, the greater the number of points, the more powerful the scale is in discriminating, but respondents soon become unable to make fine distinctions and round off. Thus, on 100 point thermometer scales, most answers are at 25, 50 and 75 or end in zero (e.g., 30, 40, 60).

Some researchers continue to use scales with an even number of points to force an answer, but the work of Schuman and Presser (1981) suggests that the distributions of positive–negative are similar in odd and even scales and suggests that the middle category does provide information on those who are really undecided.

2.3.4 The Questionnaire

Questions are not asked in isolation but are grouped together in a questionnaire. Sometimes a questionnaire contains questions more or less on the same topic. Other times questionnaires are made up of groups of questions on quite unrelated topics, a so-called omnibus questionnaire. As noted earlier, the interpretation of questions often is influenced by the context in which they are asked. Other questions that have been asked provide what may be one of the more powerful contexts for interpreting the meaning of questions. Thus it is not surprising that the study of question order effects is one of the most active areas of methodological research. Smith in Chapter 4 (this volume) in this section points out, however, that context effects are fairly rare in real-world surveys.

One of the more well established findings of research on order effects is that when a general question and a more specific question on the same topic are asked together, the general question may be affected by its placement, but that the specific question is not. An example is:

> Taking things all together, how would you describe your marriage? Would you say that your marriage is very happy, pretty happy, or not too happy?
>
> Taken all together, how would you say things are these days? Would you say that you are very happy, pretty happy, or not too happy?

Answers to the general happiness question are affected by its placement, but answers to the marriage happiness question are not. There are still a number of mysteries about this type of effect, and we are not yet able to reproduce it reliably. It does seem likely, however, that the effects that do occur come at the interpretation stage; that is, the interpretation of the general questions under some circumstances, which are still poorly known, can be altered by the preceding questions.

Most studies of context effects have been with questionnaires administered by interviewers in which respondents do not know what subsequent questions will be. With mail and other self-administered questionnaires, however, where respondents can know all the questions before they start answering, this knowledge has an impact on context effects, which is discussed in this section by Dillman and Tarnai in Chapter 5.

Answering questions on one topic may prime memories that make them more accessible for retrieval when answering subsequent questions. Recent work by Tourangeau and his colleagues (1989) suggests that preceding questions can prime schemata — a set of closely related arguments — that lead to different responses on subsequent questions. For example, asking attitude questions about welfare cheating, as contrasted with questions about helping the poor, before asking questions about support for welfare programs affects the responses to the latter question.

One of the mysteries of these types of order effects is their differing direction. Sometimes the effect is one of assimilation, that is subsequent responses are in the same direction as the preceding ones, and sometimes the effect is one of contrast, that is, subsequent responses are in the direction opposite to the preceding ones. Both assimilation and contrast effects have been reported in the literature. (See Chapter 3.)

Order effects may also arise from judgments about the logical consistency of answers to related questions. A good example of this type of effect was given by Cantril (1944), who found that support for allowing Americans to fight in the German army at the beginning of World War II (before the United States became involved) was affected by whether the question came before or after a related question about attitudes toward Americans fighting in the British army. When the German question came after the British question, there was greater support for allowing Americans in the German army than when it came before the British question. These effects seem to occur when two questions are asked together about the same policy but in two different applications which have different levels of support. The item that has the least support is affected by its placement relative to the more popular item. When it comes after the more popular item, it receives more support than when it comes before the more popular item.

In addition to the cognitive effects of order, placement in the questionnaire may affect item responses because of changes that have gone on during the interview. Typically interviewers develop greater rapport with respondents as the interview progresses, and respondents may be motivated to be more open or work somewhat harder when good rapport has been established. Questionnaire designers usually put more sensitive questions toward the end of the interview, by which time, they believe, good rapport will have been developed.

On the other hand, if the interview is long, fatigue, both for the respondents and the interviewers, may be a factor in reducing the willingness to respond fully or to undertake difficult cognitive work such as extensive behavioral reports on activities in the distant past. Fatigue effects are difficult to demonstrate because there are so many other factors involved in complex studies with long interviews that might affect the responses. In long questionnaires, however, it is a factor to keep in mind when deciding on the placement of questions.

2.4 THE FUTURE OF RESEARCH ON QUESTIONNAIRES

One need not have a crystal ball to predict that cognitive psychological theories and insights will continue to play a major role during the next decade and beyond in questionnaire research. It is unrealistic, however, to think that we shall discover perfect methods of asking difficult questions that require major cognitive effort. A more modest goal should be achievable — to understand better how respondents answer questions so that the cognitive difficulty of questions can be reduced and the magnitude of the remaining measurement errors reasonably well estimated.

The key problem in questionnaire design will remain the enormous complexity of human language. For a long time we have hoped that linguistic research would provide new insights into the question and answer process, but these hopes have not been fulfilled. As linguistics continues to develop, we continue to hope that there will arise the same fruitful cross-pollinization between linguistic and questionnaire research that has already occurred with the cognitive psychologists.

CHAPTER 3

RESPONSE ALTERNATIVES: THE IMPACT OF THEIR CHOICE AND PRESENTATION ORDER

Norbert Schwarz and Hans-Jürgen Hippler
Zentrum für Umfragen, Methoden, und Analysen (ZUMA)

3.1 INTRODUCTION

That the choice and presentation order of response alternatives in a survey can greatly influence the obtained results is no news and has been widely documented in the survey literature (cf. Payne, 1951; Sudman and Bradburn, 1974; Molenaar, 1982; Hippler and Schwarz, 1987). However, the underlying cognitive and communicative processes are not well understood, which makes it difficult to predict which effects may be expected under various conditions. In the present chapter, we will review a cognitive research program that explores the psychological processes that mediate the impact of response alternatives on respondents' reports. In addition to summarizing parts of our own research, we provide a selective review and conceptual integration of the available literature on the effects of choice and presentation order of response alternatives.

We begin with a comparison of open and closed question formats, focusing on the information that respondents extract from different types of response alternatives provided to them. Subsequently we explore the

The reported research was supported by grants Schw278/2 and Str264/2 from the Deutsche Forschungsgemeinschaft to N. Schwarz and F. Strack, grant SWF0044 – 6 from the Bundesminister für Forschung und Technologie of the Federal Republic of Germany to N. Schwarz, and ZUMA's "Cognition and Survey Research" program. We thank George Bishop for helpful comments on a previous draft.

Measurement Errors in Surveys.
Edited by Biemer, Groves, Lyberg, Mathiowetz, and Sudman.

ISBN: 0-471-53405-6

impact of the order in which response alternatives are presented and outline a cognitive model of response order effects.

3.2 RESPONSE ALTERNATIVES: WHAT THEY MAY TELL YOUR RESPONDENTS

The advantages and disadvantages of open- and closed-response formats have been the topic of considerable debate in survey research since its early days (e.g., Lazarsfeld, 1944; Krech and Crutchfield, 1948). As Converse (1984) noted, the debates were primarily based on "enlightened common sense, in-house experience, and practical constraints" (p. 279), rather than on systematic experimental research. As a result, the methodological debate lacked a coherent theoretical framework and the actual data base is still surprisingly small, given the importance of the issue.

Survey researchers typically assume that response alternatives constitute a "measurement device" that respondents use to report their answer. According to this assumption, respondents recall an opinion from memory, or compute it when asked, and select a response alternative to communicate their opinion to the researcher. As long as the response alternatives allow them to communicate their opinion, they are assumed to have little impact on the obtained results. Systematic bias is only expected if the response alternatives are too constrained, thus providing no opportunity to communicate the "true" answer. This assumption is captured in the concept of "question constraint" (see Schuman, 1985), which holds that respondents assume that they have to work within the set of response alternatives provided to them.

From a cognitive point of view, the concept of "question constraint" addresses only parts of the relevant processes. The survey interview is best considered as an ongoing conversation that includes the intertwined tasks of question comprehension, recall of information from memory, computation of an answer, and reporting of this answer to an interviewer (see Feldman, 1991; Strack and Martin, 1987; Strack and Schwarz, in press; Tourangeau and Rasinski, 1988, for more detailed discussions). Much as in other forms of social discourse in everyday life, all contributions of the participants to the conversation may influence each of these stages of the question answering process. In the survey interview, the contributions of the interviewer/researcher include the response alternatives provided to respondents, and respondents treat these contributions as they treat any other contribution to an ongoing conversation. That is, they proceed on the basis of the cooperativeness principle that governs the conduct of conversation in everyday life (cf. Grice, 1975; Clark, 1985). This principle holds that every contribution should be relevant to the aim of the ongoing conversation, and that speakers should not provide information that is irrelevant to the task at hand. Moreover, speakers are required to make their contributions informative, that is, to provide information that the recipient needs, rather than information that the recipient already has — or may take for granted anyway. Conforming to these conversational norms requires a considerable degree of inference to determine which information is "informative" in the specific context given (cf. Schwarz, Strack, and Mai, 1991; Strack and Schwarz, in press). In the survey interview, this context is, in part, constituted by the response alternatives. Accordingly, response alternatives not only serve to record respondents' answers, but also to define respondents' substantive tasks. They therefore influence respondents' interpretation of the questions and determine which information they use in making a judgment and which responses they consider appropriate to report.

Whereas researchers in experimental (social) psychology pay close attention to the information that their subjects may extract from the research procedures used (cf. Wyer, 1974), the informative functions of apparently formal features of questionnaires have received little attention in social and psychological research. It is these informative functions that are of key interest in the first part of this chapter.

3.3 KNOWLEDGE AND OPINION QUESTIONS

3.3.1 Open- vs. Closed-Question Formats

Experimental studies on the impact of open- and closed-response formats in the domain of knowledge and opinion questions converge on the finding that open- and closed-response formats may yield considerable differences in the marginal distribution as well as in the ranking of items (e.g., Bishop, et al. 1988; Schuman, and Presser, 1977). On the one hand, any given opinion is less likely to be volunteered in an open-response format than to be endorsed in a closed-response format if presented. On the other hand, opinions that are omitted from the set of response alternatives in a closed format are unlikely to be reported at all, even if an "other" category is explicitly offered, which respondents in general rarely use (Bradburn, 1983; Molenaar, 1982). Several processes are likely to contribute to these findings.

First, precoded response alternatives may remind respondents of options that they may otherwise not consider. From a cognitive perspective, open-response formats present a free-recall task to respondents, whereas closed formats present a recognition task. As has been found in other domains of research, recognition tasks have been found to result in higher degrees of recall (cf. Smyth, et al. 1987). For example, when respondents are asked, "What are the most important problems facing the country today?", a precoded list of response alternatives may direct respondents' attention to issues that may otherwise not have come to mind. When reminded of it, respondents may endorse the issue as important. Accordingly, Schuman and Presser (1977) observed, for example, that "security" was mentioned 13 percent less often as an important feature of a job when respondents were asked in an open rather than a closed format. Thus, open-response formats are more adequate when the investigator is interested in the salience of an issue, where the order in which a respondent retrieves different issues and the total number of respondents who retrieve a particular issue are of primary interest (cf. Bodenhausen and Wyer, 1987). Closed formats, on the other hand, are more appropriate when the investigator is interested in a fairly complete evaluation of a large set of issues to determine their relative importance.

Second, respondents are unlikely to report spontaneously, in an open-answer format, information that seems self-evident or irrelevant. In refraining from these responses they follow the conversational maxim that an utterance should be informative, as discussed above. This results in an underreporting of presumably self-evident information that is eliminated by closed-response formats, where the explicit presentation of the proper response alternative indicates the investigator's interest in this information.

In addition, respondents may frequently be uncertain if information that comes to mind does or does not belong to the domain of information the investigator is interested in. Again, closed-response formats may reduce this uncertainty, resulting in higher responses. This differential complexity of open and closed formats is reflected in higher nonresponse rates in the open-response form, in particular among less educated respondents (Schuman and Presser, 1981).

In combination, these processes are likely to result in an "underreporting" of opinions and knowledge in an open-response format as compared to a closed format. For behavioral reports, on the other hand, the pattern may reverse under some conditions. For example, Blair et al. (1977) found pronounced underreporting in response to threatening behavioral questions with a closed-answer format. This effect is most likely mediated by the informative function of response alternatives, as will be discussed below.

As a first conclusion, we note that open- and closed-response formats pose different cognitive tasks and result in different patterns of responses, thus undermining the comparability of the data obtained under open- and closed-response format conditions. Next we consider specific types of response alternatives that are frequently used in survey research.

3.3.2 Middle Alternatives

Opinion questions often include a middle alternative between two extreme responses. This practice has been the topic of some debate. In general, the data indicate that explicitly offering the logically possible middle alternative will produce a considerable increment in the percentage of respondents who will endorse it (see Molenaar, 1982, for a review), as compared to conditions where the middle alternative has to be volunteered. Moreover, offering a middle alternative is likely to reduce the rate of "don't know" responses. Regarding the impact of middle alternatives on the endorsement of polar opposites, the findings are mixed. While Schuman and Presser (1981) found that the introduction of middle alternatives did not change the substantive conclusions drawn from the endorsement of the opposites, Bishop (1987) reported a series of studies where the introduction of middle alternatives affected the substantive conclusions. The conditions that determine whether middle alternatives draw similar or dissimilar numbers of respondents from both sides of the attitude continuum, thus affecting or not affecting the substantive conclusions, are a promising area for future research.

3.3.3 "Don't Know" Options and the Use of Filter Questions

Survey researchers frequently use "no opinion" filters to screen out respondents who may not have an opinion on the issue under investigation. This is either accomplished by offering a "no opinion" option as part of a set of precoded response alternatives (often referred to as a *quasi filter*), or by asking respondents if they have an opinion on the issue before the question proper is asked (often referred to as a *full filter*).

Research on the use of these filters (see Bishop et al. 1983; Schuman and Presser, 1981; Sudman and Bradburn, 1974, for reviews) indicates that respondents are more likely to endorse a "no opinion" option if it is explicitly offered than if they have to volunteer it. Moreover, they are more likely to report not having an opinion on the issue if a full filter is used than if a "no opinion" option is offered as part of the response alternatives — and the more so, the more strongly the full filter is worded. For example, asking respondents if they have an opinion on the issue results in fewer "no opinion" responses than asking them if they "have thought enough about the issue to have an opinion on it" (Bishop et al. 1983).

Consistent with the informative functions approach offered here, a series of experiments demonstrated that filter questions influence respondents' perception of their task (Hippler and Schwarz, 1989): The more strongly the filter question is worded, the more respondents assume that they will have to answer difficult questions and that they may not have the required knowledge. Most importantly, full filters seem to suggest to respondents that they will *not* be asked a global opinion question, in contrast to actual survey practice. Accordingly, strongly worded filter questions discourage respondents from offering any global opinions that they may hold by suggesting a more demanding task than is actually in store. In line with this assumption, all respondents who reported not having an opinion in response to a filter question in one of our studies (Hippler and Schwarz, 1989, experiment 3) subsequently provided substantive responses on a global opinion question — presumably because the global question asked was less demanding than expected on the basis of the filter. As a result, the use of full filters seems likely to result in an underreporting of opinions.

3.3.4 Rating Scales

Like discrete response categories, rating scales are ubiquitous in social research, particularly in attitude measurement. Dawes and Smith (1985) provide a careful discussion of their properties and of the empirical and psychological justifications for their use. Leaving concerns about their psychometric properties aside (cf. Nunnally, 1978), rating scales with labeled endpoints do not seem to be very controversial. Respondents are able to use these scales consistently, even in telephone interviews without visual aids (cf. Hormuth and Brückner, 1985). Seven-point scales seem to be best in terms of reliability, percentage of undecided respondents, and respondents' ability to discriminate between the scale values (Cox, 1980). Thus, seven plus or minus two is the usual recommendation. Moreover, scales which provide verbal labels for each scale value seem more reliable than scales with labeled endpoints only (Krosnick and Berent, 1990).

Researchers should be aware, however, that the terms used to label the endpoints, and the terms used to designate the separate values of verbal rating scales, affect the obtained distribution (Rohrmann, 1978;

Wegner et al. 1982; Wildt and Mazis, 1978). Moreover, Schwarz, Knäuper et al. (in press) observed that respondents may use the specific numeric values provided by the researcher to interpret the meaning of the scale's labels. A representative sample of German adults were asked, "How successful would you say you have been in life?". The question was accompanied by an 11-point rating scale, ranging from "not at all successful" to "extremely successful." However, in one condition the numeric values of the rating scale ranged from 0 ("not at all successful") to 10 ("extremely successful"), whereas in the other condition they ranged from −5 ("not at all successful") to +5 ("extremely successful"). The results showed a dramatic impact of the numeric labels. Whereas 34 percent of the respondents endorsed a value between 0 and 5 on the 0 to 10 scale, only 13 percent endorsed one of the formally equivalent values between −5 and 0 on the −5 to +5 scale. Subsequent experiments indicated that this difference reflects differential interpretations of the term "not at all successful." When this label is combined with the numeric value 0, respondents interpret it to reflect the absence of success. However, when the same label is combined with the numeric value −5, they interpret it to reflect the presence of failure.

This differential interpretation of the same term as a function of its accompanying numeric value is also reflected in inferences that judges draw on the basis of a report given along a rating scale. For example, in one of our experiments, a fictitious student reported his academic success along one of the above scales, checking either a −4 or a 2. As expected, judges who were asked to estimate how often this student had failed an exam assumed that he failed twice as often when he checked a −4 than when he checked a 2, although both values are formally equivalent along 11-point rating scales of the type described above.

3.4 BEHAVIORAL QUESTIONS: WHAT RESPONDENTS LEARN FROM SCALES

So far, we have seen how the set of response alternatives provided by the researcher helps respondents to define their task. This, however, does not exhaust the informative functions of response alternatives. As a considerable body of research has demonstrated, response alternatives also inform respondents about the researcher's knowledge of, or assumptions about, the range of opinions or behaviors in the population, thus providing information about the "real world," which respondents may use in computing a judgment.

For example, research on the use of response alternatives in the assessment of behavioral frequency reports (see Schwarz, 1990a,b;

Table 3.1. Reported Daily TV Viewing as a Function of Response Alternatives

Reported Daily TV Viewing			
Low Frequency Alternatives		High Frequency Alternatives	
Up to 1/2 h	7.4%	Up to 2 1/2h	62.5%
1/2 h to 1h	17.7%	2 1/2h to 3h	23.4%
1h to 1 1/2h	26.5%	3h to 3 1/2h	7.8%
1 1/2h to 2h	14.7%	3 1/2h to 4h	4.7%
2h to 2 1/2h	17.7%	4h to 4 1/2h	1.6%
More than 2 1/2h	16.2%	More than 4 1/2h	0.0%

Note. $N = 132$. Adapted from Schwarz et al. (1985), "Response Scales: Effects of Category Range on Reported Behavior and Comparative Judgments," *Public Opinion Quarterly*, 49, 388–395. Reprinted by permission.

Schwarz and Hippler, 1987, for reviews) indicates that respondents assume that the range of the response alternatives reflects the researcher's knowledge about the distribution of the behavior. Specifically, values in the middle range of the scale are assumed to reflect the "average" or "typical" behavior, whereas the extremes of the scale are assumed to correspond to the extremes of the distribution. These assumptions influence respondents' behavioral reports as well as related judgments in various ways.

3.4.1 Behavioral Reports

First, respondents use the range of the response alternatives as a frame of reference in estimating their own behavioral frequencies and report higher frequencies on scales that present high rather than low frequency response alternatives. The results of a study on TV viewing, shown in Table 3.1, illustrate this effect (Schwarz, et al. 1985). Whereas 37.5 percent of the respondents reported a daily viewing of 2 1/2 hours or more if presented the high frequency response alternatives, only 16.2 percent of the respondents did so if presented the low frequency response alternatives.

This reflects that individual instances of mundane and frequent behaviors, such as watching TV, are not separately represented in memory (see Bradburn et al. 1987; Schwarz, 1990a; and Strube, 1987, for reviews). Rather, individual episodes tend to blend into one generic representation, thus making it difficult to determine their frequency on the basis of a "recall the episodes and count their number" procedure (see Blair and Burton, 1987; Sudman and Schwarz, 1989). Accordingly, respondents have to rely on an estimation strategy, for which they will use any information that seems helpful (Bradburn et al., 1987). One piece of information that is highly salient in the interview context is the range of the response alternatives provided. Respondents are therefore likely to use this range as a salient frame of reference in estimating their own behavioral frequency, resulting in higher estimates on high frequency compared to low frequency scales. The impact of this salient frame of reference is more pronounced when the episodic information that respondents can recall from memory is less. Accordingly, it is more pronounced for proxy reports than for self-reports (Schwarz and Bienias, 1990, experiments 1 and 2), and is eliminated when respondents have a chance to refresh their memory, e.g., by browsing through a TV program guide.

3.4.2 Comparative Judgments

The impact of the response alternatives is not limited to the behavioral frequency question with which they are provided. Given the assumption that the scale reflects the distribution of the behavior, checking a response alternative is the same as locating one's own position in the distribution. Accordingly, respondents extract comparison information from their own location on the response scale and use this information in making subsequent comparative judgments (e.g., Schwarz et al., 1985; Schwarz and Scheuring, 1988).

For example, respondents of the TV viewing study described above reported that TV plays a more important role in their leisure time (experiment 1), and described themselves as less satisfied with the variety of things they do in their leisure time (experiment 2), when they had to report their TV viewing on the low rather than on the high frequency scale. This reflects that checking a value in the upper range of the low frequency scale suggested to them that they watch *more* TV than "average," whereas their location on the high frequency scale suggested to them that they watch *less* TV than "average" (cf. Table 3.1).

Interestingly, the use of scale location in comparative judgments is not limited to respondents, but does affect the users of their reports as well. For example, experienced physicians were more likely to consider that having a given physical symptom "twice a week" reflected a serious medical condition if that frequency were reported on a low rather than a high frequency scale (Schwarz, Bless et al., 1991).

3.4.3 Question Interpretation

Finally, if the target behavior is open to interpretation, as is often the case when subjective experiences are assessed, respondents may use the frequency range of the response alternatives to determine the exact reference of the question. For example, respondents who were asked to report how frequently they feel "really annoyed" on a low frequency scale subsequently reported more extreme annoyances as typical for their experience than respondents who had to give their frequency report on a high frequency scale (Schwarz et al. 1988). This suggests that respondents used the frequency range of the response alternatives to determine what the researcher meant by "really annoyed," asking themselves "Does that refer to major or to minor annoyances?" Given that minor annoyances are frequent, whereas major ones are rare, the frequency range of the response alternatives helped to clarify the meaning of the question. Accordingly, the same behavioral question in combination with different frequency alternatives is likely to assess different experiences.

3.5 SUMMARY: THE INFORMATION FUNCTION OF QUESTIONNAIRES

We conclude from these and related findings that respondents actively use apparently "formal" features of the questionnaire as a source of information to determine what is expected of them and to solve the cognitive tasks posed in the survey interview (see Schwarz, 1990a; Schwarz and Strack, 1991; Strack and Martin, 1987; and Strack and Schwarz, in press, for more extended discussions). Accordingly, response alternatives are not only technical measurement devices. They are better conceptualized as part of an ongoing conversation between the researcher and the respondent. Although survey respondents are aware that they are supposed to answer rather than to ask questions, they still bring many assumptions to the survey interview that govern the conduct of conversations in everyday life (cf. Grice, 1975). Most importantly, they expect all participants to provide only information that is relevant to the issue at hand, and they assume that the selection and presentation of response alternatives comply with that norm. As a result, they conduct their own share of thinking, judging and communicating within the framework of the ongoing conversation, much as they would be expected to do in everyday life — except that in the survey interview that context is prestructured by the questionnaire. For that very reason, we have to pay close attention to the context that we set up in devising the questionnaire, unless we want to assess what we evoked in the first place.

3.6 THE IMPACT OF RESPONSE ORDER: ACCOUNTING FOR PRIMACY AND RECENCY

Once the researcher decides which response alternatives should be selected, he or she needs to determine the order in which they are presented to respondents. Again, the survey literature indicates that the presentation order may strongly influence the obtained results. Theoretically, *primacy effects*, that is, higher endorsements of items presented early in the list, as well as *recency effects*, that is, higher endorsements of items presented late in the list, may be obtained. While response order effects have occasionally been reported when the response alternatives present an ordered set of categories that constitute a verbal rating scale (e.g., excellent, very good, good, fair, poor), they are rare under these conditions (see Mingay and Greenwell, 1989). In contrast, response order effects have frequently been obtained when each response alternative presents a different opinion on an issue, and respondents are asked to select the one that best represents their own position. Several processes are likely to contribute to these findings, as discussed below. Moreover, these processes may result in complex interaction effects, depending upon the specific conditions. Accordingly, this area of research is characterized by a complex set of apparently contradictory findings.

3.6.1 Elaboration Processes

As one heuristic framework for understanding the nature of response order effects, we suggest that each response alternative may be portrayed as a single persuasive argument (Schwarz, Hippler et al. 1991; see also Krosnick, 1991a, and Krosnick and Alwin, 1987, for a related conceptualization, emphasizing memory limitations). Borrowing from research on the processing of persuasive communications (see Petty and Cacioppo, 1986 for a detailed review), we assume that a given item is more likely to be endorsed the more positive cognitive responses it elicits, that is, the more agreeing thoughts the respondent generates. Conversely, a given item should be less likely to be endorsed the more disagreeing thoughts it elicits. The number of thoughts generated, however, is not only a function of the content of the item per se, but also a function of the degree of cognitive elaboration that a given mode of data collection permits.

Suppose, for example, that a long list of response alternatives is presented to respondents on a show card as part of a face to face interview, or in a self-administered questionnaire. Under these conditions, "items presented early in a list are likely to be subjected to deeper cognitive processing," as Krosnick and Alwin (1987, p. 213) noted. "By the time a respondent considers the later alternatives, his or her mind is likely to be cluttered with thoughts about previous alternatives that inhibit extensive consideration of later ones." Accordingly, a given response alternative is more likely to be endorsed if presented early rather than late in the list, provided that it is "plausible" to the respondent, thus eliciting agreeing thoughts. Conversely, an "implausible" response alternative that elicits disagreeing thoughts is less likely to be endorsed if presented early.

Note that this analysis implies that order effects are likely to depend on respondents' attitudes, which determine what is "plausible" for a given respondent. For example, a "liberal" item is likely to elicit agreeing thoughts from liberal respondents, but disagreeing thoughts from conservative respondents. Accordingly, liberal respondents should be *more* likely to endorse the item if presented early rather than late on a list, whereas conservative respondents should be *less* likely to endorse it under these conditions. As a result, primacy and recency effects in subsamples may cancel one another, resulting in the apparent absence of order effects in the sample as a whole. Hippler, et al. (1990) report data supporting this hypothesis.

Assume, however, that the items are not presented visually, but are read to respondents by the interviewer. In this case, respondents have little opportunity to elaborate on the items presented early in the list, because the time that is available for processing each item is restricted by the speed with which the interviewer moves on to read the next one. "Under these circumstances, respondents are able to devote most processing time to the *final* item(s) read, since interviewers usually pause most after reading them" (Krosnick and Alwin, 1987, p. 203). In addition, respondents may find it difficult to keep all response alternatives in mind without visual help. Accordingly, items that elicit agreeing thoughts from a given respondent should be more likely to be endorsed if presented late rather than early in the list, resulting in recency effects under auditory presentation formats. Again, the reverse holds for items that elicit disagreeing thoughts.

In summary, response order effects are assumed to depend on the items' serial position, their plausibility for a given respondent, and the administration mode used. If the response alternatives are presented on show cards or in a self-administered questionnaire, items presented early in the list are more likely to be extensively processed than items presented later, resulting in *primacy effects*, provided that the item is plausible to the respondent (i.e., that it elicits agreeing thoughts). In contrast, if the items are read to respondents, the last response alternatives are more likely to be extensively processed and recalled than the first ones, resulting in *recency effects*, again assuming plausibility of the items for the respondent. Given that the likelihood of endorsement may be expected to decrease as more extensive processing uncovers flaws in implausible items, the reverse predictions hold for items that lack plausibility for the respondent (i.e., items that elicit disagreeing thoughts; see Schwarz, Strack, and Mai, 1991, for further discussion).

The predicted interaction of serial position and administration mode for plausible items has been supported by secondary analyses of a large number of split-ballot experiments with representative samples of the adult population in West Germany, which were originally conducted by the Allensbach Institute under the direction of Elizabeth Noelle-Neumann, since the early 1950s (see Schwarz, Hippler, and Noelle-Neumann, 1991, for a review). Additional analyses of these archival data, using respondents' attitudes as an approximation for the likely plausibility that a given item may have for them, provided preliminary support for the assumed role of item plausibility (Hippler et al. 1990). We emphasize, however, that our conclusions are based on secondary analyses involving different questions under different presentation formats. While the consistency of the data patterns across widely different questions suggests that the conclusions are likely to be valid, more tightly controlled experiments using the same questions and comparable samples under all conditions are definitely needed.

3.6.2 Memory Limitations

Not surprisingly, response order effects have frequently been attributed to respondents' memory limitations, and Krosnick and Alwin (1987; see also Krosnick, 1991a) suggested that the interaction of serial position and administration mode described above may reflect memory processes. Despite their popularity in the survey literature, however, memory based accounts are difficult to reconcile with the available data.

As psychological research on the learning of long lists of verbal expressions indicates, the recall of verbal material depends on its serial position in the list and the time delay between learning and testing (see Smyth, et al., 1987, for a review). Material that is presented at the beginning of the list is more likely to enter long-term memory than material that is presented later, because the first few items "suffer less competition for time and space in immediate memory from other items" (Smyth, et al., 1987, p. 123). On the other hand, material that is presented at the end of the list may still be in short-term memory if recall follows learning without much delay. This results in an interaction effect of serial position and the delay between learning and recall: Without delay, material presented at the end of the list is much more likely to be recalled than material presented at the beginning, reflecting that the later items can be recalled from short-term memory. Accordingly, recency effects are typically obtained under no-delay conditions, and they are more pronounced if the material is read to subjects rather than presented visually (Murdock and Walker, 1969). If recall is delayed, however, material presented at the beginning of the list is more likely to be remembered than material presented at the end, reflecting that the early items entered long-term memory, whereas the later items can no longer be recalled from short-term memory. This results in primacy effects under delayed recall conditions. Material that is presented in the middle of the list is least likely to be recalled under any conditions.

How do these findings bear on response order effects in survey measurement? Given that respondents report their answers immediately after exposure to the response alternatives, no delay between "learning" and "recall" is introduced. Accordingly, response alternatives presented at the end of the list should be easily accessible in short-term memory. Because recall of late items from short-term memory is better than recall of early items from long-term memory under no-delay conditions, this should result in pronounced recency effects. Primacy effects should only be obtained if a delay is introduced between exposure to the response alternatives and respondents' reports. This is typically *not* the case in survey interviews. Nevertheless, *primacy* rather than recency effects have typically been reported in survey experiments with long lists of response alternatives (e.g., Payne, 1951; Mueller, 1970; Ring, 1975; Krosnick and Alwin, 1987), suggesting that memory limitations are not the primary source of response order effects in survey measurement.

In fact, long lists of response alternatives are usually presented on show cards that remain available until respondents report their answer, thus placing little burden on their memory to begin with. Moreover, response order effects have consistently been observed on questions that present only two or three response alternatives (e.g., Payne, 1951; Schuman and Presser, 1981; Schwarz, Hippler, et al. in press). This limited number of response alternatives, however, should be easily accessible in short-term memory under the no-delay condition of survey interviews, thus rendering the emergence of memory-based order effects unlikely unless the alternatives are overly complex. We conclude from these inconsistencies that memory limitations are *not* the dominant source of response order effects in survey measurement; whether the cognitive response approach that we offered above, building on Krosnick and Alwin's (1987) discussion, fares much better, on the other hand, remains to be seen.

3.6.3 Contrast Effects

Finally, the likelihood that a given response alternative is endorsed does not only depend on its serial position per se, which may influence its cognitive elaboration or memorability, but also on the nature of the preceding response alternatives. This possibility has received little attention in survey research. Specifically, if a given item is preceded by an item that is more extreme on the dimension of judgment, a *contrast effect* may emerge, provided that all items are to be judged along the same dimension (cf. Ostrom and Upshaw, 1968).

Suppose, for example, that respondents are asked to select persons that they like well, and that an extremely well-liked person is presented in the middle of a list. If so, a moderately liked person will seem *less* likable if he or she is presented in the second half of the list, following the extreme stimulus, than if he or she is presented at the beginning of the list, preceding the extreme stimulus. If we compared the two orders of this list, this judgmental contrast effect would lead us to conclude that a pronounced primacy effect emerged. On the other hand, if the person presented in the middle of the list were extremely dislikable, the same mechanism of judgmental contrast would increase the endorsement of moderately liked persons presented in the second half of the list. In that case, a comparison of both order conditions would lead us to conclude that a pronounced recency effect emerged. Note, however, that the underlying cognitive process of judgmental contrast is quite different from the cognitive elaboration and memory processes discussed above.

A classic example of such a contrast effect was reported by Noelle-Neumann (1970). Specifically, respondents were presented a list of food items and were asked to select the ones that are typically "German." Respondents were more likely to consider a number of food items, such as noodles or potatoes, as typically "German" when they were preceded by rice than when they were not. Thus, introducing rice as the first item resulted in pronounced contrast effects in the perception of the other food items. Finally, the evaluation of rice itself was unaffected by order manipulations.

Contrast effects of this type are a function of the items' extremity on the underlying dimension of judgment. Introducing a more extreme item results in a wider "perspective" regarding the set of stimuli, thus affecting the evaluation of moderate stimuli as described in Ostrom and Upshaw's (1968) perspective theory. Accordingly, these effects do also emerge under conditions where each item is likely to receive about the same degree of attention and elaboration, for example, because *each* item has to be rated along some scale. For example, Schwarz et al. (1990) observed that subjects evaluated a number of drinks as being more "typically German" when they were preceded by "vodka" (an extremely atypical drink) than when they were preceded by "beer" (an extremely typical drink).

Moreover, contrast effects of this type do *not* require that the items are presented on the same list. Rather, they may also emerge if the extreme item is presented as part of a preceding question, provided that this question taps the same dimension of judgment. For example, in some conditions of the Schwarz, et al. (1990) study, we asked some respondents to estimate the percentage of Germans who drink vodka, and others to estimate the percentage of Germans who drink beer, before they rated the typicality of other drinks. As expected, subjects who estimated the percentage of Germans who drink vodka rated subsequent drinks as more typically German than subjects who estimated how many Germans drink beer, replicating the contrast effects obtained when all stimuli were presented on the same list. Other subjects, however, were asked as part of the preceding questions to estimate the caloric content, rather than the consumption, of vodka or beer. While this question also serves to render these drinks highly salient in the interview context, it does not tap the typicality dimension that underlies estimates of the consumption of these drinks. Accordingly, estimating their caloric content did *not* influence subsequent typicality ratings. Thus, we conclude that contrast effects can emerge as a function of preceding questions *if* these questions tap the same underlying dimension of judgment.

This emergence of contrast effects bears in important ways on the emergence of primacy and recency effects in general: If an extremely *positive* item is presented as part of the stimulus set, it will *decrease* the endorsement of subsequent moderate items. If an extremely *negative* item is presented, on the other hand, it will *increase* the endorsement of subsequent moderate items. These judgmental effects may lead the researcher to conclude that the data show pronounced recency or primacy effects. Accordingly, the phenomenon of judgmental contrast may dilute the emergence of memory and elaboration phenomena.

3.6.4 Summary of Response Order Effects

As the preceding discussion illustrates, response order effects may be a function of the serial position of the item, the item's plausibility for a given respondent, the extremity of adjoining items, and the administration mode used. The effects become complex when several of these factors are simultaneously present, as discussed in detail by Schwarz, Hippler, and Noelle-Neumann (1991). The absence of significant response order effects may indicate that none of the factors causing them was sufficiently powerful in a given case, but it may also reflect the different effects canceling each other. Because the relative power of the factors involved is difficult to evaluate a priori, the emergence of response order effects is likely to remain a surprise in many specific cases, although the underlying processes seem systematic and their effects replicable.

CHAPTER 4

CONTEXT EFFECTS IN THE GENERAL SOCIAL SURVEY

Tom W. Smith
NORC and University of Chicago

4.1 INTRODUCTION

After a half-century of study, question order remains one of the least developed and one of the most problematic aspects of survey research. As Schuman and Presser remark in their work on survey methodology (1981, p. 77), "Overall, order effects ... constitute one of the most important areas of methodological research. They can be very large [and] are difficult to predict." This perplexity is shared by Bradburn (1983, p. 302) who observes, "No topic in questionnaire construction is more vexing or resistant to easy generalization than that of question order" and by Groves (1989, p. 479) who notes, "(T)here seems to be no general theory that predicts when such effects are to be expected and when they should not be expected."

In part because of our limited ability to predict their occurrence, there is some disagreement in the literature over how common context effects are. Tourangeau et al. conclude that "The literature on survey context effects may create the impression that such effects are relatively rare, involving items on a few scattered issues. These results here indicate otherwise ... (Tourangeau et al. 1988, pp. 22–23)." This impression of pervasiveness is supported by numerous instances in which

This research was done for the General Social Survey project directed by James A. Davis and Tom W. Smith. It revises and expands upon GSS Methodological Report No. 55. The project is funded by the National Science Foundation Grant SES-8747227.

Measurement Errors in Surveys.
Edited by Biemer, Groves, Lyberg, Mathiowetz, and Sudman.

ISBN: 0-471-53405-6

changes in question order have upset time series or caused other undesired measurement variations (Smith, 1986; Smith, 1988d; Astin, et al., 1988; Cowan, et al. 1978; Roper, 1990; Johnston and Bachman, 1980; Turner and Martin, 1984; and Gibson, et al. 1978).

Schuman and Presser, on the other hand, reach a conclusion that at least differs in emphasis — "Question order-effects are evidently not pervasive . . . but there are enough instances to show they are not rare either (1981, p. 74)." This non-pervasive impression is supported by numerous failures to produce context effects in experiments designed to do so (Schuman and Presser, 1981; Smith, 1983a; Turner and Martin, 1984) and by the ability of different survey organizations to produce similar marginals when the same questions, but different question content (as well as other variations), existed (Turner and Martin, 1984; Smith, 1978; Smith, 1982a).

To date only one study has conducted a general search for context effects.[1] Schuman and Presser (1981) examined the 1971 Detroit Area Study (DAS). The DAS used split ballots in order to accommodate various experiments in either question order or wording. They looked at 113 attitude items that were not the designed objects of these experiments, but appeared after the experiments and thus varied in context due to the prior experiments. Apparently using simple random sampling (SRS) assumptions, they found eight significant differences at the 0.05 level, just two above what chance would predict. Their inspection of these eight suggested that three probably represented real effects and the rest were due to sample variation.

In this chapter we conduct a similar analysis using the 1988 and 1989 General Social Surveys (GSS).[2]

While the survey methodologist would usually think of context as creating variation in the target question, most substantive analysts usually characterize context effects as creating bias or distortion. This latter perspective rests on the notion that there is a true value for the target item free from context (see Introduction and Chapter 1). One way to think about context effects is to imagine a sample of all possible contexts with the resulting distributions on the target variable ranging from low to high. We presume that something resembling a normal distribution would arise and that one might consider the contexts clustering at the middle of the distribution as neutral. Substantive analysts would prefer these contexts.

---

[1] Also see Bradburn and Mason, 1964 who tested for 14 differences in marginals across four forms and found no statistically significant variation.

[2] The GSSs are full probability samples of adults living in households in the United States. Personal interviewing is used. For complete details see Davis and Smith, 1990.

Another (and different) way of thinking of a neutral vs. biasing context would be to consider the contexts that duplicated how the issue appeared in the public arena as the "neutral" context. It would be neutral in the sense that the survey context would duplicate the real world context and not add any measurement variation. While there is no easy way to determine what survey context might duplicate the real world context and no survey context could really approximate the real world context, it still might be possible to identify some survey contexts that deviated from the real world. For example, a survey context that linked together persons A and B and lowered the ratings of B when B followed A might be considered biasing if it created a link in the respondent's mind that did not exist among the public in the real world.

4.2 THE GSS DESIGN

The GSS has monitored trends in American attitudes and behaviors since 1972. The GSS has striven to measure true change undistorted by measurement artifacts (Smith, 1986; 1988d). To avoid distortion in the time series, the project has tried to keep all facets of data collection and processing constant; using identical sampling procedures, question wordings, data processing, and so forth. When necessary changes have occurred, such as the periodic updating of the sampling frame, the GSS has utilized calibration designs or other features that allow the separation of measurement variation from true change and the maintenance of a consistent time series.

One of the measurement variations that has been most difficult to control for has been order or context (Smith, 1986). The exact replication of question order has not been possible because of several other features of the GSS:

1. Until 1988 the GSS employed a rotation design under which sets of GSS items appeared on two out of every three surveys. (For details of the rotation design see Smith, 1988c.) This meant that the same items appeared only on every fourth survey.

2. The content of the survey has changed because of the addition and deletion of items. Such changes were especially large between 1972 and 1973 when the survey length was substantially increased and between 1984 and 1985 when items were dropped or switched from permanent status to rotating in order to open up space for annual supplements.

3. Split-sample experiments on question wordings, other features, and context itself have altered the order of questions.

To minimize the unavoidable variation in order caused by these factors, the GSS has tried to (1) keep order as constant as possible within the rotation patterns so that the order of the first and fourth, second and fifth, and third and sixth years was the same, (2) maintain the order of all scale items unaltered (e.g., the seven-item abortion scale or the 15-item national spending scale), (3) place new items either at the end of the survey (as in the case of the topical and cross-national supplements) or at the end of individual scales (as when an additional abortion question, was added to the abortion scale), (4) consider the possibility of order effects when constructing the questionnaire each year, (5) check all "blips" in the time series for possible context effects and (6) study context effects via split-sample experiments both to extend our general knowledge of context effects and to test for specific effects and possible distortions to the time series suspected on the GSS (Smith, 1986; 1984; 1983b; 1983a; 1982b; 1981).

With the switch from the rotation across time design utilized by the GSS until 1988 to the split-ballot design employed since 1988, the GSS is able to test for possible context effects by comparing response distributions across ballots. (For details of the new design see Smith, 1988c.) In brief, under the new split-ballot design items that would have appeared on two of the next three surveys instead appear on two of three random sub-samples (called ballots), each with one-third of the cases in a particular year.

Retrospectively the switch means that ballot A in 1988/89 represents the 1982 and 1985 surveys, ballot B represents 1983 and 1986, and ballot C represents 1984 and 1987. (Ballots also resemble earlier years in the GSS series, but the closeness of the match decays over time due to the addition and deletion of items and other variations. The matches between the 1982 through 1987 surveys and the ballots in 1988 and 1989 are much closer than the match with earlier years.)

This switch from a rotation to a split-ballot design makes it possible to experimentally test for context effects by comparing items across ballots. And since the ballots largely duplicate the order of questions in previous years, one can usually generalize these findings to changes across past GSSs. In effect, differences across ballots on the 1988 and 1989 surveys should duplicate changes across rotation years in past GSSs. Consequently, order effects on the 1988/89 GSS identify order effects that have been distorting the time series on past GSSs.[3]

[3] In addition, over the years the GSS has had 13 planned context experiments (Davis and Smith, 1990).

Table 4.1. Statistically Significant Differences* Across Ballots by Contextual Groups (1988–1989)

	SRS		Adjusted	
	n	%	*n*	%
		1988		
Contextual Groups **(number of variables)**				
HEF/Sample Frame (9)	2		0	
Start (36)	1		0	
Religion Module (92)	9		1	
Sex Behavior Supplement (7)	0		0	
Sub-total (144)	12	8.3	1	0.7
Other (358)	33	9.2	13	3.6
Total (502)	45	9.0	14	2.8
		1989		
Contextual Groups **(number of variables)**				
HEF/Sample Frame (9)	0		0	
Start (36)	0		0	
Forbid/Allow Experiments (6)	1		1	
Sex Behavior Supplement (10)	0		0	
Sub-total (61)	1	1.6	1	1.6
Other (381)	35	9.2	13	2.4
Total (442)	36	8.1	14	3.2

* Chi Square × 0.667; SRS, simple random sampling; HEF, household enumeration form.

4.3 ANALYSIS OF CONTEXT EFFECTS

Table 4.1 shows the results of cross-tabulating the 502 variables on the 1988 GSS and the 442 variables in 1989 with ballot. Overall using SRS assumptions, 9.0% of the comparisons in 1988 and 8.1% of comparisons in 1989 are significant at the 0.05 level. Using a design effects adjustment of 0.667 lowers the percentage of significant differences to only 2.8% in 1988 and 3.2% in 1989.[4] Table 4.1 breaks the GSS variables into five categories. Two of the categories, variables from the sample frame and household enumeration form (HEF) and variables occurring at the beginning of each ballot, do not vary in context across ballots. Two other groups, variables from the topic modules and variables from the sexual behavior supplement are both internally consistent in order across ballots and

[4] Studies of design effects on the GSS and other NORC surveys typically find that design effects are mainly in the range of 1.0 to 1.5. For this reason, chi-squared tests were conducted using design effects of 1.0 and 1.5.

separated from the rest of the questionnaire by a buffer of identical questions. For these four groups the likelihood of context effects is nil to negligible. The fifth group consists of all other variables, which to a greater or lesser extent differ in their order across surveys.

In both 1988 and 1989 there are very few significant differences among the context-similar groups while more significant differences appeared among the context-different variables. For 1988 the SRS results in Table 4.1 find almost as many significant differences for the context-similar categories (8.3%) as for the context-different categories (9.2%), but the adjusted significance tests show that most of the context-similar variables had differences of only borderline significance. With adjustments only one of the context-similar variables (0.7%) showed a statistically significant difference with 13 of the context-different variables (3.6%) still registered as significantly different. For 1989 both the SRS and adjusted probability levels show few (1.6%) significant differences among the context-similar questions. Among the context-different questions 9.2% showed significant differences prior to adjustments, but only 3.4% were significant after adjustment.

4.4 ASSESSING SUBGROUP DIFFERENCES

Since fewer than 5% of the comparisons in both 1988 and 1989 were significant in the adjusted comparisons, it might be possible to argue that no true differences are occurring, that all differences might occur by chance. Rather than make this general probabilistic conclusion, we evaluated each of the 28 statistically significant differences by (a) comparing the 1988 and 1989 results both to one another and to the 1982–1987 surveys, (b) evaluating the differences in context across the ballots, and (c) searching for conditional or associational effects consistent with our proposed explanations for context effects.

Each of these evaluations involves some difficulties. The comparison of 1988 and 1989 is generally straightforward since most variables appeared both years and context was usually identical across years. The pre-1988 comparisons are more problematic however. We carried out the pre-1988 comparisons by grouping together the two most recent years representing the same rotation of ballot (1982 and 1985; 1983 and 1986; 1984 and 1987). We used only the two most recent rotation cycles because changes from the addition and deletion of items and other reasons make earlier years less comparable to the 1988 and 1989 ballots than more recent years. Since the pre-1988 comparisons are usually based on nearly 3,000 in each "ballot" condition (compared to about 500 on each ballot in 1988 or 1989), rather small differences can be statistically significant. More importantly, collapsing years into quasi ballots does not eliminate time as a factor. Not only would items varying across time coincidental with rotation create associations, but items showing secular trends also relate to rotation.

Assessing the cause of context effects is also difficult because such effects are complex and not clearly understood (Schuman and Presser, 1981; Smith, 1989; Tourangeau, 1987; Tourangeau, Rasinski, Bradburn, and D'Andrade, 1988; Tourangeau and Rasinski, 1986). While context effects usually occur between related items, the interconnection between questions can be subtle and hard to predict. In addition, for most of the context-different comparisons in Table 4.1 (and in particular the 28 significant differences) preceding context is complex, involving many questions covering numerous subjects. Although the question(s) inducing a context effect usually closely precede the affected question, this does not have to be the case. (See Bishop et al., 1982; Schuman et al., 1983 for examples of long-distant order effects.) It is effectively impossible to assess in detail the entire context in which each of the affected questions appeared. We examined and noted the four immediately preceding questions (which because of sub-questions and indices often involved many variables) and generally searched an additional 15–20 questions for possible context inducing questions.

Finally, to check the plausibility of these context explanations, we examined if any other association could be established between the hypothesized triggering variable(s) and the affected question. Context effects come in various forms, and in many instances there is no detectable connection between the two variables other than the marginal changes on the affected variable (Smith, 1982b). There are two types of context effects, however, that not only lead to a change in the marginal distribution of the affected variable, but also have other connections between the variables. First, in the case of associational order effects, the relationship between the variables is increased when the context effect is operating. This can occur when the preceding variable creates a frame of reference by which the subsequent variable is evaluated (Smith, 1981). Second, with conditional order effects the impact on the subsequent question occurs only among people giving certain responses to the prior question or, in the extreme case, the context effect is in the opposite direction, depending on how the preceding question was answered (Smith, 1982b; 1989).

We first considered whether the explanation for the context effects implied conditional and/or associational effects. In the majority of cases, the proposed explanations did not suggest such effects. In some other cases conditional or other effects were implied by our explanation, but the triggering variable appeared on only one ballot so we could not empirically evaluate our explanation. When a conditional or associational effect was predicted and it was possible to test it, we carried out the appropriate comparisons on both the pre-1988 and 1988–89 pooled data. For example, we examined whether helpfulness was more associated with

Table 4.2. Statistically Significant Differences

	Pre-1988	Dir.	1988	Dir.	1989	88–89 Pooled	Judgment
A. Pre-1988							
CAPPUN	0.0000*	O	0.240	O	0.659	0.638	No
COURTS	0.0000*	O	0.444	O	0.114	0.185	No
DRUNK	0.0000*	O	0.070	O	0.944	0.131	No
MEMCHURH	0.0000*	S	0.066	S	0.029	0.005*	Yes
Five Sats.	0.0000*	S	0.032	S	0.127	0.0009	Yes?
13 Confid.	0.0000*	S	0.035	S	0.186	0.013	Yes
B. 1988 Likely							
AGED	0.0000*	S	0.003*	S	0.033	0.0001*	Yes
ANOMIA6	0.0000*	S	0.005*	S	0.070	0.001*	Yes
HELPFUL	0.0000*	S	0.001*	S	0.239	0.001*	Yes
JOBMEANS	0.007*	S	0.003*	S	0.864	0.046	Yes?
RACCHURH#	0.0000*	S	0.000*	S	0.003*	0.0000*	Yes
SATCITY#	0.0000*	S	0.000*	S	0.003*	0.0001*	Yes
SUICIDE1	0.0000*	O	0.007*	O	0.822	0.086	No
TRUST	0.0000*	S	0.004*	S	0.000*	0.000*	Yes
C. 1988 Unlikely							
BUSING	0.0000*	S	0.008*	O	0.551	0.159	No
MEMNAT	0.007*	O	0.008*	S	0.500	0.038	?
MEMYOUTH	0.057	S	0.013*	O	0.120	0.526	No
SAVESOUL	—	—	0.005*	—	—	—	No
RACOPEN	0.026	O	0.007*	S	0.458	0.074	No?
VETYEARS	0.526	O	0.003*	O	0.720	0.056	No
D. 1989							
ABRAPE	0.002*	S	0.194	O	0.013*	0.439	No
ALLOWRAC	—	—	—	—	0.006*	—	No
ANOMIA7	0.014*	S	0.049	S	0.006*	0.001*	Yes
CONCLERG	0.0003*	S	0.025	S	0.007*	0.0004*	Yes
CONFINAN	0.0000*	S	0.302	S	0.005*	0.0048*	Yes
LIBATH	0.001	S	0.737	S	0.003*	0.021	?
LIBMIL	0.073	S	0.111	O	0.005*	0.402	No?
RELIG16	0.084	O	0.196	O	0.003*	0.0005*	No
SATFRND	0.0000*	S	0.054	S	0.004*	0.0001*	Yes
WORDD	0.902	O	0.328	S	0.0009*	0.023	No?
WRKSLF	0.959	O	0.473	O	0.011*	0.381	No

= Also on 1989 list.
* = significant at 0.05 level after adjustment for design effects.
Dir. = Direction; S = Same as 1989; O = Opposite 1989.
? = borderline.

violence questions when these questions were in close proximity (an associational effect) and whether support for home care of elderly parents was higher after the helpfulness item only among those expressing the belief that people are helpful (a conditional order effect).

Table 4.2 shows how the variables differed across ballots in 1988 and 1989 and across rotation years from 1982 to 1987 (Pre-1988) and makes a judgment on what differences represent real order effects.[5]

The table shows the 31 variables that statistically varied in 1988 or 1989 or were candidates from the pre-1988 analysis.[6]

For 13 of the 15 variables judged not to represent true context effects, the pooled 1988–89 differences are not statistically significant (when adjusted) and the direction of the pre-1988, 1988, and 1989 differences are not consistent. In addition, context is fairly similar in each case and no probable source of an order effect was detected. For two variables, trying to convert someone to Christ (SAVESOUL) in 1988 and allowing a racist to speak in public (ALLOWRAC) in 1989, the variables appeared on only one survey and therefore no cross-survey check is possible. They are the only two context-similar variables to show significant differences and neither seems a very likely candidate for a true context effect. SAVESOUL has identical context on all three ballots in 1988 for at least the previous 28 questions and no variables closely related to it in either placement or association show signs of a context effect. Similarly, for ALLOWRAC in 1989 the immediate context is similar on both ballots and none of the five related variables in the forbid-allow experiment show any context effects. In addition, examination of the forbid-allow experiments for possible context effects from prior civil liberties questions revealed no signs of a context effect.

For two variables, we have considered the context effects as uncertain. For the first, membership in nationality groups (MEMNAT), the pattern is really similar to several that we judged not to represent real effects. We have placed these variables in the uncertain category primarily because we found another membership variable probably has a real context effect and therefore we are accepting the idea that something about the different contexts influences at least one membership item and therefore might be influencing other membership variables as well. The second uncertain variable, the item on permitting a book written by a militarist in the public library (LIBMIL), is of borderline significance for the pooled 1988–89 sample and the differences are always in the same directions. On the other hand, the immediate context is very similar across ballots.

---

[5] A summary of the questions immediately preceding each of the questions listed in Table 4.2 is available from the author. Context may also be determined by consulting Appendix B in Davis and Smith, 1990. In addition, the full question wordings may be obtained from Davis and Smith, 1990.

[6] The 1988 variables are broken into two groups based on a preliminary evaluation using only data through 1988 (Smith, 1988a). Compared to our current evaluation after using the 1989 survey, predictions for one of the 14 variables (SUICIDE1) changed.

4.5 SUSPECTED CONTEXT EFFECTS

That leaves 11 to 12 variables (11 Yes's and 1 Yes?) that are significantly different for 1988–89, in the same direction in the pre-1988, 1988, and 1989 comparisons, and have plausible, if unproven, context explanations for the differences and/or a past history of being prone to context or other method effects. Not all of these effects are equally probable or of similar magnitude, but based on the criteria listed above they are considered as probably representing real context effects.

First, belief in the helpfulness of people (HELPFUL) is higher when following items about leisure activities and lower when following items about economic standing and violence. We believe that the economic items might focus thoughts on non-helpful behavior (competition and self-interest) and/or the violence approval questions might emphasize criminal and aggressive behavior and thus reduce one's evaluation of people's helpfulness.

Second, trust in people (TRUST) is less when following three crime control variables than when following a mixture of questions on how exciting one's life is, job values, reasons for advancement, and homosexuality. As with helpfulness, we believe that the crime items might have reduced evaluations of the trustworthiness of people.

Third, there is lower satisfaction with the local community (SATCITY) when questions about the helpfulness and fairness of people and leisure time activities come first than when 15 Stouffer civil liberties questions, a question on Communism, an item on equalizing wealth, and a 13-part question on confidence in institutions come first. Past research has identified happiness and satisfaction items as susceptible to order effects (Smith, 1986; McClendon and O'Brien, 1983; Schuman and Presser, 1981). In this situation, however, the source of the impact is not readily apparent. One possibility is that a contrast effect after the Stouffer items raised positive evaluation of one's community (i.e., in contrast to the various politically or socially deviant groups asked about in the Stouffer questions, aspects of one's immediate life may be viewed more positively). We believe that the effect is greater for satisfaction with community rather than the other satisfaction items because satisfaction with community is the first satisfaction item. (For a similar first-only effect see Smith, 1981.)

Fourth, satisfaction with one's friends (SATFRND) is also apparently influenced by the same factors mentioned in the discussion of SATCITY above.

Although the other three satisfaction items (satisfaction with family, leisure time, and health — SATFAM, SATHOBBY, SATHEALT) do not show clear, individual context effects, it appears that all are being influenced in a similar way as these variables. For the five satisfaction (Five Sats.) items in the pre-1988, 1988, and 1989 comparisons, satisfaction is always greater when following the Stouffer questions than in the other contexts.

Fifth, the anomia item on it hardly being fair to bring a child into the world (ANOMIA6) is more agreed with when following the spending priority items than when coming after race relation items. (Another anomia item immediately preceded ANOMIA6 on both versions.) The no-children anomia item has been shown to be highly sensitive to context effects in the past (Smith, 1983b), but we are unsure what context in the current situation depressed or elevated anomia.

Sixth, the anomia item that states public officials do not care about citizens (ANOMIA7) is more agreed with when following the spending priority items than when coming after race relations. Since ANOMIA7 immediately followed ANOMIA6 we assume that a common context is influencing both.

Seventh, confidence in the people running banks and financial institutions (CONFINAN) is higher when the preceding context includes questions about number of children, unemployment history, work hierarchy, and occupational questions than a version that asks about social class, personal financial status, union membership, and occupation. (On both ballots 12 other confidence items appear before the financial item.) We speculate that confidence in banks and financial institutions may be lowered by the prior questions on personal financial conditions. This explanation is supported by the fact that when asked first about personal finances there is a positive correlation between personal difficulties and low institutional confidence, but when the personal financial questions did not appear first, personal financial evaluations and institutional confidence are not related. For example, when the personal financial questions came first, among those reporting improvements in their financial situation 30% had a great deal of confidence in financial institutions. Among those reporting no change in their personal condition 25.5% had a great deal of confidence and among those reporting being worse off only 20% had a great deal of confidence. When the personal financial question did not appear first, there was no association [the percentages were: confidence if personal finances better (18%), same (22%), and worse off (21%)].

Eighth, confidence in the people running organized religion (CONCLERG) is also higher in the latter condition described above. While we believe that there is a context effect affecting confidence ratings in general (see below), we do not know why significant individual effects showed only for CONFINAN and CONCLERG. They are not closely related institutions and appear at nearly opposite ends of the lists of institutions. In addition, confidence in religion does not show the conditional association to one's personal financial situation that the financial item does.

It appears however that context is impacting on the confidence scale as a whole rather than just on the two variables that show clear, individual differences. An additive confidence scale has marginally significant context differences for 1988–89 and shows the same difference in the pre-1988 data. Across the 13 confidence items in the pre-1988, 1988, and 1989 comparisons, confidence is greater in the version that omits the subjective financial questions 37 times out of 39. In addition, the confidence questions have been shown to be susceptible to other order effects (Smith, 1981).

Ninth, support for having elderly parents live with their children (AGED) is less after eight feminism questions (and items on school prayer and the UN) than after items on helpfulness and economic standing. We hypothesize that either exposure to feminism items emphasizing modern roles reduced support for traditional family care or that an expression of the helpfulness of people increased willingness to accommodate elderly parents (or both may be acting).

Tenth, more memberships in church-affiliated organizations (MEMCHURH) are reported after a battery of questions on pornography than after questions on smoking, drinking, and education. Since religious membership is preceded on both ballots by membership questions on 14 other groups, the impact apparently occurs across this buffer. Once again we are uncertain of why the context occurs. It is known that the church question is ambiguous and subject to varying interpretations and thus may be subject to contexts that activate these various understandings (Smith, 1988b). There might be a self-presentation effect of mentioning more religious affiliations after the pornography questions. Or the "sin" questions may lead people to define church-affiliated organizations in a wider fashion, thinking perhaps of their religious affiliation in general and/or perhaps drawing in morality-oriented groups in addition to religions.

Eleventh, more attendance in a racially integrated church (RACCHURH) is reported following questions about outlawing interracial marriages and support for open housing than following those about degree of neighborhood integration and approval of school integration. One possible explanation is that the neighborhood question frames the church integration question such that people tend to think of their local congregation ("church"), while in the other context more people think of their denomination ("church"). Since a local congregation is less likely to be integrated than a denomination, more integration would be reported in the former case than the latter.

Finally, selecting a "feeling of accomplishment" as an important preference in a job (JOBMEANS) is more common following questions on sexual and reproductive issues than after anomia items and racial items. (In both cases the job value question was immediately preceded by a question on whether getting ahead in life depends on luck or hard work.) This item is of only borderline plausibility. The effect may emerge from the negatively phrased anomia items' reducing mentions of the somewhat idealistic "accomplishment" preference and promoting the selection of more pragmatic selections such as high income, job security, and a chance for advancement as important preferences. However, the effect is not consistently greater among those with anomic feelings, which leaves this explanation open to question.

4.6 CONCLUSION

Overall the number of context effects created by the rotation design employed by the GSS prior to 1988 are minuscule; 11–12 probable order effects out of over 500 variables. Even this "dirty dozen" was concentrated among a few topics: anomia, misanthropy, satisfaction, and institutional confidence accounting for 8 of the 12 context effects. Distortions to the existing time series are also generally small in magnitude. Of the 12 probably real effects the average effect size is 7.5 percentage points (using pooled 1988–89 figures) with the largest being trusting people (TRUST) at 11.0 percentage points and the smallest being the less certain job values question (JOBMEANS) at 2.6 percentage points.[7]

Together with the Schuman and Presser DAS study, our GSS study suggests that unanticipated context effects might occur once out of every 40–60 questions. In one regard, however, this is probably an underestimate, since on the GSS, and presumably on the DAS, batteries of questions on one topic (e.g., the seven items on abortion or the 13-item confidence battery) were asked in a block and not varied across ballots. Since context effects are most likely to occur between closely related items, the failure to experimentally vary items within topical blocks probably underestimates the frequency of context effects. The widely discussed abortion example is a prime example of what can occur when the order within such a block is disturbed (Smith, 1989). See also Schuman and Presser's (1981) discussion of same. For other examples see

[7] Corrections to the time series due to these and other measurement variation are detailed in Smith, 1988d.

Astin, et al., 1988; Smith, 1981; and Smith, 1984. We suspect that context effects within scales are rather common, probably even typical. Since such scales tend to be replicated as units without changes to their internal order, these effects are rarely studied and whatever context effects exist generally remain fixed across administrations of the scale.

In addition to the commonness of context effects, the review of context effects on the GSS due to ballot position suggests certain characteristics of questions that make them prone to such effects.

First, many of the affected items seem to be rather general in their scope. For example, the items on helpfulness and trustworthiness ask for global judgments on the nature of man and one of the anomia items asks people to consider "the way things look for the future." All of these items theoretically make relevant nearly a lifetime of memories of the most varied and complex sort. Since the mind is neither disposed to such massive processing of memory nor to a random sampling of same, it is likely that the sampling of memories would be notably influenced by those memories and thoughts triggered by prior questions and that such priming would affect one's subsequent response to such general questions. While context effects do not always only involve such memory-overloading questions, context effects are more likely among extremely wide-ranging questions and these probably are among the more common types of context effects.

However, not all wide-ranging questions are readily susceptible to context effects. If people have a predetermined answer to the question that can be directly retrieved rather than having to undertake a general search, then biased memory sampling and therefore context effects are not likely to occur. For example, the inquiry "What is your favorite food?" in theory might require a search of a lifetime of eating experience. However, many people have consciously decided on their "favorite" food and have to report only on that judgment rather than retrieve and evaluate all their memories regarding foods.

Second, ambiguous terms may make questions more susceptible to context effects. In the case of the question on integrated churches it is likely that the context redefined respondents' understanding of the term "church" and a similar impact may have occurred on the church membership question. Or in the food example noted above people may interpret "food" as referring to main courses, desserts, or five course dinners depending on how prior context makes them understand the vague term "food." These types of context effects depend on the redefinition of question meaning and intent rather than on selective memory sampling as above (Smith, 1989).

Third, demographics seem to be relatively immune to context effects. We did not find a single example of a context effect for a demographic question and the literature reports few such effects. However, 2 of our probable 12 effects are on factual, behavioral reports and these are as likely to occur as attitudinal effects. (There are more attitudinal effects because there are more attitudinal questions on the GSS.) Demographics differ from the other items in several regards. First, they usually deal with concrete, basic facts that are well-known by respondents. Second, they deal with matters that are well understood by respondents. For most demographics people do not have to do extensive memory searches. There is also probably less misunderstanding between respondents and investigators on what is being asked for, since the information — age, marital status, etc. — is commonly used and frequently asked for. To some extent standard demographics tend to avoid the memory sampling and definition problems cited above.

This is not to say that there are no problems and no misunderstandings in demographics or that context cannot affect demographics, but rather that demographics tend to be less context sensitive than other questions. (They are relatively immune to context, because respondents understand what is being asked for and that information is readily available.) Also, the above generalizations about demographics apply to simple, basic demographics. Factual states that are difficult to report on or obscure may suffer as readily from context effects as did the items on church membership/integration.

Among the probable context effects on the GSS, we have examples of several of the different types of effects described by various taxonomies and theories of context effects (Tourangeau, et al., 1986; 1988; Smith, 1982b; 1989). These include effects acting at the interpretation, retrieval, and judgment stages. We have also identified three factors that contribute to context effects. Context effects are more likely to occur in questions that (1) require wide-ranging memory searches because the subject covers many relevant memories, (2) access memories that have not been previously organized into a summary evaluation that supplies a simple, direct answer to the question being posed, and (3) utilize ambiguous terms and/or have uncertain intent. These are clearly not the only characteristics of questions that are relevant to context effects (for an extensive list see Tourangeau and Rasinski, 1988), but may be among the most common.

CHAPTER 5

MODE EFFECTS OF COGNITIVELY DESIGNED RECALL QUESTIONS: A COMPARISON OF ANSWERS TO TELEPHONE AND MAIL SURVEYS

Don A. Dillman and John Tarnai
Washington State University

5.1 INTRODUCTION

In recent years, substantial research interest has focused on cognitive approaches to question design in an effort to obtain more accurate recall information. For example, it has been found that the ability of respondents to recall accurately their visits to doctors within particular time periods is improved by having respondents reconstruct details of the visit and by encouraging the use of benchmark dates (holidays or birthdays) (Means, et al., 1989). By probing for details such as "Who drove?" or "Who was the health care provider?" the memory of specific visits was cued or otherwise reinstated. These methods improved recall accuracy from 41 percent before the intervention to 63 percent afterwards (Jobe and Mingay, 1989). Use of an 18-month calendar on which volunteers were asked to mark important events they could accurately date, such as birthdays and holidays, resulted in subjects improving their recall from 32 percent to 60 percent of the actual number of interventions. These studies were based upon personal interviews of the respondents.

In this chapter we report results on recall of a specific behavior — the wearing of seatbelts — applied to two other survey methods, telephone and mail. To be tested is whether recall of the use of seatbelts in their most recent trip in a car is improved by first asking people to recall details

Appreciation is expressed to Richard Kulka for comments on a previous draft of this chapter.

Measurement Errors in Surveys.
Edited by Biemer, Groves, Lyberg, Mathiowetz, and Sudman.

ISBN: 0-471-53405-6

of each trip, i.e., destination, distance, and whether they were passengers or the drivers. The design of the study enables us to compare results within as well as across methods for both "quick recall" and "cognitively designed" question formats.

5.2 SOCIAL DESIRABILITY AND MODE DIFFERENCES

A small but expanding body of research on telephone vs. mail mode differences now exists. Although this research has been mostly conducted on types of questions different from those to be examined here, that research provides reasons for expecting mode differences in the current situation. However, the literature is far from conclusive on the nature and causes of mode differences.

One focus of past research on mode differences has been the tendency of respondents to offer socially desirable answers to threatening questions, which has been observed for all three methods of administration (Sudman and Bradburn, 1974). The literature is far from consistent on which method produces the most accurate answers. However, Bradburn (1983) has suggested that the more anonymous method of administration appears to work somewhat better by lowering the degree of under- or overreporting. Several studies that have made explicit mail vs. telephone comparisons support that perspective, providing evidence that socially desirable answers are more likely to be offered on the telephone than by the more anonymous (in the sense of lacking interpersonal interaction) mail method (Hochstim, 1967; Locander, et al., 1976; Siemiatycki, 1979; DeLeeuw, 1990b; and Aquilino and Losciuto, 1990). Inasmuch as seatbelt usage was required by state law at the time of this study and violators could be fined for noncompliance, it might be expected that socially desirable answers (greater use of seatbelts) would be offered by telephone respondents. However, compared to questions examined in the above studies, reporting the use of seatbelts seems somewhat less sensitive and therefore not a major reason for expecting socially desirable answers for either method.

5.3 COGNITIVE PROCESSING AND PRIMACY VS. RECENCY EFFECTS

Another focus of past research has been the influence of the order in which answer choices are presented (see Chapter 3). For example, Krosnick and Alwin (1987) have argued that respondents are more likely to choose early items (defined as a *primacy effect*) from a list when it is presented in a visual mode; selection of later items (defined as a *recency effect*) is more likely to occur when questions are presented in an auditory mode. The explanation they offer for this effect is that items presented early in the list in a visual format are subject to more extensive cognitive processing. It is argued that these early items tend to establish a frame of reference for evaluating the later items. Israel and Taylor (1990) have confirmed that early items on a mail questionnaire are more likely to be selected, but only when respondents can choose more than one answer and not when a single choice is to be made.

Schwarz, et al. (1989) and Hippler, et al. (1989, 1990) have demonstrated that the primacy–recency mode effects identified by Krosnick and Alwin (1987) are not limited to the asking of long lists of items. Even for dichotomous attitude questions, primacy effects are likely to be obtained when question choices are presented in the visual mode. Further, they have shown that a precondition for response order effects is the need for respondents to form a judgment on the spot. They demonstrate that at least in the auditory format the recency effect does not exist when respondents have a preformed judgment on the issue. Their results also demonstrate that for some subsamples, i.e., those with preformed judgments, education, traditionally thought to be an important reason for differences, has no effect.

The question examined here — the extent to which people report wearing a seatbelt during the most recent trip(s) in their cars — differs from the questions examined by Hippler, et al. (1990), and Krosnick and Alwin (1987) in two important ways. First, people were asked to report a specific behavior for which the answer choices can be considered correct or not correct, unlike attitudes which are not so unequivocally true or false. Second, instead of being given nominal answer category choices, respondents were presented with ordinal choices. Respondents were stimulated to think about an underlying scale or continuum, making the response task one of placing themselves onto that implied continuum in the most appropriate category.

5.4 EXTREMENESS AS A RESULT OF CONTEXT

In a previous study using scalar response categories, Dillman and Mason (1984) found that respondents to telephone (auditory) questionnaires were substantially more likely to choose the first mentioned category than were those who responded to the mail (visual) form, the opposite of the effect predicted by Krosnick and Alwin (1987). Visual mode respondents were significantly more likely to choose intermediate categories. The effect was especially pronounced for a series of nine items on community qualities, e.g., police protection and quality of schools. These findings from a field survey were closely replicated under laboratory conditions by Tarnai and Dillman (in press). In that study it was concluded that part of the difference was due to context effects, that is, the ability of mail survey respondents to scan the entire set of questions and see that they were using the same response categories for a number of items. Thus, a mail survey respondent might hold back on the use of extreme categories, in a sense "saving" extreme choices for potential use with later items. However, it was also demonstrated in that paper that most of the differences could not be explained by context considerations.

5.5 EXTREMENESS AS A RESULT OF A "TOP-OF-THE-HEAD" RESPONSE

One possible explanation for the remaining differences stems from the time pressures to respond, which produce a "top-of-the-head" response for the auditory mode in contrast to a more carefully reasoned response for the visual mode. Telephone interviewers tend to control the pace of the interview and may exert a sense of urgency to respond more quickly than would be the case for respondents who can proceed at their own pace to complete self-administered questionnaires. It has often been assumed that an advantage of mail questionnaires is that respondents to such questionnaires provide more thoughtful answers because of being able to go at their own pace to complete them (Dillman, 1978). Hippler and Schwarz (1987) suggest that the less time respondents have to form a judgment, the more likely they are to respond to the first piece of information that comes to mind. The effect of this "top-of-the-head" tendency for scalar questions seems likely to be to encourage selection of either end of the scale rather than one of the intermediate categories. Indirect support for this explanation of mode differences is offered by Bishop et al. (1988), in a study that shows weaker differences exist in self-administered questionnaires than in telephone surveys. However, these questions were not of the scalar type utilized by Tarnai and Dillman (1988).

If respondents are more prone to provide "top-of-the-head" answers on the telephone, we expect that (1) they would be more likely to choose the most easily remembered extreme categories at either end of the scale when asked in a quick recall format and (2) use of a cognitively designed series of questions would mitigate the top-of-the-head response and bring answers to self-administered and telephone responses closer together. The design of our present study also makes it possible to test for certain saliency or preformed judgment effects identified by Hippler, et al. (1989).

5.6 STUDY DESIGN

The current test of quick recall and cognitively designed questions was included in a 1988 study of seatbelt use in Washington State (Tarnai and Dillman, 1988). The purpose of the survey was to determine compliance with a new state law mandating the wearing of seatbelts. This law had become effective in July 1986 and in January 1987 fines were being assessed for noncompliance. Thus the issue had moderate salience to Washington drivers.

For the experiment, separate statewide mail and telephone samples were drawn randomly by Survey Sampling, Inc. from all households in the state with listed telephone numbers. The mail questionnaires were designed in a booklet format and administered according to total design method (TDM) principles but without a third follow-up (Dillman, 1978). Completed questionnaires were received from 790 respondents, or 66 percent of the originally sampled households (or 76 percent of those questionnaires not returned to sender because of bad addresses). The telephone households received a prior letter informing the respondent that he or she would be called. Interviews were completed with 76.2 percent of the 526 households contacted, or 53.7 percent of all household numbers originally drawn for the sample. (Eighty-two numbers were disconnected and 57 were still not contacted after six attempts.) Thus, cooperation rates for the two surveys were nearly identical.

In all versions of the survey, the person with the most recent birthday was asked to complete the questionnaire. The proportion of males who responded was greater for the mail survey (56 percent) than for the telephone survey (48 percent), a finding that was expected since the tendency still exists for telephones to be listed in husbands' names more frequently than vice versa. Thus, letters were more likely to be addressed to male household members.

The questions of interest to this study were embedded near the middle of a 67 item questionnaire (10 pages in the mail version). A variety of questions concerning attitudes about seatbelt use and behavior regarding seatbelts preceded as well as followed the experimental questions. In the quick recall format, respondents were asked "We would like to ask about the most recent time you drove or rode anywhere in an automobile or other vehicle such as pickup or van. During this most recent ride would you say your seatbelt was fastened: (1) all the time, that is, every minute the car was moving; (2) almost all the time; (3) most of the time; (4) about half of the time; (5) less than half of the time; (6) not at all during the time the vehicle was moving." The question was immediately repeated for the second most recent time and third most recent time.

In the cognitive format, respondents were asked the same question but in each case it was preceded by the following four questions to encourage accurate recall: A. When was the last time (or next to the last time before that) you drove or rode a motor vehicle? B. Could you tell us generally where this most recent trip began and where did it end? C. About how long was this trip? D. During this trip, were you a driver, front seat passenger, backseat passenger, or other? The formats for the telephone and mail versions of the questions are shown in Exhibit 5.1.

Asking respondents about the last three times they rode in a vehicle, rather than only their most recent trip, was done to provide a gradation of difficulty in recalling their use of seatbelts. It was reasoned that the greater the difficulty of recalling one's behavior, the greater the likelihood of a "top-of-the-head" response being substituted for carefully thought out answers. Thus, the likelihood of "top-of-the-head" responses in the quick-recall version of the questionnaire was expected to be greater for the second and still more for the third trip than for the most recent one. To the extent that cognitively designed questions make a difference, we would expect them to make less of a difference for the most recent trip than for the ones taken before that.

Hippler, Schwarz, and Noelle-Neumann have reported that cognitive elaboration on an issue in preceding questions may eliminate response effects, with respect to question order (1990). Although virtually all of the questions preceding the experimental ones concern some aspect of seatbelt use, none of them had dealt specifically with seatbelt use in a specific situation. The test questions represented the first place in the questionnaire that respondents were asked to report use of a seatbelt for a specific trip in a motor vehicle.

5.7 FINDINGS

The first question to be addressed is whether telephone respondents were more likely than mail respondents to utilize the end categories, which we initially interpreted to be a top-of-the-head response for the quick recall questions. From Table 5.1 we can see that a modest trend in this direction is evident, ranging from 1.8 to 10.3 percentage points more for the "all the time, that is, every minute" category. From 0.2 to 3.9 percentage points more of the telephone respondents used the "not at all during the time the vehicle was moving" category compared to mail respondents.

Data in this table also provide modest evidence that the top-of-the-head telephone effect is greater, as predicted, for the presumably more difficult to recall second and third most recent vehicle trips. The percentage differences for the first, second, and third trips are +1.8, +6.5, and +10.3, respectively, for telephone respondents in the "all the time" end category. No such trend is evident at the other end of the scale.

Exhibit 5.1. Experimental Seatbelt Use Question Formats

A. *Quick Recall Format:*

Q-20 Next, we would like to ask about the most recent time you drove or rode anywhere in an automobile or other vehicle such as a pickup or van. During this most recent ride, would you say that your seatbelt was fastened . . .

1 ALL THE TIME, THAT IS, EVERY MINUTE THE CAR WAS MOVING
2 ALMOST ALL THE TIME
3 MOST OF THE TIME
4 ABOUT HALF THE TIME
5 LESS THAN HALF THE TIME
6 NOT AT ALL DURING THE TIME THE VEHICLE WAS MOVING

Q-21 What about the second most recent time that you drove or rode in a motor vehicle? Would you say that your seatbelt was fastened . . .

Same categories as Q–20

Q-22 What about the time before that, that is, the third most recent time that you drove or rode in a motor vehicle. Would you say your seatbelt was fastened . . .

Same categories as Q-20

B. *Cognitively Designed Format:*

Next, we would like to ask you to please think about the last three times you drove or rode in a car or vehicle such as a pickup or van.

Q-20 First, when was the last time you drove or rode in a motor vehicle . . .

1 TODAY
2 YESTERDAY
3 SOME TIME BEFORE THAT

Q-21 Could you tell us generally where did this *most recent trip* begin and where did it end?

TRIP BEGIN______________________________
TRIP END______________________________

Q-22 About how long was this trip?

1 LESS THAN A MILE
2 ONE TO FIVE MILES
3 LONGER

Q-23 During this trip, were you the . . .

1 DRIVER
2 A FRONT SEAT PASSENGER
3 A BACK SEAT PASSENGER
4 OTHER

Q-24 During this trip, would you say your seatbelt was fastened . . .

Same categories as Q-20, Part A

Q-25 Next, we would like to ask you about the *second most recent* time that you drove or rode in a motor vehicle. When did this second most recent trip take place? (Etc. for remainder of second and third most recent trips.)

Table 5.1. Mail and Telephone Distribution for "Quick Recall" Questions on Seatbelt Use During Three Most Recent Trips

	Most Recent*			Second Most Recent**			Third Most Recent***		
	Mail ($N=401$)	Telephone ($N=165$)	Difference	Male ($N=404$)	Telephone ($N=165$)	Difference	Mail ($N=403$)	Telephone ($N=164$)	Difference
All the time, that is, every minute the car was moving	56.9%	58.7%	+1.8	56.9%	63.4%	+6.5	54.6%	64.6%	+10.3
Almost all the time	17.8	10.5	−7.3	16.4	10.5	−5.9	18.7	9.9	−8.8
Most of the time	6.5	6.4	−0.1	7.4	6.4	−1.0	5.3	4.1	−1.2
About half the time	3.5	4.0	+0.5	3.7	5.8	+2.1	4.6	2.9	−1.7
Less than half the time	3.5	4.7	−1.2	3.7	1.7	−2.0	4.1	4.1	0.0
Not at all during the time the vehicle was moving	11.8	15.7	+3.9	12.0	12.2	+0.2	12.7	14.0	+1.3

*$\chi^2=6.317$, $p=0.277$.
**$\chi^2=7.049$, $p=0.217$.
***$\chi^2=9.402$, $p=0.094$.

Overall, chi square tests of differences reveal that the telephone and mail responses are not significantly different for any of the three most recent trips. Our general conclusion is that although a slight trend is evident, these data provide no statistical support for the top-of-the-head hypothesis that mode differences exist when using the quick recall format.

The next question to be addressed is whether differences exist for cognitively designed questions across survey modes. Table 5.2 reveals no differences for any of the three most recent trips. None of the distributions exhibited significant chi squares. The only slight trend is that for the second and third most recent trips the extreme category of "all the time..." is 7.1 and 4.2 percentage points greater, respectively, for the telephone mode. Thus, we conclude that cognitively designed question sequences produce nearly the same results for the two modes.

However, a comparison of data between Tables 5.1 and 5.2 reveals a surprising finding. The percentages in the terminal categories at both ends of the scale are substantially higher for the cognitively designed questions, rather than being depressed as the top-of-the-head hypothesis would suggest. These differences between quick recall and cognitively designed questionnaires are summarized in Table 5.3. It can be seen here that the cognitively designed format "all the time..." category increased 13.1 to 16.6 percentage points for mail and 10.5 to 14.1 percentage points for telephone over the quick recall format. The other extreme category "not at all..." also increased, 3.7 to 6.0 percentage points for mail and 1.7 to 3.6 percentage points for telephone. All of the differences between question formats are significant. Surprisingly, mail responses exhibit greater change in this direction than do the telephone responses. The major effect was to drain people from the intermediate categories toward the end categories rather than the opposite, as we had predicted. Thus, although we found that the top-of-the-head effects seem to be operative under the quick recall format, i.e., telephone respondents showed a slight trend towards choosing the end categories, the cognitively designed questionnaire respondents were even more likely to make that kind of response.

One possible explanation for this seemingly contradictory finding is that the correct answer is, in fact, one of the extreme categories. Suppose that respondents see themselves as wearing seatbelts only part or most of the time in their daily lives. Under the quick recall format, the top-of-the-head phenomenon may cause respondents to think first in terms of what they usually do, but they do not have (or take time) to think about their actual most recent trip in the car. Thus, the "top-of-the-head" tendency may be to select categories closest to their usual behavior, which may or may not be one of the end categories. Then, when the cognitive question

Table 5.2. Mail and Telephone Distribution for Cognitively Designed Questions on Seatbelt Use During Three Most Recent Trips

	Most Recent*			Second Most Recent**			Third Most Recent***		
	Mail ($N=456$)	Telephone ($N=244$)	Difference	Mail ($N=456$)	Telephone ($N=244$)	Difference	Mail ($N=419$)	Telephone ($N=232$)	Difference
All the time, that is, every minute the car was moving	70.0%	70.9%	−0.9	70.4%	77.5%	+7.1	71.2%	75.4%	+4.2
Almost all the time	8.1	6.1	+2.0	8.8	5.3	−3.3	8.0	4.2	−3.8
Most of the time	2.0	1.6	+0.4	1.7	2.0	+0.3	2.7	3.3	+0.6
About half the time	1.3	1.6	−0.3	1.5	1.2	−0.3	1.1	2.5	+1.4
Less than half the time	0.9	0.4	+0.5	0.9	0.0	−0.9	0.7	0.0	−0.7
Not at all during the time the vehicle was moving	17.8	19.3	+1.5	16.7	13.9	−2.8	16.4	14.6	−1.8

*$\chi^2=6.317$, $p=0.277$.
**$\chi^2=7.049$, $p=0.217$.
***$\chi^2=9.402$, $p=0.094$.

Table 5.3. Differences Between Percentage Distributions for Quick Recall and Cognitively Designed Questions (former subtracted from the latter) on Seatbelt Use for Three Most Recent Trips

	Mail			Telephone		
	Most Recent	Second Most Recent	Third Most Recent	Most Recent	Second Most Recent	Third Most Recent
All the time, that is, every minute the car was moving	+13.1	+13.5	+16.6	+12.2	+14.1	+10.5
Almost all the time	−9.7	−7.6	−10.7	−4.4	−5.2	−5.7
Most of the time	−4.5	−5.7	−2.6	−4.8	−4.4	−0.8
About half the time	−2.2	−2.2	−3.5	−2.4	−4.6	−0.4
Less than half the time	−2.6	−2.8	−3.4	−4.3	−1.7	−4.1
Not at all during the time the vehicle was moving	+6.0	+4.7	+3.7	+3.6	+1.7	+0.6
χ^2 test for differences between question formats	$\chi^2 = 48.312$ $p < 0.001$	$\chi^2 = 43.397$ $p < 0.001$	$\chi^2 = 52.490$ $p < 0.001$	$\chi^2 = 18.294$ $p < 0.01$	$\chi^2 = 19.677$ $p < 0.01$	$\chi^2 = 9.928$ $p < 0.05$

is asked, respondents actually consider details of the most recent trip with the result that they move from this "normative" response towards giving a more accurate representation of their actual behavior on each trip.

Fortunately, a limited test of this explanation can be provided. In an earlier survey question, respondents were asked in general how often they wore their seatbelts. Answers to this question can be cross-tabulated with answers to the three most recent trips questions. If our hypothesis about a normative effect is correct, we would expect that for those who always wear *or* do not wear their seatbelts the top-of-the-head response would be to select the appropriate end category for the quick recall questions, and for the cognitively designed questions to move towards the intermediate categories. For those who wear their seatbelts only part of the time, we would expect the top-of-the-head response to be a selection of an intermediate category under the quick recall format and that respondents would migrate towards the other categories under conditions of the cognitively designed questions. Such differences might be expected based upon the Hippler, et al. findings that response order effects are more likely to occur when there is no preformed judgment, thus requiring one to think in more detail to provide an accurate answer (1990). Thus, the response category most accessible as a "top-of-the-head" answer for the most recent trips would be the answer closest to the respondent's usual seatbelt behavior.

Comparisons of the respondents' usual behaviors with reports of their most recent trips, shown in condensed form in Tables 5.4 and 5.5 for mail and telephone, respectively, provide only partial support for this expectation. Under the cognitive format, use of the end categories increases for all types of respondents, not just for those respondents whose usual behavior is to wear their seatbelts part of the time. The pattern of differences between question formats is virtually the same for both the mail and telephone modes of data collection.

The percent of "all the time" wearers of seatbelts who reported wearing their seatbelts all the time in their three most recent trips increased 9.3 to 9.4 percent for the telephone cognitive format and 9.1 to 10.1 for the mail cognitive format. The percent of the "never" wearers who reported "not at all" was 15 to 26.2 percentage points for the telephone cognitive format and 7.1 to 20 percentage points higher for the mail cognitive format versus the respective quick response formats. In all cases the percentage of "all the time" and "never" wearers who reported the same behavior for the three most recent trips was 90 percent or higher under the cognitively designed format. If the usual behavior question is a valid measure of normal behavior, it can be argued that the effect of the cognitively designed question was to increase the accuracy of respondent's answers, as found by Jobe and Mingay (1989) for doctor visits.

The changes in responses for "part of the time" users are in the direction we expected, i.e., away from the middle categories toward the

Table 5.4. Comparison of Mail Respondent's Behavior on Three Most Recent Trips with Previously Reported Usual Behavior

	HOW OFTEN SEATBELT IS USUALLY WORN*								
	All the Time			Part of the Time			Never		
Was seatbelt worn on:	Quick Recall	Cognitive Format	Difference	Quick Recall	Cognitive Format	Difference	Quick Recall	Cognitive Format	Difference
Most recent trip?									
All the time	85.9%	96.0%	+10.1	29.4%	48.7%	+19.3	2.7%	0.0%	−2.7
Part of the time	13.7	3.2	−10.5	60.0	27.9	−32.1	10.8	6.4	−4.4
Not at all	0.4	0.8	+0.4	10.6	23.4	+12.8	86.5	93.6	+7.1
Second most recent trip?									
All the time	88.9	96.0	+9.1	26.5	48.7	+22.2	5.4	2.3	−3.1
Part of the time	10.2	2.8	−7.4	62.4	32.0	−30.4	13.5	4.3	−9.2
Not at all	0.9	1.2	+0.3	11.1	19.2	+8.1	81.1	93.6	+12.5
Third most recent trip?									
All the time	86.8	97.2	+9.4	22.5	49.4	+26.9	5.4	2.1	−3.3
Part of the time	11.8	1.6	−10.2	63.8	32.7	−31.1	18.9	2.1	−16.8
Not at all	1.3	1.2	−0.9	13.8	18.0	+4.2	75.7	95.7	+20.0

*The question: Since January 1987, Washington drivers and front seat passengers may be fined for not wearing seatbelts. Since that time, how often have you worn your seatbelt: (1) Always, (2) Most of the Time, (3) Sometimes, (4) Rarely, (5) Never, (6) Don't Know.

Table 5.5. Comparison of Telephone Respondent's Behavior on Three Most Recent Trips with Previously Reported Usual Behavior

	HOW OFTEN SEATBELT IS USUALLY WORN*								
	All the Time			Part of the Time			Never		
Was seatbelt worn on:	Quick Recall	Cognitive Format	Difference	Quick Recall	Cognitive Format	Difference	Quick Recall	Cognitive Format	Difference
Most recent trip?									
All the time	86.4%	95.7%	+9.3	24.0%	47.7%	+23.7	0.0%	0.0%	0.0
Part of the time	10.7	3.6	−7.1	54.0	20.9	−33.1	25.0	5.0	−20.0
Not at all	2.9	0.7	−1.8	22.0	31.4	+9.4	75.0	95.0	+20.0
Second most recent trip?									
All the time	89.3	98.6	+9.3	30.0	59.3	+29.3	12.5	10.0	−2.5
Part of the time	10.7	0.7	−10.0	54.0	23.3	−30.7	12.5	0.0	−12.5
Not at all	0.0	0.7	+0.7	16.0	17.4	+1.4	75.0	90.0	+15.0
Third most recent trip?									
All the time	88.4	97.8	+9.4	34.7	56.6	+21.9	12.5	0.0	−12.5
Part of the time	10.7	0.7	−10.0	44.9	26.5	−18.4	18.8	5.0	−13.0
Not at all	1.0	1.5	+0.5	20.4	16.9	−3.5	68.8	95.0	+26.2

*The question: Since January 1987, Washington drivers and front seat passengers may be fined for not wearing seatbelts. Since that time, how often have you worn your seatbelt: (1) Always, (2) Most of the Time, (3) Sometimes, (4) Rarely, (5) Never, (6) Don't Know.

extreme answers. The percentage in the "all the time" categories for the most recent trip increased dramatically, from 19.3 to 26.9 percentage points higher for mail and 21.9 to 29.3 percentage points higher for telephone versus the quick recall formats. These data clearly support the idea that the quick recall format encourages a normative (i.e., usual behavior) response and the cognitive format results in a substantial shift in people's answers toward the extreme categories, particularly towards the "all the time" category.

These dramatic shifts among part-time users suggest that the top-of-the-head phenomenon may be at work, at least among some respondents. However, the results for other users suggest that some other factor, perhaps inherent to the use of ordinal scales with memorable end categories, may also be at work. Although the shift for the extreme "all the time" users was not in the expected direction, it is in the direction of suggesting greater congruence between usual behavior and most recent trip behavior.

Although there was no means available to us for measuring actual seatbelt use behavior in this study, one indirect indicator of the accuracy of people's answers may be the consistency among the three answers provided by each respondent. If respondent answers are influenced by their normative behavior and/or if less careful consideration is being given to recalling actual seatbelt use under the quick recall format, it is expected that statistical correlations between the three answers provided by each respondent will be higher than under the cognitively designed format. To test this idea, Pearson correlations were calculated among the three reported uses of seatbelts by survey mode. For the quick recall method, correlations for each of the three two-way comparisons between reported uses were 0.84, 0.81, and 0.87 for mail, and 0.75, 0.69, and 0.79 for telephone. As expected, the correlations were slightly lower under the cognitive format: 0.73, 0.76, and 0.77 for mail, and 0.71, 0.70, and 0.69 for telephone. The differences are from 0.04 to 0.11 higher for the quick recall method in all comparisons except one where a difference of 0.01 in the opposite direction was observed. We conclude that in general responses were less consistent for the cognitive method, providing limited evidence that recall was improved by the cognitively designed procedure.

5.8 POTENTIAL GENDER EFFECTS

Further analysis of the data revealed that gender was significantly related to reports of seatbelt use during the three most recent trips. Females were consistently more likely to report wearing seatbelts "all of the time" during each of their most recent three trips, the differences (female–male) being 11.8, 12.6, and 8.9 percentage points. Similarly, females were less likely to report "not at all" wearing their seatbelts (−9.8, −10.8, and −7.2 percentage points, respectively). As reported earlier, females were also more likely to respond over the telephone than by mail, 52 vs. 44 percent of the respective samples.

Therefore, it was deemed important to test for gender effects that would influence our conclusions. A linear logistic analysis was conducted with the seatbelt use questions considered as dependent variables, while mode, question form, and respondent gender were used as independent variables. For all three analyses of seatbelt use, the two-way and three-way interactions among the independent variables were nonsignificant. It was concluded that the combining of the analyses for males and females would not significantly affect the main conclusions reported above. Therefore, only the analysis including both males and females has been reported.

5.9 MODE EFFECTS FOR OTHER QUESTIONNAIRE ITEMS

As previously noted, 67 items were included in the questionnaire, only three of which were the central focus of the specific experiment reported here, and for which minimal mode effects were found to exist. A pattern of mode effects was noted for some of these remaining items, which are reported here for the contextual benefit they provide for interpreting results from the experiment.

Seventeen questionnaire items involved similar answer alternatives of *yes*, *no*, and in most cases *don't know* or *maybe*. Eight of the questions concerned reported behaviors; tests for mode differences on these items are reported in Table 5.6. The remaining nine items concerned opinions on seatbelt issues and the tests for mode differences are reported in Table 5.7.

The important finding is that only three of the eight behavioral questions exhibit mode differences that are statistically significant. In contrast, all nine of the opinion questions exhibited significant mode differences. Seven of these differences were significant at the 0.001 level. The overall pattern of results is for telephone respondents to be more likely to answer "yes" rather than to give one of the other responses. The experimental seatbelt use questions examined in this study were also behavioral in nature and thus appear consistent with these behavioral items in not exhibiting mode effects.

However, the questionnaire also included two other behavioral items which used four-choice ordinal response categories, and the responses clearly exhibited socially desirable characteristics. One of

Table 5.6. Comparison of Responses to Behavioral Questions by Mode of Administration

	Response Categories									
	Yes			No			Don't Know			Main Effects
	Mail	Tel	Diff	Mail	Tel	Diff	Mail	Tel	Diff	Chi-Square
1. In the past six months, have you heard or seen anything:										
A. in the media about the Washington safety belt law?	59.7	49.5	−10.2	30.4	43.3	12.9	9.8	7.2	−2.6	4.06*
B. about safety belts having saved someone's life or reduced injuries in an accident?	63.3	63.7	−0.6	27.3	31.1	3.8	8.3	5.2	−3.1	0.34
C. about attempts to repeal the Washington safety belt law?	14.3	14.7	0.4	76.4	83.3	6.9	9.4	1.2	−8.2	10.39**
2. Have you received brochures or other promotional materials about safety belts at any time during the past year?	12.4	12.4	0.0	76.4	80.4	4.0	11.2	7.2	−4.0	0.01
3. During the past year, can you remember having been asked or told to buckle up by a driver or passenger?	47.7	43.5	−4.2	51.5	56.5	5.0	—	—	—	0.76
4. Have you told or asked either drivers or passengers to buckle up?	77.8	76.9	−0.9	21.8	22.9	1.1	—	—	—	0.34
5. Have you, or anyone you know, ever received a ticket or warning for not wearing a safety belt?	19.1	21.6	2.5	77.9	78.1	0.2	—	—	—	4.71*
6. Are you a registered voter in Washington?	82.1	80.4	−1.7	17.3	19.7	2.4	—	—	—	1.3

$^{*}p < 0.05$; $^{**}p < 0.01$

Table 5.7. Comparison of Responses to Opinion Questions, by Mode of Administration

	Response Categories									
	Yes			No			Don't Know			Main Effects
	Mail	Tel	Diff	Mail	Tel	Diff	Mail	Tel	Diff	Chi-Square
1. In your opinion can police stop you just for not wearing a safety belt?	33.6	46.0	12.4	52.5	45.0	7.5	13.9	9.0	−4.9	20.79***
2. Next we want to know your opinions about how to increase safety belt use. Please tell us whether you think these suggestions would work in increasing safety belt use.										
A. Have police write more tickets for not wearing a belt.	19.1	21.6	2.5	77.9	78.1	0.2	—	—	—	4.71*
B. Increase the fine for not wearing belts.	37.5	47.0	9.5	35.2	34.6	−0.6	27.3	18.4	−8.9	13.12***
C. Provide more information about the effectiveness of safety belts.	27.4	32.7	5.3	50.7	57.5	6.8	21.9	9.7	−12.2	17.83***
D. Make sure local police use their belts.	71.8	78.5	6.7	9.4	15.1	5.7	18.8	6.4	−12.4	22.59***
E. Have police tug on their shoulder strap as a reminder when they see an unbuckled motorist.	69.8	86.6	16.8	15.8	9.0	−6.8	14.3	4.4	−9.9	51.62***
F. Encourage employers to have strong belt use policies for on-the-job automobile use.	56.3	82.1	25.8	20.7	10.4	−10.3	23.0	7.5	−15.5	111.69***
G. Have media publicize stories about people who were saved by using their safety belts.	74.4	87.2	12.8	11.4	7.3	−4.1	14.2	5.5	−8.7	33.21***
3. Do you favor or oppose keeping the Washington safety belt law?	69.7	78.9	9.2	23.1	16.2	−6.9	7.1	5.0	−2.1	8.23**

*$p < 0.05$; *$p < 0.01$; ***$p < 0.001$. For questions 1 and 3, the third response category was "Don't Know" rather than "Maybe."

Table 5.8. Comparison of Responses to Socially Desirable Behavioral Questions by Mode of Administration

	How often do you yourself drive after drinking alcoholic beverages		How often do you ride with other drivers after they have drunk alcoholic beverages	
	Mail $n=875$	Telephone $n=400$	Mail $n=884$	Telephone $n=401$
Frequently	1.1%	1.8%	1.4%	2.0%
Occasionally	11.1	9.0	10.7	9.0
Seldom	35.8	26.5	42.5	32.2
Never	51.8	62.5	43.7	56.3
Don't Know	0.2	0.2	1.7	0.5
	100.0	100.0	100.0	100.0
	$\chi^2 = 14.82$ $p < 0.001$		$\chi^2 = 21.43$ $p < 0.001$	

these questions asked how often the respondent drove after drinking alcoholic beverages and the other asked how often the respondent rode with other drivers after they had been drinking alcoholic beverages. The answer choices were *frequently*, *occasionally*, *seldom*, *never*, and *don't know*, presented in that order. For both questions telephone respondents were significantly more likely to respond "never" (63 percent to 52 percent and 56 percent to 44 percent, respectively); (Table 5.8).

Thus, we conclude that behavioral reports are not immune to mode effects, as already demonstrated in the literature. Although the experimental items did not demonstrate mode effects, they clearly occurred elsewhere in the study questionnaires. In addition, these effects were not limited to either the first or last listed categories.

For the analyses of the questions in Tables 5.6 and 5.7, survey mode and respondent gender were both included in the logistic model, along with an interaction term. None of the interaction terms in any of the models were statistically significant. Some gender differences were found for these questions; four significant differences in Table 5.6, and three in Table 5.7. However, the lack of significant interactions suggests no differences between males and females in the pattern of mode effects.

5.10 CONCLUSION

Consistent with earlier work by Jobe and Mingay (1989) and Means et al. (1989) done for face to face interviews, we have found that a cognitively designed question sequence produces different and presumably more accurate responses to questions about past behavior than did a quick recall format, for both mail and telephone surveys. For example, the percentage of respondents reporting wearing seatbelts all of the time on their most recent three trips were 13 to 16 percentage points for the mail and 11 to 14 percentage points for the telephone cognitively designed format.

Thus, evidence now exists that the use of cognitively designed questions is important for all three major survey modes, i.e., face to face, telephone, and mail. This finding is particularly noteworthy in the case of mail surveys. It has been reasoned in the past that mail survey respondents may take their time in filling out questionnaires, carefully thinking through questions before answering them. This ability to elicit more thoughtful answers has been deemed one of the desirable attributes of that survey method. The data we have reported suggest that this image of mail questionnaires is more a myth than reality; even though the respondents have the questionnaire under their own control, they may cognitively scan questions quickly, similar to the way they are responded to when an interviewer controls the answering process. Therefore, cognitively designing mail survey questions appears to be as important as it is for other survey methods.

The extremeness effect, whereby telephone respondents give more extreme answers than do mail respondents, reported previously by Tarnai and Dillman (in press), was not reproduced in the experimental questions examined in this study. One possible explanation is that the two studies investigated quite different phenomena. The former study examined community beliefs and attitudes for which respondents seemed less likely to possess a ready-made answer. Under these conditions, the first piece of information respondents may draw upon to formulate an answer may be the scale itself and the more memorable end categories. In responding to questions about seatbelt use in the three most recent trips, on the other hand, the most easily accessible information may be one's normal or usual behavior, which would not necessarily be an extreme category. This explanation receives some support from the dramatic shift shown by part-time users of seatbelts. However, other ("all the time" or "never") users showed similar albeit far less dramatic differences between the two formats in a similar direction towards greater use of the extreme categories. This suggests that a top-of-the-head response cannot wholely account for the differences observed here.

It remains for future research to provide a full understanding of reasons for differences between the quick recall and cognitively designed question formats. Suffice it to say for now, however, that none of the major data collection procedures can ignore cognitive design issues.

Although mode effects on the experimental questions appeared minimal, evidence of mode effects existed for other survey questions, not subjected to the cognitive design experiment. Opinion questions had a far greater tendency to exhibit mode differences than did behavioral questions, except for two questions on drinking and driving, for which socially approved answers were clearly identifiable. The socially desirable answers, chosen more frequently by telephone, were neither the first nor last answer choice presented.

Increasingly, we are convinced that such mode differences between mail and telephone cannot be explained by a single perspective. Social desirability, cognitive processing, top-of-the-head responses, and context may all play a role in explaining differences between the two methods. Their importance may depend on the type of question being asked. Much remains to be learned as we seek to develop a theory of survey mode similarities and differences.

CHAPTER 6

NONEXPERIMENTAL RESEARCH ON QUESTION WORDING EFFECTS: A CONTRIBUTION TO SOLVING THE GENERALIZABILITY PROBLEM

Nico Molenaar
Vrije Universiteit

6.1 INTRODUCTION

The immediate aim of the researcher who studies question wording effects is to know whether a particular wording aspect of a particular question has an effect on the resulting responses, and thus to be able to evaluate the quality of his or her own research data. The broader aim, however, is to develop a theory about the effects of question wording. Whatever such a theory may be substantively, it should, at least, contain sufficiently general statements about the effects of the particular wording. This chapter discusses a research strategy to reach the latter aim. In particular a nonexperimental method for investigating wording effects is proposed as a possible strategy. The emphasis is not so much on research findings or on substantive theory about question wording effects, but on the research method. The discussion will be confined to wording effects on the univariate response distributions, or, more precisely, on statistical characteristics thereof.

Before discussing the nonexperimental method, we first pay some attention to the experimental strategy for establishing general wording effects and to the individual split ballot experiment which is the basis of that strategy.

Measurement Errors in Surveys.
Edited by Biemer, Groves, Lyberg, Mathiowetz, and Sudman.

ISBN: 0-471-53405-6

6.2 THE SPLIT BALLOT EXPERIMENT

The procedure for a split ballot experiment is to formulate a question in at least two different ways, and submit the variants to equivalent samples of respondents. If the resulting response distributions differ significantly, then it can be concluded that the wording variable manipulated has influenced the responses to the question. Early examples of such experiments can readily be found in the literature. (See Blankenship, 1940; Rugg and Cantril, 1942, 1947.)

The split ballot design has many advantages. The structure is simple and clear; the design enables one to study the effects of any wording variable whatsoever. It also is a good starting point for more detailed investigations, for example to find out which categories of respondents are most sensitive to the question wording, and for testing specific substantive hypotheses. It is also a powerful design methodologically. Internal validity, defined as the extent to which the potentially disturbing variables are controlled for, is high. Some of the disturbing variables, such as the characteristics of the respondent, (most) interviewer characteristics, and, sometimes, the overall position and the context of the experimental questions in the questionnaires are controlled for by having them randomized over the wording conditions. Other disturbing variables such as the mode and the time of the data collection, sometimes the position in the questionnaire and the context, and, most important to us, the content of the question and all wording aspects of the question other than the wording aspect that is manipulated are controlled for by holding them constant.

Holding variables constant at some value is a good way to ensure internal validity, but it lowers the external validity, which is defined as the extent to which the experimental finding can be generalized over other questions and other conditions. A split ballot experiment is just a single case study, which is obviously too small a base for constructing a theory and for being relevant to opinion research practice as well since most of the time, opinion researchers use other questions in other conditions. The obvious strategy to solve the generalizability problem is the experimental approach, discussed in the next section.

6.3 SOLUTION OF THE GENERALIZABILITY PROBLEM: THE EXPERIMENTAL APPROACH

The experimental approach consists of repeating the experiment with other questions different in content, and, if possible, also with the other disturbing variables held constant at other values. From the combined

Table 6.1. Effects Observed in 17 Similar Question Wording Experiments. (Numbers within parentheses are estimated)

	Imbalanced Version		Balanced Version			
Question	n_1	y_1	n_2	y_2	t	Reference
Q 1	(485)	0.10	(480)	0.07	1.65	Rugg and Cantril, 1942
Q 2	(480)	0.25	(475)	0.33	−2.60*	Rugg and Cantril, 1947
Q 3	(480)	0.25	(470)	0.39	−4.81*	Rugg and Cantril, 1947
Q 4	444	0.62	445	0.63	−0.28	Blankenship, 1940
Q 5	449	0.32	445	0.37	−1.48	Blankenship, 1940
Q 6	480	0.87	454	0.90	−1.64	Blankenship, 1940
Q 7	487	0.21	454	0.10	5.01*	Blankenship, 1940
Q 8	471	0.54	460	0.53	0.21	Blankenship, 1940
Q 9	480	0.47	460	0.47	0.03	Blankenship, 1940
Q 10	458	0.81	394	0.87	−2.31*	Blankenship, 1940
Q 11	374	0.41	394	0.13	9.04*	Blankenship, 1940
Q 12	455	0.71	445	0.72	−0.23	Schuman and Presser, 1981
Q 13	503	0.70	485	0.70	0.03	Schuman and Presser, 1981
Q 14	507	0.38	494	0.29	2.98*	Schuman and Presser, 1981
Q 15	1368	0.46	1300	0.44	1.45	Schuman and Presser, 1981
Q 16	440	0.68	429	0.66	0.78	Schuman and Presser, 1981
Q 17	497	0.34	480	0.32	0.63	Schuman and Presser, 1981

* Significant at the 5% level; n_1 and n_2 are sample sizes; y_1 and y_2 are proportions responding "yes."

results a statement may then be derived about the effect of the wording variable concerned, that is sufficiently general for theory construction.

To illustrate this approach, and to discuss some of its features, we chose 17 experiments from the literature, all of which are sufficiently similar (not perfectly equal) as to the wording variable that is manipulated, namely: the "formal balance/imbalance" of the question. In the imbalanced version, for example, "Do you approve X?," only one verb is made explicit ("approve") with the obvious alternative ("not") left implied. In the balanced version, "Do you approve X or do you not?," both verbs are explicit (see Schuman and Presser, 1981, p. 180). All questions involved in the experiments are attitudinal questions of the yes/no type. Most experiments are carried out with face to face interviews, but some were done by telephone interviews. The experiments vary as to the time, the population of respondents, the context and the position of the question in the questionnaire and, most important of all, as to the content of the question involved and the remaining wording aspects. The results are given in Table 6.1. The data are corrected for the numbers of nonsubstantive responses.

It may be seen from Table 6.1, that the majority of the experiments did not yield a significant effect; only six effects were significant, but in

Table 6.2. Analysis of Variance of the Effects Observed in 17 Similar Wording Experiments (respondents are the units of analysis)

Source of Variation	Sum of Squares	*df*	Mean Sum of Squares	*F*
Formulation	0.643	1	0.643	0.33
Question	843.175	16	52.698	
Interaction	31.201	16	1.950	9.70*
Error	3,496.721	17,388	0.201	
Total	4,371.740	17,421		

* Significant at the 5% level.

different directions. The statement, for example, that an (obvious) verb will be chosen more frequently when it is explicit than when it is not, is not generally warranted, although the effect may be observed and even theoretically explained in an individual case. On the general level, the unconditioned effect of the balance/imbalance variable, if there is any effect at all, turns out to be unpredictable. Theoretically, one may go on to perform a multivariate meta analysis on this sort of data (see, for example, Sudman and Bradburn, 1974); the present set of experiments, however, is too small for such an analysis. One thing may be clear from Table 6.1: the effects of the present wording variable are quite diverse. For many other wording variables the situation is similar.

For an overall estimation of the effect of the balance/imbalance variable, an analysis of variance is performed on the data, using the respondents as the units of analysis. The results are presented in Table 6.2.

The main effect of the wording variable turns out to be not significant. In cases like this, where effects vary to a certain extent, the data can be analyzed by another method as well, namely by treating the questions (or the experiments, if you like) as the units of analysis rather than the respondents. The results of this alternative analysis of variance are in Table 6.3.

Again, the conclusion is that there is no significant overall effect. The latter ANOVA implies a more conservative test than the former. It is presented here to show that questions can be used as the units of analysis, as we do in the nonexperimental research design to be discussed in the coming sections.

The experimental approach is certainly a powerful strategy to ascertain systematic overall wording effects, in particular because the potential disturbance variables are controlled by randomization. Because there are so many wording variables to be studied for their possible effects, and so many different values and value combinations at

Table 6.3. Analysis of Variance of the Effects Observed in 17 Similar Wording Experiments

Source of Variation	Sum of Squares	*df*	Mean Sum of Squares	*F*
Formulation	0.0014	1	0.0014	0.34
Question	1.7667	16	0.1104	
Error	0.0654	16	0.0041	
Total	1.8335	33		

Note: The questions (experiments) are the units of analysis.

which the disturbance variables can be held constant, the experimental approach is quite laborious an undertaking; it would require a large number of experiments. The alternative nonexperimental approach seems to be more efficient.

6.4 THE NONEXPERIMENTAL APPROACH: THE RESEARCH DESIGN

The basic reasoning behind the nonexperimental design is that if a particular wording variable F has a systematic direct effect on a characteristic of the response distribution, for example the mean, then F should have, in a population of questions, a direct relationship with the means of the response distributions.

The differences compared with the experimental approach discussed in the previous section are:

1. The questions (and their response distributions) serve as the units of analysis, not the respondents and their individual responses. This means that the method is more conservative in detecting significant wording effects. The approach can be considered as a form of meta analysis where questions rather than studies are the units of analysis. The design involves a "secondary analysis" of opinion questions that were used in past surveys, and their factually obtained response distributions.

2. Because the questions are not randomized over the wordings, the potentially disturbing variables may be related to the wording variable, the effect of which one wants to estimate. That implies the necessity to make the separate disturbing variables explicit, in order to be able to control for their effects in a multivariate analysis.

3. It is impossible and senseless to include the complete, unique question texts as control variables, as could be done in the experimental approach. Instead, one has a limited number of control variables that have been identified as such. That means that the nonexperimental method provides less complete control for the potential disturbance variables in comparison with the experimental approach. This results in a relatively higher amount of error variance.

The basic procedure of the design is as follows: (1) select a number of questions and corresponding response distributions; (2) code the content and the wording of each question into a number of content variables and wording variables, respectively; these are the independent variables; (3) calculate a characteristic of each response distribution; (4) determine the partial (direct) effect of a particular wording variable on the distribution characteristic chosen, controlling for the effects of all other independent variables. The research problem corresponding to the nonexperimental design is: given the fact that characteristics of response distributions in a set of survey questions vary greatly, how much of that variance can be accounted for by the wording of the questions? And, which particular wording variables account for most of that variance?

Some years ago, I carried out nonexperimental research along these lines, reported in Molenaar (1986). This exploratory research, is described in more detail in the remaining sections.

6.5 NONEXPERIMENTAL RESEARCH ON WORDING EFFECTS

6.5.1 The Sample

The investigation involved 518 different survey questions with associated response distributions, taken from a Dutch archive, the Steinmetz Institute in Amsterdam. In the archive are stored the questionnaires, questions and the primary data of hundreds of surveys. The archived surveys were mainly conducted in the Netherlands by nongovernmental institutions and individual researchers; they were acquired from the investigators with their consent. The questions for the investigation were randomly selected among those satisfying a number of criteria. The criteria are partly set because of practical reasons (coding complexity, for example) and partly in view of the characteristics of the response distributions to be calculated.

The most important criteria were the following: (a) a national sample of at least 500 respondents age 16 years and older; (b) face to face interviews; (c) general subject matter, that is, not time-bound or topical; (d) closed; (e) asking for absolute judgments (not comparative); (f) attitudinal; (g) using answer categories that clearly indicate (degrees of) preference or no preference; (h) monotone with respect to the answer categories; (i) one dimensional; (Likert-type questions, for example, were excluded on the basis of this criterion); (j) pertaining to the areas of politics, outlook on life, marriage, sexuality or child rearing. By setting these criteria, the scope of generalization has been reduced, of course.

6.5.2 The Dependent Variables

The characteristics of the response distributions chosen were: the mean, the standard deviation, and the percentage of nonsubstantive responses ("no opinion," "don't know" and "no answer" taken together). These characteristics must be calculated in such a way that they are independent of how the questions happened to be worded, in particular independent of the number of answer categories provided. For this purpose, the mean and the standard deviation were calculated in a special way, making it possible to compare them across questions with different numbers of answer categories. To be specific, the standard deviation used in this investigation is a relative one: it is made proportional to the maximum standard deviation that is possible for the number of answer categories associated with each question.

6.5.3 The Independent Variables

The major restriction for choosing the independent variables was that their values had to be, as much as possible, explicitly traceable in the written question texts. I chose the variables by reading many question texts, asking myself in what respects they differed, and then formulating generic variables to cover those differences. The restriction implied, for example, that "content-related" characteristics of the questions, such as social desirability, threat, sensitivity and saliency, were not measured. It should be noticed that the content-related question characteristics are implicitly controlled for in the experimental approach.

The independent variables can be divided into three sets: the content variables, the wording variables, and some remaining variables. The variables of each set are listed below, without going into the details of how they were measured.

Content variables

1. The content area of the information asked in the question (global politics, specific politics, outlook on life, sexuality, marriage, child rearing).
2. The type of judgment asked for in the question (feelings/wishes, evaluations, norms, reasons, possibilities).
3. The time or time period of the event(s) to be evaluated in the question, ranging from the past to the present (and indefinite) to the future.
4. The frequency of the event(s) to be evaluated (once, several times, and indefinite).
5. The object about which an evaluation is asked, ordered in six classes according to its social proximity to the respondent, ranging from the respondent herself or himself to impersonal matters and abstract concepts.

Wording variables

1. The readability of the question as measured by a Dutch version of the reading ease formula developed by Flesch.
2. The amount of information on a "flash card," ranging from one (no card used) to answer categories plus part of the question.
3. The length of the introductory text, measured in classes of five words.
4. The strength of incentives used to stimulate the respondent to answer (e.g., ". . . Would you indicate if you agree or do not agree?" ". . . tell me if you agree or not").
5. The degree to which the text of the question emphasizes that the respondent's subjective opinion is asked for (e.g., "Do you think . . .," ". . . in your opinion . . .," ". . . according to you . . .").
6. The degree to which the set of answer categories that is provided in the introductory part of the question text is skewed or asymmetrical in the positive or negative direction with respect to an assumed midpoint. These categories, if present, are often only a subset of the answer categories that are actually provided with the question. In, for example, "I would like to know if you agree with the following statement: . . . (statement). . . Agree/Disagree," the word "agree" in the introduction is a skewed "subset" of the answer categories; the direction of the skewness depends on whether the verb in the statement is positive or negative.
7. Whether knowledge that is judged to be relevant to respondents for answering the question is transmitted in the introductory part of the question or not.
8. Whether the actual question is formulated as a declarative sentence (statement) or as an interrogative sentence.

9. Whether the question is a hypothetical question or not.
10. The question type, that is, dichotomous or multiple choice.
11. The number of subordinate clauses in the question; this variable measures an aspect of the question's linguistic complexity.
12. The degree to which the formulation of the attitudinal object of the question is biased in the positive or negative direction. This variable will be referred to as *the evaluative biasedness of object formulation.* In an opinion question about euthanasia, for example, "euthanasia" can be biased positively by emphasizing the humanitarian aspects (mild, painless death) or negatively by emphasizing the immoral or criminal aspects (killing people).
13. The degree to which the set of verbs provided in the question text is skewed or asymmetrical in a positive or negative direction with respect to an assumed midpoint.
14. The degree to which opinions of others, in constructions like "Some people. . . other people. . .," are given unequal weight in the positive or the negative direction. For example, in the construction "Some people say (. . . something positive). What do you think?" the positive side of the possible opinions is given a stronger weight. In the construction "Some people say (. . . something positive), other people say (. . . something negative). What do you think?" the positive and the negative side are given equal weight (provided, however, that the positive and negative statements are balanced too).
15. The degree to which the question is positively or negatively loaded by the presence of particular clauses or emotionally toned words.
16. The number of substantive answer categories.
17. The evaluative value of the answer category that is mentioned first; that is, the degree to which that category reflects a positive or negative opinion.
18. The number of times a neutral middle alternative is mentioned among the answer categories or within the text.
19. The completeness of the answer category labels: as full, well-formed sentences, as repetitions of the predicates provided in the question text, or as short terms that are not comprehensible without reverting to the question text (yes/no, agree/don't agree, and the like).
20. The absence or presence of the "no opinion" or "don't know" option.
21. The absence or presence of the category "no answer."
22. The degree to which the formulation of the object, the verbs and the answer categories are logically correct.

Remaining variables

1. The year in which the question had been posed to respondents, ranging from 1965 up to and including 1981.
2. The overall position of the question in the questionnaire.

Table 6.4. Variances Accounted for by All Predictor Variables Together, and the Proportional (Partial) Contributions by the Respective Sets of Predictors

	Distribution characteristic		
	Mean	Standard Deviation	% NSR
All predictors	0.260* (0.203)	0.450* (0.408)	0.513* (0.475)
Proportional contribution by:			
Content Variables	0.056* (0.032)	0.040 (0.016)	0.183* (0.163)
Wording Variables	0.188* (0.151)	0.394* (0.366)	0.238* (0.203)
Position Variables	0.026* (0.020)	0.013 (0.007)	0.034* (0.028)

* Significant at the 5% level. NSR, nonsubstantive responses. Entries are R-square values, with adjusted R-squares within parentheses. $N = 518$ questions.

3. The sequential number of the question within a series of questions using identical answer categories.

These three remaining variables will be referred to as "position variables."

In all, 37 independent variables (including dummy variables) were included in the research: 12 content variables, 22 wording variables and three position variables. The questions were coded into the variables on the basis of highly structured protocols. Most work was done by university students. To assess the reliability of the coding process, 231 randomly chosen texts were coded for a second time. The mean reliability of the codings was measured as Kappa = 0.90.

6.6 ANALYSIS AND RESULTS

The data were analyzed with the use of multiple correlation and regression for each distribution characteristic separately. First of all, it is interesting to know how much of the variance of each distribution characteristic could be accounted for by all 37 predictor variables together and how much the respective sets of variables contribute to that variance. The latter is an indication of the relative importance of the separate sets. These summary results are presented in Table 6.4.

With all predictors included, the percentage of nonsubstantive responses (% NSR, for short) and the standard deviation appear the most predictable distribution characteristics, as measured by the amount of their variance that could be accounted for. The wording variables turn out to be the most important set, even when corrected (adjusted) for the different numbers of variables in the respective sets; the position variables contribute least to the variances explained; the content variables are somewhere in between.

It is more interesting, however, to know which particular wording variables are most important from among the set of 22 wording variables. To determine that, two steps were taken. The first step was to include all content and position variables in the equations; then the 22 wording variables were selectively added, using a stepwise regression procedure. By this procedure a particular wording variable is selected for inclusion if it meets some specified statistical criteria. In this manner four wording variables were selected with respect to the mean, seven with respect to the standard deviation, and five with respect to the percentage NSR. The proportion of the variance that is accounted for by all predictors together (which is a smaller number now) decreased about 0.025 to 0.036 in comparison with the figures in Table 6.4; the proportional contributions by the (smaller number of) wording variables decreased 0.030 to 0.040. Almost all of the variance that is attributable to wording appeared to be accounted for by only four to seven wording variables. From the wording variables thus selected, we further considered, in the second step, a variable to be relatively important, if its own contribution to the explained variance exceeded one percent. Table 6.5 displays the finally selected variables and their estimated effects. The numbers in front of the variables correspond to the identification numbers in the lists in Section 6.5.3.

The following statements describe the results in Table 6.5.

1. If the "no opinion" option is absent, the percentage of nonsubstantive responses is lower than when that option is present. The most likely explanation of the effect is that the interviewers are more willing to accept the "no opinion" response when that category is provided for than when it is not.
2. Yes/no questions yield higher percentages of NSR than do multiple choice questions. The effect is strongest for questions about politics with only two answer categories; that is, when the questions are difficult to answer because of their subject matter and when the range of possible answers is very narrow.
3. The size of the standard deviation decreases when the number of answer categories increases. To explain this rather strange effect, it should be remembered that the standard deviation used in the investigation is a relative one: it is proportional to the maximum standard deviation that is possible for the number of answer categories in the question. The effect reflects the tendency for respondents to avoid answer categories that are further removed from the middle of the scale (as measured by the number of intermediate categories) and that are in that sense, more extreme. If the "normal" standard deviations are calculated, the relationship is reversed: the standard deviation increases when the number of answer categories increases.

Table 6.5. Effects of the Wording Variables That Are Selected by the Stepwise Procedure, and Whose Contribution to the Variance Explained (R-square) Is More Than 1%

		Distribution Characteristic		
Wording Variable	Statistic	Mean	S.D.	% NSR
20. 'No opinion' option (absent → present)	r	−0.02	0.01	0.41*
	B			0.22*
	R^2			0.034
10. Question type (choice → yes/no)	r	−0.09*	0.51*	0.24*
	B		0.25*	0.30*
	R^2		0.060	0.070
16. Number of substantive answer categories (few → many)	r	0.07	−0.52*	−0.17*
	B		−0.28*	
	R^2		0.223	
2. Amount of information on card (none → much)	r	0.02	−0.26*	−0.10*
	B		−0.19*	
	R^2		0.016	
12. Evaluative biasedness of object formulation (negative → positive)	r	0.19*	−0.11*	0.11*
	B	0.14*		
	R^2	0.019		
14. Evaluative biasedness of *other's* opinions (negative → positive)	r	0.19*	−0.15*	0.00
	B	0.15*		
	R^2	0.034		
15. Evaluative biasedness of question text (negative → positive)	r	0.25*	0.04	−0.03
	B	0.27*		
	R^2	0.074		

* Significant at the 5% level. S.D., standard deviation; r is the zero-order correlation, B is the standardized partial regression coefficient; $N = 515$ questions.

That has a rather obvious explanation: differences in opinions among the respondents are expressed when more answer categories are provided for.

4. Yes/no questions yield response distributions with a higher "relative" standard deviation than choice questions do, especially when the questions contain more than two answer categories. With respect to yes/no questions one can think of the sequence "agree/depends/disagree" and, with respect to choice questions, of the sequence "never/sometimes/often." The effect means that in yes/no questions in comparison with choice questions, the category (categories) in the middle of the scale is chosen less frequently and the border categories are chosen more often. A possible explanation is that category scales

of yes/no questions are organized in a bipolar fashion by the respondents, whereas the category scales of choice questions are organized unipolarly. In a bipolar scale the middle answer category appears as the neutral dividing point between the positive side of the scale and the negative side; it reflects a situation of undecidedness and evokes feelings of dissatisfaction (Schönpflug and Büch, 1970). This is usually not the case with respect to the middle category in a unipolar scale.

5. The size of the "relative" standard deviation is inversely related to the amount of information given to the respondents on a card. This effect appears to be strongest when the card at least contains the answer categories. An explanation is that when the answer categories are written on a card, the respondent can literally "see" the structure of the set of answer categories, the ends of the scale, and the extremity of the border categories. This is less likely when the categories are conveyed to the respondent orally. The tendency to avoid the border categories is therefore more likely to occur when the categories are provided on a card.
6. If the object of the question is formulated in more positive terms, respondents are more likely to choose the answer categories that express positive judgments. This effect should be attributed to the emotional or social pressure on the respondents to choose an answer that is in accordance with the suggested direction.
7. If the positive opinions of others are emphasized more strongly in constructions such as "Some people. . . other people. . .," the respondents show a greater preference for such opinions. The explanation is similar to the explanation of the previous effect.
8. Respondents are more likely to provide positive answers, as indicated by higher means of the response distribution, if there is a greater degree of positive suggestion resulting from the clauses and/or the emotionally loaded terms in the question. The explanation is, again, similar to the explanation of the sixth statement.

6.7 QUALITY ASPECTS OF THE NONEXPERIMENTAL RESEARCH DESIGN

6.7.1 Limitations of Nonexperimental Designs

In order to be able to determine unambiguously the characteristics of the response distributions and to compare them across questions, homogeneity among the questions on a number of dimensions is required. To that end, I introduced a number of criteria to select the questions; as a consequence, the results of the nonexperimental investigation will apply only to certain types of questions. There were also practical reasons for taking a subjective sample of questions, for example to avoid extremely complex coding procedures.

A second limitation is that only the possible effects of wordings normally used in current survey practice can be investigated with the nonexperimental design. With experiments the effects of any wording variable can be investigated, even the effects of wordings which never occur in current practice. In my sample of questions, for example, there was no question that contained a real "no opinion" filter, which is a variable that has been studied for its effects time and again in experiments.

Furthermore, it is not possible or sensible to take all differences in the content and the formulation of the questions into account by constructing separate variables. The wording and the content variables to be included in the research have to be generic variables from the very beginning. In single experiments you can indeed study any individual and unique wording for its effect, but when it comes to summarizing a series of experiments, you also have to define generic variables.

Finally, to keep the research manageable, respondent and interviewer characteristics were not included. For similar reasons, characteristics of the context of the questions were also not included; their inclusion would have doubled the number of variables. Content-related characteristics of the questions (social desirability, threat, sensitivity, and the like) were also not included in the research described above, but that is not an inherent limitation of the nonexperimental design.

6.7.2 Stability of the Response Distributions

The nonexperimental method examines wording effects on the basis of the assumption that the response distribution obtained with a particular question is fairly "typical" for that question; or, to put it in another way, that response distributions obtained with a particular question are relatively stable. If the response distributions of identical questions can be proven identical, or stable, the results from the nonexperimental study can be considered more reliable.

In the process of selecting the questions for the investigation, 122 different sets of identical questions with corresponding response distributions were available as a by-product. More precisely, the questions within each set were identical except with respect to the three previously mentioned position variables. One question from each set had actually been included in the sample of the nonexperimental research, on a

Table 6.6. Analysis of Variance of the Distribution Characteristics Performed on 122 Sets of Identical Questions ($N = 515$ Questions)

Characteristic	Source of Variation	Sum of Squares	*df*	Mean Sum of Squares	*F*	Omega Square
Mean	Between sets	1,100.2	121	9.09	58.22	0.931
	Within sets	61.4	393	0.16		
	Total	1,161.6	514			
Standard Deviation	Between sets	152.5	121	1.26	49.42	0.916
	Within sets	10.0	393	0.03		
	Total	162.5	514			
% NSR	Between sets	17,896.6	121	147.91	30.71	0.875
	Within sets	1892.9	393	4.82		
	Total	19,789.5	514			

random basis. The size of the sets varied from two to 18 questions. If the "stability assumption" holds, the distribution characteristics for questions within a set should have a much lower variance than the distribution characteristics for questions that are from different sets. The stability assumption was tested by carrying out analyses of variance on these 122 sets of questions. The results are presented in Table 6.6.

It can be concluded from the omega-square values (a measure of explained variance comparable to eta-square; see Hays, 1973, p. 680) that about 90 percent of the variance of each of our distribution characteristics can be attributed to questions that are different from one another (between-set variance), and 10 percent to questions that are identical in content and wording (within-set variance). That leads to the conclusion that response distributions obtained with identical questions are relatively stable, and, accordingly, that the analyses in the main nonexperimental investigation were probably performed on distribution characteristics which were, on the whole, reasonably stable.

6.7.3 Internal Validity of Nonexperimental Research Design

As has been said, in the experimental approach the overall direct effect of a particular wording variable can be established while almost all potential disturbance variables are controlled for without having them identified separately, namely by the factor Question. In a nonexperimental approach, however, the wording effects are established while control-

Table 6.7. Summary Results of Regression Analyses of the Distribution Characteristics Upon Three Position Variables, After a Quasi Experimental Control (QEC) and a Nonexperimental Control (NEC)

	Mean		Standard Deviation		% NSR	
Control Procedure	QEC	NEC	QEC	NEC	QEC	NEC
Total Accounted Variance	0.952	0.635	0.943	0.707	0.928	0.724
Variance Accounted for by the Control Variables	0.948	0.618	0.938	0.695	0.914	0.719
Proportional Contribution by the Position Variables	0.077	0.046	0.081	0.039	0.163	0.018

Entries are R-square values. $N = 383$ questions. All estimates are significant at the 5% level.

ling for only a limited number of disturbance variables, namely the content variables and the wording variables that have been identified as such.

These variables can be viewed as a subset of the variables that comprised the factor Question. As was indicated above, the difference between experimental and nonexperimental approaches is manifested in higher error variances in the latter case.

The material described in the previous section provided the opportunity to investigate the effects of the three position variables (year of administration, overall position in the questionnaire, and series number) on the distribution characteristics in two ways. First, the effects can be measured while using the complete, unique questions as the control variable. This control variable is similar to the factor Question in the experimental approach, with 121 levels now (or 121 dummy variables), each of which represents a complete, unique question. This control procedure may be called the quasi experimental control procedure (QEC). Secondly, the effects can be established while controlling for the 34 content and wording variables that were involved in the nonexperimental investigation. This control procedure may be called the nonexperimental control procedure (NEC). Because of the limited capacities of the computer program, however, the analyses were carried out on 95 randomly chosen sets, composed of 383 questions. The results are presented in Table 6.7.

The analyses reveal that the complete, unique question texts (QEC) account for 90 to 95 percent of the variance of each distribution characteristic, whereas the 34 content and wording variables (NEC) account for about 60 to 70 percent. This indicates the systematically higher amounts of error variances in NEC in comparison with QEC, as was expected. When corrected (adjusted) for the number of predictors involved in the respective equations, the R-square values in the table are about 0.01 to 0.04 lower.

Table 6.7 also reveals that of the variance left unexplained by the control variables, a small proportion can be further accounted for by the three position variables; two to five percent in the NEC condition, and systematically somewhat more in the QEC condition. We may ask, then, which individual position variable(s) would be the most significant, within either control procedure. For that purpose, the three position variables were selectively added to the respective equations, using a stepwise procedure. With the additional criterion that the specific contribution of a selected variable to the variance explained should be at least one percent, one rather small effect is left in the QEC condition; the effect of the overall position of the question in the questionnaire upon the percentage NSR. ($B = 0.14$ (significant at the 5 percent level) with R-square $= 0.010$.) The comparable effect in the NEC condition was $B = 0.08$ (also significant at the 5 percent level) with R-square $= 0.004$. Obviously, the later a question is posed in the interview, the higher is the percentage of nonsubstantive responses.

This evidence leads to the cautious conclusion that the nonexperimental method provides less complete control for the potentially disturbing sources of variance, as noted from the higher amount of error variance. The method appears to control for the major disturbing sources of variance, as may be noted from both approaches indicating about the same effects. The evidence, however, is not conclusive.

6.8 CONCLUDING REMARKS

The nonexperimental research design is not to be considered the panacea for studying wording effects. It is proposed here as an alternative but more efficient research design to arrive at general statements about unconditioned wording effects. For immediate aims, the split ballot experiments remain the best design.

Like most experimental approaches, the nonexperimental approach does not provide an answer to the legitimate question: "How should I formulate my question in order to get responses of good quality?". For answering that question, additional information would be necessary, which for attitudinal questions is difficult to obtain. The multitrait multimethod design, recently developed by Andrews (1984), may offer a new perspective to this validity problem.

CHAPTER 7

MEASUREMENT ERRORS IN BUSINESS SURVEYS

Solomon Dutka and Lester R. Frankel
Audits and Surveys, Inc.

7.1 INTRODUCTION

Many measurement techniques used in business and establishment surveys are similar to those used in surveys of individuals and consumers (see Chapter 12). All involve interviewer–respondent, question–answer and stimulus–response procedures. In business surveys, however, information is also gathered by the retrieval of records and by direct observation. It would appear that by using these procedures, the net effect of measurement errors in business surveys would be minimal compared with consumer surveys. Nevertheless, experience has shown that even with the use of sophisticated electronic retrieval systems, errors in business surveys do occur that have major consequences for survey estimates.

In any system of measurement there are essentially three causes of error: (1) measurement of the wrong variable, (2) incorrect or biased measurement and (3) variability due to measuring instruments. It is the purpose of this chapter to examine these measurement errors that occur in business surveys and to suggest methods to reduce or eliminate them.

The term "business surveys" as used in this chapter connotes those surveys in which the sampling and enumeration unit, generally outside the home, may be (1) an entity such as a business establishment, a retail store, a restaurant, an institution, a physician's practice, etc. or (2) an individual in his or her capacity as a member of a business with a title such as owner, president, purchasing agent, computer operator, chef, etc.

Measurement Errors in Surveys.
Edited by Biemer, Groves, Lyberg, Mathiowetz, and Sudman.

ISBN: 0-471-53405-6

7.2 TAXONOMY OF BUSINESS SURVEYS

Business surveys can generally be categorized as either enumerative or analytical, and by whether the objective is to study the business unit itself or obtain other relevant information associated with the unit.

Enumerative surveys of business units measure the characteristics of establishments such as sales and number of establishments by size or by Standard Industrial Classification (SIC) groupings. They may also measure the number of users of different types and brands of equipment, corporate estimates of capital expenditures, and future economic outlook.

Analytical surveys of business units include product and brand satisfaction studies, measurement of receptivity toward new business products, satisfaction with services received, and attitude towards price variation by suppliers.

Many kinds of establishments, such as retailers and wholesalers, function as conduits in the flow of consumer products from the producer to household or individual users. They represent a major source of business information. Thus, for enumerative studies, they provide information on the movement of products and brands, both nationally and locally, at the retail level, from wholesalers and from manufacturers. They also provide level of product inventory by brands at the retail level, nationally, and within individual sales areas; and they provide information on the prevalence of promotion and advertising material in retail outlets and restaurants, in waiting areas such as doctors' and dentists' offices, and in airports and train terminals.

Analytic surveys on the effect of advertising and promotion, the acceptance of new products, and package designs are conducted in controlled test market experiments where effectiveness is measured based on product movement at the retail level.

7.3 UNITS INTERVIEWED AND MEASURED IN BUSINESS SURVEYS

It is impossible to list all the kinds of establishments that are sampled in business surveys, but some recent examples of the types of establishments or offices from whom data have been collected by our company include: business users of the Yellow Pages (telephone directories of businesses), dentists and physicians, users of copying machines, users of mainframe computers, users of communication equipment, hospitals, libraries, school systems, country clubs and pro shops, retailers and retail chains, wholesalers and warehouses, and automotive dealers.

7.4 MEASUREMENT TECHNIQUES IN BUSINESS SURVEYS

Interviewing methods in business surveys include the standard techniques involving face to face and telephone interviewing and self-administered mail questionnaires. Recently the use of computer inputs by respondents using either telephone modems or floppy disks has increased.

A significant factor in business surveys is the retrieval of records such as invoices, purchase orders, sales, and prescriptions written and filled. This may involve hand transcription or the collection and processing of computer output from wholesalers, mass merchandisers, chain headquarters and manufacturers. There has been a significant growth in the use of scanner devices in retail outlets (see Chapter 18).

Observational methods play an important role in business surveys. These include observations of static phenomena such as distribution of brands in retail outlets and service operations, such as hotels, theatres and sporting arenas; checking shelf facings in supermarkets; and monitoring the incidence of specific outdoor and indoor advertising.

Time sampling of dynamic phenomena is also important. This includes measures of the movement of people or customers entering or leaving an establishment; the movement of objects such as sales of perishable foods, sales from vending machines, reading of specific magazine issues in waiting rooms; traffic counts to determine usage of different car makes; and behavior of sales and other service personnel.

7.5 TYPE III ERROR

Statisticians are familiar with Type I and Type II errors in testing of hypotheses. In business surveys, where data are used for decision-making, Type III errors may be even more important (Loebl, 1990). Type III error occurs when the survey specifications are not met in the interviewing or measurement process and irrelevant information is obtained. For example, in business surveys this can come about when collecting information on the the wrong topic and/or interviewing the wrong person in a firm.

Type III error is less of a problem in household surveys or surveys of small establishments. In these studies, the description of the study objectives is more easily communicated, and the specification of eligible respondents is clearcut. It is perhaps for these reasons that this type of error has not received much attention in the survey literature. In conducting surveys in large companies, however, Type III errrors may be disastrous and could invalidate the entire survey. Thus, for surveys designed to measure the acceptance of a new technical product, there may be many people involved in the decision process, including a departmental manager, the technical director and the purchasing agent. Without prior knowledge or pretesting to determine whom to interview, Type III errors may occur.

Many business surveys in large companies are designed to measure brand awareness, image and the reputation of suppliers of industrial equipment and services. Others are concerned with satisfaction and experience with specific products currently in use in the establishment. It is essential that the questionnaire cover the relevant products and that the appropriate person or persons be interviewed.

Before a survey questionnaire is administered in a large business establishment, it may be necessary to use a separate screening questionnaire or series of questions to determine the types and functions of persons who should be interviewed. For example, in a series of studies dealing with marketing information systems in large companies, we use a preliminary questioning procedure to obtain the names and titles of relevant company personnel at the sampled location. We then ask to speak to the person who, for example, is responsible for organizing, implementing and overseeing the computerized management or marketing information system (MIS). After making contact with this person, we verify his or her functions and point out that we are not referring to the data processing operations manager, but to the MIS director or manager. After being assured by further questioning that we have reached the key person, the interview is conducted.

When asking about pieces of equipment or operations, it is necessary to be sure that the respondent is referring to the same things that the survey organization is concerned with. Thus we have used phrases like "workstations, not stand-alone PC's nor dedicated word-processors." In studies dealing with document retrieval we indicate that we are talking about the retrieval of an actual paper document or its image that had been previously stored on microfilm or optical disk. "Simply retrieving information from a computer data file, however, is not what me mean by document retrieval."

Of all the possible measurement errors that can occur in surveys, Type III errrors are the most unobtrusive. These errors, however, are preventable through the use of subject matter specialists, through care in the preparation of the questionnaire, and by vigorous pretesting and probing by experienced interviewers.

7.6 MEASUREMENT BIAS

The discussion of measurement bias is in three parts. We first describe measurement biases resulting from interviewing methods. This is followed by a discussion of the biases from record retrieval. The section ends with a discussion of possible biases from observational methods.

7.6.1 Interviewing Methods

Measurement error in business surveys occurs when, in response to a question or inquiry, the respondent unintentionally or deliberately furnishes information other than the true value sought. (See the Introduction and Chapter 1 for a discussion of true value.) The net difference between the true value and what is reported is the bias. Such bias can often be detected through a direct on-site observation (Dutka and Frankel, 1975). For example, in a survey to obtain estimates of the brands and types of electrical appliances in use in small restaurants, that is, those with less than 100 seats, telephone interviews were conducted with the owner or operator. The items of interest were ovens, microwaves, grills, refrigerators, freezers, dishwashers, toasters and blenders. A pretest was conducted to determine the accuracy of response. Interviews were conducted by telephone with a sample of 100 establishments and brand data for all appliances were recorded. Follow-up visits were completed with 92 restaurants and brand information was obtained by personal observation for 580 appliances. Of these, 552 or 95.2 percent of the brands coincided with information obtained from the interview. The overall disagreement rate was 4.8 percent and the rate varied by type of equipment. The implication of these findings is that the estimate of the percent owning a specific brand may be subject to some bias.

The seriousness of such biases depends upon the size of the sampling error since the total survey error is a combination of bias and sampling error. For example, suppose that the market share for a particular brand is overstated by 4.8 percent, regardless of the size of the brand share. As the brand share increases, the total mean squared error (MSE) of the estimated percent also increases as a result of response bias. With a 50% brand share, the MSE for a sample size of, say, 500 is 0.0011 (computed as $(0.5)(0.5)/500 + [0.05(0.048)]^2$). Here the squared bias is about 53% of the MSE. For an 80% brand share, the MSE of the estimate is 0.0018 or about 1.6 times the MSE at 50%. Further, response bias contributes 82% to the MSE.

Cost considerations have important implications on the choice of sample designs and the allocation procedures. For example, assume that the cost of a telephone interview is $5.00 and that the cost of an observation is $12.50. For a fixed budget of $2,500, one could opt for 500 telephone interviews and accept the net bias of 4.8 percent or one could use observations, which we here assume have no bias, but for a smaller sample of 200. In Figure 7.1, the total survey error for each design is shown, expressed as 100 times the square root of MSE.

For a brand share of 56% or less, the telephone interview design will

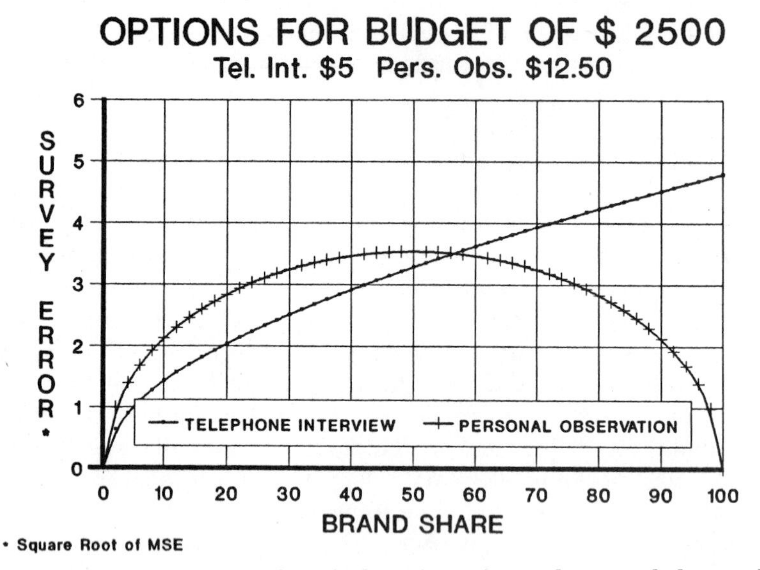

Figure 7.1 Total survey error for telephone interview and personal observation.

be more cost-efficient than observations. However, if it is expected that the leading brand has a share of about 65% and the others fluctuate about 15%, the design can more efficiently incorporate both modes.

Sometimes in business surveys, biases may be purposely introduced by the respondent. This may occur when the respondent, knowing the sponsor of a study, may slant his or her responses for or against the sponsor or deliberately supply other kinds of misleading answers.

We confronted this situation at our company in a study conducted for *Women's Day* magazine, "Women's Influence in the Purchase of a Car." This was a survey among automobile dealers to determine the features and car characteristics that are important to a woman in the showroom when she is shopping alone or with a male companion. Interviews were conducted with the owner, manager, or head salesman of the dealership.

When pretests were conducted by highly experienced interviewers, they sensed that some respondents were evasive, and apparently many suspected that we were working for another dealer or for their own manufacturer. Subsequently, we obtained the cooperation of NADA, the National Association of Automobile Dealers by offering them a chance to add questions for the benefit of the association. By announcing their sponsorship through an advance letter, we were able to obtain excellent cooperation and more candid responses.

7.6.2 Record Retrieval

It is often believed that measurement obtained from records or so-called hard data are devoid of error. In many business surveys this may not be the case, as indicated in the examples given below.

a. Out-of-scope data. Measuring warehouse withdrawals is one method for obtaining estimates of sales of branded merchandise to consumers through retail outlets. This procedure seems to be very efficient for high turnover grocery items with their large variety of brands and package sizes since these data are generally already on computer tapes.

A bias may exist if one assumes that all the shipments of wholesalers go to retail outlets and then on to consumers. Some items are shipped by wholesalers to restaurants, hotels, steamship companies, schools, army bases or other institutions and are thus out of scope. Using total warehouse shipments results in an overstatement of the consumer market.

For consumer durables, the overstatement for a specific time period may be even greater since shipments to retailers are not matched by sales to consumers. For example, since sales of automobile tires and accessories are low at certain times of the year, manufacturer shipments may overstate consumer sales for some time periods and understate them at other times.

b. Misleading records. It is believed that manufacturer shipments provide an impeccable source of consumer sales estimates over an extended period of time, but since recordkeeping practices vary among manufacturers, it is difficult to use shipment data to reflect brand shares or units sold in domestic markets. Some shipments of products, such as toasters or detergents, go to commercial users. Other shipments are exported. Some manufacturers franchise the use of their brand names in other countries. These brands are manufactured abroad and are distributed in this country — the "gray" market.

The following biases came to light when our audit data for consumer purchases were compared with our client's shipment data. In one case, it was claimed that our estimates of consumer purchases in the New England area for a product were too low compared with company shipments, even after allowing for the product in the pipeline. It was discovered that, unknown to the manufacturer, his North East distributor was shipping goods to the Philippines. In another case, it was believed that our retail estimates of fourth quarter sales compared to shipments for a newly introduced brand were too low. It turned out that the sales manager had "borrowed" shipments from January sales to show a successful prior year.

c. Contamination. In retail audits, an outlet is visited by an auditor (interviewer) periodically to count inventory and purchase invoices in order to obtain sales to consumers during the period. Occasionally, when the dealer is running short of an item, he or she may buy some for cash from another dealer, and no regular purchase invoice is kept. As a consequence, the retail sales estimate of such items to consumers will be understated. The effect may be detected by comparing current sales with those of the preceding period. In the form used by auditors the sales for the previous period are indicated. The auditor is required to follow up any unusual discrepancy and to note the reason for the follow-up on the form.

7.6.3 Observation

A major difference between business and household surveys is in the extensive business use of observational methods. These are possible because the activity occurs in a public place such as a store or mall. When done well, observational methods are not subject to the kinds of problems that we discussed earlier for interviewing methods. There are, however, possible biases in these procedures also.

Observational methods have been used to make traffic counts, to measure the movement of perishable foods, to determine the frequency of magazine pickups in waiting rooms, to audit newspaper sales by street vendors, and to measure the use of particular interstate highways by recreational vehicles (Dutka and Frankel, 1960; Houseman and Lipstein, 1960).

Observational procedures require that a sample of time periods be selected. The frequencies obtained are usually expressed in terms of an hourly rate and then projected to total hours. This technique is sometimes referred to as *time interval sampling* (Suen and Ary, 1988). If these time periods are selected using probability methods no biases result. Sometimes, however, time periods may be selected for the convenience of the store owner who says "don't do your observations on the days that my store is busiest." Obviously, this results in understatements of activity as well as biases in the kinds of customers observed.

With some exceptions, observational studies are conducted on private property and require the cooperation of the owner or manager. While the initial sample design may be unbiased, low cooperation of places where observation is to be conducted raises the possibility of serious sample biases. These biases can sometimes be reduced by stratification or post-stratification, but in some cases the sample biases may exceed the response biases in interviewing methods.

7.7 RESPONSE VARIATION

As with bias, response variation can be examined for interviewing methods, record retrieval and observational methods. Again, there are some unique features for business surveys.

7.7.1 Interviewer Variance

On the whole, it is probably the case that interviewers are less a source of variance in business than in household surveys. Interviewer effects arise when the interviewer is asking the questions, recording the answers or sometimes because of the characteristics of the interviewer.

Since business respondents are, almost by definition, more knowledgeable about the topic of the study than are the interviewers, they are less likely to need or receive interviewer interpretations of the question, thereby eliminating one cause of interviewer variance. Also respondents recognizing that the interviewer is not necessarily an expert will give more complete answers that are easier to code.

Interviewers in business surveys are not likely to have prior expectations about the answers and are more likely to record what the respondent said. Also, business surveys usually deal with more factual topics where interviewers are better able to record the given answers without summarizing or interpreting them.

Interviewer characteristics may affect the answer if they are seen as related to the questions, e.g., race of interviewer is important if questions deal with racial issues. It is highly unlikely that any interviewer characteristic will be seen as related to a business survey.

Finally, many business surveys are done using self-administered questionnaires, which completely eliminate the possibility of interviewer effects.

7.7.2 Record Retrieval

The retail store audit is a method that has been used with great success over the past fifty years to estimate retail sales of branded merchandise within a fixed time period. At the beginning of the period, inventory counts are made for all brands being measured for each store in the representative sample.

At the end of the period, the same stores are revisited and counts of closing inventories are made. At that time, all of the purchase invoices during the period are examined and recorded. For a particular brand, the

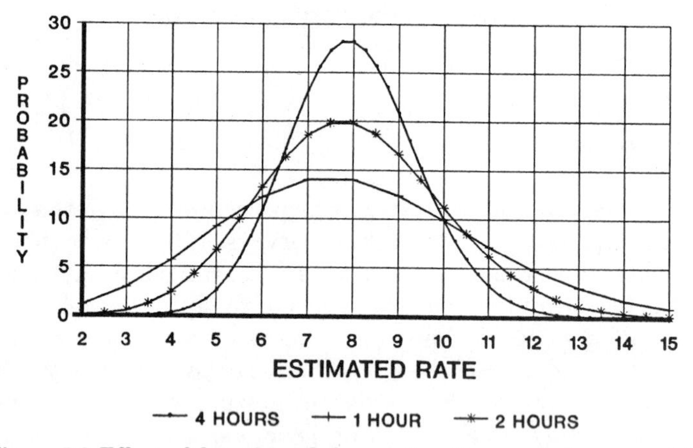

Figure 7.2 Effect of duration of observation time periods on precision.

retail sales are computed by taking the dealer purchases and adding or subtracting the difference between opening and closing inventory. A less costly, and in some cases a more efficient procedure, is also used to estimate retail sales. This estimate is based solely on the use of purchase invoices, and inventories are not taken. It is assumed that the difference in inventories between the two time periods is zero.

Depending on the turnover of sales during an audit period and the number of outlets in the sample, the impact of variation of inventory can be determined. In general, for high turnover products such as soft drinks, gasoline, and cigarettes, this source of variation can, for most practical purposes, be ignored. For a test audit, however, where the observation period is short, this source of variation should not be ignored (Frankel, 1969).

7.7.3 Observational Methods

The reliability of observational methods generally depends on two factors — the length of the time interval and the number of sites that are used from which time intervals are selected (Deming, 1950). From the measurement viewpoint, the longer the time interval, the more precise the estimate of the hourly rate. Further, the activity being observed may occur very infrequently so that the hourly rate of occurrence can be modeled by a Poisson distribution.

The effect of observation time periods on the precision of the hourly rate estimate is shown in Figure 7.2. The distributions of the estimates based on one, two and four-hour periods is plotted in a population where, on average, eight events occur in an hour.

7.8 SUMMARY

In this overview, we have tried to outline the major kinds of business surveys and the measurement errors that need to be considered. Measurement errors in business surveys are similar to those in household and individual surveys. However, their effects may be more critical in business surveys since many industries are dominated by a few very large establishments. Measurement errors in these large establishments can significantly increase total survey error.

A larger variety of methods, including record retrieval and observation, are typical in business surveys. These create special problems. The use of newer technologies, while helpful, does not eliminate measurement error and may create additional ones. Type III error, measuring the wrong thing, is more critical and difficult to avoid in business surveys.

Measurement errors in business surveys are preventable and/or controllable subject to cost–benefit considerations. Choices may include the option of living with the measurement error, the use of alternative measurement techniques, the use of mixed mode measurements, and the use of extensive pretesting. A close ongoing relationship with the client is needed to insure that measurement errors do not destroy the usefulness of the results for decision-making.

SECTION B

RESPONDENTS AND RESPONSES

CHAPTER 8

RECALL ERROR: SOURCES AND BIAS REDUCTION TECHNIQUES

Donna Eisenhower
Westat, Inc.

Nancy A. Mathiowetz
U.S. Bureau of the Census

David Morganstein
Westat, Inc.

8.1 INTRODUCTION

The survey measurement process, as previously noted, is fraught with many potential sources of error. This chapter will examine theories of memory and the process of retrospective recall and discuss a means for reducing error through the sample design, the data collection instruments, the interviewing methodology, and post-survey statistical adjustment. Within this context, the respondent will be examined as a source of invalid information.

The types of questions respondents are asked in surveys are quite diverse, ranging from questions on their attitudes and opinions about current events to questions which focus on quantitative facts about events or behavior in a respondent's past. Although there is substantial evidence of response problems associated with attitude questions (e.g., Turner and Martin, 1984) this review examines only those response errors associated with the recall of autobiographical memories.

The opinions expressed in this chapter are those of the authors and do not represent the views or policies of the U.S. Bureau of the Census or Westat, Inc.

Measurement Errors in Surveys.
Edited by Biemer, Groves, Lyberg, Mathiowetz, and Sudman.

ISBN: 0-471-53405-6

The field of survey research to a large extent is dependent upon the retrospective recall ability of respondents. Whether the task is one of recalling the details about a specific episode or estimating the frequency of an activity or event, the respondent must search his or her memory for a response that corresponds to the cue or question. To the extent that an individual is not able to recall the occurrence of an event or details about past events, or is affected by his or her present psychological state or environment, the quality of the data becomes questionable.

How is it possible for survey respondents to forget events, especially events that others would tend to classify as important, salient or traumatic? To begin to understand the answer to this question, we need to examine the interview process. Tourangeau (1984) outlined the respondent's task from a cognitive perspective as a four-step process: comprehension, retrieval, judgment, and response. The potential for measurement error actually begins with a process which precedes these, the encoding of information. This interview process has been evaluated from several disciplines including cognitive psychology (with respect to theories of memory, retrieval, and judgment), linguistics (theories of communication and shared meaning of words) and survey methodology (the documentation of respondent characteristics and task characteristics related to error that is not predictable from the other two literatures). We will be drawing from these various literatures to hypothesize and explain various errors related to the respondent in the survey process.

8.2 ENCODING PROCESS: KNOWLEDGE TO ANSWER SURVEY QUESTIONS

Memory has been defined as "the dynamic process associated with holding a memory trace, the product of an experience, over time and the retrieval of information about it" (Crowder, 1976). Tulving (1975) makes the distinction between two types of memory, episodic and semantic. Episodic memory is concerned with the storage and retrieval of "temporally dated, spatially located, and personally experienced events and episodes and the temporal/spatial relationship among these events." In contrast, "semantic memory is the system concerned with the storage and utilization of knowledge about words and concepts, their meanings and interrelationship."

Theories of episodic memory need to account for three phenomena of memory: accuracy, incompleteness, and distortion.[1] One class of memory theories, "schemata," have emphasized the incompleteness and distortions of memory. The development of schema theories is based on Bartlett's rejection of the notion that memory representations consist of accurate traces that are stable over long durations (Bartlett, 1932). The global interpretation of schema theories is that they focus on the general knowledge a person possesses about a particular domain. Current schema theorists propose that what is encoded, or stored in memory, is heavily determined by a guiding schema or knowledge framework that selects and actively modifies experience in order to arrive at a coherent, unified, expectation-confirming and knowledge-consistent representation of an experience (Alba and Hasher, 1983).

According to schema theory, knowledge, beliefs, and attitudes that are stored are a highly selected subset of all that has been presented to an individual. The organized cognitive structure that guides the selection process is the schema. The theory suggests that not only is memory selective in what is stored, but it is also abstractive. Rarely, if ever, is a verbatim record stored in memory; rather meaning appears to have the highest priority for memory storage. Memory also appears to have the property of being interpretative, as is evidenced by individuals filling in missing details or distorting information so as to be schema-consistent. The schematic approach to memory postulates that memory has the property of being integrative — incoming information is integrated with other related stored information about an episode to create a single representation of a complex event.

How well do schema theories address the three properties of memory — accuracy, incompleteness, and distortion? The four central assumptions of schema theories offer adequate explanations for why memory is incomplete or distorted. The schema theory approach to memory provides little theoretical basis for understanding the third aspect of memory, accuracy. Research indicates that memory for complex events is far more detailed than schema theory suggests.

Once encoded, are memories permanent? Two alternative theories about memory and the permanency of information stored in memory are relevant:

1. Information once acquired by the memory system is unchangeable

[1]For the purpose of this work, *accurate* or *accuracy* will refer to those responses or recollections which conform exactly to "truth"; *incomplete* is defined as not containing all elements; and *distortion* is the process of misrepresentation of facts.

and errors in memory result either from an inability to find stored information or from errors made at the time of the original perception.

2. Stored information is highly malleable and subject to change and distortion by events occurring during the retention stage.

The theory of permanently stored information is appealing in its ability to explain accurate recall after long periods of time during which there has been little or no rehearsal. People's experience in occasionally recovering an idea or memory that had not been thought of for an extended period of time often provides individuals with the evidence they need to support a theory of "permanence." However, there appears to be far more evidence that memory is malleable and subject to change with the introduction of new material over time. Loftus' research (1975, 1977, 1980) has indicated that reports of real-world, complex events can undergo systematic and predictable distortions. However, it is not inconceivable that some aspects of the same event may be permanent while other aspects may be malleable, depending upon the particular type of memory and the encoding process for that memory.

8.2.1. The Effects of Respondent Rules

One survey design factor related to the encoding process is the choice of a respondent rule, that is, whether to require that all information be collected from the individual of interest (self-reporters) or whether to accept proxy reports as part of the design. Many surveys are designed such that individuals within a household may report for other household members, resulting in a mix of self and proxy responses. To the extent that information is encoded and eventually retrieved differently when reporting for oneself than when reporting for others, respondent rules may contribute to overall levels of response error. The cognitive tasks and the social psychological influences of reporting for oneself are hypothesized to be quite different from those affecting proxy reporting. Memories concerning oneself may be organized differently than memories concerning others, in part due to the fact that visual images that are possible in observing others are usually not possible in self descriptions. For behavior that is not observed by the respondent, the visual image is that of the spouse or child reporting the event, whereas a self-reporter will draw on images of the event itself. Similarly, the internal evaluations related to a particular event will differ for self and proxy reports; for the self-reporter the evaluation will be directly related to the event, for the proxy reporter, the evaluation is in part a response to the person to whom the event occurred. Social desirability influences may more strongly affect self responses, as the respondent attempts to portray him- or herself in a positive light to the interviewer. Although many studies have compared self and proxy responses, the findings from this literature are questionable due to the confounding of respondent rule with characteristics of the individual (those home at the time of the interview are defined as self respondents; Moore, 1988). Often characteristics of interest to the analyst (e.g., health, labor force participation, etc.) are correlated with the availability of an individual and thus their ability to serve as a self-respondent.

Thus, the encoding of information and subsequently the means by which to retrieve that information as well as the willingness to report may all be affected by respondent rules. Chapter 9 examines the difference in self and proxy reporting from an information processing perspective. However, further research is needed to understand the differences in encoding and retrieval processes for different types of respondents; until then the survey designer has little information by which to determine the best means for reducing errors from this source.

8.3 COMPREHENSION

Comprehension of the interview question is the "point of entry" to the response process. Does the question convey the concept(s) of interest? Is there a shared meaning among the researcher, the interviewer, and the respondent with respect to each of the words as well as the question as a whole? The comprehension of a question involves not only knowledge of the particular words and phrases used in the questionnaire, but also the respondent's impression of the purpose of the interview, the context of a particular question, and the interviewer's behavior in the delivery of the question. Cognitive research in comprehension has emphasized the importance of prior knowledge in understanding words and groups of words, the semantic memory for specific concepts. Questions that are written so as to limit the interpretation made on the part of the respondent will be more effective in directing the retrieval process. The context of the questions and the details provided by the questions will guide the respondent's selection of the appropriate script.

The psycholinguistic view emphasizes that language use does not have to do with words and what they mean; it has to do with people and what they mean, that is, the speaker's intentions. The question–answer nature of the interview, in which interviewers are trained to be noninteractive with the respondent, is based on the belief that the researcher, interviewer, and respondent share a "common ground" (Clark and Schober, 1990). The implication for the survey researcher is that each respondent brings a set of information and experiences which comprises their cultural and personal common ground. Cultural common ground is that information common to the culture and shared by the groups to which the person belongs; personal common ground consists of those experiences jointly shared by individuals engaging in conversation. Survey questions are written to maximize the cultural common ground and minimize the use of personal common ground between the interviewer and the respondent.

Sound questionnaire design, interviewer training, and comprehensive training manuals help to assure adequate respondent comprehension. Since comprehension has more to do with the questionnaire and the choice of question wording than with the respondent, the reader is directed to Chapter 2.

8.4 RETRIEVAL OF INFORMATION

To address questions concerning past behavior or events of interest, the process of comprehension is followed by the respondent's attempt to retrieve information from long term memory. Although a distinction is often made between the storage of information and retrieval of that information, for most life events (as opposed to laboratory tasks), the distinction is not very useful. We know little about what is stored or encoded in memory apart from what is retrieved. With respect to the measurement of response error, the distinction is somewhat irrelevant; both types of forgetting will lead to incomplete or inaccurate information. Response error may indicate that the information was never stored in long term memory, that the original means for accessing the information ("the memory trace") no longer exists, and/or that the retrieval cue was not adequate. However, with respect to the reduction of response errors the distinction is important since trace-dependent forgetting cannot be affected by the interview situation, whereas cue-dependent forgetting implies a possible reduction in response error through the use of well-formulated questions, aids to recall, or providing the respondent with sufficient time to retrieve the information.

Under what circumstances are we able to recall accurately? Any learning task occurs in a particular context which is defined by a combination of the external environment and the learner's own internal environment, for example, mood or emotional state. As stated by Tulving (1975), the interpretation, storage, and retrieval of an event is affected by the surroundings in which that event occurred. Evidence from laboratory settings indicates that retention is higher when both learning and recall take place in the same environment and when the subject's mood is similar at learning and testing. Recall may also be affected by the respondent's motivation to perform the task (see Cannell, et al., 1981). Within a survey context, Fisher and Quigley (1990) found that recall of specific food items doubled when respondents were interviewed using a "cognitive" interview, where the respondent was initially encouraged to "think about the environmental context and ... the relevant psychological context" at the beginning of the interview to guide the data collection. A similar cognitively based questionnaire, discussed in Chapter 10, to improve reporting of smoking history was also successful in improving recall.

Experimental work completed in the field of cognitive psychology coupled with the research related to nonsampling errors in survey research has generated a list of factors which are correlated with errors in retrospective recall. These factors include:

1. Interference: no ability to distinguish between similar events and the introduction of new and possibly conflicting information;
2. Length of time between the occurrence of an event and the recall of that event;
3. The salience of the event;
4. The respondent's psychological state—for example, his or her mood.

8.4.1 Intervening Events

Classical interference and information-processing theories suggest that as the number of similar or related events occurring to an individual increases, the probability of recalling any one of those events declines. An individual may lose the ability to distinguish between related events, resulting in an increase in the rate of errors of omission. Inaccuracy concerning the details of any one event may also increase as the respondent makes use of schema related to a class of events for reconstructing the specifics of a particular occurrence. When faced with the task of recalling specifics among a set of similar or related words or events, subjects (in psychology experiments) have tended to exhibit two patterns of accurate recall. The first pattern, "proactive inhibition," refers to the situation in which exposure or the occurrence of an earlier event X inhibits the recall of subsequent events. "Retroactive inhibition" models describe the other pattern that has been evidenced, specifically that event X will be less well recalled if it is followed by a related event Y than it would have been had Y not occurred. In general, interference theory suggests that forgetting is a function of both the number and temporal pattern of related events in long-term memory.

The effect of "intervening events" on recall is not limited to the occurrence of similar events. Interference can also be the result of new information related to the original event or learning the outcome or result of an event. An example of the latter type of interference is the error in retrospective recall of voting behavior and preference following the outcome of an election (Fischhoff, 1975).

The optimum solution to minimizing bias imposed by intervening events is to interview the person as close as possible to the occurrence of the event of interest. This admittedly is normally not possible. The next best solution is to try to recreate the mood and context that the respondent was experiencing at the time of the event. Most simply this can be done by asking the respondent to think back to the time of the event. Statements can be made to encourage this "thinking back" process such as "we know the outcome of the election is now known, but thinking back prior to this what did you think, feel or do."

If the number of intervening events is a particular problem in a given survey, or if the outcome of an event is well known, then it might be worth asking fewer questions and allowing respondents to take their time to think back. The pace of the interview is often too fast to allow respondents to sort out intervening events from the event itself. The use of memory aids and dividing the period into shorter, more manageable periods for the respondent may also help to reduce bias due to the introduction of intervening events.

8.4.2 Length of Recall

Intuitively, the most obvious aspect of forgetting is that we recall more and more poorly with the passage of time. McGeoch (1932) was the first to suggest that the hypothesis that memories fade because of the passage of time is theoretically sterile and wrong. He showed that with the amount of time since learning held constant, it was possible to experimentally vary the amount of forgetting by manipulating what went on during the intervening time. His work forced researchers to reevaluate the existence of a correlation between time and memory as an indication that time is the causal factor in memory loss.

However, research findings confirm that the greater the length of the period being recalled, the greater the expected bias due to respondent recall error. This relationship has been supported by empirical data from studies of consumer expenditures and earnings (Mahalanobis and Sen, 1953; Neter and Waksberg, 1964); studies of hospital data and medical records (Woolsey, 1953; Cannell, 1965; U.S. National Center for Health Statistics, 1965 and 1967); studies of motor vehicle accidents (Cash and Moss, 1969; Waksberg and Valliant, 1978); studies of crime (Murphy and Cowan, 1976); and studies of recreation (Gems, Ghosh, and Hitlin, 1982; Ghosh, 1978).

By extending the recall period, a larger number of events may be observed, resulting in smaller sampling errors, but with potential increases in bias. Since most research has indicated decreased bias correlated with shorter recall periods, an obvious approach is to use a shorter recall period. However, too short a recall period may result in bias due to telescoping of events into the reference period. For example, if you ask a respondent about a large, expensive purchase such as a refrigerator or about a critical health event such as a surgery, and the reference period is too short and the relatively rare event actually occurred in the recent past, then research has shown that the respondent will typically forward telescope these events and answer accordingly. A short recall period is used most effectively when asking the respondent to recall frequent, routine events. This situation is a classic one in designing a survey, that is, selecting a recall period which minimizes the total mean squared error by controlling the contributions from possible biases and sampling error.

Response errors which relate to the length of the recall period are typically classified as either *telescoping error*, that is the tendency of respondents to report events as occurring earlier or later than they actually occurred, or *recall decay*, the inability of the respondent to recall the relevant events occurring in the past (errors of omission). Telescoping generally is thought to dominate errors of recall when the reference period for the questions is of short duration, while recall delay is more likely to have a major effect when the reference period is of long duration.

One means by which to reduce recall error associated with longer recall periods is to use varying recall lengths for different respondents, depending upon their expected level of activity. Respondents who are expected to have many events to recall are asked to recall them for a shorter period, thereby reducing the potential recall bias for their contribution to the overall estimate. For example, those who have chronic health conditions requiring frequent trips to the doctor may be asked to recall these over a shorter period than those being asked to recall less frequent trips to the doctor for unanticipated acute conditions. Where possible respondents are stratified in advance so that the interviewer can be instructed as to which instrument (which length of recall) is appropriate for which respondent. The lack of perfect stratification does not eliminate the rationale for using varying recall lengths, it only diminishes the effectiveness of the method for reducing potential biases.

As noted above, one type of error associated with length of recall is telescoping events either into or out of the recall period. Nonroutine events which occurred prior to the recall period may be telescoped in and be reported erroneously by the respondent while more frequent, routine events may suffer from memory decay. To reduce the effect of telescoping, a bounding or preliminary interview has been used to anchor the initial point of reference in the respondent's memory.

Bounded recall procedures were developed by Neter and Waksberg (1964) in a study of recall of consumer expenditures. The general methodology involves completing an initial interview which is unbounded; often the data from this interview are not used for estimation but are solely used as a means for reminding respondents in subsequent interviews about the behaviors that have already been reported. Bounded procedures work most naturally in longitudinal survey designs, where bounding can be built into the survey without substantially increasing costs. The major drawback of the bounded interview design is that it requires multiple interviews and is therefore costly. To reduce costs, some have experimented with using bounded recall within a single interview. Sudman, et al. (1984) tried this by asking respondents in a single interview to recall health events in a previous month and then to recall events for the current month.

Other methods for reducing telescoping error are the use of significant dates, such as a holiday, or other key events relevant to the respondent as anchors for reference periods. Loftus and Marburger (1983) used the eruption of Mt. St. Helens as the landmark event for two studies; the result was significantly reduced incidence of forward telescoping of crime victimizations. Respondents' provisions of their own landmark event, New Year's Day, also substantially reduced forward telescoping. However, the use of a key event does not always reduce bias due to the length of the recall period. Chu, et al. (1990) found that dividing the year into four holiday periods did not improve the accuracy of recall of hunting and fishing trips when compared to ordinary annual recall. Perhaps the holiday periods were not salient to the seasonal context in which these respondents engaged in hunting and fishing activities.

Among the earliest methods developed for reducing the possible bias related to length of recall are the use of memory aids. These techniques are used to jog the memory of the respondent by suggesting related items or topics. These aids may include calendars, maps, or, when a number of events occur not only over a period of time but simultaneously as well, a diary. The use of diaries is an alternative procedure which reduces the reliance on recall and the dependence on a reference period. However, Bishop, et al. (1975a) found that diaries require more time and effort from both the respondent and investigators and, in their study, 90 percent of the information recorded in the diary was obtained on a retrospective questionnaire. Similarly, Sudman (1964) found severe recall bias in diaries of consumer expenditures. He compared expenditures in weekly diaries for grocery products and monthly diaries for nongrocery products to manufacturers shipment data and found an underreporting rate of 45 percent for grocery products and 45–65 percent for nongrocery products.

When a relatively small number of events happen simultaneously, a log may be a suitable recording device for assisting the respondent. The log serves the purpose described earlier of organizing the period of recall into a chronologically ordered list, facilitating recall. This will apply even if the respondent fails to maintain the log on a regular basis. Various formats have been tested for improved respondent use aimed at obtaining more accurate recording.

A calendar used as a memory device can jog memory and structure it in time. As such, it is useful in helping with problems related to memory decay. When provided with calendars, respondents often recall additional events. This point is confirmed by many instances in the literature where respondents were asked a question about frequency of an event within a time frame, an answer was recorded, and then the respondent was shown a calendar and asked the same question, which resulted in the respondent reporting a greater number of events.

When the survey variables include geographic locations, for example sites visited or trips taken, maps will serve to remind the respondent of activities associated with specific places. The use of a map as a memory aid provides assistance in two ways. It facilitates the recall of specific events that are location oriented and also offers a way of organizing events, in this case spatially and temporally, which facilitates recall. When questions are related to locational issues, the use of maps may prevent double-counting frequent, routine events.

The use of photographs may assist the respondent in recalling events more accurately when recall and recognition of a large number of items, especially when some of these may be unfamiliar to the respondent, are involved. For example, recall bias was reduced by the use of color photographs of products, when asking respondents whether and how often they used an array of over twenty different household and automotive products containing methylene chloride (Eisenhower, et al., 1986). Photographs can also be used to facilitate self-assessment of features such as skin tone and assessment of other physical features of objects of questions such as weave of cloth, etc.

8.4.3 Salience of an Event

Salience, as defined by Webster, refers to those items or events which have "prominence" or have "emphasis." In both survey and psychological research, saliency of an event is often measured by the "emotion" related to the event or the extent to which the event is anomalous with respect to related schema or a person's life in general. Linton (1982), in assessing her own ability to recall events over a period of several years, suggests that the events best remembered are those that: (1) were emotional at the time of occurrence; (2) marked a transition point in a person's life; (3) remained relatively unique. Sudman (1975) states that there are three dimensions that distinguish between events which are more or less salient to respondents: the uniqueness of the event, its economic or social costs/benefits, and the continuing consequences.

The salience or importance of an event is hypothesized to affect the strength of a memory trace. The more salient the event, the stronger the memory trace and subsequently, the less effort or search of memory necessary to retrieve the information. This suggests that salient events are less subject to errors of recall decay than less salient events. Cannell and Henson (1974) found that events which were important to the individual were reported more completely and accurately than those of less importance. Psychology experiments have reported that tested recall of students under conditions of emotionally charged material was more accurate as compared to emotionally neutral material (Strongman and Russell, 1986). However, more salient events may also be subject to greater error in estimation of frequency or telescoping, due to the prominence of the event in memory. Chase and Harada (1984) discovered that the importance of swimming as a leisure activity to the respondent was moderately and positively associated with the percentage of error in estimation; the more important the activity, the more likely estimates of participation were overreported.

Waksberg and Valliant (1978) examined recall bias in reporting product-related injuries monthly at the end of a six-month period. The interviews were "rolled" so that the end of the six-month period varied seasonally for different respondents. There was a steady decrease in number of injuries reported as the length of time between the injury and the interview increased. Most of this decrease was attributed to forgetting of events, although some telescoping of more critical events may also have been present. The attrition in reporting of injuries was much greater for minor injuries than for those that require medical care. Elapsed time had virtually no effect on the reporting of injuries that were medically attended and caused restrictions in the person's activities for six days or more. Conversely, there was a drop of over 50 percent in the reporting of injuries which were not treated in emergency rooms and did not restrict the person's usual activities.

The saliency of the respondent's mood at the time of the event or about a category of events also influences the respondent's recall ability. Rholes, et al. (1987) did a study to determine whether the impact of emotional states on recall varies when subjects read statements expressing positive or negative self-evaluative ideas or when they describe somatic states that frequently accompany positive or negative mood states. They thus manipulated subjects' mood states. Although somatic and self-evaluative statements generally had equally strong effects on subjects' mood, self-evaluative statements had a stronger impact on recall of life experiences than the somatic statements. Whereas positive somatic experiences were quickly remembered, negative ones were not. In addition, self-evaluative statements affected recall independently of their effects on subjects' mood state.

The saliency of the event or mood associated with the event can be used to enhance recall by giving the respondent time to "think back" and recreate the environmental context of the event. Instructions and wording to this effect can be built directly into the questionnaire. Interviewers can also be trained to probe for such reconstructions. This technique works best with nonroutine events, although Means, et al. (1988) enhanced respondent recall of routine medical visits for ongoing, chronic conditions by associating them with more salient events to the respondent such as birthdays and vacations.

If the saliency of the event or affective state associated with the event(s) is thought to result in overestimation then techniques must be used to neutralize the effect. Neutralization can be accomplished by building in a statement of the importance of accuracy to the study; how people respond in a variety of ways (high, medium, low occurrences), and that there are no value judgments made to the answers given other than the importance of accuracy. The environmental context can be recreated to remember events. The neutralizing techniques can be used to encourage accurate reporting.

8.4.4 Estimation

Autobiographic survey questions are not limited to questions which ask details concerning one event or memory but often ask the respondent to estimate frequencies of events or make "mathematical" judgments. Examples of estimation questions would include those which ask the respondent to report the number of times he or she has been to a doctor in the last year, the average expenditures for food per week, and the number of different magazines the respondent has read during the last month. In response to the question "How many times have you been to a doctor in the last year?" a respondent may attempt to recall all episodes or may rely on an estimation strategy. The decision as to which approach to use will depend upon both the frequency of the event and the time period of interest. If the time period is long, or the number of events frequent, estimation may provide the respondent with the most efficient means for answering the interviewer's question. For example, respondents asked to recall frequent, routine events over the past year, typically estimate what they usually do during a month or what they did last month and multiply this by twelve (months).

Judgment heuristics (Kahneman and Tversky, 1971; 1973; Tversky and Kahneman, 1973) suggest that imprecise rules are used in making frequency judgments. In making frequency judgments, it is hypothesized that the laws of probability are replaced by a limited number of heuristics to provide a response.

One of the heuristics most relevant to the retrieval of information to answer survey questions is the availability heuristic. When people judge the frequency of an event on the availability and speed with which they can recall an occurrence of the event, they are using the availability heuristic. The strength of association is used as a basis for a judgment of frequency. Factors which have an effect on the availability (the ease of recall) have a corresponding effect on the apparent frequency of repetitions. Frequency judgments are made both by assessing the availability of specific items or by a more general assessment of the strength of association. Temporal and spatial relationships among repeated occurrences have been shown to affect the estimation process. When the number of repetitions is relatively small, a person may attempt to estimate the frequency by recalling specific occurrences. These repeated occurrences are more likely to be stored and recalled as distinct units when they are spaced temporally wide apart. When the number of events are large, or closely spaced together, the respondent will most likely use heuristics to estimate the response. Based on the availability of a heuristic, the direction of response error would be predicted to be an overestimation of rare, but salient events, and an underestimation of events which are closely spaced.

Estimation strategies used by respondents will often suffice for the level of accuracy required and affordable in a given survey. Interviewers can be trained to use probing techniques and/or formal worksheets can be attached to questionnaires to ascertain what estimation strategies are actually being utilized. Where estimation procedures are expected to significantly impact the respondents' answers to a questionnaire, cognitive laboratories and formal pretests should be utilized to uncover the range of strategies in advance. To assure consistency, a preferred strategy could be selected and formalized for use in the larger survey. Respondents could be instructed to use a given estimation strategy in the absence of precise, actual numbers. Memory aids, previously discussed, can also be used to produce more accurate estimates.

8.4.5 Other Factors Affecting Recall

As noted above, the recall task is affected by several factors which the survey designer can control, including the length of the recall period or requesting that the respondent reference available records, and others which are less controllable, such as the salience of the event being reported. Other task variables which can be controlled include the question administration, for example, whether the question is open or closed and the interviewing situation, that is the mode of administration and the length of the interview. Many of these factors have been shown to be related to levels of response error and may therefore be related to the task of retrieving information. For example, question length, specifically long open-ended questions, obtain higher levels of reporting threatening behavior (e.g., consumption of liquor, sexual activities) than short, closed questions (Bradburn, Sudman, et al., 1979).

One means by which to reduce task difficulty is the organization of the questionnaire. Tourangeau (1984) stresses that the organization of a questionnaire which is logical and a natural sequence of events will help the respondent more accurately recall events and may also reduce the response burden. Reiser, et al. (1985) found that recall was greatly facilitated when the interviewer asks a respondent about a sequence of specific actions before asking about a general action, e.g., shopping in a department store (action), before asking about paying, a general action that could be applied to a number of activities.

The concept of reducing the difficulty of the respondent's recall task can be applied to a variety of survey dimensions. For example, one of the most difficult characteristics of an event to recall is the date on which the event occurred; yet many questionnaires begin series of questions concerning an event by asking the respondent to report the date.

When asking the respondent to recall several events which occurred over a period of time, research has shown that an orderly review of the time in question is preferred to an unstructured one in which the respondent recall events in a more or less haphazard fashion. There are several obvious choices for directing this review. One technique is that of retracing the time period in sequence beginning either with the most recent time boundary and working backward or beginning with the oldest time boundary and working forward (Loftus, et al., 1990).

8.5 JUDGMENT

Once information has been retrieved, the respondent may still need to integrate several pieces of information to answer a question. Integration of information is a difficult cognitive process and one in which the respondent may draw upon the use of inferences and judgment strategies to complete the reporting task. As noted in the discussion of estimation, respondents draw upon general guidelines (heuristics) when making judgments concerning frequency of events. Krosnick and Alwin (1987) suggest that respondents are "cognitive misers" in making judgments about alternative responses, attempting to use the least amount of effort or time to make decisions.

In addition to the availability heuristic discussed above, Kahneman and Tversky (1971) hypothesize that other general rules are used in formulating judgments, the "representativeness" heuristic and "anchoring and adjustment." In using the representativeness heuristic, people judge the probability of an uncertain event by the degree to which it is both similar in properties to the parent population and reflects the salient features of the process by which it is generated. Events that are judged to be more representative are also judged to be more likely. Anchoring and adjustment refers to the process in which the respondent determines an initial response (the anchor) and then adjusts, based on comparisons to the anchor. For example, Bradburn, et al. (1987) suggest that when answering questions concerning the frequency of dental visits, respondents begin with the expected norm, twice a year, and then adjust. The response categories and the order in which they are presented may also provide information to respondents about norms and appropriate anchors on which to adjust (see Chapter 3).

8.6 RESPONSE COMMUNICATION

The final task for the respondent in the interview process is to communicate to the interviewer the result of the comprehension, retrieval, and judgment tasks. Cognitive processes are central to response formulation: for closed-ended questions, the decision rules relate to a choice and for open-ended items, to the formulation of an answer. Tversky (1972) suggests that respondents make choices by eliminating those choices that lack some desirable feature. In the absence of social desirability factors affecting responses, the underlying assumption of all closed-ended or scale items is that the respondent has an ideal in mind and attempts to replicate this in his or her response. The final choice of a preferred option depends on what criteria are used for the selection process and the selection of these criteria is somewhat probabilistic.

Social psychology also provides concepts of relevance to the topic of response communication. The respondents' willingness to retrieve information will in part depend on how motivated they are and the "real" importance of surveys as perceived within the interaction of the respondent and the interviewer. The respondents' motivation will, in part, depend on how reasonable we are as survey designers. In other words, have we explained a reasonable purpose of the survey; have we logically constructed the sequence of questions on the questionnaire; and have we made the questionnaire comprehensible. The interaction and rapport with the interviewer will also enhance or reduce the respondents' willingness to assume the burden of recalling and reporting accurately.

8.7 POSTSURVEY ADJUSTMENT PROCEDURES TO REDUCE THE EFFECTS OF RECALL ERROR

The sources and methods described above should be considered as the first and most appropriate steps for reducing the possibility of recall error. According to methods of quality assurance, errors should be prevented by proper design and planning and not eliminated or reduced by detection after the fact. However, it is unlikely that any survey design can eliminate all recall error. Accordingly, there may be a need for data adjustment. A number of estimation procedures have been developed to provide estimates of recall events that have a reduced mean squared error. Two of these techniques are the use of calibrating data and composite estimators.

When the recalled responses can be checked with secondary data, correction factors can be developed which adjust responses for expected biases. Secondary data sources may be available, although often only for a subsample of interviewed respondents. Examples of such secondary data obtained from the respondent might be invoices, check registers, and receipts. Secondary data might also be obtained from sources other than the respondent, for example, a physician reporting on the number of visits made or an employer providing information on salary and benefits. Models can be developed which relate the biased recall responses to the assumed more accurate secondary data and then can be used to develop an adjustment procedure. These models may range from the very simple to complex multivariate forms (U.S. Bureau of the Census, 1978).

Consider a situation in which two recall periods can be employed, one of length t and the other of greater length, kt. Let x_t denote an estimate based upon recall period t and let x_{kt} denote an estimate based upon the longer time period, kt. In a typical situation x_t is unbiased while x_{kt} has some bias, denoted by B. Presumably, x_{kt} has smaller sampling error than x_t since it is based upon a longer exposure period. We might find that the sampling error of x_t is σ^2/t while that of x_{kt} is as small as σ^2/kt.

A composite estimator (x_c) combining these two statistics would take the form of:

$$x_c = Wx_t + (1 - W)x_{kt}.$$

An analysis involving the magnitude of the sampling errors, σ^2/kt, and of the bias, B, would yield the proper choice of W so as to minimize the mean squared error of the composite estimator s_c. The two choices for estimators to be combined can take other forms as well.

For example, the two estimates that are combined may have one with suspected biases but with relatively small sampling error. In a similar fashion, two survey estimates can be developed from the identical set of respondents, one using less expensive but potentially biased methods and one expected to have little bias but greater sampling error. Again the principle is to obtain a minimum mean squared estimate which is a composite of the two statistics.

8.8 CONCLUSION

The recent interest by survey methodologists in cognitive psychology stems from the belief that to evaluate the sources of forgetting and what to do about memory loss, it is important to understand the process and structure of memory. This endeavor requires research on each of the dimensions of the respondent's task from the encoding of information to the response formation process. The largest void in the research at the present time concerns respondent's retrieval strategies.

There is still much research to be explored with respect to the respondent as the source of error. Among the topics of interest we would include:

- The need to ask respondents about the saliency of the event. All secondary analyses of response error that have included saliency as a correlate of error have imposed the researcher's notion of what is salient. To understand the respondent's viewpoint may bring more light to the usefulness of this particular characterization of an event.

- Research related to understanding how the respondent retrieved the information. If we know what type of process lead to what types of errors, we will be better able to minimize those errors during the interviewing process.

- The need to use error-free validation information, or at least a means for assessing both the level of error in the validation records as well as the error introduced in the matching of respondent information to validation information when assessing response error.

- The expanded use of experimental design to test alternative approaches to reduce response error.

- Use of a representative sample for research so that inferences about response error will not be limited to a subset of the population.

CHAPTER 9

MEASUREMENT EFFECTS IN SELF VS. PROXY RESPONSES TO SURVEY QUESTIONS: AN INFORMATION-PROCESSING PERSPECTIVE

Johnny Blair
University of Maryland

Geeta Menon
New York University

Barbara Bickart
University of Florida

9.1 INTRODUCTION

9.1.1 The Problem and the Traditional Approach

While there is a fairly large literature on proxy reporting in surveys, the findings are not consistent (Moore, 1988). Differences between self and proxy reports have been found in some studies and not in others. Observed differences seem to be related to the relationship between the proxy and target respondent, the amount of shared information, the kinds of questions asked, and the method of survey administration. But the conditions under which these factors apply, their relative effects and possible interactions are not well understood.

The ways in which information is recalled and used to answer a survey question may differ when the question is about oneself versus someone else. In this chapter, we investigate whether information processing differs between self and proxy respondents, and whether those differences help explain proxy measurement effects.

Survey research has relied primarily on self reporting, with proxies used for reasons of cost and respondent availability. For example, in

Measurement Errors in Surveys.
Edited by Biemer, Groves, Lyberg, Mathiowetz, and Sudman.

ISBN: 0-471-53405-6

situations where a target respondent is incapable of self response — such as with children or respondents too ill to be interviewed — proxies are often used. In addition, proxies are used for respondents who are available only at substantial additional costs, such as when data on all household members are needed. Such surveys which use proxy reporters on an "as needed" basis have accounted for most, though not all (see Mathiowetz and Groves, 1985), of the methodological investigations concerning the effects of proxy reporting (Moore, 1988). The lack of random assignment to a reporting condition creates problems for interpreting observed differences. Self-selection biases or other uncontrolled variables may account for more of the differences than does the self/proxy report condition. In these studies, the aggregate self report data have usually been compared to the aggregate proxy data, since both self and proxy reports for the target respondent are not available for individual level comparisons. Several approaches to the interpretation of these aggregate data have been used. Sometimes the total frequency or amount of reporting by proxy and self reporters have been compared. For example, on sensitive questions such as drinking alcohol or voting frequency, it is often assumed that higher reported amounts or frequencies indicate more accurate reporting. In some instances, both the self and proxy aggregate reports have been compared to a validation source. Unfortunately, for many behaviors and events of interest either there are no validation records, or they are unavailable to the researcher.

Post hoc explanations of the observed differences are often proposed. For example, it has been suggested that differences between self and proxy reports are affected by different amounts of information and different willingness to report certain behaviors. Such explanations, while of some practical use, are not based on a theoretical framework and thus provide little insight into the general conditions which affect the quality of these reports. In this study, question topic, pair relationship and administrative mode were examined with respect to self/proxy differences. The results reported are confined to the effects of the question topic and of the relationship between the respondent–proxy pair. It is these two factors that we have integrated into an information-processing theoretical framework.

9.1.2 The Information-Processing Perspective

Based on information-processing theory, which suggests under what conditions proxy and self reports will converge, we developed and tested several hypotheses. This approach should provide a theoretical basis for

(a) selecting the best informant within a household for particular types of reporting; (b) designing questionnaires to encourage respondents to use effective recall and response formation strategies; and (c) measuring or controlling variables that influence the response process.

The logic of the design of an information-processing study involves three general objectives:

1. Determine the processing strategies naturally used by respondents;

2. Determine the conditions that increase the likelihood of using particular strategies; and

3. Examine reporting errors under alternative conditions.

Investigating respondents' response formation requires either testing hypotheses that make statements about what response outcomes are expected if particular processes are occurring, or having relatively direct access to respondents' information-processing strategies. While there is no consensus on whether respondents can report accurately about their cognitive processes, the literature does suggest that verbal protocols are useful both as data with which inferences about processes can be made (Simon and Kaplan 1989; Bishop 1986) and as tools to assist in hypothesis formation.

Laboratory think-aloud interview data were used primarily to test hypotheses about respondent processing strategies in answering behavioral and attitudinal questions. Both laboratory and field interviews were used to examine reporting errors.

Our primary measure of reporting accuracy is the convergence between self and proxy reports. This is an appropriate measure for attitudinal items, when one wants the proxy to report the target respondent's opinion — such as feelings about a particular political group, or assessment of health status. For factual items, such as income, or behavioral items, such as doctor visits, convergence may not be the ideal measure of choice. If both the self and proxy reports are inaccurate, high convergence would not be meaningful. Ideally, both reports should be compared to a validation source. This would allow a comparison of the conditions under which the self or the proxy report is most valid.

For some kinds of items, such as voting, known to be overreported in self reports (Presser, 1990), or a sensitive item, such as drinking behavior (Sudman and Bradburn, 1982) which is often underreported, convergence between self and proxy reports may not be the best measure of accuracy. However, for most factual and behavioral items which are not clearly threatening or socially desirable, the convergence between the self and proxy reports is an appropriate measure.

9.2 LITERATURE REVIEW

9.2.1 Why the Processes Used to Form Self versus Proxy Reports Might Differ

As noted in Chapter 8, several authors (Strack and Martin, 1987; Tourangeau, 1984; Tourangeau and Rasinski, 1988) have proposed general models of the survey response process. These models are quite similar and suggest that a respondent must: (1) interpret the question, (2) retrieve appropriate information or a prior judgment, (3) make a judgment, and (4) report a response. This section focuses on factors that could result in different processes being used for making self and proxy reports and the effects of these differences on reporting accuracy.

In most cases, knowledge about ourselves and others is acquired differently. While we learn about ourselves through our daily experiences, information about others is acquired in a number of ways. For example, you could learn about your spouse's activities through shared experiences, by discussions or by conversations with a third party. In addition, there are different degrees of shared experiences. Couples could jointly participate in an activity, could observe their spouses engaging in an activity or participate in related experiences. Both the inputs available to form a judgment and the storage of this information are likely to be affected by how the knowledge is acquired (Alba and Hutchinson, 1987).

First, the information one has about another person may vary as a function of how the information is learned. Shared experiences (either direct or related) should increase both the amount and quality of inputs available to form a judgment. In addition, the topic's importance to a respondent may affect attention and elaboration. People are likely to process a message more thoroughly when they are involved with the issue (Petty and Cacioppo, 1982). Also, people may be more likely to attend to and elaborate on another person's behavior when it is important to them. For example, self-relevant information has been shown to receive increased elaboration at encoding (Kuiper and Rogers, 1979; Rogers, et al., 1977). This is thought to create a more elaborate memory trace for the information, resulting in enhanced recall. Furthermore, information about important others is more accessible than information about less important people. For example, Bower and Gilligan (1977) showed that information about one's spouse received increased elaboration and was better recalled than information about a less important person.

Second, how information about events is stored depends on both how the information is acquired and on the frequency of occurrence and saliency of the events. Rare and outstanding events are likely to be stored as individual episodes, while frequent and mundane events are more likely to be stored in general knowledge structures or semantic stores (Tulving, 1972, 1983). When another person's behavior is *not* learned via joint participation in an event, the information may be less likely to be stored episodically. This has implications for how judgments are constructed. For example, frequency reports of mundane behaviors (those retained in semantic stores) are usually based on estimation procedures, while reports of outstanding events (those retained in episodic stores) are likely to be recalled individually and counted (Blair and Burton, 1987; Bradburn, et al., 1987; Schwarz, 1990a). Because information about others is more likely to be retained in semantic stores, proxy reports of behavioral frequencies are more likely to be based on estimation strategies, regardless of the characteristics of the event (Bickart et al., 1990). Proxies' ability to use a recall and count strategy may be related to their level of participation in the event.

If proxy reporters cannot rely on episodic inputs to form judgments, what other inputs could they use? Hoch (1987) suggests that people will use three inputs in answering a question about another person: (1) their own attitude, judgment or behavior, (2) the perceived level of similarity between themselves and the other, and (3) other relevant information, such as prior conversations or observed behavior. Supporting this model, Davis, et al., (1986) found that people anchored on their own preferences in predicting their spouses' preferences for a variety of consumer products, and then adjusted their judgment based on the spouses' perceived level of influence in the purchase decision for that product. Likewise Menon et al. (1990) found that people were likely to use general knowledge about the other person and information about themselves when making a proxy report.

Because much of the information we have about others is not acquired through shared experiences, this information is likely to be represented as episodes which relate to the occasion of learning about the event. In contrast, storage of information about ourselves is more likely to be represented as episodes related to the actual event or occasion. This has important implications for the cues that will be effective in enhancing recall. Specifically, cues which were part of the encoding context are more effective in enhancing recall (Tulving and Thomson, 1973). Thus, the cues that will be effective in enhancing retrieval of information may vary for self and proxy reporters. For example, Larsen and Plunkett (1987) found that information about reported events was accessed through the memory of the context in which the respondents learned about the event, versus cues related to the event itself. Thus, event cues may be more effective in enhancing recall for self than proxy reports.

Similarly, unless information about others is acquired through shared experiences, it may not be obtained in chronological order. Therefore, the memory representation may not include time or date information. When answering survey questions, people may use dates in several ways. First, dates are used to delimit a reference period (Brown, et al., 1985; Bradburn, et al., 1987; Loftus and Marburger, 1983). This may be more difficult for proxy reporters, and thus, may result in more errors due to telescoping and/or omission. Second, date information could affect the way people search the reference period when they must count multiple events (Loftus and Fathi, 1985). Because proxy reporters often lack date information, they may be likely to use a different strategy in searching the reference period.

Motivational differences between self and proxy reporters may also exist. This is important for several reasons. First, motivation could affect the amount of cognitive effort used by respondents to search memory for relevant information. Proxy reporters may be more likely to "satisfice," or provide quick and easy survey answers (Krosnick, 1991c). Krosnick and Alwin (1987) show that respondents who satisfice are more susceptible to certain response effects. Second, survey respondents are often motivated to present themselves in a socially acceptable way (Blair et al., 1977; Bradburn, et al., 1979; Hippler and Schwarz, 1987; Sudman and Bradburn, 1982). Self reporters may be more likely to try to present themselves favorably to the interviewer than proxy reporters, particularly on sensitive issues. When questions are sensitive, more accurate reports may actually be obtained from proxy reporters. However, there may be some cases when proxy reporters may try to present the other person in a favorable light because it reflects on themselves. For example, parents may intentionally misrepresent their teenagers' alcohol consumption.

9.2.2 Implications of These Psychological Factors for Survey Measurement

The above discussion points out some factors which may affect the convergence and accuracy of proxy reports. In designing surveys, we could measure variables thought to affect the survey response process, and then use these measures to adjust responses statistically. In addition, we could use various aspects of the questionnaire, such as question order, cues provided, or the length of the reference period to reduce measurement error.

When respondents are not easily able to retrieve relevant information from memory to form a proxy report, we may be able to use the questionnaire design to make that information more accessible. If a couple (that is, the reporter and the person of interest for the questions) is similar on a particular dimension, one's own report may be useful in making a judgment about the other, and can result in more accurate reports (Davis, et al., 1986; Hoch, 1987). Increasing the accessibility of the self report should increase the likelihood that it is used in answering the proxy question (Feldman and Lynch, 1988). This can be accomplished by changing the order in which questions are asked, or by including questions about the similarity of the couple. The use of an anchor and adjustment strategy could lead to higher convergence between self and proxy reports if the couple is, in fact, similar.

As noted above, the amount of knowledge we have about another person and how that knowledge is acquired may affect the strategy used to form a proxy report and the quality of the report. People who participate jointly in behaviors or share experiences should have similar inputs available in memory to construct a judgment. This may lead to higher convergence between reports. The importance of the behavior or topic to the respondent may affect the degree of elaboration that information receives and again, may lead to higher convergence between self and proxy reports. Couples who have lived together longer may have a greater opportunity to obtain information about each other. Thus, the length of the relationship may be positively related to convergence between reports. In Figures 9.1 and 9.2, two sets of hypotheses derived from these ideas are shown: hypotheses about processes and hypotheses about convergence and order effects.

9.3 OVERVIEW OF METHODS AND DESIGN

In both the laboratory and field, pairs of respondents were asked an identical set of questions about themselves and about the other person. The questionnaire for both the laboratory and field interviews consisted of a wide range of behavioral, attitudinal and demographic questions. Almost all of the questions had been used in numerous surveys by other organizations. In the laboratory, both concurrent and retrospective think-aloud protocols were elicited from fifty pairs of respondents. Respondents were either married or living together as married. Subjects were solicited through advertising and paid $25 per couple for the interviews, each of which averaged 45 minutes per person and was tape recorded. The sample consisted of both students and non-students from Champaign-Urbana, Illinois.

P1: **Event Cues**
Self reporters should be more likely than proxy reporters to use event cues in retrieving a specific event.

P2: **Estimation**
Proxy reporters should be more likely than self reporters to use estimation strategies (versus a recall and count strategy).

P3: **Automatic Response**
In think-aloud interviews, self reporters should be more likely than proxy reporters to report that an answer came to them automatically.

P4: **Anchor and Adjustment**
Respondents asked both self and proxy questions will tend to use a heuristic in which the first question's response provides a base for an estimate for the second question. This will occur whether the self-proxy questions are alternated or are grouped into sets. There will be a stronger order effect when each proxy report question immediately follows the self report.

P5: **Chronological Search**
In searching the reference period for an event or behavior, self reporters should be more likely than proxy reporters to use chronological sequences.

Figure 9.1 Hypotheses about processes.

C1: **Level of Participation**
The correspondence between proxy and self reports about a given behavior or issue should increase as the proxy reporter's level of participation in the behavior increases. (However, this relationship may not be linear, as there may be a sharp break in the correspondence between proxy and self reports for events that are observed versus those only heard about or otherwise learned about indirectly.)

C2: **Importance**
There should be greater convergence between proxy and self reports when the target behavior or issue is important to the respondent.

C3: **Order Effects**
The size of order effects will be inversely related to both the proxy reporter's level of participation in the event or behavior and the importance of the behavior to the respondent.

Figure 9.2 Hypotheses about convergence and order effects.

In the second phase of the study, several shorter versions of the questionnaire were used in an RDD telephone survey of 201 respondent pairs. These versions varied the interview conditions in several ways, described below. The field interviewing provided the main data to examine convergence. The sample was selected using standard RDD sampling procedures. The interview averaged 20 minutes per person.

The quantitative examination of the verbal protocol data required the development of a coding frame for alternate processing strategies. Since this type of coding is not standard, some discussion of these codes and their development is useful.

Our hypotheses required coding respondent strategies for: counting versus estimation of frequencies; memory search strategies; recall cues; and other information used to answer a question. Initially, we listed broad categories within each of these areas. The members of the research team conducted the interviews and coded them from the tapes. Then a different team member coded each interview. In discussing these results, the team refined and clarified exactly what was meant by each type of strategy, determined that subjects often used multiple strategies in responding to a single question, and found that additional coding categories were needed. Using this new list, detailed coding instructions were written and coders, who were not research team members, were trained. All of the interviews were then recoded. The coding categories are shown in Exhibit 9.1.

9.3.1 Analysis Plan

The analysis plan included: (a) the use of protocol analysis to determine naturally used strategies and strategies used under alternative conditions; (b) the analysis of the accessibility of self reports on processing strategies; and (c) analysis of reporting convergence. Our primary measure of reporting accuracy was the convergence of self and proxy reports. However, prior to the analysis of convergence, we examined the means and marginal distributions of the self and proxy reports; we found no meaningful statistical differences. We concluded that generally, for our variables, proxy reports were not biased.

Convergence was measured in two ways. The measure of convergence for continuous variables is the correlation between the self and proxy report. For noncontinuous variables, the percent agreement between the self and proxy report is used. Statistical significance was computed using Fisher's Z transformation of the correlation coefficient. Since the sample sizes in many of the table cells are very small, many of the observed differences, viewed singly, are not statistically significant. However, since the number of cells/items is large, we have also used binomial signs tests, in which the number of observations that are in the direction predicted by an hypothesis is compared to the number of observations in the opposite direction. In many of the instances discussed, although the differences are not large, the patterns are clear. The discussion of results is based mainly on the signs tests.

Exhibit 9.1. Verbal Protocol Coding Scheme

Code	Description
	10's: Automatic Response
11	Automatic
12	Event did not occur/Nonevent
13	Retrieval of prior judgment
14	No probe, therefore no protocol
15	Don't know
16	Not applicable (for skips)
	20's: Counting Strategies
21	General recall and count
22	by domain
23	by observation
24	with adjustment for uncertainty
25	with expression of uncertainty (no adjustment)
26	by domain with adjustment for uncertainty
27	by domain with expression of uncertainty (no adjustment)
	30's: Rate-Based Estimates
31	General rate-based estimation
32	by domain
33	with adjustment based on specific incident
34	with adjustment based on general knowledge
35	with adjustment based on nonoccurrence
36	with adjustment for uncertainty
37	with expression of uncertainty (no adjustment)
38	by domain with adjustment for uncertainty
39	by domain with expression of uncertainty (no adjustment)
	40's: Enumeration-Based Estimates
41	General enumeration-based estimation
42	by domain
43	with adjustment based on specific incident
44	with adjustment based on general knowledge
45	with adjustment based on nonoccurrence
46	with adjustment for uncertainty
47	with expression of uncertainty (no adjustment)

Exhibit 9.1. *Continued*

Code	Description
	50's: Anchoring Strategies
51	Same as self
52	Based on prior answer
53	Anchor on self and adjust
54	Anchor on norm and adjust
55	Anchor on another specific person and adjust
56	Anchor on proxy and adjust
	60's: Misc. Codes for Attitude Questions
61	Based on specific behavior/event
62	Based on discussions with other
63	Based on general knowledge
64	Based on attitude toward issue
	70's: Search Strategies
71	No order/search
72	Forward search
73	Backward search
74	Anchor on date and forward search
75	Anchor on date and backward search
76	Search by domain
77	Based on another event
78	Based on regularity of behavior
	80's: Event Cues (for counting strategies)
81	Person mentioned
82	Place mentioned
83	Emotional reaction to event mentioned
84	Time of event occurrence mentioned
85	Characteristic of event mentioned
86	Based on prior response (one answer triggers off another)
87	Based on cues used from question
	90's: Reference Period
91	Anchor date on public event
92	Anchor date on personal event
93	Anchor date on season of the year
	General characteristic of event/person (for estimation strategies):
94	"Always..."
95	"Never..."
96	"Nowadays..."/"Usually..."

Table 9.1. Total Number of Event Cues Used by Self and Proxy Respondents ($n = 94$ pairs)

Behavior	Self	Proxy	Self/ Proxy
Newspapers	18	6	3.0
Magazines	13	5	2.6
Books for work	4	4	1.0
Books for fun	13	10	1.3
TV during the week	18	1	18.0
TV during the weekend	20	4	5.0
Crime	2	1	2.0
Illness	5	1	5.0
Doctor visits	4	2	2.0
Beer	10	10	1.0
Liquor	3	5	0.6
Drunkenness	6	2	3.0
TOTAL	116	51	2.4

9.4 RESULTS

9.4.1 Hypotheses About Processes

We used the results from the coded verbal protocols of the face to face interviews to examine hypotheses P1 through P3 (Figure 9.1).

Since information about oneself is hypothesized to be elaborated in greater detail and encoded more deeply than information about others, we expected self reporters to use more event cues than proxy reporters in retrieving information about behavioral frequencies (P1). Table 9.1 provides support for this hypothesis. The "Self" and "Proxy" columns in this table present the absolute number of event cues (such as a person, place, time or characteristic of the event) that were used by respondents in reporting how many times they (self report) or someone else (proxy report) had engaged in a particular behavior. The "Self/Proxy" column presents the ratio of usage of event cues in making a self and a proxy report. A ratio greater than one indicates more usage of event cues in making a self rather than a proxy report.

It is apparent that more event cues are used — and used more frequently — when one makes a self report (the overall ratio is 2.4). This pattern holds for every behavior except drinking liquor. In general, we conclude that respondents have more accessibility to specific event information related to themselves and use this information in making a self report. Such information about other people is less accessible, as is shown by the use of fewer event cues in the proxy reports.

Similarly, the use of a counting strategy should be more predominant in self versus proxy reports about behavior frequencies, since detailed information is more readily accessible for oneself. Proxy reporters may have to rely on more general information, and hence be more likely to use estimation strategies (P2). Table 9.2 shows the percentages of respondents using a count or an estimation strategy in making frequency judgments about themselves and about another person on different behaviors. The last column of Table 9.2 shows the ratios of self to proxy usage of estimation strategies. Again, if the ratio exceeds one, estimation is a more common strategy for self reports. The data indicate that in general, respondents used the estimation strategy more for a proxy report than for a self report (the overall self/proxy ratio for use of estimation strategies is 0.78). If we look at each of the individual behaviors, however, the pattern is not as clear. In four out of the twelve behaviors, estimation was used more frequently in formulating self rather than proxy reports.

We hypothesized that respondents should mention that a behavioral frequency report came to their minds "automatically" more often when they make a self versus a proxy report (P3). Since the number of mentions of an automatic formulation of a response is fairly small by topic, we aggregated the mentions across the different topics on which frequency judgments were elicited. We found that there were 33 mentions of an automatic process when a self report was made, as opposed to 26 mentions when a proxy report was made (the ratio of self/proxy is 1.27).

We also examined whether respondents would be more likely to consider the reference time period mentioned in the question in the case of self versus proxy reports of a behavioral frequency. Though we did not have a direct measure to look at this, we used an indirect dependent measure. If respondents use estimation strategies rather than counting strategies in arriving at a frequency report, we can infer that the reference period specified in the question was less likely to have been considered. This is because estimation strategies are used in a decontextualized general manner, where a rate for a general time period is used as a basis to arrive at a frequency judgment for the target time period. On the other hand, the use of a counting strategy requires the consideration of the reference period. The support we obtained for hypothesis P2, therefore, also provides support for this reference period issue since estimation was used as a strategy more in proxy than in self reports.

For the laboratory interviews, four basic questionnaire versions were used in a balanced 2×2 experimental design, in which there were two independent variables, priming and buffering. The priming variable was manipulated at two levels, presence of a prime and absence of a prime. The prime consisted of asking respondents immediately preceding the substantive item how similar they and their partners were on the

Table 9.2. Percentage of Self and Proxy Counting and Estimation Strategies[a]

Topic	Counting Strategy		Estimation Self/Proxy Ratio[b]
	Self	Proxy	
Newspapers	51.2% (84)	42.5% (73)	0.849
Magazines	88.2% (85)	75.0% (72)	0.472
Books for work	92.5%[c] (40)	77.1%[c] (35)	0.327
Books for fun	87.2% (47)	83.0% (47)	0.753
TV during the week	24.6%[c] (81)	16.9%[c] (72)	0.895
TV during the weekend	75.7% (74)	69.1% (49)	0.794
Number of relatives	96.1% (147)	94.0% (133)	0.650
Illness	74.0% (73)	76.3% (59)	1.097
Doctor visits	97.2%[c] (71)	83.1%[c] (71)	0.166
Beer drank	55.8% (43)	65.0% (40)	1.262
Liquor drank	75.6% (33)	66.7% (77)	0.732
Drunkenness	76.7% (30)	86.9% (23)	1.778
TOTAL	76.6% (856)	70.0% (728)	0.780

[a] Number in parentheses represents the number of reports. The number of self and proxy reports differ due to the use of other strategies.
[b] Indicates the ratio of self to proxy usage of *estimation* strategies. Ratios greater than 1.0 indicate estimation used more for self than proxy reports.
[c] Significant at $p \leq 0.05$.

substantive item. The buffer was also manipulated at two levels: a low level in which the proxy question was asked shortly, if not immediately, after the self report; and a high level, in which all the self-report questions were asked, and then all the proxy report questions. Ninety-four pairs of respondents participated in the laboratory interviews.

Hypothesis P4 suggests that proxy reporters will often use an anchor and adjustment strategy, and that the frequency of use of this strategy will be directly related to the accessibility of the self report. If this

Table 9.3. Processes used in Reporting Attitudes Towards Political Groups: Summary Table

			Proxy Reports			
			Low Buffer		High Buffer	
Process[a]:	Total Self Reports	Total Proxy Reports	No Prime	Prime	No Prime	Prime
Anchor on self	—	19.1%	16.1%	30.8%	16.4%	12.7%
Based on general knowledge about self/proxy	14.0*	37.5*	38.5	39.2	36.5	35.5
Based on general knowledge about the political group	42.1*	11.1*	6.9	14.1	12.2	11.8
Based on discussions with other persons	—	7.7	7.6	9.1	6.0	8.1
Based on a specific behavior or event	6.9	6.7	3.84	10.0	7.8	5.4
TOTAL N	94	94	26	24	22	22

* Significant at $p < 0.05$.
[a] Percentages do not sum to 100 because the number of processes reported in a respondents' protocol on a single question could range from zero to three.

hypothesis is correct, respondents should be more likely to use an anchor and adjustment strategy when the self report is obtained immediately before the proxy report, and less likely when the reports are separated by other items. We also expect that questions asking respondents to compare their partners' behavior or attitude to their own should increase the accessibility of the self report.

Two verbal protocol codes provided evidence that anchor and adjustment strategies were used: "same as self" and "anchor on self and adjust" (see Exhibit 9.1). The frequency of these codes by treatment condition is shown in Table 9.3. For attitudes about political groups, 19.1% of respondents across all experimental conditions reported using an anchor and adjustment strategy to make a proxy report. These results indicate that the use of anchor and adjustment may be less common than assumed previously (e.g., Davis et al. 1986; Hoch 1987). Respondents are much more likely to rely on general knowledge about the other person (37.5%) when making a proxy report.

Proxy respondents who were primed in the low buffer condition were almost twice as likely (30.8%) to use anchor and adjustment than in any other condition. However, when there was a buffer, there were no differences between priming conditions. Thus, priming a respondent

Type of Prime	*Low Buffer* Self/Proxy Questions Alternated	*High Buffer* All Self, Then All Proxy Questions
Similarity	52	50
Difference	50	50
No Prime[a]		
how different	24	24
how alike	26	26
Norm[b]		
preceding self	24	26
preceding proxy	26	24
Total	202	200

[a] In this control condition, the question about how alike or different is the pair was asked after the substantive question about the behavior or event.
[b] Comparison on self or other to a population norm; results not included in this chapter.

Figure 9.3 Sample allocation to questionnaire versions.

about similarity with the other after the self report increases the use of anchor and adjustment. As the time between the self and proxy report increases, proxies are less likely to use an anchor and adjustment strategy (also see Menon et al., 1990).

In the field experiments, we again manipulated the accessibility of a self report in two ways. First, the order of self and proxy questions was varied. In a random half of the sample, all self reports were obtained before the proxy reports, creating a time buffer between self and proxy reports. In the other half of the sample, the self and proxy questions were asked together for each topic (low buffer condition). Second, respondents were asked priming questions, whose intent was to heighten or prime the accessibility of information about the self. Figure 9.3 describes the 2x4 experimental design, with two levels of questionnaire format: high buffer and low buffer; and four priming conditions: (1) control or "no prime," (2) similar to self, (3) different from self, and (4) comparison to a norm.

Two measures were used to examine anchoring and adjustment: (1) verbal protocols from the think-aloud interviews, and (2) correlations between one's self report and the proxy report. We will refer to this correlation as "perceived similarity" (see Hoch 1987), which indicates the extent to which respondents see themselves as similar to their partners and may thus use their self report as a basis for making the proxy report. This must be considered along with the extent to which the couples' self reports actually agree, or their "actual convergence." If perceived similarity is greater than actual convergence, this may be taken as evidence that a respondent is relying on his or her self report more than he or she should, and thus is using an anchor and adjustment strategy.

Table 9.4. Actual Agreement Compared to Perceived Agreement by Type of Prime and Buffer Condition[a]

	Actual Agreement		Perceived Agreement					
			Type of Prime					
			None		Similarity		Difference	
	Low Buffer	High Buffer	Low Buffer	High Buffer	Low Buffer	High Buffer	Low Buffer	High Buffer
TV: Monday–Friday	0.28	0.18	0.77	0.32	0.42	0.66	0.61	0.63
TV: Weekends	0.14	0.49	0.55	0.51	0.26	0.76	0.62	0.70
# Times Beer	−0.30	0.26	0.80	0.16	−0.11	0.25	−0.01	0.89
# Bottles Beer	−0.01	0.11	0.34	0.63	0.06	0.36	0.77	0.87
Health Rating	0.15	0.18	0.25	0.17	0.15	0.44	0.23	0.45
Days of Work Missed	−0.02	0.05	0.60	0.19	−0.07	0.03	−0.05	0.21
# Doctor Visits	0.01	0.16	−0.13	−0.01	−0.38	−0.15	−0.13	−0.01
Attitude–Drug Control	0.21	0.18	0.88	0.62	0.88	0.68	0.86	0.68
Attitude–Abortion	0.50	0.53	0.83	0.80	0.79	0.84	0.68	0.65

[a] The estimates are correlation coefficients.

In the field study, one of the three priming questions preceded items about five other topics: hours of television watched; drinking beer; health; attitudes toward drug policies; and attitudes toward abortion and family planning. The priming questions asked respondents to assess either the similarity between their partners and themselves, the differences between the pair, or the similarity of the partner to a norm. In Table 9.4, the correlation between each respondent's self report and proxy report (perceived similarity), and the correlation between the pair's self reports (actual agreement) are shown for each treatment condition. Although the size of the differences between correlations varies across topics, a consistent pattern is evident.

Perceived similarity relative to actual similarity is taken as an indication of anchoring. Several comparisons can be made. First, when no priming question was used (the control condition), an anchoring strategy was more likely to be used in the low buffer condition. This was expected, as the self report would be more accessible when the self and proxy items are asked sequentially. Second, in most cases, respondents were more likely to anchor on their self reports when similarities were primed in the high buffer condition. A similar pattern occurred when differences were primed, although it was much weaker. This runs counter to the results reported above. It is possible that asking respondents to compare their partners to themselves in the low buffer condition actually increases the salience of self report. Thus, they are more likely to consider their self reports when making the proxy reports, but they may actually be discounting the self reports. This could be due to the way they interpret the question or to avoid appearing biased in their responses.

It should be noted that the perceived agreement between responses was always higher than the actual agreement, indicating that in all cases respondents rely to some extent on their self reports. In addition, there was no difference in convergence between reports across experimental conditions. Thus, at this point it appears that an anchoring strategy is no more accurate than other strategies.

Finally, for the items about drug control, there appears to be a main effect of question format, such that perceived agreement is higher in the low buffer condition, regardless of the priming question. It is possible that people did not have well-formed attitudes on this issue, and thus used their own response whenever it was accessible.

9.4.2 Hypotheses About Convergence

The examination of the hypotheses about convergence is based on the field data. For each behavior item, a follow-up question about how much the respondent participated in the target behavior was asked. Additionally, in a set of post-interview questions, respondents were asked how important each behavior is to them. In Table 9.5, the responses to the participation and importance questions have been dichotomized into "more or less" participation or importance.

There are consistently higher correlations when the level of participation is greater, as the first part of hypothesis C1 suggests, though the differences are not large. Our methods do not allow a clear test of the second part of the hypothesis about events that are observed versus those only heard about. One might speculate that an item such as "books read for work" might be more likely to be heard about rather than observed. But without asking directly about observation, we cannot be sure.

The hypothesis about importance (C2) was not supported by the data. We have suggested elsewhere (Sudman et al., 1990) that this may have to do with self-representation effects. It may also have to do with the concept of importance being a difficult one to measure in a single question in a way that makes sense across a range of behavioral and attitudinal variables. In either case, respondents' assessment of importance was not found to be a good predictor of self–proxy report convergence.

The number of years that a couple has lived together is another variable that one might reasonably expect to be correlated with the level of participation, and hence a predictor of convergence. We dichotomized

Table 9.5. Correlations between Self and Proxy Reports by Level of Participation/Discussion and Importance

	Participation Discussion		Importance	
	More	Less	More	Less
Hours of TV watched:				
Weekdays	0.52	0.42	0.29	0.51
Weekends	0.66	0.46*	0.52	0.62
Times drank in past month:				
Beer	0.72	0.58*	0.58	0.97
Liquor	0.93	0.69*	—	—
Number of times drunk in past year	0.28	0.24	−0.09	0.16
Average number of daily snacks	0.62	0.48	—	—
Number of books read for work/school	0.17	0.08	0.08	0.30
Number of books read for pleasure	0.54	0.42	0.78	0.56
Number of newspapers read	0.36	0.68*	—	—
Days missed from work/school in year	0.48	0.40	—	—
Times seen doctor	0.50	0.30*	—	—
Seriousness of condition	0.39	0.20*	—	—
General health rating	0.61	0.44*	—	—

* Difference between correlations significant at $p < 0.05$.

years living together into two years or less, and more than two years. While there was no support for our expectation for behavioral variables, there is for attitudinal variables. As can be seen in Table 9.6, the correlations between self and proxy reports for attitudes toward labor unions, the National Rifle Association (NRA), the Ku Klux Klan (KKK), and women's groups all are significantly higher for couples that have lived together more than two years. The correlations on the environmental question did not show any difference. The percent agreement on the question about government support for family planning is in a direction opposite to our expectation.

Hypothesis C3 suggests that the effects of questionnaire structure on proxy reports should be moderated by a respondent's level of participation in the behavior or the amount of discussion on the topic. It was expected that the order manipulation (high buffer/low buffer) would have a larger effect for those couples with lower levels of participation or discussion. Specifically, respondents should be more likely to use an anchoring strategy in the low buffer condition. However, as shown in Table 9.7, the opposite appears to be the case.

Table 9.6. Correlations (or percent agreement) between Self and Proxy Attitude Reports, by Years Living Together

	2 years or less	More than 2 years
Labor unions	0.53	0.60
NRA	0.60	0.70
Women's groups	0.28	0.42
Environment	0.35	0.32
Ku Klux Klan	0.09	0.18
Govt. support for family planning	82.5%	76.0%
Abortion for family planning	81.5%	84.5%
Abortion for birth defect	60.0%	69.5%

The use of an anchoring strategy for behavioral questions was not affected by question order or by level of discussion. Respondents were able to retrieve other inputs from memory to answer these questions. However, as shown in Table 9.8, level of participation did affect the perceived similarity of the pair. Perceived similarity was higher when level of participation was high, regardless of question order. Similarly, respondents tended to rely on their self reports more for answering attitude questions when the level of discussion between the pair was high (see Table 9.7). People who discuss a topic frequently might be more likely to agree, or at least know how the other person stands relative to one's own position, and thus may be more likely to use their self reports.

9.5 SUMMARY

In this research, we have used a variety of techniques for examining hypotheses about differences between self and proxy survey reporting. These included face to face interviews eliciting both concurrent and retrospective verbal protocols and using postinterview questions, as well as telephone interviews. In both settings, alternative questionnaires were used in experimental treatments to test effects of priming and accessibility. We also used a wide range of behavioral, attitudinal and factual questions. The consistency across these techniques and question items lends support to our findings.

As we hypothesized, respondents do not use the same reporting strategies for self and proxy reports. Most clearly, estimation is used much more frequently in proxy than in self reporting. And, while

Table 9.7. Correlations between Self and Proxy Reports by Level of Discussion and Buffer Conditions

	Less Discussion		More Discussion	
	Low Buffer	High Buffer	Low Buffer	High Buffer
Books read for work/school	0.03	−0.02	0.64	0.05
Books read for fun	−0.01	0.22	0.89	0.22
Health rating	0.40	0.08	0.16	0.36
Work loss days	0.19	0.55	−0.04	0.00
Doctor visits	−0.03	−0.10	−0.01	0.08
Favorability				
KKK	0.41	0.73	0.08	0.85
Labor unions	0.60	0.72	0.65	0.60
NRA	0.62	0.71	0.83	0.86
Woman's groups	0.59	0.67	0.64	0.99
Environment groups	0.61	0.85	0.65	0.96
Job President is doing				
Inflation	0.55	0.43	0.73	0.67
Unemployment	0.60	0.53	0.73	0.70
Trade deficit	0.56	0.64	0.65	0.64
Economy	0.67	0.49	0.72	0.70
Foreign competition	0.66	0.28	0.62	0.72
Honesty of				
Politicians	0.63	0.60	0.74	0.60
President	0.62	0.46	0.78	0.72
Senator Simon	0.61	0.57	0.68	0.80
Gov. Thompson	0.56	0.70	0.77	0.78
Senator Helms	0.69	0.74	0.79	0.72
Gov. Dukakis	0.62	0.48	0.69	0.80
Senator Kennedy	0.59	0.52	0.78	0.73
Gary Hart	0.81	0.90	0.93	0.95
Effectiveness of government drug abuse program	0.86	0.61	0.88	0.7
Average	0.62	0.57	0.70	0.76

characteristics of the questionnaire, such as priming and question order, affect the strategies used, actual convergence is more difficult to increase. Characteristics of the respondent pairs — such as years living together — appear to affect convergence, though there is some inconsistency in our findings.

The concept of level of participation begins to provide a model for understanding a large part of the difference between self and proxy reporting. It is a somewhat more complex concept than we have so far allowed for and encompasses dimensions beyond those we have examined to date. The literature on social distance and attention may provide a direction for further development and testing of this concept.

Table 9.8. Correlations between Self and Proxy Reports, by Level of Participation and Questionnaire Form

	Lower Participation		Higher Participation	
Behavior	Low Buffer	High Buffer	Low Buffer	High Buffer
Hr. of TV watched				
Weekdays	0.66	0.54	0.60	0.67
Weekends	0.44	0.65	0.82	0.78
Times drank beer in month	0.17	0.19	0.67	0.80
Times drank liquor in month	0.30	0.07	0.99	0.95
Average	0.35	0.32	0.75	0.81

There are several implications of our results for improving survey design. Certainly the findings for level of participation, years living together, and importance of a topic area should help in designing screening questions for deciding whether to use an available proxy respondent. The findings on the effects of accessibility of self reports will be useful in suggesting situations where one might encourage the use of the anchor and adjustment strategy, and how, operationally, that can be done.

CHAPTER 10

AN ALTERNATIVE APPROACH TO OBTAINING PERSONAL HISTORY DATA

Barbara Means and Gary E. Swan
SRI International

Jared B. Jobe and James L. Esposito
U.S. National Center for Health Statistics

10.1 TWO CHALLENGES TO CONVENTIONAL SURVEY METHODOLOGY

In recent years we have seen two challenges to conventional wisdom about how survey interviews should be conducted — one from the perspective of sociolinguistics and the other from that of cognitive psychology. Sociolinguists, examining the survey interview as a form of interpersonal interaction, have questioned the validity of the assumption that survey items can be regarded as fixed, context-independent measures. Further, they argue that by violating respondents' expectations for social interchange, the standardized interview introduces misunderstandings that result in inaccurate data. Cognitive psychologists have been concerned with somewhat different issues. They examine the memory demands posed by survey items and consider the interview as a method for eliciting recall from autobiographical memory. From this perspective, questions arise concerning the compatibility between commonly used survey structures and the way that human memory functions.

These two challenges to standard survey interview techniques

[1]This chapter is based in part on research supported by DHHS Contract No.200-88-7074 to SRI International from the U.S. National Center for Health Statistics. We wish to express our thanks to Katherine Habina, Alison Dame, and Lisa Jack for their diligent work in conducting interviews, coding data, and running analyses.

Measurement Errors in Surveys.
Edited by Biemer, Groves, Lyberg, Mathiowetz, and Sudman.

ISBN: 0-471-53405-6

prompted us to develop an experimental alternative approach to obtaining personal history data. This approach is being tested in a study of long-term recall of events related to smoking and smoking cessation. The availability of a sample of current and former smokers who participated in a 1982–83 study of smoking behavior is making it possible to compare the accuracy of retrospective reports obtained through this experimental technique with those obtained through standard survey methods. The latter part of the chapter reports preliminary results from this study and discusses their implications for survey methodology.

10.1.1 The Sociological Critique

Suchman and Jordan (1990) assert that survey interviews are fundamentally social interactions and as such cannot really be neutral measurement instruments. Many factors, such as age, culture, gender, education, and personal experiences of the respondent all affect the interpretation of a questionnaire item, regardless of the standardization in interviewer behavior. Chapter 19 examines these sociolinguistic issues, including an extensive comparison of standardized interviewing and conversational norms. Table 10.1 summarizes these contrasts between the rules of conversation and those of standardized survey interviews. Unlike conversation, the survey interview is structured in a way that precludes use of the mechanisms employed in everyday personal communications to clarify each participant's message and repair misunderstandings.

Table 10.1 Interactions in Conversations Versus Those in Survey Interviews

Conversations	Survey Interviews
Participants identify a topic of mutual interest	Topics are identified in advance by survey designer
Topics evolve dynamically but follow rules of continuity	Topics may switch suddenly, with no logical continuity
Questioned person can reject or reframe the question	Respondent must answer question as it is written in advance, whether or not it fits individual circumstances
Participant's talk is adapted to information the other participant has provided	Interviewer uses standard wording so that respondent's prior responses have no effect
Procedures are used to obtain clarification when ambiguity is encountered	Requests for clarification are responded to with repetition of the same question

10.1.2 The Psychological Critique

Our own research has focused on limitations of standard survey interviews from another perspective — that of cognitive studies of memory phenomena (Jobe and Mingay, 1990; Means and Loftus, 1991). In looking at national surveys, one of the first things that strikes the memory researcher is the tremendous demands that are made on the respondent's autobiographical memory. People are asked to report the number of days they missed work over the last year for a specific health condition, how often they've been to a restaurant in the past month, and all the clothing purchases they have made in the last 90 days. Such questions presuppose that the respondent is not only able to remember each relevant event, but also to place events in time accurately enough to eliminate any recalled incident that did not fall within the specified reference period.

Given that surveys are going to include such difficult memory tasks, the typical survey interview does not appear to be structured in a way that is most conducive to recall. It is well established that a person's ability to retrieve an event from long-term memory is heavily dependent on the cues available at the time of attempted recall. As noted in Chapter 9, the greater the similarity between cues present when the person tries to recall an event and the cues present when the event was encoded, the more likely the subject is to be successful in retrieving the event from long-term memory (Tulving, 1983). Necessarily, survey interviews are generally administered at a time and place quite unlike those surrounding the events being asked about. Additional impediments to retrieval of past events are raised by the standard survey's stricture against "irrelevant" discussion of the topic of the survey item. This prohibition discourages the respondent from reconstructing the context in which the event occurred — an activity that is important for generating useful retrieval cues.

In addition, the order in which surveys pose questions about a topic is often less than optimal for stimulating event retrieval. In a study of memory for use of health facilities during the past year, Means, et al. (1989) found that the reason motivating a visit was the best remembered aspect of it. Subject protocols suggested that the preferred method of recalling events was in terms of a narrative structure, with an identifiable problem, actions, and a resolution. Whereas the problem or reason for an event appears to be a good recall cue, the time when it occurred is a poor one. In a study of longer term autobiographical memory, Wagenaar (1986) found that information about *when* an event occurred was poorer than information about *what* was done, *where* it was done, or *who* did it as an initial cue for retrieving other information about the event.

Table 10.2 Conditions Facilitating Recall Versus Those in Survey Interview

Conditions Facilitating Recall	Conditions in Survey Interviews
Overlap between cues available at recall and those at time event was encoded	Context bears few cues similar to those present at time event was encoded
Presence of *why* cues more helpful than *when* cues	Topic introduced with questions about *when* and *how often*
Use of a narrative structure	Questions shift focus from one event to another without providing for well-formed descriptions

This memory research raises questions about the efficacy of the order in which event topics are addressed in many survey interviews. In the smoking supplement for the U.S. National Health Interview Survey (NHIS), for example, the interviewer opens the topic by asking about frequency and dating of periods of smoking and smoking abstinence.

- How many times within the last year have you made a serious attempt to stop smoking cigarettes?
- When did you last make a serious attempt to quit smoking?
- When you tried to quit, how long did you stay off cigarettes?
- Of all the times you have tried to quit smoking, what was the longest period you stayed off cigarettes?
- When did you make the serious attempt to quit smoking?

Only after answering these questions are respondents asked about how and why they changed their smoking behavior. A very different question order would be recommended based on the research literature on autobiographical memory.

A second feature of the typical survey interview that appears less than optimal from a cognitive psychology viewpoint is that the subject is repeatedly asked to shift focus. In the portion of the smoking history section of the NHIS shown above, for example, the respondent is asked first to think about all attempts to stop smoking, then about the most recent attempt to stop smoking, and then about the longest cessation period. The survey's use of shifting reference periods can be expected to disrupt natural retrieval strategies, if we are correct in our interpretation that an event story or narrative form, in which the subject develops a coherent narrative of a single event, is preferred for autobiographical recall, much as a story structure facilitates recall of prose material (Stein and Nezworski, 1978). Table 10.2 summarizes the differences between the conditions that laboratory research suggests are most conducive to recall of past events and the conditions that apply in most standard survey interviews.

10.2 THE CURRENT STUDY: LONG-TERM RECALL OF SMOKING AND SMOKING CESSATION

These criticisms from the viewpoints of sociolinguistics and cognitive psychology suggest that survey interview techniques are flawed as methods for obtaining valid information about people's past lives and behavior. The study to be reported here explores an alternative approach to obtaining personal history information. The study is designed to have applications for the design of interviews dealing with personal history over an extensive time frame. Although the specific application used in this study concerns smoking history, the same techniques could be used in surveys concerning long-term residential patterns, employment history, or any other category of significant life event.

In many studies of smoking behavior, researchers want to measure the varying patterns of smoking engaged in by one individual historically, differences in smoking patterns across individuals, the effect of smoking cessation efforts on smoking patterns, and the resultant effect of these patterns on illnesses such as lung cancer and heart disease (U.S. Public Health Service, 1985). Often surveys ask respondents to indicate (1) when they started smoking, (2) when, if ever, they finally stopped, and (3) what intervening periods of abstinence occurred. Clearly, answering such questions can be very difficult for a middle-aged person who has stopped and started smoking repeatedly over a course of three or four decades.

In our current study, participants are all individuals who were involved in a formal program to stop smoking sometime in 1981 or 1982. After stopping smoking as part of that program and staying off cigarettes for three months, they joined an SRI study of factors contributing to smoking relapse. Some subsequently resumed smoking while others remained abstinent. Our interview questions deal with their smoking patterns just prior to participating in the stop-smoking program and with prior and subsequent attempts to stop smoking.

In designing the experimental interviews, we ordered interview questions in a way that would be compatible with memory retrieval processes — first asking questions about the reasons *why* the subject did something, followed by *how* they did it, the *result*, and, finally, *when* the event occurred. In this way, the subject is allowed to describe the essentials of an episode before being asked to place the episode in time.

In addition, in our experimental condition we tried to adhere to conversational axioms without sacrificing objectivity. The interviewer first seeks open-ended responses to questions such as "Why did you try to stop smoking at that time?" and then asks the subject to look at a list of standard responses and select those that are relevant. The interviewer acknowledges information that the subject has already provided and uses that information to form individualized probes. For example, a subject

Table 10.3 Contrasts Between Experimental and Standard Interview Conditions

Feature	Experimental Interview	Standard Interview
Question order:	*Why* tried to stop smoking, *how* tried to stop smoking, *why* restarted, *when* stopped and started	*How many* times tried to stop, *when* was most recent stop, duration of longest stop, *how* tried to stop smoking, *why* ever tried to stop, *why* ever restarted
Continuity:	Each episode described completely before respondent is questioned about other episodes	Focus of questions shifts as on NHIS
Standardization:	Balance sought between conversational axioms and standardization; questions about reasons and methods asked first in open-ended format followed by provision of standard response options	Questions completely standardized; subjects given standard response set for all items
Relevance:	Subjects encouraged to talk about any details that might help them remember the event	Subjects discouraged from rambling

who said he could not remember when a quit attempt occurred might be probed "Well, you said you are pretty sure you smoked at least up until spring of 1979 when you were still working as a gardener, and you're pretty sure you participated in the SRI project in February of 1982. That leaves a period of three years . . . . What about when you were singing in the choir, were you smoking then?" In this way, we encourage subjects to talk about the details of their past lives that have connections to changes in their smoking behavior and that can serve as retrieval cues to help them remember the events in their smoking history.

This alternative interviewing approach is being contrasted with a standard survey interview approach in which questions are sequenced in an order patterned after the NHIS smoking supplement and the interviewer adheres strictly to the interview protocol. In this standard condition, subjects are not asked about topics with open-ended questions before the standard response options are given. Subject digressions are discouraged. If a subject says that he cannot remember something such as the date of an attempt to stop smoking, he is simply urged "well, try to give me your best estimate." Table 10.3 summarizes the differences between the two interview conditions.

This study is unusual because we are able to compare these two conditions in terms of the accuracy of the information obtained from subjects. This is possible because the subjects were all participants in

SRI's 1982–83 study of smoking behavior and relapse (Swan, et al., 1988; Swan and Denk, 1987), funded by the National Heart Lung and Blood Institute (NHLBI). As part of that study, extensive smoking histories were obtained from all participants. Subjects completed a written questionnaire asking about their current smoking habits and all periods of smoking cessation and attempts to quit during the prior ten years. These subjects were then followed up for a 12-month period, during which they were regularly asked about their smoking status and had saliva samples taken and analyzed for tobacco byproducts. The records from the earlier study provide a standard to which we can compare subject retrospective reports in our interviews.

10.2.1 Subjects

Seventy-six individuals, all of whom participated in the 1982–83 study of smoking relapse, served as subjects. Of these, 20 were interviewed in 1988–89 during a pilot phase of our research, and 56 were interviewed between 1989 and 1990 during the main data collection. Much, but not all, of the interview protocol remained the same between the two study phases, permitting pooling of data for some items. Subjects were randomly assigned to interview groups.

10.2.2 Procedures

All subjects were interviewed individually in their homes, places of work, or other settings convenient to them. Four researchers served as interviewers, each conducting approximately equal numbers of experimental and control interviews. For the experimental group, the interviewer began by asking the subject about the quit attempt immediately before joining SRI's 1982–83 smoking relapse study. These subjects were first asked about *why* they wanted to stop smoking at that time in their life. They were then asked questions about how much they had been smoking prior to joining the stop-smoking program, the number of family members and friends who smoked at that time, and the extent to which these people made it easier or harder to stop smoking. Only after addressing all of these questions about this attempt to stop smoking were they asked to try to recall the month and year they had stopped smoking as part of this program. These questions were followed by less detailed inquiries about why and how they had tried to stop smoking on any other occasions between 1977 and 1982. For each period of smoking cessation, they were first asked why they had tried to stop, followed by questions about methods used, and then dates. Finally, they were asked about changes in smoking status after participating in the SRI study.

In the standard interview condition, subjects were first asked to estimate the month and year when they had stopped smoking prior to participating in the SRI smoking relapse study. They were then questioned about the details surrounding their participation in the stop-smoking program. After answering these questions, they were asked to consider all their other smoking cessations between 1977 and 1982. They were first asked how many cessations had occurred and then the month and year when each period of cessation started and stopped. Next, they were asked to report all the methods they had ever used to try to quit smoking during this period and all of the reasons why they had ever resumed smoking after trying to stop.

The form of the questions used with the standard interview group (asking about methods and reasons for all of their 1977–82 quit attempts at once) was identical to that used on the questionnaire subjects filled out in 1982. In contrast, the experimental group was asked to consider the motives and methods for each event individually. In addition, the experimental subjects were first asked to describe their reasons or methods in their own words and only subsequently given the standardized response categories that had been used on the 1982 questionnaire. (Responses were aggregated across quit attempts for comparison with their 1982 responses.) Thus, any variance related to similarity of method should favor the standard interview group. In both conditions, the interview concluded with questions about the subject's current smoking status and the amount smoked presently or date of final quit, whichever was relevant.

10.2.3 Results and Discussion

Subject responses during their individual interviews were compared with data recorded concerning their smoking cessation in 1981 or 1982 and compared to 6- and 12-month follow-up data obtained as part of the earlier SRI study by Swan, et al. (1988).

Memory for When the Quit Attempt Occurred

The experimental interview technique was designed to help subjects generate a context for remembering incidents in their smoking histories. A key prediction was that subjects would be better able to place an event in time if they first had the opportunity to recall the reason for wanting to stop smoking and the context in which the event occurred rather than being asked first about when the event had happened. Memory accuracy

Table 10.4 Deviations Between Recalled and Actual Date of Stopping Smoking

	N	Mean Absolute Deviation (in months)	Standard Error
Total Sample	56	7.47	1.50
Experimental Interview	28	4.99	1.29
Never smoked again	13	3.95	2.02
Smoked again	15	5.90	1.68
Standard Interview	28	10.74	2.56
Never smoked again	14	3.58	1.17
Smoked again	14	17.91	3.98

Note: A value equal to the total sample mean plus one standard deviation was imputed for two subjects who reported that they had no idea when the event had occurred.

for the month and date in which the subject stopped smoking before joining the SRI study was estimated for the 56 subjects in the main part of the study. (In the pilot phase, subjects were told the year in which this stop-smoking attempt occurred.) For each subject, the remembered month and year when he or she stopped smoking before joining SRI's study was contrasted with the date on record. The mean absolute deviations between the two dates for the experimental and standard interview subjects are shown in Table 10.4.

On average, subjects were able to date this attempt to stop smoking (which had occurred some six to nine years prior to the interview) within 7.47 months of its actual occurrence. Dating errors usually involved subjects' remembering the event as occurring more recently than it actually did ("forward telescoping"), with the average date remembered being 4.23 months too recent.

There were large differences between the two interview groups in the precision with which this event was dated. Subjects in the experimental interview condition erred by an average of just 4.99 months (absolute deviation) compared to an average of 10.74 for those in the standard interview condition. Thus, the experimental technique appeared highly effective in helping subjects to place this event in time, $t(40) = 2.01$, $p < 0.05$, cutting the dating error by more than half.

Another important factor in determining the accuracy with which subjects could date this event was whether this attempt to quit smoking was ultimately successful (i.e., the subject never smoked again). It is quite reasonable to assume that the date of *the* final cessation of smoking is more likely to be discussed or otherwise mentally rehearsed than the date of a quit attempt that was not sustained. This is clearly the case: The

Table 10.5 Consistency Between Interview Smoking History and 1982-83 Follow-up Data

	First Follow-up		Second Follow-up	
	N	% Agree	*N*	% Agree
Experimental Interview	33	0.94 (0.04)	34	0.97 (0.03)
Standard Interview	24	0.75 (0.09)	28	0.75 (0.08)

Note: Follow-up data were not available for some subjects.
Standard errors appear in parentheses.

mean absolute deviation between the remembered date and the actual date for the attempt was 3.76 months if it turned out to be a "final" attempt and 11.70 if it did not, $t(39) = 2.94, p < 0.01$. A general linear model test, with both condition and finality of the quit as factors, found a significant effect for the interaction between these two variables, $F(1,52) = 5.87, p < 0.05$, as well as significant main effects [$F(1,52) = 5.19$, $p < 0.05$ for condition and $F(1,52) = 10.16$, $p < 0.01$ for finality of quit]. Subjects in the experimental interview group were good at dating the event regardless of whether it turned out to be a final quit (3.95 and 5.90 month deviations for final and nonfinal attempts, respectively). In contrast, subjects in the standard interview condition were quite good at dating final quit attempts (mean deviation of 3.58 months) but very poor at dating the smoking cessation if it turned out not to be ultimately successful (mean deviation of 17.91 months).

Memory for Smoking History After the Quit Program

After participating in the program to stop smoking in 1981 or 1982, a majority of subjects remained abstinent, according to both records from the follow-up conducted for approximately one year and their own accounts in 1989 or 1990. There were 37 subjects (49 percent), however, who resumed smoking, although a majority of these have since quit. The smoking history provided by each subject in our interview was compared to the recorded dates and follow-up data collected roughly 6 months and 12 months after quitting smoking in 1981 or 1982 for the earlier study. In this way, we were able to assess subjects' memory for their smoking behavior in the year following their 1981 or 1982 smoking cessation. Table 10.5 compares the two interview conditions in terms of the consistency of reports with the recorded follow-up data. Experimental interview subjects tended to give more accurate histories for the period after the stop-smoking program. Their reported smoking status at the

Table 10.6 Consistency in Description of Social Environment

	Experimental n = 38	Standard n = 37[a]
Objective Social Environment Items[b]		
Number Possible Matches[c]	4.32	4.38
	(0.15)	(0.13)
Matches with 1982–83 Questionnaire	3.11	2.92
	(0.20)	(0.17)
% Matches	0.71	0.66
	(0.04)	(0.03)
Subjective Social Environment Items[d]		
Number Possible Matches	2.55	2.54
	(0.10)	(0.09)
Matches with 1982–83 Questionnaire	0.97	1.19
	(0.15)	(0.14)
% Matches	0.40	0.48
	(0.06)	(0.05)

[a] Does not include one subject whose 1982 questionnaire could not be located.
[b] Includes whether lived alone, number of adults in household, number of smokers in household, proportion of friends who smoked, proportion of coworkers who smoked.
[c] Number possible varies depending upon whether the individual lived alone and whether or not the individual was employed at the time the questionnaire was completed.
[d] Includes ratings of degree of support received from household members, friends, and coworkers, each on a four-point scale.
Note: Standard errors appear in parentheses.

time of the first follow-up matched the follow-up record data in 94 percent of the cases compared to 75 percent for standard interview subjects, $Z = 1.95$, $p < 0.10$. The advantage for experimental subjects was clearer for the second (12-month) follow-up. The smoking status reported in the interview for the time of the second follow-up matched that on record for 97 percent of experimental interview subjects compared to 75 percent of those in the standard condition, $Z = 2.75$, $p < 0.05$.

Memory for Events Surrounding the Quit Attempt
Each subject's responses in the individual interview were compared also to his or her responses on a written questionnaire administered during the 1982-83 smoking relapse study. Differences between the two interview groups were not pronounced. Memory for the type of cigarette smoked at that time (i.e., its length, whether or not it was mentholated) was similarly good for both groups (71 percent recall for the experimental interview group and 69 percent for the standard interview).

Table 10.6 shows the data concerning memory for the social context within which this attempt to stop smoking took place. Some of the items about social context concerned objective data, such as the number of people in the household and the number of them who smoked, the proportion of friends and coworkers who smoked, and so on. Other items requested subjective judgments — for example, ratings of the extent to which friends were helpful in the attempt to stop smoking. Not surprisingly, memory was better for objective aspects of the social environment surrounding the stop smoking attempt. On this type of item, responses of the experimental group during our interview matched those on their 1982–83 questionnaire 71 percent of the time and those of the standard condition subjects matched 66 percent of the time (a nonsignificant difference). For the subjective items, experimental interview subjects' responses matched 40 percent of the time compared to 48 percent among standard interview subjects. The trend toward greater consistency in the responses to subjective items among the standard interview group was not statistically significant, but does suggest that these subjects may have an advantage in terms of similarity of the strategies used to answer the question on the two occasions. Either a standard interview or a written questionnaire is likely to elicit a rough estimate ("gut reaction") when a complex question such as "How supportive were your friends when you tried to stop smoking?" is posed. The slower paced, more detailed discussion elicited in the experimental interview was designed to lead subjects to actually recall specific people and how they responded to the attempt to stop smoking. To the extent that subjects did so and actually tried to mentally average the degree of support over various friends or family members, they may have tended to come up with a different answer than the estimate provided in 1982 after a more superficial consideration of the item on the questionnaire (see Krosnick, 1991c).

Another item on the 1982 stop-smoking attempt concerned the reasons for wanting to stop smoking at that time. The interview used the 12 response options provided on the 1982 questionnaire, asking subjects to indicate all that applied. Subjects in the experimental interview responded to these options after describing the reasons for wanting to quit smoking in their own words. Subjects in the standard interview merely responded to the set of options.

Figure 10.1 shows some examples of the reasons experimental interview subjects provided in response to the open-ended question, and the response options selected by the same subjects. Subjects in this condition clearly selected some options that were quite different from anything they had talked about in response to the open-ended query, but preceding the standard question with the open-ended item may have made them more conservative in selecting options: They were less likely than subjects in the standard interview condition to select options that they had not marked on the 1982 questionnaire (mean of 0.92 versus 1.20

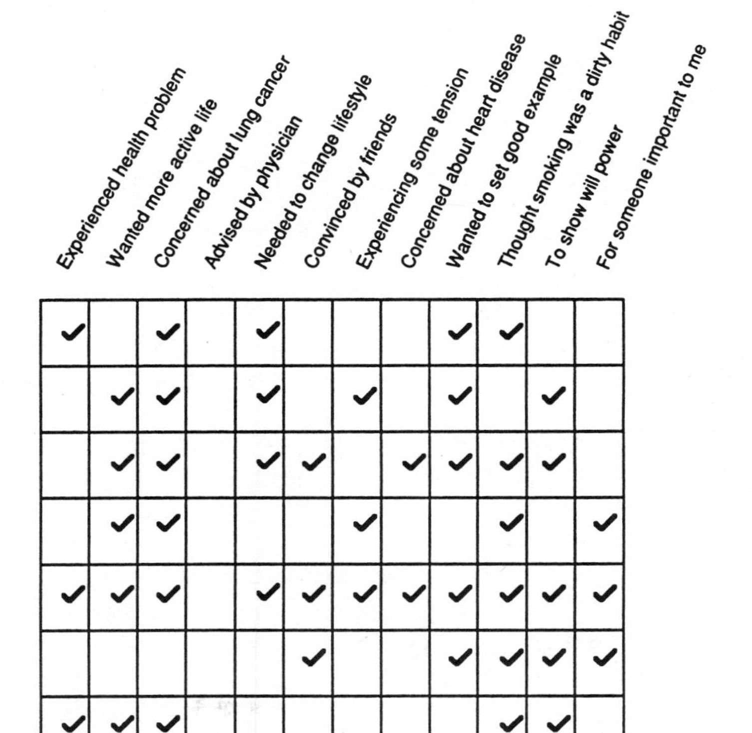

Open-Ended Response	Experienced health problem	Wanted more active life	Concerned about lung cancer	Advised by physician	Needed to change lifestyle	Convinced by friends	Experiencing some tension	Concerned about heart disease	Wanted to set good example	Thought smoking was a dirty habit	To show will power	For someone important to me
S#7: *Mother had recently died of cancer...I was smoking a lot and not enjoying it.*	✓		✓		✓				✓	✓		
S#12: *We had recently moved from Oregon and were planning for children...We wanted to be nonsmokers...*		✓	✓		✓		✓		✓		✓	
S#23: *I wanted to get healthier...A guy I was dating didn't smoke. Also a friend was in this stop-smoking program...*		✓	✓		✓	✓		✓	✓	✓	✓	
S#29: *Dad died and mother was concerned about people who smoke in general...It was really devotion to my mother...*		✓	✓				✓			✓		✓
S#35: *I had recently quit drugs and alcohol so I thought I should quit smoking too.*	✓	✓	✓		✓	✓	✓	✓	✓	✓	✓	✓
S#48: *The real motivator was my wife. She really wanted to quit and I just sort of went along.*						✓			✓	✓	✓	✓
S#51: *I was anticipating a divorce and wanted to take control of my life...*	✓	✓	✓							✓	✓	

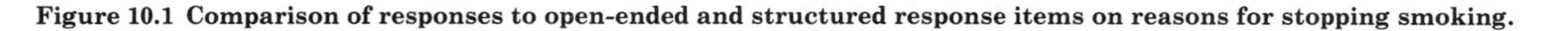

Figure 10.1 Comparison of responses to open-ended and structured response items on reasons for stopping smoking.

Table 10.7 Consistency Between 1982 Questionnaire and 1989 Interview Items on Reasons for Quitting Smoking in 1981/82, by Condition

	Experimental $n=38$	Standard $n=37$[a]
Reasons for Quitting Smoking		
Number Possible[b]	7.66	7.05
	(0.34)	(0.33)
Hits	4.71	4.54
	(0.28)	(0.33)
False Alarms	0.92	1.20
	(0.18)	(0.22)
Proportion Hits	0.64	0.64
	(0.04)	(0.03)
(Hits-False Alarms)/Possible	0.48	0.44
	(0.04)	(0.05)

[a] Does not include one subject whose 1982 questionnaire could not be located.
[b] Number possible varies depending upon number of items marked on 1982 survey.
Note: Standard errors appear in parentheses.

per subject). When the proportion of matches between the interview and the 1982 questionnaire responses is corrected for these "false alarms," the experimental interview group had a score of 48 percent compared to 44 percent for those in the standard interview condition, a difference that was not statistically significant. (See Table 10.7.)

Other Attempts to Quit Smoking

Table 10.8 compares the two interview groups on items concerning attempts to quit smoking prior to the attempt just before joining the SRI smoking relapse study. One of these items concerned reasons for starting to smoke again after prior quit attempts. This item was similar to that described above concerning reasons for trying to stop smoking. Subjects in the experimental interview condition saw the 12 response options only after describing reasons for relapsing in their own terms. Here, there was an even clearer indication that the experimental procedure reduces the number of response options selected. Experimental group subjects chose a smaller number of options ($\bar{X}=1.96$) than did standard interview subjects ($\bar{X}=3.44$). As a result, they had both a lower proportion of hits (responses that matched those on their 1982 questionnaire) and a lower proportion of false alarms. Both groups performed little better than chance in selecting reasons for smoking relapses that matched those they had given in 1982–83 (corrected recall rates of 6 and 20 percent for experimental and standard interview groups, respectively).

A second item concerning prior quit attempts was more objective; it concerned the methods that had been used to try to stop smoking.

Table 10.8 Consistency Between 1982 Questionnaire and 1989 Interview Items on Reasons for Prior Relapses and Methods Used, by Condition

	Experimental $n=38$	Standard $n=37$[a]
Reasons for Prior Relapses		
Number Possible[b]	4.26 (0.30)	3.76 (0.37)
Hits	1.32 (0.23)	2.14 (0.27)
False Alarms	0.64 (0.17)	1.30 (0.24)
Proportion Hits	0.28 (0.05)	0.61 (0.06)
(Hits-False Alarms)/Possible	0.06 (0.10)	0.20 (0.08)
Methods Used Prior to '82 Quit		
Number Possible[b]	1.79 (0.19)	1.92 (0.18)
Hits	0.79 (0.10)	0.75 (0.13)
False Alarms	0.25 (0.08)	0.26 0.11)
Proportion Hits	0.53 (0.07)	0.40 (0.07)
(Hits-False Alarms)/Possible	0.34 (0.11)	0.24 (0.13)

[a] Does not include one subject whose 1982 questionnaire could not be located.
[b] Number possible varies depending upon number of items marked on 1982 survey.
Note: Standard errors appear in parentheses.

Performance on this item was somewhat better. There were no large differences between the interview groups in the number of options selected (1.04 for experimental versus 1.01 for standard subjects). As shown in Table 10.8, the experimental interview group had a corrected recall score of 34 percent compared to 24 percent for the standard group, a nonsignificant difference.

10.3 IMPLICATIONS FOR SURVEY RESEARCH

The experimental interview technique used in this study appeared effective in helping subjects recall an accurate smoking history for the period surrounding a smoking cessation that had occurred some seven to nine years earlier. Both memory for the date when the smoking cessation had occurred and the report of smoking status a year afterward were more accurate among experimental interview subjects.

At the same time, the data also indicate that another factor that needs to be considered is the salience of the information to be remembered. For those subjects whose attempt to quit smoking in 1982 or 1983 was successful (who never resumed smoking), memory for the date of this event was highly accurate regardless of the interviewing technique. The date for such an event is likely to become one of the "landmarks" in autobiographical memory — that is, one of the relatively few events for which an exact date is stored, and thus a potential aid in reconstructing dates for other personal events. The advantage of the experimental interviewing technique was found for less salient events — i.e., quit attempts that were not ultimately successful. Since many of the events asked about in standardized interviews are not necessarily of high salience for the respondents, this finding is important. At the same time, it should be acknowledged that the memory for smoking behaviors going back as many as nine years displayed by our subjects was almost certainly enhanced by the fact that they had the unusual event of participating in the 1982–83 SRI smoking relapse study as a cue for recalling other events in their smoking histories.

Although the aspects of their smoking histories that could be validated by project records were more accurate for experimental subjects, there was no comparable advantage for consistency between responses to more subjective survey items administered on the two occasions seven to nine years apart. When interview responses were compared to responses on a questionnaire completed in 1982 or 1983, none of the differences between experimental groups was statistically significant. In part this may reflect the relatively small size of our sample, but it also appears to reflect the instability of the type of response called for on the questionnaire (e.g., responses to 12 feasible justifications for trying to stop smoking).

Overall, the data suggest that, to the extent that the survey designer seeks to get accurate retrospective reports of specific medium-salience events, the techniques explored here have promise, especially when there is a requirement for placing the events in time. At this stage in our research, however, our data do not permit us to pinpoint the locus of our experimental interview's positive effect: it could be the change in question order to one more compatible with autobiographical memory. Alternatively, the change in the nature of the interaction to one in which personal details are elicited and the subject is encouraged to relate the target event to other things going on in his or her life at that time may be the critical factor (see Means and Loftus, 1991). These two aspects of the intervention need to be separately tested and evaluated as potential tools for national surveys.

Before such techniques are considered for implementation on a wider scale, we would want to identify the contribution of these individual factors to increases in validity as well as their costs in terms of increased time to complete the interview. While a change in question order could be implemented in national surveys without greatly increasing interview time, the more interactive interview technique and the request for subjects to recall major life incidents that might serve as cues for recalling the target information would be likely to have major implications for time requirements and interviewer training.

In our study, experimental interviews averaged three times as long as control interviews (33 versus 11 minutes in a sample of 10 subject pairs matched on number of quit attempts remembered). This is probably an extreme case because we are comparing subjects questioned about each incident separately (the experimental group) with those who responded once for a whole class of incidents (standard group). Comparing the two approaches for survey items concerned with specific events, rather than classes of events, would presumably show smaller differences in the time required to conduct the two types of interviews. Alternatively, there may be more efficient ways of providing a meaningful context for subjects' attempts at recall, perhaps by providing a piece of general or personal information that can serve as a date marker and a retrieval cue (see Loftus and Marburger, 1983).

Interviewer selection and training in our study was more extensive than in many large-scale surveys, but not as costly as one might suppose. The training was conducted in 3–4 spaced sessions of several hours each over a period of three weeks, and included modeling and coaching in the experimental techniques, using videotaped practice interviews. The majority of interviews were conducted by research assistants with bachelor's degrees and interview experience but no advanced training in social science or research methodology.

Certainly a much broader research base is needed to support decisions concerning whether or not the additional training and implementation time for cognitive interviewing techniques are warranted for various applications. Given such information, individual survey designers can weigh increased costs against the value they place on the estimated increment in the validity of individual responses. To the extent that the latter is a serious concern, we believe that alternative interview methods designed to facilitate respondents' recall and to clarify communication are worth investigating.

CHAPTER 11

THE ITEM COUNT TECHNIQUE AS A METHOD OF INDIRECT QUESTIONING: A REVIEW OF ITS DEVELOPMENT AND A CASE STUDY APPLICATION

Judith Droitcour
U.S. General Accounting Office

Rachel A. Caspar, Michael L. Hubbard,
Teresa L. Parsley, and Wendy Visscher
Research Triangle Institute

Trena M. Ezzati
U.S. National Center for Health Statistics

Research on social desirability suggests that asking a person to openly admit to a socially undesirable behavior will elicit responses that minimize self-stigmatization. Some researchers (e.g., DeMaio, 1984; Gove and Geerken, 1977) have distinguished between the personality trait of "need for approval" (giving socially desirable answers to all questions) and "trait desirability" (certain traits or behaviors are undesirable to everyone). Persons with high need for approval are likely to give socially desirable answers to any form of question. Jones et al. (1984) suggest that in social interactions, people are motivated to avoid stigmatization in order to protect their self- and social identities.

With this in mind, researchers have checked respondents' answers to survey items against administrative records, information from intake interviews at drug clinics, and even the results of lie detector tests. These

The views expressed in this chapter do not reflect the positions of the U.S. General Accounting Office. Primary responsibility for Sections 11.1 and 11.2 belongs to Judith Droitcour; the remaining sections were prepared by the other authors.

Measurement Errors in Surveys.
Edited by Biemer, Groves, Lyberg, Mathiowetz, and Sudman.

ISBN: 0-471-53405-6

validity studies have indicated a pattern of underreporting for socially disapproved items (Hyman, 1944; Cannell and Fowler, 1963; Clark and Tifft, 1966; Phillips and Clancy, 1970; Cisin and Parry, 1980; Locander, et al., 1976) and exaggeration of socially desirable items (Parry and Crossley 1950–51; Weiss, 1968; Locander, et al., 1976). Many of the studies also showed varied levels of response distortion by socioeconomic status or other variables (Hyman, 1944; Parry and Crossley, 1950–51; Weiss, 1968).

To combat respondent underreporting on socially disapproved items, investigators have tried various techniques such as loading questions to assume each respondent has engaged in a disapproved behavior (e.g., Kinsey, et al., 1948), rigging answer categories to imply that many persons have frequently engaged in the behavior (the "displaced norm," see Cisin, 1963; Parton, 1950), "warming up" respondents with introductory items (see Warwick and Lininger, 1975), and increasing response privacy via a confidential interview location (Gold, 1966) or sealed "secret ballots." Some of these methods were used in the studies cited above — studies that documented response distortion. For example, a drug survey questionnaire using most of these stratagems failed to obtain admission of heroin use by many of those who had once indicated use of the drug in applying to a treatment clinic. Thus, more powerful techniques are needed.

11.1 METHODS OF INDIRECT ESTIMATION

Beginning in 1965, a variety of *indirect* survey-based estimation techniques have emerged. These techniques reduce the level of self-disclosure that a truthful answer entails. An indirect question eliminates the need for an individual respondent to reveal whether he/she has actually engaged in the behavior. Yet when responses are tallied across large numbers of respondents, the information collected will provide a statistically unbiased prevalence estimate for the population as a whole — and selected subgroups. Several methods have been devised, including randomized response, aggregated response, and nominative techniques.

In the first version of randomized response (Warner, 1965), the respondent operates a random device and obtains a secret instruction to give either a truthful or a false response to a "yes-no" question. The interviewer is unaware of the instruction the respondent received. Provided that the probabilities for true and false answers were known and had been set at unequal values (i.e., not 0.50/0.50), a statistically unbiased estimate could be derived.

In a subsequent version — the "unrelated questions method" (Horvitz, et al., 1967; Greenberg, et al., 1969) — the respondent is instructed to randomly answer *either* a sensitive question *or* an unrelated innocuous question such as "Were you born in April?" Estimation is based on knowing the probabilities of answering each question.

Randomized response methods generate large variances. For example, if the probability of selecting the sensitive question is 0.5, then a randomized response estimate would have a variance of *at least* four times that of a corresponding conventional estimate — and sometimes considerably higher (Pollack and Bek, 1976; Miller and Cisin, 1980).

Turning to aggregated response, the first version (Warner, 1971) instructed the respondent to add a (secret) random number to his or her quantitative answer. In a subsequent version (Boruch and Cecil, 1979) respondents in one subsample add answers to two quantitative questions, whereas respondents in the other subsample subtract one answer from the other. Combining the means of the two subsamples yields an estimate for the population mean:

$$\text{Mean}\,(X+Y) + \text{Mean}\,(X-Y) = (\bar{X}+\bar{Y}) + (\bar{X}-\bar{Y}) = 2\,\bar{X}.$$

Thus, it is easy to solve for the means of the separate X and Y questions. This method results in comparatively lower variance than previously described randomized response techniques (Miller and Cisin, 1980).

A very different approach, the nominative technique, derives from the multiplicity methods developed by Sirken (1975). In this technique, respondents serve as informants by providing reports about *other persons'* experiences. The first attempts to measure sensitive behavior in this way (Sudman, et al., 1977a) simply asked respondents to report either what percentage of their friends had used marijuana or how many of their three best friends had used the drug. A more statistically defensible version (Rittenhouse and Sirken, 1980; Fishburne, 1980; Miller, 1985) asked the respondent to report, first, how many of his or her friends had used heroin, and second, how many *other friends* of each reported (or "nominated") heroin user also knew about his or her heroin use — and thus could report that user. (The potential for bias is present to the extent that there is error in reporting the number of friends.) The latter information allowed each "nominated heroin user" to be weighted inversely to his/her chance of being reported. Variances may be very low, but are dependent upon the distribution of these weights (see Cisin, 1980).

The various indirect techniques each have advantages and disadvantages. First, with respect to reduced self-disclosure and the social desirability of responses, both randomized response and aggregated response reduce the level of self-disclosure required of respondents.

Respondents who engaged in the disapproved behavior are thereby encouraged to provide more honest answers. However, these methods may have the *reverse* effect on *other* respondents who did *not* engage in the disapproved behavior and are now asked to voice answers that may seem to convey the false impression that they did engage in that behavior (Miller, 1981). The social desirability issue is not completely avoided by such techniques. Various stratagems have been devised for minimizing response distortion on the part of respondents not participating in the disapproved behavior; one example would be to combine socially *positive* items with the socially disapproved item (see Zdep and Rhodes, 1976–77). Only the nominative technique eliminates the self-disclosure/social desirability issue.

Second, with respect to respondent understanding, randomized response requires respondents to perform complex tasks for which interviewers must provide "mind-boggling" explanations, thereby risking respondent suspicion (Greenberg, et al., 1969), confusion (Wiseman, et al., 1975–76; Shimizu and Bonham, 1978) and uncertainty as to the level of disclosure that a truthful answer actually entails (Tracy and Fox, 1981). Aggregated response requires simpler tasks, but again, these must be explained to respondents. By contrast, the nominative technique is unobtrusive and does not require interviewer explanations.

Third, with respect to respondent ability to provide the required information, randomized response and aggregated response present no problems, but unfortunately the nominative technique requires respondents to provide information about other persons that they often do not possess.[1]

11.2 THE ITEM COUNT METHOD

The item count method (Miller, 1984) provides each respondent with a short list of items describing behaviors and asks him or her to count and report *how many* he or she has engaged in — not which ones. In the simplest version of item count, a random subsample (Subsample A) is shown a four-item list that includes the socially disapproved behavior item; the remaining respondents (Subsample B) are shown an identical (three-item) list from which the disapproved item has been removed. By comparing responses from the two subsamples, an estimate is obtained.

[1] About 20 percent of the respondents who knew a heroin user could not estimate how many of the user's other friends also knew about his or her heroin use.

Specifically, to derive an item count estimate, the mean number of behaviors ($\bar{X}_{4A}$) reported by Subsample A on the four-item list is calculated; then, the mean number of behaviors ($\bar{X}_{3B}$) engaged in by Subsample B on the corresponding three-item list is calculated. The difference between the two list means provides the item count estimate of the socially disapproved behavior, as shown in Appendix 11.1. The variance associated with this estimate is quite high, partly because responses that include the disapproved behavior item are required only of persons in Subsample A (not those in Subsample B).

The *double-lists version of item count* increases the efficiency of the estimator. The double lists generate *two* separate item count estimates, which are then averaged to provide a single estimate. The first estimate is provided by lists X_4 and X_3, say, as described above. The second estimate is provided by (new) lists, Y_4 and Y_3, say. Respondents in Subsample A respond to lists X_4 and Y_3, whereas respondents in Subsample B respond to lists Y_4 and X_3. The two estimates are averaged such that:

$$EST\,(P) = \hat{p} = 1/2\ [(\bar{X}_{4A} - \bar{X}_{3B}) + (\bar{Y}_{4B} - \bar{Y}_{3A})]$$
$$= 1/2\ [(\bar{X}_{4A} - \bar{Y}_{3A}) + (\bar{Y}_{4B} - \bar{X}_{3B})]$$

This method cuts the variance in half, because reports including the disapproved behavior item have been obtained from twice as many respondents (i.e., the entire sample, not just a random half):

$$\begin{aligned} VAR(p) &= 1/4\ [VAR(\bar{X}_{4A} - \bar{Y}_{3A}) + VAR(\bar{Y}_{4B} - \bar{X}_{3B})] \\ &= 1/4\ [VAR(\bar{X}_{4A}) + VAR(\bar{Y}_{3A}) - 2r_{X4\cdot Y3}\sqrt{VAR(\bar{X}_{4A})}\,\sqrt{VAR(\bar{Y}_{3A})} \\ &\quad + VAR(\bar{Y}_{4B}) + VAR(\bar{X}_{3B}) - 2r_{Y4\cdot X3}\sqrt{VAR(\bar{Y}_{4B})}\,\sqrt{VAR(\bar{X}_{3B})}] \end{aligned}$$

where r_{XY} denotes correlation between X and Y.

The item count approach is closely related to previously developed techniques. First, like the randomized response unrelated questions version, item count obtains an estimate via administering a combination of innocuous and socially disapproved items to one subsample and administering the innocuous item(s) only to the other subsample; in fact, the double-lists version of item count is very similar to a variant of unrelated questions version of randomized response. As with randomized response (and aggregated response), the level of self-disclosure is reduced; respondents who have engaged in the disapproved behavior are therefore likely to provide more truthful answers, but the reverse effect is risked for respondents who have *not* engaged in the behavior. Specifically, some item count respondents may be reluctant to voice a non-zero answer to a list that includes the socially disapproved item. Strategies for minimizing distortion on the part of nondeviants include using (1) innocuous items that are low-prevalence,[2] (2) innocuous items that are socially positive, and (3) "traditional" methods of encouraging honest answers, such as private self-administered answer sheets for item count responses.

Second, like aggregated response, item count uses a simple respondent task: combining answers to multiple questions. Since item count represents the simplest possible version of this task, there seems to be virtually no uncertainty about the operant level of self-disclosure, and very little (if any) risk of suspicion and confusion deriving from the task itself.

Third, although item count can be unobtrusive, it is unlike the nominative technique in that it asks only about the respondent's own behavior and thus avoids the problem of the respondent's lacking information about other persons.

Earlier field tests of the item count technique used an unobtrusive format. In the area of drug use, item count estimates were obtained in the early to mid-1980s (Miller et al., 1986b), using a theme of "risk taking behaviors" (with items such as "married someone I'd known less than one year" or "tried to help someone who was drowning or whose boat was in trouble"). The item count estimates were then compared to self-report data. In a test of heroin use among young men, item count resulted in substantially higher estimates (i.e., among males aged 18 to 34 years, the heroin ever-use estimate was 7.3% for item count, and 3.6% for direct self-reports); and among those who had the opportunity to try heroin, the item count estimate of actual heroin use was 23 percent. In subsequent tests to estimate past-year use of marijuana and cocaine, the estimates were very similar for item count and for self-report (i.e., with either method, the estimates for young adult males were about 40% for past-year marijuana use and 20% for past-year cocaine use).

Another field test used a "social exchange" theme to estimate criminal behavior involving property crimes (Miller et al., 1986a). Here, introductory lists were used to "warm up" respondents, and their first task was to count how many items on a list were things that had *happened to them*; these introductory lists were used to estimate being a victim of

[2] In an item count field test (described below) to estimate numbers of young men who had engaged in property crime, the item count double-lists method used one list of low-prevalence items and another of somewhat higher prevalence items. The lower prevalence list generated a relatively high criminal-behavior estimate, whereas the list with higher prevalence items generated a near-zero estimate. (See Miller et al., 1986a.) When using low-prevalence items on estimation lists, it may be advisable to use high-prevalence items on additional introductory lists. This gives all respondents a chance to report some behaviors that they have engaged in.

property crime. Then, near-identical item lists were used and respondents were asked to count how many items were things they had *done to someone else*. This approach yielded an estimate that 4.3 percent of males aged 18 to 34 had "broken into some else's house or car" within the past five years. Varying lists used in this field test also indicated that lists with *low-prevalence* innocuous or socially valued behavior are superior to lists with higher prevalence behaviors.

To summarize, the item count method combines many strengths of the earlier methods of indirect questioning. The respondent's task is straightforward and burden to the interviewer is minimal. Special care must be taken in developing the lists to maximize response validity, however. The major drawback of the item count method is the potentially high variance that results. The development of the double lists design increases the efficiency of the estimator, but the problem of large variances will continue to be problematic for rare behaviors and/or low-risk population groups.

11.3 THE NATIONAL HOUSEHOLD SEROPREVALENCE SURVEY PRETEST

The primary objective of the National Household Seroprevalence Survey (NHSS) was to estimate the prevalence of human immunodeficiency virus (HIV) infection in the U.S. noninstitutionalized civilian population, aged 18 to 54 years. While many studies estimate HIV prevalence in certain population subgroups, most include self-selected rather than randomly selected individuals, and the results cannot be generalized to the total U.S. population. Phase 1 of the NHSS was a feasibility study that involved a pilot study and a pretest. The pilot study was conducted in Allegheny County (Pittsburgh), PA, and involved 263 respondents. The pretest was conducted in Dallas County, TX, and involved 1,449 respondents. For the pretest, segments were classified into one of three HIV risk strata (high, medium, or low), based on eight key HIV risk indicators. These indicators included information such as: AIDS morbidity rates by Census tract, hepatitis B morbidity rates by Census tract, and the relative number of persons admitting to intravenous (IV) drug use by Census block. The three risk strata were balanced on race/ethnicity so that their distribution in percent black, percent Hispanic, and percent white were approximately equal to the overall race/ethnicity percentages for Dallas County.

Information about HIV infection and the behaviors that put an individual at increased risk for this infection are extremely sensitive, so there is considerable concern about the potential for both nonresponse and response bias in a household-survey-based HIV estimate. Anonymity was an essential component of both the pilot study and the pretest and all procedures were designed to ensure the anonymity of and protect the privacy of all participants. The anonymous approach to data collection was designed to maximize the participation of individuals contacted for the survey, regardless of their level of HIV risk. It was hoped that this approach would decrease the nonresponse bias that occurs if respondents at higher risk for HIV infection participate in the survey at lower rates. Following the pilot study, the issue of estimating the impact of this response bias was raised. Plans were made to include an indirect questioning method as part of the NHSS pretest. What follows is a discussion of the research and policy considerations which led to the use of the item count technique as part of this pretest.

11.4 PRELIMINARY TESTING: FOCUS GROUPS AND INDIVIDUAL COGNITIVE INTERVIEWS

11.4.1 Unresolved Issues About Form and Content of Item Count Questions

As discussed previously, respondent acceptance of item count questions is highest when the items in the lists "fit together" or make sense to the respondent. In the NHSS pretest, this meant they should make sense in a self-administered questionnaire about HIV infection. Items should also be nonverifiable behaviors and it should be highly unlikely that any respondent would have performed all of the behaviors described by the items. These features would be important in an embedded form of item count, where no attention is drawn to the technique. It would be important to know whether or not respondents see the technique as unusual, and whether or not they see it as providing enhanced privacy. Care must be taken not to make an honest answer to the item count question more stigmatizing than answering a single question, thus making an embedded question no better than a direct one (Clark and Tifft, 1966). However, if the question does not completely hide the sensitive behavior, then it may produce suspicion or reactance (Brehm and Brehm, 1981; Webb, et al., 1966). Yet, a separate set of item count questions may suggest to respondents that their answers to direct questions are not as private as the item count answers, thereby arousing a more general suspicion about all of the survey techniques the respondent encounters.

11.4.2 Focus Groups

Our initial testing centered on respondents' reactions to the randomized response and item count techniques. We conducted two focus groups to assess the range of reactions and to identify specific factors which affected those reactions. These groups were told the purpose of the indirect questioning techniques and received demonstrations of the item count method and several variants of randomized response (with different random selection devices and non-key questions). Respondents in these groups were RTI employees — one group of non-college-educated data entry employees, and one group of employees with college or post-graduate degrees. Neither group was related to the NHSS project in any way.

Respondents' reactions to the indirect questioning techniques were mixed. Overall, randomized response was not well received. All respondents saw the technique as highly obtrusive and potentially disruptive. Generally, the response to answering randomized response questions was laughter. Using random selection devices like flipping coins or drawing marbles was seen as highly incongruent with a serious survey. Despite knowing the purpose of indirect questioning, they still found the randomized response technique highly unusual.

Item count questions were less troublesome to respondents. They understood how to answer them and their comments centered around the content of the non-key items. Items completely different in content from the key item caused respondents to be suspicious of why they were being asked. Because respondents were aware that the purpose of the questions was to learn about HIV risk behaviors, they felt that admitting to one or more innocuous or positive behaviors on the lists might be misconstrued as admissions of the more negative key behavior. Moreover, a number of respondents wanted to know how information could be gained by asking questions with the item count method.

11.4.3 Individual Cognitive Interviews

To further investigate those issues raised in the focus groups, fourteen cognitive interviews were conducted using RTI employees and homosexual men from the surrounding community as respondents. Respondents were given one of several versions of the self-administered materials to complete and asked to talk about their thought processes in answering the questions. These materials consisted of a questionnaire containing direct questions about HIV risk behavior and a section containing item count questions. Due to the sensitive nature of the questions, respondents completed the questionnaire alone and tape recorded their thoughts as they worked. After the questionnaire was completed, the interviewer returned and conducted a debriefing interview.

We did not draw respondents' attention to the item count questions as they completed the questionnaire. The item count questions had minimal instruction for how to complete the questions and the questions were embedded within the larger questionnaire. The item count questions were placed in a section of the questionnaire with a heading of "Risk Questions." They included items which we saw as related to HIV infection and which were perceived by respondents as fitting in with a questionnaire designed to measure the prevalence of HIV infection. The non-key items were behaviors involving travel to places where AIDS is more prevalent, blood contact, contact with AIDS patients, sexual practices, and sexually transmitted diseases.

These interviews and recordings indicated that respondents found the item count questions embedded in the larger questionnaire answerable and unobtrusive. They experienced no difficulties in reading and understanding written instructions for the self-administered item count questions and did not make mistakes in marking answers. There was even some evidence that at least one respondent felt more comfortable answering an item count version of a question than a direct question asking about the same content.

Comments elicited during debriefing interviews were similar to those from the focus groups. Respondents wanted to know how information could be gained by asking questions in this way. Respondents voiced no objections to the general content of the item count questions (although they did comment on particular items).

The item count questions were taken before the Dallas County Community Advisory Panel, an oversight committee composed of local leaders who were to approve all materials to be used in the pretest phase of the study. This group had serious reservations about the content of the lists. They believed that including items which described behaviors such as traveling, blood contact, and contact with AIDS patients would perpetuate myths about modes of HIV transmission. Because the NHSS was to be a government-sponsored survey about HIV infection, they felt that asking about these activities would suggest to respondents that these behaviors are risk behaviors, when in fact they are not.

Several RTI researchers also had concerns about this version of item count. They thought that to work correctly, item count required a clear perception on the part of respondents that their privacy is being protected. They proposed that the way to enhance perceived privacy was to explicitly state how and why item count questions protect privacy. Simply having respondents answer embedded item count questions would be no different than answering direct questions since the respondent would be unaware of any additional privacy protection.

These considerations meant that we would have to move to a format in which there was an enhanced description of how item count protected privacy and the items did not relate to HIV risk. This also meant we would have to develop explanations for interviewers to give to respondents regarding how the item count technique "works." Additional research was needed to find out which characteristics of the item count lists were associated with the greatest perceived privacy protection.

11.5 CONTROLLED EXPERIMENTS

11.5.1 Privacy Ratings of Item Count

In an experiment addressing perceived privacy levels, respondents were given different lists of item count questions and asked to indicate which ones best protected their answers. We constructed item count lists which varied on two dimensions: the *type* of non-key items (either stigmatizing or innocuous and, among the stigmatizing items, either similar or dissimilar in content to the key item) and the *number* of non-key items in the question (ranging from one to five low-prevalence items). In addition, we tested lists in which a single, nonstigmatizing, *high-prevalence* item was included with the various types of low-prevalence items. Such a list might be useful for two reasons. First, it would make respondents feel that admitting to at least one behavior on the list was normal — thus forestalling the perception that admitting to any behavior in the list was tantamount to an admission of the risk behavior. Second, it could serve as a useful manipulation check: If a behavior was included which 99 percent of respondents would have done over a given time period, then we could see whether or not respondents were correctly completing the item count questions by seeing if most respondents gave counts of at least one for the item count questions.

In the first stage of testing, we constructed a 6 (number of items) $\times$ 3 (type of items) $\times$ 2 (replications) factorial experiment in which 22 respondents made ratings of the protection given by the questions embodying the various combinations of number and type of items.

Respondents rated the protection provided by each of 36 different item count lists (4 warm-up lists were discarded from the analyses) on a scale of 0 = "No Protection," 1 = "Poor," 2 = "Moderate," 3 = "Good," and 4 = "Complete Protection." In the second stage, 23 respondents made paired comparisons of the protection given by the 15 most protective questions from Stage 1. A random subset of all 15! pairings was made in order to reduce respondent burden. Questions were paired which differed only on a single dimension. Respondents for these experiments were RTI employees and members of the local community.

Significant main effects were found for the number of items ($F(5,71) = 3.59$, $p < 0.05$) and types of items ($F(2,40) = 3.28$, $p < 0.05$). Respondents preferred longer lists (four or five non-key items) over short ones (one or two items) and questions containing all innocuous items over those containing any number of stigmatizing items (both comparisons significant at $p < 0.05$, Tukey's HSD test). The difference between lists including a single high-prevalence item and those with all low-prevalence items was not significant. The paired comparison ratings from the second stage confirmed the preference for innocuous items over stigmatizing items (all $p < 0.05$); respondents did not strongly prefer having lists with one high-prevalence item over lists with all low-prevalence items.

11.5.2 Tests of Instructions Using Final Item Count Lists

The final test was a four-way factorial field experiment using respondents who were more representative of the general population. The three item count factors were:

- *Item Content*: lists of four low-prevalence items vs. lists of one high-prevalence and three low-prevalence items (all items described innocuous behaviors),
- *Instruction Content*: short instructions (containing a brief statement that item count protects privacy and an explanation of how to complete the questions) vs. long instructions (containing an additional explanation of how item count protects privacy),
- *Instruction Form*: written (respondent-administered) vs. oral (interviewer-administered) instructions.

While not a random sample of the local community, respondents for this experiment did provide heterogeneity in terms of race, sex, educational attainment, marital status, and age. We made special efforts to include respondents with lower reading comprehension levels.

Each respondent was asked to complete a questionnaire which included a series of direct questions and one version of the item count questions, presented as separate documents. Respondents were then asked whether they preferred answering item count questions or direct questions about risk behaviors. Error patterns, written answers to a question about respondents' impressions of item count, and oral comments made during post-experimental debriefings were also examined.

Of a total of 61 respondents completing the item count questions, 27 preferred direct questions, 16 preferred item count questions and 21 expressed no preference. There was a stronger preference for direct questions with oral instructions than with written instructions (21 vs. 6, $\chi^2_{(2)} = 19.11$, $p < 0.001$). Preference for direct questions over item count was not associated with item content or with instruction content.

Item count errors were (a) unanswered questions, (b) questions where individual items were marked, rather than indicating the number of items in the space provided for answers, and (c) answers interpretable as indicating the performance of a given behavior rather than a count (for instance, when the number checked as the answer corresponds to the ordinal position of a high-prevalence item, which was always last in the list).

Eleven respondents had at least one error in completing the item count questions. More errors were made in the written instructions condition than in the oral instructions condition (10 vs. 1, respectively, $\chi^2_{(1)} = 8.23$, $p < 0.05$). More respondents made errors when given lists containing a high prevalence item than when given all low prevalence lists (7 vs. 4) and with short vs. long instructions (6 vs. 5). However, these differences were not statistically significant. Nevertheless, we considered the higher number of errors with lists of high-prevalence items unacceptable, and we chose to use only low-prevalence items in the final version of the item count questions. Our analysis of errors was restricted to those that could be clearly detected from the written form. To the extent that respondents made errors which are not discernable, our error counts will be underestimates.

While respondents generally understood how to answer item count questions, several respondents commented in oral debriefings that they failed to see the technique as a privacy protection, despite explicit instructions emphasizing that fact. Most respondents thought that item count was unnecessary and that anyone who agreed to complete the questionnaire would be motivated to tell the truth and would prefer direct questions.

Other comments dealt with the items themselves, the technique in general, and respondents' preferences for direct questions over item count questions:

- Four respondents indicated that they did not like answering questions about non-AIDS activities.

- Four respondents wondered how anything could be learned by asking questions in this way.

- Six comments by respondents were expressions of preference for direct questions or general criticism of item count (item count is "tricky," "dumb," "too long," "beating around the bush").

Because they did not understand the purpose of item count, respondents sometimes attempted to impose some meaning of their own on the task. Several thought it might be a way of ascertaining social class. One respondent thought item count might be a personality test. Several people also thought that saying that they had done any behavior in a list with a key item in it might be misconstrued as an admission of the key behavior, making them trust the question less, not more.

Not all comments about item count indicated incomprehension or rejection. Nine respondents commented that item count hides answers. These comments can be interpreted as indicating that respondents recognize that item count protects an individual's privacy. However, these comments were usually couched as criticisms (item count is "imprecise," or "insensitive"; "You can't know which question is being answered"). Six of these nine respondents still preferred direct questions in their preference ratings. This preference for direct questions was not significantly different from the preferences of those who did not see item count as hiding answers. Finally, three respondents commented that they did see item count as providing additional privacy which could make people with something to hide feel safer.

11.6 FINAL VERSION OF ITEM COUNT USED IN THE NHSS PRETEST

A two-part Sample Person Questionnaire was used in the NHSS Pretest to obtain HIV risk information from respondents. Part 1 included questions related to demographic characteristics, health conditions, and factors or behaviors that may increase a person's risk for HIV infection. Part 2 contained the item count questions. Part 1 and Part 2 were separate documents. The item count questions were always administered after the direct questions were completed. It is important to note that the item count technique had never been incorporated into a questionnaire that also included a direct question for the item of interest. The possibility of an interaction effect between the two forms of the key questions led us to prefer a design in which a random half of the sample completed the item count questions prior to the direct questions, with the other half completing them following the direct questions. However, study sponsors were unwilling to implement such a design because of the possibility it would compromise the quality of the estimates based on the direct questions.

The setting for questionnaire completion was, in most cases, a completely private location within the household. After the respondent had reviewed an introductory letter, watched the project videotape, signed the consent form, and given a blood sample (if required), the interviewer played the audio portion of the project videotape that stressed the importance of providing truthful answers on the questionnaire. The respondent then completed Part 1 of the Sample Person Questionnaire, either by reading the questions and marking the answers, or by listening to the questions and answer categories as the interviewer read them and then marking the answers. In either case, the interviewer never saw the respondent's answers. When Part 1 was completed, the respondent taped the edges with tamperproof tape and placed it in a mailing envelope.

The instructions for the item count instrument consisted of a script read verbatim by the interviewer to the respondent before completing the questionnaire. A showcard with a sample item count question was given to the respondent to look at while the interviewer read the instructions (see Appendix 11.2). The instructions explained exactly how the sample question should be answered and how the item count format protects the respondent's privacy. The instructions also stressed the importance of marking only the answer box and not the individual items, and of answering the questions honestly. After receiving instructions, the respondent completed the item count questions using the same method of administration (that is, with or without interviewer assistance) that had been used for Part 1. Again, the interviewer never saw the respondent's answers. The respondent taped the edges of Part 2 with tamperproof tape, then sealed it into the mailing envelope with Part 1. The interviewer invited the respondent to accompany him or her to the nearest mailbox to mail the envelope.

Two versions of the item count instrument, the Blue Form and the Yellow Form, were developed for the NHSS Pretest to implement the double-lists design. The time period of interest in all the item count questions was approximately 10 years — "Since January 1978."

The first question in the item count instrument was a warm-up question containing five innocuous items. It included a very high prevalence item (getting a haircut) which was designed to ensure that respondents should report at least one item on this list. The warm-up question was identical in the Blue Form and the Yellow Form. Following this were the four item count lists used to estimate the prevalence of intravenous drug use and receptive anal intercourse. The final two questions in the item count instrument were comment questions. A closed-ended question asked whether the respondent believed that direct questions or item count questions were best for getting honest answers. The final question was an open-ended one that asked the respondent to comment about the item count method for asking personal questions.

11.7 RESULTS FROM THE NHSS PRETEST

11.7.1 General Survey Results

The screening rate for the NHSS pretest was 98 percent. No significant differences were seen in the screening rates across the three risk strata. In the regular survey, both a blood sample and a questionnaire were required for participation. In a followup study of nonrespondents, individuals were given the option of completing a questionnaire without providing a blood sample. Therefore, two interview response rates were obtained for the pretest, one for persons who completed the questionnaire, and another for persons who also provided a blood sample. These response rates were 90 percent and 84 percent, respectively. Again, no differences were observed in either response rate across the three risk strata. The overall response rates for the NHSS pretest were 88 percent for questionnaire completion and 82 percent for blood and questionnaire response. These rates are the products of the screening response rate and the associated sample person interview rate.

For the purposes of this discussion, missing data includes both blank items and unusable responses. Unusable responses include illegible, out-of-range, and multiple responses, as well as refusals to answer items. For Part 1 of the questionnaire, which contained the direct questions, 83 percent of completed questionnaires had no blank items or unusable responses. Lower missing data rates were seen for Part 2 of the questionnaire, which contained the item count questions. Of these completed questionnaires, 92 percent had no blank items or unusable responses. The rates for the individual item count questions ranged from one to two percent. However, more than six percent of respondents did not answer the question which asked the person's opinion on the best questioning method.

11.7.2 Item Count Results and Evaluation

At the end of the field work, the 47 interviewers who worked on the NHSS pretest were asked to complete a debriefing questionnaire concerning how well the procedures for implementing both Parts 1 and 2 of the pretest questionnaire had worked. Procedures for Part 1 worked well; 30 interviewers (70%) felt the procedures worked very well, while the remaining 13 interviewers (30%) felt the procedures worked fairly well. In contrast, only 22 interviewers (51%) felt the procedures for item count worked very well. An additional 28 percent felt the procedures worked fairly well, and 21 percent felt the procedures did not work very well. The most common problems noted with the item count technique were that the interviewers' script and instructions were too long, and that the script and instructions insulted the respondent's intelligence and made the interviewer appear foolish. Despite these flaws, however, nearly all the interviewers (95%) felt that the majority of their respondents understood how to respond properly to the item count questions. Interviewers who conducted interviews in Spanish noted no problems with the translated item count questions.

We were concerned about the extent to which respondents would understand the instructions for marking their answers (and thereby yield usable data) on the item count questionnaires. Therefore, we manually reviewed each item count questionnaire and tallied the frequency of identifiable recording errors. Errors were rare. Only eight respondents (0.6%) marked more than one answer box, making their responses unusable. Only 23 respondents (1.6%) failed to complete one or more of the five questions, of whom five (0.3%) returned the questionnaire entirely blank. Twenty-one respondents (1.5%) marked the individual behaviors in the lists which applied to them. (All but three also marked the appropriate answer box.) These responses were not unusable, but merely required editing.

A second check on respondents' comprehension of the item count questions, however, is not so favorable. Of the 62 persons who reported intravenous (IV) drug use on the direct question, at least seven of them (11.3%) then failed to report this in the item count questionnaire — that is, they reported having done none of the things in the list containing the drug use item. For the receptive anal intercourse question, 27 of 196 people (13.8%) giving a positive response to the direct question reported "none" in the corresponding item count question. Surprisingly, the rate of inconsistent reporting on this item is higher for women (15.0%) than for men (9.3%).[3] There were no differences in the percentages of respondents making these errors by education, age, or language translation of the questionnaire.

In examining the distributions of item count responses, we find a larger than expected number of respondents reporting two, three, or even four behaviors in the lists of items. This is alarming because it increases

[3]The number of men reporting receptive anal intercourse in the direct question is low (43), so this comparison is based on a smaller case base than for women, among whom there were 153 positive responses to the direct question.

the variance of the item count estimates. In our preliminary testing of the item count procedures, we noted the tendency of several respondents to mark the response in the answer box that corresponded to the ordering of the item in the list that was applicable to them. For example, persons who had done only the third item in the list were sometimes inclined to mark the "3" answer box rather than the "1." We suspect that some of the high reports in the item count lists may be attributable to this error, but we have no way to verify this. It is possible that the items in the lists, in conjunction with each other, are simply not as rare as we had thought.

In conclusion, we found little evidence in the hardcopy questionnaires that respondents had difficulty answering the item count questions properly. Other indications, however, imply that there may be "invisible" errors in the data resulting from respondents' lack of understanding the procedures.

At the end of the item count questionnaire, respondents were asked which of the two questioning methods (direct or item count) they thought was "the *best* way to ask personal questions so that most people will give an honest answer." Since one of the primary strengths of the item count technique is that it does not require the respondent to directly reveal his or her participation in a sensitive behavior, we thought that respondents who had more to "hide" might be more likely to perceive this as the best method. In addition, we expected that respondents in interviews which were not completely private and self-administered might report that item count was the better technique. Finally, we suspected that preferences might differ by educational level and other demographic characteristics. Table 11.1 reveals that, while not always statistically significant, the direction of these findings provides support for each of the hypotheses.

Overall, only 30 percent of respondents said they think the item count technique is best. Those who were HIV positive and those in the higher risk strata were more likely to prefer the item count method, though the differences were not statistically significant. Somewhat surprisingly, persons openly reporting IV drug use on the direct question were less likely than those with negative responses to report that the item count technique is better.[4] Men reporting receptive anal intercourse in the direct question were somewhat more likely to report item count as the best method, but there was virtually no difference for women. Again, however, none of these differences are statistically significant.

As we had expected, respondents interviewed when another household member was present were significantly more likely to report item

[4]This finding is consistent with anecdotal knowledge of social scientists studying drug use and the comment of one respondent that "If you shoot drugs you usually don't mind admitting it."

count as the better technique. Similarly, item count was more likely to be perceived as the better approach by respondents for whom the questions had to be read by the interviewer rather than using a purely self-administered approach. Finally, persons using the Spanish translation of the questionnaire were slightly more likely to report item count as the better technique than persons completing the interview in English, but this difference was not statistically significant.

Education shows a clear negative linear relationship in the percentage of respondents reporting that the item count method is better (linear contrast $p < 0.05$). Men were significantly more likely to report item count as the best questioning method perhaps because receptive anal intercourse is viewed as a more sensitive topic among men than among women. Marital status is not strongly associated with perceptions of the best technique. (All independent contrasts $p > 0.05$.) White respondents were significantly less likely than persons of other racial groups to believe that the item count technique is best, and persons of Hispanic origin were more likely to report this method as best. There is no clear relationship between age and which method of questioning is perceived as best.

While one question asks respondents for their opinion on the "best" technique, it does not ask them their personal preference in questioning method. This fact was highlighted numerous times in the open-ended comments provided by respondents. Many wrote phrases such as "Personally, the direct approach appeals to me most. However, I can see how the latter method would work best for some people." Another theme among the comments was that conditional upon true confidentiality, there is no reason *not* to ask the direct question. Other respondents commented that the item count technique seemed "dumb" or a waste of time or money, or that they did not understand how responses to the questions could possibly be of any use. Several likened the item count method to "beating around the bush" and commented that AIDS is too serious an issue not to address it directly.

11.7.3 Prevalence Estimates

In comparing the estimates from the two questioning methods, we first examined the item missing data rates. For IV drug use, these rates were low for both methods — 0.7 percent for the direct question vs. 1.0 percent for the item count lists. For receptive anal intercourse, the missing data rate was higher (2.5%), primarily due to men apparently not understanding that they were to answer this question. (The item missing data rate

Table 11.1. Questioning Method Perceived as Best by Various Characteristics of the Respondent and the Interview Situation

	n	% Reporting Item Count as Best	% Reporting Direct Question as Best	Tests of Statistical Significance
Total Sample	1,352[a]	30.0	70.0	
Risk Stratum				
High	372	32.8	67.2	$\chi^2 = 2.074$
Medium	537	29.6	70.4	$p = 0.13$
Low	443	28.0	72.0	
HIV Blood Test Result				
Positive	15	33.3	66.7	$\chi^2 = 0.075$
Negative	1,255	30.0	70.0	$p = 0.79$
Answer to Direct Question on IV Drug Use				
Yes	58	24.1	75.9	$\chi^2 = 1.195$
No	1,288	30.1	69.9	$p = 0.28$
Answer to Direct Question on Anal Sex				
Men:				
Yes	40	37.5	62.5	$\chi^2 = 0.408$
No	591	32.5	67.5	$p = 0.52$
Women:				
Yes	148	28.4	71.6	$\chi^2 = 0.066$
No	559	27.4	72.6	$p = 0.80$
Another Household Member Present During Interview				
Yes	94	41.5	58.5	$\chi^2 = 6.698$
No	1,220	29.3	70.7	$p = 0.01$
Questions Read Aloud by Interviewer				
Yes	70	42.9	57.1	$\chi^2 = 4.832$
No	1,246	29.3	70.7	$p = 0.03$
Spanish Questionnaire Used				
Yes	60	38.7	61.3	$\chi^2 = 1.304$
No	1,288	29.6	70.4	$p = 0.26$
Education				
Less than 9th grade	72	44.4	55.6	$\chi^2 = 18.061$
Some high school	157	36.3	63.7	$p = 0.000$
Finished high school or GED	322	35.4	64.6	
Some college	403	27.8	72.2	
Finished college	391	22.0	78.0	
Gender				
Male	639	32.9	67.1	$\chi^2 = 5.482$
Female	713	27.4	72.6	$p = 0.02$

Table 11.1. (Continued)

	n	% Reporting Item Count as Best	% Reporting Direct Question as Best	Tests of Statistical Significance
Marital Status				
Married	526	29.7	70.3	$\chi^2 = 5.263$
Widowed	23	21.7	78.3	$p = 0.001$
Divorced	229	34.9	65.1	
Separated	69	30.4	69.6	
Never married	505	28.3	71.7	
Race				
Black	351	35.9	64.1	$\chi^2 = 9.087$
White	861	26.7	73.3	$p = 0.000$
Other	140	35.0	65.0	
Hispanic Origin				
Yes	185	37.3	62.7	$\chi^2 = 4.245$
No	1,167	28.8	71.2	$p = 0.04$
Age				
18–30	635	30.4	69.6	$\chi^2 = 2.908$
31–45	539	28.2	71.8	$p = 0.06$
46–54	167	33.5	66.5	

[a]Six respondents (0.4%) said "both"; 86 (6.0%) responded "don't know" or left the item blank. These persons are omitted from further analysis.

was 3.1% for men vs. 1.9% for women.) The item count lists fared better for this question, with a missing data rate of 0.9 percent.

Before turning to the prevalence estimates from the two techniques, we first note the high levels of reporting on the item count lists *without* the key behaviors. The unweighted percentage of respondents engaging in at least one of the non-key behaviors ranged from 7.5 percent to 23.4 percent on the four lists without the key items. From a variance point of view, for the item count technique to work efficiently, these values need to be very close to zero. However, to avoid bias it may be best for items to have values greater than zero. Unfortunately, our selection of these items was inhibited by factors described above, and we were unable to be as true as we would have liked to this requirement.

As Table 11.2 shows, our prevalence estimates suffered high variance estimates as a result of this weakness. The prevalence estimates show few consistent trends in the comparison of the two methods, and we believe this is a result of the instability of the estimates. Due to the high variance of the item count estimates, there are no statistically significant differences between any of these prevalence rates and the corresponding direct estimates. Prevalence rates are presented for the total sample for IV drug use, and for men and women separately for receptive anal intercourse. We also calculated estimates separately for the three risk strata. Although not statistically significant, there is a trend of higher reporting of drug use and receptive anal intercourse using the item count method among respondents in the high risk strata. The trend suggests that the item count method, if not plagued by problems related to high levels of reporting for sensitive behaviors, may be useful for eliciting reports of sensitive behaviors. We also calculated estimates for the subset of persons remaining after omitting persons who had already admitted to the sensitive behavior in the direct question. The item count estimates for these groups would provide an indication of the extent of "lying" about the sensitive behavior. This, however, is apparently very minimal as evidenced by the negative prevalence estimates resulting. As Miller, et al. (1986b) point out, negative item count estimates can result when the true prevalence of the sensitive behavior is extremely low so that it is outweighed by differences between subsamples with respect to the nonsensitive behaviors in the list.

11.8 CONCLUSIONS

As researchers use surveys to probe further into highly sensitive behaviors and issues, the problem of eliciting honest answers from respondents is likely to increase in magnitude. The list of issues considered sensitive (such as income, voting behavior or religious beliefs) has been expanded to include topics such as sexual practices, illicit drug usage, and other illegal activities. Developing methods of indirect questioning which are easy to administer, are accepted readily by respondents, and which do not result in highly inflated variance estimates should become a priority. The item count technique is a step in this direction.

The experimental evidence from the NHSS project suggests that the technique can be administered easily and that, for the most part, respondents appear to understand how to answer the questions. Although not always statistically significant, the data support the hypothesis that respondents who are more likely to have engaged in the sensitive behavior, as well as those who answer the questions in a less protected format, are more likely to prefer the item count method. The changes made to the technique for the NHSS pretest (necessitated by policy concerns) appear to have detracted from the overall simplicity of the technique, and caused greater variance in the prevalence estimates. The requirement that respondents be informed about the purpose of the

Table 11.2. Prevalence Estimates for Key HIV Risk Behaviors for Direct Questions and Item Count Method, by Risk Stratum

	IV Drug Use		Receptive Anal Intercourse			
			Men		Women	
	Direct Question[a]	Item Count[b]	Direct Question[a]	Item Count[b]	Direct Question[a]	Item Count[b]
Total Sample	3.9%	0.2%	4.3%	1.1%	19.4%	17.4%
	1,435	1,428	679	690	730	737
	0.8%	2.1%	1.5%	5.0%	2.2%	3.6%
Risk Stratum:						
High	3.9%	10.4%	15.8%	24.5%	17.6%	24.0%
	400	394	202	208	185	187
	0.8%	3.8%	4.2%	5.3%	3.5%	10.3%
Medium	4.2%	0.3%	5.5%	5.3%	16.0%	14.7%
	575	573	273	274	296	299
	0.8%	2.9%	1.6%	4.0%	1.9%	8.8%
Low	3.9%	−1.2%	2.5%	−2.7%	20.7%	18.6%
	460	461	204	208	249	151
	1.1%	3.0%	2.0%	7.2%	3.1%	4.4%
Answer to Direct Question = "No" or Missing Data		−2.6%		−3.0%		−1.1%
		1,367		647		585
		2.0%		5.1%		3.0%

Note: Numbers in cells refer to the estimate, the number of cases, and the standard error, respectively. Estimates are weighted to reflect unequal probabilities of selection and are adjusted for nonresponse. Standard errors reflect the complex sample design and unequal weighting.

[a]Base = Number of cases with nonmissing data on direct question.

[b]Base = Number of cases with nonmissing data on lists with and without key item.

item count questions may, in the end, have defeated the purpose of including the questions. The NHSS pretest experience also points to the importance of the selection of the non-key items used in the item count lists. Researchers using this technique should be warned to carefully scrutinize the relationship between the suspected prevalence of the key item and the prevalence of the other items in the list. Results from the NHSS pretest indicate that the percentage of respondents who reported having done at least one thing from a list of non-key items was much higher than had been anticipated. This problem was compounded by the fact that the estimated prevalence of HIV infection for Dallas County was even lower than had been expected (0.42% as calculated from the NHSS pretest blood results and adjusted for nonresponse). It may be that for certain sensitive behaviors which have extremely low prevalence rates the item count technique cannot be adequately operationalized because the non-key items would not appear to provide any protection to the respondent. For example, an acceptable non-key item for the NHSS pretest from a variance standpoint might have been, "I have climbed to the top of Mount Everest." The item fulfills the criteria of being both low prevalence and nonstigmatizing. However, such an item might not appear to provide much protection to the respondent since the likelihood of an individual climbing to the top of Mount Everest is so extremely low. In addition, this item would be out of place in a questionnaire on HIV risk behaviors. Finding rare, nonstigmatizing behaviors related to the key item may prove to be one drawback of the item count technique since, in many cases, what makes a behavior sensitive is the fact that not many individuals engage in it. Thus, while the item count technique has many attractive features for studying sensitive behaviors, care must be taken in developing the non-key items. And, in cases where the behavior of interest is very rare, item count may not be the method of choice.

ACKNOWLEDGMENT

The authors would like to acknowledge the following individuals who contributed their ideas and advice to this research: Ira Cisin, Kathryn Dowd, Ralph Folsom, Anne Freemon, Charles Haley, Adele Harrell, Daniel Horvitz, Richard Kulka, Judith Lessler, James Massey, Michael Weeks, Emelita Delon Wong, and Donald Zimmerman.

APPENDIX 11.1 THE SIMPLEST FORM OF THE ITEM COUNT/PAIRED ESTIMATOR

Let P denote the population parameter being estimated, e.g., the proportion who engaged in the deviant behavior. Then an unbiased estimator of P is

$$\hat{p} = \bar{X}_{4A} - \bar{X}_{3B}$$

where $\bar{X}_{4A}$ is the mean number of behaviors on the four item list engaged in by persons in Subsample A and $\bar{X}_{3B}$ is the mean number of items on the three item (deviant behavior omitted) list engaged in by persons in subsample B.

Proof:

$$E(\bar{X}_{4A}) = E(\bar{X}_{3A}) + E(P_A)$$

where P_A is the proportion of subsample A who engage in the deviant behavior and $\bar{X}_{3A}$ is the mean for list A defined in analogy to $\bar{X}_{3B}$. Since, subsample A is a random sample of the population, $E(P_A) = P$.

APPENDIX 11.2 SHOWCARD WITH SAMPLE ITEM

EXAMPLE

Question: How many of the following things have you done *since January, 1978*?

- I have bought a new car.
- I have travelled to England.
- I have donated blood.
- I have gotten a speeding ticket.
- I have visited a shopping mall.

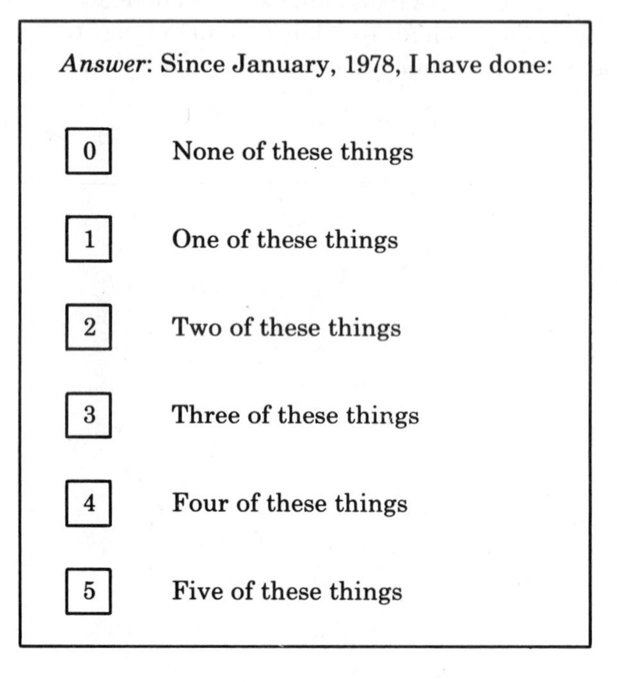

Answer: Since January, 1978, I have done:

[0] None of these things

[1] One of these things

[2] Two of these things

[3] Three of these things

[4] Four of these things

[5] Five of these things

CHAPTER 12

TOWARD A RESPONSE MODEL IN ESTABLISHMENT SURVEYS

W. Sherman Edwards and David Cantor
Westat, Inc.

12.1 INTRODUCTION

12.1.1 Purpose and Scope of This Chapter

The methodological literature on data quality in establishment surveys is not extensive (Federal Committee on Statistical Methodology, 1988). The published research (e.g., Cox, et al., 1989; Chapter 7) does, however, demonstrate special data quality problems in establishment surveys. These include problems of frame coverage, idiosyncratic patterns of participation and item nonresponse, difficulty in identifying appropriate respondents, and so on. Compared with the relatively rich literature on measurement error associated with respondent characteristics and response processes of individuals in household surveys, however, little research has focused on comparable issues in establishment surveys. Although some recent design and methodological work in establishment surveys (Cox, et al., 1990; Palmisano, 1988; Gower and Zylstra, 1990) has used techniques borrowed by household survey methodologists from cognitive science (see Chapter 20 for a description of these methods used in household surveys), it is not clear how applicable models of survey response processes developed from research on household surveys are to establishment surveys.

The purpose of this chapter is to develop a model of survey response processes for establishment surveys, drawing from both the survey methods literature and from the literature dealing with organizational behavior. Features of the model will be supported primarily by information from the authors' qualitative research on the Occupational Safety

Measurement Errors in Surveys.
Edited by Biemer, Groves, Lyberg, Mathiowetz, and Sudman.

ISBN: 0-471-53405-6

and Health System (OSHS), conducted by the U.S. Bureau of Labor Statistics, as well as by examples from other establishment surveys. Both establishment and household surveys considered here will largely be national surveys conducted by or for the U.S. Federal government. The chapter is not intended to make definitive statements about measurement error in the OSHS survey or any other survey, but to begin building a conceptual framework for the study of measurement error in establishment surveys generally.

12.1.2 Definitions

An *establishment* is usually defined as "an economic unit, generally at a single physical location, where business is conducted or industrial operations are performed" (Federal Committee on Statistical Methodology, 1988). Establishments include for-profit businesses, nonprofit organizations, government agencies, educational institutions, and so on.

We can then define an *establishment survey* as a census or sample survey whose sources of information are public or private businesses, agencies, or other nonhousehold organizations, or individuals acting as representatives of them. An establishment survey may ask for information about the establishment itself or about a more aggregated (corporation, agency) or less aggregated (division, department) level of organization. Examples in the U.S. include the Current Employment Statistics Survey and the Census of Economic Establishments. Although not usually included in the definition of establishment surveys, surveys which contact establishments to ask about specific individuals associated with the establishment (employees, members, customers, or patients, for example) may suffer from response errors similar to establishment surveys. Examples are verification surveys where the individuals and their relationships with establishments are predetermined, often in a household survey, such as the Medical Provider Survey and Health Insurance Plans Survey components of the National Medical Expenditure Survey, and verification surveys associated with the Survey of Income and Program Participation.

12.2 COMPONENTS OF A MODEL OF SURVEY RESPONSE FOR ESTABLISHMENT SURVEYS

Households and establishments are clearly not mutually exclusive. Family farms, for example, would meet either definition, and thus could be included in either an establishment or a household survey. Household surveys and establishment surveys have other common ground: both usually involve a respondent answering questions in a questionnaire.

Obvious differences between households and establishments come quickly to mind, as do differences between household surveys and establishment surveys. Establishments have a greater range of size than households; individuals' roles and inter-relationships within establishments are very different from those within households. Establishments have very different relationships with their environments than do households. With respect to surveys, Federal household surveys are typically conducted by an interviewer over the telephone or face to face, while establishment surveys are often conducted by mail, with or without follow-up by an interviewer. (This difference may mean that interview mode effects are important in comparing the two kinds of surveys. See Chapters 5 and 13.)

Despite the great areas of difference, we shall take the areas of overlap as justification for beginning our search for a response model for establishment surveys by examining such a model developed for household surveys.

12.2.1 Cognitive Model of Household Survey Response

Chapter 8 describes five stages of the survey response process in household surveys: encoding, comprehension, retrieval, judgment and communication of an answer. This set of processes is the result of an interaction between a respondent and a questionnaire, typically mediated by an interviewer. The respondent, the questionnaire, and the interviewer all have characteristics affecting the response process, potentially introducing response error. Characteristics of the respondent include his or her motivation, cognitive skills, and values and attitudes. As noted in Section A, several aspects of the questionnaire affect the response process, including question wording, context, and sequencing. The interviewer in his or her behavior, values and attitudes, and personality can also affect the response process. A good deal of survey methodological research has attempted to measure these effects and to design interventions to reduce the resulting response error.

A particularly important respondent characteristic, in that it may be influenced by survey design and procedures, is the respondent's *motivation* to perform the cognitive tasks involved (Cannell, et al., 1981). For example, a respondent well-motivated to give complete and accurate information is more likely to do the work necessary to understand complicated questions, retrieve complex data, or manipulate data encoded in ways not conforming to the structure of the question.

We have also included as a respondent characteristic the term "cognitive skills," by which we mean (in this model) the respondent's innate or learned abilities to perform the cognitive tasks required. For example, some respondents are more skilled at recall than others, and some are able to make more complex and more accurate judgments than others. Since these skills are largely not influenced or measured by the survey design, the survey methods literature has little to say about the effect of differences in response skills on measurement error.

Other influences on the response process are typically not physically present during the interview. These include the household or family of the respondent, the survey sponsor or administering organization, and the larger environment in which these all reside.

12.2.2 Similarities and Differences Between Household and Establishment Surveys

Let us now examine the first-level model of factors affecting the response process for establishment surveys.

The Respondent and the Questionnaire
The most basic similarity between household and establishment surveys is the usual presence of an interaction between questionnaire and respondent, although the establishment "questionnaire" may be more likely to have labeled answer spaces than questions. (See also Chapter 17 for a discussion of measurement error in establishment surveys without respondents.) Because U.S. Federal establishment surveys tend to be conducted by mail, the interviewer is omitted from the model, but could as well be added. As with the household survey model, both the respondent and the questionnaire have certain characteristics affecting the response process.

The Information System(s)
The major difference between household and establishment response models is the latter's inclusion of an "information system." The information system consists of the collection procedures, the processing procedures, and the quality control for information maintained by the establishment. Not all establishment surveys focus on information available from records — for example, questions may focus on decision making processes, employee morale, or job satisfaction. However, the use of information systems or records is far more prevalent in establishment than household surveys. Both household and establishment surveys often ask the respondent to report information for or related to others besides himself or herself (although there are both one-person households and one-person establishments). However, the relationship between the respondent and the others for whom he or she is providing information is generally quite different between households and establishments. Although some household surveys encourage respondents to consult records like calendars or checkbooks, household survey respon-

dents acting as proxies typically rely on their memories of shared experiences or conversations with other household members to answer survey questions. Because of differences in size and the nature of interpersonal relationships between households and establishments, respondents to establishment surveys may be less likely to have the kinds of relevant personal memories household survey respondents have. In the absence of these kinds of memories, establishment survey respondents often rely on paper or electronic records to answer survey questions. The Federal Committee on Statistical Methodology (1988) made the following observations:

> Establishment surveys typically seek hard data for which records are available. This is a central characteristic which both simplifies the collection and complicates the interpretation of the data. The collection is simplified because there are hard data on record from which the data of interest are extracted, rather than relying on the memory, opinions, or interpretations of the respondents as is often the case for household surveys.
>
> However, in establishing the concepts and definitions to be used in the surveys, special care must be taken to consider carefully the establishments' recordkeeping systems, definitions, and data availability to avoid introducing specification error into the data.

In the opinion of the cited report, establishment surveys trade one set of sources of measurement error (those associated with a respondent in a household survey) for another set (those associated with establishments' records). While this view is oversimplified, it indicates that characteristics of the information system(s) (records, automated systems, informal systems) used by survey respondents can affect the response process and introduce error.

The presence of this additional player in the first-level model has important implications for the respondent's role. Whereas the interviewer acted as a mediator between the questionnaire and the respondent in the household survey model, the respondent is now in the role of a mediator between the questionnaire and the information system(s) in the establishment survey model. As we shall describe in Section 12.3 this expanded role implies an expanded set of "response skills" required by the respondent.

Second-level Influences on the Response Process

The respondent's task in an establishment survey is to answer for the establishment, as the establishment's representative, rather than answering for himself or herself and as a proxy for other individuals, which is more the norm in household surveys. Thus, characteristics of the establishment and of the respondent's role in the establishment may be important potential sources of measurement error in establishment surveys.

Just as household survey respondents' answers to survey questions may be influenced by "society," an establishment's "environment" also can have profound effects in its responses to survey items. One widely used approach to organizational theory views organizations as "open systems." Buckley (1967) described the meaning of an "open system" as follows:

> That a system is *open* means, not simply that it engages in interchanges with the environment, but that this interchange is *an essential factor* underlying the system's viability, its reproductive ability or continuity, and its ability to change.

Katz and Kahn (1966) warn of the dangers of research on organizations that views them as closed, rather than open, systems. The organization is continually dependent upon its environment for input, and so it is continually adapting in response to changes in the environment. "Irregularities in the functioning" of an organization that may be attributed to environmental effects must be considered in any model of the organization's behavior. Thus, an establishment's environment may have even more effect on the survey response process than does the household respondent's environment on his or her survey responses. We also note that the survey sponsor or administrator may play an important role in an establishment's environment, as a regulator or advocate, for example, confounding the effects of the environment in general.

12.3 THE RESPONSE PROCESS IN ESTABLISHMENT SURVEYS

In this section, we shall explore the direct or analogous applicability of each of the components of the cognitive model of survey response processes described in Chapter 8 to establishment surveys. Since many of our observations will be supported by research on the Occupational Safety and Health System (OSHS), Section 12.3.1 describes that data system. Given the discussion in Section 12.2, the principal source of difference in response processes between household and establishment surveys would seem to be that household respondents are assumed to answer from memory while establishment survey respondents often rely on an information system for answers to survey questions. Recalling the overlap between "establishments" and "households" and the similarities of establishment and household surveys, it seems reasonable to assume that some establishment survey respondents answer from memory for at least some survey questions. In fact, we know this to be the case. Thus, it seems likely that the household response process model applies *in toto* at least some of the time in establishment surveys. It is appealing to conceive of the response process when information systems are involved as having activities analogous to the cognitive steps of encoding, comprehension, etc. The balance of this section will examine each of these cognitive steps for its applicability to establishment surveys when records are used.

12.3.1 The Occupational Safety and Health System (OSHS)

Since 1972, the U.S. Bureau of Labor Statistics (BLS) has conducted an Annual Survey of Occupational Injuries and Illnesses. It is conducted by mail from a probability sample of about 280,000 establishments in private industry. The Annual Survey is based upon a record-keeping system required of all employers in specified categories who have 11 or more employees, and of a sample of small employers and employers in other categories. The system involves maintaining Occupational Safety and Health Administration (OSHA) Logs and Supplemental Records, on which are recorded information about work-related injuries and illnesses. This system, mandated by the Occupational Safety and Health Act, was intended as a regulatory mechanism rather than simply for statistical uses. Inspectors from the OSHA may visit establishments and review their records at any time. Establishments not in full compliance with the record-keeping requirements are subject to fines. Despite such sanctions, research (Eisenberg and McDonald, 1988; National Research Council, 1987) has indicated considerable error, both under- and over-reporting as well as noncompliance, in the OSHS.

The survey currently collects only summary totals from the OSHA Log. BLS is redesigning the OSHS program to increase the amount of information available from the annual survey, and BLS and OSHA are working to streamline the record-keeping process. BLS has designed and implemented a series of pilot tests of proposed revisions in forms and procedures. The primary data cited in this report are based on interviews with record-keepers participating in three of these pilot studies. The purposes of this research were (1) to learn about the processes used by record-keepers and establishments in keeping OSHA records and (2) to obtain reactions to test forms.

The sample comprised 24 establishments in five states; each establishment had agreed to participate in a pilot test, completed the first quarter's forms for the test, and reported one or more injuries or illnesses during the quarter. The sample was stratified to include both large and small establishments and to include both establishments that would routinely be required to complete OSHA forms ("nonprenotified" establishments) and those that would only be required to complete the forms if they were selected for the Annual Survey ("prenotified" establishments). Since prenotification establishments often reported no injuries or illnesses, it was more difficult to identify an adequate sample of such establishments in all of the states. Data collection included review of the test forms submitted by sample establishments and semi-structured in-person interviews with the record-keepers completing the forms.

12.3.2 Encoding of Information

Regardless of whether the establishment survey respondent answers from memory or uses an information system, *some* cognitive encoding is relevant to the response process. At the very least, the respondent must have encoded information about how to get the answer to a question. If the response is obtained from an information system, some set of steps must have occurred for the information to be in the system. One may think of this as analogous to cognitive encoding; we have called the steps *record formation.*

Because record formation is more visible and perhaps more uniform across subjects than cognitive encoding, measurement errors attributable to differences in record formation may be more easily preventable or measurable than those attributable to encoding. Further, because encoding of relevant information and access to information systems may vary widely within an establishment, respondent selection in establishment surveys is almost certainly more critical than in household surveys — one would not expect a mailroom clerk or a receptionist to have access to payroll records, for example. Thus, encoding and its analog, record formation, may be of more interest to establishment survey designers and methodologists than to those involved with household surveys.

The Record Formation Process

An extensive literature of systems analysis and design examines the introduction of error into systems within establishments, and another literature in organizational behavior examines the effects of systems on behavior within establishments, including the behavior being measured by the system. A simple, very general model of processing systems (Knight and McDaniel, 1979) is presented in Figure 12.1. A set of inputs are subjected to one or more processes, producing outputs. Inputs,

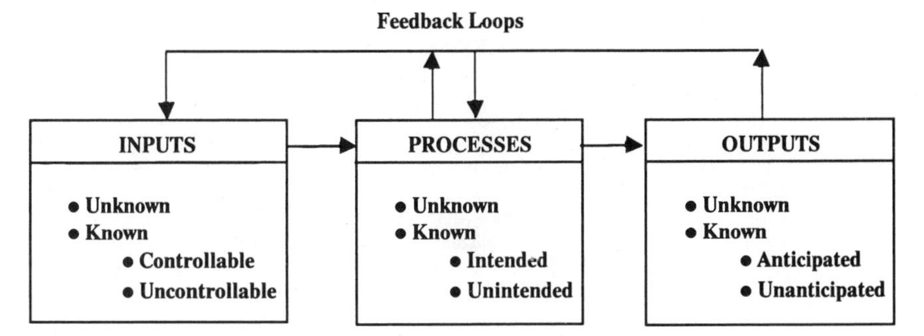

Figure 12.1 General model of information systems. Adapted from Knight and McDaniel (1979).

processes, and outputs each include known and unknown elements. Among known inputs, some are controllable and some are not; among known processes, some are intended and some are not; among known outputs, some are anticipated and some are not. By comparing actual outputs with intended outputs, for example, feedback and control can be built into a system. In general, two kinds of measurement error can occur in systems: errors of selectivity (inappropriate inputs are accepted or subjected to certain processes, or some appropriate inputs are not accepted or subjected to certain processes) and errors of distortion (processing steps make inappropriate changes or fail to make appropriate changes). The purpose(s) and corresponding specifications of systems in individual establishments or groups of establishments may be identified and studied, with the goal of identifying and mitigating or measuring the error that the systems may introduce into survey items.

Less easily identified and studied are behaviors associated with information systems within establishments, which may also contribute to measurement error. Since information systems in establishments are intended to measure and often control the behavior of individuals within the organization, each system affects the behavior of every contributor, depending in part upon how the contributor feels the system's outputs will affect him or her. Lawler and Rhode (1976) identify a number of aspects of systems and the kinds of dysfunctional behavior that can be induced, including rigid bureaucratic responses, gamesmanship, and deliberate distortion or withholding of information. This phenomenon is well known in the survey methods literature, as well as in social psychology and other disciplines, and is incorporated in the "response" step of the response process model. However, it can be a source of error in the record formation process as well.

All establishments contacted for the OSHS pilot study research

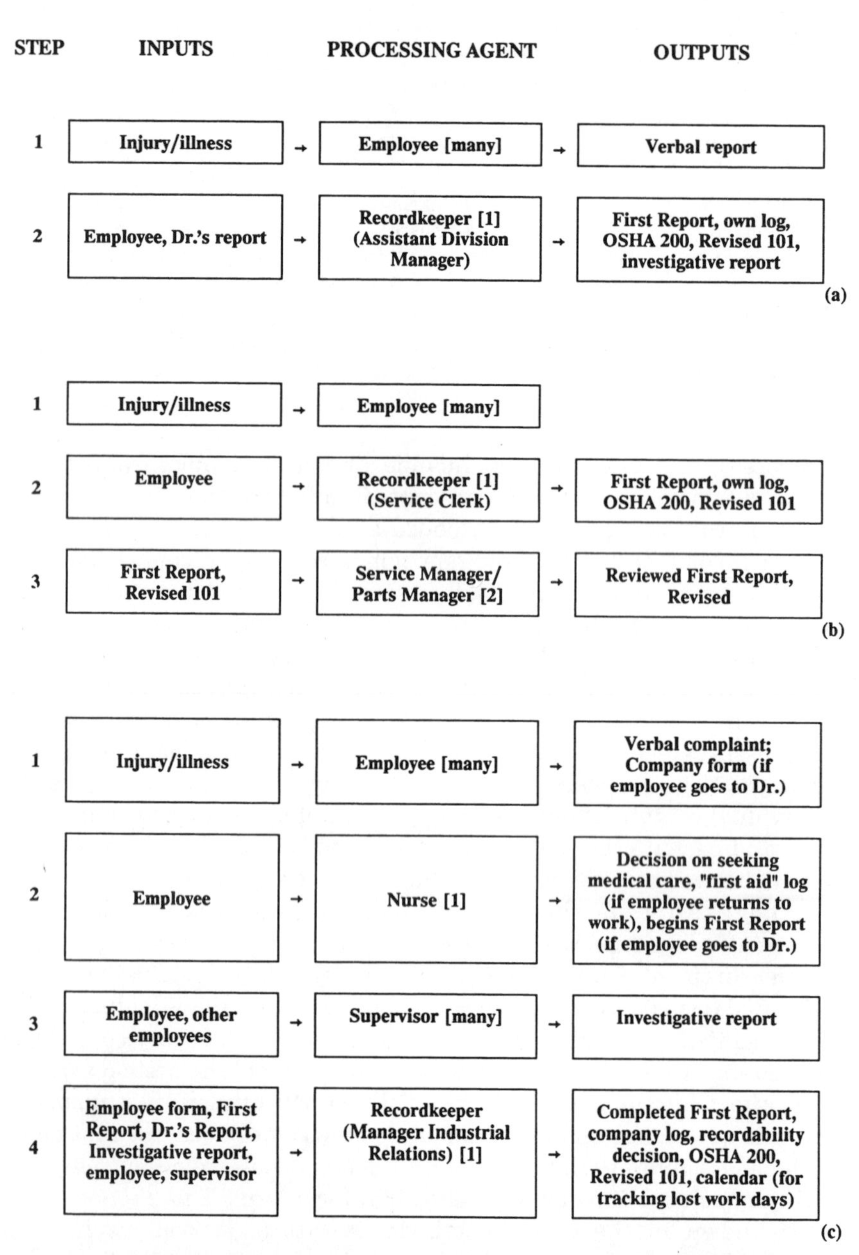

Figure 12.2 Sample flow models of record-keeping systems from interviews with OSHA record-keepers. a: typical system in small establishment. Single record-keeper completes all required forms. b: single record-keeper has output reviewed by managers. c: record-keeping system of medium-sized manufacturing firm. Number in brackets = how many people process information.

combined OSHA record-keeping with that required for Worker's Compensation. In all of the states visited, the First Report of Injury (a state or insurance carrier form for documenting work-related injuries) was an acceptable substitute for one of the OSHA forms, the Supplemental Record. Figure 12.2 presents three diagrams as examples of systems used to record injuries and illnesses observed in the OSHS research. The number and form of inputs, processing agents, and, to some extent, outputs vary across establishments. Figure 12.2a presents a typical system from a small establishment, also common in larger ones that experience relatively few work-related injuries and illnesses: all employees injured on the job go to the Assistant Division Manager, who is responsible for coordinating medical care and completing all required forms. Figure 12.2b presents a variation, in which the single record-keeper has his or her output reviewed by managers — a feedback loop. In most of the establishments interviewed, however, no such feedback loop was part of the system.

Larger establishments tend to have more complicated systems, with more processing agents. Figure 12.2c represents the record-keeping system of a medium-sized manufacturing firm. All injured or ill employees are sent to the company nurse. If the employee is then sent to the doctor, both the employee and the nurse complete forms. The employee's supervisor is also responsible for completing a report on the incident. The OSHA record-keeper, the manager of industrial relations, then completes the OSHA and state forms, using the written input from the employee, nurse, and supervisor.

The number and sequence of processing steps by each agent varied considerably. The variations were due to procedures imposed by the establishment, by parent companies, by insurance carriers, and by the record-keepers themselves. Generally, the more complex systems in terms of the number of processing agents also had more processing steps for which the record-keeper (or record-keepers — one of the 24 establishments divided the record-keeping tasks between two employees) was responsible.

Because the more complex systems have more processing steps and the record-keeper often relies on secondary information to complete OSHA forms, one might expect such systems to be more subject to distortion errors (using the systems term). However, in the simpler systems injuries are relatively infrequent and record-keepers have many other duties. The relative lack of experience with the forms in such systems led to far more visible distortion errors (we did not validate reports of injuries and illnesses) found in our review of the forms completed by OSHA pilot test record-keepers.

In all the establishments we visited, the outputs included both OSHA and state forms, and in some cases also included insurance forms or company-specific forms, usually for the purpose of safety investigation. While the OSHA regulations acknowledge the overlap of OSHA and state requirements (they allow substitution of a state First Report of Injury for the OSHA Supplemental Record, which gives details of each injury or illness), the specifications for the two forms are different. This difference led to the most significant source of error we observed — a kind of "selectivity error" in systems terms — recording all injuries on OSHA forms for which a state First Report of Injury was completed. In fact, according to the OSHA definition, many of these injuries should have been classified as "first aid" and not recorded on the OSHA Log. Other research, which has validated reported work-related injuries and illnesses and/or deaths, has found both under- and over-reporting of injuries and illnesses (Eisenberg and McDonald, 1988) and considerable under-reporting of deaths from work-related causes (National Research Council, 1987).

The OSHS research supports the hypothesis that the record formation process is potentially an important source of measurement error for establishment surveys. To the extent that the survey respondent is involved in record formation, as was always the case in the OSHS research, the respondent is one source. The establishment, in defining the purpose of the system and imposing procedures on it, is also a source of error. Finally, sources external to the establishment — in the OSHS research, state worker's compensation laws, insurance companies, and parent organizations — may also affect record formation and the potential for measurement error in an establishment survey.

Respondent Selection

An extensive literature on respondent selection focuses on differences between self-reports and proxy reports (Moore, 1988). Very little of this literature goes on to examine differences between different categories of proxy respondents. A validation study of reports of hospitalizations (Cannell, 1965) demonstrates some differences in agreement between proxy interview reports and record checks, depending upon the relationship of the proxy to the subject.

The most important consideration in respondent selection for household surveys of behavior is *knowledge* — does the selected respondent have access to the information requested? Typically, this "access" is through memory or some informed estimation process. If a respondent never encoded requested information (or relevant information for making an estimate), the remaining response processes become irrelevant. Some household surveys for which one person serves as household informant incorporate procedures to identify the "most knowledgeable" person in the household about the survey's subject matter. The "most knowledgeable" person is often the one who fulfills a particular role — the main meal preparer, for example, for a survey about food safety.

Functional role in establishment. Role specialization is obviously widespread in establishments. Thus, identification of the "most knowledgeable" respondent for an establishment survey is particularly important in limiting response error. The respondent should logically be someone involved in the relevant function: someone involved in administering employee benefits for a survey asking about health insurance coverage of employees; an accounting staff member for a survey asking for financial information. Such a respondent would be likely to understand the survey items (comprehension) and have knowledge of or access to records concerning the information requested (information retrieval). This is a common-sense notion incorporated into most establishment surveys, even if in no other way than how the envelope containing the questionnaire is addressed: "Personnel Manager, XYZ Organization, etc."

Functional role and level of authority. Who actually completes the questionnaire within an establishment is partially under the control of the survey, but is ultimately decided within the establishment. Because establishments are organized differently and assign different priorities to particular activities, the levels and functional roles of respondents to an establishment survey may vary widely. In 15 of the 24 establishments selected for the OSHS research, the primary record-keeper was a manager or business professional. The remaining nine were clerical staff. One of the managers and one of the clerical staff were "line" employees, that is, their non-OSHA responsibilities related to the establishment's manufacturing or service function. Nine of the managers and six of the clerical staff were administrative (personnel, human resources, safety), while five of the managers and two of the clerical staff had both line and administrative functions. All of the record-keepers with line responsibilities were in establishments with fewer than 100 employees.

A similar distribution was found among respondents to a survey of employer-sponsored health benefits conducted by Westat for the Health Insurance Association of America (HIAA). The job titles of 1,143 respondents were coded according to functional role and level. Overall, about two-thirds of the respondents were supervisory or management and one-third technical or clerical; about three-fifths were in personnel functions, including benefits, insurance claims, human resources, and so on, while two-fifths were in nonpersonnel roles, including accounting, general management, and service or production roles.

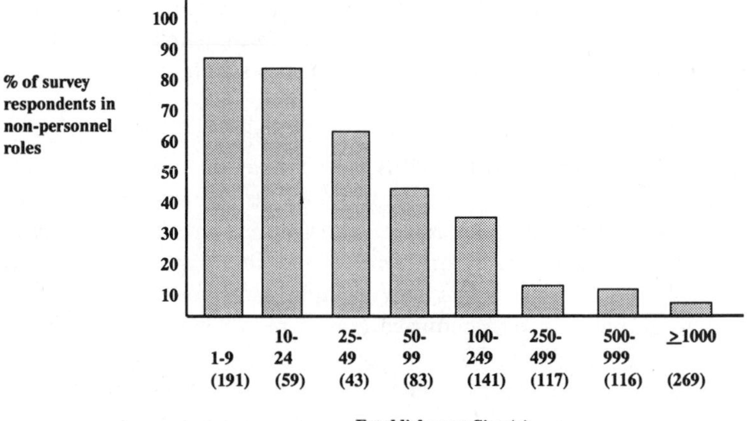

Figure 12.3 Proportion of survey respondents in nonpersonnel roles by establishment size. Source: HIAA 1990 Profile of Employer-sponsored Health Benefits.

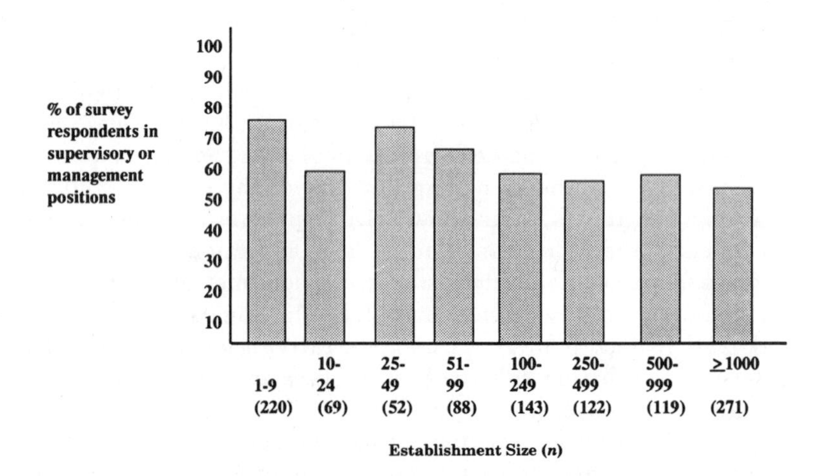

Figure 12.4 Proportion of survey respondents in supervisory or management positions. Source, HIAA 1990 Profile of Employer-sponsored Health Benefits.

Figure 12.3 depicts the proportion of respondents to the HIAA survey in nonpersonnel roles by establishment size. This distribution is consistent with the intuitive expectation that smaller organizations have less role specialization. The distribution of organizational level of respondent by establishment size (Figure 12.4), on the other hand, is fairly uniform across establishment size categories, with the exception of establishments of fewer than ten employees. This distribution suggests that the significance accorded the survey or its subject by responding establishments was not particularly related to the size of the establishment.

Position relative to information system. Another potentially relevant classification of establishment survey respondents is their relationship to the information system(s) serving as the source of data. Lawler and Rhode (1976) identify three organizational positions with respect to information or control systems: persons who are measured and controlled; persons responsible for system maintenance; and decision makers using the system's output. A given individual may occupy more than one position relative to a particular system (or may have no relation to a particular system). Among OSHS record-keepers interviewed, all were responsible for system maintenance, and all presumably would have been subject to measurement had they been injured (one record-keeper interviewed in fact had been injured during the pilot period). Generally, managers who kept the OSHA records were also in some sense decision makers with regard to the data in the system. Many had formal safety review responsibilities, either self-defined or imposed by the establishment.

Individuals typically occupy different positions in the Lawler and Rhode schema with regard to different systems. The OSHA forms require record-keepers to enter the number of work loss days or restricted activity days associated with each injury or illness. In many of the establishments interviewed (generally the larger ones), the OSHA record-keeper was not responsible for the system necessary for this tracking. For such respondents, obtaining this information was quite burdensome, particularly since it required return trips to the forms at some interval after most of the information had been entered. In other establishment surveys that request information from different systems, more than one respondent may be required to complete the survey form(s). The identification and enlistment of multiple respondents may be an ad hoc process or may be planned as part of routine survey procedures.

Summary. Respondent selection certainly should be guided by potential respondents' knowledge of survey subject matter. Other factors that might be considered include the respondent's functional role in the establishment, his or her level of authority, and his or her relationship to the information system(s) from which answers to survey questions will be obtained; still other factors may be relevant as well.

The factors to be considered will depend in part on the kinds of cognitive demands the survey will make on the person responding and the establishment, as described in the following sections.

12.3.3 Comprehension

This step in the response process model encompasses the respondent's interpretation of the intent of a survey question. If the respondent's interpretation matches the survey designer's intent, no error is introduced in this process. If the interpretation does not match the intent, "specification error" (Federal Committee on Statistical Methodology, 1988) may result. In an establishment survey, the respondent is responsible for interpreting the survey question or item, just as in a household survey. The respondent's interpretation may be affected by other factors in the establishment model, such as trade jargon (environment) or a particular use of a term in the establishment itself. One specification problem discovered in the OSHS research was that the state First Reports of Injury defined one item slightly differently from the OSHA forms.

A long and extensive literature on preparing survey questions (e.g., Payne, 1951; Sudman and Bradburn, 1982) sets forth guidelines for avoiding specification error. It may be that, on average, establishment survey respondents have better developed skills for comprehending survey items than household respondents, since many routinely deal with forms and requests for information in their jobs. Such skills might, for example, include the ability to read and understand complex definitions of terms used in survey questions. It also seems reasonable to hypothesize that the ability of establishment survey respondents to comprehend questions may be related to their levels of authority and to the salience of the survey topic to their roles in the establishment.

12.3.4 Information Retrieval

Once the survey respondent has comprehended a question, he or she must make a judgment about how to go about answering the question. This judgment may be influenced by the interviewer or by an instruction in the questionnaire. In household surveys, the direction of this step is usually taken for granted: the respondent will attempt to recall the answer. In establishment surveys, the direction is often less clear.

Retrieval of information from memory requires effort and may be assisted by memory cues or suggested retrieval strategies. "Retrieval" from an information system, which we will call "record look-up," also requires effort — usually physical effort to locate and mental effort to identify the information being requested. We make the assumption that greater effort required in retrieval results in greater likelihood of measurement error. An inappropriate retrieval process may lead to measurement error, as may a respondent's lack of motivation to perform the required effort. Krosnick and Alwin (1987) suggest that survey respondents are "cognitive misers," making minimal effort to come up with an acceptable response. They invoke Simon's (1957) principle of "satisficing," a concept often employed in organizational behavior theory.

The Record Look-up Process

Respondent's relationship to system. As we have noted, citing Lawler and Rhode (1976), establishment survey respondents have differing relationships to the systems of records they may need to access. If the respondent must request information from or through someone else in the organization, this additional step presents another chance for introducing measurement error. The respondent must interpret the survey for the person who maintains the information and rely on that person's look-up skills. Thus, the complexities of interpersonal and interdepartmental relationships in the establishment may affect measurement error in the record look-up process.

Timing of record look-up. The timing of record look-up activities may also be related to their relative ease and accuracy. For example, billing data from some medical providers is optimally available after posting and before moving paper records to microfilm or storage. Thus, it would be possible to request the information too early or too late. Surveys may also request different pieces of data that are optimally available at different times. For example, the OSHA regulations require forms to be completed within one week of a work-related injury or illness, and also reporting of the number of work loss or restricted activity days resulting from an injury or illness. The latter information may not be available to a record-keeper until a payroll is run; employees may also be away from work for a considerable period of time after an injury or illness occurs. Thus, OSHA record-keepers must often return to forms to complete them.

Rehearsal of record look-up process. Regardless of who actually looks up the information, one element that appears to cross the boundaries of respondents reporting from recall and from records is the notion of rehearsal. Alba and Hasher (1983) identify "differences in the number and timing of rehearsals" to be an important factor in the accuracy of recall. Knight and McDaniel (1979) identify two important (and related) characteristics of requests for information from systems, the frequency of the request (is it routine?) and whether the request is continuous or noncontinuous. Using this classification, Knight and McDaniel suggest that routine, continuous requests are less subject to error than unusual, one-time or sporadic requests. They also suggest that requests for information will be treated as routine (whether appropriately or not) unless there is some extraordinary aspect to the request, again bringing in the notion of "satisficing" in the survey response process.

In the OSHS, the information request is routine and continuous for many ("nonprenotified") establishments, but neither routine nor continuous for "prenotified" establishments, i.e., those selected for the Annual Survey who are otherwise not required to maintain OSHA records. For some nonprenotified establishments, the OSHS request may not be routine for some period of time because of personnel, ownership, or other changes. An OSHS study of establishments in Massachusetts and Missouri (Eisenberg and McDonald, 1988) found that many establishments cited for record-keeping violations offered an excuse related to recent losses of staff or organizational restructuring. Review of completed OSHS pilot test forms also indicated that establishments with few injuries or illnesses in the reporting period were more likely to make errors on the test forms than those with many incidents.

Respondent Motivation

Although included in the section on record look-up, respondent motivation is relevant to each of the processes of survey response — motivation is required for comprehension of complex questions, for the effort of making judgments about responses, and for overcoming tendencies to edit responses for social desirability or other reasons. Discussing a model of survey response assumes some level of motivation, since the respondent has at least agreed to participate. Traditionally, survey design features to motivate respondents have focused on persuading them to participate. More recently, work by Cannell et al. (1981) has focused on survey design and interviewer behavior to motivate respondents during the interview.

In establishment surveys, the decision to participate may be made by an individual, or more than one individual, representing the establishment. While altruistic or personal motives for participation may be present, it seems likely that the decision to participate often involves some calculation of the potential benefit to the establishment against the cost and potential risk of completing the survey. For example, some of the establishments interviewed in the OSHS research used the OSHA record-keeping forms for their own investigation of accidents and injuries, with the goal of improved job safety. Once the decision to participate is made, the decision-maker(s) may then delegate the actual completion of the survey form to another member of the establishment. Thus, even in voluntary surveys, the actual respondent may or may not be completing the survey out of his or her own choice. It seems reasonable to assume that this is more often true in larger establishments, which generally have more levels of authority.

The OSHA record-keeping is unusual for an establishment survey, in that establishments are required to maintain the forms and respond to the Annual Survey. "Motivation" is encouraged by OSHA inspections, which can lead to fines for failure to maintain accurate records. Record-keepers we interviewed cited other motivating factors: insurance companies would only reimburse establishments for medical care provided to workers injured on the job if the appropriate forms were completed, and establishments were protecting themselves against the potential for lawsuits brought by injured employees. All of these may be considered "environmental" factors affecting motivation.

12.3.5 Judgment

Because information recalled from memories or looked up in records often does not fall neatly into the proper answer slots in a survey questionnaire, respondents must make judgments about the information to formulate a response. We suggest that information source is largely irrelevant to classifying this judgment step in the response process model; that is, the cognitive task is the same for household and establishment surveys.

We have already noted the observed tendency of OSHA record-keepers to equate injuries and illnesses that are recordable by OSHA guidelines with those that are reportable by state workers' compensation laws, when in fact the "OSHA recordable" injuries and illnesses are a subset of the "state reportable" ones. In order to record only the appropriate injuries and illnesses, the OSHA record-keeper must understand the BLS guidelines and apply them to the injuries and illnesses on state First Reports of Injury. The second step is one of judgment.

Among the 24 record-keepers interviewed for the OSHS research, only seven (29 percent) clearly knew and applied the BLS guidelines. The remaining 17 record-keepers applied some inappropriate rule to decide whether to record an injury or illness on the OSHA forms. These rules were all under-selective; that is, they would potentially result in the recording of more injuries and illnesses than would be appropriate.

All seven of the record-keepers correctly applying the BLS guidelines for recordability were managers and were record-keepers in "nonprenotified" establishments (that is, establishments required to maintain OSHA records continuously). Five of the eight record-keeping managers not correctly applying the guidelines were in "prenotified" establishments (whose first experience with OSHA forms was the pilot test). Although all record-keepers could not be classified according to their relationship to the system (the Lawler and Rhode taxonomy), the seven making correct judgments generally used the system to make decisions (about safety procedures, etc.), while those making incorrect judgments were generally only system maintainers.

The relatively better performance of managers as compared with clerical staff, of nonprenotified as compared with prenotified, and of decision-makers as compared with system maintainers all make intuitive sense and also hold out some hope for the ability of establishment survey respondents to perform difficult judgments.

12.3.6 Communication

As defined in the response process model in Chapter 8, communication of a response retrieved from memory and judged appropriate to the question consists of (1) filtering the response according to personal standards, societal norms, or other characteristics and of (2) providing the answer to survey administrators. As with the judgment step, we suggest that communication is essentially the same cognitive process in an establishment survey, although the influences on it may be somewhat different in the establishment setting.

The notion of "social desirability" is often cited as a reason that respondents adjust responses to survey questions (e.g., Sudman and Bradburn, 1974). To enhance one's image in the eyes of an interviewer or unseen reviewer of a self-administered questionnaire, or perhaps as a form of self-deception, survey respondents often over-report behaviors perceived as socially desirable such as voting and under-report behaviors perceived as socially undesirable such as drinking alcoholic beverages. When a survey respondent is reporting the behavior of an establishment with which he or she has some affiliation, a similar tendency may cause him or her to respond so as to present the establishment in a more favorable light.

Beyond the respondent's own perceptions of how survey responses will make the establishment appear, the establishment may have explicit or implicit policies on the divulging of information about itself. These policies may be aimed toward enhancing public image, protecting competitive advantages in the marketplace, concealing illegal or unethical behavior, or influencing the opinions of financial markets, for example. This hypothesis is consistent with the "open system" concept of Katz and Kahn described earlier — the respondent's behavior is influenced by the establishment's policies, which are in turn influenced by factors in the establishment's environment.

Deliberate under-reporting of work-related injuries and illnesses is apparently widespread, although our research (not surprisingly) did not identify any such instances and under-reporting cited by OSHA inspectors tends to be explained by establishments as due to unintentional oversight. Despite the potential for inspection and fines, the incentives for under-reporting are clear: OSHA targets industries and individual establishments for inspection in part based upon the number of injuries and illnesses reported, and establishments may fear lawsuits by injured employees or the potential of expensive alterations to procedures or physical plants for safety improvements. A similar situation, for example, exists with regard to the police-reported crime rate (e.g., U.S. Department of Justice, 1989), which is collected by individual police departments reporting the number of offenses in their respective jurisdictions. A number of studies have found individual jurisdictions, or even precincts within jurisdictions, that substantially under-report crimes.

One factor that may influence the communication of survey responses is the presence of quality control procedures within an information system or specifically with regard to a survey. One would expect that the more individuals involved in preparing or reviewing the information, the less likely any particular individual would be to alter it. Of course, a quality control process within an establishment could be guided by explicit or implicit establishment policies that would result in agreement on alterations in information.

12.4 DISCUSSION

In terms of the response process, the major difference between establishment surveys and household surveys appears to be the establishment survey respondent's use of information systems instead of recall to answer the survey questions. However, establishment survey respondents often use recall, and household survey respondents may use their own information systems, either as recall aids or to supply information directly. The difference is one of degree. It thus seems reasonable that a response process model developed for household surveys would also be applicable to establishment surveys, at least when recall is used, and that, conversely, a response process model developed for establishment surveys would be applicable to household surveys when records are used.

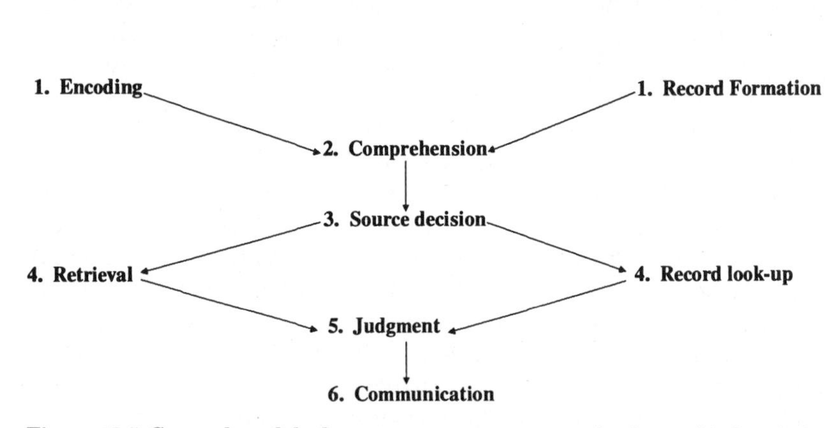

Figure 12.5 General model of survey response process for factual information.

We have examined the applicability of a cognitive model of the response process developed for household surveys to establishment surveys, suggesting that the cognitive activities of comprehension, judgment, and communication are directly applicable. We have also suggested that respondents who use information systems employ steps analogous to the cognitive activities of encoding and retrieval, and have called them record formation and record look-up. Finally, we have suggested that all survey respondents perform one additional cognitive activity in the response process: making a decision about the source of information to answer a survey question. This set of suggestions is summarized in Figure 12.5, a comprehensive survey response process model applicable to respondents using either recall or information systems.

Beyond the response process itself, the model we have presented includes some of the important factors influencing process. As described in Section 12.3, the establishment survey is subject to some influences that are quite different from those affecting a household survey. Perhaps the most important influence on the response process, and consequently on the potential for response error, is the characteristics of the respondent, in particular his or her relationship to the information system(s) used for answering survey questions. Other important influences seem to include the establishment's other uses for the information system, the system's specifications, establishment policies about release of information, and environmental influences relevant to the survey's content.

These influences are relevant to questionnaire design and development of survey procedures. For example, how well does a question match with customary specifications for the system(s) that are expected to supply answers? It may be necessary to include questions about system specifications to interpret responses to some questions. For a simple example, the Current Employment Statistics Survey asks for total payroll in dollars, and also asks for the payroll time period. It may also be helpful to ask questions about system specifications to help evaluate measurement error in less straightforward cases.

As another example, consider respondent selection procedures. Depending upon the survey content and the level of judgment the respondent is expected to exercise, the respondent's relationship to information systems needed and possibly the respondent's level in the establishment may be important in limiting measurement error. If survey procedures do not ensure that the proper respondent is selected (and they almost never can), it may be helpful to include questions about the respondent, such as his or her position in the establishment and relationship to the information systems used, to help evaluate measurement error.

As we noted in the introduction, the methodological literature on establishment surveys is very limited. The model we have presented suggests a number of potentially interesting hypotheses for methodological research:

To what extent do respondents to an establishment survey use recall or information systems? What are the implications for measurement error in the survey as a whole?

What is the distribution of respondent's level within the establishments and relationship to information systems used to answer survey questions? What are the implications for measurement error?

What is the correlation between response effects and independent variables of interest to survey analysts, such as establishment size? What about dependent variables?

Issues such as these can be informed by the disciplines of survey methods, information systems, and management science. Establishment survey data provide a wide array of statistics of interest to government and the business community. Continued methodological research is needed to evaluate and improve the quality of this data.

SECTION C

INTERVIEWERS AND OTHER MEANS OF DATA COLLECTION

CHAPTER 13

DATA COLLECTION METHODS AND MEASUREMENT ERROR: AN OVERVIEW

Lars Lyberg
Statistics Sweden

Daniel Kasprzyk
National Center for Education Statistics

13.1 INTRODUCTION

During the last twenty years, the collection of information in surveys has undergone a transformation in the means by which data are gathered. Mail surveys and face to face interviews provided the primary mechanisms for collecting data during the 1940–1970 period. The increased availability of and access to telephones, improved sampling methods through random digit dialing, and improved technology have resulted in a substantial increase in the use of the telephone as the means by which information is gathered in sample surveys. In countries where telephone coverage is extensive, telephone interviewing has become the dominant mode of data collection, at least in some government agencies and large survey research firms. However, in 1987 only eleven countries had telephone coverage rates exceeding 90 percent (Trewin and Lee, 1988). If it were possible to determine a worldwide dominant data collection mode, we believe it would be face to face interviewing.

The increased availability of the telephone has, it seems, stimulated the discussion of the relative advantages and disadvantages of the various ways in which data are collected and has contributed to the observation that the selection of the data collection mode is a complex decision related to the goals of the individual study. A number of factors enter into the decision process. These factors include the expected quality of the data, funds available for the study, the questionnaire

Measurement Errors in Surveys.
Edited by Biemer, Groves, Lyberg, Mathiowetz, and Sudman.

ISBN: 0-471-53405-6

content, the population under study, and the administrative and staff resources available to the survey organization. Knowledge of the methodological goals of the study and consideration of the various factors affecting the choice of the mode are essential in deciding which of the following most common methods of data collection is most appropriate: face to face interviewing; telephone interviewing (centralized or decentralized); self-administered methods (for example, mail and diary surveys); direct observation; administrative records; various combinations of methods.

Design decisions related to the execution of the data collection operation are associated with each method of collecting data. These decisions, usually determined by the researcher, include sample coverage, questionnaire content and length, respondent rules, specification of callback and refusal conversion rules, timeliness of results, extent of interviewer training, etc. Often decisions are easily made because the alternatives are unrealistic or not practical for a particular study. For example, the literature widely recognizes the use of a diary method to collect many types of household expenditure data in a developed country. As a consequence, alternatives to this data collection method are not seriously considered. Similarly, data on crop yields are collected using direct observation. In these cases, discussions concerning the mode focus on other issues such as the period the diary is to cover (for e.g., one week or two weeks) or whether the direct observation should come from a combination of two or more sources.

In cases where alternatives exist, the choice between modes is usually guided by factors such as the estimated data collection cost, predicted nonresponse rates, expected level of measurement error, and length of the data collection period. These are important issues in the selection of mode; however, in this chapter the emphasis of the discussion on alternative modes is solely on measurement error issues.

Thus, when alternatives exist the choice of mode is a complex decision. This is especially true when measurement errors are exceptionally important. Cost estimates can usually be made and nonresponse rates can often be predicted relatively accurately. The marginal measurement error mode effect is, however, not easily established. The only way of quantifying the true mode effect on measurement errors is to conduct mode comparison experiments controlling on all other design parameters. Since each mode has its own set of design constraints — for example, the number of allowable callbacks for a telephone or mail survey is likely to be much larger than for a face to face interview — the experimental comparisons are likely to be confounded. This difficulty leads to analysis of differences in results generated by alternative modes with realistic design parameters. Biemer (1988) and Groves (1989) provide useful reviews of the design issues associated with mode comparison studies.

In general, the literature on mode comparisons has not shown large differences in estimates between modes. This may just mean, however, that the sum of measurement errors associated with one mode is similar to that of the other mode. The common situation is that the choice of mode depends on a number of factors, some of which we have identified above, where measurement errors is just one. Given a mode with its specific set of design parameters, we may identify factors affecting measurement errors and in some cases estimate these errors. The issue of mode and measurement errors then is more concerned with errors associated with each mode and how they can be corrected or eliminated rather than with a choice between alternative modes.

The section of this volume on mode and measurement error contains contributions on interviewer errors (Chapters 14 and 15), measurement errors in diary surveys (Chapter 16), measurement errors associated with direct observation in agricultural surveys (Chapter 17), and the use of electronic data collection devices in retail surveys conducted by the A. C. Neilsen Company (Chapter 18).

This review chapter provides a discussion of a number of different methods of data collection, the sources of measurement error associated with each one, and the difficulties associated with mode comparison experiments (Sections 13.2 and 13.3). Special attention will be given to the nature and effects of interviewer errors, as well as the control of interviewer errors (Section 13.4). We also provide a discussion of recent developments in data collection technology (Section 13.5).

13.2 THE MEANS OF DATA COLLECTION IN SURVEYS

This section reviews a number of different methods of collecting data in surveys and the sources of error associated with the methods. Books have been written about these methods; consequently, our task in this section is only to provide a brief overview.

13.2.1 Face to Face Interviewing

Face to face interviewing is the mode in which an interviewer administers a structured questionnaire to a respondent within a limited time period and in the presence of the respondent. Face to face interviewing has been a popular, demanding, and expensive method of data collection. Research during the last decade has challenged the once strongly held view that face to face interviewing obtains the best quality data. Face to face interviewing, however, provides a data collection environment that allows a wide degree of flexibility. The flexibility allows a potentially longer, more complex interview to be conducted. The interviewer, by virtue of his or her presence, may motivate and encourage respondents to answer the survey questions as well as probe for more complete and accurate responses. The "personal" aspect of this data collection method can be conducive to establishing trust and building rapport to complete the response task successfully. Face to face interviewing also allows the use of a wide variety of visual aids to help the respondent answer the questions.

A risk exists, however, that the quality of data obtained in face to face interviews can be affected by the presence of members of the household who are not part of the study. This is particularly true when the topic of the survey is viewed as sensitive by the respondent. Another risk, labeled *social desirability bias*, may exist in this data collection environment. DeMaio (1984) notes that social desirability is difficult to define, but it seems to share two elements — the notion that some things are good and others are bad, and the notion that respondents want to appear "good" and answer questions to appear that way. Thus measurement error may occur because of the reluctance of respondents to report socially undesirable traits or acts. De Leeuw and van der Zouwen (1988) reviewed a number of studies and in their meta-analysis discuss social desirability in the telephone and face to face interview settings.

The biggest difficulty with face to face interviewing and measurement error is the interviewers themselves. Interviewers affect respondents' answers in a way similar to the clustering effect in cluster sampling. Interviewers affect responses through their individual performance patterns when conducting interviews. They each ask the questions in their own style and at their own pace; each may have his or her own way of recording the information, particularly with regard to open-ended questions. Each may probe and elicit responses differently. Of course, each interviewer may interact and engage the respondents in the response tasks in a wide variety of ways. Differences in the way the interviewers do their jobs lead to situations where some interviewers appear to obtain different answers compared to others.

Interviewer effects or the extent to which interviewers influence data occur because of a complex combination of personality and behavioral traits, which manifest themselves in an interview setting. The interviewer effect increases the total variances of the statistics under study. Most interview surveys use estimation methods that do not take the interviewer effect into account; the true variances of statistics obtained in these surveys will be underestimated. There are two ways of addressing this source of error — through reduction and measurement. Reduction of interviewer variance can occur through a program to "standardize" interviewers' performances so that they complete their tasks in a uniform way. This is accomplished through a training, observation, and monitoring program. Fowler and Mangione (1990) discuss standardized survey interviewing and methods to minimize interviewer-related error. Fowler's contribution to this volume, Chapter 14, also focuses the discussion on reducing interviewer-related error through training, supervision, questionnaire design, and assignment size. Measurement of interviewer effects can occur through sample designs that allow for the calculation of the variance attributed to interviewers. Groves (1989) reports on the magnitude of interviewer effects across a number of studies. Other examples of studies studying the size of interviewer effects include Kish (1962), Bailey, et al. (1978), the U.S. Bureau of the Census (1979), and Groves and Magilavy (1986).

13.2.2 Telephone Interviewing

Since the interviewer, the medium by which information is transmitted from the respondent to the questionnaire, plays a central role in both face to face and telephone interviewing, it is not surprising to find that many of the same problems exist in both methods of collecting data; in particular, the most disturbing component of measurement error in both methods is the interviewer variance component. Telephone interviewing provides less flexibility than face to face interviewing. Visual aids cannot be used during the interview unless they are mailed to the respondent in advance. While this procedure has been implemented successfully (Bradburn, 1983), it is not typically done. Complicated and open-ended questions are more difficult to administer over the telephone and many consider lengthy telephone interviews to be impractical. Soliciting information on topics requiring the use of records and/or requiring respondent calculations may be difficult and awkward in a telephone interview setting. The anonymity of the interviewer may improve reporting on sensitive topics; however, it is not advantageous for developing rapport and increasing the motivation and interest of the respondent.

Surveys which use telephone interviewing are less expensive and often proceed more rapidly than face to face interviewing. These surveys can often be implemented rapidly with a smaller staff than face to face interviewing. If interviewing is centralized, supervision and quality control can be improved through direct observation and monitoring of the assignment. Interviewer effects can be considerable as the individual workload size will increase substantially in a telephone interview setting. A discussion of a variety of issues related to telephone interview surveys can be found in Dillman (1978), Groves, et al. (1988), and Groves (1989).

13.2.3 Self-Administered Mail Surveys

Self-administered mail surveys offer a less flexible method of data collection when compared to telephone or face to face interviewing. A mail questionnaire is sent to a sample respondent and introduced by means of an introductory letter. The data collection organization has no control over the response process. Consequently, contamination of responses by individuals other than the intended respondent can easily occur. Mail questionnaires should be visually appealing to stimulate and motivate the interest of the respondent and should be neither of excessive length nor complexity. Especially good questionnaire design and formatting are essential since a wide variation exists in the reading and writing ability of the population. A large effort must be made to minimize the misreading and misinterpretation of questions by the respondent.

Self-administered mail surveys are often not considered as a means of data collection because of the traditionally high nonresponse rates associated with them. Dillman (1978, 1983), however, argues that despite limitations in a mail questionnaire it is possible to overcome historical limitations of this method and increase response rates to acceptable levels. In addition, mail surveys are usually underrated from the measurement error point of view. For example, all topics are possible to investigate, even sensitive topics. There is no interviewer effect, there is more time and room for considered answers, as well as the potential for asking questions with many response categories. Respondents can set the pace of the process, completing the form at a leisurely and considered pace. If properly motivated, respondents may also seek out records and documents to accurately complete the questionnaire. In fact, using meta-analysis on a wide variety of mode comparison studies, De Leeuw (1990a) concluded that mail surveys performed best regarding the measurement error criterion when compared with face to face and telephone interviewing; these differences were small, however.

The absence of an interviewer with this data collection method should reduce the risk of social desirability effects. Dillman (1978), reviewing several studies, concluded that mail surveys have lowest probability of producing socially desirable responses when compared with telephone and face to face interviewing. Order effects may also be reduced because there are reduced serial-order constraints. Ayidiya and

McClendon (1990) show that response order and question order effects are more or less eliminated in mail surveys.

13.2.4 Diary Surveys

Diary surveys are generally self-administered forms on topics difficult to study using other modes of data collection. The structured questionnaire is replaced by a diary where respondents enter information about frequent events such as expenditures, time use, and television viewing. To avoid recall errors the respondents are asked to record information soon after the events have occurred. Often this means that respondents should record information on a daily basis and sometimes even more often than that. Thus, successful completion of a diary requires the respondent to take a very active role in recording information. Diary surveys differ from mail surveys because interviewers are usually needed to contact the respondent at least two times. At the first contact, the interviewer delivers the diary, gains the respondent's cooperation, and explains the data recording procedures. At the second contact, the interviewer collects the diary and, if it is not completed, assists the respondent in its completion. During the recording period other visits may be made by the interviewer to assist the respondent in completing the diary, to answer questions about recording procedures, and encourage cooperation in completing the task. One major limitation of diaries as a means of collecting data is the commitment required of the respondent to complete the task. Because of the need for a high level of commitment, diary reporting periods are fairly short, typically varying in length from one day to two weeks.

In theory, the main advantage of the diary method is its lack of reliance on memory in the data gathering process. If properly implemented, information about an event should be recorded soon after its occurrence. In practice, this does not always occur and respondents fail to record events or incidents when fresh in their memories. Other problems exist with measurement errors in diary surveys. First, the reporting or recording pattern may change over time. Research has shown that reporting levels start decreasing early during the reporting period (see Chapter 16). The short recording period implies that the diary method cannot be used for estimating the occurrence of rare events. Consequently, the diary method is often complemented with another form of data collection.

A second problem with diary surveys concerns the design of the diary itself. The structure and complexity of the diary can present significant practical difficulties for the respondent. The diary must be capable of completion by the general population, in which a wide range of educational levels are present as well as varying amounts of experience with completing forms.

A third problem with measurement errors in diary surveys is that respondents may change their behavior as a result of using a diary. It is easy to see how the detailed recording of purchases in a diary survey can influence responding individuals by reminding them of their purchasing behavior. Similarly, the recording of time use may alter respondents' behaviors.

A fourth problem in diary surveys is the insufficient attention given by the respondent to recording the activity or purchase in the diary. This may occur in several ways, resulting in underreporting of the behavior or activity being measured. For example, the recording of events in a diary may not occur in a regular or systematic way, say for instance at the time of purchase. This may result in either inadequate detail concerning the event being recorded or forgeting about the event altogether. In addition, it is difficult to obtain accurate record keeping for all individuals in a household, yet the use of one diary per household often occurs in practice. This also results in underreporting due to the use of proxy reporting.

A fifth problem is one that resembles a nonresponse problem. In some diary surveys, an initial nonrespondent, as a result of a conversion to a respondent, is allowed to report later in the survey period than initially planned. This leads to several kinds of biases, such as fewer respondents participating during the vacation period and more respondents than expected at the end of the survey period (Lyberg, 1980). The underlying assumption with this practice is that the reductions in nonresponse error outweigh any increase in measurement error.

Pearl (1968) provides a comprehensive review of expenditure survey methods, while Neter (1970) and Tucker and Bennett (1988) discuss measurement errors in expenditure surveys and, in particular, the diary survey aspect of such surveys. Chapter 16 provides a discussion and overview of measurement errors associated with the U.S. Consumer Expenditure Survey, analyzing performance characteristics such as the degree of diaries' completion and the variation of estimates of expenditures by day and week. Chapter 16 also provides a comparative analysis of expenditure estimates derived from two methods of data collection — the face to face interview and the diary method. Time-use studies often use a diary method approach; Lyberg (1989) discusses measurement issues in a Swedish time-use study and compares estimates obtained using two methods of data collection. Plewis, et al. (1990) also discuss measurement error in time-use studies and conclude that approaches to collecting time-use data that include time budget interviews, self-completion diaries or both are preferable to retrospective recall.

13.2.5 Administrative Records

Administrative records are records which contain information used in making decisions or determinations, or for taking actions affecting individual subjects of the records. The term usually refers to records about persons, although other entities, such as corporations, may be treated by law as legal "persons," about whom decisions and actions are taken (U.S. Office of Management and Budget, 1980). The actions include functions like licensing, insuring, taxing, regulating, charging, paying or conveying benefits or penalties.

In the statistical use of administrative records, the interest is in studying the characteristics and attributes of groups of individual entities as opposed to identifying individual entities and taking some action. The use of administrative records for statistical purposes has a long history as a means of gathering data. Early censuses had administrative purposes; the vital record systems in Europe date back to the seventeenth century. Recently, advances in computer science, computer technology, and record linkage methods along with progress in the systems used to maintain administrative records have made this method of collecting statistical data more attractive to statisticians.

Errors in data collected from administrative records are of the same type observed in other modes of data collection — namely, coverage, response, and processing errors. Issues of comparability, a concern of many data collections, also become evident in the statistical uses of administrative records. Quality problems specifically associated with administrative records include:

i. Rarely does the statistician have control over the kinds of data on the record; nor does the statistician have much to say about updating and coding as these areas are in the province of the keeper of the record system.

ii. Information important to statistical uses of administrative records, but only marginally useful for administrative purposes, may often be imperfectly recorded, processed, and checked.

iii. The administrative purpose of the record system is the reason the system exists; consequently, the production of statistical data will always be of secondary importance. Thus, the availability of timely data will always be a problem that must be addressed.

iv. Conceptual differences usually exist between the statistical application and the administrative use.

v. Coverage problems may exist depending on the administrative source.

vi. The quality of data in administrative systems is difficult to assess.

Using administrative records to produce register-based statistics is very appealing. The records and data exist and are relatively inexpensive to use. Respondent burden is nonexistent if data can be collected without contacting the respondent. Register-based censuses have been proposed repeatedly by most Nordic countries; in fact, Finland and Denmark now have register-based censuses. Administrative records are more commonly used to provide small area data, to augment data collected in a survey, to generate information for the estimation process, and as a basis for building statistical registers.

Trade-offs between reduced data collection costs and decreased respondent burden and the loss of control by the statistician of the measurement method are serious matters that have emerged as the main theme for research on the statistical uses of administrative records (Kilss and Alvey, 1984; Redfern, 1986; Leyes, 1991).

13.2.6 Direct Observation

Direct observation as a systematic method of data collection is the recording of measurements and events using an observer's senses (vision, hearing, taste) or physical measurement devices. It differs from other methods of data collection in that data are collected directly rather than through the reports of a respondent; the observer or measurement device, in essence, becomes the respondent. Some examples of applications include: auditing of stocks in market research, providing visual estimates of crop yield, or the counting of the number of drivers not using seat belts. Direct observation through the use of measurement devices is seen in studies requiring the reading and interpreting of aerial photographs, scales in agricultural surveys, or the recording of human activity patterns.

Direct observation really constitutes several modes of data collection, and applications exist in many disciplines. In anthropology, for instance, participant observer studies are common. With this method the researcher lives with and joins in the daily activities of the subjects under study. In psychology, observers might collect data by observing their subjects through a one-way mirror. In market research, an observer may be replaced by an electronic measurement device to record television viewing, and in environmental studies pollution may be monitored and measured through specialized instrumentation.

Direct observation can have several advantages over methods of data collection that require the involvement of a respondent. If the information desired is likely to be unreliable because of either recall difficulties or practical difficulties in obtaining the cooperation of a respondent to do complex and arduous tasks, then direct observation may be desirable.

However, measurement errors may be introduced into the recording of observations by observers in ways similar to the errors introduced by interviewers. Observers may misunderstand concepts and misperceive the information to be recorded; observers may change their behaviors and patterns of recording information over time because of complacency or fatigue. It is just as difficult to standardize observer behavior as it is to standardize interviewer behavior. Observers, like interviewers, exhibit between- and within-observer variability. Bias and variability can be kept under control through training and supervision programs.

Measurement error associated with direct observation may also be related to the instrument or device used to gather information. In this situation the instrument may be providing biased readings because it has not been calibrated. Studies using instruments or devices require a continuing program to validate and recalibrate to reduce systematic error in estimates.

Large scale data collections using direct observation methods are found in most agricultural surveys. Chapter 17 discusses errors of direct observation in crop yield surveys. Data are collected in crop yield surveys through an observer's visual estimates, observations of weather events and growing conditions, and sample measurements of growing plants. Much of the work on nonsampling errors originated in agricultural surveys and the associated observational errors. Mahalanobis (1946) was prominent in modeling and measuring these kinds of errors; Sukhatme and Seth (1952) also provide a survey model for observational errors.

The control and measurement of errors of direct observation resembles that of other methods of data collection. Training programs, observer monitoring and observation programs are important to the standardization of procedures and the consequent reduction of measurement errors. Various forms of process control have begun to be used in agricultural surveys (Hanuschak, 1991).

13.2.7 Mixed Modes of Data Collection

The methods of data collection described above are the principal modes used in sample surveys. In theory, if a particular method of data collection is thought "best" to study a topic, one would expect its use throughout the study. In practice, however, few studies rely solely on one mode for data collection. Cost problems, incomplete coverage, nonresponse and measurement errors almost always result in the use of two or more modes for collecting data. Some common combinations of modes include face to face interviewing and telephone follow-up, mail and telephone follow-up, administrative records and mail follow-up, and diaries and face to face interviews.

Applications of mixed mode designs are of two types. One type is a mixed mode application using two or more sampling frames. This design strategy is useful in overcoming the incomplete coverage of random digit dialing (RDD) to study the entire population. An area sample with face to face interviewing is combined with an RDD sample and telephone interviews to improve population coverage. Designs of this kind are characterized by a prespecified sample allocation to the two frames based on information about errors and costs. Applications are provided by Casady, et al. (1981) and Groves and Lepkowski (1985).

A second type of mixed mode application can be viewed as a mixed mode strategy. This can occur in several ways. Typically, one main data collection method is chosen for a study, but to achieve acceptable response rates or reduce study costs a second or third mode is used. The primary mode is used to its maximum potential with certain constraints, such as a prespecified number of callbacks; then another mode is adopted for the principal purpose of increasing response rates. While this is a practical strategy, it is clearly a strategy of compromise. If a particular design and mode is most appropriate for the subject under study, then any deviation should result in data of less quality. For instance, some questions asked in a face to face interview may be inappropriate in a telephone interview. Similarly, sensitive questions asked in a mail survey may result in different response error distributions when asked in a telephone interview. Some shortcomings of this strategy can be addressed in the planning phase of the survey; however, oftentimes an ad hoc strategy with minimal control is adopted.

A second application of a mixed mode strategy may occur in the use of one mode to screen a population for a subpopulation with rare attributes, and the use of a second mode for individuals who agree to participate in the study; for example, a telephone sample may be appropriate to identify individuals with unusual characteristics, but the follow-up interview could be completed by means of a mail survey or face to face interviewing.

A third application of a mixed mode strategy frequently occurs in panel surveys, where the purpose of the strategy is usually to reduce costs. In the Current Population Survey, a nationally representative panel survey conducted by the U.S. Bureau of the Census, a mixture of face to face and telephone interviews is conducted (U.S. Bureau of the Census, 1978). The National Medical Care Expenditure Survey (NMCES) used telephone interviews in the third and fourth rounds of the panel and face to face interviews otherwise (Horvitz and Folsom, 1980). Mixed mode strategies in panel surveys try to take advantage of the best features of each mode. Face to face interviewing in the first round of a panel survey improves on the incomplete coverage associated with nontelephone households while simultaneously establishing a relationship with the household to maintain cooperation and response in future rounds of the panel (Kalton, et al., 1989).

13.3 MODE EFFECTS ON MEASUREMENT ERRORS

13.3.1 The Basic Problem

An unconfounded study comparing modes of data collection, that is a study comparing the marginal effect of the data collection mode, is difficult to design. The aim of such a design is to compare estimates between modes with equivalent samples and equivalent design parameters. While this may be a useful intellectual and empirical exercise, its significance to a survey practioner is limited. As mentioned earlier, associated with each mode is a set of decisions intended to take full advantage of the benefits of the mode. Thus, designing the study with equivalent design parameters must by definition result in one of the modes having design parameters with little practical significance.

Mode comparison studies should compare data collection systems developed for the two modes. Mode comparison studies typically use a set of evaluation criteria consisting of data quality and costs. For example, some criteria are unit and item response rates, completeness of reporting, similarity of response distributions, the absence of social desirability responses, and validity. In this situation we are not concerned with the marginal effects of the data collection method, but whether the two systems produce equivalent results given all the differences in data collection using two modes.

13.3.2 Mode Comparison Results

A relatively extensive literature exists on mode comparisons where outcomes are compared. A number of them have been studied in meta-analyses conducted by De Leeuw and Van der Zouwen (1988) and De

Leeuw (1990a). In the former, the authors provide evidence of small differences between telephone and face to face interviewing. This is especially true of sensitive or threatening questions. In the latter, the author reviews analyses of face to face, telephone, and mail surveys and concludes that the differences are, once again, small. Other examples of mode comparison studies are found in Groves and Kahn (1979), where few significant differences in univariate and bivariate response distributions were found when comparing telephone and face to face interviewing modes. The U.S. National Center for Health Statistics (1987), Sykes and Collins (1988), and Bishop, et al. (1988) also discuss mode comparison studies as does Chapter 16. In the latter, an interview–diary comparison, a complex picture emerges in which a surprising number of items result in equivalent results; however, the diary does appear to improve the reporting of smaller, less salient purchases, whereas the interview yields better data on less frequent or more salient purchases.

The results of mode comparison studies are affected by many decisions in the development of a data collection system; the topic and population under study are two aspects that may influence the results. Examples exist in which comparisons with preferred procedures result in evidence on, for example, underreporting when a particular mode was used. These findings can guide researchers in the selection of a data collection method.

13.4 INTERVIEWER ERRORS

13.4.1 The Nature and Effects of Interviewer Errors

The interviewer is the principal agent by which information is gathered in many studies. As such, one might expect considerable concern to be expressed that the interviewer is a potentially large source of error in surveys. Interviewers, of course, do make errors and these errors can occur because of misunderstanding or carelessness resulting from erroneous interviewer manuals and insufficient training. On average these errors may cancel for each interviewer, resulting in small interviewer biases. Individual interviewer errors, however, may be large and noncompensating, resulting in large errors for individual interviewers. To the extent that these errors are large and in the same direction, a bias, as measured in the mean squared error of the estimate, will result.

Interviewers may make uninformed errors as a result of misconceptions, bad habits, or lack of knowledge. They may incorrectly follow or skip instructions or not ask questions as worded. There is evidence that some interviewers deliberately falsify questionnaires from time to time by fabricating responses to selected items and even entire questionnaires. Biemer and Stokes (1989) describe a study conducted at the U.S. Bureau of the Census where, during 1982–1985, 140 interviewers were discovered cheating and another 31 were suspected of cheating. Interviewers who deliberately falsified data were both newly recruited and experienced interviewers.

Another way in which interviewers may contribute to errors in estimates is through their complex personal interactions with respondents. Many believe the interviewers contribute a personal bias to the data collected, the bias resulting from a mixture of personal characteristics and behavior. An extensive literature exists on how interviewers' demographic and socioeconomic characteristics may influence respondents' answers. Individual studies exist which indicate that specific interviewer characteristics such as age, race, and sex may influence some estimates. Social desirability effects have been documented by Schuman and Converse (1971), Anderson, et al. (1988a, 1988b), and Reese, et al. (1986) in studies of interviewer race and ethnicity effects. Fowler (1966) studied the effect of respondent education on the way respondents and interviewers felt about each other and how they interacted during the interview. Weiss (1968) studied a sample of welfare mothers in New York City and validated the accuracy of several items on the survey and found that similarity between interviewer and respondent with respect to age, education, and socioeconomic status was not necessarily conducive to more valid reporting. There were instances, however, in which better reporting occurred when the interviewer had a higher level of education than the respondent. Groves and Fultz (1985) studied interviewer gender differences as a source of error and found that on a variety of factual items no effects of interviewer gender were observed, although on attitudinal items greater optimism was expressed when the interviewer was male. Groves (1989) reviews a number of such studies and concludes, in general, that demographic effects appear to apply when the measurements are related to the characteristic but not otherwise. Fowler (1990) in his review, however, concludes that despite considerable research the relationship between interviewer characteristics and interviewer-related error has seldom been established.

Other interviewer factors may also play a role in interviewer-related error, such as voice characteristics and interviewer expectations. Their importance is difficult to ascertain. Oksenberg and Cannell (1988) studied vocal characteristics of telephone interviewers and their relationship to refusal rates. Hyman, et al. (1954) first raised the issue of interviewer expectations and the effect of these expectations on survey data. Sudman, et al. (1977b) studied interviewer expectations about the difficulty of obtaining sensitive information and observed weak effects on the relationship between expectations of difficulties in interviewing and actual difficulties encountered.

Under a proper design the interviewers' contribution to the total variance of the estimate can be obtained through the correlated response variance component. Such designs are discussed in Chapters 14 and 23. These chapters demonstrate that the effect of correlated interviewer error is to increase the variance of estimators in proportion to the average size of the interviewers' assignments. Sometimes these increases can be quite dramatic, leading to severe underestimates of the true variance, if regular variance estimators are used.

13.4.2 The Control of Interviewer Errors

Three different means to control interviewer errors are: training, supervision or monitoring, and workload manipulation. It is believed that standardization of the measurement process especially as it relates to interviewers' tasks leads to a decrease in interviewer effects. Fowler (1990 and Chapter 14) reviews the need for standardization of survey work. One way to accomplish standardization is through a training program of sufficient length to cover interviewing skills and techniques as well as information on the specific survey — its goals, questionnaire, and procedures. The purpose of interviewer training is to inform the interviewers of the correct methods and procedures and reinforce the correct implementation of their tasks so that interviewer variation is minimized.

Typical training programs for new interviewers have several components. The first component consists of home and classroom study, on the job training, additional home study, and a final review test. The home study may consist of basic information about the survey being conducted and the job of interviewing; classroom training expands on the home study through lectures, audio-visual presentations, written and oral exercises, and practice interviews. For new interviewers, training in connection with their first several assignments provides the experience needed to build confidence in their ability to complete their tasks. The second component of training occurs after the interviewers have obtained some interviewing experience in connection with their regular assignments. For example, additional home studies are possible as well as feedback associated with reinterview programs (see Chapter 15) and field observations. For continuing surveys, refresher classroom training is also held periodically. The third component of training may be considered supplemental training and is given to interviewers weak in certain aspects of the survey, such as response and accuracy rates. Jabine, et al. (1990) briefly describe the training and supervision program for the Survey of Income and Program Participation, a continuing national household survey program of the U.S. Bureau of the Census.

Initial interviewer training is necessary, at least to the extent that basic concepts and procedures are learned. The literature on the effects of various lengths of interviewer training is scarce. Fowler (1990), Fowler and Mangione (1986), and Fowler (Chapter 14) conducted such experiments and concluded that additional training after some point does not necessarily result in reduced interviewer effects. One reason for this lack of knowledge may be that it is not known what interviewer behaviors or interviewer errors contribute most to interviewer variance. Therefore, developing new training programs is difficult without knowing exactly where the emphasis ought to be placed; lengthening the program may, as demonstrated, result in only marginal effects. Groves, et al. (1981) describe attempts to link behavior with interviewer effects, but the results are inconclusive.

Controlling the quality of the survey is important, and supervision and performance monitoring are essential ingredients of a quality control system. The objectives of such a system is to monitor interviewer performance through observation and performance statistics and identify problematic questions. The extent of supervision and monitoring varies among surveys and organizations. The U.S. Bureau of the Census maintains a relatively thorough program for many of its surveys. Reinterview programs and field observations are conducted to evaluate individual interviewer performance. Monthly performance statistics, such as error rates, production ratios, and nonresponse rates are given to the interviewer as well as feedback on problems identified in completed questionnaires through the clerical edits conducted at regional offices. Interviewers not meeting acceptable quality and productivity standards take remedial training. If this does not help, they are relieved of their positions. Examples of these systems are found in U.S. Bureau of the Census (1978), Jabine, et al. (1990), and Forsman and Schreiner (Chapter 15). Weinberg (1983) offers views on this topic from the vantage point of a large nongovernment data collection organization.

Observations in the field are conducted using extensive coding lists or detailed observers' guides where the supervisor or monitor checks whether the procedures are properly followed. For instance, the observation can include the interviewer's grooming and conduct, his or her introduction of himself or herself and of the survey, the manner in which the questions are asked and answers recorded, the use of flashcards and neutral probes, and the proper use of the interviewers' manual. In other instances, the observer tapes the interview and codes interviewer behavior. Monitoring and coding are discussed in detail by the U.S. Bureau of the Census (1978), Collins (1980), Sykes and Morton-Williams (1987), and Cannell and Oksenberg (1988).

A third way to control interviewer effects is to change the average workload; as mentioned above, interviewer variance increases as average workload increases. If average workload were reduced to one respondent per interviewer, the correlated component of response error would vanish. This, however, is an impractical solution to the problem, because it is simply not cost-effective. Such a severe reduction in workload would likely result in the occurrence of other errors, such as a higher nonresponse rate, because interviewers would never really learn the survey or feel comfortable with the respondents. In addition, the lack of a correlated component of response error does not mean that all variation is gone; some is now transferred to the simple response variance. The issue is, therefore, to find optimal average workloads. Groves and Magilavy (1986) discuss optimal workload as a function of interviewer hiring and training costs, interview costs, and size of the intra-interviewer correlation. Determining optimal workloads is a difficult task because the value of the intrainterviewer correlation varies between statistics in the same survey. Workloads in large national household survey programs at the U.S. Bureau of the Census, like the Survey of Income and Program Participation and the American Housing Survey, range between 60 and 80 households. There are, however, examples of large surveys, some with considerably less and some with considerably larger average workloads.

Interviewer errors and interviewer variability can be measured in various ways. Basically, different systems for reinterviews (replication) and interpenetration are used. These methods are treated in detail both in this volume (Chapters 15, 23, and 27) and in Mahalanobis (1946), Kish (1962), Hansen, et al. (1964), Bailar and Dalenius (1969), Fellegi (1974), U.S. Bureau of the Census (1975, 1978), Bailey, et al. (1978), Groves and Kahn (1979), Collins (1980), and Groves and Magilavy (1986).

13.5 RECENT DEVELOPMENTS AND NEW CHALLENGES

Advances in computer technology have changed the methods of data collection. As observed earlier, the widespread availability of telephones has affected the manner in which researchers approach the question of how best to gather data. The availability of microcomputers has resulted in changes in how data are gathered over the telephone via computer assisted telephone interviewing (CATI), and now the advent of small, light weight personal computers has resulted in innovations in how face to face interviewing is conducted via computer assisted personal interviewing (CAPI). The use of other technology, such as electronic meters and scanners, while not particularly widespread at this time, will likely gain greater acceptance in the future. The use of new technologies will open numerous research areas in the measurement error associated with these technologies.

13.5.1 Computer Assisted Data Collection

Computer assisted data collection refers to a wide variety of data collection activities featuring the use of computer technology to obtain information from respondents. The automation of telephone interviewing (CATI) has become a standard collection methodology during the last 15 years. It is not difficult to argue that this method of data collection should now be viewed as a "basic" data collection method rather than a new development. Nevertheless, we include it as part of the discussion of computer assisted collection methods.

The experiences with CATI are generally favorable (Nicholls and Groves, 1986; Nicholls, 1987). CATI replaces the paper and pencil method of recording responses; the interviewer asks respondents questions as displayed on a computer monitor and records responses directly into the computer. As most CATI systems are centrally administered, greater control and supervision of interviewers is possible. CATI has been adopted by literally hundreds of survey organizations as the preferred method of data collection.

During the last ten years or so, other computerized data collection methods have evolved. Computer Assisted Personal Interviewing (CAPI) is a personal interview conducted either at the home or at the business of the respondent using a portable personal computer. It differs from CATI in that it usually implies the interviewer and respondent are physically together. CAPI has been used to collect supplement data in the U. S. National Health Interview Survey (Rice, et al., 1988) and has been used in a survey to assess nutritional needs of the population of the United States (Rothschild and Wilson, 1988); in the Netherlands, CAPI has been used in the labor force survey (van Bastelaer, et al., 1988). At this time, the implementation of large-scale use of CAPI is somewhat constrained because of the need for lighter computers with long-life batteries and the need for more user-friendly software.

Other examples of new data collection methodology concern several forms of computer assisted self-interviewing (CASI) in which the respondent answers questions without the benefit of an interviewer. Three examples, as described in a report prepared by a subcommittee of the U.S. Federal Committee on Statistical Methodology (U.S. Office of Management and Budget, 1990), stand out. Touchtone data entry (TDE) allows respondents to answer computer generated questions using the keypad of their touchtone telephones. TDE is relatively common in applications outside the survey field; for example, it is used for banking by telephone, for rental car reservations and college class registration. The U.S. Bureau of Labor Statistics uses TDE in its Current Employment Statistics Program (Werking, et al., 1988). The data collection method is suitable for surveys containing few and simple items; its principal limitation is that the respondent must have a touchtone telephone, a limitation that will decrease in the future. Phipps and Tupek (1990) report on measurement errors in a TDE system; they find few serious problems with TDE as a mode of data collection and note that many errors can probably be reduced with experience.

Voice recognition entry (VRE) allows respondents to answer questions by speaking directly into the telephone. It is similar to TDE, but rather than using the telephone keypad for transmitting responses the respondent speaks into the telephone. The computer translates the respondent's answers into text for verification with the respondent and then stores the text into a data base. This is a very "natural" way of collecting data, but the main limitation is the current state of technology in voice recognition. Dialects and accents can provide formidable difficulties in accurate voice recognition. The U.S. Bureau of Labor Statistics is testing VRE, but there has been no practical application to date (Clayton and Winter, 1989; Byford, 1990).

A third example of CASI is called prepared data entry (PDE). This form of data collection allows respondents with compatible microcomputers or terminals to access and complete a questionnaire on their computers. Respondents can enter data in response to preprogrammed floppy disks and then mail the disks to the collecting agency, even electronically. The technique has found applications at the U.S. Internal Revenue Service for electronic tax return transmission and at the U.S. Energy Information Administration for the collection of monthly refinery data from petroleum companies (U.S. Office of Management and Budget, 1990). Saris (1989), in a review of recent technological innovations, reports on a slight variation to the above described methods of data collection; he describes a system called "tele-interview." In this system every person in the sample is provided with a computer and modem. With these, a television, and a telephone, an interview is sent from the central computer to the home computer of the respondent. The respondent answers all questions on the personal computer and sends the results back to the central computer.

The advantages of automated data collection are improved data quality, improved timeliness, improved survey management, and increased flexibility in general. These methods offer improvement in control by reducing manual intervention, promoting standardized procedures, and providing for on-line editing capability. The technologies described above, with the exception of CATI, are in various stages of development, and, therefore, there are almost no studies on measurement errors related to them. Even CATI has surprisingly few studies on measurement errors. Groves and Nicholls (1986) and Groves (1989) review CATI and the issue of data quality.

CATI as well as CAPI offer advantages not present in paper and pencil interviewing; automatic altering of question wording, automatic question skips, arithmetic feedback, and range and edit checks are possible in a CATI and CAPI environment. Research on the measurement error of these features is rare. While interviewer falsification in a CATI environment is almost impossible, other CATI specific errors are possible, such as the misrecording of responses because interviewers are not skilled typists or a negative respondent reaction because of slowness in recording responses. Few if any studies exist on such topics.

There is an increasing use of special electronic instruments in data collection. Meters are used for measuring media ratings and electronic scanners are used in many stores to collect data on retail sales. Donmyer, et al. (Chapter 18) discuss the use of scanners and their associated measurement errors in A.C. Neilsen Company's continuing surveys of the grocery retail trade.

13.6 ENDNOTE

A number of factors enter the decision to select one method of data collection over another. Budgets, speed of implementation, resources and staff of the data collection organization, sample coverage, expected response rates (unit and item), length of the questionnaire, types of questions asked (open versus closed, sensitive versus nonsensitive, complex versus simple), need for visual aids, expected social desirability bias, expected risk of contamination of responses, and interviewer effects are factors that enter the complex decision of selecting a method of data collection appropriate for a study. The benefits and limitations of each method ought to be weighed. This chapter has sought to review the various methods of data collection and their associated measurement errors. Despite the ever increasing knowledge of errors in surveys, it is apparent that much more has yet to be learned. In fact, while technological advances have in some ways made the data collection task easier, these advances have opened up new areas for research on measurement error.

CHAPTER 14

REDUCING INTERVIEWER-RELATED ERROR THROUGH INTERVIEWER TRAINING, SUPERVISION, AND OTHER MEANS

Floyd J. Fowler, Jr.
Center for Survey Research,
University of Massachusetts, Boston

14.1 OVERVIEW OF INTERVIEWER-RELATED ERROR

Interviewers in social surveys can have an important effect on the answers to the questions they ask. The goal of this chapter is to describe ways of reducing error in surveys that can be attributed to the interviewers.

Interviewers play several roles in most surveys, typically including:

1. Making contact with those chosen to be interviewed and enlisting their cooperation.

2. Orienting respondents to their tasks and responsibilities in the interview.

3. Managing the question and answer process, including asking the questions, clarifying and probing respondent answers as needed to make sure that questions' objectives are met and recording answers.

With each of these areas of responsibility comes the potential to affect survey error. For example, interviewers affect the representativeness of the sample in their role of enlisting cooperation. However, the focus of this chapter is on the interviewers' contribution to response error: the extent to which the answers given by respondents do not correspond with the "true value" of what researchers are trying to measure.

There are various reasons why the answers given by respondents are not "accurate": respondents are not able to provide the information

Measurement Errors in Surveys.
Edited by Biemer, Groves, Lyberg, Mathiowetz, and Sudman.

ISBN: 0-471-53405-6

requested, respondents are not willing to provide the information requested, and respondents misinterpret the questions and, in essence, answer the wrong questions.

Problems that occur across all interviewers are not the responsibility of the interviewer. Responsibility for question wording lies with the researcher. However, interviewers can contribute to error because of their personal characteristics or job performance.

A key defining criterion for good measurement in surveys is "standardization." Standardization in social surveys, like other scientific measurement, means that the procedures are consistent each time an observation is made and data are collected. To say that a staff of interviewers should be standardized in order to minimize error is tautological. To the extent that individual interviewers, or groups of interviewers, can be associated with the answers that result, then the interviewers are not standardized (they are not the same in ways that affect the data) and they are related to error in the estimates.

A goal of any survey researcher should be to minimize the extent to which interviewers differ among themselves or across respondents in ways that influence answers. Available general strategies include the procedures interviewers are given, the selection, training and supervision of interviewers, the questions interviewers ask, and the size of interviewer assignments. This chapter reviews the theory and evidence for the importance and effectiveness of these various options for minimizing interviewer-related error. See also Chapters 19, 21, and 22. Let us first digress briefly to describe the ways in which interviewer error is studied.

14.2 DETECTING INTERVIEWER-RELATED ERROR

In most surveys, the interviewers' contribution to the total error cannot be calculated. The reason is that the respondents interviewed by a particular interviewer usually are assigned haphazardly or on the basis of convenience. Variation in answers due to interviewers cannot be distinguished from variation due to sampling or response variance. Also, if the answers obtained by an interviewer are distinctive, it is not possible to dissociate the effects of the interviewer from the idiosyncrasies of that interviewer's sample.

In order to estimate interviewer-related error, it is necessary to control interviewer assignments. The ideal design would have each interviewer interviewing a representative subsample of a total survey sample. Such designs, however, are difficult to implement in practice. For a survey of households using face to face interviewing procedures, cost considerations would normally preclude giving interviewers assignments without regard to proximity. Even for telephone surveys, it is common for interviewers to prefer to work during certain time shifts, but not on others, which means that respondents interviewed by any given interviewer may not be representative of the entire sample.

A variation that also permits an estimate of interviewer effects entails creating interviewer assignments by splitting subsets of sampled units systematically between at least two interviewers. With such so-called interpenetrated designs, when differences in the answers produced from two or more interviewers' assignments are greater than one would expect due to normal sampling variability, interviewer-related error can be inferred and estimated.

The most commonly used statistic for estimating interviewer-related error is the intraclass correlation, typically referred to as rho (ρ_{int}). The strategies for calculating ρ_{int}, and the complexities thereof, have been extensively discussed by Fellegi (1964), Kish (1962), Hansen, et al. (1961), Groves and Magilavy (1986), Stokes and Yeh (1988), and Groves (1989). Of the various approaches, Groves (1989) recommends the one proposed by Kish (1962), which is slightly different from that used by Hansen, et al. The advantages of the Kish approach include the fact that the value of ρ is independent of the number of interviews taken per interviewer. Therefore, it is possible to compare results across surveys. It also is comparatively easy to compute, since a standard analysis of variance program will produce the numbers needed to calculate the value. However, an adjustment is needed if the number of interviews per interviewer vary markedly for a particular survey (see Groves and Magilavy, 1980). Readers also are referred to Groves (1989) for a more thorough discussion of the calculation issues.

The basic approach to estimating ρ_{int} is:

$$\hat{\rho}_{int} = \frac{\dfrac{V_a - V_b}{m}}{\dfrac{V_a - V_b}{m} + V_b} \tag{14.1}$$

where V_a is the between mean square in a one-way analysis of variance with interviewer as the factor, V_b is the within mean square in the analysis of variance, and m is the total number of interviews conducted by an interviewer.

Values of ρ_{int} range from 0 to 1.0. A value of 0 means that answers to a particular question are unrelated to the interviewer; in our terms, the interviewers are perfectly standardized. Higher values of ρ_{int} indicate questions for which the impact of interviewers on the answers and the amount of interviewer-related error is correspondingly higher.

Although the value of ρ_{int} is unrelated to the assignment size of the interviewers, the significance of ρ_{int} for the total error that is interviewer-related is proportionate to the average number of interviews taken per interviewer. The equation for calculating that is as follows:

$$\text{deft} = \sqrt{1 + (n - 1)\rho_{int}} \tag{14.2}$$

where deft is the extent to which standard errors are increased beyond the effects of sampling due to the effect of interviewers and n is the average number of interviews per interviewer.

The values of ρ_{int} are specific to each question in each survey. Groves (1989) reports the distribution of values of ρ_{int} across numerous studies. Although the distributions vary from study to study, he calculates the mean value of ρ_{int} based on 300 estimates to be about 0.01, with a standard error of 0.001. Even though most values are 0.02 or less, if an interviewer takes 50 interviews on average, the effect of a ρ_{int} of 0.02 is to increase standard errors by 40 percent over what they would be with no interviewer-related error.

Although using the intraclass correlation as a way of measuring interviewer-related error is perhaps the most common in the survey literature, there are at least three other kinds of designs that have been used to study these phenomena.

1. Associating interviewer characteristics with the answers obtained. For example, if male and female interviewers have comparable samples, the resulting data should be similar. To the extent that differences exceed sampling errors, one can infer that the gender of the interviewer was related to the answers obtained.

2. There are a few studies in which survey answers were evaluated against records as a way of assessing error associated with each interviewer.

3. There is a set of studies that do not look directly at the resulting data, but rather look at the way interviewers perform their tasks. Although such studies do not always provide direct links to the quality of data, in some cases they are germane to our understanding of the relationship between interviewers and the error in the data they produce.

The following sections present the existing evidence regarding the importance of various interviewer characteristics and behaviors in generating survey error and the efficacy of options available to researchers for minimizing such error.

14.3 STRATEGIES FOR REDUCING INTERVIEWER-RELATED ERROR

14.3.1 Introduction

On the survey team, the researchers and interviewer supervisors are responsible for managing interviewers so that they are as consistent as possible across all interviews. The idea of creating a consistent measurement experience for a survey involving tens of interviewers and hundreds of respondents may seem, on the surface, to be impossible. Some government surveys routinely employ hundreds of interviewers to conduct thousands of interviews. Indeed, critics such as Suchman and Jordan have argued that standardized interaction for large surveys is an unrealistic ideal (Suchman and Jordan, 1990). Chapter 19 also presents a discussion of the difficulty of making interviews consistent.

Each interviewer brings a number of clearly evident characteristics to the interview that might affect the way respondents answer questions. Interviewers establish some kind of relationship with each respondent, and, inevitably, there are different dynamics in each relationship. Although interviewers are given the same questions to ask, interviewers sometimes must explain and clarify questions in order to obtain answers; what they have to explain or clarify will differ from one respondent to another.

So Suchman and Jordan are correct that each interview is unique. The fact that values of ρ_{int} are low for many items is evidence that interviewers often are consistent "enough" not to influence answers. The fact that interviewers do affect some answers to a significant degree defines the problem for researchers: to minimize the variation in interviewers that affects answers, and to minimize the effect of interviewer-related error on the total error in their data. Researchers have at least five ways in which they can attempt to reduce the effect of interviewer-related error on survey data:

1. Give interviewers procedures to follow that will minimize the extent to which they will distinctively influence the answers that are given.
2. Attempt to control interviewer behavior through training and supervision.
3. Select interviewers least likely to affect answers.
4. Design questions that can be administered consistently.
5. Reduce the size of the interviewer assignment, which will reduce the

impact of whatever error is generated by interviewers on the total error of survey estimates.

14.3.2 Interviewer Procedures

There are four rules given almost universally to interviewers about how to be consistent with other interviewers and reduce their individual effects on data:

1. Read questions exactly as written.
2. In the event that a respondent fails to answer a question adequately in response to the initial reading, follow-up probes and questions should be nondirective.
3. Answers should be recorded without interviewer interpretation or editing.
4. Interviewers should maintain a professional, neutral relationship with the respondent, which minimizes any sense of evaluation of the content of the answers produced.

Giving interviewers well-developed scripts, so that they can ask all respondents questions exactly as written, was one of the early innovations in improving the quality of survey research. When interviewers were simply given a set of topics to be covered in an interview, they were free to make up their own questions. There was a lengthy debate about how much flexibility to give interviewers in the wording of questions, which Converse (1987) describes very well. Even today, critics such as Suchman and Jordan (1990) and Mishler (1986a) question whether the standard of having interviewers use the same wording for all respondents is best. Perhaps, they argue, interviewers can improve questions by individualizing them for respondents.

In fact, studies of interviewer behavior show that interviewers do not always read questions the way they are written. Observational studies show a range of from 5 to over 60 percent of questions are at least slightly reworded by interviewers (Bradburn, Sudman, et al., 1979; Cannell, et al., 1968; Fowler and Mangione, 1990; Cannell and Oksenberg, 1988; Groves, 1989). The literature is less clear on how much minor rewording matters with respect to the data that result.

Efforts to show relationships between how careful interviewers are about question wording and their contributions to error have not consistently shown positive results. Groves and Magilavy (1986) monitored interviewer behavior and related the results to estimates of the intraclass correlation associated with interviewers. Interviewers were rated on how well they read questions (including the clarity of their reading), and this rating was not clearly related to interviewer-related error. Fowler and Mangione (1990) report a similar study, also with inconclusive results.

Another place to look for evidence of the importance of reading questions exactly as worded is the literature on question wording. Studies reported in this literature are not focused on interviewing *per se*, but they shed light on the way that question wording affects answers, which in turn helps to assess the effect of changes in question wording. For example, Schuman and Presser (1981) report on a series of studies comparing the results of slight changes in question wording in interviews with comparable samples. In those studies, they show numerous clear examples of small wording differences that produce large effects.

For example, Schuman and Presser report on the importance of including a "don't know" option in a question. Consider the following: "In general do you favor or oppose the Agricultural Reform Act of 1978, or do you have no opinion on that?"

A common error made by interviewers in reading questions is to omit a response alternative, particularly one that may seem superfluous to some interviewers. Since the Agricultural Reform Act was fictitious, the "no opinion" response should have been common. Schuman and Presser show that omitting "or do you have no opinion on that" changed the number of people who said they had no opinion from 90 percent to 62 percent.

Collins (1980) gives an example that helps to relate this finding to ρ_{int}. Interviewers were given discretion about whether or not to read a neutral response category. Very high levels of interviewer-related error occurred, which Collins attributes to differential use of the neutral option.

Fowler (1989) compared the results of two forms of a question about exercise. One form explicitly stated that walking was exercise, while the other form did not mention walking explicitly. Significantly more people said they exercised when the inclusion of walking was explicit. That finding implies that changing the question by leaving out two words, "including walking," would directly affect the data.

Some wording changes have little effect on the distribution of answers. For example, substituting "end a pregnancy" for "abortion" had no impact on answers in one series of tests (Schuman and Presser, 1981). Yet Rasinski (1989) found that respondents were much more likely to say government spending was "too little" for "assistance to the poor" (over 60 percent agreed) than when asked a parallel question about spending on "welfare" (less than 25 percent said it was "too little").

The "exercise" example cited above shows the value of having interviewers ask exactly the same questions in order to minimize interviewer-related error. However, some wording changes affect answers, others do not, and we lack good generalizations about which changes matter most.

The way interviewers probe inadequate answers has long been thought to be a critical part of the way that interviewers influence answers. Interviewers can err in two different ways. First, they can fail to probe when an answer requires elaboration or clarification. Second, they can use what is called a directive probe, a probe that suggests an answer or increases the likelihood of one answer over others.

Hyman, et al. (1954) found that interviewers were likely to influence answers through their probing in the following way: if respondents provided an answer that was consistent with interviewer expectations, interviewers were less likely to probe it than if an answer was given that was inconsistent with the kind of answer interviewers expected to get.

Fowler and Mangione (1990) found that probing was the task at which interviewers felt least confident. Studies of interviewer behavior show that poor probing skills are most likely to show up in interviewers who are not well trained (Fowler and Mangione, 1990; Billiet and Loosveldt, 1988). Cannell, et al. (1968) found that nearly 40 percent of the probes used by Census interviewers in a health survey were directive probes. About 20 percent of probes used were judged to be inappropriate in studies by Rustemeyer (1977) and Oksenberg (1981). Brenner (1982) reports over 30 percent of questions produced leading or directive probes. Most important, as will be discussed later, the rate at which questions require probing is a major correlate of ρ_{int}. However, although such a link seems likely, we lack direct evidence of the relationship between probing performance and the amount of interviewer-related error.

With respect to recording answers, interviewers have different tasks, depending on the characteristics of the question. When respondents are asked to choose one of a fixed set of answers, the recording task is simply to check a box or circle a number that corresponds with the answer given by the respondent. When respondents are to answer questions in their own words, interviewers are required to record answers as close to verbatim as possible in order to minimize their effects on the data.

Hyman, et al. (1954) again is a source of information that interviewers are inconsistent in the way they record. Interviewer expectations were found to influence what interviewers included or left out of their supposedly verbatim recordings; interviewers were more likely to record the statements that were consistent with their expectations, and more likely to omit statements that were inconsistent with their expectations. Fowler and Mangione (1990) found that interviewers were very good at recording fixed response questions, but that recording open-ended answers was much more problematic. They do not provide an estimate of how much imperfect recording contributed to total error in survey data, but they found that questions that produced high rates of recording errors also produced high interviewer-related error. A study by Rustemeyer (1977) found interviewers made recording errors that affected the resulting data in 10 to 14 percent of the answers they recorded.

Failure to establish an interpersonally neutral and nonjudgmental relationship with respondents is probably a less important source of interviewer-related error. Hyman, et al. (1954) found that respondents are very unlikely to accurately guess the views of interviewers about the topics being covered in an interview. Sudman, et al. (1977b) looked in vain for relationships between interviewers' own expectations and answers given by respondents (although there was some tendency for interviewers to have higher rates of "no answer" to questions they thought would be troublesome to respondents). Collins (1980) also reports negative results in studies of the effect of interviewer opinions on interviewer-related error.

The tone of the relationship interviewers establish with respondents is a related issue. Interviewers have two potentially conflicting goals: to establish a relationship in which communication is comfortable for respondents (the traditional term for this is *rapport*) and to establish a professional, task-oriented relationship in which reporting accurately takes precedence for respondents over concerns about presenting a good image.

Interviewers do differ in the extent to which they appear friendly and encourage conversation that is not strictly task-related (Cannell, et al., 1968). The data on the significance of a friendly style are inconclusive.

Fowler and Mangione (1990) report a study that showed that interviewers who initiated some humor had more health events reported by respondents who had not completed high school. They also found some evidence that interviewers whom respondents rated high on "friendliness" tended to obtain answers that were less biased. Henson (1973) manipulated interviewer style and, overall, found no significant differences in accuracy of reporting related to whether the interviewers behaved in a more interpersonal or a more professional style (although the more professional style produced better data for one subclass of event reports). In contrast, Weiss (1968) found that interviewers who rated rapport with respondents high obtained less accurate data (as determined when answers were checked against records).

Definitive studies of the effect of interviewer style on interviewer-related errors have not been done. It seems likely, based on what we know, that a variety of interpersonal styles will work. Probably a much more important issue than whether or not an interviewer has a warm, friendly style is the content of the feedback interviewers give to respondents about what they are expected to do.

Few interviewer manuals address the role of the interviewer in training the respondent and setting goals and expectations for the respondent. Yet, clearly, it is the responsibility of the interviewers to orient respondents to the task of being respondents.

Fowler and Mangione (1990) emphasized the importance of training the respondents how to participate in the question-and-answer process. They argue that when respondents do not understand their roles, how they are supposed to behave, or what the interviewer is supposed to do, it makes the task of the interviewer harder. There are, however, no data so far that directly document the value of having the interviewer train the respondent for reducing the error in data.

Another aspect of the orientation of the respondent, the goals and expectations that are set for respondent performance, clearly have been found to affect survey data.

For example, Cannell and associates found that interviewers differed in their effectiveness in motivating respondents to report accurately (e.g., Cannell and Fowler, 1964). They also found that interviewers often reinforced counterproductive behavior, as well as encouraging the kinds of thoughtful behaviors that would lead to accurate reporting (Cannell, et al., 1968). To correct this, Cannell and his associates designed a series of experiments to try to make interviewers more consistent in setting the goals of accuracy and completeness of reporting for their respondents. Five different techniques have been tried at different points:

a. Slowing the pace of the interview.

b. Providing an audiotape model of a good respondent at the beginning of an interview (Henson, 1973).

c. Giving interviewers standardized instructions to work hard and be accurate.

d. Building in a systematic program of reinforcement of thoughtful answers, together with criticism of quick and apparently thoughtless answers (e.g., Marquis, et al., 1972).

e. Having respondents sign, or verbally agree to, a commitment to provide accurate and complete answers.

Although the effectiveness of the procedures has varied somewhat from study to study, and across subgroups of the population, in general all of these techniques have proven to be effective ways of improving the quality of data (Cannell, et al., 1977b, 1987). There has been only one experiment that permitted study of the effect of these techniques on the error that can be correlated with interviewers, and the results were suggestive but inconclusive (Cannell, et al., 1987; Groves, 1989).

In conclusion, the links between the typically prescribed interviewer behaviors and interviewer-related error are not well established. The rules for good interviewers deserve scrutiny and are in need of further research. Present knowledge and future directions may be summarized as follows:

a. Interviewers often do not read questions as written. The wording of questions can have a major effect on answers, but we lack good generalizations about when that is and is not the case.

b. Nondirective probing is the most difficult interviewer task, and it almost certainly is one key to interviewer-related error. Designing questions that require a minimum of probing is probably one important way to reduce interviewer-related error. However, there is controversy about what kind of help interviewers should give when respondents have problems understanding or answering questions. Those most concerned about standardization argue for minimal intervention, while others argue that respondents need more help, and more flexible help, in order to meet question objectives (Suchman and Jordan, 1990). Both positions are held more on theoretical than empirical grounds.

c. Recording answers accurately is probably not a controversial guideline.

d. Guidelines for how interviewers should relate to respondents are perhaps the least well developed. Being more sociable may facilitate communication for some respondents, while it also increases chances of distinctive interviewer effects on answers. The evidence is clear that the goals interviewers communicate affect the data, and giving interviewers specific protocols for orienting respondents improves reporting. There still is much need for research on the best ways for interviewers to train respondents.

14.3.3 Controlling Interviewer Behavior Through Training and Supervision

The five ways of reducing error listed in Section 14.3.1 and the rules in Section 14.3.2 outline a desirable set of behaviors for interviewers. The next question is how to manage interviewers so they behave in desired ways.

There are two basic ways to do that: training and supervision.

Interviewer Training

Almost all survey organizations provide some training to interviewers. Training protocols vary markedly in the degree to which they are structured, their length, and the mix of lectures, demonstration, self-study, discussion, and supervised practice. Based on feedback from interviewers and supervisors, the areas that require the most attention are probing and enlisting respondent cooperation. Involving interviewers actively in the training process, through discussion and especially supervised practice (in addition to more passive activities like lectures and demonstration) probably is a valuable component of a good training program (Mockovak, 1989; Fowler and Mangione, 1990).

At least two studies in the survey research literature examine the effect of training on interviewer skills. The most extensive of those was carried out by Fowler and Mangione (1990). They examined the effect of four training programs that differed in length and in content:

- A half-day program consisting only of lectures and a demonstration, with no supervised practice.

- Two- and five-day programs, which included a small amount and a great deal of supervised practice, respectively, in addition to more extended opportunities for discussion and lectures.

- A ten-day training program, which greatly exceeds the training that is routinely given in any survey organization.

Interviewers were then assigned a sample of households at which to conduct 30-minute face to face interviews about health-related topics. One feature of the design included tape recording interviews done by a sample of the interviewers. These tape recordings were later evaluated regarding the extent to which interviewers asked questions as worded, probed appropriately, recorded answers appropriately, and handled the relationship with the respondent in a neutral, nonbiasing fashion. The evaluations included both counts of specified behaviors and ratings. Raters, of course, were blind to the kind of training that interviewers had received.

Table 14.1 presents the relationship between the results of these evaluations and the training program to which interviewers were assigned. It is quite clear that those interviewers who received the shortest training program were much less adept at the skills thought to promote standardized data collection. Those skills were markedly improved by being exposed to the two-day training program that included some supervised practice. With respect to the particular skills examined in Table 14.1, however, most of the skills did not further improve with the longer training sessions. However, probing open-ended questions, which

Table 14.1. Selected Measures of Interviewer Behaviors from Coding Taped Interviews, by Training Program

Interviewer Behaviors	Length of Training Program				
	< 1 day	2 days	5 days	10 days	*p*
	Average Number Interviewer Behaviors per Interview				
Questions Read Incorrectly	21	7	14	6	< 0.01
Directive Probes	8	5	5	3	< 0.01
Failed to Probe Inadequate Answers	8	6	5	5	< 0.01
Inaccurate Recording of Closed Question Answers	1	1	1	*	0.05
Inaccurate Recording of Open Question Answers	4	2	2	2	< 0.01
Inappropriate Feedback	2	*	*	*	< 0.01
	Percentage of Interviews Rated Excellent or Satisfactory				
Reading Questions as Worded	30	83	72	84	< 0.01
Probing Closed Questions	48	67	72	80	< 0.01
Probing Open Questions	16	44	52	69	< 0.01
Recording Answers to Closed Questions	88	88	89	93	0.74
Recording Answers to Open Questions	55	80	67	83	< 0.01
Nonbiasing Interpersonal Behavior	66	95	85	90	< 0.01

* Less than 0.5 times per interview
Source: Fowler and Mangione (1986).

is undoubtedly the most difficult task for interviewers, continued to improve with the increasingly long programs of training.

In a very similar study, Billiet and Loosveldt (1988) compared the performance of interviewers who had little training (3 hours) with those exposed to a three-day training program. They, too, found that probing skills were markedly affected by training.

Both of the above-cited studies also provided some evidence that training was associated with error reduction in surveys. Billiet and Loosveldt found the effect of training primarily on questions that required probing. Fowler and Mangione found that the value of training interacted with the way in which interviewers were supervised in order to produce an effect on the quality of data.

With respect to the content of training, Fowler and Mangione found that the inclusion of supervised practice, beyond lectures, reading manuals, and demonstrations seemed to make a critical difference in interviewer skills. Their ten-day training program was an attempt to go beyond teaching concrete skills to provide a more general understanding of the reasons that interviewers create error and how they can avoid affecting answers. They found minimal evidence that skills were improved (as indicated in Table 14.1) though those with ten days of training were usually as good as any. In analyses of how training related to resulting data, they also found some evidence that the longer, more in-depth training was counterproductive, in some cases seeming to produce more interviewer-related error (Fowler and Mangione, 1986, 1990).

Interviewer Supervision

A great deal of what passes for supervision of interviewers consists of monitoring their costs, productivity, and individual response rates. Although these are obviously important concerns for any research organization, none of these has anything to do with response error. Reviewing completed interviews provides information about whether or not codable answers are recorded, but not about interviewing techniques or error in the data.

There are two strategies that directly assess the way interviewers collect the data and the resulting quality of the data: directly monitoring interviews, either through observation or tape-recording, and reinterview programs.

Reinterviews as a method of quality control are discussed in detail in Chapter 15. As noted there, although reinterview programs have the potential for measuring interviewer-related error, it is unusual for reinterview programs to serve that function.

The best way to find out how interviewers are conducting interviews is to observe them. When interviews are done face to face, an observer can be on the scene. More practically, interviews can be audiotape-recorded, and supervisors can review the results. When interviews are done from centralized telephone facilities, supervisors can directly monitor how interviewers carry out interviews.

Of course, giving interviewers evaluative feedback needs to be done in a systematic way. Cannell and Oksenberg (1988) and Fowler and Mangione (1990) give guidelines for using systematic coding of interviewer behavior to supervise interviewers. The training and supervision of supervisors seems likely to be significant in how well such programs work, but there are no such studies.

The critical objective of interviewer supervision is to directly monitor the way that interviewers conduct interviews. Both studies cited above, Billiet and Loosveldt (1988) and Fowler and Mangione (1990), found that a program using tape-recording to supervise personal interviews had a positive effect on interviewer-related error in surveys. Billiet and Loosveldt found that questions that require probing were more influenced by interviewers when they were not tape-recorded. Fowler and Mangione used two indices of data quality: the intraclass correlation ρ_{int}, and a measure of the extent to which answers on average were likely to be biased (due either to underreporting or social desirability bias). According to both those indices, if interviewers had received more than the minimum training, interviewers who were tape-recorded produced less error. It is also important to note that Fowler and Mangione found no evidence of adverse effects of the tape-recording process either on response rates or on respondents' post-interview reports of how they felt about the interview experience.

In summary, the existing data support the generalizations that at least two or three days of training, including some supervised practice, is necessary to give interviewers minimally adequate skills, and that supervision through tape-recording or direct monitoring is needed in order to make sure that interviewers use those skills.

14.3.4 Interviewer Selection

Choosing people who are good interviewers, or who have skills likely to make them good interviewers, is another obvious way to improve the quality of interviewing. However, despite numerous efforts to identify good interviewers, beyond minimally adequate reading and writing skills, there are no consistent correlates in the literature between interviewer characteristics and the quality of interviewing.

Interviewers do bring characteristics to interviews, such as their race, age, sex, and general social class, that are obvious to respondents and do differ from interviewer to interviewer. The hope for a standardized data collection process is that such characteristics will not affect the data that result from an interview. For the most part, the research literature tends to indicate that that indeed is the case. For example, Groves and Magilavy (1986) found interviewer age, gender, and education to be unrelated to the value of ρ_{int}. In addition, in most surveys in which it has been studied, interviewer demographic characteristics have been found to be unrelated to the data that result.

There are some exceptions. Schuman and Converse (1971) compared the answers of white respondents when interviewed by black and white interviewers. They found that on about 10 percent of the items in the survey, those most related to how white respondents felt about black people, the race of the interviewer affected the answers.

Robinson and Rhode (1946) (religion), Hyman, et al. (1954) (race and gender), and Pettigrew (1964) and Williams (1964) (race) also report some differences in answers related to race, religion, or gender of the interviewer.

Most of the positive findings of interviewer demographic characteristics affecting answers tend to be consistent with the generalization that respondents tend to avoid answers that they believe will make the relationship to the interviewer uncomfortable. The vast majority of survey questions in combination with most interviewer demographic characteristics do not fall in that category and will be unaffected by demographic characteristics of interviewers.

Although standardization has been our criterion and typically is the only way we can identify interviewer-related error, there are two studies that should be noted because they enable researchers not only to detect differences between interviewers but to assess the relative accuracy of answers. Weiss (1968) compared responses about income from Welfare gathered by interviewers recruited from low income neighborhoods (who were presumably more similar to and might be more trusted by respondents) with results from the regular middle class interviewers that typically are used by survey organizations. When she compared the survey answers to the records, she found not only that the two staffs differed, but that the professional interviewers obtained more accurate information.

On a different topic, Anderson, et al. (1988a) studied reports of voter participation of black respondents interviewed by black and white interviewers. Again, they found differences. Moreover, when reports of voting were compared with voting records, they found that blacks overreported voting participation to black interviewers; that is, there was more error obtained by black interviewers from black respondents than by white interviewers.

Both of these studies serve to warn researchers that when responses are associated with interviewer characteristics, it is not always easy to predict which answers are best. As Hyman, et al. (1954) observed long ago, similarity can be beneficial to rapport and, at the same time, potentially detrimental to accuracy.

In conclusion, for most surveys, choosing particular interviewer characteristics is not an important way to reduce interviewer-related error. However, when questions are asked that are very relevant to obvious interviewer characteristics, researchers may be well advised to consider systematically trying to measure the amount of error associated with interviewer demographics, rather than (or in addition to) trying to control it.

14.3.5 The Role of Question Design

Although training and supervision seem to have an important effect on how interviewers carry out their tasks, probably the most important step a researcher can take to minimize interviewer-related error is to give interviewers questions that can be used in a standardized way.

As noted earlier, Groves (1989) tabulated various studies that calculated intraclass correlations for interviewers. The distributions differ, but in all studies at least some questions were significantly influenced by the interviewer who asked the question. It is reasonable to ask which questions are most likely to be influenced by interviewers.

There is a history of looking for the questions most subject to interviewer effects. For example, there are some analyses in the literature that suggest that factual items are less likely to be affected by interviewers than questions about attitudes or opinions (Hansen, et al., 1961; Fellegi, 1964; O'Muircheartaigh, 1976). However, Fowler and Mangione (1990) and Groves (1989) report analyses of much larger samples of questions that, like the results reported by Kish (1962), indicate no consistent differences in interviewer-related error between attitudinal and factual questions.

Open-ended questions (as opposed to fixed-choice questions) have also been targeted as susceptible to interviewer effects. Hyman, et al. (1954) found that interviewers selectively probed open-ended questions. Groves and Magilavy (1986) replicated an earlier finding by Gray (1956) that the number of answers given in response to open-ended questions was affected by interviewers. Fowler and Mangione (1990) and Groves and Magilavy (1986) found that whether or not an answer was given at all to open-ended questions was related to interviewers. However, in both the latter analyses, the content of answers to open-ended questions was no more related to the interviewer than the answers to closed questions.

Sudman, et al. (1977b) looked at questions that interviewers consider to be sensitive, reasoning that interviewer effects might be most evident among such questions. They found no such effects; Fowler and Mangione (1990) replicated that result.

Finally, Fowler and Mangione (1990) systematically compared questions rated likely to be difficult to answer with those rated easy and found no significant association with values of ρ_{int}.

In short, there is some evidence that certain aspects of questions that are answered in an open-ended form may be more susceptible to interviewer effects than questions asking respondents to choose from a set of fixed alternatives. However, on average, there is very little evidence that either the content or the form of the question *per se* is a good predictor of where interviewer-related error is most likely to be found.

After failing to find question characteristics that strongly predicted where interviewer-related error would be found, Fowler and Mangione (1990) went back to tape-recordings of the interactions between interviewers and respondents to see what stood out as distinctive characteristics of questions with high intraclass correlation values. Using a coding scheme similar to that reported by Morton-Williams and Sykes (1984) and Cannell and Oksenberg (1988), their finding was that questions that required a distinctive amount of probing on the part of interviewers in order to obtain an adequate answer were most susceptible to interviewer effects. Probing is the skill that requires the most interviewer judgment. Hence, perhaps it is not surprising that questions that consistently require probing are those where interviewers are most likely to behave differently in ways that will affect the answers. They also found that questions subject to recording error (where the recording task was not easy) were subject to higher than average interviewer effects, as well.

In the last six years, there has emerged an integration of cognitive psychology and survey methodology to design a protocol for evaluating questions prior to pretesting them in the field. "Think aloud" interviews and in-depth interviews, such as those described by Jabine, et al. (1984), Lessler (1987), Willis, et al. (1989), and Royston (1989) lay out protocols for identifying terms that respondents do not consistently understand and tasks that respondents cannot consistently perform. See also Chapter 20. Studies of respondent comprehension find considerable ambiguity in most survey instruments (e.g., Belson, 1981). When interviewers are confronted with a need for definitions that are not written into questions or a need to help respondents with questions that are difficult to answer, they have to innovate. Innovation by interviewers leads to inconsistent interviewing and interviewer-related error.

Beyond laboratory testing of questions, problem questions can be identified through better pretesting. Cannell, et al. (1989) and Sykes and Morton-Williams (1987) have shown that pretest interviews can be tape-recorded and the behaviors coded to produce reliable and meaningful information about question problems. There are three key behaviors to tabulate: 1) the rate at which interviewers have to probe, or the mirror image of that, the rate at which respondents give inadequate answers; 2) the rate at which respondents ask for clarification of questions; and 3) the rate at which interviewers misread or rephrase the questions. When any of these behaviors is found in a significant number of interviews, say 15 per cent or more, it indicates a question that warrants attention and that probably needs to be revised in order to make it better for respondents and interviewers.

14.3.6 The Size of Interviewer Assignments

A fifth way of reducing interviewer-related error is by using more interviewers to carry out a fixed number of interviews, thereby reducing the average number of interviews per interviewer. The significance of this strategy is of two types: reducing the impact of interviewer effects on the total survey error and reducing the potential for bias due to interviewer fatigue.

On the first point, as noted previously, for a given value of ρ_{int}, the extent to which interviewers affect answers is proportionate to the number of interviews taken per interviewer. Although the interviewer-related error cannot be estimated unless assignments to interviewers are interpenetrated, when it can be estimated, larger interviewer assignments will produce larger estimated standard errors for any given intraclass correlation value.

There obviously is a potential cost trade-off that needs to be considered. It costs more to train and, perhaps, to supervise a larger staff of interviewers. Conceivably, a larger staff, on average, may be less capable or less experienced, which in turn may produce more error of other types, such as higher nonresponse. However, given the fact that at least some lack of standardization is inevitable, there is potential for reduction of interviewer-related error by using more interviewers and thereby reducing the assignment size.

In addition, there is some evidence in the literature of interviewer fatigue and reduction in quality over time. Cannell, et al. (1977a) reports results of a record-check study in which it was found that the interviewers with larger assignments had a much lower proportion of known hospitalizations reported to them in health interviews. The correlation was −0.72.

In a second study, newly trained interviewers were given equivalent assignments over each of five weeks. It was found that the percentage of known hospitalizations reported declined monotonically from the first week to the last week (Cannell, et al., 1977a).

Although the evidence is not definitive, over time interviewers may become less conscientious, more willing to go quickly, and less concerned about accuracy. That may be one additional reason for considering strategies to limit the size of interviewer assignments.

14.4 CONCLUSION

Almost certainly, interviewer-related error is underappreciated. Although it is easy to ignore, because it cannot be estimated without special designs, whenever researchers have looked for interviewer-related error, there has always been an abundance of it available.

The most widely accepted approach to reducing survey error probably is to attack sampling error through increasing sample size. However, steps to reduce interviewer-related error often may be more cost-effective than sample redesigns as ways to reduce error. In this final section, we try to put the options for reducing interviewer-related error in the context of total survey design.

Respondents need to be oriented in how a standardized interview is to be carried out. It is clear that teaching the respondent what is expected, and what the interviewer has to do, will make it easier for the interviewer to do the job in a standardized way. It also is well documented that interviewers convey different expectations for respondent performance and for standards in the interview (e.g., Fowler and Mangione, 1990). The value of techniques developed and tested by Cannell and his associates (Cannell, et al., 1977b) to consistently convey to respondents what is expected and the importance of reporting accurately and completely is well demonstrated. These steps cost virtually nothing, they work to reduce bias in estimates, and, although it has not been well documented, they probably work to reduce interviewer-related error as well.

Interviewer training and supervision clearly matter. Relatively untrained interviewers, who have been trained for less than a day, generally will not have mastered the basic techniques for standardized interviewing. Beyond that, once they have received at least minimally adequate training, say two or three days, whether or not they use the skills they have learned will depend in part on how they are supervised. Monitoring telephone interviewers and tape-recording face to face interviews add to the total cost of a survey. Not all questions in surveys are significantly affected by interviewers. However, for questions in a survey that are likely to be significantly affected by interviewers, training and monitoring are very efficient measures.

Choosing interviewers with particular characteristics probably is not an important way to control interviewer-related error. For most surveys, getting the best trained and supervised interviewers possible, and giving them good instruments, is probably the best path to reducing interviewer-related error.

One of the most important ways to minimize interviewer-related error is to give interviewers questions that can be administered in a standardized way. The key to doing this is better question evaluation and pretesting prior to the inception of a survey. Particular targets of evaluation should be the extent to which questions require probing or clarification. Although better question evaluation requires some time, as well as expense, better evaluation of questions in the laboratory and during pretesting is probably the most cost-effective way to reduce interviewer-related error. At the same time, it is very likely to reduce response error for other reasons as well.

Finally, researchers should seriously consider limiting the number of interviews per interviewer in any survey. For those items that are most susceptible to interviewer effects, the resulting reduction in total error can be significant. In addition, limiting assignment sizes can reduce the effects of interviewer fatigue.

CHAPTER 15

THE DESIGN AND ANALYSIS OF REINTERVIEW: AN OVERVIEW

Gösta Forsman
University of Linköping

Irwin Schreiner
U.S. Bureau of the Census

15.1 INTRODUCTION

Reinterview is an important tool for estimating and reducing response errors in interview surveys. In this chapter we attempt to describe the state of the art in reinterview. Various aspects on reinterview are illustrated by examples from the U.S. Bureau of the Census, which has played a leading role in reinterview research.

The term *response error* is used here for any error that occurs during the data collection stage in an interview survey. Response error may depend on, for example, imperfect instructions to the interviewers, badly designed questions or questionnaires, coincidental factors that affect the interviewer or the respondent during the interview, or deliberate errors by the respondent. Other response errors are closely connected to the interviewer. For example, interviewers may administer the questionnaires in different ways or they may assist the respondents in different ways. Also, interviewer demographic and socioeconomic characteristics can affect the behavior of respondents, thereby resulting in variation in responses. A special source of response error is the deliberate falsification of interview results by interviewers.

Survey models have been developed to meet the need for an integrated treatment of sampling errors and response errors. Throughout this chapter we shall use concepts defined according to the U.S. Bureau of the Census' survey model (Hansen, et al., 1951, 1961, 1964). See

Measurement Errors in Surveys.
Edited by Biemer, Groves, Lyberg, Mathiowetz, and Sudman.

ISBN: 0-471-53405-6

also Chapters 14, 23, 24, and 27. In this model the mean squared error is decomposed in the following general terms: sampling variance, simple response variance, correlated response variance, covariance of the response and sampling deviations, and the squared bias. Two major schemes are available for measuring the survey model's response error components, both developed in the late 1930's and the early 1940's and highly influenced by the development of the theory of statistical experiments. The method of *interpenetrated subsamples* (Mahalanobis, 1946, Bailar, 1983, and Kish, 1962) is particularly designed for estimating the correlated component of the response variance. The method of *replicated measurement on the same units* is designed for estimating simple response variance and response bias.

In interview surveys, replicated measurement is called *reinterview*. Reinterview methodology was developed in the United States and India during the 1940's and has been used in a number of countries since then. Today, the most extensive reinterview programs are conducted at the U.S. Bureau of the Census and at Statistics Canada within their large-scale recurring surveys and population censuses.

By reinterview we mean a new interview where the questions of the original interview (or a subset of them) are repeated. If better data quality is required in the reinterview, several questions may replace the original question. The reinterview always refers to the same time point as the original interview. Thus, our definition does not include interviews of the same person in two or more waves of a panel survey, since different waves concern different time points.

Section 15.2 contains a general discussion of various purposes and designs of reinterview studies. Section 15.3 deals with design issues; in particular, with the reinterview questionnaire and the evaluation of differences between original interview and reinterview (reconciliation). The analysis of reinterview data is discussed in Section 15.4 (evaluating field work) and Section 15.5 (model-based analysis). In Section 15.6 we discuss the use of computer-assisted interviewing in reinterviews. Section 15.7 contains some concluding remarks.

15.2 PURPOSE AND DESIGN OF REINTERVIEW

15.2.1 Purposes of Reinterview

There are two major purposes for conducting reinterviews: (1) to evaluate field work; and (2) to estimate error components in a survey model. In the first area, reinterview is used to identify interviewers who (1a) are falsifying data, and (1b) misunderstand procedures and require remedial

Table 15.1. Cross-classification of an Individual Population Characteristic

		Original Interview		
		$y_{u1}=1$	$y_{u1}=0$	Total
Reinterview	$y_{u2}=1$	a	b	$a+b$
	$y_{u2}=0$	c	d	$c+d$
	Total	$a+c$	$b+d$	$n=a+b+c+d$

training. In the second area reinterview is used to (2a) estimate simple response variance, and (2b) estimate response bias. In this chapter, we refer to these four purposes of reinterview as 1a, 1b, 2a, and 2b, respectively.

In the first area, evaluating field work, the use of reinterview depends on the kind of survey conducted. For one-time surveys and censuses, where there is little time to provide feedback to the interviewers, the reinterview serves primarily to identify falsification of data. For recurring surveys reinterview is used, not only to detect falsification, but also to provide feedback to the interviewers regarding their performances so that they can be improved in the future. Tolerance tables (see Section 15.4.3) can be applied to determine if the number of errors meets or exceeds the acceptable standard. For those interviewers who fail to meet the standard some form of remedial training can be provided to improve their performances.

For the second area (purposes 2a and 2b), a major difficulty in the reinterview is to meet the survey model assumptions necessary for estimation of error components. For purpose 2a it is important that the reinterview be an independent replication of the original interview. This is difficult to achieve since the respondent might remember his or her first answer when the reinterview is conducted (conditioning). For purpose 2b the reinterview should be conducted in whatever manner produces the "true," or correct response.

We shall illustrate estimates of simple response variance and response bias for a categorical variable. Consider Table 15.1, the cross-classification of two measurements of an individual population characteristic (e.g., whether a person is in the labor force or not) obtained at two occasions ($t=1$ or 2) from a single sample of individuals, $u=1, \ldots, n$. Let $y_{ut}=1$ if Yes is recorded for the unit u in the trial t ($t=1$ for original interview, $t=2$ for reinterview) and $y_{ut}=0$ otherwise.

Assuming that the two measurements y_{u1} and y_{u2} are independent and identically distributed (i.i.d.), the gross difference rate, g, divided by two, that is,

Table 15.2. Cross-classification of Answers to the Question: "Does your (house/apartment) have a clothes washer (----/in the apartment)?'

		Original Interview Yes	No	Total
Reinterview	Yes	961	23	984
	No	53	385	438
	Total	1,014	408	1,422

Source: Smith (1987)

$$g/2 = \sum_{u=1}^{n} (y_{u1} - y_{u2})^2/2n = (b + c)/2n \tag{15.1}$$

is an unbiased estimator of the simple response variance (see Section 24.4.1). Replacing the i.i.d. assumption with the assumption that the reinterview provides true values, an unbiased estimator of the overall response bias of the first measurement is given as the net difference rate,

$$\bar{e} = (b - c)/n. \tag{15.2}$$

To illustrate the above formulas, as well as the trial to trial variability in response that is present even for factual items, consider Table 15.2. The table shows the interview/reinterview cross-classification of answers to the question "Does your (house/apartment) have a clothes washer (----/in the apartment)?" in the 1984 American Housing Survey – Metropolitan Sample (see Smith, 1987).

The gross difference rate, g, is $(53 + 23)/1422 = 0.053$ and the net difference rate is $-30/1422 = -0.021$. Note that the interpretation of these measures in terms of error components depends on the design of the reinterview, that is, which of the above model assumptions are met. In our case, the reinterview was designed for estimating simple response variance. Thus, $g/2 = 0.027$ is an estimator of the simple response variance, while the net difference rate should not be regarded as a measure of response bias.

15.2.2 Design of the Reinterview

The design of a reinterview depends upon how it will be used. The four uses 1a, 1b, 2a, and 2b each have their own optimal designs, which we will refer to as the four *basic reinterview designs*. Each basic design is characterized by the following six factors:

1. *The method of reinterview sample selection.* The reinterview sample can be a one-stage sample of respondents, households, or clusters

of households (such a cluster may consist of, e.g., four neighboring households). The reinterview sample can also be a two-stage sample, where the original interviewers are primary sampling units, and respondents (or households or clusters) within interviewers are secondary sampling units (ssu). Such a two-stage sample permits a proper allocation of ssu's over interviewers.

2. *The choice of reinterviewer.* The reinterviewers can be selected from the same pool of interviewers as the original interviewers. They may also be selected from among the most experienced interviewers in this pool. A third option is to select the reinterviewers from a group of supervisors. By "best" interviewers we mean the most experienced interviewers or supervisors.

3. *The choice of respondent.* The respondent can be the same as in the original interview; he or she can be chosen according to the same procedure as in the original interview ("original respondent rule"); the respondent might be the most knowledgeable person in the household, or each person could respond for himself or herself ("self-response").

4. *The design of the reinterview questionnaire.* The reinterview questionnaire may be exactly the same as the original questionnaire, or may contain a subset of the original questions. To achieve "true" values, the reinterview questionnaire may contain probing questions.

5. *Whether or not to conduct reconciliation.* When the responses obtained during the reinterview differ from those obtained in the original interview the differences are evaluated through a process called *reconciliation.* During reconciliation the respondent is provided with the information received in both interviews and asked to determine what is the correct information. We will return to this subject in Section 15.3.2.

6. *The choice of mode.* The choice is between telephone and face to face interviews. See the discussion in Section 15.3.4.

Table 15.3 shows the optimal requirements for each of the four basic designs, 1a, 1b, 2a, and 2b, of reinterview.

15.2.3 Reinterview Programs at the U.S. Bureau of the Census

The policy at the U.S. Bureau of the Census is to insist on some kind of check in each survey to ensure that the interviewers conduct the interviews. However, the measurement of response error is only pursued when the sponsor is interested and willing to foot the costs. Thus, designs 1a and 1b are quite common while designs 2a and 2b are used more rarely.

Table 15.3. Overview of Optimal Requirements for Various Reinterview Designs

	Purpose			
Factor	Detect Falsification 1a	Evaluate Field Work 1b	Estimate Response Variance 2a	Estimate Response Bias, 2b
Sample Selection	Two stage[1]	Two stage[1]	One stage[3]	One stage[3]
Choice of Reinterviewer	Supervisors		Same pool as original	Best
Choice of Respondent	Original interview respondent		Original respondent rule	Most knowledgeable or self
Design of Questionnaire	Sample of questions[2]	Sample or all	Exactly the same	Probing questions
Reconciliation	Generally	Yes	No	Yes
Choice of Mode	Telephone if possible		Same as original	Face to face

[1] Ideally, parts of all interviewer's assignments shall be reinterviewed, that is, a one-stage reinterview sample shall be used. This is, however, usually too expensive in practice.
[2] The sample of questions should be spread throughout the questionnaire to ensure that all parts are covered in the reinterview questionnaire.
[3] Since error components are to be estimated from designs 2a and 2b, the reinterview subsamples must be representative subsamples of the original survey sample.

The reinterview programs for the U.S. decennial census use design 1a, designs 2a and/or 2b. Design 1a is done during the enumerator follow-up of the mail nonresponses. Designs 2a and/or 2b are completed later.

The *regular reinterview programs* use several designs and are conducted continually.

In this chapter, we report research from the reinterview program for the *Current Population Survey* (CPS). The program is conducted monthly and attempts to meet the objectives of all four designs within one design. As a result, certain sacrifices in efficiency or in meeting the assumptions are made because of the incompatibility of the four designs.

Two subsamples exist within the CPS reinterview sample itself; a 75 percent subsample in which differences between the original interview and reinterview are reconciled, and a 25 percent subsample in which they are not. Estimates of response bias have been obtained from the former and estimates of response variance have been obtained from the latter.

The large subsample is used to evaluate the performance of the interviewing staff.

The CPS design works relatively well for evaluating interviewer performance. However, there are several criticisms of the reinterview program with regard to the measurement of response error:

- For response variance estimation, the assumption that the reinterview is an independent replication of the original interview is violated. The reinterviewers are supervisory personnel, who are supposedly the "best" interviewers. Also, because the results are used to evaluate the interviewers, the reinterviewers try to contact the original respondent instead of using the "original respondent rule" as discussed in Section 15.2.2 above.

- For response bias estimation, the assumption that the reconciled reinterview yields the truth is also violated. Violation occurs primarily by the use of household respondents (i.e., one respondent responds for all persons in a household) and because the reinterviewer conducts the reconciliation, making it possible to change the reinterviews. Other factors, such as *recall lapse* and *falsification*, may also play a role.

Most reinterview programs for other major recurring surveys at the U.S. Bureau of the Census currently focus exclusively on evaluating interviewer performance (design 1b). There are a few surveys where the primary aim of the reinterview is to measure simple response variance (design 2a) and several one-time surveys where the sole purpose of the reinterview is to ensure against falsification of data (design 1a).

15.3 REINTERVIEW DESIGN ISSUES

15.3.1 The Reinterview Questionnaire

Designs 1a and 1b

When the purpose is to evaluate interviewer performance, certain criteria apply in the selection of questions for the reinterview:

1. *The questions should be factual where the same answers would be expected in the interview and reinterview.* This criterion aims at eliminating "soft" or opinion type questions, which may have high response variance. Such questions are not useful in evaluating the performance of an interviewer.

2. *The questions should be ones whose distribution of responses is not heavily skewed in one direction.* This criterion aims at selecting questions

which the interviewer must ask. It should not be possible for the interviewer to easily guess the correct answers.

3. *The questions should be asked of all or most respondents and not skipped for a sizeable portion of the sample.* This criterion aims at ensuring enough responses to evaluate how well the interviewers understand the proper procedures. With the reinterview sample proportion being rather small (around 5 percent) this becomes a critical issue.

4. *The questions should be scattered throughout the questionnaire.* This criterion aims at ensuring that the interviewers are asking all parts of the questionnaire. Surveys which require lengthy interviews may result in interviewers conducting "shortened" interviews. In these instances the interviewer does not conduct a complete interview, but instead uses the results from part of the interview to complete the rest.

Design 2a
When the purpose of the reinterview is to estimate response variance, then the original interview questions must be repeated in their exact forms. The only exception is that some questions may have to be reworded to apply to the appropriate time period because of the time lag between the two interviews.

If the original interview is fairly lengthy, it may be possible to include only a subset of the original survey questions. Then, careful consideration must be given to ensuring the questions are not taken out of context. Earlier questions may have to be inserted to ensure the appropriate sequence of questions are asked as in the original interview.

Design 2b
When the purpose is to estimate response bias, then

(1) the original question may be repeated and differences between the two responses reconciled, or

(2) there may be a series of questions replacing the original question in an effort to obtain the "true" or correct response.

Note: When a reinterview is designed to meet several purposes within the same design, the reinterview questionnaire will contain separate questions to satisfy each particular purpose.

15.3.2 Reconciliation

In order to evaluate interviewer performance, that is, purposes 1a and 1b, reconciliation is performed. It is then not only important to determine the correct response, but also determine who is responsible for the difference in responses. When the purpose of the reinterview is to measure response bias, purpose 2b, reconciliation is also almost always conducted.

Reconciliation may be done during the same contact as the reinterview or in a third or separate contact with the household. In the latter situation the reinterviewer is not provided with the original interview responses.

When reconciliation is done as part of the reinterview process, provision must be made for making available the original responses to the reinterviewer. This information may be placed within the reinterview questionnaire or in a separate document. The reinterviewer is instructed not to look at the original responses until the reinterview is completed.

When the reinterview document is not lengthy, the original information is usually added at the end of the questionnaire. This procedure has its drawbacks because the reinterviewer must flip back and forth to compare and determine whether the original interview and reinterview responses are different. Using a separate document is somewhat easier; however, the reinterviewer must work with both forms at the same time and, if there is no place to put them down, reconciliation can still be rather cumbersome.

Currently at the U.S. Bureau of the Census a new approach to the design of a combined reinterview and reconciliation form has been implemented for the Survey of Income and Program Participation. It involves lining up the original and reinterview responses to a particular question so they face each other on opposite pages. The original responses are covered by a blank page which can be opened up during reconciliation.

When reinterview includes reconciliation, the form must provide space for (1) recording the correct information; (2) determining who is responsible for the difference (if used for interviewer evaluation); and (3) determining what caused the difference to occur. For the latter, space is also provided to obtain both the interviewers' and the respondents' explanations.

Research on reconciliation has been conducted in connection with the CPS. Comparisons have been made on the estimates of simple response variance (the gross difference rate) from the two reinterview subsamples. Recall the CPS reinterview sample is subdivided; in one part of the sample differences are reconciled and in the other part they are not. In the latter the reinterviewer is not provided with the original interview responses. Theoretically, the estimates from both subsamples should be the same. However, U.S. Bureau of the Census (1963) and O'Muircheartaigh (1986) show there are in fact substantial differences in

Table 15.4. Gross Difference Rates in CPS Reinterview: Unreconciled Subsample and Reconciled Subsample (Before Reconciliation), 1958–1961 and 1982–1984

	Average Gross Difference Rate			
	1958–1961		1982–1984	
Category	Unreconciled Subsample	Reconciled Subsample	Unreconciled Subsample	Reconciled Subsample
Civilian				
Labor Force	4.1	2.3	4.9	2.5
Employed	3.9	2.1	3.8	1.9
Unemployed	1.9	1.1	3.1	1.5
Nonagriculture	3.5	1.9	3.7	1.9
Full-Time	4.4	2.3	4.3	2.5
Part-Time	4.6	2.6	5.0	2.7

Source: U.S. Bureau of the Census (1963) and O'Muircheartaigh (1986).

estimates from the two subsamples. As is shown in Table 15.4, this difference is of the same magnitude over the years 1958–1961 and 1982–1984.

The U.S. Bureau of the Census (1963) concluded "evidence is unequivocal that reported differences before reconciliation are not independent of original CPS results." This report points to the difficulty of conducting an independent reconciliation, if the reinterviewers are to reconcile.

Within the CPS framework, Schreiner (1980) reports an independent reconciliation experiment, conducted from April 1978 through September 1979. During this period differences in the unreconciled subsample were reconciled independently, that is, by a third contact with the household. The study did not reveal any significant differences between the net error rates for the two reinterview subsamples, i.e., no difference in the reconciliation methods could be found. Another conclusion was that differences attributable to not only the reinterviewers but also to the interviewers and the respondents were being suppressed when the reinterviewers conducted the reconciliation.

Limitations of Reconciliation

The most obvious limitation of a reinterview program using reconciliation to measure validity is that a respondent may knowingly report false information. If he or she is consistent in the reporting of that information then there is no way in which the reinterview will yield the "true" information. In a study of the quality of the CPS reinterview data, Biemer and Forsman (1990) concluded that up to 50 percent of the errors in the original interview are not detected in the reconciled reinterview.

Even when the respondent chooses to cooperate and provide correct information, reconciliation may still be less than a perfect instrument for arriving at the truth; for example, the respondent's attitudes about which of the responses is correct. If the respondent is inclined to support the original response in order not to look bad, then we will not be getting at the truth. This is more likely to occur when reconciliation is conducted during the same contact as the reinterview.

There are other problems with reconciliation. If the respondent does not understand the question, an erroneous response is likely. Who the respondent is also comes into play. Unless the person conducting the reconciliation does it with the most knowledgeable respondent, the information obtained may not be completely valid.

Finally, reconciliation as normally conducted at the U.S. Bureau of the Census relies heavily on the reinterviewer to ask the appropriate probes. All that is provided in the way of instruction is to present the respondent with the different responses and then ask "What is the correct information?" How to probe for the correct information is left to the reinterviewer's discretion.

15.3.3 Time Lag Between Original Interview and Reinterview

The time lag between the original interview and the reinterview varies between a few days to several months. We distinguish two situations:

For design 1a, that is, if the purpose of the reinterview is to detect fabrication of interviews, the time lag should be as short as possible. The important point in this case is that the respondent remembers whether the interview has taken place or not.

For designs 1b, 2a, and 2b, that is, if the purpose is to provide feedback to the interviewers on their performances or to measure error components in survey models, time lag is a critical issue. The time lag should not be so great as to cause the respondent to forget what actually happened, nor so short that the respondent merely remembers the original response and repeats it. How long the time lag can be is dependent upon the nature of the data being collected. The more the data are subject to variation, the shorter the time lag.

Ideally, for designs 1b, 2a, and 2b, the two interviews should be independent. This is a necessary assumption for the estimation of simple response variance (design 2a). Independent interviews are also important when the reinterview aims to find the "truth," since dependence can lead to repetition of erroneous answers in which case no reconciliation is conducted.

Usually, a specific time lag is adopted in each survey depending on administrative factors, the purpose(s) of the reinterview and the type of variables of study. For example, in the CPS, dealing with labor force items, a one-week time lag is used as a compromise between different aims of the reinterview program. In the Content Reinterview Studies of the U.S. Population Census the time lags are usually several months. The variables of study are stable ones, such as race, sex, education, etc. In the World Fertility Survey, conducted in several developing countries during the time period 1972–1983, the reinterview program dealt with relatively stable demographic items, and time lags up to several months were used.

Research on optimal time lags in different reinterview situations is rare. In a pilot survey, Wikman (1980) asked the reinterview respondents if they remembered the original answers to about 100 questions regarding living conditions. He found that none of the 20 respondents remembered their answers after about four weeks. This result led to the use of a three-week time lag in a sequence of five reinterview studies in the Swedish Survey on Living Conditions conducted during the 1980's. O'Muircheartaigh (1982) suggests a regression type analysis to test independence between the two interviews. Using measures based on the Census Bureau's survey model, Bailar (1968) compares results from three-month and six-month time lags in the 1960 U.S. Population Census Post Enumeration Survey. She found that, for some items, the shorter time lag should be preferred. For the majority of the items, no differences could be detected with the sample sizes used.

The problem of dependence between the two interviews may in certain situations be dealt with by using different respondents in the interviews. This can be done by using proxy interviews or, as was the case in the 1960 U.S. Population Census Post Enumeration Survey, asking the respondent's employer about labor force items in the reinterview.

15.3.4 Field Implementation Issues

One issue associated with the field work is *who should conduct the reinterview*, interviewers or supervisors. For designs 1a and 1b, the most appropriate choice is the supervisor. For design 2a the choice rests with the interviewer and for design 2b it should be the "best" interviewers, which in some cases may be the supervisors.

The emphasis of most reinterview programs at the U.S. Bureau of the Census is to control the quality of interviewing; consequently, the usual choice is to use the supervisors as reinterviewers. This limits the use of the reinterview as a vehicle for measuring response variance.

A second issue associated with the field work is the *mode of the*

Table 15.5. Reinterview Costs, Expressed as Percentage of the Total Field Costs, for Four U.S. Census Bureau Surveys. October 1988 Through September 1989

Survey	Percentage of Total Field Costs	Reinterview Sampling Percentage
Current Population Survey (CPS)	2.4	3.3 (1 in 30)
National Crime Survey (NCS)	2.4	5.6 (1 in 18)
Survey on Income and Program Participation (SIPP)	0.8	5.6 (1 in 18)
Consumer Expenditure Survey (CES)	0.5	5.6 (1 in 18)

reinterview, telephone or face to face interviewing. The geographical spread of the reinterview sample generally dictates that it be conducted by telephone, if at all feasible. Even with a two stage reinterview sampling design, with interviewers as primary sampling units, there is still considerable cost due to supervisory staff traveling.

When the original survey interview is conducted using face to face interviewing, this can have a significant impact on the use of the telephone for conducting reinterview. Flash cards and other visual aids cannot be used over the telephone and this may severely restrict the content of the reinterview.

The use of the telephone for reinterview causes problems for both evaluating interviewer performance and measuring response error. First, in evaluating interviewer performance only certain parts of the questionnaire can be checked. Second, in measuring response variance, the assumption that the reinterview is an independent replication of the original interview may be violated. For response bias it may be more difficult to determine the truth.

For the reinterview programs conducted at the U.S Bureau of the Census the use of the telephone is maximized. This results in considerable cost savings. Table 15.5 shows the costs for four reinterview surveys represented as a percentage of the total field cost.

The reinterview sampling rate is included in the table to illustrate that costs per unit are less in the reinterview than in the survey itself. The differential in costs between the CPS/NCS and the SIPP/CES relates to the ratio of telephone in the original survey to telephone in the reinterview and also to the amount spent on the listing check operation (normally included as part of the reinterview operation).

A third field work issue is how to introduce the reinterview to the respondent in order to obtain cooperation. This is usually done by

Table 15.6. CPS Reinterview Nonresponse Rates (1982–1988)

Year	Original Interviews	Reinterview Nonresponse	
		Number	Rate (%)
1982	36,074	3,097	8.6
1983	36,004	3,554	9.9
1984	34,625	3,733	10.8
1985	34,580	4,255	12.3
1986	24,917	2,824	11.3
1987	21,646	2,785	12.9
1988	19,569	2,676	13.7

explaining to the respondent that the purpose is to check on the quality of the interviewer work or the data, not to check on the respondent.

The use of the reinterview to measure response error can be severely restricted if the reinterview nonresponse rate is too high. At the U.S. Bureau of the Census the goal is to limit to 10 percent the nonresponse rate for households that were interviewed in the original survey. In recent years for CPS, as shown in Table 15.6, staffing and budget reductions have hampered this effort.

A fourth issue is how to limit the reinterview respondent burden both in length and frequency. For surveys where households or persons are interviewed numerous times over a fairly short time period this is particularly important.

In the CPS, where households are in sample four consecutive months, retired for eight, and then back in for four more consecutive months, a reinterview is conducted only once per household. This creates a problem for the measurement of response error because the result is a bias in the reinterview sample (see Bailar, 1974).

Thelen (1979) investigated the impact of reinterview on the basic CPS. She concluded that the overall effect of the reinterview on the subsequent month's nonresponse rate was very minor (1.7 percent). She also found that the third contact taken in the independent reconciliation experiment (Section 15.3.2) did not have any adverse effect on the subsequent month's nonresponse rate.

15.4 EVALUATING INTERVIEWER PERFORMANCE

The two purposes, identification of false data (1a) and the evaluation of interviewer performance (1b) will be discussed separately in this section.

15.4.1 Methods to Detect Interviewer Falsification

Deliberate falsification of interview results by interviewers is one of the most serious problems leading to response error. As pointed out by Biemer and Stokes (1989), interviewer cheating can take different forms: the interviewer may fabricate answers to all or a subset of the questions in the questionnaire. An interviewer may deliberately deviate from prescribed procedures, such as conducting a telephone interview where a face to face interview was indicated, or conducting the interview with a willing but inappropriate respondent. The significance of the cheating problem is illustrated by the fact that the U.S. Bureau of the Census, during the years 1982 to 1987, identified 205 interviewers who falsified interviews.

When designing a sampling plan for detecting interviewer falsification in recurring surveys, there are certain factors to consider. These include: (i) how often a specific interviewer's assignment should be reinterviewed; (ii) the optimal allocation between respondents (or households/segments of housing units) over interviewers; and, (iii) the time for reinterview (it should not be predictable).

Detecting interviewers who falsify data requires a two-stage reinterview sampling plan (Table 15.3). One problem in sampling for a 1a and 1b reinterview design is to avoid the possibility that an interviewer can predict the time of the reinterview. For example, in an earlier CPS design, interviewers were placed in six separate groups and were selected twice a year for reinterview; once from January to June and a second time from July to December. Thus, an experienced interviewer whose assignment was reinterviewed in January knew he or she was "safe" until July. In the present CPS design, the interviewers are placed in 15 groups and each group is randomly assigned to a particular month within a 15-month period. In addition, in each month a second group is randomly selected for reinterview from one of the remaining 14 groups. In this way unpredictability is introduced into the design. In order to limit the number of times an interviewer is selected for reinterview over a 15-month period, an upper limit is set at four times.

Schreiner, et al. (1988) report a Census Bureau study aimed at designing an optimal reinterview sampling strategy for detecting interviewer falsification. The research areas included:

- Whether interviewer falsification is clustered; for example, interviewers falsifying data for all households in a cluster of neighboring housing units. If it is, then resources can best be used by spreading the reinterview sample over as many clusters as possible rather than selecting entire clusters.

Table 15.7. Rate of Interviewer Falsification for Three Surveys Conducted at the U.S. Bureau of the Census

Survey	Rate (%)
Current Population Survey (CPS)	0.4
National Crime Survey (NCS)	0.4
New York City Housing Vacancy Survey (NYC-HVS)	6.5

Source: Schreiner et al. (1988)

• The patterns of falsification. If, for example, interviewers who falsify data tend to do so infrequently, but on a large proportion of their assignments, then the best strategy is to reinterview a small amount of their assignments fairly often. On the other hand, if interviewers who falsify data tend to do so on a small proportion of their assignments quite often, then the best strategy is to reinterview as much of their assignments as possible less frequently.

The study resulted in three major findings:

1. The shorter the length of service, the more likely an interviewer will falsify data.

2. Certain differences occur in the falsification patterns of new (less than one year of service) and experienced interviewers (a year or more of service). Experienced interviewers tend to have a more sophisticated falsification pattern than new interviewers; e.g., experienced interviewers cheat less frequently and usually avoid falsifying data the first time a respondent is interviewed in a panel survey.

3. There is little or no clustering of falsification in neighboring housing units.

The implications of these findings are:

1. It may be useful to reinterview work of newer interviewers more frequently;

2. For experienced interviewers an effective reinterview sample ought to be concentrated on respondents for which cheating is most likely; and

3. Selecting clusters of neighboring housing units appears to be a viable approach.

Another result from the study is displayed in Table 15.7. The falsification rate is derived as the proportion of falsifying interviewers detected through the regular reinterview among all interviewers. (For details on the definitions of falsification rate, see Schreiner, et al., 1988.)

The reinterview rate depends on the kind of interviewer staff used. CPS and NCS are recurring surveys with a relatively stable staff of interviewers. By contrast, the NYC-HVS is done once every three to four years and is carried out by a temporary staff of interviewers.

Biemer and Stokes (1989) propose a statistical model for describing dishonest interviewer behavior. This model assumes cheating is a random event governed by a probability distribution whose parameters depend on the interviewer.

15.4.2 Administrative Fabrication System

The traditional approach to the design of a reinterview program to detect interviewers who are falsifying data involves a random selection process in which all interviewers have an equal probability of selection. Another method of identifying falsifying interviewers is to stratify the sample of interviewers, using certain performance variables to define strata. For example, an interviewer that turns in a high proportion of unemployed persons in an area which has low unemployment might be suspected of falsifying data. Other performance variables can be production rate (cases completed per day), mileage, and vacancy rate. At the U.S. Bureau of the Census the approach of using performance variables to stratify the sample of interviewers is referred to as an "administrative fabrication system." Such administrative samples were used as a part of the reinterview operation during nonresponse followup in the 1990 U.S. Decennial Census as well as in the 1988 Dress Rehearsal for the 1990 Census.

15.4.3 Tolerance Levels

The vast majority of the interviewers are honest but some of them might have a genuine misunderstanding of the proper procedures. What is called for here is some kind of quality control procedure that determines whether interviewer performance is acceptable or not.

Tolerance tables, based on statistical quality control theory, are used to determine if the number of differences after reconciliation exceeds a certain acceptable limit. If they do, the interviewer's performance is considered unacceptable. Some form of remedial training is then provided to bring the interviewer's performance to a satisfactory level.

The tolerance limits vary somewhat and are set based upon the nature of the data being gathered. This is because, while the attempt is made to select "factual" questions, the levels of respondent error for these questions differ from survey to survey. For the reinterview programs at the U.S. Bureau of the Census, the tolerance limits applied to the differences in the content of the questionnaire are based upon a theoretically acceptable quality level ranging between six and ten percent. Mersch and Dyke (1978) discuss the construction and use of quality control tables.

For the CPS, interviewer failure rates concerning the content of the questionnaire are based upon an acceptable quality level of six percent. The data gathered pertain to the labor force status of individuals within the household. Over the ten year period of 1979 through 1988, the percentage of interviewers failing in this area averaged 1.5 percent (see Leonard, 1990).

For the CES, interviewer failure rates concerning the content of the questionnaire are based upon an acceptable quality level of ten percent. The data gathered pertain to consumer expenditures that are subject to higher respondent variation than labor force data; therefore, the interviewer is allowed a higher level of error in the CES. The reinterview program for this survey has been in place since November 1979. Since the onset of the reinterview program through December 1987 the percentage of interviewers failing in the recheck of the content of the questionnaire has averaged 3.5 percent (see Pennie, 1989).

15.5 MODEL-BASED ANALYSIS OF REINTERVIEW DATA

This section reviews the analysis of data from the basic reinterview designs that estimate response variance (design 2a) and response bias (design 2b).

15.5.1 Simple Measures of Consistency

If we regard the interview and the reinterview as two separate surveys, the most simple measure of the deviation between the interview and the reinterview is obtained by comparing the summary measures of the two resulting distributions. A drawback with direct comparison of statistics computed from the two data sets is, however, that it would itself provide estimates of precision based on only one degree of freedom. Instead, we will discuss here measures of consistency based on the *individual* response deviations.

Several measures of consistency have been developed outside statistics, to be applied when two measurements are conducted on each unit. Perhaps the most simple one is the proportion of cases identically classified in the two measurements, the *index of crude agreement.* This measure does not take into account the fact that some agreement will occur by chance even if the measurement is completely random. To avoid this and other drawbacks, Cohen (1960, 1968) elaborated the index. These indices and other measures of consistency are further discussed in Landis and Koch (1976) and in O'Muircheartaigh and Marckwardt (1980).

15.5.2 Survey Models

Although the above measures reflect the variability in a measurement procedure, they do not meet the need for an integrated treatment of different errors in surveys. Measures defined under a survey model can be interpreted in terms of components of the total error of the survey estimates. Some of these components can be estimated by means of reinterview data. Critical assumptions usually made then are (1) the reinterview is independent of the first interview and (2) the two interviews are conducted with the same data collection procedure and under the same general conditions, or, alternatively, (2′) the reinterview and a reconciliation provide "true" values. If the assumptions are met, measures like the net difference rate (Eq. 15.2) can be interpreted in error terms. Often, however, it is hard to verify whether the assumptions are met or not.

The Census Bureau survey model, described in Chapter 24, Section 24.2.1 is the most commonly used survey model. Important study schemes for estimating error components of this model by means of reinterview data were presented by Fellegi (1964) and Bailar and Dalenius (1969). Bailar (1968) and Biemer and Forsman (1990) discuss various problems when the basic assumptions of the model are not met. A model similar to the Census Bureau model was used to analyze reinterview data in the World Fertility Survey. See, for example, O'Muircheartaigh and Marckwardt (1980). For general discussions of survey models, including the analysis of reinterview data, see Lessler (1984) and Forsman (1989).

15.5.3 Alternative Model-Based Approaches for Analyzing Reinterview Data

Different approaches exist to extend the traditional survey models based on error decomposition by including characteristics on, for example, the study individuals, the interviewers, or other aspects on the survey environment. The purpose is to identify characteristics of error-prone groups of respondents and interviewers.

Forsman (1987) used a logistic regression model for reinterview data from design 2b. The 0–1 response variable reflected consistent/inconsistent responses between two interviews. Since true values were assumed in the reinterview, an inconsistency could be interpreted as an error. Using a more general approach, Hill (Chapter 23) studied the number of inconsistencies for a respondent over not only one but several questions. If inconsistencies occur independently over a fairly large number of different questions, the number of inconsistencies in a particular questionnaire may be modeled by a Poisson distribution.

The Hill approach is suitable when there are a number of questions in a questionnaire that are equally important. For surveys with one dominating variable, like labor force status in a labor force survey, logistic regression is more appropriate.

In recent years, models using latent variables have been used or suggested for analyzing reinterview data. Johnson and Woltman (1987) used the Rasch model, a special case of the latent trait model. The advantage of the Rasch model in this context is that it provides a conceptual framework for evaluating errors that is more relevant for variables that cannot easily be conceived of as having a true state corresponding to one of the response categories of a question, than is the framework used in conventional survey models. Another class of models that may be used for analyzing reinterview data is structural equation models, especially the LISREL model. See Munck (Chapter 29).

15.6 USE OF COMPUTER ASSISTED INTERVIEWING FOR REINTERVIEW

Computer Assisted Telephone Interviewing (CATI)

Under CATI, the perspective on reinterview changes since interviewing conducted at a centralized telephone facility can be monitored. The interviewers in a centralized setting know they can be monitored, but not when. Monitoring can be used both to deter falsification (design 1a) and also to provide feedback to the interviewers if they need help (design 1b). This allows the focus of reinterview to be exclusively on the estimation of error components (designs 2a and 2b).

CATI offers the opportunity for designing and conducting reinterview in a manner that satisfies the major requirements for obtaining uncontaminated estimates of response error.

For both the measurement of response variance and response bias there are certain advantages associated with a CATI environment. For response variance, it becomes more feasible to:

1. Use regular interviewers to conduct reinterview rather than supervisory personnel.

2. Instruct the interviewers to follow the original respondent rule rather than trying to reinterview the original respondent.

For response bias, it is now possible to use the best interviewers to conduct reinterview and conduct the reinterview with the most knowledgeable respondent or the respondents themselves.

These changes can easily be implemented because the reinterview is not being used as a vehicle for evaluating the performance of the interviewers.

For both response variance and response bias there is much more flexibility as to when the reinterview can be conducted (time lag). In a centralized telephone facility, reinterviews can be conducted much more quickly (if this is what is desirable) than in a decentralized field operation.

Finally, for response bias, there are some specific advantages concerning reconciliation of differences.

1. The reinterviewer has no access to the original interview data until after the reinterview is completed.

2. It is not possible for the reinterviewer to alter the reinterview response once the reinterview is completed.

3. Recognition of when there is a difference and what constitutes a difference can be programmed into the instrument. The computer makes the comparisons rather than the reinterviewer.

Research on CATI Reinterviews at the U.S. Bureau of the Census

A CATI reinterview study to measure simple response variance for the labor force data from the CPS was conducted in 1988 at the U.S. Census Bureau's Hagerstown Telephone Facility (HTC). See Schreiner (1989).

The reinterview sample was drawn from cases originally interviewed at the HTC. The sample was an unclustered sample of one out of every 18 households.

The reinterview procedures were set up as much as was feasible to replicate the original interview scenario: (i) the regular staff of HTC interviewers were used to conduct the reinterview, (ii) the reinterview was conducted as soon after the original interview as feasible (generally, reinterview began two or three days after original interviewing started), and (iii) the original respondent rule was applied to the reinterview. This meant asking to speak to the household respondent that completed the interview the previous month since this was the person the interviewer initially asked for at the time of the original interview.

Although the study was not designed to be directly comparable to the measures of simple response variance obtained from the regular ongoing CPS field reinterview program, it is interesting to note the close agreement between the two methodologies displayed in Table 15.8.

Table 15.8. Indices of Inconsistency, in Percent, for People Employed, Unemployed, and Not in the Labor Force for Matched Persons: CPS/CATI Reinterview ($n = 1454$) and Regular CPS Reinterview ($n = 8217$) for January–December 1988

Category	Reinterview Methodology	Index of Inconsistency[1]	90% Confidence Limits
Employed	CPS/CATI	7	5 to 9
	Regular CPS	8	7 to 9
Unemployed	CPS/CATI	38	29 to 50
	Regular CPS	38	34 to 43
Not in the Labor Force	CPS/CATI	9	7 to 11
	Regular CPS	10	9 to 11

[1] Index of inconsistency is the ratio of the simple response variance to the total variance of individual responses multiplied by 100.
Source: Schreiner (1989).

The 1990 Content Reinterview Study (CRS) of the 1990 Census was also conducted using CATI. The CRS was designed to measure response bias or variance for a selected group of the 1990 Census long form sample items. The CRS attempted to maximize self-response, and the regular staff of interviewers conducted the reinterviews.

The questions for which a response variance measure was required were identical to those of the 1990 Census presentation. A series of probing questions were used for those questions in which a response bias measurement was desired. No reconciliation was performed.

Future CATI plans for CPS at the U.S. Bureau of the Census call for developing a reconciled reinterview instrument to measure response bias. Initial plans include the use of supervisors to conduct the reinterview. Reinterview will most likely begin the day following interview closeout and the reinterviewers will attempt to obtain the reinterview from the original household respondent (the one who was considered the most knowledgeable).

Computer Assisted Personal Interviewing (CAPI)

The use of laptop computers in interviewing offers certain possible advantages for reinterview compared to paper and pencil interviews. These advantages concern the reconciliation and are the same as those mentioned above for CATI reconciliation. We disregard here the possibility for an interviewer to manipulate the computer. Since CAPI is conducted in a decentralized setting, it does not offer the possibility of unobtrusively monitoring the interviewing.

15.7 CONCLUDING REMARKS

When designed properly, reinterview can be an extremely useful tool for estimating and reducing measurement errors. According to information available in the form of published reports, it is, however, used by just a few statistical agencies. This is probably because the method is considered fairly expensive. We have shown in Table 15.5, however, that with the use of the telephone, field reinterviews can actually cost less per case than original survey interviews. Since reinterview is effective when conducted on a subsample of the original survey sample, the costs can be moderate. Also, with the advent of computer assisted interviewing in a centralized setting, costs can be kept minimal while the usefulness of the reinterview is increased.

Within the chapter we point out certain limitations regarding the use of reinterview to estimate response error. Concerning estimates of response bias, a respondent may report consistent but invalid information. Concerning estimates of simple response variance, a respondent may report consistent information, but may be merely repeating what he or she reported in the initial interview (conditioning). In the former situation the result is an underestimate of the response bias and in the latter an underestimate of the simple response variance. In spite of this, reinterviews have often discovered substantial levels of response error.

The above conclusion leads us to one major reason why reinterview is not always considered a popular endeavor. It can be the bearer of bad news. However, if one is interested in reducing response error and improving the quality of the survey, a reinterview is warranted. Oftentimes, though, budget constraints and short term goals result in little, if any, effort in this regard. This is foolhardy. Short term savings can result in poorer data quality. While the development and implementation of a reinterview may initially raise costs, over the long run the results, if heeded, can lead to a significant improvement in data quality.

CHAPTER 16

EXPENDITURE DIARY SURVEYS AND THEIR ASSOCIATED ERRORS

Adriana R. Silberstein and Stuart Scott
U.S. Bureau of Labor Statistics

16.1 INTRODUCTION

Diary surveys are designed to obtain detailed recordings of events when or soon after they occur, thus potentially eliminating recall problems typical of interview surveys. Diary methods are used in many fields of data collection, such as marketing, health, time use, and nutrition. Most household expenditure (or budget) surveys use multiple collection methods, typically diary and interview, as discussed in many studies (e.g., Pearl, 1968; Neter, 1970; Sudman and Ferber, 1974; Kemsley, et al., 1980; Stanton and Tucci, 1982; Redpath, 1987). The U.S. Consumer Expenditure Survey (CE) is a national survey including two components with separate samples and methodologies: the diary, completed by respondents for two consecutive weeks, and the interview, with five quarterly interviews. Data from this survey are used to derive a picture of current living conditions, develop expenditure weights for the Consumer Price Index (CPI), and supplement some categories of the National Accounts. Methodological aspects of the CE have been summarized in recent years (Dippo, et al., 1977; Pearl, 1979; Jacobs, et al., 1989), and several studies have reported on measurement error effects (Silberstein and Jacobs,

---

This chapter is dedicated to the memory of Curtis A. Jacobs. The authors are grateful to Joni Ancona for statistical and computing contributions in Section 16.4. Opinions expressed are those of the authors and do not necessarily represent the official position of the Bureau of Labor Statistics. Due to special purposes of the study, estimates presented here differ from official estimates.

Measurement Errors in Surveys.
Edited by Biemer, Groves, Lyberg, Mathiowetz, and Sudman.

ISBN: 0-471-53405-6

1989; Silberstein, 1989 and 1990; and Tucker, 1990). These effects are further examined in this chapter, with emphasis on the diary method.

A review of the type of measurement errors prevalent in expenditure surveys is presented in the next section, followed by an analysis of measurement errors based on data from the 1987 CE. The performance of the diary method is analyzed in terms of degree of completion (Section 16.3) and reporting length effects (Section 16.4). The comparability of diary and interview reports for the same types of expenditures is analyzed in terms of specificity of reports (Section 16.5), and reporting levels and distributions (Section 16.6). Summary and conclusions can be found in Section 16.7.

16.2 MEASUREMENT ERRORS IN DIARY SURVEYS

16.2.1 Differing Effects in Diaries and Interviews

The collection mode affects the relative importance of survey errors, although error sources are similar across mode. Respondent conditioning, inadequate recall, and proxy reporting are potential sources of error at the data collection stage. A more general source is insufficient cooperation, in the form of poor or incomplete responses (Krosnick, 1991c). Respondent burden is often cited as a contributing factor to measurement errors, and long questionnaires are a primary cause, regardless of collection mode. A fast interview pace tends to aggravate this factor in an interview survey (Cannell, 1987). Diary surveys present a different type of burden, since respondents must take an active role in recording data for an extended time period.

Respondent cooperation tends to decrease and conditioning tends to increase with longer reporting periods in a diary and with a greater number of cycles of participation in panel surveys. Conditioning, however, is difficult to measure. The largest changes are typically seen very early in the diarykeeping process. One manifestation of conditioning is a tendency for respondents' behavior to change as a result of participating in a survey, although the conclusion drawn from the British Family Budget Survey experiments is that behavioral changes are not uniform nor systematic in one direction or another (Kemsley, et al., 1980, p. 70). Another manifestation of conditioning is a change in data quality, which may be positive when related to the respondents' learning curve, or negative when stemming from declining interest as time increases (Silberstein and Jacobs, 1989). A reporting period of 14 days is regarded as optimal in expenditure diary surveys (Kemsley, et al., 1980), but four-week diaries have also been suggested as very desirable (Rao and

Vidwans, 1989). Different lengths of diarykeeping may be appropriate for other subject matters; for instance, diaries as short as 24 hours are preferred when collecting time-use data (Juster, 1986; Lyberg, 1989). Systematic declines in expenditure estimates by diary day and week have been observed in the 1972/73 U.S. Diary Expenditure Survey (Dippo, et al., 1977; Pearl, 1979), in the British Family Expenditure Survey (Kemsley and Nicholson, 1960; Kemsley, 1961; Turner, 1961), and in other diary surveys. Additional evidence of these effects is found in this study of the current CE.

Recall errors in interview surveys are produced by the partial recollection of past events, increasing with longer reference periods and less salient events (Silberstein and Jacobs, 1989). Recall errors may affect diary reports as well, since lapses in diarykeeping often occur. In addition, some diaries may be completed through interviewer-coached recall at diary pickup. Some aspects of this procedure, allowed in the CE diary, are documented here. Salient events can be erroneously included in survey responses (or telescoped) at higher rates than less important events (Neter and Waksberg, 1965), thus distorting estimates. Telescoping errors affect interview reports more than diaries and can be greatly reduced by *bounding* techniques in repeated interviews (Silberstein, 1990). CE diary procedures tend to limit the extent of telescoping, as noted by Tucker (1990): first-week reporting tends to be bounded by the use of interviewers' examples from respondents' previous-week purchases, in explaining diarykeeping, and a natural bounding of the second week is accomplished by the first-week diary pickup.

Proxy reporting refers to reporting events about persons other than the respondent, a process that affects both diary and interview responses. Personal diaries eliminate proxy reporting errors, provided all eligible persons participate. Interestingly, large households, which would profit most from this procedure, were found to be the least likely to participate fully in the 1979–1980 Hong Kong Household Expenditure Survey (Grootaert, 1986). Several countries use multiple household expenditure diaries, among them France and the United Kingdom, whereas only one diary is provided to respondents in the United States. Certain demographic characteristics of respondents, especially lack of education, greatly reduce the ability to complete diaries (Rao and Vidwans, 1989), but they affect to a lesser degree the task of answering interviewers' questions. Social desirability effects are added distortions in both modes, but are more acute in face to face interviews. Diaries were found, for example, to give more accurate information about alcohol consumption than questionnaires in a small-scale study conducted in Finland (Poikolainen and Karkkainen, 1983).

Among other factors affecting data quality, questionnaire design is of greater importance in guiding diary respondents than interview respondents. The diary can be cumbersome or impractical, making the recording task difficult. The type of wording and examples can be sources of error since they are interpreted as cues, reminding respondents of certain items or events and not others. Different expenditure diary formats have been adopted in each country, ranging from a blank page, as the 1989 *Budget de Famille* in France, to a diary form with preprinted entries throughout, as the 1985 *Rilevazione sui Consumi delle Famiglie* in Italy. The effectiveness of different types of diary forms, especially journal versus product formats, has been investigated by Sudman and Ferber (1971). A small-scale Swedish experiment, conducted in 1983, compared an expenditure diary with preprinted headings to a diary with no headings; the diary with preprinted headings was considered favorable, even though it only increased reported expenses by five percent (Näsholm, et al., 1989). The current CE diary form is of the journal type with two pages per day, although increasing the number of pages is under consideration. A modification already adopted pertains to greater structure in the diary pages, suggested by results from the 1985 Diary Operational Test. This test compared two diary formats that differed by the extent of structure and preprinted entries (Tucker and Bennett, 1988; Vitrano, et al., 1988; Tucker, et al., 1989; Tucker, 1990).

16.2.2 Diary Methods in Expenditure Surveys

A basic design topic in household expenditure surveys is the choice of expenses for a mode of data collection (Lamale, 1959; Pearl, 1968; Kemsley, et al., 1980). A related topic is whether to limit a survey or mode to a single group of expenses in order to reduce respondent burden. The relative difficulty of the response task must also be considered in selecting the collection method. Diary coverage restricted to purchases of food resulted in higher recording levels in a small-scale experiment reported by Kemsley, et al. (1980, p. 65). Several expenditure surveys limit diaries to food and a few other types of expenses (e.g., the surveys in Mexico, Canada, and France). Other countries include all expenses in a diary, but limit the interview to larger, more infrequent, expenses. This method is used with data collection by interview prior to the diary, as, for instance, in the United Kingdom and Sweden, or with an interview after the diary, as in Italy and other countries. The Dutch Budget Survey offers an example of two diary components, the first an intense report for two weeks, the second a recordkeeping for a whole year, restricted to expenses above a certain amount.

Another design topic is whether the survey is ongoing or intermittent. A long interval between surveys leads to difficulties in building expertise in field personnel (Näsholm, et al., 1989) and may miss changes in economic conditions. Several countries (e.g., the United Kingdom and the United States) have continuing surveys, but most countries conduct only intermittent surveys, every five or ten years. A design adopted by some countries (e.g., France) includes a general survey at intermittent dates and a continuing survey dedicated to rotating groups of expenses, such as apparel. The CE has been ongoing since 1980 with two components, collected by the U.S. Bureau of the Census under contract to the U.S. Bureau of Labor Statistics (BLS). The interview is a panel survey with five waves, the first used primarily for bounding purposes. There are about 6,000 respondents per year, who are asked to report all expenses incurred during the previous three months. Expenses for groceries are reported as weekly *usual* amounts. The diary includes about 5,000 respondents per year, who are asked to record all expenses for two one-week periods. The overlap between the two data collection modes is justified on the grounds that the number of larger items is usually small in a one-week diary (Pearl, 1968). The interview, by its broad scope of data collection, permits microlevel studies of annual family spending patterns.

Estimation of annual expenditure means varies across countries and may be complex, due to different time periods and expenses covered by multiple surveys. CE *integrated* annual estimates are derived using both components (U.S. Bureau of Labor Statistics, 1990). Except for certain expense groups that require one or the other survey as the source (e.g., diary for grocery expenses, and interview for housing expenses), the method of integration involves detailed comparisons of estimates from the two sources, considering the frequency of reports and the estimated mean squared error (MSE). Estimates of personal consumption from the National Accounts are used to develop MSE estimates. While these comparisons indicate that both components underestimate certain commodities (e.g., apparel), detailed comparisons within these commodities lead to choices that are not always consistent with a priori expectations (Dippo, et al., 1977; Pearl, 1979). Comparisons of expenditure means and additional characteristics of the expenditure distributions, presented in this chapter, further investigate this topic.

The remainder of this section explains additional features of the current CE and describes the data used in the analysis. The diary and the interview, although independent, have the same multistage sample selection design with 109 primary sampling units. The unit of analysis is the *consumer unit* (CU), which comprises those members of a household who are related and/or pool their income to make joint expenditure decisions. Multiple CU's at the same address are asked to report separately, as in cases of students living in university-sponsored housing or other dwellings with unrelated persons economically independent from each other. Response rates have averaged 85 percent in both survey components. The diary sample design allows continuous data collection throughout the year, with a double sample size during the last six weeks in the year. This technique is designed to capture the increased spending for that season, but findings from this study raise doubts about its effectiveness. The issue of holiday spending has been addressed in other countries; the Netherlands, for example, has designed a special diary for this purpose.

The set of expenses analyzed pertains to commodities ordinarily chosen from the diary, such as food and beverages, and selected other commodities for which the source varies depending on the individual item. These include categories of apparel, home furnishings, entertainment, personal services, and transportation. Certain types of expenses, for which diary day variations and mode comparisons are not meaningful, are excluded from the study: major expenses (e.g., furniture, airline fares, automobiles), expenses with regular payments (e.g., insurance payments), and infrequent expenses (e.g., legal fees).

16.3 DIARY PLACEMENT AND COMPLETION

16.3.1 Diary Placement

Timing constraints for diary placement are used to insure coverage of all shopping days. The CE design requires that diary weeks be contiguous, and each sample case be assigned a date for the first-week diary placement, the *earliest placement date*. A seven-day leeway is allowed after this date, but diaries that cannot be placed within this deadline are, in most cases, considered noncontacts. A postponement of up to a month is allowed in the Swedish Family Expenditure Surveys in order to increase response rates, a procedure that can introduce bias because of respondents' self-selection; a 1978 study indicated that 90 percent of the respondents started the diary at the requested time (Näsholm, et al., 1989).

In the CE, diary start and placement have different procedures for each of the two diary weeks. Sample housing units (addresses) are first notified by a letter informing the occupants about the purpose of the survey and the upcoming visit by the interviewer. This visit, with no appointment, is used to obtain cooperation, record the family roster, and place the first week diary. There are two other visits: a middle visit to pick up the first-week diary and place the second-week diary, and a third visit to pick up the second-week diary. Income and employment information for each adult member is collected at the end of the third visit. Recordkeeping for the first-week diary is supposed to start on the day after placement. This procedure is followed in most cases (92%), whereas 5 percent of the diaries start on the day the diary is placed. The remaining diaries start two or more days after placement (some of these are Saturday placements for Monday scheduled starts). The second-week diary is to be placed on the day after the end of the first week, but, in practice, is sometimes placed together with the first-week diary (12%), other times on the last day of the first week (15%), and yet other times later than prescribed, requiring data collection by recall (11%). The percentage of late second-week placements is relatively small, considering the difficulty in arranging three household visits within a two-week period. An obvious explanation is that interviewers place some of the second-week diaries early, when it appears that there might be a problem in making a timely visit.

The design calls for an even placement of diaries by day of the week. The daily distribution is fairly even, except that few diaries start on Monday (5%). This is probably due to limitations in weekend placements, and, as a result, more than the average number of diaries start on Tuesday (21%).

16.3.2 Diary Completion

Followup procedures after diary placement are usually included in diary surveys, in order to maintain cooperation and improve reporting. The use of these procedures is not easy to enforce, however, and different guidelines may be established informally within the same survey operation, introducing regional office and interviewer effects in the survey. An example of high degree of followup is the 1985 Expenditure Survey in Singapore. Interviewers were instructed to visit the households daily to insure that diaries were properly filled out, and for households requiring interviewers' assistance, daily visits were timed, as far as possible, just after the housewives returned from markets to avoid memory losses (Majulan Singapura, 1985, p. 10). In several countries, interviewers' assessment of diary completion are used, resulting in descriptive codes for each respondent. Diaries not completed for the whole period and/or by each eligible family member are not considered responses in the United Kingdom. Screening of diary data may be implemented to exclude poorly completed diaries. In the United States, for instance, minimal acceptance criteria are used to reclassify as nonresponses about 4 percent of the diaries, through computer screening.

At diary pickup, CE interviewers check the diary pages and enter additional expenses recalled by the respondent. A number of explicit questions are asked for certain items (e.g., alcoholic beverages, gasoline, and mass transportation), if they do not appear on the diary pages. The interviewer assigns a *recall status* code at the end of the visit, summarizing the extent of completion by recall: 1) diary completed by respondent (referred to here as *no recall*), 2) *partial recall*, or 3) *total recall*. On average, three-fourths of the diaries are completed by respondents. Completion rates refer to this group of diaries, although these are not necessarily complete or do not include recalled expenses, since recall may be used during the week and in late placements, as discussed earlier.

Most respondents (90%) keep the diary both weeks, but only two-thirds complete it both weeks without recall at pickup. In most years, 10 percent of the diaries have partial recall, and 15 percent have total recall. These diaries exhibit lower expenditure levels than diaries completed by respondents: means are as large as 30 percent lower for diaries with partial recall, and as large as 60 percent lower for diaries with total recall, depending on the type of expenditure. These differences may indicate that few expenses are made by these respondents or that cooperation and/or ability to report are lacking. Completion rates are considerably lower (62%) for older respondents (65 and over) living alone, although higher rates (71%) are experienced by these respondents when not living alone. Respondents' education influences diary completion; not surprisingly, completion rates are significantly lower for those with less education. Regional variation is also apparent.

Diarykeeping patterns were examined in 1984 during the second-week pickup of the CE. In this study of 1,200 cases, Tucker (1986) found that 71 percent of CE respondents made entries immediately after expenses occurred or at the end of the day. The average completion rate was 81 percent. Respondent feelings about the diary ranged from positive (42%), to neutral (23%), e.g., diary is boring, or format is awkward, to negative (35%), e.g., requires too much effort. Interviewers answered that over two-thirds of the respondents seemed conscientious. The 1978 study reported by Näsholm, et al. (1989) found that 50 percent of Swedish respondents recorded expenses on the day of purchase.

16.4 DAY AND WEEK EFFECTS

16.4.1 Intraweek and Interweek Variation

Differences in reporting are usually observed within a diary period, typically with a decline by day and week. This phenomenon is best

Table 16.1. Diary Day and Week Effects: Percent Ratios of Expenditure Means by Day to Overall Mean

	Food at Home			Food Away from Home 1987		Apparel 1987		Other Diary Expenses 1987	
	1972–73	1987							
	r	*r*	*s*	*r*	*s*	*r*	*s*	*r*	*s*
TOTAL		100	—	100	—	100	—	100	—
WEEK 1		105	0.9	104	1.3	110	3.0	105	1.4
Day 1	148	147	5.3	121	4.7	138	12.7	134	5.6
Day 2	106	107	6.4	111	3.7	109	12.9	106	3.6
Day 3	102	102	3.4	103	2.7	103	11.3	113	6.5
Day 4	94	93	4.4	105	4.3	114	22.2	94	5.1
Day 5	101	98	3.9	106	7.5	98	10.4	103	5.1
Day 6	92	96	4.9	99	3.6	99	12.9	94	4.2
Day 7	87	96	4.6	87	2.9	106	10.1	89	6.2
WEEK 2		95	0.9	96	1.3	90	3.0	95	1.4
Day 1	116	110	5.0	104	3.5	104	11.3	103	4.1
Day 2	96	94	4.3	96	3.3	91	9.2	99	5.9
Day 3	91	90	4.0	102	6.0	86	7.1	96	6.8
Day 4	98	93	4.8	95	3.6	97	8.0	94	3.8
Day 5	91	95	4.2	95	3.3	71	7.3	88	4.9
Day 6	87	85	3.8	91	3.4	97	12.9	89	3.4
Day 7	90	95	4.0	87	4.2	86	7.8	98	7.2

Notes: *r*, ratios; *s*, standard error of ratios. The 1972–73 ratios reproduced from Pearl (1979, p. 49).

depicted when attention is restricted to diaries completed solely by respondents. Table 16.1 displays percent ratios of expenditure means, estimated using diaries with no recall in both weeks ($n = 3{,}935$). The source of this and subsequent tables is the 1987 CE. Each of the four expense groups shows a decline by diary day, with the largest decline between day 1 and day 2, especially in the first week. For food at home, it is possible to compare the 1987 results with the 1972–73 CE, also from diaries completed by respondents (Pearl, 1979). The similarity is striking, both surveys showing the first day of week 1 almost 50 percent greater than the overall estimate and 40 percent greater than the mean for week 1. The first day of the second week for food at home is 10 percent greater than the overall mean and 16 percent greater than the mean for week 2. The mean expenditure of the second diary week is 20 percent lower than the mean of the first week for apparel, and 10 percent lower for the other three groups.

Apart from daily fluctuations in shopping trips, both within an individual household and across households, a number of factors tend to affect reporting levels during the diary period. A peak on the first day, sometimes extended to the second and third day, is an expected pattern in diary surveys, and several hypotheses have been advanced. Of course, there is the possibility that some respondents neglect to turn pages of the diary booklet. A more likely explanation is a fatigue effect or reduced interest after the first day and, even more so, for the second week. This notion is supported by time-in-sample effects in panel surveys, and by the fact that many expenditure categories, including food, are considered underreported in the CE (Gieseman, 1987). Another viewpoint is that the first day is affected by external telescoping errors, that is, earlier purchases are mistakenly attributed to the first day. This would tend to occur when respondents delay making entries in the diary, and then attempt to reconstruct purchases; in these cases, however, subsequent diary days would be inflated more than the first, as events often seem more recent than they really are. A novelty effect may influence diarykeepers to make more purchases than usual at the beginning of the diary period. This hypothesis does not seem plausible as a general respondent attitude, since, as noted by Pearl (1979) and pointed out by detailed analysis of the 1987 data, the decreasing pattern by day appears consistently for many types of expenses. In addition, if there are changes in spending habits, they can occur at any time during the reporting period. Small surges in reporting, noted on the seventh diary day for some expenditures, may be interpreted as increased spending or as increased reporting, stimulated by the interviewer's visit.

16.4.2 Test Results

The CE survey design distributes starting days across all days of the week, and this is kept roughly in balance in field collection, except for lower numbers of starts for Mondays. Thus, means are available for every combination of diary day and the seven days of the week (week days Sunday through Saturday). The day the recordkeeping starts determines the relationship between diary day and week day. For instance, for households with start day Wednesday, diary day 3 is necessarily week day 6 (Friday) for both weeks. Means by diary day and week are computed as equally weighted averages of means by diary day, week, and week day. Equal weighting of the 98 means that result from this classification is applied also in computing total means, and the ratios shown in Table 16.1 reflect this procedure. (The 1972–73 ratios were also calculated in this manner, except that diary day ratios were computed as averages of ratios for each week day.)

Statistical tests of day and week effects are based on Hotelling T^2

Table 16.2. Diary Day and Week Effects: Results from Hotelling T^2 Test

	Overall (df = 6)	Day 1 vs Days 2–7 (df = 1)	Differences Among Days 2–7 (df = 5)
Diary Week 1			
Food at Home	64.3***	61.5***	6.2
Food Away	77.6***	17.5***	50.1***
Apparel	7.7	5.3*	0.9
Other	55.2***	29.1***	17.6*
Diary Week 2			
Food at Home	19.8*	10.7**	6.0
Food Away	11.8	5.1*	6.5
Apparel	9.0	1.7	8.8
Other	12.6	3.3	5.7

* Significant at 5 percent level.
** Significant at 1 percent level.
*** Significant at 0.1 percent level.

and ANOVA methodologies. For the Hotelling T^2, a 12×49 contrast matrix is applied to a vector of 49 means (combinations of diary day and week day) for each week separately, along the lines discussed in Johnson and Wichern (1982, Section 6.2) and Koch and Lemeshow (1972). The first row compares diary day 1 to the other diary days, rows 2 to 6 contrast the remaining differences among diary days, and rows 7 to 12 include contrasts among week days. The covariance matrix for the T^2 statistic is estimated by the balanced repeated replication method (with 44 replicates), reflecting the complex survey design. The total of 44 degrees of freedom restricts the effects that can be tested simultaneously. For instance, main effects for diary day and week day comprise 12 degrees of freedom and their interactions 36, so these cannot be tested together. Results appear in Table 16.2. The test for food at home, points out the unique day 1 effects in both weeks, and this contrast appears to explain nearly all diary day effects. Significant contrasts are found for food away from home and for the "other expense" group, but only in the first week. Day 1 of week 1 is higher for apparel, but to a lesser degree; none of the other contrasts is significant for apparel.

The ANOVA model chosen is a $7 \times 7 \times 2$ Latin square design with repeated measures, containing effects for day of the week, start day, diary day, and week, plus some interactions. This approach permits incorporating more factors of interest into a single analysis, than the Hotelling

Table 16.3. Diary Day and Week Effects: Results from Latin Square ANOVA.

	Diary Day $F_{6,42}$	Weekday $F_{6,42}$	Week $F_{1,7}$	Diary Day*Week Interaction $F_{6,42}$
Food at Home	19.0***	63.7***	14.6**	6.3***
Food Away	8.5***	70.4***	22.8**	2.5*
Apparel	1.5	17.6***	25.9**	0.3
Other	5.6***	23.9***	9.2*	4.7***

* Significant at 5 percent level.
** Significant at 1 percent level.
*** Significant at 0.1 percent level.

T^2. The drawbacks are that the complex design effects are not taken into account and certain interaction terms are confounded with main effects. Laid out as Plan 12 in Winer (1971, Chapter 9), diary day and week day are two factors of the Latin square, and the third factor reflects the grouping of diaries according to start day. There are two Latin squares (weeks 1 and 2) with the same pattern of levels of the first and second factors. Two means are obtained for each cell of this design by splitting the sample into two parts, from the two half-samples used to form the 44 survey replicates, resulting in a total of 196 observations entering the Latin square ANOVA. Table 16.3 shows the F statistics for tests of interest. Diary day is significant, except for apparel, and so is diary week, though not too strongly for the other expense group. The interaction of diary day and week is significant as well, except for apparel, establishing a lack of parallelism between the two weeks. This interaction is marginally significant for food away from home. The strongest effects are for week day, as expected, reflecting weekly spending patterns. For instance, Saturday is traditionally a big day for grocery and other shopping. The tests for start day and the interaction of week day and diary day are not significant (results not shown), although some combinations of these factors are believed to produce mild effects.

Alternative measures for examining diary day effects are reporting rates (percent CU's reporting one or more expenditures in the category) and number of expenses reported. Both of these variables exhibit the pattern of a strong first-day effect and a general decline across the reporting period. Latin square results for reporting rates are even stronger than those for means; in particular, apparel exhibits a significant diary-day effect. The declines by diary day tend to be smoother, since the heavily right-skewed distributions for most expenditure items cause some instabilities in the means.

16.5 SPECIFICITY OF DIARY AND INTERVIEW REPORTING

16.5.1 Diary Entries

Considerable detail is required by expenditure diaries in order to account for small and frequent purchases, such as food items. The development of CPI weights is usually the motivation for the level of detail chosen. Although the brand name is not collected, unlike in marketing surveys, diary questionnaires include examples that suggest a great deal of effort should be placed on the recording task. Respondents are asked to report individual items, including packaging type and other characteristics of the purchase, for food, and information on the family member for whom purchased, for apparel.

Predictably, many diary entries do not have the desired detail or format. A single cost may be given for a list of items (*combined entries*), and items may be listed with insufficient detail (*nonspecific entries*), e.g., *meat* (type of meat not known), or *soap* (whether detergent or hand soap not known). Even less specific entries are *groceries* and *clothing*. A summary from the 1987 CE diary indicates that a small percentage of entries (7% for food at home) are reported in a nonspecific or combined manner, but these entries constitute 26 percent of total expenditures for food at home and 11 percent for apparel. (See Table 16.4.) While the coding scheme adopted determines what is considered specific, more than half of the expenses that fail this test are reported with very general entries. Pearl (1979) also noted this problem, reporting a similar value (15%) for apparel.

The frequency of reporting in a nonspecific or combined manner tends to increase with the amount of recall used, as shown in Table 16.4. Diaries completed by total recall often contain these types of entries: 56 percent of total food at home expenses are reported as groceries, and 18 percent of apparel expenses are reported as clothing (Table 16.4, col. 5). A much lower incidence of these reports occurs when diaries are completed during the week by respondents (cols. 2 and 3). An analysis by diary day shows that a greater proportion of expenses is reported specifically on the first day, although specificity does not seem to decline by day afterwards.

16.5.2 Examples of Item Imputation and Coding

Lack of specificity has implications for estimation. CE diary and interview entries are coded into a set of detailed items covering all

Table 16.4. Percent of Expenses Within Commodity from Nonspecific or Combined Diary Entries

		Two-week Diaries			
			Other Diaries by Weekly Recall Status		
	All Diaries[a]	Diaries with No Recall Both Weeks	No Recall	Partial Recall	Total Recall
Food and Beverages at Home					
Nonspecific/combined entries	26	15	24	37	66
Groceries	16	6	10	26	56
Food and Beverages Away from Home					
Nonspecific/combined entries	37	33	40	32	55
Apparel					
Nonspecific/combined entries	11	9	19	12	22
Clothing	6	4	5	8	18
Small Home Furnishings and Equipment					
Nonspecific/combined entries	7	6	5	7	10
Miscellaneous Diary Expenses					
Nonspecific/combined entries	16	16	20	15	14
Number of diary weeks[b]	12,413	7,870	613	1,090	1,796

[a] Two-week and one-week diaries, including diaries with recall status not stated.
[b] Excludes diaries with no expenses.

possible household spending, resulting in data files convenient for carrying out estimation and many analytic uses. The occurrence of nonspecific and combined entries necessitates item imputation, referred to as *allocation*. This happens less frequently in the interview, aided by the highly structured set of questions and the presence of the interviewer. On the other hand, failure to recognize that a question stated by the interviewer covers a particular item may be a factor limiting reports in an interview setting, in addition to fatigue and recall effects. Detailed reporting is the trademark of diaries, and can produce more complete reporting if timely recording is carried out by respondents. A shortcoming of open-ended diary formats is occasional coding difficulties, whereas diaries with preprinted entries have the drawback of needing summarizations by diarykeepers.

Anticipating the topic of the next section, diary and interview data are compared for two expense categories, to illustrate completeness and specificity issues. The more suitable source is selected for the official estimates at the detailed level, based on comparisons of means and variances. For a recent period, the interview has been selected for Men's Suits, Men's Sportcoats, Women's Suits, and Girls' Dresses & Suits, but the diary for Women's Tailored Jackets and Boys' Suits, Sportcoats, and Vests. One might expect the interview to be preferred for all these items, which are relatively expensive and not purchased routinely. Item imputation is partly responsible for this anomaly. With the allocation method used for the 1987 apparel data, a $30 diary entry stated as *clothing*, for a man, generates 14 proportionally *allocated records*, that is, part of the $30 creating a Men's Suits record, part a Men's Hosiery record, etc.[1] Thus, allocated records do not represent typical values: the mean for Men's Suits for allocated records may be very small, compared to the mean for similar expenses reported specifically, *nonallocated records* (e.g., $7 vs. $180, for 1987). The effects of imputation tend to diminish at aggregate levels of estimation, but means and variances for detailed categories with large portions of allocated records are distorted. This can affect mode comparisons and source selection. Furthermore, it can be argued that lack of specificity is not equally likely for all items, that is, rarer and more costly purchases such as suits are more likely to be stated as individual entries, whereas they have an equal chance in the imputation system.

For all six clothing items mentioned, the percentages of expenses reported specifically in the interview are between 70 and 80 percent, while a large majority of diary records for these items come from allocation, except for Girls' Dresses & Suits. The allocated means for the interview are similar to the nonallocated means, indicating that allocations usually involve a small number of items, not the broad range of *clothing*. For Girls' Dresses & Suits, the diary gives a higher annual mean. This seems reasonable, since, as will be seen in the next section, the diary performs well in the apparel area overall, and a girl's dress is a more common apparel expense than any of the other items in the group. Except for Girls' Dresses & Suits, however, the interview appears to be a better source for the group examined here. This can be explained both in terms of the extended time coverage and greater structure in the interview.

The next example illustrates cases where the diary yields higher estimates than the interview, at detailed levels of comparison. Table 16.5 shows a comparison for Tableware/Kitchenware items, all with low reporting frequency and low dollar value. The overall mean is higher in

[1] The set of apparel items is restricted according to the household age/sex composition. An improved method, implemented in 1990 for apparel and since 1982 for groceries, further restricts the items to a variable subset, according to a selection without replacement by probability proportional to size of expenditure shares.

the diary, mainly due to the reports in the last item, used in the diary only. This last item, coded relatively frequently, includes either combined purchases, similar to *clothing* above, or individual items not clearly assignable to one of the other codes. For instance, "salt-and-pepper shakers" is an actual diary entry coded in this item, although it could fit into three or four of the other items if information was available on what it was made of. Conversely, a respondent hearing an enumeration of houseware items in the interview might fail to think of the salt-and-pepper shakers at all. The diary appears to be the better source for Tableware/Kitchenware, since it captures more expenditures, often small in value. Along with the short reporting period, the open-ended format may contribute to increased reporting.

16.6 DATA COMPARISONS BETWEEN DIARY AND INTERVIEW

16.6.1 Choice of Collection Mode

The diary is designed to give a detailed accounting of spending, including the smallest or most trivial expenditure. The lengthy reference period of the interview increases the odds that at least one purchase of a given item will be reported and offers a more comprehensive picture of household spending. For larger purchases, memory is likely to be strong. If the interview is intended for large, salient purchases and the diary is needed for frequent purchases, especially detailed data on food, the question becomes where to draw the line. This section examines expenditure distribution statistics to assess where the two modes may be capturing expenditures differently. Others have studied this topic in the past. Pearl (1968), in particular, suggested that the interview should be the source for most home equipment and furnishings, including large and small appliances, and major clothing purchases, possibly using a dollar value to distinguish what is "major." Other than trips and vacations (which are excluded from the diary), Entertainment & Hobbies appear to be assigned to the diary.

Comparisons are made in three areas, Small Home Furnishings, Apparel, and Entertainment & Hobbies. Annual means for categories within these areas appear in Table 16.6; starred items refer to statistically significant differences, at the 5 percent level or lower. The tests are Scheffé multiple comparisons based on the Hotelling T^2 for differences between two mean vectors, with the covariance matrix derived using balanced repeated replication, as in Section 16.5 (Johnson and Wichern, 1982, Chapters 5 and 6). The categories are composed of items common to the two surveys, avoiding most of the larger items, such as major appliances, and are roughly ordered from those expected to be more

Table 16.5. Annual Means for Tableware/Kitchenware Items

Item	Diary	Interview
Total	$40.00	$32.90
Plastic Dinnerware	$ 0.50	$ 1.50
China & Other Dinnerware	6.80	11.20
Flatware	1.00	4.70
Glassware	7.70	5.00
Silver Serving Pieces	0.50	1.00
Other Serving Pieces	0.40	2.00
Nonelectric Cookware	7.80	7.50
Tableware/Nonelectric Kitchenware	15.30	—

suitable for diary collection to those less suitable. All apparel expenses are included, except for infants' apparel and apparel services.

The large number of categories for which the means are reasonably close is somewhat unexpected, although sometimes within these categories different items are being captured. In each area, the last category consists largely of salient or rarely purchased items, and the interview mean is significantly higher than the diary mean. For Small Home Furnishings, diary means are significantly higher only for Linens and Tableware/Kitchenware, although they are greater overall and in four of the seven categories. Diary means are significantly higher overall and in four categories of Apparel. Within Entertainment & Hobbies, diary means are slightly higher for Audiovisual Materials, which includes tapes, records, and film rentals, but interview means are higher in the other seven, significantly higher in four cases. It is perhaps surprising that the interview obtains more expenses for Admissions and Photographic Supplies and, in the other direction, that the diary estimates are roughly comparable to the interview for Audio Equipment and Memberships & Fees.

Dippo, et al. (1977) compared means, standard deviations, and CV's from the 1972–73 survey. For most Household Equipment and Home Furnishings expenses, the interview gave larger estimates for more salient items, except that estimates for furniture were nearly equal, as expected. In Apparel, in contrast to the 1987 results, the interview had higher estimates, except for Footwear and certain accessories. In Entertainment, results were more mixed in the 1972–73 survey.

16.6.2 Detailed Analysis of Distributions

Underreporting is the principal direction of error for both components. The source with the higher overall mean is generally accepted as the better source. Variance comparisons usually favor the interview, since its panel design and longer reference period yield a much larger number of reports. The number of diary reports can become quite small for detailed items. This leads to examining additional features of the expenditure distributions in order to study reporting characteristics by mode, without undue influence of sample size.

Two statistics to be considered are the *reporting rate*, percent respondents with a nonzero expenditure for a specific item, and the *spending mean*, the mean of the nonzero part of the distribution. As discussed in Section 16.5.2, allocated records can cause drastically lower values for means and other statistics from the nonzero distribution. This suggests examining the allocated and nonallocated distributions separately. In simplified form, we can write

$$\text{expenditure mean} = \text{spending mean} \times \text{reporting rate}$$

where these quantities are by week for the diary (each week separately) and by month for the interview. This expression is based on a simple conditional expectation argument: there is no contribution from the zero part of the distribution (i.e. respondents reporting no expenses in the category), so the expenditure mean is the conditional mean for respondents with expenses (the spending mean) times the probability that a respondent has one or more expenses, estimated by the reporting rate. The difference in reference period is an additional factor complicating comparisons. Since there are $52/12 = 4.3$ weeks per month, the interview should obtain 4.3 times the diary expenditures, assuming equally complete reporting. This can partly be realized through a higher reporting rate, that is, more households reporting a purchase over the longer reference period, and partly through a larger spending mean, representing an accumulation of purchases across the month. Another source of difference between the two collection methods pertains to the fact that out-of-town expenses are included in the interview, but not in the diary; this difference has been smoothed by the exclusion from the diary of respondents out-of-town the whole week, ordinarily included as respondents with zero expenditures.

Comparisons reported here are based on the nonallocated distribution (specific entries). For a detailed apparel item, Men's Suits, 1.1 percent of the respondents report purchases in the interview, compared to 0.3 percent in the diary; the spending means are $228 and $178 for the interview and diary, respectively, or 1.3 times higher in the interview. The product of the I/D (i.e., interview/diary) ratios for these two statistics is 4.6, suggesting better coverage for the interview. (Analysis of the allocated records supports this choice.) For Women's Dresses, 6.1 percent report purchases in the interview compared to 2.9 percent in the diary; spending means are 93.3 and 57.2, respectively. The product of I/D

Table 16.6. Annual Expenditure Means from Diary and Interview for Selected Categories.

	Diary	Interview
Small Home Furnishings	$365.7	$342.8
Decorative Items	$109.3	$ 94.9
Tableware/Kitchenware	40.0	32.9*
Linens	60.8	43.7***
Housekeeping Equipment	69.7	74.1
Interior Furnishings	38.8	32.7
Small Appliances	26.4	26.4
Curtains & Drapes	20.7	38.1***
Apparel	$1,268.9	$1,063.8***
Tops, Pants, & Skirts	$ 401.5	$ 276.9***
Sleepwear, Undergarments, Hosiery, etc.	164.5	123.7***
Footwear	190.1	129.7***
Sportswear	139.6	123.5*
Watches, Jewelry, & Luggage	134.1	141.1
Coats, Jackets, Suits, Dresses, & Uniforms	239.1	268.9*
Entertainment & Hobbies	$590.0	$702.0***
Audiovisual Materials	$ 60.7	$ 54.5
Admissions	63.6	75.3*
Photographic Supplies	31.4	44.0**
Reading Materials	131.9	138.1
Toys, Hobbies, & Pets	137.1	170.8**
Audio Equipment	18.1	19.4
Memberships & Fees	136.1	153.9
Photographic Equipment & Musical Instruments	11.1	46.0***

* Significant at 5 percent level.
** Significant at 1 percent level.
*** Significant at 0.1 percent level.

ratios for this item is 3.5, indicating stronger reporting in the diary. Table 16.7 displays the I/D ratios for the more costly and salient apparel items, with quite variable results. Using 4.3 as the critical value for selecting mode, the diary is preferred for four items, including Girls' Dresses & Suits, consistent with Section 16.5.2, while the interview is preferred for the other five items discussed there, plus four additional items.

Table 16.8 contains I/D ratios for several categories in each of the three major areas; the last category within apparel in this table and in Table 16.6 summarizes the items appearing in Table 16.7. From Tables 16.6 and 16.8, it appears that the diary captures more apparel spending overall and in four of six categories. Table 16.7, however, shows that more discrimination between modes is possible at the very detailed level.

Returning to Tableware/Kitchenware, the combined ratio of 5.0 in

Table 16.7. Reporting Rate and Spending Mean Ratios for an Apparel Category

Coats, Jackets, Suits, Dresses, and Uniforms[a]	Reporting Rate Ratio (I/D)	Spending Mean Ratio (I/D)	Product of Ratios (I/D)
Men's Suits	3.6	1.3	4.6
Men's Sportscoats	12.5	0.7	8.8
Men's Coats & Jackets	2.6	1.4	3.6
Men's Uniforms	16.3	0.6	9.9
Boys' Coats & Jackets	3.7	1.2	4.3
Boys' Suits, Sportscoats, Vests	3.6	1.3	4.7
Women's Coats & Jackets	2.2	1.7	3.7
Women's Dresses	2.1	1.6	3.5
Women's Tailored Jackets	7.0	1.4	9.9
Women's Suits	5.7	1.6	9.1
Women's Uniforms	3.7	1.3	4.9
Girls' Coats & Jackets	3.0	1.6	4.6
Girls' Dresses & Suits	1.8	1.5	2.8
Girls' Uniforms	4.6	1.4	6.4

[a] Omits "Other Clothing," coded for interview only.
Note: Ratios are based on nonallocated records only.

Table 16.8 suggests better coverage for the interview. However, for this category, when allocated records are included, many from the last item in Table 16.5, the diary provides larger estimates, as discussed in Section 16.5.2. The two ratios in Table 16.8 provide evidence that the diary and the interview capture different Tableware/Kitchenware items. The low reporting rate ratio suggests that many (small) purchases are missed in the interview; this is supported by the high spending mean ratio, in the sense that interview purchases tend to be larger. The low reporting rate in both components argues against the alternative view that the higher interview mean simply represents the accumulation of weekly purchases.

These two examples are intended to suggest additional tools for analyzing expenditure data from the two sources. There is not necessarily a downward bias in diary estimates for large or rare purchases due to the smaller number of reports, but rather a larger variance. Examining reporting rates and spending means for several years may help in the interpretation of the results. In addition, the shape and percentiles, especially the median, of the nonzero part of the distribution offer potential for analysis.

16.6.3 Comparisons by Month

Household spending exhibits strong seasonal patterns for various commodities and services, even at aggregate levels. Many items have

Table 16.8. Reporting Rate and Spending Mean Ratios for Selected Categories.

	Reporting Rate Ratio (I/D)	Spending Mean Ratio (I/D)	Product of Ratios (I/D)
Small Home Furnishings			
Decorative Items	1.7	2.4	4.1
Tableware/Kitchenware	1.0	4.9	5.0
Linens	1.3	2.3	3.0
Housekeeping Equipment	1.0	4.9	5.1
Interior Furnishings	1.8	2.1	3.7
Small Appliances	3.2	1.5	4.8
Curtains & Drapes	3.3	2.9	9.5
Apparel			
Tops, Pants, & Skirts	1.4	2.1	3.0
Sleepwear, Undergarments, Hosiery, etc.	1.4	2.1	2.9
Footwear	2.1	1.5	3.2
Sportswear	1.7	2.0	3.4
Watches, Jewelry, & Luggage	1.8	2.6	4.6
Coats, Jackets, Suits, Dresses & Uniforms	2.4	1.7	4.3
Entertainment & Hobbies			
Audiovisual Equipment	2.8	1.4	4.0
Admissions	4.0	1.4	5.6
Photographic Supplies	7.3	0.9	6.5
Reading Materials	2.1	2.7	5.6
Toys, Hobbies, & Pets	1.9	2.9	5.5
Audio Equipment	2.5	1.9	4.8
Memberships & Fees	3.1	1.7	5.1
Photographic Equipment & Musical Instruments	3.7	5.0	18.2

Note: Ratios are based on nonallocated records only.

high December or fourth quarter spending levels associated with the holidays while many others tend to have increased levels in the summer, with regional differences. The diary and interview should yield comparable seasonal patterns, but striking differences are found in certain areas. Monthly spending effects are expressed in the form of percent ratios of individual monthly means to the average monthly expenditure across all 12 months. Each diary week is assigned to the month in which it begins, and this is the basis for computing monthly means; interview data refer to the expenditure month stated by respondents. (See Table 16.9.)

Substantial increases occur in November and December for the diary for Apparel, but they are almost dwarfed by the strong December peak in the interview. Similarly, for Small Home Furnishings and Entertainment, the interview has a sharp December peak, while the peak is divided between November and December in the diary. These seasonal patterns

Table 16.9. Percent Ratios of Monthly Means to Average Monthly Expenditures in Selected Commodities.

	Jan.	Feb.	Mar.	Apr.	May	June	July	Aug.	Sept.	Oct.	Nov.	Dec.
Apparel												
Diary	66	68***	95***	102***	94	91	80	99***	98	94*	154***	156***
Interview	65	58	79	87	92	88	81	113	99	86	99	251
Small Home Furnishings												
Diary	70	63***	85	109	109**	77**	88	109**	107***	91	142***	149***
Interview	71	80	81	113	126	91	93	93	82	85	99	184
Entertainment												
Diary	96**	86	107***	91	75***	84***	92	110***	89	97	125***	145***
Interview	84	84	84	94	91	99	96	92	92	104	103	177

* Significant at 5 percent level.
** Significant at 1 percent level.
*** Significant at 0.1 percent level.

suggest different measurement error effects in the two components. For the diary, possible explanations are: (1) greater fatigue effects, with respondents being busier than usual during December; (2) disclosure problems, since respondents are not likely to write down gift items for other family members; and (3) space limitations of the diary pages, in which most of the space is earmarked for food. For the interview, December data come from interviews conducted in January, February, and March, usually times of less activity and pressure at home. After the holidays, remembering gifts purchased may be a pleasurable part of the exchange between respondent and interviewer. Internal telescoping errors may be present in December, again due to holiday associations. This is supported by the fact that November ratios in all three areas (Apparel, Small Home Furnishings, and Entertainment) are lower for the interview than for the diary. Another cause for low November levels could be fatigue, since some of these expenses are reported in December with time pressures mentioned above.

Purchases of gifts, such as jewelry and toys, are known to increase in December, lending support to the notion that disclosure concerns may contribute to low diary reporting. Summary retail sales show that spending is 1.7 times higher in apparel stores and 2.8 times higher in jewelry stores in December and up by about 10 percent in November, compared to previous months (U.S. Bureau of the Census, 1990). These figures suggest that the seasonal effects are somewhere between diary and interview values for apparel. Further investigation of the presence and causes of differences in seasonality appears to be a fruitful area of research on measurement errors in expenditure surveys. Seasonality has some interest in its own right, since quarterly as well as annual data are published.

16.7 SUMMARY AND CONCLUSIONS

This study is based on data from the 1987 U.S. Consumer Expenditure Survey. Declines in reporting after the first diary day and in the second diary week are found to be sizable. Estimates derived from diaries completed solely by respondents show that the first day of the first week exhibits means 35 percent greater than means for the combined two weeks. First-day estimates for the second week are 5 percent higher than the two-week average. Estimates for the second week are 11 percent lower than first-week estimates. These effects have been noted in other expenditure diary surveys, and are in remarkable agreement with findings from the 1972-73 U.S. Consumer Expenditure Survey. They help explain some of the underreporting in the diary.

About one-quarter of the diaries depart from the intended diary method, in the sense that some or all of the expenditure data are recalled with the aid of the interviewer at pickup time. This procedure appears to stem from lower respondent motivation or, in some cases, inability to perform the recordkeeping task. These diaries have lower means and substantially higher percentages of combined or nonspecific entries. Further research is needed to examine to what extent these effects are associated with respondent characteristics.

Many diary entries lack specificity and this contributes to a substantial portion of expenditures, 26 percent for food at home, 37 percent for food away from home, and 11 percent for apparel. Especially for combined entries such as clothing, the lack of specificity creates the need for item imputation, with possible distortions. This causes difficulty only at rather detailed levels, and is less a problem for the interview, due to its structured questionnaire and the presence of the interviewer.

Certain commodities (e.g., apparel) are known to be underreported in both CE components, and it appears that one method alone may not capture sufficient data. The findings tend to support the general assumption that for routine items the diary captures smaller expenses to a greater degree than the interview, whereas for less frequent or more salient expenses the interview yields better data. Useful clues about diary performance are provided by examining spending means and reporting rates, taking into account the effects of item imputation or allocation. The diary captures higher levels of spending in most of the apparel and small home furnishings categories studied, while the interview is higher overall in the entertainment and hobby area. The two modes of data collection show remarkable similarities in several areas.

Multiple modes of data collection are used in several countries, according to expense type and amount. This study has shown some advantages of selecting the mode at very detailed item levels, made possible by the overlap in the CE components. However, sound reasons can be advanced for identifying the most appropriate mode at broader levels. Respondent burden is reduced if entire groups of expenditures are not required, and, conceptually, it may be more satisfying and consistent to design one component for estimating expenses in a given area.

Comparison of the seasonal patterns of the two components shows significant differences, symptomatic of underreporting. Findings are presented for three commodities with the greatest differences: apparel, small home furnishings, and entertainment & hobbies. Seasonal patterns differ for the month of December, with higher values in the interview. It is speculated that diary reporting is low in December due in part to fatigue effects and disclosure concerns related to holiday gifts. In addition, interview reporting for December may be better than for other months, especially for apparel. Further research is needed to shed light on the causes of this phenomenon.

CHAPTER 17

A REVIEW OF ERRORS OF DIRECT OBSERVATION IN CROP YIELD SURVEYS

Ron Fecso
U. S. Department of Agriculture

17.1 INTRODUCTION

Direct observation is a form of data collection which is useful and often superior to respondent questioning. For example, interviewers may find direct observation faster or more comfortable for items such as sex or race. Direct observation is useful in such diverse fields as anthropology, traffic safety, biology, and physics. Education, psychology, and sociology researchers use direct observation methods to study vision, hearing, behavior disorders, task analysis, learning, cognition, and group interaction.

Discussions of direct observation found in these applications often describe modes of observation. For example, in the participant observer mode, the researcher could live or work with the subject(s) of study. In a nonparticipant mode of observation, the researcher could be behind a screen, a one-way mirror, sit in the rear of a classroom, or use audio and/or video recording. Direct observations may use senses other than visual, such as hearing a pause or dialect in telephone interviews, touching to determine texture, or tasting. Use of measurement instruments, counting, and record examination are also direct observations.

Here, the proposed definition of direct observation is the recognition and noting of a fact or occurrence using the observer's senses or physical measurement devices designed to quantify the characteristics or events. Thus, direct observation error (DOE) is the difference between the true value and the value obtained from the direct observation. As with response error, which has a similar definition, cognitive processes play a

Measurement Errors in Surveys.
Edited by Biemer, Groves, Lyberg, Mathiowetz, and Sudman.

ISBN: 0-471-53405-6

role in DOE. Examples include an observer forgetting categories to observe, misunderstanding concepts, or altering the observed event. Rounding, complexity of measurement, and fatigue also affect DOE.

This chapter discusses the direct observation techniques used to measure crop yields. Crop yield, which varies considerably from year to year, is an economic variable of great interest. Yield fluctuations directly affect prices on the agricultural commodity trading exchanges, allocation of transportation services, agricultural credit, purchasing programs of commodity users, manufacturers of farm capital goods, national income, taxes, trade balances, and agricultural price support programs.

Surveys which measure yield rely on direct observation of the crop or local area environment. Visually and tactually observed variables include crop condition, maturity, pest or disease infestation, counts, and environmental monitoring. The three principal survey approaches are: (1) visual or "eye" observation of growing crops; (2) sample measurements of growing plants (called objective measurement or crop cutting surveys); and (3) collecting environmental observations (growing conditions) such as soil moisture or weather events. A report by the Food and Agricultural Organization (United Nations, 1982) indicated that some combination of these methods was in use in at least 87 countries.

Eye estimates reported by individuals from the crop growing regions provide one of the basic methods of data collection. In the United States, this method dates back to 1862 when the *American Agriculturist* asked subscribers to send monthly crop reports during the growing season (Sanderson, 1954). Then as now, the emphasis was on obtaining timely and accurate data on acreage and yield.

Doubts about the accuracy of eye estimates made by local inhabitants led to the development of surveys with increasing objectivity. For example, route sampling was a method in which enumerators drove a predefined route and systematically selected fields for more detailed counts and observations (Hendricks, 1942). This method understated actual yields by about five percent. Again, eye estimates, rating the size of ears of corn compared to a standard ear, were believed to be a major source of error. Another suspected contributor to error was incorrect identification of the crop. The history of attempts to find better methodology includes the use of experienced eye estimators, administrative records, incorporating weather, field harvesting, small plot crop cuttings, and using new technologies (photography to X-rays and Landsat satellite imagery).

This chapter presents some historic background on the use of direct observation methods in crop yield surveys conducted by the National Agricultural Statistics Service (NASS) of the U. S. Department of

Agriculture (USDA). The basic forms of yield measurement described include reported condition and yield, direct crop measurement (cutting), and some technological methods for measurement. Finally, the control and measurement of direct observational errors in NASS's yield surveys will be reviewed.

17.2 NASS SURVEYS

Estimates of harvested acreage and production by the USDA for most major field crops originated in 1866. Current surveys which produce yield statistics for the crop cycle begin with reports on farmers' planting intentions (acres), followed by estimates of acreage actually planted, forecasts of yield per acre during the growing season, and, finally, estimates of harvested yield. Forecasts and estimates represent two distinct concepts in NASS. Forecasts relate to an expected future occurrence, such as crop yields expected during the growing season. Estimates refer to an accomplished fact, such as crop yield after harvest.

Gathering data for NASS crop estimates is the responsibility of the State Statistical Offices (SSOs). There are 45 SSOs, one per state, except only one in the six-state New England region. The SSOs conduct surveys using various combinations of mail questionnaires, telephone interviews, face to face interviews, and field observations. Survey statisticians in the SSOs manage the survey activities. They conduct training sessions for interviewers and enumerators, supervise the field supervisors, review survey questionnaires manually, run computer edits, and correct data before summarization. Statisticians in Washington, D.C. assure that the edits are complete, summarize the data to state, regional, and national levels, and review the summaries.

The field staff involved in these surveys consists of about 1,300 enumerators and 230 supervisory enumerators distributed throughout most counties. Field staff conduct face to face and telephone interviews. About 550 of the field enumerators and 150 of the field supervisors collect objective measurement samples. In addition, nearly 700 telephone enumerators and 50 supervisors work in the SSOs.

17.2.1 Yield Forecasts and Estimates

NASS makes monthly yield forecasts during the growing season and end of season estimates using farmers' reported data and/or measurements of the crops taken in fields by trained enumerators. In past years, monthly nonprobability sample surveys of crop reporters (farmers, ranchers, and other people knowledgeable of local agricultural conditions) provided subjective appraisals of local crop conditions and probable crop yields. The crops survey program is currently undergoing a transition as probability designs replace the nonprobability surveys. The new surveys request yield reports rather than reported condition.

Reported Condition

The concept of "normal condition" was first used for reporting of condition of crops during the growing season around 1880. Normal condition was designated as 100 percent. The concept, developed to provide forecasts during early crop development, allowed farmers to evaluate potential output without having to consider future impacts of weather, insects, and other influences.

Disappointment with forecasts led to several changes in the condition concepts and question wording. For example, in the early 1950's, the questionnaire mailed to crop reporters in the first month requested a "REPORT FOR YOUR LOCALITY" and asked "What was the average date that [insert crop] was up to a stand this year?" In subsequent months, the questionnaire asked the respondents to "REPORT FOR YOUR IMMEDIATE LOCALITY" the "Condition of [insert crop] [insert month] 1 as compared to normal." The instructions stated "A 'normal' or 'full' yield, represented by 100 percent condition, is that yield per acre which is expected when the season is favorable and insects and disease have caused little or no damage. 'Normal' (100 percent condition) is higher than that of most seasons but may be exceeded in an unusually favorable year. If the percent condition is 10 percent above normal, the answer should be filled in as 110; if 5 percent below normal, it should be reported as 95, etc. In estimating the condition of [the crop], consider not only the growth and appearance of plants, but also all other factors which influence the probable yield per acre."

Recent nonprobability surveys instructed crop reporters to "Report the condition of crops now, as compared with the normal growth and vitality you would expect at this time, if there had been no damage from unfavorable weather, insects, pests, etc. Let 100 percent represent a normal condition for field crops." The reporters then answered, for their localities, the question: "[CROP], condition in — PERCENT."

Reported Yields

After the first month of crop growth, these questionnaires asked reporters to provide a value for average yield along with reported condition. During the 1950's, the wording was "Probable yield per acre in your locality this season." More recently, the question read "(CROP), probable yield per acre this year in (weight unit: bushels, pounds, etc.)," with instructions to answer "for your locality." The future basis of yield reports removes locality or immediate locality concepts which have varied somewhat in use over time. Farmers will be asked a more specific question for a more restricted area. "What is the expected yield per acre for (CROP) to be harvested on the total acres you operate?" Note that the words "probable" or "expected" are not defined or calibrated for the respondent in any of these questionnaires, while "condition" was defined more fully.

Objective Yield Surveys

Studies of crop cutting methods to further enhance the quality of yield data began on a regular basis in the U.S. during the 1930's, but stopped during World War II. Sanderson (1954) reviews the early experiences. Reintroduction of the method (sometimes called objective yield or objective measurement surveys) for several crops occurred from the 1950's to the early 1980's.

NASS conducts objective yield surveys for major crops in the major producing states. Enumerators visit approximately 1,900 corn, 1,600 cotton, 1,900 soybean, and 2,500 wheat fields to make objective yield counts on two small plots in each field. Plots comprise only a small fraction of the total area in a field. Counts, vegetative measurements, fruit measurements, and weight of fruit are direct observations collected from the plots. Near harvest, part of the plot is harvested (crop cutting). The design allows forecasts based on data taken from the monthly measurements and the crop cutting. After harvest, two new plots are located on a half-sample of fields to determine harvesting loss (gleanings). Biological (gross) yield minus estimated harvest loss determines net yield per acre. Due to the uniqueness of each crop, efficient estimation requires the use of different plot sizes and measurement of different plant characteristics. A comparison of the survey designs for several crops is in Matthews (1985). Further detail about the design and associated estimators is in Francisco, et al. (1987). Reiser, et al. (1987) discuss forecasting methods.

17.3 ERRORS IN EYE, OBJECTIVE, AND TECHNOLOGICAL MEASUREMENTS

17.3.1 Eye Estimates

Eye estimates of condition, yield, and some objective yield observations are examples of magnitude scaling. Anderson, et al. (1983) mention that magnitude scaling is useful since people can often make finer distinctions than category scales permit and they can indicate extreme responses better. Yield survey data provides some examples of magnitude scaling, other observational methods, and their errors. As pointed out by Converse and Presser (1986), there is, however, a lack of methodological work concerning magnitude scaling.

Condition

Reporters judge the "condition" of the crop based on appearance. Scales used range from "very good, good, average, poor, very poor" to "percent of normal for the area." Conditions reported as a percentage may use a "standard" or "normal" condition defined by crop appearance or a specific yield, such as yield last year. Condition may also be based on an average yield over a specified number of prior years. Used more vaguely, reporters must form their own concepts of "normal" or "full" crop.

The 100 percent of normal condition, regardless of the definition, tends to equate to a higher than average yield prospect. Normal tends to be conceptualized as a full crop, forgetting years of crop failure in the appraisal. Further, the stability of the normal crop concept presents a problem when improved farming practices, abandonment of marginal acreage, or variety changes occur. The reliance on "normal" vegetative appearance results in a failure to reflect technologically induced yield increases as a percentage greater than 100.

A reporting problem may occur at the beginning of a period of adverse environmental conditions such as drought or excessive heat. The crop condition may appear good, even though the reporter knows that continuation of the adverse environment would seriously reduce the crop's potential. Further, the reporter generally does not reduce condition reports to reflect the zero yields of abandoned acres, while officials may not have the information to reduce the acreage. Thus, in years of acreage reduction, production forecasts based solely on condition may be high.

Reporting condition as a percentage of normal yield presents an estimation problem because normal yield will vary by locale. Also, reporters have a tendency to underreport conditions as a percentage of average long term yield and to underestimate year to year variation. Sarle (1932) identified these biases almost 60 years ago and concern for the lack of objectivity remains today.

Sanderson (1954) reports several other results. Condition reports for wheat may understate yield prospects in dry years and overstate prospects in wet years, an indication that additional factors may be neglected by the reporters. Reports may underestimate the ability of the crop to recover from unfavorable effects. Finally, the reporter neglected nonvisible but important yield determinants for some crops, such as subsoil moisture and beginning stages of diseases. Although there are no recent studies of these error sources, the problems are not likely to have diminished very much. Soil moisture and disease may be an exception, as modern farmers are better trained to be aware of these factors and take actions to produce in a profitable manner.

Despite these problems, condition reports can provide useful information for crop forecasting. (A comparison of the yields derived from the alternative methods in the U.S. is provided later.) The reliability of condition reporting as an indicator of crop yield is best for crops with a high correlation between vegetative appearance and yield. Crops have growth periods during which weather effects are critical. Crops with short critical periods, such as corn, tend to have more reliable condition reports after the critical period. Continuous fruiting crops, such as cotton and soybeans, and tuber crops, where the yield is underground, have less accurate forecasts. The best indications usually occur after fruit, such as heads of wheat or ears of corn, are visible. In general, early season reports are less dependable when there is an increased chance of future extreme weather conditions which would influence crop development.

In summary, the appeal of condition reporting was the simplicity of basing reports on the appearance of the crop, relieving reporters of the task of judgment about future impacts upon crop yield. Although data users may find condition as reported in percent interesting, they usually want it converted to an official value for yield. Yield can be estimated from condition reports using ratio and regression methods. Yet current sentiment favors reports of yield values.

Yield

Various eye estimators can report yield. The U.S. selects farmer-reporters randomly each year. In contrast, panels of volunteers or trained reporters can be used. For example, in the People's Republic of China, groups of involved local inhabitants inspect fields visually. Using historic yield information for each field, the group derives a consensus yield (Fecso, 1984).

Although eye estimation has worked reasonably well, several deficiencies exist. The major problem is the error of early season yield forecasts. Larger early season errors are a problem for all methods. The main limitation is the inability to predict future weather. A forecast of prospective yield on a given date makes assumptions about weather conditions and damage from insects, diseases, or other causes for the remainder of the growing season. Even when reporters can observe current conditions accurately, weather, disease, insects, or other condition changes may significantly alter the final estimate from the earlier forecast.

Reporter yield observations undoubtedly are influenced by error sources beyond the act of direct observation. For example, the reporter most likely is not looking at the crop when replying by telephone or mail. Thus, recall problems may occur. Eye estimates also may be subject to reported biases which could include: desire to affect prices (underreporting), desire to impress (overreporting), and concerns that the report will be linked with the reporter's participation in other government programs (reporting the value used in the other program). Sample variances may also be biased. Eye estimates of the yield value for a farmer's field were found to have a lower sample variance than actual harvested yields (Irwin, et al., 1938). This condition occurs when errors of estimation tend to be in the direction of the mean.

Yield concepts may differ for reporters using different harvesting equipment or methods and for differing crop uses. Feed crops, especially those fed to livestock without threshing, have larger errors of observation since reporters have less historic yield information upon which to judge the current crop. Mixing several crops in an area presents problems for eye estimation of both acreage and yield in the field. Kiregyera (1987) discusses similar problems found in surveys in Africa. Although mixed cropping is rare in the United States, a similar problem does occur when parts of the field are unplanted or contain different varieties. Little has been done recently to measure the contribution of these individual error sources.

The reporter's yield definition is influenced by many concepts. The acreage used to determine yield per acre must be clear. The use of planted, harvested, contracted, allotment, or interplanted acres needs definition. Precisely defined weights or counts are needed. When yield is reported in containers, such as boxes or bushels, the reporter must know the weight per container. Weights in husks, on stalks, or in shells must be distinguished from the weight of the fruit fully removed from other plant matter. Finally, the delivery point can affect yield concepts, since harvest at the field often differs from delivery to market. Amounts consumed on the farm, used for seed, destroyed, lost during harvest, or lost in shipment also affect yield reporting.

17.3.2 Objective Measurements

The desire for more reliable data led to the design of crop cutting surveys. Mahalanobis (1946) outlines the extensive early work, starting with studies done during the 1920's. Excellent references for the breadth of methodology are Zarkovich (1966), Huddleston (1978) and United Nations (1982). This section reviews research conducted at NASS related to direct observations in the objective yield surveys. The discussion will generally follow the chronology of the data collection process through the season.

Enumeration begins with the placement of the unit in a field. Units not located in a random place in the field, for example, tossing hoops and some pacing methods of placement, probably favor local good areas of the crop, resulting in upward biases. Bias in these procedures has always been a concern. In the U.S., plots are located by stepping an assigned random number of paces into the field from an edge point, also located randomly. This process raises concern that the enumerators, seeing the condition of the crop in the direction of pacing, may consciously or unconsciously alter their normal pace to arrive at a desired location. Influences include seeing an easier area to work or an area perceived as looking more typical of the field or vicinity.

Plot biases can be examined by testing the assumption of equal distributions of the two plots in a field. It was suspected that the second unit done in a field may be less subject to an enumerator's subconscious desire to make the plot representative of the field. Fatigue may also cause a selection of an easier plot. In a study of two states in one year, Wood (1972) found no end-of-season differences between the two plots in sampled wheat fields, although measurements of some early season characteristics did differ. No reference to the power of the test was made. Matthews (1986) looked at all states for three years and found differences varied by year and state. His study used a split sample experiment to test for the effect of a procedural change thought to reduce bias. In the new procedure the plot starts after a five-foot buffer distance which begins at the end of pacing. Although the power was too low to find significant differences by state, lower counts occurred in the buffered plots in almost all the states, indicating that some enumerator effects changed with the buffer.

In contrast, Tremblay and Fecso (1988) studied the randomness of plot location in sunflower yield surveys and questioned the effectiveness of a five-foot buffer. Measurement of the plot using a tape was also examined. Since the desired 15-foot plot length converts plot observations into yield per acre values, errors in plot length are inversely related to errors in computed yield. In what seems to be a simple measurement process, over half the samples had remeasurements of the 15 foot plot length which were not 15 feet (Fig. 17.1). The fairly normal looking distribution, although slightly skewed above 15 feet, suggests measurement error sources such as slack tape, rounding, or pulling anchor pins.

Battaglia (1985) also found problems with the placement of soybean

Measured Plot Length (FEET)	Frequency	Cumulated frequency	%	Cumulated percentage
12.0|	1	7	0.15	1.06
13.5|	1	8	0.15	1.21
13.6|	1	9	0.15	1.36
13.8|	1	10	0.15	1.52
14.0|	2	12	0.30	1.82
14.1|	2	14	0.30	2.12
14.2|	1	15	0.15	2.27
14.3|	4	19	0.61	2.88
14.4|	5	24	0.76	3.64
14.5|*	8	32	1.21	4.85
14.6|	6	38	0.91	5.76
14.7|*	14	52	2.12	7.88
14.8|**	28	80	4.24	12.12
14.9|*****	68	148	10.30	22.42
15.0|*************************	325	473	49.24	71.67
15.1|*******	95	568	14.39	86.06
15.2|***	43	611	6.52	92.58
15.3|*	16	627	2.42	95.00
15.4|*	10	637	1.52	96.52
15.5|	5	642	0.76	97.27
15.6|	3	645	0.45	97.73
15.7|	1	646	0.15	97.88
15.8|	1	647	0.15	98.03
17.6|	1	648	0.15	98.18
19.3|	1	649	0.15	98.33
19.8|	1	650	0.15	98.48
19.9|	2	652	0.30	98.79
20.0|	5	657	0.76	99.55
20.1|	2	659	0.30	99.85
20.2|	1	660	0.15	100.00

10 20 30 40 50
Percentage

Figure 17.1. Distribution of 15-foot plot length. 1984 sunflower research.

plots. Enumerators have difficulties identifying rows in narrowly spaced soybeans. Before this study, enumerators were allowed to use a rectangular-shaped plot rather than rows when they encountered difficulties. The change, although an operational aid, caused an upward bias due to over-inclusion of plants. Procedures now stress not to use rectangular plots unless rows cannot be distinguished after careful observation.

The search for plant characteristics which improve early forecasts often results in field procedures which are very difficult for the enumerator to follow. Battaglia found that counts on the first plant were generally higher than counts on the following plants. Differences increased for subsequent plants, especially for difficult counts (nodes and lateral branches) and large counts (blooms and pods). Differences often exceeded 15 percent. All plants should have a similar distribution. Thus, the possibility that enumerators pace to a better-yielding starting point, or experience fatigue making the large counts (sometimes 100 pods per plant) under very hot and humid conditions, was suspected. Another problem involves the correct inclusion or exclusion of plants on the edges of the plot. Shape of the plot can interact with cultivation practice to produce a systematic placement bias. Also, smaller units are more likely to overestimate bias due to inclusion of border plants (including plants rather than excluding plants near the unit's edge). The bias approaches zero with plots over 50 square feet, but larger cut units can produce loss of grain during harvesting, threshing, drying, etc. (Zarkovich, 1966).

Counting is subject to large errors, even for simple counts like the count of the letter "e" in this paragraph (Faulkenberry and Tortora, 1979). In general, undercounts are a problem in objective yield surveys, especially when counts are high or enumeration conditions are difficult. For example, Wigton and Kibler (1972) found undercountings of filberts on small branches in a tree survey. The undercounting was associated with overlooking nuts hidden by leaves. Stripping all the nuts from the terminal limb and counting them on the ground reduced undercounting from eight percent to three percent. These destructive counting techniques are useful to reduce counting error, but only during the last visit, since forecasting reliability decreases when models do not use data collected from the same place across the season (Nelson, 1980, Pense, 1981, and Reiser, et al., 1989). On the other hand, handling damage from previous observations of the unit reduced counts at the end of the season (Nelson, 1980 and Pense, 1981).

Bigsby (1989) found measurements which show promise for improving forecasts of corn ear weight. Operational procedures obtain the length of the cob over the husk for ears in the unit without removing the ear or pulling back the husk. Additionally, a destructive sample of five ears outside of but near the sample unit is selected and husked. The enumerator selects a representative row of kernels and measures it for each ear. This research also collected over-the-husk measurements, length of a representative kernel row (in contrast to length of cob) and diameter of the cob two inches from the tip and one inch from the butt of the ear. These measurements represent dimensions related to the volume of the cob and grain and the bearing surface of a corn cob. Bigsby cited difficulties in finding end points of a row over the husk and in the subjective determination of a "representative" row. Models using a volume or surface area variable (computed from the diameter measurements and the kernel row length) had mean squared errors of 30 to 50 percent lower than the prior models. Although some of the improvement may come from the more reliably measured values, transforming the data to a "volume" variable seems to be the major improvement. Over-the-husk measures of row length did not produce models which forecast better than those based on the easier to observe kernel row length. Since experimental error should be lower when making the observations on the same ears, measurement errors when trying to feel through the husk to measure row length must be large enough to negate the hoped for efficiency.

The work by Bigsby corroborates the findings of Nieto de Pascual (1959). In his work, kernel weight per square inch of bearing surface of the cob was nearly constant across years. Thus, measurement of the bearing surface area, or number of kernels, is a key measurement. The number of rows of kernels in an ear was counted and multiplied by the number of kernels counted in a row "considered typical of the ear at hand." This product estimates the total number of kernels in the ear. Figure 17.2 compares the rather subjective, eye derived estimate of kernels to the actual count on a subsample of 35 ears. Nieto de Pascual studied the visual errors and found that the eye estimate is best in the center of the distribution (from 550 to 800). At higher numbers of kernels, the eye estimate tends to underestimate actual kernels. Overestimates occur at lower values. Irregular kernel patterns were cited as a difficulty for the person who is eye-estimating. A suspected cause of overestimation was the observer representing these ears by a more regular kernel pattern.

Steele (1987) looked at the enumerator's ability to determine the maturity category of corn ears. In the operational procedure, enumerators husk ears outside one of the plots to determine maturity stage. In a research measurement, enumerators examined the ears inside each sample plot, without damaging the ears. A multinomial model tested the hypothesis of no systematic patterns of misclassification by the two procedures. In August and September, significant departures from the null hypothesis were found in almost all states. As the crop matured, in October and November, it should be easier to feel through the husk and the hypothesis was viable.

Bond (1985) studied procedures used to determine grain weight in the wheat objective yield survey. Measurement problems included counts of heads differing between field and laboratory over half the time, occasionally by over six percent. Minor procedural changes, such as recounting in the field, were made to help reduce errors. Moisture meter estimation of dry matter, rather than oven drying, resulted in a slight, but statistically significant difference in dry matter. Finally, grain moisture level influenced meter accuracy slightly.

A recent study measured some direct observational errors in the corn objective yield survey (Reiser, et al., 1989). Existing forecast models treat

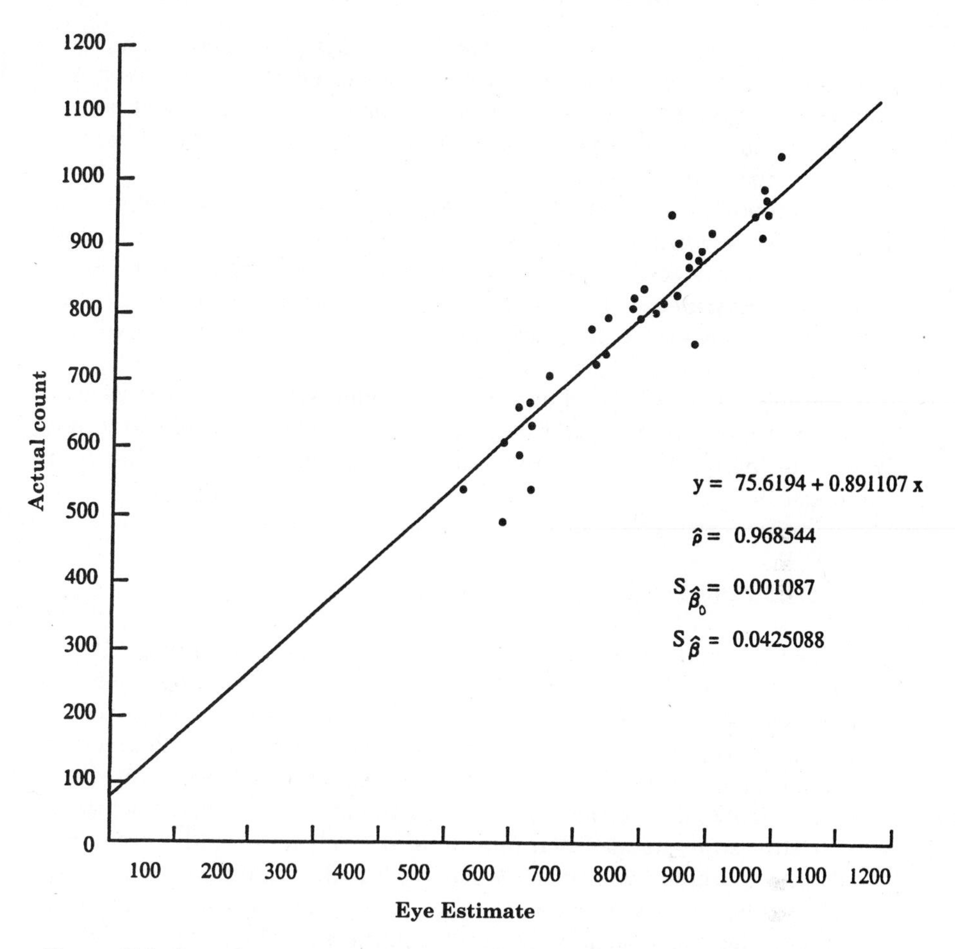

Figure 17.2. Actual corn kernel count versus eye estimate, August 15. From Nieto de Pascual (1959).

the data, collected for several months during the growing season, as essentially cross sectional. That is, each month has a separate forecast model. Panel models incorporated both the cross sectional and time series aspects of the survey. Various forms of panel models, familiar in the social sciences (Wiley, 1973), were examined and reliability ratios, the ratio of the variance of the true measure to the total variance, were computed. Easy counts were measured virtually without counting errors in all months. For example, the number of ears with kernels had a 0.99 reliability ratio. Length of cob over husk had reasonably high reliability (0.75), and appeared preferable to length of kernel row (0.16), possibly because measurements were on the same ears over time. The low reliability of the length of kernel row reflects between-ear, over-time variability inherent in this destructive sample. Using a model which identified the annual repetitions of the survey, reliability of length values varied across years.

The report identified two uses for the models. First, reliability scores, used to monitor month-to-month and year-to-year changes, could serve as a quality control procedure to warn of changing enumerator or training effects. Secondly, identifying measurement error in the prediction variables would help assess the impact of measurement error on forecasts.

These panel models and nested error models which treat the year to year changes in the errors (Reiser, et al., 1987) can help provide direction in forecasting research for both model specification and data collection problems.

17.3.3 Technological Observations

The desire to improve survey accuracy or to reduce cost motivated crop yield researchers to attempt to use sensing devices to supplement or replace visual observations. The 1970's had researchers interested in using photography in fruit tree surveys. Vogel (1970) investigated the use of photographs in tart cherry sampling, hoping that fruit counts from photographs would improve estimates of the number of fruit per tree.

In this study, a metal frame divided each side of the tree into quadrants. A photograph was taken of each of the eight divisions during the early season survey and again at harvest time. Different counters often obtained considerably different counts from the photographs. A replicated design tested for between-counter differences. Although no significant difference was found, the paper did not mention power. Photo counts also were compared to expanded counts from a sample of limbs. Green fruit counts from the photos averaged only 18 percent of the limb count expansions. Counts of ripe fruit on photographs were somewhat better, but averaged only 34 percent of the limb count expansions. Although no mention of validity and reliability was made, the report contains indications that both validity and reliability problems exist in using the photo counting technique for both green and ripe fruit.

Similarly, Allen (1977) found that counts of citrus fruit from aerial photographs using an adjustable stereoscope were poor before the fruit changed color, while counts by direct visual observation from the same distance correlated with actual fruit. There was general agreement between counters. Counts were also made using optical density readings with a densitometer. Here, light passes through the developed film of photographed trees to measure spectral readings and identify fruit by their spectral signature. Strong correlations were found between these counts and actual fruit counts, but not consistently over months. These techniques never progressed to an operational use.

The use of satellite "observation" (such as LANDSAT or SPOT) to estimate area and forecast yield has received research attention for over 20 years. Despite many encouraging results reported in the studies, to date there are no ongoing operational uses in the U.S. Especially for yield data, these techniques are not cost effective compared to traditional methods (Allen, 1990).

17.4 CONTROL AND MEASUREMENT

The studies just mentioned indicate that a variety of opportunities for errors of direct observation exist when determining yield statistics. Commonly assumed error sources include the observer's inability to grasp the complexity of the data, lack of experience in the task, lack of experience in historic comparisons, and fatigue, as well as the task difficulty and variability inherent in the measurement devices. The control and measurement of errors of direct observation can use methodology similar to that for response or measurement errors. Since these techniques are described in detail in other chapters of this book, they are not described here. Ongoing activities to control and/or measure DOE in yield statistics and ideas for future measurements are examined below.

17.4.1 Ongoing Activities

Data Collection Activities

Control and measurement activities used to assure and/or improve the general quality of the yield surveys at NASS evolved from the history of research discussed earlier. Control of the farmer-reported data consists primarily of checks for outliers during computer editing. Values outside a predetermined range are output for review. Values outside a critical range (a larger range) are set to missing values if not changed. Critical edit failures are usually keypunching errors or reporting of production rather than yield. Currently, there are no direct measures of the effect of observational error on these data.

There is some concern about recent changes to the reporter methods. Such concerns are not new. Sanderson (1954) warns against the tendency to change definitions and scales frequently, cautioning that mere changes do not necessarily improve data quality. Discontinuing condition reporting may be the ultimate change, but should probably be viewed as the long delayed outcome of the development of objective yield surveys. Limited bridging studies (using both methods for a year in a few states) have not exhibited gross differences in results. There are concerns about the power of these bridging studies and the loss of the panel aspects of the prior design. Yet, the overall impact can be monitored as shown in the next section.

Control of the data collection activities for the objective yield surveys is more involved, using training, supervision, and editing. Training of the objective yield enumerators is done by each state office. Usually, there are 10 to 15 enumerators in a state. Nationally, about 15 percent of the enumerators are new to the survey each year, but they almost always have worked on other NASS surveys. All new enumerators work several sample fields with an experienced enumerator after the training school. Thus, using experienced enumerators may be the main form of control.

Training averages about one and one-half days in length and usually includes a field exercise. The training of enumerators is done locally by state agricultural statisticians. Training these trainers occurs at a national training session. In 1990, budget limitations forced the cancellation of the training of the trainers for objective yield surveys. Limited training was done at a training session for another earlier survey and national statisticians provided assistance in states where the trainer lacked experience. The pyramid style of training raises questions about consistency of training between states. Further, the reduced trainer instruction is an error source we will need to watch, especially if this lower level of training continues. Since many characteristics are not present at the time of training, color slides were used to show future measurements and color photo inserts were enclosed in enumerators' working materials. These practices became another victim of budget tightness as supplies were exhausted. Currently, there is no formal evaluation of the effectiveness of the objective yield training. Participants take quizzes during the training, but they are not scored. There are some ongoing efforts to develop evaluation of the training in other surveys. Hopefully, these efforts will be useful and expand to the objective yield program.

The objective yield survey has a quality control program designed to aid in the supervision of enumerators, detect faulty equipment, and detect outlier data. The immediate supervisor spends several hours with each enumerator near the beginning of the survey to observe sampling work and assure the use of correct procedures. Later in the survey, the state office selects at least one sample per enumerator per crop for a revisit by the supervisor. Whenever possible, the enumerator accompanies the supervisor to the check field. The supervisor checks the plot layout, counts and observations, then discusses differences with the enumerator, resolves the differences and documents reasons for the differences. These checks are not regularly summarized. Small sample sizes and some nonrandom selection would produce measures of questionable use. Also, the length of time to the revisit can produce valid differences in some direct observations such as maturity category and number of fruit. Control of equipment used to determine weight or moisture content consists of regular calibration. Finally, hand and computer editing during the survey provides some opportunities to revisit fields to verify data. At this time, computer edits are simple range checks and printouts of the ten most extreme values for each data item. NASS is exploring the use of portable data recorders to provide in-field resolution of many edit failures.

Measurement Activities

Several indirect measures of nonsampling error are produced for the reported and objective yield surveys. For example, nonresponse rates, usually under ten percent in yield surveys, can indicate the potential for problems with coverage bias. Completion rates by mode of interview for reported data, edit failure counts, and lists of outliers are typical of other indirect measures. None of these measures specifically identify direct observational errors, instead they may provide a warning of the need for more specific study.

Measuring the error of all the data used in the various yield surveys might seem like an ideal goal. In reality, it is not likely that enough personnel or financial resources would be available to study all the data. As a result, the search for the possibility of serious error begins with measurement at an aggregate level. For example, yield may be available from marketing or processing organizations, from assessments or censuses made by other governmental groups, or from expert opinion. These estimates provide a check on NASS methods, recognizing that they are subject to their own sources of error, such as administrative error or definitional error.

In situations where data are available from multiple survey sources across time, errors are often detectable. In the U.S., the yields reported by a volunteer "panel" of farmers and by a probability sample of farmers were available for corn for several years. The nonprobability sample yield averaged six percent below the probability sample yield. In another series, farmer-reported yield for the same farmers as the objective yield survey averaged three to four percent below the objective method.

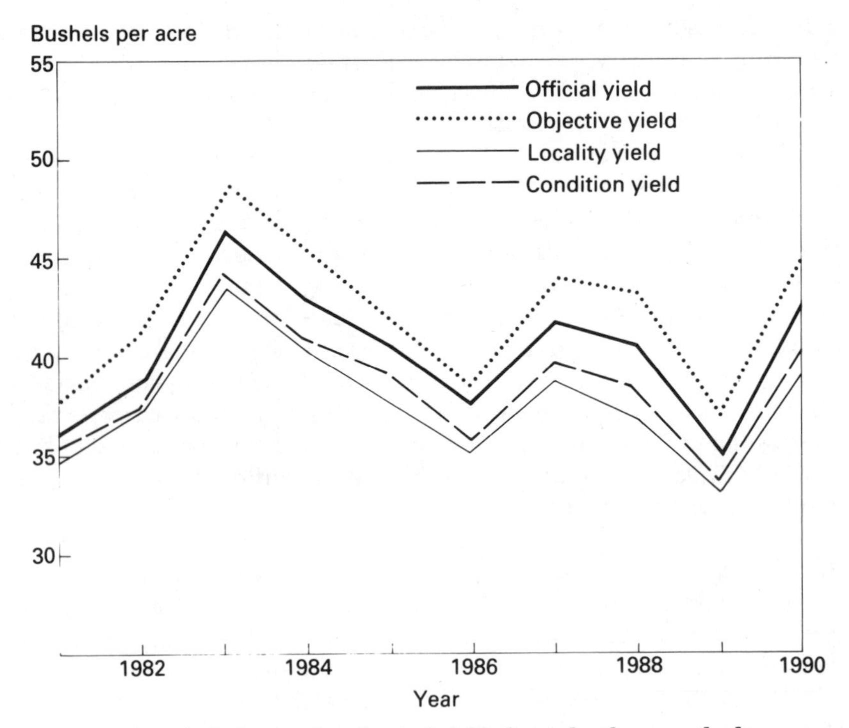

Figure 17.3. A typical time series chart of yields from the three methods compared to the official yield.

Figure 17.3 presents survey information for a typical crop. Yield derived from farmer reported condition and yield, and objective yield are compared. These time series charts indicate the ability of each method to identify year to year change and thus provide useful information. The figure depicts the general pattern found for most crops. Objective yield is higher than reported yield and reported condition, and there is reasonable stability across years in the biases. Secondly, the relatively small variability between methods within year compared to the variability of the official yield indicates the ability of the methods to provide useful data. Development of more statistical evaluations of these time series is an area of possible research. The most important question is which level corresponds to the true yield. That is, where should the board value be set? Validation surveys help answer this question.

Validation

Objective yield surveys are a rare example where a value very close to the "truth" is available at reasonable cost. This true yield is the weight of the harvest from the field taken at a nearby grain elevator divided by field acres. Validation studies compare field yield estimated by objective yield methods to true field yield. NASS conducts periodic validation studies to determine whether changes in varieties or cropping practices alter sampling error or bias. Changing errors require modifications of survey procedures or a change in the use of the survey data.

Historically, validation studies were conducted on one crop in several (but not all) states in a given year. Each state was assigned enough subsample fields and additional plots within fields to produce precise estimates of the difference between sample procedures and harvested yield for the study area. The validation surveys identified a wide range of nonsampling errors. Warren (1985) presents a comprehensive history of validation studies for the corn objective yield survey. Although measurement of DOE was not a specific goal in these studies, many of the errors identified are the result of observational tasks. Although results vary by state or year, a relatively consistent upward bias of between two and five percent was evident since 1966. In another study, Warren (1986) found that objective yield estimates were 11 percent higher than farmer reported yield for matched fields in the North Central States (a major growing region). Farmer-reported yield was lower for fields not going to a weight station. Objective yields were higher than yields from the Census of Agriculture and weighed yields. Ears selected to go to the laboratory were heavier than others in the unit. Furthermore, harvesting more than seven days after the crop cutting lowers the estimated yield.

The general consensus was that objective methods were about five percent above "true" yield while reported methods were about five percent low. Similar results occurred in wheat and soybean validation studies. As a result of several internal reviews of the yield surveys and their quality control activities, the validation survey methodology was changed.

The validation program was redesigned as a double sampling approach to measure bias in the estimate for all states in the survey (Fecso, 1986). Redesigned studies have been done since 1987 for soybeans. A relatively consistent bias of about six percent between the objective survey yield and weighed harvest yield has been estimated. The average bias estimates are 2.2, 2.3, and 3.2 bushels per acre for three years of study, with a standard error of the estimate consistently near 0.9 bushels per acre. The bias was significantly different from zero for all three years. Preliminary data from the 1990 study show bias down to 0.4 bushels. This decline from the average of the three prior years is not yet explained, but the only known change is the reduced training mentioned earlier.

These results are encouraging because, previously, studies did not have sufficient power to detect specific error sources. The soybean validation provides NASS with a relatively inexpensive methodology to assure the accuracy of the official yield statistic. There is increasing interest in expanding the methodology to other crops.

17.4.2 Future Directions

Although the validation approach provides a good and affordable overall measure of error sources and a control of the official estimate, improvements to survey operations or further research into specific causes should not be ruled out. Ideas which should be developed include: measurement and evaluation of training methods, exploring panel approaches to yield reporting, and developing more timely and statistically based edits. NASS staff are also developing improved cost tracking and reporting of indirect measures of error to aid management decision making.

In the future, basic research into the nature of DOE in yield surveys could rely on more cross disciplinary research. Concepts from psychological research, educational measurement, sociological research, and engineering can be applied to yield surveys to identify, understand, and measure DOE and to develop better training and field procedures.

Since the system of surveys produces satisfactory statistics and NASS has increasing demands for additional statistical output in other areas, it is not likely that funds will become available for extensive research in these areas. Even though the ad hoc methods for yield surveys have worked reasonably well, yield surveys need a formal statistical control system to assure the validity of the time series of estimates and to optimize operating and research resources.

17.5 ENDNOTE

Much has been learned in the long history of yield measurement surveys. Yet, this catalogue of errors may present quite a pessimistic impression. Actually, the quality of these surveys parallels that of most other large surveys; the information meets the expectations of most users while the providers struggle to improve accuracy with declining resources. In fact, more is probably known about errors in yield estimates over time than most other survey estimates because it is possible to measure consistency between several methods of estimation. Hopefully, this error listing provides a foundation for understanding and improving yield survey quality, and some insights into problems with direct observation which may occur in other applications.

CHAPTER 18

MEASUREMENT ERROR IN CONTINUING SURVEYS OF THE GROCERY RETAIL TRADE USING ELECTRONIC DATA COLLECTION METHODS

John E. Donmyer, Frank W. Piotrowski, and Kirk M. Wolter
A.C. Nielsen Company

18.1 INTRODUCTION

Electronic data collection technology has had a dramatic impact on surveys of the grocery retail trade in the United States. Historically, manual field audits were the primary source for store-level retail trade data, collecting inventories and shipments (or purchases) into the store, among other things. Sales were derived from purchases and inventory turnover. Measurement error, for the most part, was a function of the expertise of the field auditor. Today, in-store optical scanners are the primary source for data collection. Generally, scanner technologies have provided more timely and detailed survey data and have changed the composition of measurement errors.

In Section 18.2, we trace some of the history of surveys of retail stores, and discuss the methods and technologies used for data collection in such surveys. The attending structure of any measurement error that may occur is presented in Section 18.3, including a model for the error together with a description of possible sources of error. Section 18.4 presents error detection or editing strategies and possibilities for *ex post* evaluation or validation of data quality. The chapter closes with a brief summary in Section 18.5.

The entire discussion focuses on surveys of grocery stores, the principal sector in which the A. C. Nielsen Company does business, yet the models, error structure, conclusions, and findings also apply to other retail outlet types, such as drug stores and mass merchandisers. Nielsen

Measurement Errors in Surveys.
Edited by Biemer, Groves, Lyberg, Mathiowetz, and Sudman.

ISBN: 0-471-53405-6

grocery surveys, whether measured through in-store audits or scanning at the point of sale, collect consumer buying information on sales volume, prices, and promotions, including newspaper advertisements, in-store displays, and manufacturer- or retailer-sponsored coupons. In the balance of this chapter, to simplify the discussion, we shall focus principally on the sales volume and price variables.

18.2 STORE SURVEY

18.2.1 Background and History

The A. C. Nielsen Company launched the marketing research industry in the early 1930s, with the introduction of the Nielsen Drug Index and the Nielsen Food Index. These indexes provided customers bimonthly tabulations of total sales volumes and prices for many items then sold in drug and grocery stores. In the early years of these indexes, the data were collected by means of manual store audits. On a bimonthly basis, Nielsen field personnel visited a sample of stores, taking a complete physical inventory of many categories of items. The inventory data were collected only for those item categories for which Nielsen had paying customers. The inventory change from the previous audit period plus any shipments into the store during the period (ascertained by review of all warehouse and purchase order records at the store) was the derived estimate of sales volume for the period.

For the most part, these early surveys were paper-driven. Individual store data were sent to a central production facility for aggregation and projection into geographic region and total United States reports. This production process was clerical in nature, taking advantage of the technology of the day, which was essentially the mechanical adding machine.

With these early retail indexes, Nielsen offered its manufacturing customers a new statistic for measuring the success of their marketing efforts. This innovation was the "market share" statistic, or the percent of sales of a given product category attributable to a particular product, brand, or manufacturer within the category. Market share quickly became a standard tool for fulfilling the manufacturer's objective of creating and maintaining customers (Peckham, 1978).

Audit-based surveys remained the primary tracking vehicle for consumer nondurable goods for the next forty years. Then, in the early 1970s, the technology of collecting consumer sales data at the store level began to change. The Uniform Code Council was established to create industry standards to facilitate the use of optical scanners in retail stores in the United States. The Council, which is sponsored by several grocery industry associations, provides print specifications (Uniform Code Council, *UPC Symbol Specification Manual*, 1986) for the machine-readable symbol known as the Universal Product Code (UPC). The UPC is an eleven-digit, all numeric code that identifies the consumer package and/or the shipping container. The code consists of one character that identifies item type, a five-digit manufacturer identifier, and a five-digit item code. The item-type character distinguishes among random-weight items (e.g., meat and produce), fixed-weight food items, drug items, coupons, etc. While the UPC Council assigns manufacturer numbers and provides guidelines concerning UPC assignment, item codes are actually assigned by each manufacturer, who may or may not rigorously adhere to the guidelines.

The first test installation of a retail scanning system occurred in a Kroger store in Cincinnati, Ohio in 1972. Nielsen participated in evaluating the results of this test by comparing scanner-generated sales with manually audited figures. Scan sales compared well for many items, but many problems were found also. Certain packaging types resisted scanning due to materials used, shape, or color density and contrast of the UPC symbol. Also, some manufacturers failed to use unique UPCs for different flavors or colors of a given product.

Over the ensuing eighteen-year period, scanning was adopted by a majority of the large grocery organizations across the U.S.; manufacturers assigned UPCs to all, or nearly all, consumer package goods sold in the U.S.; and Nielsen developed new tracking services based upon samples of scanning supermarkets. Today, 56 percent of all U.S. supermarkets (a supermarket is defined as a grocery store with $2 million or more annual all commodity volume or ACV) scan merchandise at the point of sale, representing 75 percent of total supermarket ACV.

Our scanning samples evolved over this eighteen-year period, beginning with the launch in 1979 of scanning samples in two metropolitan areas (or markets), and the launch in 1980 of the first national scanning sample, based upon only 100 stores. Today's scanning sample is composed of 3,000 supermarkets and produces weekly sales tabulations for the total United States, 50 local markets, and 150 individual grocery organizations (or chains) within markets. This sample is supplemented by an audit-based sample of 800 smaller grocery stores, representing the stratum of stores below $2 million in annual ACV.

18.2.2 Current Methods and Technology

Currently, there are approximately 170,000 retail establishments in the United States classified as grocery stores. See U.S. Office of Management and Budget (1972); in particular, Standard Industrial Classification code

541. We cover this universe using the stratified probability sample of 3,800 sample stores described above. Scanning data (including UPC, sales, and price) are collected from respondents in the large store stratum, whereas in the smaller stores, where the use of checkout scanners are often prohibited by cost, all data are collected by means of traditional field audits. For the large grocery stores, the explanatory or causal variables, such as presence of promotional displays, are collected by Nielsen field personnel using hand-held computers with wand scanners.

Over 900 data collection points, representing the 3,000 scan sample stores, send Nielsen weekly scanning data once per week, using magnetic tapes, data cartridges, data cassettes, and floppy disks. The retailers use all known equipment vendors and model types (NCR, IBM, SWEDA, etc.), and all known point-of-sale systems and data formats (ISD, BASS, PC MASTER, etc.). Many stores forward their data directly to Nielsen, while others first preprocess the data on a host computer at chain headquarters before forwarding it. In a few cases, Nielsen obtains continuous sales data in real time by direct, electronic data-capture methods, and aggregates the data to the weekly level. For causal data collection, the field worker scans the appropriate UPC by way of the wand, and key-enters other required information through on-screen prompts.

Many stores use retailer-specific product codes, known as locally-assigned or price-look-up codes (PLU) in place of scanning the UPC. This practice is normally reserved for promoted or difficult-to-scan merchandise, such as 50-pound bags of dog food. The in-store checkout scanners collect the selling price and total sales, while field workers collect the set of UPCs associated with the PLU codes at the time of the weekly causal data collection. In the case where a given PLU code covers two or more UPCs, the total sales are apportioned to the UPCs prior to report tabulation using historical relationships.

An important, and perhaps unique, feature of our survey is that sales data are reported only for those products which sell during a seven-day period. Within retailer, there are approximately 17,000 packaged-good items stocked in a typical store, and only 12,000 or so sell in any given week. Between retailers, however, there are roughly 900,000 unique UPCs, and all of these have appeared, at least once, in a Nielsen sample store.

18.3 MEASUREMENT ERROR MODEL

In what follows, the fundamental measurements of interest are sales volume, as expressed either in physical units or dollar value, and price of individual items or UPCs. Let Y_{hri} denote a specific variable of interest for the i-th item, the r-th store, in the h-th stratum. In our applications, we obtain, process and produce weekly data for the large store stratum and bimonthly sales data for the small store stratum. Thus, Y_{hri} should be interpreted as reported sales or price for a given week or bimonth, as appropriate. Where necessary, we append the subscript, t, signifying the time period.

Thissen (1990) presents a comprehensive discussion of measurement problems that can occur in scan data, Y_{hri}, and some recommendations for reducing or eliminating the problems. We shall summarize a few of the more common problems, as well as some of the important audit-data problems, in the context of a modification of the classical error model articulated, for example, by Hansen, et al. (1961, 1964), and Koch (1973). Our modified measurement error model is expressed by

$$Y_{hri} = \sum_{i' \varepsilon I_{hr}} \Delta_{hri'} \Gamma_{hri'i} (y_{hri'} + e_h + u_{hr} + v_{hri'}) \qquad (18.1)$$

where $\Delta_{hri'}$ is an indicator variable signifying whether or not the store reports the sales of the item; $\Gamma_{hri'i}$ is an indicator variable signifying whether or not the store classifies the sales of item i as item i'; I_{hr} denotes the set of items available for sale by store (h,r); $y_{hri'}$ denotes the true value of item i' in store (h,r); and e_h, u_{hr}, and $v_{hri'}$ are components of reporting error, given $\Delta_{hri'} = 1$. Each of the components of this model will be defined later.

As for the classical measurement error model, all of the terms in (18.1) should be regarded as conditioned upon the general survey conditions, say G. In the case of the large store stratum, G includes but is not limited to choices from the following categories: type of scanning equipment; use of a centralized host computer by the chain organization, or not; scanning policies within the store; and manufacturers' policies regarding packaging. In the case of the small store stratum, G includes counting, keying, and classification practices by the field auditor; filing practices and completeness by the grocery store concerning records of shipments or purchases; and Nielsen's processing and follow-up activities. Overall, the realized G is a composite of the specific realized conditions affecting each of the stores in the universe.

The various components of model (18.1) attempt to represent many of the real reporting problems that arise in surveys of consumer packaged goods collected by electronic means. The indicator variable Δ signifies whether the item sales are reported or not. In most sample surveys, the survey organization controls the items to be collected by way of the questionnaire, and missing data arises because a targeted respondent refuses to cooperate, or supplies faulty data, and so on. On the other hand, for the sample of large stores we receive scan data by tape or by direct transmission in whatever format is convenient for the sample store or store chain, and have only limited means of knowing or checking whether there are items not reported.

For the large store stratum, errors can and do occur in the grocery chain's host computer prior to our receiving their data. Occasionally, some of the sales movement data is dropped, not written to our tape, and thus $\Delta = 0$. If a tape is unreadable, or is only partly unreadable, data may be dropped in Nielsen processing, and thus $\Delta = 0$. In some cases, the checkout clerk may consistently enter the price, but not the UPC, and thus $\Delta = 0$. For the small store stratum, the field auditor may overlook an item or may overlook inventory in the back room. The retailer may misfile or destroy purchase records. All of these circumstances may lead to nonreporting, or in other words, $\Delta = 0$.

The second indicator, Γ, signifies classification error, arising from the fact that even though the item-specific sales volumes may be reported, they may be misclassified under wrong UPCs. For the large store stratum, PLU codes cause classification problems as do new UPC codes. On the order of 5,000 new UPC codes are created and discovered each week. The PLU codes, the new UPC codes, as well as any ambiguous existing UPC codes require clerical research, and in many cases callbacks to the grocery store so as to achieve the correct product classification. Errors can and do occur in this research process.

In some cases, data is keyed, not scanned, at the point of sale, leading to the potential of keying error or omitted sales movement. In other cases, the checkout clerk at the grocery store may gather several similar items and scan one of them repeatedly to represent the others. This practice may, e.g., record 10 bags of orange powdered soft drink instead of one orange, one cherry, one strawberry, and so on.

For the small store stratum, field auditors may misclassify one item as another, and the grocery stores may misclassify their purchase records, thus leading to classification error.

Finally, the e-, u-, and v-components span all of the measurement errors that may occur to sales volume or price, given that the item is reported. The e-component reflects a stratum or store-chain effect. This may arise if a host computer at the chain headquarters collects and processes all data from all affiliated stores before sending the data to Nielsen. This effect may be zero in instances where no host intermediary is used, with data sent directly from the sample stores to Nielsen. The u-component reflects a store-level effect. The premise here is that scanning and management practices utilized within the store will serve to impart a common effect to all items within the store. The v-component is an item/store-specific effect, which varies from item to item and store to store.

For the large store stratum, there are numerous e-, u-, and v-effects. Occasionally, a store or store chain may supply year-to-date data mistakenly, instead of the current week's data. Data for slightly less or more than a full seven-day week may be sent, as might data for a seven-day reporting period that is offset by a day or two from the actual reference week. Stores or store chains sometimes send erroneously the same data two weeks in a row. Credits or returns or voids by the store may not be handled properly.

Frequently, there are problems with multipack items; depending on the item's packaging and scanning practices at the store level, the same UPC code may identify a single unit (e.g., a can of soft drink), a six-pack or an eight-pack. Clerical staff or computer edits must identify how each multipack item is scanned at each store, so that unit sales of comparable items will be aggregated across stores, and prices of items across stores will refer to the same unit definition.

A similar problem in product definition occurs both for special promotional packaging, e.g., razor blades are added on to a can of shaving foam, and for reused UPCs where a manufacturer suddenly begins using a UPC for a new product, even though it had been assigned, heretofore, to another product. For both situations, comparable products across stores must be identified so that data will be aggregated properly.

Occasionally, the sales movement and price data that are sent to us are misaligned, representing different reference weeks. This error can vary from week to week and store to store. For some chain organizations who employ a central host computer as an intermediary, the price and sales data may be misaligned chain-wide.

For the small store stratum, counting mistakes or keying errors can be made by the field auditor. Purchase records kept by the grocery store are sometimes wrong or incomplete. Error in the measurement of inventories or purchases leads to error in the derived sales volume.

It may be reasonable to assume that the e_h, the u_{hr}, and the v_{hri} are mutually independent random variables. Thus, measurement errors are correlated within a supermarket chain by virtue of e_h and within an individual supermarket by virtue of u_{hr}. Alternatively, we believe other correlation structures may be important in certain circumstances. For example, similar and problematic product design and packaging may result in scanning errors, and thus correlation, across all UPCs within a given brand or manufacturer. Looking at data across time, we can also imagine correlations, e.g., between e_{th} and $e_{t-1,h}$, u_{thr} and $u_{t-1,hr}$, and v_{thri} and $v_{t-1,hri}$.

For large stores measured by scanning, the between-store or between-item correlations are induced mainly by the policies and practices of the store or of the store's organization. For small stores measured by audit, these same correlations will arise principally from the phenomenon usually called interviewer (in this case auditor) variance.

Furthermore, it may be reasonable to assume that classification errors, Γ, and the errors in reporting sales volume are uncorrelated, given that sales volume is reported, i.e., $\Delta = 1$. Given these various assumptions on the correlation structure of Δ, Γ, e, u, and v, it is easy to write down an expression for the total mean squared error of survey estimators, and to identify its major components. The detailed derivation is not provided here due to space limitations.

We have only recently begun development of measurement-error models like (18.1), and have done so to provide a principled, structured means of classifying the various types of measurement error and of describing their impacts on Nielsen tabulations. We have begun to use (18.1) in the basic sense of classifying types of errors and recording their frequencies. Some of this work is reported next in Section 18.4. Model (18.1) has also proven to be of value to us in assessing the kinds of data edits that may be helpful in screening our input data and in developing the logical and statistical forms of the edits.

18.4 ERROR DETECTION AND ERROR VALIDATION

18.4.1 Error Measurements and Validations

Management and control of measurement error are critical to the success of any survey. This is particularly true for continuing market research surveys of the type conducted by Nielsen, because without proper management and control of these errors, they will, over the course of time, begin to outweigh the other components of survey error and significantly increase the total mean squared error. It is important that these errors are properly monitored, measured and validated.

Measurements. One of the most effective tools to monitor and measure the quality of any system is the statistical process control chart. In the 1920s, Shewhart first developed control charts to help partition special- from common-cause variation. When used properly, these charts are powerful tools to monitor and improve a process (Shainin, 1990). Simple control chart tolerance limits, like a 3-sigma rule, can be used. Figure 18.1 presents a typical control chart for the rate of measurement error, as defined in model (18.1). This measurement error rate is computed by dividing the total number of faulty response observations by the total number of observations with a response, then multiplying by 100. Not only is this rate easily measurable through an analysis of the results of the data editing systems, it also provides timely and actionable information to correct unforeseen problems. When the rate of measurement error falls outside the acceptable control limits, an unexpected or

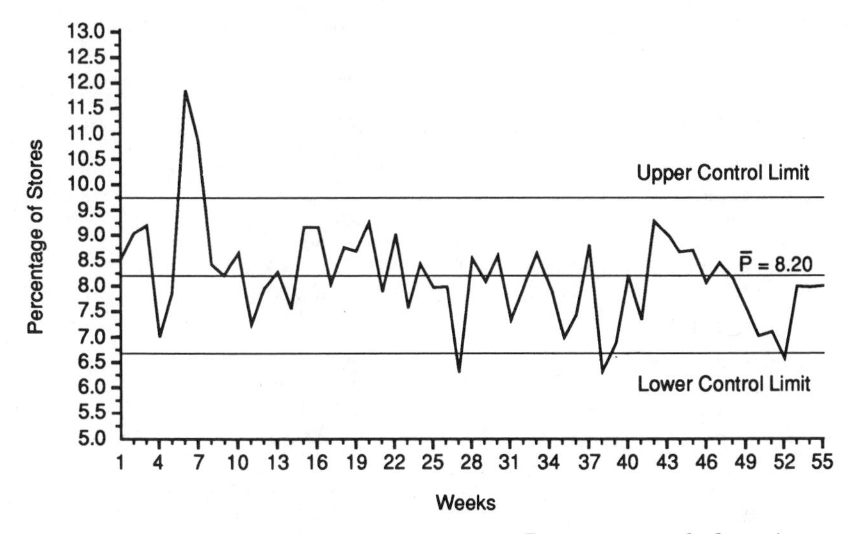

Figure 18.1. Rate of total measurement error. Process control chart (current week = 55).

new error contaminant has affected the survey data. These errors can take the form of errors of measurement or Type II errors in the edit system. In either case, the data analysts react to these out-of-control signals and identify the source of the error and recommend corrective action.

The process mean used in the control charts is generated by summing the measurement errors quantified by the edit system (Section 18.4.2) for each component in model (18.1). Table 18.1 displays the results and also quantifies the conditions which generate these errors. The error rates shown are based on 26 weeks of data in 1990.

The u-component (all items in a store) contributes the largest error rate. Approximately 6 percent of all data received is affected by this common store error condition. The e-component has many of the same error conditions as the u-component, but with one significant difference. Not only have the error contaminants affected all items in a store, but also *all stores* in a stratum. Typically, the e-component error is related to a data processing problem at the retailer's headquarters, which in turn affects all stores from that organization.

The final major component of error in model (18.1) is the v-component, or item within store-specific errors. The major contributors to this component are items that are not scanned by all stores. In some cases, retailers choose not to scan (or use item-specific PLUs) for convenience or efficiency. For example, due to the physical size of many

Table 18.1. Measurement Error Rates Weekly Summary of All Response Data

Data Conditions	UPCs per Week		Error Rates (%)
All Response Data	30,000,000		8.2
Correct Data	27,544,000		—
Erroneous Data:			
u-component of error	1,860,000		6.2
• tape problems		301,300	
• duplicate or accumulated data		619,400	
• incorrect data		939,300	
e-component of error	180,000		0.6
v-component of error	416,000		1.4
• incorrect unit sales		3,200	
• incorrect sale prices		75,000	
• difficult-to-scan items, such as bags of dog food and cigarettes		337,800	

large bags of dog food, retailers may find it convenient to enter the price of the item using a general grocery key rather than scanning the item's UPC. Cigarettes are examples of items that a retailer may elect not to scan for efficiency reasons. A typical supermarket will carry over 200 different cigarette items; however, these items will have only four or five unique prices (pack versus carton, king size versus regular). Rather than maintain all the individual items on their scanning price file, it may be more efficient (and use less disk space) to record sales by price entering or using a general PLU for all cigarette items with the same price.

Figure 18.2 presents the *u*-component error rate for selected categories with incomplete scanning data. As this figure indicates, a significant number of stores contribute to this component of error, for one or more product categories. Overall, the affected items represent 1.4 percent of all response data.

Validation. In continuing large scale surveys of the kind discussed here, an effective data validation program must contain two operational phases. The first phase is the initial validation of all survey data prior to any use of the data in final tabulations. The second phase is the validation of tabulated data. In both phases, statistical techniques such as reinterviews, cross validation, and record checks are required to validate data effectively.

Reinterviews are primarily used to validate audit data. An experienced field auditor will be sent to selected stores to rework the previously completed audit for selected items. Any differences found must be reconciled, and if necessary, audit procedures adjusted. The reinter-

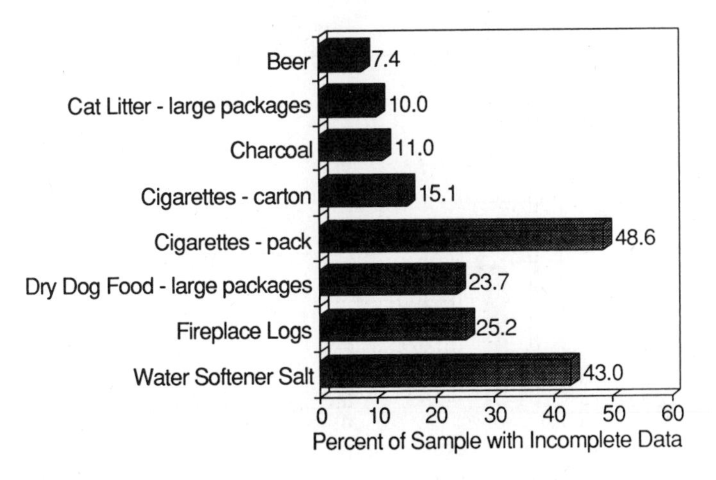

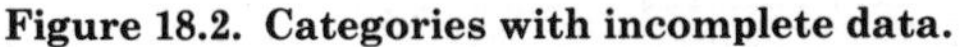

Figure 18.2. Categories with incomplete data.

views are performed on a scheduled basis at the close of each survey period, and on demand, to resolve a specific challenge to the survey data.

Cross validations, where the survey variables are compared with each other or to an expectation, are primarily utilized in phase one of the data validation program. This technique is used extensively to determine which stores provide "complete" scanning data for the difficult-to-scan categories. For these categories, scanned dollar sales and number of UPCs received from the retailer are compared to in-store observations made by Nielsen auditors. These observations include: number of items in stock, location of items in store, and sales recording practice observed for the category (scanned, price-entered, PLU-entered, etc.). For each category with a scanning problem, a criterion is established for appraising stores as either "complete" or "incomplete." For example, one of the rules for a store to be "complete" for large package dog food is that at least 80 percent of the items observed in stock must appear on the retailer's tape (20 percent are allowed to be slow movers with no sales for a given week, based on empirical research). If this and all other rules are true, the store's data is acceptable for tabulation.

Figure 18.3 provides a distribution of sales rates for "complete" and "incomplete" stores for large package dry dog food. The sales rate is an index of dollar sales for the category per million dollars of all commodity volume. It should be noted that while low sales rates are typically associated with "incomplete" stores, some "incomplete" stores can have

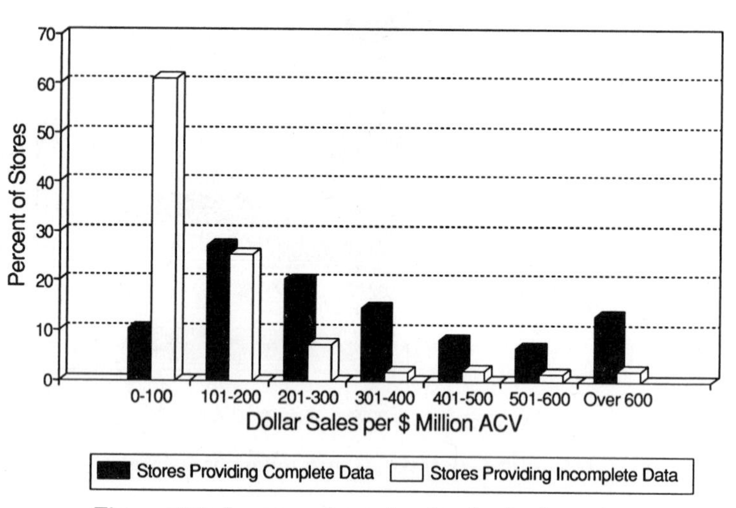

Figure 18.3. Large package dry dog food sales rates.

relatively high sales rates, indicating that only a small portion of data is missing.

While phase-one validations are designed primarily to minimize the reporting of incorrect data, the primary function of phase two is to detect Type II edit failures and estimate the magnitude of the error in order to improve the overall edit and imputation process. Typically, record checks are most often used to validate data in this phase. This procedure compares the observed survey data, Y_{hri}, to data, $\hat{Y}_{hri}$, thought to be true, obtained from independent sources external to the survey process. These independent sources of data include retailer financial data (separate from scanning data) or manufacturer shipment data. A recent example of record checks occurred within the carbonated beverage category. During the summer of 1990, a popular promotion was a parking lot sale. Customers meeting the store's requirement (presenting coupon, or total sales over a certain amount) could purchase the promoted item at a discount. The physical location of the item, however, was outside the store, so the majority of item sales were not scanned. In each known case, the store manager was contacted to obtain the true level of item sales. Figure 18.4 summarizes the scanned data versus manager's data for these events.

Once these measurement errors were fully understood and mea-

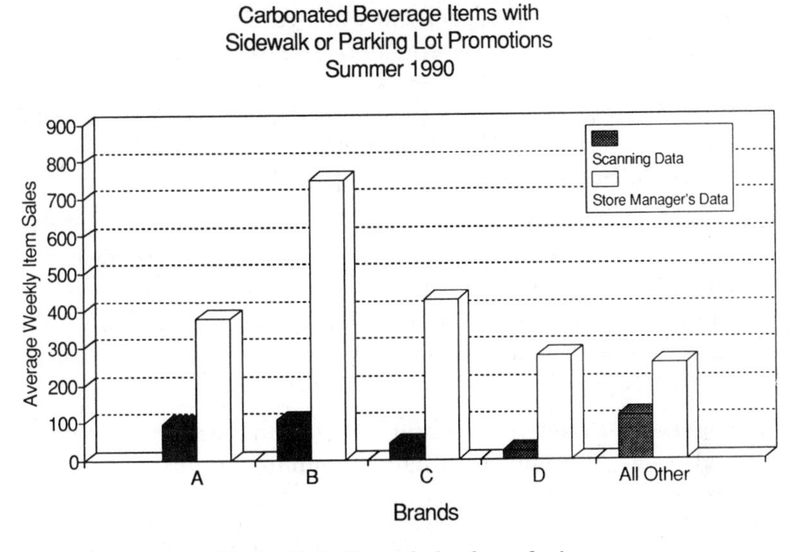

Figure 18.4. Record check analysis.

sured, appropriate edit and imputation procedures were established to estimate consumer sales. Nielsen auditors, during their weekly store visits, record all UPCs which are not scanned due to special promotions of this type. During the weekly edit process (Section 18.4.2), these UPCs are manually reviewed and compared to the manager's information, if available, or to historical data for the store. Consumer sales are then imputed for all UPCs with incomplete sales data.

18.4.2 Data Editing

Traditionally, most editing systems have involved considerable manual review of the data. These systems are obviously costly and time-consuming. In most editing systems today, there is a balance between manual review and computer automation [Greenberg and Petkunas (1987); Little and Smith (1987)]. Given the volume of data collected — in the large store stratum alone, 30,000,000 records are processed weekly — and the need for timely reporting, automation must be a primary objective of any editing system. It should not, however, be exhaustive. Analyst review is still required for selected error situations.

Prior to the design of an edit system, each error component of the measurement-error model must be understood. Typically, estimates of each component are made using data from past or comparable surveys, pilot studies, or subsamples specifically designed to evaluate data errors. The overall edit structure is dependent on the frequency and size of these errors.

Our editing system is designed to detect each type of error identified in model (18.1), and is divided into three stages which are applied sequentially to the incoming data for a particular time period:

(1) Store Level Stage: The largest error component of model (18.1) is the u-component, which imparts a common error effect to all items in a store. The initial edit stage is performed on aggregated store level data to determine if all items have a common error effect. If a store fails this stage, all items are classified as errors. Otherwise, the data are subject to stages two and three.

(2) Item Level Stage: The υ- and Δ-error components, i.e., errors to individual items, are identified using computer-intensive item-specific edit routines.

(3) Processing Level Stage: Classification and tabulation errors, the Γ-component, are detected at the end of the process, after all stores have been processed, edited in stages one and two, and tabulated.

It should be noted that the e-component of error is not addressed statistically with edits, but rather operationally with analyst review guidelines. For example, analyst review of the edit output is deferred until all stores in a chain are processed, to help identify a common store-chain error effect.

Before we detail more of the specifics of this edit system, we would like to make a few comments on imputation. Normally, editing and imputation are discussed together, since most effective data processing systems have these procedures intertwined. There is a very fine line, for example, between nonreceipt of data, which can be defined as nonresponse error, and receipt of totally contaminated or unreadable data, which can be defined as either nonresponse or measurement error. We have purposely omitted discussions on imputation, due to the constraints of this book.

18.4.2.1 Store Level

The most fundamental question the edit system must answer is whether all data from a particular store have a common "error" effect. Since a typical large grocery store may handle 10,000 to 15,000 different items, it would not be efficient to test each item to determine the store's overall usability. Instead, items are aggregated into homogeneous groups and tested, using group tolerance bands. The tolerances are determined from previous estimates of the groups' sales distributions. Group definitions are a function of product type, product use, and stocking practice.

It should be noted that during holiday weeks, the tolerance bands are adjusted to reflect increased group sales. Also, all group aggregations are not treated equally. Item-specific knowledge and understanding of error-generating mechanisms suggest, for example, to weight carbonated beverages, and health and beauty aids higher than, say, breakfast cereals. The former categories are prone to multipack- and UPC-type errors, whereas cereal is not.

Once a store is classified as usable, i.e., there is no common error effect, its data are filtered through computer-intensive item-level edits (discussed in the following section). Utility programs utilizing artificial intelligence techniques are also used to classify nonusability conditions such as: fewer days of data than specified; more days of data than specified; accumulated data from previous periods; truncated transmissions of data; and duplicate data from a previous period. In some situations (such as a day's problem) these utilities suggest corrective action to adjust nonusable data into usable data.

18.4.2.2 Item Level

Once a store is determined to be usable for the survey period, each item must be subjected to item-level edits to minimize the v- and Δ-component in error model (18.1). Two separate edit processes are used for inspection of sales units and item price.

Sales Units. The test statistic, P, weights the potential error by the size of the item and hence gives greater importance to more frequently purchased items. Thus,

$$P = B_0\left[(S_{rit} - S'_{rit})/\frac{(S_{rit} + S'_{rit})}{2}\right]^2 \times \left[\frac{S_{rit} + S'_{rit}}{2}\right].$$

S_{rit} denotes unit sales for store r, item i, time t; and S'_{rit} denotes the expected sales for the store, item as computed through baseline time-series techniques. The constant, B_0, reflects the type of item (e.g., multipack or high-variance item) and the statistic can be reduced to:

$$P = B_0(S_{rit} - S'_{rit})^2/\left[\frac{S_{rit} + S'_{rit}}{2}\right].$$

If we let $S'_{rit} = S_{rit,1}$ and $B_0 = 2$, this statistic is also the first order series approximation of Strobel's (1982, 1984) information measure D.

This P statistic not only has great theoretical appeal, as evident above, it also has great intuitive appeal. In practice, many analysts mentally weigh relative change by the magnitude of absolute change when determining significance, and can easily understand the application of this statistic. Both the P statistic and the information measure have been used in editing and exception reporting of retail surveys. Moreover, information theory has been useful in a variety of fields, including marketing (e.g., Hauser, 1978; Herniter, 1973).

One example of use of the P statistic is in the manual audits of small stores. For this segment, the survey procedure of dependent interviewing is used. The auditor is provided the results from the previous two audits while conducting the current audit in-store. The P statistic helps the auditor reconcile the exception items prior to transmission of the data. Roughly 10 percent of all items are identified as exceptions and warrant a recheck of the audit.

It should be noted that dependent interviewing is used primarily to assist the auditor and improve data quality. However, it can also be used by the auditor to falsify or fabricate the audit. This type of false audit is known as "curbstoning," "fictitious auditing," or, at Nielsen, "plugging a store." The penalty for "plugging" is termination of employment; therefore, the occurrences are very rare. At any rate, both the data edits and validation procedures must recognize this possibility. For example, the P statistic can also be used to detect data sets that are suspiciously too smooth.

Price Edits. Unlike sales-movement edits which are trend-based, price edits must be cross-sectional in nature. A given price for a given item in a given store may require a continuous adjustment for the life of the survey. This is not due to an error in the retailer's reporting of price, but rather an inconsistency in the price basis. For example, retailer A may record a six-pack of 12-ounce cans of cola as one unit sold for $2.40. Retailer B may count the same item as six units sold for $0.40 each. Both methods are correct. However, in tabulating the data, our surveys must handle the item consistently for all stores. The most common way to handle these multipack alignment problems is through tolerance range tests. Using all reported prices received, a tolerance band of acceptable prices is generated. Since the upper and lower tail extremes are potentially contaminated data, a trimmed measure of spread is used. All prices which fall outside this acceptable range are corrected; either adjusted to the same multipack basis, or corrected to an expectation.

For scanning stores, it has been our experience that approximately one percent of all prices need to be adjusted. Of these adjustments, about

75 percent are multipack realignments. The remainder are true price errors such as missing price multiples (e.g., 3 for $1.00, where the 3 is missing) or incorrect outliers.

18.4.2.3 Processing Level

Other sources of measurement error are classification errors, such as incorrectly describing a 6 1/2-ounce bag of potato chips as a 7-ounce bag, and tabulation error, such as incorrect placement of a new or changed UPC in a client's report. Classification errors are normally detected at the input or individual-store level, while tabulation errors are detected at the end of the survey process after all stores have been processed and tabulated.

Classification errors are most effectively identified using logic checks on the incoming data. To ensure that all UPCs are properly described and classified on a continuous basis, current UPC activity and descriptions for a store are logically compared to previous data. A change in incoming description or observed activity after a long period of inactivity (e.g., six months) initiates a recheck of the UPC classifications.

Unlike most measurement errors, tabulation errors most often surface at the end of the process and after final reports have been produced. Two edit systems are used to detect tabulation errors. One incorporates ratio-edit theory, the other a true outlier adjustment procedure.

A typical tabulation of the scanning data consists of category, brand, and UPC estimation for total U.S. and 50 major markets. The time frame is normally eight weeks with four individual "current" weeks and four previous weeks. These relatively short time series are edited for extreme weekly variation using sequential ratio edits. Greenberg and Surdi (1984) discuss ratio edits at the U.S. Bureau of the Census. For each observation in the series, a ratio is computed and tested using all other observations. If any ratio fails, manual inspection of data is initiated to resolve the failed series.

A true outlier adjustment system can also be used to control measurement errors. The statistical literature is rich with discussions and theories on outlier observations. Typically, most systems are designed to reduce the mean squared error of the estimate. In these systems, an adjustment is applied to the outlier observation, such that the reduction in sampling variance is not offset by the increase in bias. Fuller (1970) proposes a series of censored estimators for the mean of the Weibull distribution and an outlier test on the value of the shape parameter. These estimators work well with most distributions which exhibit "tail" characteristics of the Weibull distribution. A system of this type has been employed by the A. C. Nielsen Company since 1980.

18.5 SUMMARY

In this chapter, we have discussed changes to the structure of measurement error in continuing surveys of the grocery retail trade, due to the use of electronic data collection methods. We have presented a generalized measurement error model appropriate for this new technology. Currently, this model is being used by A. C. Nielsen to identify error sources, and to establish a classification system to measurement error. The chapter also described data editing techniques and data validation procedures in use by Nielsen. While the cost of data quality can be measured in terms of resources required, such as data analysis, computer programmers, and CPU time, the true cost must also consider the potential lost revenues from clients. If not carefully monitored and controlled, measurement error in these surveys may lead to clients making wrong business decisions which ultimately leads to lost clients.

SECTION D

MEASUREMENT ERRORS IN THE INTERVIEW PROCESS

CHAPTER 19

CONVERSATION WITH A PURPOSE — OR CONVERSATION? INTERACTION IN THE STANDARDIZED INTERVIEW

Nora Cate Schaeffer
University of Wisconsin, Madison

19.1 INTRODUCTION

All three components of the survey interview — interviewer, respondent, and questionnaire — shape what happens in the interview and the resulting data. Strategies for improving survey data that focus on only one component may neglect ways in which they interact. The chapters in this section contribute to the study of the joint effects of these components of the interview. The first two chapters propose frameworks for examining the joint effects of the interviewer, respondent, and questionnaire (Chapter 19) and of the respondent and questionnaire (Chapter 20). The remaining three chapters present empirical studies (Chapters 21–23).

The classic characterization of the interview as a "conversation with a purpose" (Bingham and Moore, 1924, cited by Cannell and Kahn, 1968) has been used by both survey practitioners and critics. One recent critical examination concludes that ways in which the standardized interview is not a conversation may undermine its purpose (Suchman and Jordan, 1990). Survey researchers might respond that although the standardized interview resembles a conversation and uses participants' conversational skills, it is not a conversation, nor is it meant to be. The interaction between the interviewer and respondent is, however, one potential source of observational errors. Moreover, measurement-by-talk imposes constraints that require complicating theories of measurement. Understanding how talk in interviews differs from talk in other situations is needed to improve measurement theory and practice.

Measurement Errors in Surveys.
Edited by Biemer, Groves, Lyberg, Mathiowetz, and Sudman.

ISBN: 0-471-53405-6

Critics are correct that locating where standardization fails and suggesting solutions require a closer look at interaction in the interview. Overlooking the problems of standardization poses the risk of canonizing current failures. But reforms that ignore the justification for standardization run the risk of repeating old mistakes.

19.2 THE STANDARDIZED INTERVIEW: THE CASE FOR AND AGAINST

Underlying competing assessments of the survey interview are different views of three interrelated topics: the appropriate goals of research, the role of the interviewer, and validity and error. Issues of validity (reviewed in Chapter 1) motivate concern with the first two issues, and together they suggest taking a closer look at the site of data collection.

19.2.1 The Case for the Standardized Interview . . .

The standardized survey interview takes measurements on a large number of persons, for a set of topics selected by a researcher, in such a way that measurements can be aggregated to describe a population. Aggregating measurements is only straightforward if the same question is asked of all respondents; each respondent has one of a set of mutually exclusive answers; and no respondent–answer combinations are empty (Galtung, 1967). Strictly speaking, answers are only comparable if all respondents are also asked to accept or reject the same choices (see Schaeffer and Thomson, in press). Ideally, errors are taken into account by theories of construct validity and models of response errors (see Chapter 24).

Survey researchers train hundreds of interviewers of varying abilities to collect thousands of interviews. Differences between interviewers in the way they understand the investigator's purpose, administer questions, and probe answers increase interviewers' contributions to variable survey errors and bias. These in turn undermine the goal of obtaining comparable answers from all respondents. Two standardizing procedures designed to reduce between-interviewer variation and meet the demands of comparability are using highly scripted closed questions and restricting the ways interviewers may deviate from this script. By training interviewers to read questions as written and to use neutral probes, and by turning certain requests back to the respondent, survey researchers attempt to hold key behaviors constant. The relationship between the interviewer and respondent is also constrained so that the demands of the personal relationship will not outweigh those of measurement. Ironically, perhaps, it is because their research tradition recognizes the power of the interviewer that survey researchers have made the standardized interviewer almost powerless over a broad range of decisions, in the judgment of some observers.

That reporting context affects answers is implicit in the notion of a "classical true score" (see Chapter 1 and Bohrnstedt, 1983). Standardization takes seriously the effects of context on validity by attempting to hold crucial aspects of context constant, although this is overlooked by arguments that standardization typically sacrifices validity for reliability (e.g., Cicourel, 1964, pp. 76-81).

19.2.2 . . . And the Case Against It

Some critics argue that the goal of mass measurement may be fine in principle, but impossible in practice. The researcher's agenda may be unrelated to the lives of respondents (Cicourel, 1964, 1974; Mishler, 1986a). Partly as a result, answers in an interview lack ecological validity — the meaning conferred by grounding in the everyday (Cicourel, 1982). Because they do not examine what actually happens in the standardized interview, survey researchers pressure respondents into categories that fail to capture their experiences. By sweeping problematic answers into categories such as "don't know" or "refused," survey researchers ignore signals that negotiations about meaning have broken down. These related concerns focus on what survey answers do and do not mean, and together implicate the validity of responses in the standardized interview.

One solution emerging from such criticisms is to "empower" respondents and interviewers. At one extreme are proposals that the respondent should be invited to collaborate in defining the research question.[1] This goal entails revising the power asymmetry of the interview, so that interviewers become "learners" and "advocates" (Mishler, 1986a; see also, Paget, 1983). Validity would be improved if interviewers could pursue important issues and respondents could describe their experiences using culturally "natural" forms of talk.

---

[1]This recommendation attempts to resolve some of the political issues inherent in research (e.g., Briggs, 1986, pp. 120–125; Mishler, 1986a, pp. 122–132), but raises problems of its own. Community residents may be unaware of issues that the researcher judges important to their lives; or communities may comprise factions with competing goals, among which the researcher must choose.

Respondents' stories could then be analyzed using the tools of narrative analysis (Mishler, 1986a).

A less extreme proposal is to reformulate the role of the interviewer to minimize the damage standardization does to validity. Respondents present unpredictable situations, and attempting to script the interview in advance deprives interactants of interactional resources to use in negotiating the unexpected (Suchman and Jordan, 1990). Giving the interviewer flexibility in diagnosing and correcting misunderstandings could increase validity: question wording might not be standardized, but *meaning*, which is more important, would be (see Briggs, 1986, p. 24; Suchman and Jordan, 1990, p. 240).

19.2.3 Talk — But Not a Conversation

Some of the critics' illustrations include what most survey practitioners would recognize as "bad" questions (e.g., Mishler, 1986a, p. 60) or rare situations (e.g., Suchman and Jordan, 1990, p. 237). But practitioners themselves sometimes produce questions that in retrospect are clearly bad and encounter situations that require reconsidering the assumptions of their questions. Criticisms run the risk of being ignored, however, when they fail to confront the general goals of survey research. Simply rejecting surveys may mask an unwillingness to face similar problems posed by other styles of research (e.g., Briggs, 1986, p. 10; Hyman, et al., 1975 [1954], pp. 4–9). Three interrelated difficulties survey researchers encounter in attempting to benefit from these criticisms are the use of "everyday conversation" as a standard (e.g., Mishler, 1986a; Suchman and Jordan, 1990), a focus on the structure rather than the content of the interview (e.g., Mishler, 1986a), and an implicitly deterministic model of the interview (e.g., Mishler, 1986a; Suchman and Jordan, 1990).

Use of Conversation as a Standard

Objections that the standardized interview is not a conversation — at least not an "ordinary" conversation — play on the ambiguity in the word "conversation." Sometimes conversation is used synonymously with "talk in interaction," which could include interviews, and sometimes it is used more restrictively. It seems useful to recognize "interview" and "conversation" as distinct classes of talk in interaction (e.g., Schegloff, 1990), which can be compared with each other. This requires specifying what is distinctive about "conversation" and what is added or subtracted by "ordinary." Almost any interview (see Werner and Schoepfle, 1987, pp. 306–310; Button, 1987) is different from conversation — speech-exchange in which the "order, lengths, and contents of turns are specified neither by prior social prescriptions nor by persons with socially prescribed responsibility for managing the interaction" (Wilson et al., 1984, p. 160). But the inequality in control over topics and turns at talk that characterizes interviews also occurs in other native speech events (e.g., a parental lecture on good conduct), some of which might be considered conversations.

The contrast between standardized interviews and conversation is most instructive in that respondents may apply rules learned in other speech events and reject attempts to impose interview rules (e.g., Briggs, 1986, and the discussion below). Interviewers may begin to discard interview rules in response (see Chapter 21). Although norms with which interviewer and respondent are more familiar may never be completely suppressed, it is an open question whether systematically giving participants more access to "normal" conversational resources would improve the quality of the interaction or the resulting data. Any movement to remove constraints on the interviewer must pay attention to effects on interviewer variability.

There is no simple criterion to use in evaluating how the specialized nature of the standardized interview affects validity. The artificiality of the standardized interview does not necessarily invalidate responses. After all, both the "true self" we reveal with friends and the "polite self" we present to strangers are social constructions, as is the sense that the former is a more "real" self (Lofland, 1969). People express important things about themselves in situations even odder than the survey interview, and keep parts of themselves secret in "everyday life." The relationship of answers produced in a standardized interview — with the kind of validity they have to offer — to other kinds of data — with the kind of validity they have to offer — requires more investigation than it usually receives. Recognizing the distinctive character of interviews is important to understanding the relationship between respondents' expressions in interviews and in other contexts.

Focus on Structure

In his discussion of the research interview, Mishler (1986a) advocates less structured interviews and the techniques of narrative analysis. Such proposals entail two problems for survey researchers. First, there is the problem of how to combine and compare the answers of different respondents, particularly when interviewer variability may be substantial. Second, the analytic techniques described seem more suitable for understanding the process and structure of interviewing than for analyzing the *content* of answers across respondents. This content is central for survey researchers, who worry that a respondent might produce different stories in different unstandardized interviews. The criticisms do, however, emphasize that the structure of interaction is one source of its meaning and that the line between "structure" and "content" is often fine. "Nonattitudes" (Converse, 1970) illustrate this close connection between structure and content in standardized interviewing.

Deterministic Models of Interaction

Discussions of interaction — for example, ones that describe how a respondent understood a particular question in a particular interaction — appear to use an implicitly deterministic model (for an exception, see Churchill, 1978; see also the discussion of turn-taking models in Wilson, et al., 1984). The "validity" of a response is situated — it is simply that respondent's answer in that conversation. Analyzing validity would require seeking criteria that would be constructed in other interactions. Emphasizing a specific interaction as a source of meaning is also compatible with a focus on structure and a lack of concern for rules for combining stories and generalizing about their contents. At its most extreme, such determinism neglects rules for making generalizations about content or implies that they are impossible. This determinism is, however, a reminder that responses are always negotiated in specific cases and there may be systematic causes for what appears as "error" in aggregate models.

19.3 SURVEY RESEARCHERS EXAMINE INTERACTION IN THE INTERVIEW

The models that survey researchers use to study issues related to interaction in the interview accept the goals outlined earlier. Sudman's and Bradburn's (1974, pp. 418) model emphasizes the structure of the interview. Variables of interest are associated with the interviewer role (role behavior and extra-role characteristics), the respondent role (role behavior), and the structure of the task (question form, mode of administration, etc.). Cannell and Kahn (1968) propose a "motivational model which treats the interview as a social process and regards the interview product as a social outcome" (1968, p. 538). Although they also characterize the role structure of the interview, Cannell and Kahn underscore the importance of motivational and cognitive factors, such as the respondent's perceptions of the purpose of the interview, and the participant's actual behavior. A third class of models combines these elements somewhat differently and focuses on the process by which the respondent arrives at and reports an answer (Cannell, et al., 1977a; Cannell, et al., 1981, p. 393; Tourangeau, 1984).

These models can be seen as nested and complementary. The structural model identifies central elements of the interview; the social-psychological model suggests that they must be subjectively transformed to be relevant to participants; the cognitive model shows how they enter into the response process. Similarly, much of the research relevant to understanding interaction in the interview can be roughly classified according to whether it measures structural features of the interview (e.g., interviewer race, question form, etc.) or participants' perceptions of these features and their behavior in the interview. Here I focus on the latter.[2]

19.3.1 Participants' Orientations and Definitions

The importance of a "phenomenology" of the interview, which would capture the participants' shifting definitions of the interview, has long been recognized (Hyman, et al., 1975 [1954]). Studying participants' perceptions of the interview, however, requires either constantly interrupting the interview to ask about current perceptions or undertaking a second interview (see Chapter 20). In the first case, the intervention is likely to affect the perceptions asked about; the second approach adds problems of memory, dissonance reduction,[3] and social desirability. Consequently, there are few studies of participants' perceptions of the interview.

An exception is interviewers' expectations, some of which can be measured before interviewing begins. Hyman, et al. (1975 [1954]) distinguished among expectations based on interviewers' theories about attitude structure, about respondents' social roles, and about the probability distribution of responses in the population. They concluded that interviewers' expectations about respondents have more influence on what interviewers record than the interviewer's own beliefs, partly because interviewers are trained not to communicate their beliefs. Later studies have distinguished among general expectations about a study

---

[2]Research that examines structural features of the interview is discussed elsewhere in this volume (see Chapters 13, 14, 22, and 23). Studies of interviewer effects and variability describe the impact of the respondent-interviewer interaction without studying it directly (see Hyman, 1975 [1954]; and reviews in Groves, 1989; Hagenaars and Heinen, 1982a; and Lessler, 1984). I also do not discuss studies of interviewers' vocal characteristics (Barath and Cannell, 1976; Blair, 1977–78; Oksenberg, et al., 1986; Oksenberg and Cannell, 1988).

[3]For example, respondents overwhelmingly prefer that the interviewer behave on the second interview just as she did on the first (Cannell, et al., 1968, p. 12) and also prefer the mode of administration to which they were first exposed (Groves, 1979).

and those for specific questions and among effects on cooperation, item nonresponse, and responses (e.g., Bailar, et al., 1977; Singer and Kohnke-Aguirre, 1979; Singer, et al., 1983; Sudman, et al., 1977b).

Respondents' global perceptions of the interview have received little attention recently. One reason is probably the conclusion of early investigations that participants' behavior had a greater impact on the course of the interview than did their orientations to the interview (e.g., Cannell, et al., 1968). An important exception is that the respondent's definition of the task appears to affect reporting (Cannell, et al., 1968, pp. 31–32; see also Chapter 20 this book).

19.3.2 Behavior in the Interview

A series of systematic studies has provided an increasingly complex view of interaction in the interview. In one of the earliest studies, interviewers' and respondents' different orientations to the interview were suggested by the much greater frequency with which respondents initiated talk unrelated to the interview (Cannell, et al., 1968). The finding that feedback from interviewers affected the amount of information respondents provided (Cannell, et al., 1977a; Marquis and Cannell, 1969, cited in Cannell, et al., 1981) led to devising feedback that reinforces appropriate role behavior by respondents (Cannell, et al., 1981) and to recommendations for more scripting of the standardized interview (e.g., Fowler and Mangione, 1990).

These studies have been important in developing ways to identify and record features of interaction. Without such methods, the components of traditional concepts such as rapport — which captures a range of qualities — remain unanalyzed (see Goudy and Potter, 1975–76; Hyman, et al., 1975 [1954]). Rapport has recently been replaced with the more neutral concept "interviewing style." Behavior coding systems are essential for comparing professional or formal styles, on the one hand, and interpersonal, socio-emotional, or personal styles, on the other.

Coding Behavior in the Interview

Studying interaction requires classifying a continuous stream of behavior. Three related systems for coding interaction have evolved. The earliest of these systems (Marquis, 1971b) associates a series of codes for interviewer and respondent behaviors with each questionnaire item. Codes indicate, for example, whether the interviewer modifies the question wording or the respondent requests clarification. The coding system has been used in evaluating interviewer performance (Cannell, et al., 1975; Cannell and Oksenberg, 1988; Mathiowetz and Cannell, 1980) and in pretesting instruments (Cannell and Robison, 1971; Morton-Williams, 1979; Morton-Williams and Sykes, 1984; Oksenberg, et al., 1989).

A different approach begins with a detailed analysis of the rules of standardized interviewing and the role requirements and actions they imply (Brenner, 1981c). The resulting coding system classifies each action by whether it meets the actor's role requirements or how it fails to do so (Brenner, 1981b, 1982). The behaviors can then be compared across interviewers (Brenner, 1982) or types of questions (Brenner, 1981b), using tree diagrams that show the sequence of actions that begins with asking a questionnaire item. Dijkstra, et al. (1985) extended previous coding systems for an experiment in which interviewers were trained to behave in either a formal or a personal style (see also Chapter 21). Each speech act is coded for four variables which indicate the actor (interviewer or respondent), speech category (e.g., question or answer), orientation (task or personal), and adequacy with respect to the prescriptions of standardized interviewing (see also the elaboration in Dijkstra, et al., 1988, and Chapter 21).

Issues in Studying Interaction in the Interview

Studies of interaction in the interview raise several issues. First, all the interviewers in these comparisons are trained to be standardized (though this is less true in Brenner's studies), and interviewers' "personal" behaviors are highly restricted. Thus, the comparison is not between standardized and unstructured interviewing. The dangers of not controlling interviewer behavior are suggested by the results of an experiment in which the interviewer expressed an opinion about whether transportation or recreational facilities would be important to her in deciding where to move. To a later question about what would be important in a move, respondents in both styles were equally likely to echo the interviewer's opinion (Dijkstra, 1987, p. 327). A second issue is that interviewers are usually charged with persuading respondents to participate. If studies do not control how interviewers recruit respondents and success in recruitment is correlated with interviewing style, results could reflect differences among the respondents that each type of interviewer persuaded.

As these coding systems illustrate, decisions about what to record determine which features of the interview are visible and how they can be analyzed. Coding interaction requires identifying meaningful actions and their boundaries. Cannell and Kahn (1968, p. 572) identify a basic cycle within the interview: the interviewer reads a question, the respondent answers, and the interviewer improvises an indication of acceptance or rejection. This basic structure is prescribed in standardized interviewing, but it can be augmented by participants. Brenner briefly discusses the difficulties encountered in segmenting interaction (1981a, pp. 27–28; 1981b, pp. 138–139), such as overlapping speech and incomplete utterances (see also Chapter 21).

These coding schemes are evaluative because survey researchers are concerned with measurement issues, and they entail compromise because the material in any interview is overwhelming. But continuing experimentation with ways of segmenting and coding interaction is needed. Otherwise we risk concluding that features of interaction are inconsequential when we have only failed to identify what is important. For example, only when the coding system records respondents' interruptions of the initial question reading can this be revealed as a problem in question design.[4] Similarly, respondents often provide more material than a question specifically requests (Cannell, et al., 1968, p. 19). The implicit request for additional material characterizes some questions in other social situations (e.g., Clark, 1985; Francik and Clark, 1985), and respondents may assume that survey questions are intended similarly. Extended answers may occur more frequently in the first of a series of related survey questions, and the spontaneously provided information may affect the interaction for an entire series of questions (Cannell, personal communication). Although available coding systems could be used to analyze some features of such phenomena, they cannot yet describe the impact of the *content* of respondents' spontaneous offerings through time or how interviewers respond to that content.

The data produced by coding interaction pose difficult statistical and methodological problems. Analyses must consider that multiple question-and-answer sequences are provided by each respondent, questions and answers are developed through time, and respondents are nested within a small number of interviewers (Chapter 22, this volume). Statistical tests should consider these features. Furthermore, patterns of error correlations under different interviewing styles need attention.

Finally, *how* the style of interviewing affects respondents' answers must also be examined. Interviewing style might affect whether respondents attempt to be accurate or to ingratiate themselves with the interviewer (Dijkstra, 1987). Mediating processes could also include effects on the respondent's role expectations (e.g., attending to interviewer cues), the interviewer's task behavior (e.g., providing inadequate

---

[4]Marquis (1971b) codes a behavior called "blocking" (p. 235). It is unclear whether this code would be used if the respondent interrupted the reading of a question with an answer. The accompanying analysis of questions does not appear to use the code (Cannell and Robison, 1971). Oksenberg, et al. (1989) use a separate code if the respondent interrupts the initial question reading with an answer.

cues), reinforcement patterns in the interview, and the respondent's motivation or mood (Dijkstra and van der Zouwen, 1987, p. 201; see also Chapter 21). This list should also include effects on how the respondent understands the *point* of a question and uses this understanding to formulate an answer. The sequential structure of the tasks presented by different questions must be analyzed and explicitly considered in analyses of interviewing style (Chapter 21 begins this task). The apparent contradiction that a professional interviewing style yielded more accurate reports about accidents as the time since the accident increased (Henson, et al., 1976) but a personal interviewing style resulted in more accurate maps of the neighborhood (Dijkstra, 1987) suggests that the difficulty of the task must also be considered. A personal style may better motivate respondents, but make it difficult for interviewers to impose the demands of difficult recall tasks. Lastly, because the actions of respondents vary with style of interviewing, the effect of the respondent on the interviewer may also so vary.

19.4 STRATEGIES FOR STUDYING INTERACTION

Studies by survey researchers primarily focus on whether behavior meets the criteria of standardization. In this section I draw on research outside the survey tradition to consider (1) conceptualizations of the context of interaction, (2) interaction as a medium of comprehension, and (3) detailed studies of interaction that can be used in studying the survey interview. The first two motivate the third. The need to broaden measurement theory to consider the context and practice of measurement motivates all three.

The measurement models that survey researchers typically work with are more refined versions of one we use every day: If I ask someone how she is, I recognize that social factors as well as her true state influence her self-report. Similarly, measurement in surveys views self-reports as functions of underlying unobservables. And at their best, measurement models include other influences on self-reports, and studies either measure these influences (e.g., interviewer race) or hold them constant (e.g., question wording).[5] Improving the "measurement" conducted in interviews requires understanding how social and psychologi-

[5]Although investigators attempt to eliminate some sources of bias (e.g., leading questions), this is not the principal technique, as critics sometimes suggest (e.g., Briggs 1986, p. 24). Even using neutral probes implements the more general strategy of holding potentially important influences on answers constant across respondents.

cal processes are used in constructing self-reports. Just as the findings of psychologists require incorporating heuristics into measurement models (e.g., Nisbett and Ross, 1980; Tversky and Kahneman, 1973), studies of behavior in the interview can identify interactional influences on responses. An understanding of the range and power of these influences is needed to appreciate the kind of knowledge self-reports have to offer. Social context both constrains self-reports and establishes what they mean. Specifying when a relationship between variables holds (Mishler, 1979) requires refining our conceptualization of the social context in which self-reports are produced.

"Context" is one way of referring to, among other things, the global structure of the interview as a speech event, the social situation surrounding the interview, or the role relationships (both inside and outside the interview) among the participants. Context can be defined at a very local level, for example as the interpretive framework resulting from a series of turns at talk or as the immediately preceding sentence which provides the antecedent for a pronoun. Context can also be described from the point of view of goal-directed participants or from that of an observer. Finally, discussions of context can focus on the content or structure of talk or show how they are interrelated.

19.4.1 Global Context: The Interview as Speech Event

A speech event, such as an interview, is an activity "directly governed by rules or norms for the use of speech" (Hymes, 1972, p. 56), and is made up of speech acts. Briggs (1986, pp. 40–60) presents a model of the interview that directs attention to (a) the frame supplied by the social situation and type of communicative event constituted by the interview; (b) the interviewer and respondent, their respective roles, and interactional goals; and (c) the referents, message forms, channels, and codes of communication. Explicitly viewing the interview as a speech event stresses the role of communicative norms in guiding talk and interaction. The distinction between the interview as a speech event and its constitutive speech acts (e.g., questions and answers, stories, side sequences) emphasizes that context is both global and local — participants orient both to their perceptions of the purpose and structure of the interview as a whole and to their immediate talk, and both figure in participants' understandings.

Global Definitions of the Speech Event

As evidence that respondents recognize the interview as a speech event, Wolfson (1976) cites the existence of the word "interview" and respondent's exasperation when the interviewer, in an effort to impose less structure, fails to ask questions. The prescriptions respondents ascribe to the cognitive "world" of the interview (Agar and Hobbs, 1982) may vary. A respondent's global definition of the interview may be an intimate conversation or therapy session, an exercise in filling out a form, a referendum, a test, or a game (Turner and Martin, 1984, pp. 262–270), a "talk show" (Suchman and Jordan, 1990, p. 236), or an opportunity to construct a coherent positive identity (Mishler, 1986b).

People telling narratives select those details that are relevant to the point they are trying to make (Moerman, 1973), so that defining what the point of the story is to be becomes important. Similarly, a respondent's global definition of the interview's purpose influences his contribution to the interaction, and hence the resulting data. Global perceptions may motivate the respondent's participation, but subvert the interviewer's control of the situation or undermine the goal of candid and accurate responses. Or a respondent can use global perceptions of the interview's purpose to compensate for an interviewer's failures: Agar and Hobbs (1983) describe an interview with an experienced informant, Jack, to learn why he had moved from Chicago to New York. When that story proved shorter than their typical interview, Jack demonstrated orientation to a "norm" that an interview should produce at least one side of tape, which he apparently inferred from their past interviews. The interview rambled after its initial purpose was completed, because the interviewer was not prepared for another topic. But Jack attempted to produce material relevant to the research topic — he talked about his first arrest and sexual experiences in prison — illustrating his orientation to what he perceived as the interviewer's goals, at a level that subsumed the specific interview.

Global Rules Governing Talk

Talk in interviews is subject to formality constraints that have yet to be described. Formality may help ensure that the interactional tasks of the interview are accomplished (Atkinson, 1982) and may affect the use of language. Compared to language in other social situations, formal talk may be more highly structured and more consistent. Speakers may invoke their positional identities or roles more readily and use language to maintain focus on the situation (Irvine, 1979). The formal relationship between interview participants also undoubtedly constrains what can be discussed and how. For example, Jefferson (1985) compares a mother's conversations about her son with two different neighbors. Statements the mother makes in one conversation "gloss," or refer to, something that is "unpackaged" in a later conversation with a more receptive conversational partner. Only when a gloss is unpackaged, either spontaneously or in response to direct questioning, is its status as a gloss revealed. Jefferson's discussion suggests examining more closely how the interviewer collaborates in providing a social context in which formal "glosses" can be recognized and "unpackaged." (One could also argue, of course, that it is just this variability by social context that justifies standardization.)

In addition to being guided by their global definitions of the interview and rules about formal interaction, respondents' talk uses rules learned in other talk. For example, respondents may use conversational maxims not to be redundant (Grice, 1975), as shown by answers to a series of closed questions (Schwarz and Strack, 1988; Schwarz, Strack, and Mai, 1991). Similarly, respondents who are asked a series of open questions refer to previous comments and rely on the sequence of answers to clarify what they mean (Bobo, personal communication).

Studying the Context Supplied by the Interview

The work of survey researchers suggests that respondents' definitions of the interview have important consequences, but that their understanding of the researcher's goals may remain vague even after the interview (Cannell, et al., 1968). Briggs (1986, pp. 95, 106–107) suggests analyzing the structure of actual interviews and paying particular attention to comments participants make about the interview itself to uncover global perceptions of the interview.

Our understanding of the social constraints in interviews would be helped by comparative studies of the types of formality that characterize talk between strangers in a variety of roles on a variety of topics. Identifying appropriate comparisons requires an analysis of the dimensions of the social context of talk (see above), and may prove difficult. There are many studies of talk between doctors and patients (see references in Maynard and Clayman, 1991), for example, but the actors are often not strangers, and the patient typically seeks the contact and wants something from the doctor. Similarly, we need comparative studies of how statements respondents make in the standardized interview are related to statements they make in other social situations and the contextual sources of similarities and differences (see Turner and Martin, 1984, pp. 86–89; Dean and Whyte, 1958; Cicourel, 1982). Such studies would take seriously that there is no "context-free" measurement and aid the theoretical task of understanding the expression of "unobservables" in different social contexts.

19.4.2 Local Comprehension: Questions and Responses

Because surveys prescribe the questions interviewers ask and how they may react to respondents' answers, studies of interaction in the interview must attend to characteristics of questions, both as scripted and as enacted, and the rules guiding interviewers' behavior. Because questions and answers are enacted in talk, the participants' mutual comprehension is fundamentally interactive. The coordination between interviewer and respondent uses context as it appears in their "common ground" (Clark and Marshall, 1981). Common ground includes everything said in an encounter (linguistic evidence), the common experience of the participants (perceptual evidence), and things that participants believe known to others in their communities (community membership). The common ground thus changes over the course of talk. Questions and answers "become" problematic when they introduce interactional "troubles" into the common ground.

Improving Common Ground: Questions

Simply by being asked, almost any question conveys that it is intelligible, salient, answerable, and that the person asked is capable of providing an answer. Some questions also suggest what is feasible, viable, or appropriate, or what does or does not exist (May, 1989). The suggestions contained in questions are problematic for all interviews. In addition, poor question design — particularly problems of question relevance and meaning — can impair comprehension of questions and cause interactional problems. The "correct" answer to a question that makes a false assumption, for example, is rejecting the assumption (Kaplan, 1981). Such rejections require the interviewer's intervention, probably increasing interviewer variability. Researchers repeatedly face the problem that what a question turns out to have been — for example, a simple request or an admonition — depends on what the respondent does with it (e.g., McHoul, 1987).

Careful question design can address some problems in comprehension and relevance that become problems in interaction. Studies of conventional questions and spontaneous answers suggest that question writers could better exploit respondents' tendency to answer the question behind the question. For example, when restaurant owners were asked if they accept credit cards, half spontaneously listed the credit cards they accept (Clark, 1979). Werner and Schoepfle (1987, pp. 319–321) illustrate the use of ethnographic techniques to "negotiate the question." An informant is presented with an "answer" and asked to provide the question that would elicit it. The informant successively formulates a series of questions and answers to his own questions. The process shows both how questions restrict possible answers and how a desired answer restricts possible questions. Ethnographic and focused interviews used with individuals or groups (e.g., Merton, et al., 1990) may illuminate matters as "simple" as the appropriate vocabulary (e.g., Bauman and Adair, 1989) and as complex as the kinds of uncertainty respondents experience (e.g., Schaeffer and Thomson, in press). Such material can improve the relevance and intelligibility of questions by providing investigators a description of the "cognitive worlds" of potential respondents.

Improving Common Ground: Responses
To infer what a question asks for, respondents may apply rules of deletion or selection, generalization, and construction (van Dijk, 1980, pp. 46–50) to the question as it appears in the common ground. Some comprehension problems in interviews arise when participants apply these rules using community-membership-based assumptions about common ground that differ from the researcher's assumptions. A respondent may assume, for example, that no one would ask about "unimportant" visits to the doctor. When read a list of health practitioners to consider when calculating the number of medical visits, the respondent may then consider some practitioners on the list "irrelevant details" (deletion/selection). From the practitioners remaining on the list, the respondent may generalize that the question asks about visits to specialists. If the interviewer previously accepted a rough estimate as an answer, the respondent may also draw on this common ground to conclude that the interviewer wants a rough estimate of visits to specialists.

Responses are a standard mechanism by which speakers display both understanding and acceptance of prior talk. In some social situations, for example, a listener to a story may indicate how she interpreted the story by commenting or by telling a similar story (Moerman, 1973). This response allows the original story teller to determine how he has been understood. By providing fixed response categories, standardized interviewing attempts to minimize the need for interviewers to check their understanding of a respondent's answer. But standardization provides interviewers few resources to use in diagnosing and correcting problems of comprehension when this is needed (see example in Suchman and Jordan, 1990, pp. 239–240). Some methods of displaying understanding must be used cautiously in any interviewing and are inappropriate for large-scale surveys. But interviewers need neutral methods of verifying their understanding with respondents. One approach is to distinguish between scripted questions and "side-sequences" (see below) in which the interviewer may use prescribed techniques for offering clarifications and pursuing ambiguities.

19.4.3 Enacting Local Context: Analysis of Speech Acts

The survey data we analyze are products of talk: Interviewers enact questions and negotiate answers with respondents. Studies of talk in other contexts can contribute to understanding interview data in two ways — by illuminating persistent problems in surveys and by providing concepts and methods for basic research on interaction in the interview. Because all talk occurs in some social context, both strategies are inherently comparative. Comparative studies are needed to identify interactional influences on survey answers for inclusion in measurement models where appropriate. To capture the close connection between structure and meaning in talk, such studies must examine details of interaction.

Conversation analytic (CA) studies offer results, concepts, and methods useful in analyzing the sequential organization of talk and the "interactional substrate" of the interview (Marlaire and Maynard, 1990; Maynard and Marlaire, 1987). The interactional substrate consists of the interactional skills that interviewers and respondents use to arrive at "accountable" survey answers (see Garfinkel and Sacks, 1970; Button, 1987; Groves, et al., 1990). An accountable answer is one an interviewer can justify as having been properly arrived at. For example, an interviewer using a neutral probe to help a respondent select an answer is applying interviewing rules to cooperatively produce an answer that researchers using the data can then regard as "objective." CA studies also contribute to a comparative foundation for studies of interaction in the interview, but they must be used carefully. Many CA studies do not describe the role relationship between the speakers whose talk is being analyzed or the situation in which the speech fragment occurred (see Cicourel, 1987). Thus, it is not always clear how the social context might be similar to or different from the interview (see Goffman, 1981, pp. 29–34).

Applying Results from Conversation Analytic Studies

Some results of conversation analysis can show how common conversational practices might affect the quality of data collected by surveys. I present two examples, one dealing with a nagging theoretical issue, the other with a practical problem that can increase interviewer variability.

Acquiescence. Acquiescence, the tendency of respondents to agree with questions or statements using an agree/disagree format, has been described as a "quagmire" (Schuman and Presser, 1981). This swamp may be everywhere, because one way people talking together structure minimal affiliation is to design questions or assessments for agreement (Sacks, 1987; Pomerantz, 1984a). The relative preponderance of "yes" over "no" answers in talk suggests speakers often proceed in this way (Sacks, 1987, p. 57). But speakers know that listeners will not always agree with them, and need a mechanism to engineer agreement. One

Exhibit 19.1. Agreement and Disagreement in Talk

A. Delayed Disagreement

A: Yuh comin down early?
B: Well, I got a lot of things to do before gettin cleared up tomorrow. I don't know. I w – probably won't be too early.

B. Agreement Followed by Disagreement

A: How about friends. Have you friends?
B: I have friends. So called friends. I had friends. Let me put it that way.

C. Strong Agreement Followed by Disagreement

A: 'N they haven't heard a word huh?
B: Not a word. uh-uh. Not – Not a word. Not at all. Except Neville's mother got a call . . .

D. Question Reformulated for Agreement

A: They have a good cook there?
((pause))
A: Nothing special?
B: No, everybody takes their turns.

E. Agreement to Both Parts of Forced Choice Question

A: ((clears throat)) Was yer lenience in yer opinion because of the fact thatchu uh thought this would be a better way of controlling uh the members
[
B: Yes because
A: or was it because of the fact thatchu wanted them to like you.
B: Well, partly it was because I wanted them to like me and they I mean they were all my friends. Everybody in that dorm had known me fer years. An –

Source: Simplified from Sacks, 1987, pp. 58, 62–64. Panel E conforms to Sacks' comments in the text. Reprinted with permission from Multilingual Matters, Ltd, Bank House, 8a Hill Rd., Clevedon, Avon BS21 7HH, England.

mechanism is the pattern "question$_1$, pause, question$_2$, answer." The question recipient supplies a pause to let the speaker redesign the question for agreement. The pause serves as a signal, because disagreeing answers tend to be preceded by delay (Exhibit 19.1, Panels A and D). When disagreement does occur, the question recipient may first agree and then disagree, or coparticipants may negotiate by proposing candidate formulations until agreement is reached (Exhibit 19.1, Panels B, C, and D).

Drawing on their experiences in other social contexts, some respondents may assume that many survey items have been formulated for agreement. Respondents who disagree may initially signal their responses as they might in other situations; they then find that the interview does not allow them to negotiate agreement by modifying the question or statement. Because "neutral" probes do not recognize signals of disagreement, respondents may also perceive probes as requesting "agreement." Forced-choice questions may offer less escape

Exhibit 19.2. Downgraded Conventional Response Followed by Trouble

Bob:	[How are you] feeling now
Jayne:	Oh? pretty good I guess
	[
Bob:	Not so hot?
	(0.8 second pause)
Jayne:	I'm just sort of waking up,
Bob:	Hmm,
	(3.6 second pause)
Jayne:	Muh ((hiccup)) (0.9 second pause) My (ear),
Bob:	Huh?
Jayne:	My (ear) doesn't hurt (0.4 second pause) my head feels ((brief pause)) better,

Source: Simplified from Jefferson, 1980, p. 153–154.

than is sometimes hoped (Converse and Presser, 1986, p. 38). Both parts of the question may evoke agreement (Exhibit 19.1, Panel E), and respondents may then appear ambivalent (see Schaeffer and Thomson, in press). A similar problem may affect other questions. For example, a "downgraded conventional response" such as "I'm pretty good" may announce that I am not so good and invite further inquires (Exhibit 19.2).

Standardized interviewing often implicitly assumes that a respondent's first answer indicates salience or attitude strength (Kane and Schuman, 1991). Accepting the respondent's first answer without pursuing it minimizes negotiations between respondent and interviewer (and opportunities for the interviewer to influence the answer), and is thus also part of constituting the answer as produced by the respondent (see Button, 1987). Some practices depart from this model (e.g., warning interviewers to probe initial "don't know" answers, Survey Research Center, 1976, p. 17). By generally preferring the respondent's first answer, however, practice does not consider that respondents may first agree and then disagree (see also Suchman and Jordan, 1990, p. 236). Interviewer variability probably increases when respondents spontaneously develop a second answer, because interviewers' responses may vary.

CA studies suggest that acquiescence sometimes inheres in self-reports, but also that some statements, such as self-deprecations, are formulated for disagreement. (See Pomerantz, 1984a. This may contribute to race-of-interviewer effects.) Background expectations about survey questions probably vary over respondents and questions, and may dominate when the respondent does not feel strongly about an issue (Schuman and Presser, 1981). Or such expectations may be more important in talk between strangers, where affiliation depends on the immediate interaction.

This discussion suggests that we study closely how respondents agree and disagree, as well as "don't know" answers, other possible disagreements, and subsequent probes and resolutions. CA studies of talk in other contexts provide comparisons for research on acquiescence, particularly if the participants' relationship and the interaction's level of formality can be considered. Moreover, studying how people talk in other situations may help investigators face the difficult problem of deciding which answer they want, when respondents have more than one to offer, and the implications of that choice for measurement. Interviewing techniques—particularly for subjective items—that let respondents talk about a topic before presenting them with fully standardized questions may allow respondents who have several answers to resolve them.

Turn-Taking and a Simple Issue in Question Design. Part of displaying competence in everyday talk is appropriately projecting and using possible transition points for turns (Sacks et al., 1984). A question projects a transition point, and speakers appear to prefer placing questions at the end of turns (Sacks, 1987). Hearers project the beginning of their turn by recognizing the upcoming completion of the question. Writers of standardized questions frequently violate the rule of placing the actual "question" at the end of the interviewer's turn. This may happen because the topic of a questionnaire item is indicated in a question, which is then followed by restrictions, definitions, or response categories (Exhibit 19.3). Although some interviewers succeed in holding their turn with a "rush-through" (Schegloff, 1982, p. 76), the resulting confusion for respondents is evident in the frequency with which they "interrupt" the initial question once they recognize its point (see examples in Cannell, et al., 1989; Jefferson, 1973 and 1986 for discussions of overlapping talk at transitions). The interviewer may have to cut off the respondent's "premature" answer or repeat definitions or categories until the respondent learns that response categories *may* follow a question. Replacing a long item with several questions or placing the actual question at the end of an item greatly reduces the frequency with which respondents interrupt (Exhibit 19.3).

Basic Research Using Conversational Analytic Techniques

Some details of interaction, such as feedback by interviewers, are known to affect respondents' answers. But further studies of interaction may reveal previously unnoticed features as important. CA studies have identified important components of the interactional substrate of talk generally. Basic research on the interview can compare how these components appear in interviews and other interactions, in different

Exhibit 19.3. A Frequently Interrupted Question and Its Revision

First Pretest	Second Pretest
During the past 12 months, since January 1, 1987, how many times have you seen or talked with a doctor or assistant about your health? Do not count any times you might have seen a doctor while you were a patient in a hospital, but count all the other times you actually saw or talked to a medical doctor of any kind about your health.	Have you been a patient in a hospital overnight in the past 12 months, since July 1, 1987?
	(Not counting times when you were in a hospital overnight,) During the past 12 months, since July 1, 1987, how many times did you actually *see* any medical doctor about your own health?
	During the past 12 months since July 1, 1987, were there any times when you didn't actually *see* the doctor, but saw a nurse or other medical assistant working for the doctor?
	(IF YES) How many times?
	During the past 12 months since July 1, 1987, did you get any medical advice, prescriptions, or results of tests over the telephone from a medical doctor, nurse, or medical assistant?
	(IF YES) How many times?

Source: Adapted from Cannell, et al., 1989, Appendix A.

styles of interviewing, and in interviews that "work" and those that do not. Such basic research could improve questionnaire design and interviewer training, identify interactional sources of meaning, and specify how the effects of race-of-interviewer or different styles of interviewing are accomplished (see Marquis, 1971a; Chapter 21).

Side-Sequences and Interview Structure. The flow of the standardized interview may be interrupted by respondents' requests for clarifications or comments. These side-sequences may be structured as insertion sequences in which a preliminary question must be resolved before the initial question can be answered (Jefferson, 1972; Psathas, 1986; Schegloff, 1972). Less often, such a side-sequence may be initiated by an interviewer who infers that a respondent has misunderstood a question. As in other kinds of talk, something only "becomes" a misapprehension if a coparticipant makes it so (see Jefferson, 1972, p. 305). The principal types of side-sequences that occur in standardized interviews are probably personal business or "irrelevant" talk, meta-talk, respondents' requests for clarification, and repairs initiated by the respondent or interviewer (primarily probing). We know little about the structure or content of these side-sequences, but they become part of the common ground of the interview, and may have effects beyond the question-answer sequence in which they occur.

Repair. Repair occurs when one party corrects her own contribution to the talk or attempts to correct another party's (see Schegloff, et al., 1977). A common repair side-sequence in interviews is interviewer-initiated probing. Probing is frequent (85 percent of interviewers' task behavior in Cannell, et al., 1968, p. 18), and it may be the most important indication that a question is susceptible to substantial interviewer variability (Fowler and Mangione, 1985, p. 136).

Probing often begins after an answer that is perfectly adequate in other talk, but not acceptable in the standardized interview. For example, a respondent may answer, but not select a response category. When the respondent finishes answering, the interviewer can use this transition-point to display that the respondent's answer was adequate by continuing to the next question (Jefferson, 1985, p. 458) or initiate repair, usually by probing. In some kinds of talk, a speaker seeking to confirm or disconfirm an assertion that initially receives no response may clarify the assertion, review knowledge shared with the hearer, or modify his position (Pomerantz, 1984b). Speakers asking questions that receive inadequate responses appear to use similar strategies (suggested by transcripts in Sacks, 1987). Probing in the standardized interview, however, is largely restricted to repeating the question or response categories (Fowler and Mangione, 1990). If producing a remedy for a complaint legitimizes it (Jefferson, 1972, p. 310), a respondent may feel that an interviewer who simply repeats the response categories is delegitimizing his complaint.

What actually happens during probing has received little study. These negotiations, however, have considerable impact on the social atmosphere of the interview and the reliability and validity of the resulting data. In addition, interviewers self-repair question delivery, and respondents self-repair their own answers (see e.g., Jefferson, 1974). Studying respondents' self-repairs, in particular, could tell us how respondents find their way into standardized categories and what information is lost in the process.

Response Tokens. Response tokens, such as "mm hmm" or "uh huh," are highly flexible. They can be used to invite a speaker to continue talking or to indicate that an alternative response, such as laughter, is being withheld. As an invitation to continue, a response token may be seen as passing an opportunity to initiate a repair, such as asking for clarification, and thereby communicate that the speaker was understood (Schegloff, 1982, pp. 86, 88). Speakers use the token "yeah" to change or curtail a topic, and "oh" to signal receipt of new information (see Heritage, 1984, and 1985, p. 5). Given the limited range of responses available to interviewers, they may use tokens in distinctive ways (e.g., Jefferson, 1984b), particularly in telephone interviews. The way interviewers use tokens may be difficult for respondents to interpret, may inadvertently reinforce some behaviors of the respondent, or may be perceived as nonresponsive. Survey researchers have recognized the importance of response tokens (e.g., research to that date reviewed in Cannell and Kahn, 1968, pp. 581–582). But distinctions among tokens are needed to identify the interactional problems they are used to solve (as in the coding categories developed by Smit, van der Zouwen, personal communication).

Laughter. Proffering, joining, and withholding laughter are all patterned uses of this conversational resource. A speaker who laughs is regularly joined by her coparticipant (Jefferson, 1979). Laughter is also used to affiliate with a speaker or is withheld to disaffiliate or control escalation of a topic (Jefferson et al., 1987; see also the discussion of laugh tokens in Jefferson, 1972, pp. 299–301, 449). In talk about troubles, the troubles teller may laugh as a display of his ability to manage the trouble, while the recipient responds to the content of the statement, rather than to the laughter (Jefferson, 1984a, pp. 348, 358). When both parties laugh together in talk about troubles, it may indicate a "time-out" (Jefferson, 1984a, p. 356). The little evidence available suggests that a large proportion of respondents' talk unrelated to the interview is laughter, and that interviewers laugh much less frequently than respondents (Cannell, et al., 1968, pp. 19–20; see also Bradburn, Sudman, et al., 1979). Although questions distinctively susceptible to interviewer effects are not distinctive for eliciting laughter, interviewer laughter is significantly correlated with some question characteristics: difficulty (but not sensitivity), opinion questions, and open questions (Fowler and Mangione, 1985, pp. 137–139).

The work laughter does for either participant in an interview has not been explored. Given the uneven distribution of laughter in the interview, we can expect that work to differ for respondents and interviewers. Interviewers may use laughter to manage problems in their roles (such as problems caused by inappropriate questions). Laughter may also help establish and maintain affiliation (or disaffiliation) with the respondent, and hence the character of rapport and the reciprocity that underpins the interview. When a respondent greets a question with laughter, it may be the occasion for the interviewer's collaborating in or avoiding a side-sequence (Jefferson, 1972) or time-out (Jefferson et al., 1987) in which the interviewer steps out of role and then must negotiate a return to the interview (e.g., Goffman, 1981, footnote on pp. 45–46).

19.5 CONCLUSION

All interviewing — standardized or not — relies on implicit or explicit models of the meaning of self-reports, just as observational studies rely on models of the meaning of behavior. The meanings interactionally created and recorded during data collection require the attention of those who rely on the spoken words of others to describe the world. The research reviewed and proposed here suggests closer study of how conversational practices are enacted within the prescribed structure of the interview to create data, and comparisons between talk in interviews and talk in other situations. Such studies potentially have both practical and theoretical results: improvements in question design and interviewing technique, as well as increased understanding of the substantive implications of measurement based on talk.

The problem we face in improving measurement in surveys was summarized by Hyman, et al. (1975 [1954]) in words that presaged those of critics of the standardized interview:

> The twin goals of reduction of inter-interviewer variation and an increase in the validity of the results must always be kept in mind (p. 20) .... Evaluations oriented purely to the reliability problem ... run the danger of conservatism because the standard against which any interviewer's performance is appraised is that of another current interviewer, or that of all current interviewers. . . .It is only as we have as a norm a form of interviewing that approximates close to valid results, that we become radical and experimental. It must be the neglect of this latter concept of interviewer error that accounts for the rarity of innovation (p. 21).

Hyman's reflection on survey interviewing 20 years later puts the burden where it belongs: "The fault sometimes is not in our interviewers, but in ourselves" (Hyman, et al., 1975 [1954], p. xviii).

ACKNOWLEDGMENTS

This chapter benefited from a 1987 workshop on the role of the interviewer sponsored by the Social Science Research Council (SSRC) and another in 1989 sponsored by the SSRC and the Russell Sage Foundation. Discussions with Douglas W. Maynard and Harold Garfinkel contributed to the ideas developed here in ways for which they bear no responsibility. Robert Hauser, Douglas W. Maynard, Howard Schuman, and Johannes van der Zouwen provided very helpful comments on earlier drafts. My thanks to Corey Keyes for his help with the tables.

CHAPTER 20

COGNITIVE LABORATORY METHODS: A TAXONOMY

Barbara H. Forsyth and Judith T. Lessler
Research Triangle Institute

20.1 INTRODUCTION

Some survey questions are easier to answer than others. Consider the following sets of questions that could be asked of a woman who reported having a long-term relationship with John Smith:

- How many issue were there from your union with John Smith?

- Did you and John Smith have any children together? IF YES ASK: How many?

- How many children did you give birth to who were fathered by John Smith?

- How many times, if any, did you become pregnant by John Smith? IF ONE OR MORE PREGNANCIES, ASK: How many of these [INSERT NUMBER OF PREGNANCIES] pregnancies resulted in a live birth?

Which of these question sets is best? All of them, if answered accurately, would give us information about the number of issue from the woman's union with John Smith, and the last set of questions would yield additional information about the number of pregnancies that did not result in a live birth. Few of us would ask about "issue from a union" because we do not think respondents would know what we were talking about. However, what of the other alternatives? Which is better? Will women be willing to report pregnancies that did not result in live births? Does asking this additional question affect their responses to the follow-up question on the number of live births? How does referring to a baby as a live birth — a term used by professionals, but not in wide use in everyday life — affect respondent thinking? Past research has shown that women often fail to report births of children who died soon after they were born. Do any of these questions do better at preventing this error, or do we need to use a specific probe to cue memories for these births? And, most important, how can we best determine the impact that a particular formulation of a question has on the response process?

In their review of response effects (measurement errors) in surveys, Sudman and Bradburn (1974) concluded that "questionnaire construction and question formulation lie at the heart of the problem of response effects." Therefore, if we are to understand the sources of survey measurement error and find ways of reducing it, we must understand how errors arise during the question-answering process. This will allow us to develop better questions that will yield more accurate answers.

The primary objective of cognitive laboratory research methods is not to merely study the response process, but through careful analysis to identify questioning strategies that will yield more accurate answers. To that end, this chapter focuses on the thought processes that respondents use to interpret and answer survey questions.

A number of response process models exist (Oksenberg and Cannell, 1977; Tourangeau, 1984). Each of these models assumes a basic sequence that respondents follow when confronted by a question. A respondent to a survey question must first comprehend the question, of course, then translate it into a memory retrieval task, figure out a way of recalling the information, as well as make some decision as to what to report to an interviewer. The way in which this basic sequence is followed will vary according to what is asked and what the respondent knows about what is asked. Thus, a number of characteristics of the question can affect the strategy used to complete the question-answering process.

The phrase "cognitive laboratory methods" in this chapter refers to a set of tools for studying these thought processes and for identifying the errors that may be introduced during the response process. In our view, the cognitive laboratory represents an approach for understanding survey response error; as an approach, it may or may not be implemented in specialized settings. The goals of cognitive laboratory research are (a) to understand the thought processes used to answer survey items, and (b) to use this knowledge to find better ways of constructing, formulating, and asking survey questions. Because a respondent's cognitive processes are not directly observable, however, we need to use research methods that give valid data on these thought processes. Several different methods are being used in cognitive laboratories to study how respondents answer survey questions (e.g., Jobe and Mingay, 1989; Lessler and Sirken, 1985). Given the variety of available methods, it is natural to wonder whether some methods are better than others at uncovering the cognitive processes and the question characteristics that may affect response accuracy.

We began investigating method usefulness by (a) reviewing the literature on the cognitive research methods that have been used to study the survey question-answering process, and (b) conducting informal discussions with measurement staff at four fairly representative organizations in the United States that have done cognitive laboratory research. The four organizations are the Bureau of the Census, National Center for Health Statistics (NCHS), Bureau of Labor Statistics (BLS), and Westat, Incorporated. Of course, this chapter also reflects our own work at Research Triangle Institute (RTI).

We carried out this review and these discussions with several goals in mind; namely, we wanted to find out what methods were being used, which methods were used most often, whether some methods were judged as more useful than others and whether this assessment of usefulness depended on factors such as respondent characteristics, questionnaire content, stage of questionnaire development, as well as whether the different organizations have guidelines for selecting a research method.

As was anticipated, a great variety of opinion and practice flourishes in these organizations. We found some consensus as to the goals of a particular method. However, we also found potentially important differences in how the methods are implemented and used across settings. In addition, preferences for particular methods differed across agencies, and criteria for selecting one research method over another were not available. A consensus was also absent regarding procedural details on what constituted a particular method; perhaps implementation details detrimentally affect the perceived usefulness of alternate research methods at these organizations. Based on these informal discussions and our literature review, we concluded that no guidelines currently are available for preferring one cognitive research method over another. Other factors such as cost, time to carry out the research, staff preferences, and so on had a major influence on the choice of methods.

We believe that this lack of consensus may be due, in part, to a lack of theoretical and empirical work that explores how methodological details can affect cognitive laboratory results. Therefore, in Section 20.3, we present a theoretical framework that we hope will help researchers develop general hypotheses about the usefulness of available methods. First however in Section 20.2, we briefly review the methods currently being used.

Our theoretical framework assumes that specific laboratory procedures affect how respondents direct their attention toward survey questions. A method's success (i.e., utility) depends on matching the focus of attention in the laboratory to a focusing of attention in survey-interview settings. Our hypothesis on attention generates a taxonomy of methods that we use to draw more specific conclusions about method utility. The taxonomy in turn identifies important research issues that we address to develop stronger inferences about method utilities under various research conditions.

20.2 COGNITIVE LABORATORY METHODS

20.2.1 Overview

Several methods are being used in the United States to investigate the question–answering process. In Table 20.1, we have identified four general sets of these methods — expert evaluation methods that involve no interaction with respondents, expanded interview methods in which interviewers administer survey questions along with more elaborate questions about how respondents perceive the survey items and how they go about answering them, targeted methods where survey items are simply used as stimulus materials for other tasks such as free-sort classification or rating judgments, and group interview and discussion methods.

20.2.2 Expert Evaluation

The expert evaluation methods include behavior coding, expert analysis, and cognitive forms appraisal methods.

Interactional behavior coding involves observers coding interactions between interviewers and respondents during the question-answering process (Cannell, et al., 1989; also see Chapters 19 and 21). This procedure is directed at identifying questions that are difficult to ask or to understand, lack of a common understanding of terms and concepts, and difficulties in processing the requested information.

Specific behaviors believed to reflect problems that respondents have with questions are observed. The coders specifically observe the question-answering process either as it occurs or review audio or video recordings of the interview. For example, Cannell, et al. (1989) used 10 codes: two focused on interviewer behavior and eight focused on respondent behaviors. The technique permits a researcher to assign quantitative scores to questions in terms of the numbers of problems

Table 20.1. Cognitive Laboratory Research Methods Currently Being Used in the United States to Study the Question-Answering Process.

General Type of Method	Specific Method[a]
Expert evaluation	Interactional behavior coding Cognitive forms appraisal Expert analysis
Expanded interviews	Concurrent think-aloud interviews Follow-up probes Memory cue tasks Retrospective think-alouds and probe questions
Targeted methods	Paraphrasing Free-sort classification tasks Dimensional-sort classification tasks Vignette classifications Rating tasks Response latency Qualitative timing
Group methods	Focus groups Group interviews Group experiments

[a] Methods were identified during a literature review as well as informal discussions with cognitive laboratory research measurement staff at the Bureau of the Census, NCHS, BLS, and Westat, Inc., as well as RTI.

coded for a question and the percentage of respondents experiencing each problem.

Informal observation is a variation of behavior coding in which an observer notes aspects of respondent behavior that seem interesting or unusual to the observer.

In expert analysis, a researcher (who may or may not have originally constructed or formulated the questions to be asked) reviews a questionnaire to gather an understanding of the response task and to note potential problems. The questionnaire's characteristics are described, including procedures for moving through the questionnaire, type of respondent (self versus proxy), and the nature of the tasks required of the respondents. Lists of problems that respondents might have and types of mistakes they might make are made. The researcher/appraiser then classifies the observations to get a general understanding of the questionnaire and specific points where difficulties may occur. Expert analysis differs from the next method (cognitive forms appraisal) in that no formal coding scheme is applied.

Cognitive forms appraisal analyzes questions under an assumed model for the response process (Forsyth, Hubbard, and Lessler, 1990). Questions are examined and assigned codes that describe the response process and are directed at identifying problems. Sources of problems include (but are not limited to) difficulties that respondents have in understanding questions, recalling information, and formulating responses, as well as the sensitivity of questions and so on. The resulting codes are then used to identify sets of questions that are similar in terms of the response task, select questions for further study using other laboratory methods, and guide questionnaire revision.

20.2.3 Expanded Interviews

Expanded interview methods include concurrent think-aloud interviews, follow-up probe questions, memory cue tasks, and retrospective think-aloud and probe methods.

In concurrent think-aloud interviews, respondents are instructed to report their thoughts while they answer survey questions (Bishop, 1989). Such interviews are conducted during one-on-one sessions with the respondents using either self-administered or interviewer-administered questionnaires. Before the interview, respondents are given instructions about "thinking aloud." During the interview, respondents are reminded to report their thoughts through neutral probes such as "Tell me what you are thinking...", "Remember to report your thoughts...," and/or "Say more about that...." Typically, interview sessions are recorded, and the sessions are analyzed either by reviewing the audio or video tapes or by examining transcripts of the sessions.

Think-aloud results have been used to identify difficulties in question comprehension, perceptions of the response task, memory recall strategies, difficulties in selecting a response, interpretations of question reference periods, and reactions to sensitive questions.

Think-aloud interview procedures can incorporate detailed probe questions that focus respondents' attention on particular aspects of the question or on the whole question-answering process. For example, a respondent may be asked to comment on the procedures he or she used to recall some information, thus directing his or her attention toward the retrieval component of the response process. A central thesis of the model we present in Section 20.3 is that focusing a respondent's attention on a particular aspect of the question-answering process can have an important impact on whether the results of the laboratory procedure provide a valid description of the question-answering process that goes on under the general conditions of the survey. Therefore, we consider the use of detailed probe questions as a separate method.

Follow-up probing is a variation of the think-aloud method. Interviews for specific surveys are conducted under similar instructions; however, a researcher who makes follow-up probes has identified a set of focal issues based on analyses of the question-response task. For example, a researcher may be interested in how respondents interpret technical terms, how they make choices among provided response alternatives, or what their approaches are to memory retrieval when questions cover long recall periods. If general think-aloud responses do not address the pre-identified issues, then an interviewer can ask specific probe questions that do.

Memory cue tasks are used to assess recall errors due to a respondent's failure to remember events during an interview. Psychological research has demonstrated that cued recall is generally better than free recall (Loftus and Marburger, 1983). Memory cue tasks have been used to assess the potential for reducing recall error by providing cues during the survey interview.

The effects of memory cues have been studied by (a) providing respondents with examples of situations or events that may cue their memory for the events in question (Lessler, et al., 1989), and (b) using a special interviewing technique that assists respondents in coming up with their own recall cues (Means, et al., 1989). In the former study, more dental visits were reported by respondents who were randomly assigned to receive cues than were reported by those who did not receive the cues. Means, et al. (1989) found that respondents who were asked questions using the special interviewing technique did a much better job of remembering recurring similar visits for health care.

Retrospective think-alouds and probe questions have respondents first complete an interview under conditions that approximate the actual survey. Then the survey responses are reviewed, and respondents are asked about how they arrived at their answers. This may be done by using more or less nondirective probing ("Tell me what you were thinking...") or with specific probes ("Why did you report that you were prescribed a flea collar to treat your 'thrombophlebitis'?"). In one variation of the procedure, the initial survey interview is audio- or video-recorded, and the respondent and the researcher review the recording during the followup probing.

Retrospective think-alouds and probe questions are used to address the same kinds of issues as the first two methods — understanding, recall, response formulation, and so on. Because think-aloud interviews are time-consuming and because the think-aloud techniques, if used early in the questionnaire, may change a respondent's behavior on subsequent questions, retrospective techniques permit researchers to use relatively unobtrusive methods to observe a survey interview. Subsequently, more directed methods can be used to focus on issues that are identified during the observation or through prior analysis.

20.2.4 Targeted Methods

Targeted laboratory methods include paraphrasing, free-sort and dimensional-sort classification tasks, rating tasks, vignette classifications, response latency research, and qualitative timing. We call these methods *targeted* because they are generally focused on a particular component of the response process.

Paraphrasing is used to determine whether respondents understand a question and whether respondents' interpretations of the response task are congruent with that intended by the researcher. Simply stated, respondents are asked to repeat questions in their own words.

Free-sort classification tasks are used to gain an understanding of how respondents conceptionalize the topics covered by a questionnaire and to provide insight into the organization of memory. Respondents are given a set of cards that list items of interest, then instructed to sort the cards into groups. Each respondent is free to choose which criteria he or she uses to sort the cards.

In dimensional-sort classifications, respondents are asked to sort items according to gradations of some characteristic, such as their sensitivity in reporting a particular behavior, difficulty in recalling the information, or the seriousness of certain medical conditions. As with the free-sort method, the goal is to determine how respondents think about the topics of interest. However, a dimensional-sort task is more constrained; that is, a respondent's attention is directed at particular characteristics selected by the researcher.

Vignette classifications are similar to dimensional-sort tasks; however, more complex materials are used. Respondents are asked to read short descriptions of situations ("vignettes") and select category labels that best describe the situations. For example, a respondent may be given a vignette of an event and asked (a) if this is the type of event that should be reported in the survey, and (b) to identify the response category into which the situation should be classified, and so on.

A variant is the hypothetical vignette method, in which respondents are asked to imagine a person in a vignette situation. Then respondents rate the hypothetical vignette using category rating scales that refer to imagined perceptions or behaviors of the person in the vignette. Thus, respondents are not asked what their thoughts or reactions are; rather, they are asked to speculate on what other people's reactions will be. The technique has been used by Nathan, et al. (1990) to explore the willingness of respondents to answer sensitive questions.

In confidence ratings and other rating tasks, respondents answer the survey questions and then are asked to rate the degree of confidence that they have in their answers. The procedure is used to identify questions that respondents find difficult to answer. The difficulty may arise, for example, from problems with recalling information or lack of knowledge by proxy respondents. Other types of ratings may also be used. For example, in an investigation of item-count procedures to gather information on sensitive questions, Hubbard, et al. (1989) asked respondents to rate the sensitivity of questions that address various behaviors and the degree to which different item-count lists protected their privacy.

Response latency research involves measuring the time elapsed between the presentation of a question and the indication of a response. Typically, special equipment is needed to measure such response latency. The technique is partially based on a theory about how elapsed time is related to a respondent's cognitive processes. For example, we can assume that questions requiring more memory searching have longer response latencies. Thus, we could use response latencies to study whether it is easier for respondents to answer questions that are grouped according to alternative presumed memory structures. This technique was used by Tourangeau, et al. (1990) to study the cognitive structure of attitudes. Respondents answered questions that were displayed on computer display terminals. The response latency consisted of the time between the appearance of the question and the time that a respondent indicated an answer.

Qualitative timing is a similar technique; however, special equipment is not used to measure the actual response time. Rather, either at the time of an interview or when reviewing a taped interview, an observer codes the interval between the question and the response into categories (e.g., short, medium, long).

20.2.5 Group Methods

A focus group interview brings several people together to discuss selected topics. Generally, groups are constructed to represent subgroups of a survey's target population, and several groups may be necessary to represent different types of people because individual groups are often selected to be relatively homogeneous. Generally, the group sessions last for 1 to 2 hours, are recorded, and are led by one or two moderators who guide the discussion by focusing it on particular topics of interest. Focus groups are used in a variety of ways to address the question-answering process. One approach is to have respondents complete a questionnaire — usually self-administered — and then discuss their experiences. Under this approach, there are two sources of data: (a) observations of the response process during questionnaire administration, and (b) the succeeding discussion. Researchers also use focus groups to gather background information before they construct a questionnaire. DeMaio (1983) notes that this technique is especially useful for obtaining information on how people think about specific issues and their attitudes and opinions related to a survey's topics.

A variety of group interview formats also exist. Some of the important variables distinguishing different group formats include group size, degree of structure, and nature of group leadership. These variables highlight the fact that important social factors distinguish all group tasks from other laboratory tasks. Throughout the rest of our discussion, we will use the generic term "group interview" to refer to the entire set of group interview methods, including but not restricted to the focus group interview method.

Group experiments constitute a special type of self-administered survey in which respondents complete experimental versions of a questionnaire, all of which are constructed using principles of experimental design.

20.2.6 Some Similarities and Differences Among Methods

All of the methods entail having a respondent or an observer carry out some task that provides information on the question-answering process. The result usually is more information than can be obtained by merely asking the survey questions and recording the answers. However, these methods differ in several ways.

Most obviously, they vary according to their timing in relation to a respondent's thought processes. Some methods are used to gather information on the cognitive processes inherent in the question-answering process as these processes are occurring (e.g., concurrent think-alouds). Other methods attempt to look back at what has already occurred (e.g., follow-up probes). Still others are done independently of any actual observations of survey respondents (e.g., cognitive forms appraisals).

The methods also vary according to the amount of control a

researcher has over what is observed. In some of these methods, the respondent decides what information will be observed; for example, in think-alouds. When using other cognitive laboratory methods, a researcher exerts some control over which particular aspect of the response process the method focuses on. For example, when using dimensional-sort classification tasks, a researcher may ask a respondent to sort survey items according to the difficulty of recalling the information; the researcher may ask the respondent to focus attention on this particular aspect of his or her response process. In other cases, response process data are completely controlled by the researcher and are independent of any decisions on the part of the respondent. For example, in behavior coding, behaviors that are assumed to be reflective of problems in the question-answering process are preselected by a researcher, and respondents are not aware of what types of behaviors are being used as indicators of response difficulties.

These types of differences among research methods form the basis for the taxonomy of methods that we present in the next section.

20.3 GENERAL THEORETICAL FRAMEWORK AND TAXONOMY

Several of the cognitive laboratory methods ask respondents to report information about their thoughts and thought processes. Ericsson and Simon (1980) presented a taxonomy of these verbal report methods, and they used their taxonomy to generate hypotheses about the usefulness (utility) of the verbal report methods. We take the Ericsson-Simon taxonomy as a starting point, then we introduce our attention hypothesis and use it to extend the Ericsson-Simon taxonomy to cover the broader set of cognitive laboratory methods. We close this section by illustrating some of the taxonomy's methodological implications.

20.3.1 The Ericsson-Simon Taxonomy

When respondents are asked to report their thoughts as they carry out some activity, they must do two things: the assigned activity and the reporting task. Ericsson and Simon (1980) assumed that two factors determine the accuracy of verbal reports of thought processes and, thus, method utility. First, information about his or her own thought processes must be attended to by a respondent. Unattended information is simply unavailable for further processing or translation into overt response. Second, assuming information is attended to, the accuracy with which it is reported depends on its availability to the cognitive processes that are responsible for generating the report of the thought processes.

Ericsson and Simon hypothesized that information availability depends on the timing of the reporting task in relation to the timing of the thoughts that are used when carrying out the assigned activity. For example, they posit that methods requiring an immediate report of recalled information will generally yield more accurate results than will methods that involve a delay between the survey activity and reports about it. Thus, if respondents are to be able to report information concerning their thought processes during the completion of an activity, they need to have noticed what they were thinking. An example of an activity that people do without much attention to their thought processes is driving home from work. More than one person who has recently moved to a new house suddenly becomes aware of approaching his or her previous residence after some extended driving time — a rather dramatic example of a series of unattended thought processes. In contrast, imagine a woman who is trying to solve a complicated math problem and therefore is paying very close attention to her thoughts. As a person engaged in such an activity, she may even write down each thought or step in the process in order to increase her control of her focus of attention.

Second, assuming that information on one's thoughts is attended to while the activity is being completed, the accuracy with which it is reported depends on its availability to the process responsible for generating the report. Ericsson and Simon (1980) hypothesized that information availability depends on the time elapsed between the thoughts and the report of the thoughts. For example, if I asked you what you were thinking the first time you proved the Pythagorean theorem or when you got to work last July third, you probably could not tell me. Thus, cognitive laboratory methods that require an immediate report of recalled information will generally yield more accurate results than will methods that involve a delay between recall and response.

Thus, under the Ericsson-Simon taxonomy, we expect that concurrent think-aloud methods will be more useful than retrospective think-aloud methods for tracing the thought processes involved when respondents answer survey questions. Similarly, the accuracy and utility of follow-up probes will depend on whether the probed information was originally attended to and, if so, whether the information is still available at the time the probe question is asked. For example, suppose we are observing a respondent completing a self-administered questionnaire that contained skip instructions written in italics. We query the respondent on any difficulties in following the skip instructions. If the respondent has not noticed the skip instructions or has not given any responses that required skipping other questions, his or her answer may not reflect what was actually happening as he or she completed the questionnaire activity. In addition, if we used this probe after the respondents had completed all of the questions, they may not remember what they were thinking as they read the skip instructions.

20.3.2 Attention Hypothesis

Following Ericsson and Simon (1980), we assume that task timing is one factor affecting report accuracy and method utility. We extend their taxonomy by noting that laboratory method procedures may affect how respondents direct their attention. Thus, cognitive laboratory methods may alter the types of information available for reporting.

To illustrate potential attention effects, imagine a respondent asked to report the amount of money he spent on groceries in the past 30 days *and* to rate his level of confidence in the accuracy of his grocery expenditure response. The probe for a confidence rating may lead the respondent to think about his expenditures more carefully than he would in a survey interview and to recall purchases that he would not normally recall if he were not asked to focus on the accuracy of the response. The confidence-rating task may encourage him to recall or attend to information that would not be considered without the confidence rating instruction. Such an example causes us to wonder whether the cognitive research method (confidence ratings) provides valid information on the question-answering activity.

The relationships between attention and method utility may be complicated. In the foregoing example, the additional information recalled because of the confidence task instructions may actually enhance the usefulness of the confidence rating task. In general, we hypothesize that method utility will depend on the relationships among attentional strategies specified by a particular method, attentional strategies under standard survey conditions, and laboratory research goals. Section 20.3.3 describes the extended taxonomy we have developed, and Section 20.3.4 uses the taxonomy to generate more specific hypotheses about method utility.

20.3.3 Extended Taxonomy

The extended taxonomy, represented in Table 20.2, distinguishes the cognitive laboratory methods along two dimensions — task timing and attention control. Task timing refers to the timing of the reporting or data-gathering task in relation to the question-answering activities we wish to study. Attention control refers to whether a researcher has used a method that explicitly focuses on some particular aspect of the multiple thought processes that can occur during the question-answering process.

In the remainder of this chapter, we distinguish the different laboratory procedures according to their use of what we call "attention filters." Attention filters explicitly direct a respondent's attention to certain aspects of his or her thought processes, thus allowing a researcher to observe only some components of the cognitive processes.

Thus, task timing reflects the availability of information that has been attended to during the question-answering activity, and attention filter refers to the structural relationship between the types of information attended to by the respondent as he or she completes that question-answering activity and the types of information gathered under each laboratory reporting or observation task. Recall our example of the woman who is writing down her thoughts in order to increase her control over her focus of attention as she solves a complicated math problem. She employs a filter for recording thoughts and only chooses those that are relevant to solving the problem. She will not write down such thoughts as: "I wonder if Jim will remember to pick up the cleaning?" or "The Brown's cat must be in heat again; it's screaming like a banshee."

Task Timing

We distinguish four levels of task timing:

- Concurrent: Laboratory observations are gathered at the same time that the respondent is doing the thinking required to complete the question-answering activity.

- Immediate: Laboratory observations are gathered from the respondent while information on his or her thought processes is still in his or her short-term memory.

- Delayed: Laboratory observations are made when information related to the thought processes is no longer available in short-term memory.

- Unrelated: Laboratory data are not collected along with any ongoing survey response activity.

As shown in Table 20.2, concurrent research methods include concurrent think-aloud methods, memory cue tasks, survey response latency methods, and behavior coding. Immediate laboratory tasks include paraphrasing, free-sort classifications, follow-up probe ques-

Table 20.2. Laboratory Method Taxonomy

	Attention Control			
Task Timing	Unrestricted	Directed	External	Dynamic
Concurrent	Think-alouds	Memory cue tasks Response latency	Behavior coding Qualitative timing	
Immediate	Paraphrasing Freesort classification tasks	Follow-up probes Dimensional-sort classification tasks Rating tasks (including confidence-rating tasks) Vignette classifications		
Delayed	Retrospective think-alouds	Retrospective probes		
Unrelated			Cognitive forms appraisal	Group interview formats Intensive interviews

tions, dimensional-sort classifications, and confidence and other rating tasks. Delayed laboratory research methods include the retrospective think-alouds and retrospective probes. Unrelated timing methods include the cognitive forms appraisal and other formal task analysis methods, as well as some forms of focus group interview methods.

Attention Filters

In general, we hypothesize that data from cognitive laboratory methods will give useful information about survey response processes when the laboratory research procedure and survey response activities create similar attentional sets. We distinguish four types of attention filters: unrestricted, directed, external, and dynamic.

Laboratory methods that employ unrestricted attention filters place few, if any, constraints on the types of information about the thoughts that respondents attend to under a laboratory method. For example, concurrent think-aloud tasks incorporate unrestricted filters because the respondent determines what information to pay attention to — perhaps consciously or perhaps through habit or other unconscious processes.

Tasks that use directed filters include those that specify the types of information on which respondents should focus. For example, the confidence-rating method incorporates an explicit attentional instruction that asks respondents to pay special attention to the strength of the evidence on which a survey response is based. Cognitive laboratory methods that use directed filters require a respondent to interpret and implement the attention filter and thereby constrain the set of acceptable reports on the response process. For example, if we asked respondents to rate the confidence of their answers, we would not expect them to reply that they felt afraid when they reported their recent use of cocaine and were reluctant to provide the information.

Some laboratory tasks make use of external attention filters that are implemented independently of a respondent. Interactional behavior coding is an example. Under interactional behavior coding, specific constraints are placed on the types of information that are relevant to the data collection task; these constraints, moreover, are implemented by observers. A few laboratory tasks use instructions that encourage variable or dynamic shifts in attention based on preceding responses, interviewer interpretation, or respondent perceptions of the situation, among other things. For example, focus group and intensive interview methods may involve combinations of unrestricted and directed filters that are highly dependent on the individuals involved and the course of conversation.

As shown in Table 20.2, methods using attention-unrestricted filters include concurrent and retrospective think-aloud methods, free-sort classifications, and paraphrasing. Tasks presenting directed or explicit instructions that point a respondent's attention toward certain aspects of the question-answering process include follow-up probe questions, tasks incorporating memory cue aids, dimensional-sort and vignette classifications, confidence and other rating tasks, and tasks that involve ratings or measures of response latencies. Methods that use external filters include interactional behavior coding, cognitive forms appraisal, qualitative timing, and other formal task analyses. Methods involving dynamic filters include focus group and intensive interviewing methods.

It is important to note that we assume that all methods incorporate some type of attention filter and that we distinguish the methods by the kinds of filters imposed. This is an important point because it emphasizes that it is always possible for researchers, when implementing a particular laboratory method, to manipulate a respondent's focus of attention by modifying the procedures that are used.

The next sections describe some hypotheses that follow from the taxonomy presented in Table 20.2. Section 20.3.4 focuses on the relationships among the methods that are highlighted by the taxonomy. Section 20.3.5 uses the taxonomy to develop hypotheses about the conditions in which various cognitive laboratory methods will be more or less useful.

20.3.4 Relationships Among Methods

Our taxonomy highlights four particularly interesting task features:

- There are several pairs of analogous verbal and nonverbal laboratory research methods tasks.
- Directed filters that attempt to focus explicitly on specific aspects of the response process may rely on active attention processes or on more passive, automatic memory processes.
- Directed filters vary in specificity.
- Relatively few tasks involve external attention filters.

Verbal and Nonverbal Tasks

Several of the verbal report tasks discussed by Ericsson and Simon (1980) have analogous nonverbal tasks. Under our taxonomy, the paraphrase and the free-sort methods are both immediate and both employ unrestricted filters. The methods differ in that the former requires a verbal response while the latter does not. When asked to paraphrase a question, a respondent must first understand the question and then is free to choose which aspects of the question to mention during the paraphrase. When completing a free-sort classification task, a respondent again must assign meaning to the items and then is free to choose some aspect of the items to focus on when making the sort.

We would expect to make similar observations using these methods. For example, suppose we have a survey that contains some items with a 1-month reference period and others with a 3-month reference period. The items might refer to different types of expenditures — food, clothing, utilities, and so on. Respondents who attended to both the reference period and the type of item purchased might mention both when paraphrasing the question; they might say something like, "Did I buy any clothes during the last month?" Similarly, a respondent who sorted the items might first sort them according to the reference period and then according to the type of item purchased. If a respondent does not attend to the reference period when comprehending the question (as it sometimes seems to survey researchers beset by telescoping), he or she might say something like, "Have I bought any clothes?" or, if sorting the items, he or she might choose to sort them only by type of item purchased.

The follow-up probes and the dimensional-sorting tasks form a second pair of analogous tasks: Both are immediate, and both use directed attention filters. An example of analogous tasks under these two methods are (a) follow-up probes asking, "Were you reluctant to report your income?" and (b) dimensional-sorts directing the respondent to "Sort the items according to how willing you would be to report the information in a personal interview survey." The retrospective think-aloud method lacks an analogous nonverbal method. This leads us to propose an additional sorting procedure — retrospective sorting. Under a retrospective sorting procedure, respondents would sort survey items only after answering all of them. Ericsson and Simon (1980) cautioned against using retrospective verbal report methods because respondents may have difficulty distinguishing memories generated by different items. Typically, nonverbal sorting tasks are used to explore general perceptions, and identifying particular episodic memories is less important. Thus, retrospective sorting tasks may be a useful alternative method when research conditions require collecting retrospective data.

Active and Passive Attention

The methods that employ directed filters rely on two procedures for establishing a focus of attention: cueing and probing. Cognitive theory suggests that these two procedures involve different attention processes (Lachman, et al., 1979). Memory cue procedures rely on memory structure and memory retrieval processes to direct attention. These memory-based attention processes may be more passive than the attention processes that are evoked when probes are used. Probes probably involve some "passive" memory cueing, but direct probe instructions also encourage active focusing and selection of attention filters after information has been retrieved from memory. Based on this distinction, we expect that cueing methods would be useful to explore how memory structure affects survey response. Probing methods, on the other hand, would be expected to be useful for exploring how information selection and evaluation affect survey response.

In Table 20.2, all cueing procedures are concurrent, and all probing methods are immediate or delayed. We could eliminate the correlation between task timing and type of attention filters by developing concurrent probing tasks and immediate or delayed cueing procedures. For example, a concurrent probing task would involve "warning" respondents of the content of follow-up probes. We might direct them, for example, to comment on the sensitivity or difficulty of recalling information as they answer the survey questions. The warning might encourage active filtering processes at the time of survey response. Immediate and delayed cueing procedures involve presenting memory cues after respondents generate initial survey responses, and possibly involve asking for a modified survey response after cue processing.

We expect that cueing methods will generate information about memory structure and that probing methods will generate information about strategic information selection and evaluation processes. Expanding the set of cueing and probing tasks would give richer information on how memory structure and information evaluation processes affect survey responses; it would also provide more data on how we might revise questionnaire items to take advantage of memory and selection and thereby to enhance response accuracy.

Directed Filter Specificity

Under our taxonomy, the confidence-rating task is a special type of follow-up probe that involves a specific focus of attention on the strength of belief in response accuracy. Following Ericsson and Simon (1980), we expect that reports on the question-answering activities in response to specific probes will be most useful when a probe asks for information that respondents attend to as they complete the survey task.

The extended taxonomy also highlights the potential utility of expanding the set of rating procedures. For example, under some conditions it might be useful to elicit ratings of item sensitivity, item difficulty, or item "interest value." The set of rating procedures would provide a scale-based analog to the classification-based dimensional-sorting procedures. Then we would expect the utilities of the sorting and rating procedures to depend on compatibility between respondents' perceptions and the methods used to elicit those perceptions (i.e., through ratings or through classifications).

External Attention Filters
Relatively few of the methods in our taxonomy make use of external filters. Under most of the cognitive laboratory research methods, respondents exercise control over the focus of attention. This feature of the taxonomy leads us to emphasize the importance of developing and using testable models to assess the validity of laboratory observations. In the absence of "objective" knowledge of the attentional focus selected by respondents when they complete the question-answering activity, descriptive models are required to interpret the data gathered under a particular research method.

We expect that methods using external attention filters will give different kinds of information than that generated by the other laboratory tasks. Moreover, aspects of the question-answering process that can be studied using external attention filters are different in kind from those that can be studied by respondent-controlled attention filters. Therefore, we expect that information about the utility of methods employing specific external filters will be useful when we select sets of research methods that give complementary or convergent information about survey response processes.

In summary, our taxonomy highlights several features of the methods that may be related to their utility under varying survey conditions. In the next section, we use the taxonomy to develop explicit hypotheses about method utilities.

20.3.5 Method Utility

Survey researchers make use of cognitive laboratory methods for a variety of purposes, such as testing scientific theories, reducing survey error, and developing or revising survey instruments. The multiplicity of goals means that we cannot identify methods that are consistently most useful and least useful. Rather, our goal is to delineate the conditions for which each set of methods is particularly useful.

Timing of Observations
The timing of our observations is important because it can affect information availability; however, the relationship between timing and availability may vary depending on our research goals. For example, survey researchers interested in the response strategies used by respondents as they answer survey questions may find that concurrent methods

are most useful because strategy selection processes are quick, and as such do not usually get much attention from respondents. In contrast, researchers who are interested in respondents' general perceptions of survey procedures, item difficulty, or sensitivity may find that delayed methods are more useful because they can be used to assess perceptions that develop over time. Immediate procedures may be most appropriate for understanding and revising particular items and procedures, while unrelated methods may be particularly well-suited for linking general theory and specific practice. We believe insufficient information is available to generate specific hypotheses about the effects of timing and research goals on information availability. Instead, we offer a general hypothesis on the utility of procedures that employ different levels of timing in relation to the question-answering activity:

Hypothesis 1: Relationships between method timing and method utility will depend on the type of information requested.

Our extended taxonomy is based on the premise that method utility depends on the attentional focus encouraged by a method. The next subsection describes our hypotheses about method utility for each type of attention filter.

Unrestricted Attention Filters

Under methods that use unrestricted attention filters, respondents select the focus of attention. The utility of these methods depends on the match between the focus of attention encouraged by the laboratory research method and the focus encouraged under more standard survey conditions. We expect that three factors will affect the match: (a) presence of a salient structure for the question-answering activity, (b) availability of alternative strategies for focusing attention, and (c) correspondence between attention strategies and response mechanisms. We present three additional hypotheses in the following paragraphs to elaborate on these factors.

Hypothesis 2: Generally, methods that use unrestricted filters will be most useful when the processes involved in the survey activity have a salient structure. They will be less useful when salient structure is absent.

When a survey activity has a salient structure, an attentional focus is obvious to the respondent, and selecting a focus involves relatively little effort. In the absence of salient structure, the respondent may use deliberative processes to "uncover" or construct a structure, leading respondents to focus attention on details that would normally go unnoticed as they complete the question-answering activity. For example, if respondents are instructed to freely sort a heterogeneous set of items, the sort responses may reflect relatively unimportant item differences that would go unnoticed when the same items are administered in a survey interview.

Suppose we were studying questions used in a survey that had a household respondent who answered questions about the household, himself or herself, and other members of the household. We might have questions about the household, about the respondent, and about other members of the household. Some of the questions might be about current characteristics, others about past events. The questions about past events might cover items that are relatively unique (birth of a child), and some might focus on repeated events (consumption of milk). Some might be on sensitive topics (income, drug use), and others might not be sensitive (number of rooms in the house). If we gave a respondent such a set of items to sort, each of which differed on these dimensions, the structure of the items would not be salient, and the respondent would have difficulty grouping the items into categories.

> Hypothesis 3: Generally, using unrestricted filters will be most useful when survey questions contain enough structure to permit multiple interpretations. Unrestricted filters will be less useful when the survey questions present a single, transparent interpretation.

There are two reasons why we expect less useful results from using unrestricted filters when the questions encourage one obvious interpretation. First, the results may not generalize to broader or even slightly different combinations of survey questions. Second, results from the observations may give little new or unexpected information about respondents' perceptions of survey questions.

As an extreme example, imagine respondents performing free-sort classifications of survey items that deal with frequencies of substance use. If the items cover only a few substances, then respondents may choose to sort the items according to substance because the sorting is obvious and easy. Other less obvious dimensions, such as sensitivity or difficulty of the item, may be ignored because they are more difficult to detect and implement. However, if some of the items addressed a respondent's own use of drugs, others addressed use by other family members, some were about prescription drugs, others were about illegal drugs, and the time periods for reporting varied, then it would be more obvious that a number of choices were available as to how the items could be sorted. Moreover, we would obtain more information on how respondents thought about the question-answering activity by observing the different ways they choose to sort the items.

> Hypothesis 4: Generally, the utility of using unrestricted filters will depend on how easily the relevant information about the question-answering activity can be expressed under the protocol designated by the laboratory research method.

As noted above, respondents who participate in a cognitive research study are often required to complete both the question-answering activity and the task of reporting on their thought processes as they complete this activity. The procedures that are specified for giving these reports may affect how respondents choose to focus their attention when some aspects of the question-answering process are difficult to express using the designated research procedures. For example, respondents may choose to sort items with respect to perceived sensitivity, but they may not report their reactions related to the sensitivity in a think-aloud because it is more difficult to verbalize such thoughts.

Directed Attention Filters

Cognitive research protocols that employ directed filters give guidance for focusing attention. We present three hypotheses concerning the utility of methods that employ directed filters.

> Hypothesis 5: In general, research methods that directly focus a respondent's attention on some aspect of the response process will be most useful when the protocol that directs the respondent's attention has one single, unambiguous interpretation. These methods will be less useful when instructions concerning which aspect of the question-answering process the respondents should focus their attention on permit ambiguous or respondent-specific interpretation.

Alternative interpretations can be minimized through clear, unambiguous instruction; however, in some cases, the information of interest is inherently vague. For example, instructions to rate or sort items according to sensitivity may permit differences in interpretation because respondents may have different interpretations as to what it means for an item to be sensitive. However, we might make the instructions less ambiguous by asking respondents to sort items into several categories — those they would be willing to report during a personal interview survey with another member of their family in the room, those they would be willing to report in an anonymous self-administered survey, and those they would never report.

Even if we reduce or eliminate ambiguity in the instructions, we may find that the way in which we structure the task of reporting on the question-answering activity affects a respondent's interpretation more effectively. For example, we may be able to control how respondents interpret ambiguous concepts to some extent by using cueing rather than probing tasks because memory cues often invoke in respondents automatic memory processes that are less amenable to direct respondent control.

> Hypothesis 6: Explicit filter procedures are more useful when the information requested matches working memory content; they are less useful when there is a mismatch, so that respondents are directed to focus attention on information that is not available in working memory.

Respondents may react in at least two ways to mismatches between our specifications as to where to focus their attention and what their working memory content is. First, they may revise our instructions to correspond to their working memory content. Alternatively, respondents may alter or redirect subsequent question-answering activities so that the information requested is attended to and available. As an example, suppose we were interested in how respondents established the boundaries of reference periods. Such an issue is important in surveys like the U.S. Consumer Expenditure Survey, which asks respondents to report their expenditures since a preceding interview and to report the month in which a purchase was made. We might use probing procedures to investigate how respondents placed purchases in time. After a male respondent answered a question on the purchase of shoes, we might ask, "How did you decide that your purchase of running shoes was in March?" The respondent may not have made such a decision and may merely have recalled purchasing the shoes "some time not too long ago." If he revised our attention filter, he might tell us in response to the probe how he recalled the fact of purchase ("I knew I got new shoes for the Chicken Bridge Run"), which would tell us something about what he could recall about his thoughts. If he used this as a filter to redirect his subsequent question-answering activity, he might, when answering subsequent questions, focus his attention on trying to decide in which month he made his purchases.

Informally, it seems to us that data generated through revised filters will be more useful than data generated via redirected processes because the redirected question-answering processes that we observe in the laboratory will not yield information to describe what goes on in an actual survey setting. Therefore, it seems important to identify ways to detect both filter revision and activity redirection. We may be able to detect filter revision by comparing results from memory cueing and follow-up probes. We also may be able to detect activity redirection by comparing pre-probe and follow-up probe responses or by determining whether survey responses differ depending on whether explicit instructions as to the focus of attention are presented.

Finally, the risks of a researcher's misinterpreting data from methods that employ restricted filters are high. The benefits from correct interpretation are also high. Therefore, because it is important to monitor and validate information from these research methods, we suggest the following hypothesis:

> Hypothesis 7: The specific procedures we adopt for implementing the use of directed filters may affect how the respondents interpret the instruction concerning their focus of attention and how they implement the method for reporting on their thoughts as they complete the survey activity.

External Attention Filters

Methods that employ external filters use theories of respondent behavior to develop coding systems that reflect important aspects of survey interviews, survey procedures, and survey materials. These methods eliminate uncertainties introduced when respondents are responsible for focusing their attention. Therefore, the utility of methods that employ external focuses of attention depends on the soundness of the theories those methods use. To maintain high utility, coding schemes must be updated with theoretical advances, and they must be subject to review by the scientific community. We view the development of useful external filter methods as a goal of research on survey response processes.

Dynamic Attention Filters

Under these procedures, the focus of attention is dynamic and may shift rapidly. Such procedures are characterized by at least two features:

- Multiple focuses of attention may be in effect due to the presence of at least two active participants.
- Attention may shift across time through the interaction of research participants.

The interaction that characterizes these dynamic shifts in focus of attention makes it difficult to interpret data precisely. The literature on focus groups stresses the need for using trained moderators who can control and direct the discussion. Our analysis supports this prescription. An experienced moderator establishes and maintains a consistent set of attention focuses and directs the discussion among participants in the context of these filters. In addition, a good moderator has a theoretical understanding of important aspects of the survey response process so that when certain aspects of that process are spontaneously mentioned during a discussion, he or she can direct the attention of other members of the focus group to these aspects. However, because of these rapid shifts in the focus of attention, our taxonomy suggests that the dynamic methods may be most useful as part of an iterative set of methods geared toward generating hypotheses about survey procedures, survey methods, and survey instruments.

20.4 CONCLUSIONS

The overall goal of laboratory studies is to identify those aspects of survey questions that promote accurate response. Our assessment of the current state-of-the-art is that insufficient attention has been given to developing theoretical models that assess the validity of various cognitive research methods. In addition, our analysis indicates that the results from a particular implementation of a research method may be sensitive to procedural variations in implementing the method. Without a sufficiently rich theory of assessing how procedures affect results, it may appear that findings are ad hoc and idiosyncratic. We need to continually evaluate how well our research methods identify those components of the response process that affect data quality. We also need to study how these processes vary with different questioning strategies. Thus, we need to be sure that the methods we use to study the response process are valid indicators of that process.

The central premise of our chapter has been that an evaluation of cognitive laboratory methods is best guided by a theoretical framework. Such a framework can provide a structure to evaluate the relationship between the observations gathered using particular cognitive laboratory methods and the underlying response process. We believe that the taxonomy we have presented can provide such a structure, and we fully expect that the structure will be modified by continuing research.

Our framework has permitted us to identify new cognitive research methods and to generate a series of hypotheses regarding the usefulness of various research methods. Thus, it is a structure for examining a cognitive research program and beginning an assessment of the utility of alternative methods. It also provides a mechanism for reducing the ad hoc nature of the selection and application of research methods. Perhaps most important, the hypotheses outlined in our last section provide a way to begin testing our theoretical framework. Thus, our framework gives the impetus for applying scientific procedures to assess the utility and validity of cognitive laboratory methods.

CHAPTER 21

STUDYING RESPONDENT—INTERVIEWER INTERACTION: THE RELATIONSHIP BETWEEN INTERVIEWING STYLE, INTERVIEWER BEHAVIOR, AND RESPONSE BEHAVIOR

Johannes van der Zouwen, Wil Dijkstra, and Johannes H. Smit
Vrije Universiteit

21.1 RESPONDENT—INTERVIEWER INTERACTION: THE INTERVIEWING STYLE DEBATE

There are two important classes of behavior of interviewers and respondents in survey interviews. First, behavior related to the exchange of information; and second, behavior related to the relationship between interviewer and respondent (Back and Gergen, 1963). The latter type of behavior affects the adequacy of the information exchange process. In the early days of studying response effects, this recognition led to a debate whether "rapport" increases or decreases response bias (Weiss, 1970; Goudy and Potter, 1975–76).

At present this discussion mainly concerns *interviewing style*: should interviewers behave in a personal, understanding manner, or should they confine themselves to strictly task-oriented, businesslike behavior? In a study by Henson, et al. (1976), the style of interviewing was experimentally varied among interviewers by training them in both styles. Part of the sample was interviewed in a personal style, and another part in an impersonal or formal style. No clear differences were found and Henson, et al. therefore concluded that a personal style is not a prerequisite for accurate reporting.

However, in a field experiment conducted by Dijkstra (1987), it appeared that interviewers trained in a personal style obtained more accurate information than interviewers trained in a formal style.

Measurement Errors in Surveys.
Edited by Biemer, Groves, Lyberg, Mathiowetz, and Sudman.

ISBN: 0-471-53405-6

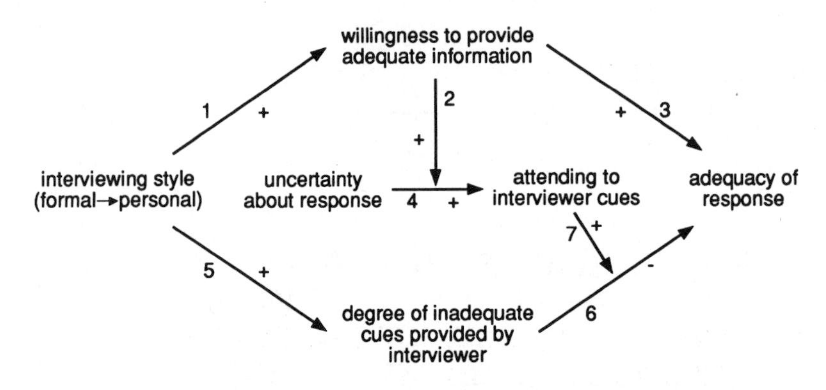

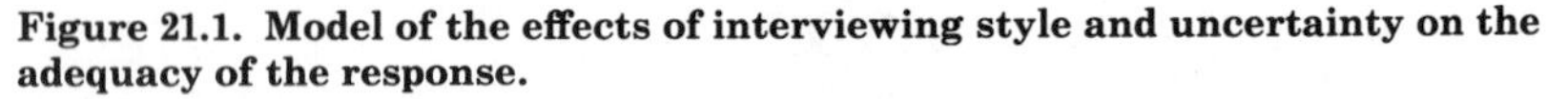
Figure 21.1. Model of the effects of interviewing style and uncertainty on the adequacy of the response.

Subsequent analysis of these data (Dijkstra and van der Zouwen, 1988) indicated that interviewers trained in a personal style showed more inadequate interviewer behavior, for example, posing questions in a suggestive manner and posing questions unrelated to the items of the questionnaire. Moreover, interviewers showing such inadequate behavior obtained more biased responses.

In this chapter we will further explore the findings regarding the effects of interviewing style on the adequacy of the response, starting with a tentative model described in the next section. The validation of this model requires the selection and the development of an appropriate method for studying the interactions between respondents and interviewers. This will be treated in Section 21.3. The data stem from a field experiment, briefly described in Section 21.4. In the remaining sections we will confront the predictions derived from our model with the data and look for the consequences of these results for further research on "style effects."

21.2 A TENTATIVE MODEL FOR THE EXPLANATION OF STYLE EFFECTS

It is hypothesized (cf. Dijkstra, 1987) that a personal style of interviewing enhances the willingness of respondents to provide information perceived by them as adequate. This causal relationship is indicated by arrow number 1 in Figure 21.1. Generally, this will lead to more adequate responses (arrow 3). However, if respondents are uncertain about their responses, *and* are eager to provide adequate information they will, according to the "cue-search model" of the interview as developed by Cannell, et al. (1977a), attend to cues provided by the interviewer about what constitutes an adequate response (arrows 2 and 4). The important point here is that cues provided by the interviewer may very well be inadequate; for example, consisting of leading probes (arrows 6 and 7).

Therefore, according to this model, *if* respondents are rather certain about their responses, and *if* interviewers are not giving inadequate cues, a personal style of interviewing will lead to better responding. However, for respondents who are uncertain about their responses, a personal style will enhance the willingness to provide adequate information, causing them to look for cues provided by the interviewer. To the extent that these cues are inadequate, this may lead to more biased responding. And, as already mentioned, from research by Dijkstra and van der Zouwen (1988), it appears that interviewers trained in a personal style of interviewing posed, on average, more questions in a suggestive manner (arrow 5).

21.3 METHODS FOR STUDYING RESPONDENT–INTERVIEWER INTERACTIONS

21.3.1 A Classification of Methods

To test this tentative model, we need methods for studying the interaction between respondent and interviewer. The existing methods for studying this interaction, described in Chapters 19 and 20, differ in three aspects:

1. The *type* of behavior studied: (1a) verbal behavior, (1b) paralinguistic behavior, and (1c) nonverbal behavior.

2. The degree to which the analysis aims at an *evaluation* of (elements of) the interaction considered; ranging from (2a) a purely descriptive analysis to (2b) a systematic evaluation of the adequacy of the role behavior of the actors.

3. The degree of *decomposition* of the interactions analyzed; ranging from (3a) a decomposition of the interview into several parts, each belonging to a particular question of the questionnaire, via (3b) a decomposition of these parts of the interview protocol into series of "turns," to (3c) a decomposition into series of "speech acts" or "utterances" of the actors, commonly indicated by the term "question-answer sequences" ("sequences" for short).

21.3.2 The Selection of the Method

For our analysis we need methods aimed at the detection of interaction characteristics, producing measurement errors in survey interviews.

Measurement errors in interviews might emerge when the transfer of information is hampered or distorted, for example by misunderstanding of the question, by using leading probes, by incorrectly interpreting answers, etc. Therefore, the emphasis in these methods will be on *content*, that is, on verbal behavior (1a).

The concept of measurement error is rather difficult to apply in the case of questions about "attitudinal" information (Sudman and Bradburn, 1974). However, it seems reasonable to expect that answers to these questions, given in interviews in which both interviewers and respondents perform their roles adequately (seen from the viewpoint of the researcher), contain relatively few measurement errors. The utilization of this assumption (in case of attitudinal questions), and the test of this assumption (with factual questions, where validating information is available) require a method with an *evaluative* component (2b).

It may be necessary to analyze the audio-taped interviews in greater detail; especially when more global descriptions of interactions, for example in terms of "rapport," turn out to be unreliable (Goudy and Potter, 1975–76), or difficult to relate to data quality (Weiss, 1970). This means that the data produced and analyzed with this latter type of interaction analysis concern either "sequences," or the "turns" or "utterances" of which they consist (decomposition level 3b or 3c). For the present analysis we will need methods for studying respondent–interviewer interactions with different decomposition levels ranging from type 3a through 3c.

21.4 A FIELD EXPERIMENT ON STYLE EFFECTS

The model described in Section 21.2 leads to a number of predictions. The data used for the test of these predictions come from a field experiment in which a personal style of interviewing was compared with a formal style. The description of this field experiment in this section is abstracted from the one presented by Dijkstra (1987).

21.4.1 The Respondents

Eight interviewers (four male and four female) were assigned to the personal interviewing style condition. Eight other interviewers (four male and four female) were assigned to the formal style condition. Each interviewer interviewed 12 male and 12 female respondents. Thus a total of 384 interviews were conducted. All assignments, i.e., interviewers to style, and respondents to interviewers, were made randomly. The results presented in this chapter are based on a stratified random sample of these interviews: 288 interviews, conducted by the 16 interviewers.

The respondents in this field experiment were randomly selected from the adult inhabitants of a recently developed urban area near Amsterdam. The interview was introduced to the respondents as a study of housing satisfaction. They received a letter from the university informing them about the alleged purpose of the study. Later an interviewer telephoned them to make an appointment. All the interviews were conducted in the respondents' homes and were audio-taped.

21.4.2 The Interviewers

The interviewers, with ages ranging between 21 and 33 years, were informed only that the study concerned housing satisfaction. They were not aware that two different style groups were used. To eliminate the threat that existing interviewing habits would interfere with the training on interviewing style, only those individuals were selected who had little prior interviewer experience. To overcome the disadvantages of relatively unexperienced interviewers, much attention was given to adequate interviewer behavior during the training.

The training of the interviewers took place in four-person groups, during five sessions, each lasting half a day. The first two sessions were devoted to adequate interviewer performance and were similar for all groups. For example, interviewers were taught to read questions as worded in the questionnaire, to probe nondirectively, to give positive feedback if the respondent's answer was adequate (e.g., "Thank you"), or to repeat the answer, and not to answer on behalf of the respondent.

The other three sessions differed between the interviewing groups. The interviewers assigned to the *personal style* were told that a good relationship with the respondent was a prerequisite for the respondent's willingness to provide the information asked for. They were trained to express a supportive and understanding attitude toward the respondent. They were taught to use personal statements such as: "I understand what moving to this house meant for you"; "How nice for you!," and so on. Special care was taken that these interviewers did not engage in inadequate behavior. Although interpretations of the respondents' answers can help to show that the interviewer understands and is interested in the respondents, such behavior was not allowed. Essentially, these interviewers were permitted only to repeat or summarize the respondent's answers, and were encouraged to express their sympathy toward the respondent.

The interviewers assigned to the *formal style* condition were trained to attend primarily to the information-gathering aspect of the interview, and to act in a polite but essentially neutral fashion. They were told that if too much attention was paid to personal matters, respondents would be distracted from their task, leading to useless or inadequate responses. They were taught to employ such emphatic, person-oriented behavior only at a minimum level of social acceptability. Like the interviewers in the personal style, these interviewers were allowed to repeat or summarize the respondent's answers, as this is generally viewed as adequate interviewer behavior. Most of the training in this formal style in fact was devoted to unlearning "natural" person-oriented behavior.

During the "style training" care was taken that interviewers in both styles did not differ with respect to task-oriented behavior, posing the questions as worded in the questionnaire or clarifying questions (the task-oriented behavior of interviewers is more fully described in Chapter 14).

21.4.3 The Questionnaire

The questionnaire involved the following topics: (a) satisfaction with housing and the neighborhood, (b) relationships with neighbors, and (c) feelings of social deprivation. The present study is mainly based on the verbatim transcriptions of three questions addressing neighbor relationships, and the responses to these questions. These questions were all designed in the same way, as "closed" questions involving the use of response alternatives and "open" follow-up questions.

Therefore, the question-answer sequences under investigation here have roughly the same structure. First (phase 1) the respondent was requested to choose one alternative from a set of response-alternatives (e.g., "yes" or "no"). The selection of an alternative by the respondent was included in this phase. Next (phase 2) the respondent was asked to motivate this choice, whereupon (in phase 3) the interviewer requested the respondent to further elucidate this motivation, if the motivation was considered insufficiently clear.

21.4.4 The Coding Procedure

The verbal behavior of interviewer and respondent was decomposed into "speech acts." A speech act was defined as any meaningful question, answer, or other meaningful verbal utterance (e.g., "feedback"). The average question–answer sequence involved approximately six speech acts by the interviewer and seven by the respondent.

These speech acts were coded according to a number of variables regarding their form and content. The coding procedure used was based on schemes developed by Cannell, et al. (1975) and Brenner (1982). The very detailed coding system enabled us to identify a large number of different types of verbal behavior (see Dijkstra and van der Zouwen, 1988). A central issue in our analyses relates to inadequate interviewer behavior. The coding procedure made it possible to locate precisely where in the sequence inadequacies occurred and to identify behavior that preceded and followed this inadequate behavior.

The most important variables used to code each speech act concerned: the actor; the direction of the information exchange, i.e., questions, answers, or other reactions; the phase of the sequence (Section 21.4.3); and the direction of the content of the speech acts, i.e., suggesting or giving a particular answer. Other variables concerned for example the degree of agreement between the wording of questions as stated in the questionnaire and the wording used by the interviewer, and whether factual or attitudinal information was concerned.

Coders were thoroughly trained in the use of the coding procedure but were kept unaware of the experimental conditions. Part of the sequences were coded independently by a different coder, yielding reliability coefficients (kappa) of over 0.80 for each variable.

An example (Exhibit 21.1) may clarify the coding procedure. Consider the following part of a question-answer sequence concerning question 37. For illustrative purposes, the codes presented here are a somewhat simplified version of the coding scheme actually used.

The coding procedure resulted in strings of coded speech acts, each string representing a sequence belonging to a particular question from the questionnaire posed to a particular respondent. Each string begins with numbers identifying the interviewer, the respondent, the question, etc. For example:

1 05 14 37 1 IQ10 RA1N IQ2N RA2N IR2U ... (etc),

where the first number (1) represents the style condition, the second (05) the number of the interviewer, the third (14) the number of the respondent belonging to this interviewer, the fourth (37) the question number, and the fifth (1) the respondent's score for this particular question as filled in by the interviewer on the questionnaire. A computer program enabled us to perform a number of analyses on these data, e.g., counting the number of sequences containing suggestive questions in each style, or detecting discrepancies between the respondent's answers and the respondent's score as written down on the questionnaire.

Essentially, the sequence serves as unit of analysis. However, as the same interviewer is involved in various sequences, these sequences are not completely independent. Although interviewers were randomly

Exhibit 21.1. Coding Question–Answer Scheme

Codes	Speech acts	Explanation
IQ10	"After moving to this neighborhood, did you try to make contact with people living here?"	The interviewer (I) poses a question (Q) from the questionnaire (phase 1), worded nondirectively (0).
RA1N	"No, not at all."	The respondent (R) answers (A) a question from the questionnaire (phase 1) with a denial (N).
IQ2N	"Is that because you are too busy with your work?"	The interviewer (I) poses a question (Q). This question is a probe on respondent's answer and suggests a specific motivation (phase 2) in line with respondent's previous answer "no" (N).
RA2N	"Yes indeed, I'm much too busy."	The respondent (R) answers (A) the probe (2) with an agreement in line with his or her previous answer (N).
IR2U	"Yes, I can imagine that."	The interviewer (I) reacts (R) to this agreement (2) with a remark showing understanding (U).

assigned to each style condition, style differences may be partly due to the fact that some interviewers contributed disproportionately to these differences. The fact that sequences are not completely independent and that some style effects may be interviewer-related makes us very cautious with regard to the application of procedures from inferential statistics. The domain of our conclusions is in fact restricted to the set of data analyzed, but tables presented in this chapter contain sample sizes useful for estimating sampling errors based on simple random sample assumptions.

21.4.5 Manipulation Checks

The coded sequences enabled us to count how many times each interviewer made "personal" utterances. Personal utterances were expressions of understanding and sympathy for the respondent (see Section 21.4.2). In the personal style condition, each interviewer made on average 0.8 such utterances per sequence; in the formal style condition this average was only 0.1 ($t = 3.71$, $\mathrm{df} = 14$, $p < 0.01$).

After the interview, the respondents completed a fourteen-item rating scale with items like "The interviewer was very understanding" or "The interviewer acted very personally." Interviewers in both style groups were judged very positive (i.e., "understanding," "personal"), but the respondents in the personal style condition expressed the most favorable judgments (the average scores were 79.8 vs. 73.9; $t = 3.48$; $df = 14$; $p < 0.01$).

It was concluded that interviewers acted according to the style requirements and that the respondents perceived the difference as intended.

21.5 THE EFFECTS OF INTERVIEWING STYLE ON WILLINGNESS AND RESPONSE

In Section 21.2 we hypothesized that a personal style of interviewing will enhance the willingness of respondents to provide information perceived by them as adequate (arrow 1 in Figure 21.1). Generally, more "willingness" will lead to responses of good quality (arrow 3). In order to test the prediction derived from the hypotheses 1 and 3 that, *ceteris paribus*, a personal style of interviewing will lead to more adequate responses, one should have indications about the quality of the responses. However, such information is usually difficult or even impossible to obtain.

However, if one is willing to accept the auxiliary hypothesis that a high proportion of "socially undesirable" responses forms an indication of more honest responses, it is nevertheless possible to test this prediction with the available data. The three questions involved in the analysis (Q37, Q38, and Q39) all seem to have response alternatives which differ from each other with respect to their "social desirability." For each of the three questions mentioned, we analyzed the sequences belonging to 144 respondents interviewed in the personal style, and 144 sequences from respondents interviewed in the formal style. In 606 of these sequences, that is in 70 percent of the 864 ($= 3 \times 2 \times 144$) sequences observed, no suggestive speech acts of the interviewers occurred. In order to cancel out as much as possible the negative effects of inadequate cues provided by interviewers (arrows 5 and 6), we focused our analysis on these 606 sequences.

From Table 21.1 it appears that for questions 37 and 38 the highest proportion of socially undesirable responses is obtained when interviewers are trained in a personal style of interviewing. With respect to question 39 there is no difference. Using the above mentioned auxiliary hypothesis that a high proportion of socially undesirable responses is an indicator of response validity, one may conclude that, at least *if the interviewers do not pose leading questions, a personal style of interviewing tends to result in more valid information than a formal interviewing style.*

Another implication of hypotheses 1 and 3 is that respondents interviewed in a personal style will try hard to give a substantive answer

Table 21.1. Distribution of the Responses to Three Questions, per Style of Interviewing, for Sequences Without Suggestive Speech Acts

	Question	Response	Formal Style (%)	Personal Style (%)
Q37:	After moving to this neighborhood, did you try to make contact with people living here?	socially desirable (yes)	36	33
		socially undesirable (no)	46	64
		do not know	18	3
		Total	100	100
			($n = 96$)	($n = 89$)
Q38:	If other people in this neighborhood try to make contact with you, do you generally comply with such an effort?	socially desirable (yes)	75	83
		socially undesirable (no)	3	8
		do not know	22	9
		Total	100	100
			($n = 115$)	($n = 100$)
Q39:	Are there people living in this neighborhood who you do not like?	socially desirable (no)	57	61
		socially undesirable (yes)	33	34
		do not know	10	5
		Total	100	100
			($n = 106$)	($n = 100$)

and to avoid nonsubstantive responses (e.g., "don't know") as much as possible. This leads to the prediction that a personal style of interviewing will produce more substantive answers than a formal style of interviewing. The data presented in Table 21.1 are in line with this prediction: *the proportion of "don't know" responses obtained using a personal style of interviewing is less than half of those obtained with interviews in a formal style.*

21.6 DIRECTION OF, AND REACTIONS TO, SUGGESTIONS OF THE INTERVIEWERS

As mentioned in Section 21.4.2, all our interviewers were instructed to refrain from posing questions in a suggestive manner. A detailed analysis of the question-answer sequences shows that suggestive speech acts hardly ever occur at the beginning of the sequence, that is, when the interviewer reads the question from the questionnaire. In this part of the sequence (phase 1), the respondent is also requested to choose one alternative from a set of response alternatives.

In the second phase of the sequence, the respondent is asked to motivate this choice. The interviewers are required to formulate this probing question themselves, partly on the basis of the preceding answer(s) of the respondent. It turned out to be rather difficult for the interviewers to avoid directive probes like, "is that because of X?": in 136 of the 864 sequences analyzed (16 percent) the interviewer posed the

Table 21.2. Direction of Suggested Reason, Offered by the Interviewer, per Style of Interviewing.

Direction	Formal Style (%)	Personal Style (%)
Accepting	63	84
Challenging	23	16
Other	13	0
Total	99	100
	($n = 60$)	($n = 76$)

phase 2 probing question in a suggestive manner. As already predicted by arrow 5 of the tentative model, *leading probes are more often used by interviewers trained in the personal style, than by interviewers trained in the formal style* (18 versus 14 percent).

A suggestion may be "in line" with the previous response (e.g., the respondent says "No," and the interviewer suggests a reason for answering "No"); this occurs in 102 sequences. But sometimes, in 26 sequences, the interviewer "challenges" the initial response, by suggesting reasons for another choice. In the eight remaining sequences, the direction of the suggestions presented by the interviewer cannot easily be described as either "accepting" or "challenging"; these few sequences are therefore coded as "other." This occurs mainly when respondents gave a "don't know" response. An example of an "accepting" probe of the interviewer, following a "No" response of the respondent to question Q37 ("After moving to this neighborhood, did you try to make contact with people living here?"), is: "Is that because you are too busy with your work?". An example of a "challenging" probe in this case would be: "Not even with the people next door?."

In Table 21.2 the 136 sequences in which an initial response is followed by a leading probe of the interviewers are cross-classified with the direction of the suggestion, and the style of interviewing.

From Table 21.2 it also appears that *the suggested reasons are mostly in line with the respondent's previous answer, thereby accepting the initial response. And this again is especially so for personal style interviews.* These results may be attributable to the fact that interviewers trained in the personal style are instructed to show understanding and sympathy with the respondent; and it is conceivable that they, mistakenly, view presenting an "accepting" leading probe as an adequate way of showing understanding and sympathy.

The tentative model predicts that respondents interviewed in the personal style attend more closely to cues provided by the interviewer than those interviewed in the formal style (arrows 1, 2, and 4). When the interviewers present leading probes, the respondents may, or may not,

Table 21.3. Reaction of the Respondent to the Suggested Reason, per Style of Interviewing

Reaction	Formal Style (%)	Personal Style (%)
Acceptance	75	84
Denial	18	9
Other	7	7
Total	100	100
	($n=60$)	($n=76$)

accept them. Again, a few reactions of respondents to suggestions presented by interviewers cannot be characterized as either "accepting" or "denying"; they are coded as "other" reactions.

The data from Table 21.3 show that *an "accepting" reaction of the respondent to a suggestion offered by the interviewer occurs far more often than a "denial," and this is especially so for respondents interviewed in the personal style*, an outcome that fits well with the prediction just mentioned.

To investigate the combined effect of the direction of leading probes by the interviewers, and the direction of the reactions of the respondents to these probes, we analyzed a larger part of the question-answer sequences. The results are presented in Figure 21.2.

Sequences in which the interviewer presents an accepting leading probe, followed by a reaction of acceptance by the respondent (+ + sequences for short) occur most frequently; others (e.g., the interviewer challenges the response, whereupon the respondent denies the suggestion of the interviewer: − − sequences) are rather scarce. When we look more closely at the "interaction-tree," we see differences between the interviewing styles. In sequences belonging to the personal style, + + sequences are over-represented (54 of the 76 sequences; that is 71 percent versus 48 percent in the formal style). On the other hand, sequences in which the respondent denies the suggestion of the interviewer, whatever the direction, more often occur in the formal style (11 of the 60 sequences; that is 18 percent) than in the personal style (10 percent). It seems that *the personal style of interviewing affects the relationship between interviewer and respondent in such a way that, compared with the formal style, the interviewer shows less "confronting" behavior, and the respondent is more apt to accept the suggestions offered by the interviewer.*

21.7 STYLE EFFECTS OVER TIME

The observed interactions of respondent and interviewer are ordered in time: they take place from the first phase of the sequence to the last phase,

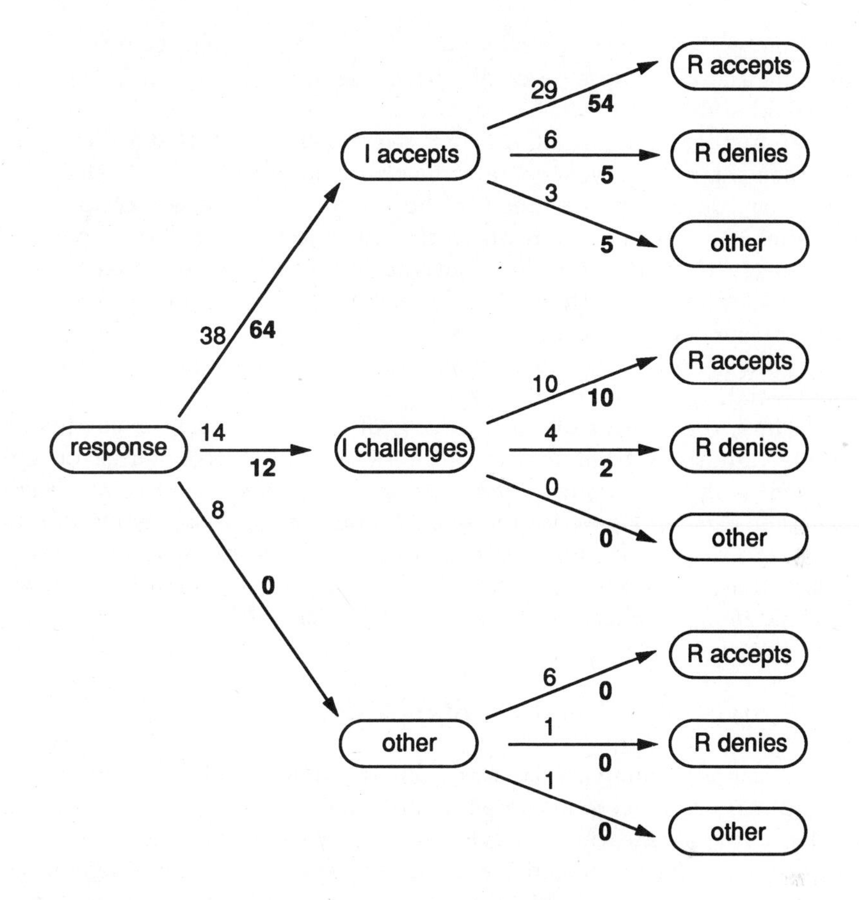

Figure 21.2. Interviewer-respondent interaction; bold numerals represent the personal style, plain numerals the formal style of interviewing.

from the first question of the questionnaire to the last question, from the first interview conducted by a particular interviewer, to the last one. It is, therefore, possible to investigate whether the effects of interviewing style change in the course of these three different time intervals.

21.7.1 Style Effect and Phase of the Sequence

In Section 21.6 we mentioned that differences between interviewing styles hardly ever occurred in phase one of the sequence: interviewers trained in both styles posed the questions of the questionnaire in about the same manner. In phase two, the interviewers are required to formulate probing questions mainly by themselves, and here we saw clear differences between the two styles. For example, significantly more probing questions were posed directively by interviewers trained in the personal style.

A further analysis of the 864 sequences shows that differences between interview styles can also be observed in the third phase of the sequence: interviewers trained in the personal style pose more phase-3 questions (i.e., questions in which the interviewers request the respondents to further elucidate their motivation for the choice of a particular response alternative) than interviewers trained in the formal style (26 percent versus 18 percent). Moreover, the proportion of suggestively posed phase-3 questions is somewhat larger for the personal style than for the formal style (54 percent versus 50 percent).

Thus, the impact of the training of interviewers in a particular "style" on their behavior, is different for the three phases of the sequence. *In the first phase there are hardly any differences in interviewer behavior between styles. In the second phase the (obligatory) probing is done more directively in the personal style. With regard to the third phase, interviewers trained in the personal style engage more frequently in further probing than interviewers trained in the formal style.*

21.7.2 Style Effect and Order of the Question

The foregoing analysis is based on sequences belonging to three questions, which were all posed in the middle of the questionnaire. In order to investigate whether style effects are related to the location of the question in the questionnaire, it is necessary to analyze sequences belonging to other questions as well. For this particular analysis we will use the coded transcripts belonging to the three questions just mentioned, in combination with the transcripts of four other questions, two of which are posed at the beginning of the questionnaire, and two others in the middle. All seven questions are designed as closed questions, involving the use of response alternatives, and open "follow-up" questions; the question-answer sequences proceed according to the three phases described in Section 21.4.3. The transcripts originate from the same 288 interviews mentioned in Section 21.4. This means that, in principle, data were used stemming from ($7 \times 288 =$) 2,016 sequences. Because the information about 34 sequences is incomplete due to recording failures, the analysis reported in this section concerns 1,982 sequences.

We focused the analysis on that part of the sequence where style effects are likely to occur: in phase two, when the interviewer is inquiring into the motivation for the choice of a particular response alternative by the respondent. To what degree is this probing done directively, and are

Table 21.4. Percentage of the Sequences in Which the (Phase 2) Probing Question Is Posed Directively, per Style of Interviewing, and Location of the Question in the Questionnaire

Location of the Question	Formal Style (%)	Personal Style (%)
At the beginning	7 ($n = 285$)	15 ($n = 287$)
In the middle	21 ($n = 706$)	25 ($n = 704$)

there differences in this respect between both styles, and between questions posed at the beginning of the interview, and questions posed in the middle? Table 21.4 gives information pertaining to this issue.

On the basis of these data one might conclude that *during the first half of the interview, probing has become more directive. This increase of inadequate interviewer behavior is largest for the formal style: the differences in this respect between the two styles become smaller in the course of the interview.*

Although the seven probing questions involved in the analysis are alike in the sense that they all ask for the motivation of the selection of a particular response alternative by the respondent, it is still possible that the differences in interviewer behavior found between questions asked at the beginning, and in the middle of the questionnaire are partly determined by the differences regarding the topics of the questions. The first two questions are related to satisfaction with housing, the others, with the relations with neighbors. Thus, we have to be cautious in drawing the conclusion that the positive effects of interviewer training "erode" during the interview. But, as we will see below, there is also another indication for such an "erosion effect."

21.7.3 Style Effect and Interview Number

It is conceivable that interviewers change their behavior during the interview period. As the number of completed interviews increases, the experience of the interviewers grows, and their recollection of the interview training might begin to fade. In order to see whether this indeed leads to changes in interview behavior, we divided the 18 interviews conducted by each interviewer in two parts: numbers 01–09 and numbers 10–18. For each half we computed the proportion of leading probes. The results for both styles of interviewing are presented in Table 21.5. We caution that no randomized order of contacting and interviewing sample cases was used, and thus respondents interviewed late may differ from those interviewed early.

As the interview campaign proceeds, interviewers in both styles become

Table 21.5. Percentage of the Sequences in Which the (Phase-2) Probing Question is Posed Directively, per Interview Number, and for Both Styles of Interviewing

Interview Number	Formal Style (%)	Personal Style (%)
01–09	11 ($n=480$)	18 ($n=467$)
10–18	22 ($n=511$)	24 ($n=524$)

more and more alike, at least with regard to the probability that they formulate probing questions in a directive manner.

21.8 IGNORING AND INTERPRETING RESPONSES

In survey research practice, one assumes that the response given by the respondent, and the answer recorded on the questionnaire by the interviewer, are largely identical, or at least congruent. The availability of verbatim protocols of large parts of survey interviews makes it possible for us to test this common assumption.

We distinguished two types of discrepancies between response and score. The first one is "ignoring": the respondent has produced a substantive response in this sequence, but the interviewer fills in "don't know" or "no response" on the questionnaire. We analyzed the 359 sequences in the formal style, in which the respondent gave a substantive response, and found 37 (about ten percent) instances of interviewers "ignoring" these responses. This percentage is much smaller for the personal style (8/364 = 2%).

The other form of discrepancy between response and score is called "interpreting": the respondent does *not* choose one of the (substantive response) categories mentioned in the questionnaire, but the interviewer still checks one of the codes belonging to a substantive response, probably by (correctly?) interpreting other information provided by the respondent. Interpreting responses by interviewers trained in the formal style occurs in 44 of the 73 sequences (60 percent) where the respondent did *not* select one of the substantive response alternatives offered. This proportion is higher for the personal style, namely 81 percent (55/68).

So we may conclude that there is a clear effect of "interviewing style" on the transformation of the responses into scores on the questionnaire: *in comparison with interviewers trained in the formal style, interviewers trained in the personal style ignore the information contained in the responses less often, and interpret this information more often.*

21.9 SUMMARY AND DISCUSSION

The results of our investigation of style effects can be summarized in the following way:

The personal style of interviewing stimulates the respondents to give adequate responses: that is, on average more honest and more substantive responses (Section 21.5). But it also gives interviewers more freedom and opportunities to engage in inadequate role behavior, like posing suggestive (probing) questions (21.6) and interpreting incomplete responses (21.8). Thus the personal style of interviewing is only superior to the formal style if we can train interviewers to show sympathy with the respondents without adding their personal views to the question-answer process.

The formal style has in principle the advantage that interviewers keep close to the questionnaire, thereby avoiding irrelevant and/or suggestive questions. However, most of the training of interviewers in this particular style was, as mentioned in Section 21.4.3, devoted to unlearning "natural" person-oriented behavior. And it looks as if this "unlearning" has only a temporary effect: in the course of an interview (Section 21.7.2), and during the interview campaign (Section 21.7.3), the interviewers trained in the formal style tend to behave more and more like interviewers trained in the personal style. Hence a training of interviewers in the formal style has to be followed by close supervision, and probably by additional retraining sessions.

The predictions derived from the tentative model for the explanation of style effects were in general in line with the outcomes of the interaction analysis. However, this model of course requires further testing with data for various types of questions. These questions then should preferably differ from each other with respect to two essential variables of the model:

1. The degree to which respondents are *uncertain* about their responses; or "easy" versus "difficult" questions.

2. The possibility for interviewers to provide *inadequate cues;* especially the possibility of providing suggestions for an answer.

Dichotomizing these two variables leads to four types of questions, A, B, C, and D, as indicated in Table 21.6.

According to the model, we would predict style effects to be different for each of these four types of questions.

Type A: The task of responding to "easy" questions (for example the factual question "How old are you") does not require much willingness

Table 21.6. The Four Types of Questions

	Suggestions for an Answer Hardly Possible	Suggestions for an Answer Possible
Easy question	A	C
Difficult question	B	D

on the part of the respondent and there is little need to look for cues provided by the interviewer. Moreover, the interviewer here has hardly any possibility of suggesting an answer. So we expect that for these questions it does not really make a difference which interview style is used: both styles yield about the same information.

Type B: Responding to a "difficult" question (for example the retrospective question "What was the party preference of your father when you were young?") requires much more willingness on the part of the respondent to come up with a substantive answer. Although respondents may feel uncertain about their responses and search for cues, in this case, the interviewers are unable to provide suggestions for answers. On the basis of our model, we expect that for Type B questions, the personal style of interviewing leads to more adequate (i.e., substantive) responses than the formal style.

Type C: Responding to these rather easy questions (for example the simple question about an actual opinion: "Are you satisfied with your house?") does not require much willingness on the part of the respondent. However, suggestions provided by interviewers are both possible and undesirable. We therefore expect that in this case the formal style yields better information than the personal style, with its higher risk of leading questions.

Type D: These questions (for example, retrospective questions about complex attitudes like "What were your main motives when you decided to have a baby?") are difficult to answer, and therefore require much willingness on the part of the respondents. Possibilities for interviewers to come up with suggestions for answers to these questions also exist. This type of questions makes the choice of a particular style of interviewing difficult. What we need here is a way of interviewing which increases respondents' willingness, and at the same time reduces the risk that the interviewer provides inadequate cues. In other words, a personal style of interviewing, combined with close supervision of the interviewers.

These predictions can be summarized as follows: for Type A questions the style of interviewing will not make much difference; for Type B questions the personal style will be the best option; for Type C questions, on the contrary, the formal style has to be preferred; while for the difficult Type D questions, a personal style needs to be supplemented with close supervision over the interviewers.

However, a questionnaire usually consists of questions belonging to different types. Which style will give the best results then depends on the particular "mixture" of these types. This dependency of style effects on the composition of the questionnaire offers an explanation for the contradictory results of research into style effects, mentioned at the beginning of this chapter: which style turns out to be "the best" partly depends on the content of the questionnaire used for the comparison of interview styles.

As stated at the beginning of Section 21.7, the interactions of respondent and interviewer are ordered in time. A personal style of interviewing needs some time to affect respondents' willingness to provide adequate information. And suggestions for answers, based on preceding responses, can only be "inferred" by interviewers when relevant information already has been given by the respondent.

A combination of these two rather trivial statements leads to two recommendations for researchers who have decided to opt for the personal style of interviewing, and who want to design their questionnaires such that the beneficial aspects of this style are exploited, and the risks minimized:

1. Easy questions (Types A and C) should precede difficult questions (Types B and D).

2. Questions with a high risk of suggestions by interviewers (Types C and D) should precede questions with a low risk (Types A and B).

These two recommendations are partially contradictory, and therefore do not lead to a strict order over the four types of questions. But the ordering C/A then D/B looks like a satisfactory compromise.

The aim of this chapter was to give an explanation for the contradictory findings regarding the effects of interviewing style on response adequacy. We constructed a model for the explanation of these style effects. This model passed a preliminary empirical test. A further test will require a combination of detailed interaction analysis with an experimental design, in which the type of question is varied systematically.

CHAPTER 22

THE EFFECT OF INTERVIEWER AND RESPONDENT CHARACTERISTICS ON THE QUALITY OF SURVEY DATA: A MULTILEVEL MODEL

Joop J. Hox
University of Amsterdam

Edith D. de Leeuw
Vrije Universiteit

Ita G. G. Kreft
University of California at Los Angeles

The impact of interviewers on survey results has been a matter of concern for several decades. As early as 1929 Rice reported on the influence of interviewers' opinions and in 1955 one of the first overviews on interviewers as a source of error was written by Boyd and Westfall. Recently, Groves (1989) in his monograph on survey errors and survey costs gives considerable attention to the interviewer as a source of survey measurement error.

In the literature on interviewer effects three research strategies can be distinguished (cf. Singer and Presser, 1989b). The first research strategy investigates to what extent specific interviewer characteristics influence responses. The second approach concentrates on the estimation of the response variability attributable to the interviewers. The third investigates the technique of questioning by interaction coding (Chapters 19 and 21). In this chapter we concentrate on the first and to a lesser degree on the second research strategy. We will critically discuss previous findings and present an appropriate model for analyzing interviewer effects. The main emphasis is on the nature of the multilevel estimation model we use, which is different from the models used in previous studies on interviewer effects.

Measurement Errors in Surveys.
Edited by Biemer, Groves, Lyberg, Mathiowetz, and Sudman.

ISBN: 0-471-53405-6

22.1 REVIEW OF PREVIOUS RESEARCH

Even in strictly controlled telephone surveys and in face to face interviews with trained interviewers, interviewer effects have been observed. When interviewer variation occurs, the responses of different respondents who are all interviewed by the same interviewer are more alike than those of respondents who are interviewed by different interviewers. Statistically, this is often expressed as the interviewer intraclass correlation (cf. Hansen, et al., 1961; Kish, 1962). In well-conducted face to face interviews the intraclass correlation typically clusters around 0.02; in controlled telephone surveys averages below 0.01 are reported (Kish, 1962; Groves and Magilavy, 1986; Tucker, 1983). A few studies (e.g., O'Muircheartaigh and Marckwardt, 1980) have reported larger effects. As the total effect of the interviewers on the overall variance of a survey statistic is a function of both the intraclass correlation and the interviewer workload, small intraclass correlations are still worrisome for researchers conducting large surveys in which each interviewer conducts many interviews (see Chapters 1 and 24).

Estimates of interviewer variance tell us how much variance in the response is attributable to the interviewer, but they do not reveal which interviewer characteristics are responsible. Three types of interviewer characteristics have been studied, either alone or together with respondent characteristics: 1) socio-demographic variables, 2) personality traits and social skills, and 3) interviewers' opinions and attitudes (cf. Collins, 1980; Hagenaars and Heinen, 1982). Although many interesting results have been reported, as yet no clear and systematic pattern has emerged. A summary of the main findings will be given below.

22.1.1 Interviewer Characteristics

The only consistent findings concerning the sociodemographic characteristics were for interviewer race, especially when race-connected questions were asked (Schuman and Converse, 1971; Hatchet and Schuman, 1975). No systematic effect of interviewers' sex on responses has been detected (Feldman, et al., 1951; Collins, 1980), but Groves and Fultz (1985) noted a tendency of respondents to express more optimism with male interviewers.

The age of the interviewer seems to have some effect on the responses, but findings are contradictory. Reviewing the literature, Sudman and Bradburn (1974) noted a tendency for older interviewers to cause less bias and response variance. Singer, et al. (1983) reported a higher response rate with older interviewers and Berk and Bernstein (1988) found less item nonresponse when the interviewer was older, but no effect of interviewers' age on response validity. Collins (1980) did not find any clear effects of interviewers' age. In contrast with these findings, Hanson and Marks (1958) reported fewer "no answers" when the interviewer was younger and Freeman and Butler (1976) noted that older interviewers in combination with older respondents caused the highest interviewer variance. Mixed results have also been found concerning interviewer experience. Feldman, et al. (1951) reported that experience had a positive effect on the number of responses to open questions, Singer, et al. (1983) found that a higher response and better data quality were associated with more experience, but Bailar, et al., (1977) reported that experienced interviewers produced more "no answers" to income questions. Boyd and Westfall (1955), Collins (1980), and Berk and Bernstein (1988) found no apparent effect of experience on results obtained. Interviewers' education and socioeconomic status were found to have no discernible effects (Berk and Bernstein, 1988; Collins, 1980; Feldman, et al., 1951; Singer, et al., 1983).

Empirical evidence of which traits make for a successful interviewer is rather scarce. Feldman, et al. (1951) found no influence on response validity of interviewers' performance on a variety of psychological tests. Nevertheless, there is some evidence that effective interviewers (i.e., interviewers who obtained more valid answers) are less socially dependent, have more self-confidence, are more socially skilled and pay more attention to details (Steinkamp, 1966). Dijkstra, et al. (1979) found that socially skilled interviewers produced less interviewer variance. Social skills are associated with knowing and utilizing rules of behavior and communication (cf. Argyle, 1969) and should be clearly distinguished from a general social disposition or social orientation. The latter can have a negative effect. For instance, Hyman, et al. (1954) concluded that characteristics such as agreeableness or cooperativeness are somewhat negatively associated with interviewer performance. Dijkstra, et al. (1979) found that person-oriented (i.e., helping the respondent) interviewers produced more interviewer variance.

Little is known about the effect of interviewers' own opinions and attitudes. Feldman, et al. (1951) and Collins (1980) found no association between interviewers' and respondents' opinions. Hagenaars and Heinen (1982) presented the results of two Dutch studies on the influence of interviewers' opinions. In the first study on the quality of employment, small and hardly significant correlations were found. In the second study, substantial and significant correlations were found between the answers of interviewers and respondents to questions concerning women's liberation. The findings reported on the influence of interviewer expectations were more consistent. Both Sudman, et al. (1977a) and Singer and Kohnke-Aguirre (1979) reported small effects.

22.1.2 Methodological Comment

Several authors have criticized the existing studies on interviewer effects; see Hagenaars and Heinen (1982). A central criticism concerns the adequacy of the statistical models used. The structure of the data to be analyzed is hierarchical, since respondents are nested within interviewers. However, the analysis models which are typically used to relate interviewer and respondent characteristics do not take this into account (Dijkstra, 1983; Hagenaars and Heinen, 1982; Groves and Fultz, 1985). In this chapter we show what goes wrong in a traditional single-level analysis of hierarchical data and present a multilevel regression model which allows us to analyze correctly the joint effects of interviewer and respondent characteristics on data quality. Related statistical models for hierarchical data are discussed in this book by Hill (Chapter 23) and Pannekoek (Chapter 25).

Multilevel models can be applied to make statistically correct inferences about interviewers and respondent effects. We will present several examples of this use of multilevel models. Another application is to statistically control for interviewer effects, in a manner analogous to the way analysis of covariance is used to control for confounding variables. While the statistical analysis in both applications is similar, the substantive emphasis is different, and we will also present examples of this second application.

22.2 MULTILEVEL MODELS FOR INTERVIEWER EFFECTS

The design of interviewer variability studies combines both interviewer and respondent variables. Typically, both interviewer and respondent variables are combined in one ANOVA or multiple regression model. Since there are many more respondents than interviewers, this creates an obvious analysis problem. Traditionally, this problem has been solved by (dis)aggregation. One approach, which is rarely used, is to attach the mean of the respondent variables to the interviewers, and to analyze the resulting interviewer level data (for a good example see Schaeffer, 1980). This procedure will however, overestimate the standard errors, resulting in a severe loss of power (type II error). Another approach which is frequently used is to attach the interviewer variables to the respondent, implying as many interviewers as respondents. This procedure multiplies the number of degrees of freedom for the interviewers, which results in tests that show "statistically significant" results far too often (type I error).

The hierarchical data structure also creates a second, more subtle analysis problem. Since the answers of respondents interviewed by the same interviewer tend to be more alike than those given by respondents who are interviewed by different interviewers, observations within the same interviewer are dependent or correlated. This dependency results in a loss of information, which introduces uncertainty in the parameter estimates of an ordinary ANOVA or regression model. Since these traditional techniques assume random sampling, they result in standard errors that are too small (type I error).

22.2.1 The Multilevel Model

An appropriate model to analyze this type of hierarchical data is a multilevel regression model, also known as hierarchical linear model (Mason, et al., 1984; Raudenbush and Bryk, 1986) or random component or variance component model (de Leeuw and Kreft, 1986; Longford, 1990). This is equivalent to performing a multiple regression analysis: one dependent variable is predicted by a number of independent variables. The difference here is that the independent variables can be defined on different levels of the data structure. In our study, we have two levels: the highest level is that of the interviewers, and the lowest level is that of the respondents. The lowest level has the most observations and is nested within the higher level.

In a traditional multiple regression model, we assume random sampling from one level only. Thus, if there were no interviewer effects we could predict a data quality indicator Y from the respondent variable R using the regression equation:

$$Y_i = \beta_0 + \beta_1 R_i + e_i \tag{22.1}$$

where the subscript i refers to respondents, β_0 is the ordinary intercept, and β_1 is the ordinary regression coefficient (slope). The errors e_i are assumed to be independent with a variance $\text{var}(e_i) = \sigma_e^2$.

If there are interviewer effects, equation (22.1) should be changed to:

$$Y_{ij} = \beta_{0j} + \beta_{1j} R_{ij} + e_{ij} \tag{22.2}$$

with the added subscript j referring to the interviewers. In the ordinary regression equation (22.1) there is a single intercept and a single slope for all respondents. In the multilevel regression equation (22.2) it is assumed that the intercepts β_{0j} and the slopes β_{1j} vary between interviewers. This means that the regression parameters β_0 and β_1 are thought of as having a distribution over interviewers with variance σ_0^2 and σ_1^2 and covariance σ_{01}.

We may try to account for this variation by introducing explanatory variables at the interviewer level. If we introduce the interviewer variable I we can write the following two regression equations for the intercept β_0 and the slope β_1:

$$\beta_{0j} = \gamma_{00} + \gamma_{01} I_j + U_{0j} \tag{22.3}$$

$$\beta_{1j} = \gamma_{10} + \gamma_{11} I_j + U_{1j} \tag{22.4}$$

where U_{0j} and U_{1j} are the error terms with variance σ_0^2 and σ_1^2.

If we substitute equations (22.3) and (22.4) in equation (22.2) and rearrange terms we obtain the regression equation:

$$\begin{aligned} Y_{ij} = \gamma_{00} + \gamma_{10} R_{ij} + \gamma_{01} I_j + \gamma_{11} I_j R_{ij} + \ldots \\ + U_{0j} + U_{1j} R_{ij} + e_{ij} \end{aligned} \tag{22.5}$$

Thus, we end up with a multilevel regression equation (22.5) which incorporates respondent, interviewer, and interaction variables and a very complicated error term $[U_{0j} + U_{1j} R_{ij} + e_{ij}]$.

Asymptotic standard errors can be calculated for the regression coefficients γ, which allow us to conduct significance tests similar to those in ordinary multiple regression analysis. There are also asymptotic standard errors for the various variance components, which allow us to test whether the variation of the regression coefficients (intercept and slopes) between the interviewers is significant.

A multilevel analysis with no explanatory (respondent and interviewer) variables at all, the so called intercept-only model, produces σ_e^2 and σ_0^2, which can be used to estimate the familiar intraclass correlation for the interviewer effect, rho, as $\sigma_0^2/(\sigma_0^2 + \sigma_e^2)$. The intercept-only model is a useful starting point for an analysis, because it indicates how much variance there is at the interviewer level. If this variance is small, the attempts to explain it should not be expected to produce very dramatic results. Note that the error variances σ_0^2 and σ_e^2 are conditional upon the explanatory variables already included in the model. If explanatory variables are included in the model, the formula $\sigma_0^2/(\sigma_0^2 + \sigma_e^2)$ no longer estimates the intraclass correlation. Nevertheless, it is still useful to compare the variance components of the regression coefficients in different models to see how well the explanatory variables explain the between-interviewer variance of these coefficients.[1]

In the general multilevel model, the regression coefficients for the respondent level variables are random, which means that they are allowed

[1] If there are random slopes the situation is more complicated (cf. Aitkin and Longford, 1986). In practice this means one should not directly compare the intercept variance of models with different random slopes.

to vary across interviewers. In other words, there are interactions between that particular respondent level variable and one or more interviewer level variables. Subsequently, we may define interaction terms which specify which of the available interviewer variables are hypothesized to interact with the respondent variable in question.

The intercept is always assumed to be random. However, we may designate some or all lower level regression coefficients as fixed, instead of random. Thus, if a statistical test shows that the between-interviewer variance of a specific regression coefficient is not significant, we may decide to model that coefficient as fixed. This statistical decision carries a substantive meaning: if we define a regression coefficient as fixed, we assume that the effect of that respondent variable on the dependent variable is identical for all interviewers. In other words: we assume no interactions between that specific respondent variable and interviewer variables. If we define all regression coefficients as fixed, this results in a model which closely resembles the ordinary regression model. The only difference is the random intercept, which reflects the between-interviewer variance.

22.2.2 Model Search

Including all interviewer and respondent variables, and adding all possible interviewer-method and interviewer-respondent interactions, would lead to statistically unwieldy and unstable multilevel models. Consequently, some heuristic is needed for the selection of variables. Longford (1990) advises to start the model construction by determining the appropriate fixed effects, followed by a procedure to test for random effects. This leads to a search procedure, in which lower level (i.e., respondent level) variables are selected before higher (interviewer) level variables and random effects are selected.

Following Longford (1990), we select the variables and build a multilevel model in several steps. First of all, for each dependent variable, a smaller set of promising respondent variables, including interactions of interviewing method with interviewer characteristics, is selected by ordinary least squares (OLS) regression (using $p = 0.10$ as a criterion).

In the multilevel analysis, we start with a fixed effects model which includes all these independent variables. A backward selection procedure is used to arrive at a model which includes only significant fixed effects. In the next step, the regression coefficients for the respondent variables are examined for possible random effects. Finally, if a random effect is found, we try to model it with the available interviewer variables.

The respondent-level variables in this study are: interviewing method, age, sex, education, evaluation of respondent by interviewer (e.g., honest, cooperative), evaluation of interview by respondent (e.g., pleasant, threatening), the attitudinal distance between interviewer and respondent, and whether the respondent had ever refused an interview before. From all the possible interactions, two were designated on theoretical grounds as potentially interesting. The first is the (Euclidean) distance between responses of the interviewer and the respondent on four multi-item scales measuring (un)happiness, self-esteem and loneliness: this variable is the attitudinal distance between interviewer and respondent. The second is the set of interactions between interviewer characteristics and method: these reflect the hypothesis that the effect of the interviewer characteristics might be different in different modes of interview. (cf. Sykes and Collins, 1988). From this set, a smaller number of variables was selected using OLS (see above).

All potentially relevant interviewer-level variables are always included in the first stage of the multilevel model exploration (i.e., age, experience, previous training, preference for telephone vs. face to face interviews, extroversion, conscientiousness, social anxiety, ability to initiate social contacts, and ability to efficiently terminate undesirable social situations; for a description see Section 22.3). Also, four respondent-level variables are always included (age, sex, education, and interviewing method used).

22.3 DATA COLLECTION

In the autumn of 1989 a controlled field experiment was conducted in which the effects of interviewer and respondent characteristics were investigated.

The subject of the questionnaire was well-being. The questionnaire included questions on general satisfaction, on loneliness, and on (un)happiness. Furthermore a large number of background questions including questions on income and work situation were asked.

Twenty interviewers were selected, varying in previous interviewer experience and training. Important selection criteria were clarity of voice over the telephone, legible handwriting and higher education. Prior to the data collection all interviewers received a standardized interviewer training based on the Survey Research Center manual (1976). Additional training was given in telephone interviewing techniques. All interviewers conducted both face to face and telephone interviews. Ten randomly assigned interviewers started with telephone interviews and then conducted face to face interviews; the other ten started with face to face interviews.

As part of the selection procedure all interviewers completed a biographical question sheet and a specially designed self-administered form of the interview schedule. Before the interviewer training started we asked each interviewer to complete a Dutch personality inventory, which measures extroversion, friendly disposition, conscientiousness, social anxiety, and ability to efficiently terminate undesirable social situations (the reliability of the personality scales ranges from 0.69 to 0.85; Akkerman, 1977; Hox, 1978).

The respondents were a random stratified sample from the general population in the Netherlands. For the complete sample (respondents and nonrespondents) detailed background information was available based on the Dutch zip code system (Geo-marktprofiel). The Dutch zip codes form an extremely fine grid; as a consequence aggregated information was available on, for instance, socioeconomic status (SES), income, type of household, and type of community for clusters of, on average, 15 households each.

The sample was randomly split in two. One half was interviewed by telephone; the other half, face to face. In the telephone situation the respondents were randomly assigned to interviewers. In the face to face situation respondents were randomly assigned to interviewers within four geographical regions in order to avoid excessive travelling. All sample units received an advance letter. The interviewers used a standardized script in asking for respondent cooperation. In all cases (telephone and face to face situations) the request for cooperation was made by telephone. At least seven callbacks were made. No attempt was made to convert definite refusals; refusers were not called back by special interviewers.

22.4 RESULTS

In the next sections we employ a multilevel model of respondent and interviewer characteristics as described in Section 22.2, with special attention to random (i.e., between-interviewer) effects. We try to answer the following questions:

1. How do interviewers affect the nonresponse?

2. How do interviewers affect the measurement of variables?

3. How do interviewers affect the respondents' feelings about the interview?

4. How do interviewers affect substantive models?

The results are discussed in separate sections. A summary table is provided in Appendix A. To help interpret the results, all regression coefficients are reported as standardized coefficients. For comparison with previous research we also give the intraclass correlation rho (cf. Groves, 1989, chapter 8).

22.4.1 Nonresponse

In the face to face situation 243 interviews were completed and 266 were done in the telephone situation. This resulted in a response rate of 51% for the face to face situation and of 66% for the telephone situation. The difference in response rates is almost entirely due to a difference in explicit refusals: in the face to face situation 191 respondents (40%) refused cooperation, while in the telephone condition only 114 respondents (28%) refused.

Nonresponse, especially the large nonresponse of the face to face interview, is a potential source of error. Fortunately, external (zip-code based) information on both respondents and nonrespondents was available. We investigated the possibility of selective nonresponse using this auxiliary zip-code information. No differences were found between respondents and nonrespondents. Also no interaction effects of respondent characteristics and mode of data collection were observed: although the response rates differ, characteristics of respondents and nonrespondents do not differ across the modes.

When the different types of nonresponse are investigated separately, a very interesting pattern is discovered. Respondents and refusers do not differ from each other, but they do differ significantly ($p < 0.05$) from the unreachables (noncontact, ill, senile, language problem). In general the unreachables were less affluent, did not own a house and were more often found in urban areas.

Cooperation Rate

Interviewers can differ in the proportions of refused and completed interviews they had from potential respondents. The cooperation rate was estimated to assess how successful individual interviewers were in persuading those able to do the interview to comply with the request (Groves, 1989). The cooperation rate is the ratio of completed interviews to all contacted cases capable of being interviewed.

The between-interviewer variance in cooperation rate was small, resulting in an intraclass correlation for the interviewers of rho = 0.02. In the multilevel model, the cooperation rate did not depend on any of the auxiliary zip-code variables. As can be deduced from the overall response rates above, the specific request (for a telephone or a face to face interview) had a large influence (beta $=0.58$; $p<0.01$). There were no significant contributions of interviewer variables. Also, the variance component for interviewers was not significant, which indicates that interviewers did in fact not differ in their ability to persuade respondents to participate in the survey.

22.4.2 Interviewer Influence on the Measurement of Variables

Three different aspects were investigated: completeness of answer, response bias caused by acquiescence, and the psychometric reliability of a measurement. Indicators for completeness were item nonresponse and the number of different statements to open questions.

Completeness: Item Nonresponse

Missing data can pose serious problems in social research. Interviewers are therefore instructed to reinforce adequate respondent behavior and probe for an answer whenever necessary. To investigate interviewer effects on missing data, we constructed a global indicator measuring the proportion of item nonresponse over all questions.

Overall, the incidence of missing data was low (1.3 percent). The intraclass correlation for interviewers was 0.01. As expected from the research literature, older respondents had more missing data (beta $=0.26$, $p<0.01$). A respondent who had missing data was negatively evaluated by the interviewer (beta $=-0.32$, $p<0.01$). The data collection method used did not influence the amount of missing data (beta $=0.07$, $p=0.13$).

Two interviewer variables were significant: a preference for face to face interviews by the interviewer (beta $=0.13$, $p=0.03$) and the extroversion of the interviewer (beta $=-0.13$, $p=0.02$). Introverted interviewers and interviewers who, when asked, expressed a preference for the face to face method produced interviews with more missing data in both telephone and face to face interviews.

The regression coefficients for the age of the respondents and the respondent evaluation by the interviewers showed significant random components, indicating an interaction with certain interviewer characteristics. However, an exploratory analysis employing the available interviewer variables did not reveal any significant interactions.

Completeness: Open Questions

Three standard open questions were used. In these questions the respondents were asked to elaborate their previous answers. For each respondent, the mean number of answers to these three open questions was computed. For completeness, the intraclass correlation for the interviewers was substantial: rho = 0.13. The number of answers depended on two respondent variables: age (beta = 0.15, $p < 0.01$) and enjoyment of the interview (beta = 0.14, $p = 0.02$). Older respondents and respondents who reported that they had enjoyed the interview produced more answers to the open questions.

Even after introducing these two respondent variables, the between-interviewer variance component remains significant ($p < 0.01$). None of the available interviewer variables we analyzed made a significant contribution to this interviewer effect. The regression coefficient for interview enjoyment by the respondent had a significant random component ($p < 0.01$), which means that this respondent variable interacted with some interviewer characteristics. However, an exploratory analysis employing the available interviewer variables did not reveal any significant interactions. We should note that in this example all interviewer variables available for analysis concerned socio-demographic interviewer characteristics and personality traits. Process variables, such as the degree of probing, were not available for further analysis.

Response Bias: Acquiescence

In the Dutch version of the "Affect Balance Scale" all positively formulated items are exactly balanced by corresponding negatively formulated items (this is not true for the American version, see Bradburn, 1969; Hox, 1986). In a balanced scale acquiescence or yea-saying can be estimated by counting the number of "agree"-responses to all items.

The intraclass correlation for the interviewers on this scale was fairly substantial (0.06). Three respondent variables were associated with acquiescence: age (beta = −0.24, $p < 0.01$), sex (beta = 0.19, $p < 0.01$), and refusal of earlier surveys (beta = 0.14, $p < 0.01$). Younger respondents, women, and respondents who had refused other surveys but were willing to participate in this one all had a greater tendency to give an affirmative response to the questions. The regression slopes for these variables did not differ between interviewers, meaning that there were no interactions with interviewer variables.

After including the respondent variables in the model, the between-interviewer variance component still remains significant ($p = 0.03$). However, none of the interviewer variables which were available for analysis made a significant contribution to this interviewer effect.

Psychometric Reliability

To gauge the psychometric reliability, we computed for each respondent the inter-item variance for four multi-item scales (Positive affect or happiness, Negative affect, Loneliness, and Self-esteem). The mean inter-item variance is an overall indicator of the scalability of the individual responses; a high variance indicates an unstable response pattern.

The intraclass correlation for the interviewers was estimated to be zero. Nevertheless, a multilevel model revealed significant interviewer effects, including random regression coefficients for the variables Attitudinal Distance and Evaluation of Respondent by Interviewer.

A respondent with variable responses or in other words, a respondent with an unstable response pattern, was negatively evaluated by the interviewer (beta $= -0.19$, $p < 0.01$). Furthermore, interviewer extroversion is negatively related to response variability (beta $= -0.09$, $p = 0.03$), indicating that interviewers who were extroverted produced less variability and a more stable response pattern. Although the mean regression coefficient for attitudinal distance between the interviewer and the respondent was not significant (beta $= 0.10$, $p = 0.19$), the regression slopes for this variable differed significantly between interviewers, indicating an interaction with certain interviewer characteristics. An exploratory analysis employing the available interviewer variables did not reveal any significant interactions.

Since the inter-item variance was computed for four distinct multi-item scales, it is possible to analyze this data with a three-level model, with "interviewers," "respondents," and "scales" as the three levels. This enabled us to analyze the between-respondent variance. The intraclass correlation for respondents was estimated as 0.15 ($p < 0.01$), which indicated that respondents do in fact differ systematically in the degree of response variability. The intraclass correlation for interviewers was again estimated as zero. The three-level analysis selected the same explanatory variables as the two-level analysis. The between-respondent variance component remained significant ($p < 0.01$).

22.4.3 Interview Evaluation

The interview was evaluated by the respondents in two different ways. First, the respondents were asked to assess the interview on a five-item questionnaire threat scale. Second, they were asked whether they enjoyed the interview.

Questionnaire Threat

Questionnaire threat was measured by a five-item scale which asked for the respondent's assessment of the interview as a whole. The intraclass correlation was small (0.01). Perceived questionnaire threat was related to the attitudinal distance between interviewer and respondent (beta $= 0.10$, $p = 0.05$) and to the interviewer's conscientiousness (beta $= -0.14$, $p < 0.01$). Respondents assessed the questions as more threatening, and as more personal and intruding on their privacy, when there was a greater attitudinal difference between interviewer and respondent, and when interviewers were less conscientious according to their answers on a personality inventory.

The regression slope for attitudinal distance did not differ between the interviewers. After including these variables, the interviewer variance was not significant.

Interview Enjoyment

The intraclass correlation for interview enjoyment was 0.05. Respondents who were interviewed face to face enjoyed the interview more than respondents in the telephone situation (beta $= -0.26$, $p < 0.01$). Furthermore, the following respondent variables were related to enjoyment: age (beta $= 0.19$, $p < 0.01$), education (beta $= -0.12$, $p = 0.00$), and respondent evaluation by the interviewer (beta $= 0.29$, $p < 0.01$). Older respondents and respondents who were positively evaluated by the interviewer reported more enjoyment. More highly educated respondents, however, reported less interview enjoyment. The regression slopes did not differ between the interviewers, indicating that there were no interactions of these respondent variables with interviewer variables.

After including the respondent variables, the between-interviewer variance component remains significant ($p < 0.01$). None of the interviewer variables we analyzed made a significant contribution to this interviewer effect.

22.4.4 Controlling Interviewer Influence on a Substantive Model

While the intraclass correlations for most methodological quality indicators are small, their combined effect on the quality of the data as a whole may still be substantial, especially if the interviewer workload is high (cf. Groves, 1989). In our case, the mean number of interviews per interviewer was moderate (25). To illustrate both the impact of interviewer variance on a substantive model and ways to control for this effect statistically, we looked more closely at two models. The first was a loneliness model, and the second was a model for well-being.

The interview survey can be viewed as a procedure for cluster sampling with interviewers defining clusters of respondents. To control for interviewer effects one can then employ an analysis model which corrects for the effects of this "sampling" design (cf. Chapter 31; Lee, et al., 1989). However, if researchers are interested in multiple regression models, the most elegant approach is to use multilevel regression models to correct for the hierarchical nature of the data. The easiest and simplest approach is then to model only fixed regression coefficients: this is similar to an ordinary multiple regression analysis, but adjusts for the between-interviewer variance. This approach is referred to as the Hierarchical Linear Model (HLM). A more general and theoretically more interesting approach is to allow the regression coefficients for the respondents to be random at the interviewer level, and to introduce explanatory interviewer variables in the model. We will refer to this as the Random Coefficient Model (RCM).

While the HLM only gives us an estimate of a regression coefficient and its corresponding standard error, the RCM also produces an estimate of the variation due to interviewers. An intuitive way to interpret this variation is to view it as an indication of how much a researcher could influence the substantive results of a study by a judicious or malicious choice of the interviewers employed (see Wiggins, et al., 1989 for a thorough discussion). Furthermore, the RCM allows us to model interviewer effects. This, too, is of interest when analyzing a substantive model, because it effectively controls for these variables.

Loneliness Model

For our comparison, we used a loneliness model derived from De Jong-Gierveld (1987). In our model, loneliness is (negatively) determined by presence of a partner, extension of the social network, satisfaction with the social network, and self-esteem. We included the respondent's sex and age in the analysis. These variables were included in the path model used by De Jong-Gierveld, but did not directly affect loneliness. In addition to these variables, we included auxiliary respondent and interviewer variables in the random coefficient model. Two of these variables were significant: attitudinal distance between interviewer and respondent, and respondent evaluation by interviewer (see Table 22.1). The coefficients in the analyses are virtually the same. The amount of variance explained hardly differs between the three models (35%, 35%, and 37%).

In the Random Coefficient Model, none of the regression coefficients has a significant random component, which means that the regression coefficients may be considered equal for all interviewers.

Well-being Model

For our comparison, we used a model for well-being derived from Burt, et al. (1978, 1979). In this model, well-being (general satisfaction) is determined by positive affect, negative affect, and satisfaction with specific life domains (housing, health, income, and social network).

Table 22.1. Loneliness Model Regression Coefficients

	OLS		HLM		RCM	
Variable	beta	*p*	beta	*p*	beta	*p*
Age	0.03	0.52	0.03	0.65	0.00	0.98
Sex	0.00	0.99	0.00	0.99	0.00	0.95
Ext. soc. netw.	−0.13	0.00	−0.13	0.00	−0.11	0.00
Sat. soc. netw.	−0.40	0.00	−0.40	0.00	−0.38	0.00
Self-esteem	−0.26	0.00	−0.25	0.00	−0.22	0.00
No partner	0.16	0.00	0.16	0.00	0.17	0.00
Att. distance[a]					0.12	0.01
Respondent eval. by interviewer					−0.09	0.04

[a] In this analysis, attitudinal distance was computed with exclusion of the loneliness score.
OLS, Ordinary Least Squares Model; HLM, Hierarchical Linear Model; RCM, Random Coefficient Model.
Source: De Leeuw, Social Cultural Sciences Foundation (NWO) grant no. 500278008.

Again, in addition to these variables, we included auxiliary respondent and interviewer variables in the random coefficient model. Five of these variables were significant: attitudinal distance, respondent evaluation by interviewer, interviewer extroversion, interviewer social anxiety, and interviewing method used (see Table 22.2). The coefficients in the Ordinary Least Squares model and the Hierarchical Linear Model analyses are virtually the same. The Random Coefficient Model yields coefficients which are generally somewhat lower. No difference was found in the amount of explained variance between the OLS and HLM models (both were estimated as 0.26); the RCM model explains a much larger amount of variance (0.34).

The coefficient for "satisfaction with income" is larger (0.19), and random over interviewers. This means that the strength of this particular effect varies with different interviewers. To decide how much, we have to look at the variance component for the regression coefficient. In this case the variance is 0.02, which corresponds to a standard deviation of 0.14. Assuming a normal distribution of the regression coefficient, this means that a large majority of all interviewers (91%) would produce a positive relationship between "satisfaction with income" and general satisfaction. In sum, the size (0.19) and the significance ($p = 0.00$) of the random regression coefficient tells us that the average regression coefficient across interviewers is positive and statistically significant. The size of the standard deviation (the square root of the variance component) tells us that selecting different interviewers is unlikely to change this picture.

Table 22.2. Well-being Model Regression Coefficients

	OLS		HLM		RCM	
Variable	beta	p	beta	p	beta	p
Neg. affect	−0.25	0.00	−0.24	0.00	−0.21	0.00
Pos. affect	0.18	0.00	0.18	0.00	0.12	0.01
Sat. house	0.05	0.23	0.05	0.22	0.04	0.32
Sat. health	0.13	0.00	0.13	0.00	0.11	0.01
Sat. income	0.14	0.00	0.14	0.00	**0.19**	0.00
Sat. soc. netw.	0.19	0.00	0.20	0.00	0.18	0.00
Att. distance					−0.12	0.01
Respondent eval. by interviewer					0.13	0.00
Extroversion					0.12	0.01
Soc. anxiety					−0.13	0.00
Method					0.12	0.00

Bold and underscored: random coefficient.
OLS, Ordinary Least Squares Model; HLM, Hierarchical Linear Model; RCM, Random Coefficient Model.
Source: De Leeuw, Social Cultural Sciences Foundation (NWO) grant no. 500278008.

22.5 SUMMARY AND DISCUSSION

22.5.1 The Multilevel Model

Interview survey data have a hierarchical structure, and multilevel analysis methods provide efficient and flexible techniques to analyze such data. The model we present is essentially a (multilevel) multiple regression model, which has been described in a number of publications (e.g. Mason, Wong, and Entwhisle, 1984; Raudenbush and Bryk, 1986; Goldstein, 1987), and for which several computer packages are available (for a critical review see Kreft, et al., 1990; three PC-based programs are listed in Appendix B to this chapter.) The models described in this volume by Hill (Chapter 23) and Pannekoek (Chapter 25) are very similar, but some of the statistical details differ. Hill's model does not assume a normal error structure, as our model does, but allows for testing and fitting more complicated error distributions, while Pannekoek's model provides a direct analysis of the interviewer intraclass correlation. These and other developments such as extensions to binomial data (Longford, 1988; Goldstein, 1991) and covariance structure models (e.g., Muthén, 1989) are important. Nevertheless, the flexibility of the multiple regression model (cf. Cohen and Cohen, 1983) and the availability of easy-to-use software make the multilevel regression model a powerful tool for the analysis of hierarchical survey data.

It is often stressed that in interviewer effect studies, respondents should be assigned to interviewers at random (cf. Hagenaars and Heinen, 1982; Groves, 1989). In large scale face to face surveys this is expensive and difficult to organize. It is difficult to use such studies for methodological research into interviewer effects, because without interpenetrating designs the interviewer and respondent explanatory variables are correlated and the interviewer intraclass correlation no longer estimates interviewer effects only. Multilevel analysis methods offer some remedies to overcome this difficulty. If the relevant respondent variables are known, they can be put into the regression model to equate the respondent "input" between interviewers by statistical means. If, after statistical control of the respondent variables, interviewer variables are still statistically significant, we may conclude that this reflects real differences between interviewers. Thus, the appropriate analysis procedure is to test whether interviewer variables explain significant variance in addition to the relevant respondent variables. Again, this is similar to analysis of covariance, with the interviewers as the independent variable and the respondent variables as the covariates to be controlled for, but the assumptions of the multilevel model are much more realistic than those of analysis of covariance. Additionally, even researchers who are not interested in interviewer effects may find it useful to include interviewer variables in the analysis, because it shows how large their potentially disturbing effect is, and offers some means of control (cf. Goldstein, 1990).

22.5.2 Interviewer Effects

Four questions were asked about the way interviewers can influence survey results.

1. The first question concerned the nonresponse. Two remarkable results were obtained: telephone interviews resulted in a higher response rate than face to face interviews, and interviewers did not differ in their ability to persuade contacts to participate in the interview. The difference in response rates between face to face and telephone interviews probably results from the specific procedures used to persuade respondents. The telephone interview involves two steps: after asking for the designated respondent in the household, the interviewer asks her/him to answer some

questions. In the face to face interview a three-step procedure is used: the interviewer telephones and after asking for the designated respondent, tries to make an appointment for an interview at the respondent's home.

As stated in Dillman's application of social exchange theory (Dillman, 1978), the perceived costs of responding may be higher in the face to face interview, while in this case the cost of refusing is the same for the telephone and the face to face interview. In a more traditional approach, interviewers are sent directly to the respondent's home, asking for cooperation on the doorstep. This increases the social cost of refusing cooperation, resulting in a higher response rate. However, for the researcher, this also increases the financial costs of callbacks.

The absence of interviewer effects on cooperation rate may be caused by two factors: the interviewers were well-trained and used a detailed script to persuade contacts to participate. This is likely to result in little variation between interviewers at the beginning of the interview, when the decision to participate is made.

2. The second question concerned the measurement of variables. Overall, the literature reports low intraclass correlations and mixed results concerning the influence of most interviewer characteristics. We find very few interviewer variables that explain significant interviewer variance (see the Summary Table in Appendix A). This difference could very well be the result of biased significance tests in earlier research, caused by disaggregating interviewer variables.

In our present study, in two cases a relatively high intraclass correlation was observed, in combination with a significant residual interviewer variance. For acquiescence, no significant interviewer variables were found. For the number of answers to open questions, again no significant interviewer variables were found. This is more surprising, since the intraclass correlation of this variable is 0.13, the highest correlation coefficient in our results. We suspect that in this case process variables may be more important than socio-demographic interviewer characteristics or personality variables. The number of answers on open questions may be more dependent on differences in the degree of paralinguistic behavior, such as saying "mhmm-hmm"after an answer. Also interviewers may differ in the manner and perseverance of additional probing after the prescribed standard probe on open questions.

Several models showed significant random coefficients, which indicate interactions between respondent and interviewer characteristics. However, further analysis using the variables available in this study did not explain this variability. Again, we suspect that process variables may be more important than the substantial number of interviewer characteristics available in this study.

3. The third question concerned respondents' feelings about the interview. Concerning experienced questionnaire threat, a small interviewer effect is observed which is explained by interviewer conscientiousness and attitudinal distance. Conscientious interviewers and interviewers who have a small attitudinal distance to the respondents lead to less questionnaire threat. Enjoyment of the interview is influenced by interviewers (rho = 0.05), but none of the interviewer variables in this study could explain this effect. The interviewer's evaluation of the interview did concur with the respondent's judgment on what was an enjoyable interview (beta = 0.29). Furthermore, older respondents, less educated respondents and respondents who were interviewed face to face stated that they enjoyed the interview more.

4. The fourth question concerned the way interviewers affect substantive models. For the loneliness model the interviewer effects were negligible. For the well-being model small effects were found, but it is unlikely that they would have substantially influenced the model interpretation. That this is not always the case is shown by Wiggins, et al., 1989 who in an earlier analysis of a survey on physical handicaps found a large random effect. In their case, a clever choice of the "right" interviewers could even have determined the sign of the regression coefficient for the respondents' age.

22.5.3 Implications

Our results suggest some implications for research utilizing face to face or telephone interviews. First of all, while the literature suggests small method effects (De Leeuw and van der Zouwen, 1988), in our case there was virtually no direct method effect. Furthermore, with the exception of the number of answers to open questions, the interviewer effects were generally small, and did not differ between the two methods. Of course, our data come from a relatively small-scale methodological study employing a thoroughly pilot-tested questionnaire, and interviewers who were well trained in both face to face and telephone interviewing, closely supervised, and provided with scripts for difficult situations. In large-scale surveys the field conditions may be less optimal, and differences between methods or interviewers may be larger. Furthermore, since multilevel methods control type I error much better than ordinary single level analyses of multilevel data, they are expected to produce fewer significant findings. This is not a disadvantage; many of the "significant" effects found in an ordinary single level analysis of multilevel data are bound to be spurious.

At any rate, our results provide few grounds for further selection of interviewers on socio-demographic or psychological characteristics (cf. Collins, 1980). Kish (1962) suggests that interviewer effects are likely to be the result of many small differences in interviewer characteristics, respondent populations, study designs and resources. Our results point in the same direction. The implication is that controlling interviewer effects by suppressing them completely or keeping them constant will be difficult. One radical solution is to get rid of the interviewer completely, either by using mail surveys (Dillman, 1978), or by using completely computerized data collection procedures (Saris, 1989), with the risk of facing an explosion in respondent errors. When there are good reasons to employ interviewers, one feasible solution is to control for interviewer effects statistically.

ACKNOWLEDGMENTS

This study was conducted while the first two authors were at the UCLA as Fulbright and visiting scholars. They are indebted to the Department of Psychology and the Program for Social Statistics for their hospitality. The data collection was made possible by grant no. 500278008 from the Social Cultural Sciences Foundation, which is subsidized by the Netherlands Organization for the Advancement of Scientific Research (NWO).

Special thanks are due to Nick Longford for his many constructive suggestions and to Gerard Kurvers and GEO-MARKTPROFIEL for their zip-code information.

APPENDIX A

Summary Table of Standardized Regression Coefficients

Independent variable	Dependent Variable						
	Cooperation rate	Item nonresponse	Open questions	Acquiescence	Psychometric reliability	Questionnaire threat	Enjoyment of interview
Interviewer Level							
1 Age							
2 Experience							
3 Previous training							
4 Prefer face-to-face		0.13					
5 Extroversion		−0.13			−0.09		
6 Conscientious						−0.14	
7 Friendly							
8 Anxiety							
9 Terminate contacts							
Respondent level							
A Method	0.58						−0.26
B Age	na	**0.26**	0.15	−0.24			0.19
C Sex	na			0.19			
D Education	na						−0.12
E Evaluation of respondent by interviewer	na	**−0.32**			−0.19		0.29
F Enjoyment by respondent	na		**0.14**				na
G Attitudinal distance	na				**0.10**^{ns}	0.10	
H Having refused earlier surveys	na			0.14			
Interactions method × A/H							
Intraclass Correlation	0.02	0.01	0.13	0.06	0.00	0.01	0.05
Variance between interviewers	ns	ns	$p=0.00$	$p=0.03$	ns	ns	$p=0.00$

Bold and underscored = random coefficient; ns = not significant at 0.05 level; na = not applicable.
Source: De Leeuw, Social Cultural Sciences Foundation (NWO) grant no. 500278008.

APPENDIX B

Program information on the multilevel model

HLM (Hierarchical Linear Modeling):
Scientific Software, Inc.
1369 Neitzel Rd.
Mooresville, IN 46185-9312, USA

ML3 (three-level analysis)
Multilevel Models Project
Institute of Education, University of London
20 Bedford Way
London, WC1H 0AL, United Kingdom

VARCL (Variance Component Analysis)
N.T. Longford
21T Educational Testing Service
Rosedale Rd.
Princeton, NJ 08541, USA

All programs require at least a PC/AT with 640k memory and hard disk; numeric coprocessor is strongly recommended.

CHAPTER 23

INTERVIEWER, RESPONDENT, AND REGIONAL OFFICE EFFECTS ON RESPONSE VARIANCE: A STATISTICAL DECOMPOSITION

Daniel H. Hill
University of Toledo

23.1 INTRODUCTION

Rational allocation of resources to reducing measurement errors from various sources requires knowledge of both the relative magnitudes of the errors and the relative costs of remedial measures. Information on relative error magnitudes is generally not available and as a result cost considerations tend to dominate decision making. The purpose of the present chapter is to develop methods of estimating measures of relative magnitudes and to provide estimates from one large scale survey as an example.

Using interview/reinterview data from the Survey of Income and Program Participation's (SIPP) Reinterview Program we shall attempt to decompose the estimated measurement error variance into that related to attributes of respondents, subject individuals (persons whose characteristics are being reported by the respondent), interviewers, and the administering regional offices. While our problem is quite similar to that of Hox, et al. in the preceding chapter, our approach is quite different. Rather than estimate a multilevel model, we analyze the data at the individual level and reflect the clustering of the individuals at other levels in our estimation of sampling errors. As was theirs, our approach is observational rather than experimental, and the results should be

This chapter is based on research conducted under a Joint Statistical Agreement (JSA90-001) between the U.S. Bureau of the Census and the Survey Research Institute of the University of Toledo.

Measurement Errors in Surveys.
Edited by Biemer, Groves, Lyberg, Mathiowetz, and Sudman.

ISBN: 0-471-53405-6

interpreted as suggestive of relative magnitudes of effects rather than definitive.

At this point very little need be said about why interviewers and respondents should be considered important sources of measurement error. Why we should concern ourselves about how measurement error varies by characteristics of the subject individuals and regional offices, on the other hand, may require some explanation. Characteristics of subject individuals can radically affect the difficulty of the reporting task presented to respondents and are strongly correlated with characteristics of respondents (e.g., married respondents tend to report on married subject individuals). If we fail to control for subject effects, our estimates of the effects of respondent characteristics on measurement error will be contaminated. Regional offices can affect data quality because interviewer training and supervision can vary from one office to the next and because the extent and quality of clerical edits and recontact of respondents for missing information can also vary.

While there is a long and continuing tradition of observational studies such as ours in the analysis of response errors using reinterview data (see, e.g., Chapter 27), our approach differs from most other work in two fundamental ways. First, we choose to analyze the entire sequence of reinterview questions as a single process rather than on a question-by-question basis. The advantages of this approach are that it results in an empirical distribution for analysis which is better behaved than the question-by-question distributions and that it yields results which are less dependent on the specifics of any one question. It does, however, require us to confront directly the issue of question-to-question error correlation. Second, we distinguish two types of measurement error — response error and procedural error. Response errors are indicated, by our definition, when responses are recorded in both the interview and the reinterview but they differ. Such response errors have been the subject of most previous analyses. Error can also be introduced in survey measures, however, because skip sequences are not always properly followed and because people may vary from time to time in their willingness to provide responses to specific questions. Indeed, as we shall see below, such procedural errors are a far more common problem in the data we examine than are true response errors. Thus, our second measure of data quality is what we will term 'combined errors,' which is the sum of response and procedural errors.

The chapter is organized into five sections. Section 23.2 presents a set of counting-distribution regression models which are useful in analyzing sequences of interview/reinterview items under various forms of population heterogeneity and item-to-item contagion (i.e., when errors in one item affect the probability of error in a subsequent item). Section 23.3 briefly describes the SIPP reinterview data and motivates our subsequent distinction between response and procedural errors. The models developed in Section 23.2 are estimated in Section 23.4 on the SIPP data and the results discussed. Section 23.5 summarizes the results and the limitations of our methods.

23.2 A MODEL

23.2.1 The Basic Model

As with most analyses of interview/reinterview data we begin the development of our model with a variant of the Hansen, et al. (1961) response error model (HHB). The literature includes numerous parameterizations of the basic HHB model for various applications. One that is particularly well suited for our purposes is that of O'Muircheartaigh (1986). According to this model the value recorded for individual j for question i at time t is:

$$y_{ijt}^{*} = y_{ij} + \beta_{ij} + \varepsilon_{ijt} \tag{23.1}$$

where y_{ij} is the true value, β_{ij} is a fixed response error or bias, and ε_{ijt} is the variable response error. With observations at two points in time it is possible to use the paired difference $y_{ij1}^{*} - y_{ij2}^{*}$ for $\varepsilon_{ij1} - \varepsilon_{ij2}$. Assuming, for the moment, that the variance of these errors is the same for all j, and that $\sigma_{\varepsilon i1}^{2} = \sigma_{\varepsilon i2}^{2} = \sigma_{i}^{2}$ then the variance of the paired difference is:

$$\sigma_{\varepsilon i12}^{2} = 2\sigma_{\varepsilon_i}^{2}(1 - \rho_{i12}) \tag{23.2}$$

where ρ_{i12} is the between trial correlation in variable errors for question i due to respondents' memory of the interview responses. Since this correlation is generally non-negative, a nonconservative estimate of the simple response variance (SRV) is simply one-half the estimated variance of the difference in the interview and reinterview responses.

Most previous applications of the HHB model have assumed that the variance of ε_i, as well as perhaps ρ_{i12} itself, differ from one respondent to the next as a function of a vector of respondent characteristics. Interview/reinterview data are then typically analyzed on a question-by-question basis using some measure of response variance (e.g., gross difference rate [GDR]) as the dependent variable and the respondent's characteristics as independent variables.

23.2.2 Generalizations

Our approach departs from the traditional in several ways. First, we expand the set of independent variables to include characteristics of the subject individual S_j, the interviewer I_j and the administering office O_j. Second, we choose to examine the interview/reinterview data as a single experiment rather than on a question-by-question basis. The reinterview data we employ consist of the responses not to a single question but to a sequence of questions, any one of which can be either consistent or inconsistent with the corresponding question in the original interview. The count of the number of inconsistencies for each case j is not only a convenient summary of its overall data quality, but its distribution can be suggestive of the underlying stochastic processes leading to measurement errors.

To see how this is, consider first the following 'thought experiment': Respondent j is presented a single question i at time t and his or her response is recorded as y^*_{ijt} — a categorical variable. After the response is recorded the respondent's memory is wiped clean, and the process is repeated a large number of times, say T. If the response variance is low and the response errors are independent, then we would expect the number of responses inconsistent with the original response (n_{ij}) which are encountered in T trials to be distributed Poisson. Accordingly, the probability of observing exactly k_j such inconsistencies is:

$$Prob(n_{ij} = k_j) = \exp(-\lambda_{ij})\lambda_{ij}^{k_j}/k_j! \qquad (23.3)$$

The term λ_{ij} has several interpretations. According to Johnson and Kotz (1969, page 90) since $E(n_{ij}) = \lambda_{ij}$, n_{ij} "... can be said to have a Poisson distribution with *expected value* λ *(instead of parameter* λ*)* if this is more convenient."[1] In the present application this interpretation is more convenient since:

$$\lambda_{ij}/T = E(n_{ij})/T \equiv E(GDR_{ij}) \qquad (23.4)$$

and, for binary data (see e.g., O'Muircheartaigh, 1986):

$$E(GDR_{ij}) = 2\sigma^2_{\varepsilon ij}. \qquad (23.5)$$

Thus, the Poisson parameter $\lambda_{ij}/2T$ for counts of interview-reinterview inconsistencies in categorical variables is a useful measure of response variance.

To investigate the relationship of measurement errors to characteristics of the subject, respondent, interviewer and regional office, we could perform 'Poisson regression' in which we parameterize λ_{ij} as:

$$\lambda_{ij} = \exp(\lambda_i + \lambda_{Si}S_j + \lambda_{Ri}R_j + (\lambda_{Xri} + \lambda_{Xmi}R_j)X_j + \lambda_{Ii}I_j + \lambda_{Oi}O_j) \qquad (23.6)$$

---

[1]Emphasis in the original. λ is also interpretable as the variance of the number of inconsistencies.

where λ_i is the intercept and λ_S, λ_X, λ_I, and λ_O are coefficients relating deviations from λ_i to the characteristics of subjects (S), respondents (X), interviewers (I) and administrating offices (O), respectively.[2] Respondent characteristics can affect estimated response variances in either of two ways — through their actual effect on $\sigma_{\varepsilon i}^2$ or through their effect on the between trial correlation due to memory (ρ_j). Following O'Muircheartaigh (1986) we assume that these memory effects are present only when the same respondent is used in the interview and reinterview. For this reason, we include R_j (a dummy equaling 1 if and only if the same person was the respondent in both the interview and the reinterview) additively and interactively with the X_j. The coefficient on R_j is interpretable as the overall effect of memory on estimated response consistency while the coefficients for the interaction terms (λ_{Xmi}) represent the deviations from this overall effect associated with respondent characteristics.

The Poisson regression model can be estimated by substituting equation 23.6 into equation 23.3 and the result into

$$\ln(L) = \sum_{j=1}^{N} \ln[Prob(n_{ij} = k_j)] \tag{23.7}$$

and maximizing with respect to the coefficients.

So far our "thought experiment" is quite traditional as is the estimation method. Indeed, ordinary logit models of response discrepancies are exactly equivalent to a Poisson regression model with $k=0, 1$.

Now, however, let us suppose the respondent is asked a sequence of questions $i=1,...Q$. If, and only if, the number of errors in T trials for each of the Q questions is distributed Poisson with λ_i, and the errors are independent across questions, then the sum of discrepancies for the Q questions will also be Poisson with

$$\lambda^* = \sum_{i=1}^{Q} \lambda_i. \tag{23.8}$$

With actual reinterview data we only have one repetition (i.e., $T=1$), thus the average GDR for the questions within the interview-reinterview sequence would be simply λ^*/Q.

In general, of course, the item-independence assumption is not defensible. The counting-distribution literature distinguishes two types of dependencies — heterogeneity and contagion. In our case, heterogeneity would occur if certain individuals are more error prone than

[2]There are, in fact, two respondents in interview-reinterview data and it is an empirical matter as to whether the effects of their characteristics on estimated error variances are the same. The same holds for interviewer characteristics.

others, while contagion would occur if errors in one question affected the probability of errors in subsequent questions. Statisticians have developed a wide variety of counting distributions which, unlike the Poisson, allow for various combinations of heterogeneity and/or contagion. Determining which of these distributions best fit the data will be the first empirical task and will provide insights into the nature of the stochastic process which underlies the response error process.

23.2.3 Further Statistical Considerations

Choice of Distribution

Our strategy for choosing an appropriate distribution is an empirical one in which we allow the data to indicate which member of a family of distributions best describes the data. Specifically we employ the Poisson-Pascal distribution to identify which of a broad set of simpler distributions is most appropriate. This distribution (obtained by Poisson mixing of negative-binomial distributions) has been shown to subsume the Poisson, the negative binomial, the Neyman Type A, and the Polya-Aeppli distribution (see Katti and Gurland, 1961 and Johnson and Kotz, 1969). Unlike the Poisson each of the latter distributions allows for various forms of heterogeneity and/or contagion.

The probability of observing no discrepancies under the Poisson-Pascal distribution is:

$$Prob(n = 0) = P_0 = \exp[-\lambda(1 - (1 + P)^{-N})] \qquad (23.9)$$

where λ is the mean of the Poisson mixing distribution and P and N are the parameters of the negative-binomial base distribution. The probability of observing exactly k inconsistencies is:

$$P_k = e^{-\lambda}[(\delta - 1)/\delta]^k/k! \sum_{x=1}^{\infty} (Nx)^{[k]}[\lambda\delta^{-N}]^x/X! \qquad (23.10)$$

where $\delta = 1 + P$ and $h^{[k]} = [h(h+1)...(h+k-1)]$ represents the ascending factorial operator.

The log-likelihood function can be obtained by substituting these probabilities into equation 23.7 which can then be maximized with respect to λ, N, and P. While the Poisson-Pascal distribution could be used as the basis for estimating the effects of the various sources of measurement error, it would be computationally cumbersome. Instead we shall use it only as a means of identifying a more parsimonious limiting distribution and for testing implicit restrictions of that distribution.

Statistical Decomposition with Complex Samples
Once an appropriate counting distribution is identified for each particular type of measurement error in which we are interested it is a relatively straightforward exercise in numerical optimization to estimate it. Furthermore, were it reasonable to assume independence across cases, it would also be a simple matter to decompose the overall goodness of fit into those portions associated with each source of error. In this case we could, for instance, calculate the pseudo-R^2 or likelihood-ratio index:

$$\check{r}_\lambda^2 = \frac{\ln(L_o) - \ln(L_\lambda)}{\ln(L_o)} = \frac{\chi_\lambda^2}{2\ln(L_o)} \tag{23.11}$$

where L_o is the maximum likelihood value obtained when only the constants are included in the model, L_λ is that obtained when all the predictors are included, and χ_λ^2 is likelihood-ratio test statistic for the hypothesis that the predictors have no effect. The importance of the various sources of measurement error can be defined analogously as partial likelihood-ratio indices for source S, the relative sizes of which are the same as the relative sizes of $\chi_{\lambda S}^2$. Either relative likelihood-ratio indices or the associated χ^2's could, if the casewise independence assumption were valid, form the basis of our statistical decomposition of measurement error variance by source.

Unfortunately, in real data the casewise independence assumption is usually violated because several of the interviews from each interviewer are selected for reinterview. An alternative is to base our decompositions on the following generalized Wald statistics:

$$\chi_S^2 = n\boldsymbol{\lambda}_S' V_d^{-1} \boldsymbol{\lambda}_S \tag{23.12}$$

where $\boldsymbol{\lambda}_S$ is the vector of coefficients relevant to source S and V_d is the corresponding portion of the estimated variance-covariance matrix of the λ obtained by the jackknife method.[3]

In the present application the jackknife method proceeds by first randomly assigning all the interviews of each interviewer into k (approximately) equal-sized subsamples. This form of the Jackknife does not fully represent sampling variance components, but focuses on the interviewer and regional office components. The model is then estimated on the full sample to obtain λ_{all} and on k pseudo-replicate samples

[3]We are aware of the fact that the Wald statistic can behave erratically (see e.g., Hauck and Donner, 1977) and that some alternatives have been suggested at least in the context of the analysis of contingency tables (see e.g., Rao and Scott, 1981 and Fay, 1985). We rely on the design-adjusted Wald statistic here because it is uniquely suited to the situation in which the design effects of several sets of variables can differ.

constructed by omitting the jth subsample to obtain k λ^j_{all}'s. From these estimates a set of k pseudo-value vectors are constructed according to:

$$\lambda^p_{all} = k\lambda_{all} - (k - 1)\lambda^j_{all} \tag{23.13}$$

Our estimate of V_d is then obtained via:

$$V_d = \Lambda'\Lambda/(k - 1) \tag{23.14}$$

where Λ is the k by n_{dim} (the number of parameters in the full model) matrix of pseudo-values (in deviation form) constructed by stacking the k $1 \times n_{dim}$ vectors $\lambda^p{}_{all}$.

23.3 BACKGROUND FOR EMPIRICAL IMPLEMENTATION

The Survey of Income and Program Participation (SIPP) is a large panel survey of a probability sample of individuals in households conducted by the U.S. Bureau of the Census. Interviews are conducted every four months on a rotating basis.

The SIPP reinterview program is an ongoing systematic operation intended to monitor data quality by checking the interviewers' work. The sample to be reinterviewed each month is a multistage probability sample of current SIPP respondents amounting to about 250 cases per month. The reinterviews are taken approximately one week after the original interview and, although they can technically be performed "by anyone in the regional office familiar with the survey" other than the original interviewer, most are performed by supervisors on the telephone.

Our sample consists of 4,889 interview-reinterview pairs from Waves 2–7 of the 1984 panel where an interview was actually conducted both times. These interviews represent the work of approximately 450 interviewers and 50 reinterviewers from the 12 regional offices.

Figure 23.1 illustrates the question flow in the reinterview questionnaire. The questions actually asked of the respondent in both the interview and reinterview are printed in bold type, while the "office check items" which are transcribed to the reinterview questionnaire from the original appear in normal type. Unless otherwise indicated, questions are asked in sequence. In most cases, however, respondents are skipped around certain questions and these skips are indicated in the figure by lines and arrows. If, in response to question 1, for instance, the respondent said the subject person had a job for at least part of the four-month reference period ("yes" on item 1), he or she would be skipped around the questions about whether the subject had spent any time

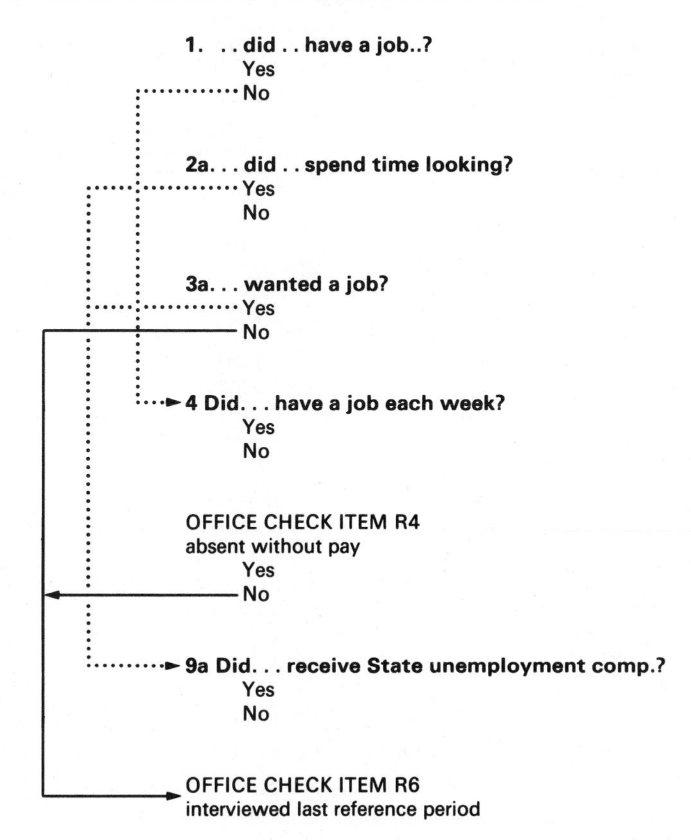

Figure 23.1. Illustration of question flow in reinterview questionnaire.

looking for a job (2a), or whether he or she wants a job (3a). Instead, the respondent would be asked about whether the subject had a job each week of the reference period (4). In Figure 23.1, a skip such as this which results from a response to a question asked in the reinterview is depicted with a dotted line. Skips for office check items, being automatic from the reinterviewer's point of view, are depicted as solid lines.

As the figure clearly shows, the skip sequences employed in the SIPP can be quite complicated. It should come as no surprise, therefore, that a number of inconsistencies will be the result of a question being skipped in either the interview or reinterview but not in both.

Table 23.1 presents the response, procedural, and combined error rates for each of the twelve substantive questions asked in the SIPP reinterview along with the mean and variance of the error counts within reinterviews. The estimated response variance ranges from a low of 0.23

Table 23.1. Estimated Response, Procedural, and Combined Error Variance, by SIPP Question

	Response Variance	Procedural Variance	Combined Variance
1. Have job?	0.81×10^{-2}	0.01×10^{-2}	0.82×10^{-2}
2a. Look for job?	1.37	0.91	1.43
3a. Want job?	1.47	1.29	1.79
4. Each week	1.78	1.18	2.23
9a. U.I. Comp?	0.98	3.69	3.82
23a. Medicare?	1.16	3.65	3.72
24. Food Stamps	0.23	1.50	1.71
26a. Mcaid now?	0.71	1.79	2.12
26b. Mcaid B4?	3.24	0.91	1.02
27a. Health Ins.?	2.14	0.64	2.75
27e. Via emplyr?	1.42	2.53	3.31
27f. Emplyr Pay?	4.04	2.24	3.87
Mean errors per interview	0.1653	0.4068	0.5721
Variance of number of errors per interview	0.1609	0.7572	1.2047

percent for the food stamp recipiency question (item 24) to a high of just over four percent for the employer payment of health insurance question (item 27f).[4] Procedural error rates also range widely from virtually zero for the first employment question (item 1) to more than three and one-half percent for the unemployment compensation question (item 9a) and the Medicare coverage question (item 23a). The combined error rates (which are the sum of the procedural rate and the response rate weighted by the fraction of the sample asked the question in both interviews) are highest for the questions on unemployment compensation and the employer payment of health insurance.

An examination of the average numbers of errors per interview reveals that procedural errors are about two and one-half times as prevalent as are actual response errors in the SIPP. Of course, part of this is due to the fact that some procedural errors will be an automatic result of prior response or procedural errors. Indeed, an analysis of this possibility reveals that roughly 60 percent of the procedural errors are automatic. This means, however, that there are at least as many spontaneous procedural errors in the SIPP data as there are response errors. While many of these procedural errors will result in item

[4]The SIPP staff dropped this question after the 1984 panel because of its low reliability.

nonresponse which will subsequently be dealt with by imputation, their high frequency is, nevertheless, indicative of serious measurement errors due to factors other than the respondent.

Before going on to our analysis of response and combined error rates, one further aspect of Table 23.1 should be noted. The mean and variance of the response errors are virtually identical whereas the variances of procedural and combined errors are much greater than their means. This suggests that while response errors may well be the result of a Poisson process, procedural errors and combined errors almost certainly are not.

23.4 EMPIRICAL IMPLEMENTATION

23.4.1 Error Count Analysis

Identification of Appropriate Distribution

Table 23.2 presents the maximum-likelihood estimates of the parameters of the Poisson-Pascal distribution and four of its special case limit distributions estimated on response errors and combined errors. As indicated by Pearson's χ^2 statistic, the Poisson-Pascal provides an acceptably good fit for both measures of error. Response errors, however, are equally well described by all the distributions including the Poisson. Indeed, the response error data are so close to those predicted by the single parameter Poisson model that technical difficulties are encountered in maximizing the log-likelihood functions for the more general distributions — the gradient does not completely vanish and the matrix of second partial derivatives is nearly singular.

In some respects the excellent fit of the ordinary Poisson is a rather remarkable result. It suggests that response errors are independent within interviews, and since the mean, λ, is small, they are rare.

The same cannot be said for combined errors, however. Here we find the independence assumption of the Poisson distribution to be seriously at odds with the data (see column 3). In this case the χ^2 exceeds 3,000. While the Poisson-Pascal provides the best fit, the Polya-Aeppli also describes the data quite adequately. The Polya-Aeppli distribution is appropriate when the events in question occur in clumps or clusters (see Johnson and Kotz, 1969). In our case such clusters would occur when an error in one item leads to errors in the rest of the sequence of questions. The shape parameter δ can be interpreted as an index of clustering, 'clumping' or contagion and is highly significant. The Polya-Aeppli is appropriate when the number of clusters encountered has a Poisson distribution while the number of errors within a cluster has a geometric

Table 23.2. Poisson-Pascal and Limiting Form Distribution of Various Measurement Errors

	Response Errors	Combined Errors
Poisson-Pascal		
θ	0.959	1.388
δ	68.72	0.300
λ	0.003	0.465
log (L)	−2,300.86	−4,940.34
χ^2	3.00	5.15
Polya-Aeppli		
δ	61.80	1.007
λ	0.003	0.567
log (L)	−2,300.88	−4,940.60
χ^2	3.00	5.73
Negative-Binomial		
δ	350.29	0.787
λ	4.059	0.450
log (L)	−2,300.69	−4,950.43
χ^2	2.74	22.76
Poisson		
λ	0.165	0.572
log (L)	−2,300.50	−5,594.29
χ^2	2.26	3,008.46
Neyman A		
δ	223.63	0.963
λ	0.0007	0.593
log (L)	−2,300.71	−4,948.99
χ^2	2.32	19.73

distribution. The stochastic process suggested is one in which primary errors occur by accident. Once such an error does occur it can result in several subsequent procedural errors — the number depending on the length of the question sequence in which the original error occurred and its position within the sequence.

Given these results, further model development will use the Poisson model for the response errors and, the Polya-Aeppli distribution for the combined errors.

Estimation of Subject, Respondent, Interviewer, and Regional Office Effects

Identification of a parsimonious specification of equation 23.6 is vital since Polya-Aeppli regression is computationally intensive and we will have to obtain jackknifed variance estimates. Our approach in doing so was first to partition the predictors into two sets upon which basis we defined two submodels — the "household" and the "organizational." These models were then estimated with a large set of predictors entered both additively and interactively. Simplifying restrictions were then

Table 23.3. Log-Likelihood Values for Various Specifications of the Household and Organizational Submodels (Polya-Aeppli Regression for Combined Errors)

Household		Organizational	
Final Model	−4,887.90 (13)	Final Model	−4,878.22 (10)
All Interview/Reinterview Interactions	−4,884.42 (21)	All Interviewer/Reinterviewer Interactions	−4,873.86 (21)
All Memory Interactions	−4,884.41 (24)		

Note: Degrees of freedom in parentheses.

imposed and tested to identify parsimonious submodels. Finally the resulting submodels were combined to form our final specification.

We based our tests of simplifying restrictions on the assumption of simple random sampling (SRS) and concentrated on the Polya-Aeppli model of combined errors. This procedure is nonconservative in the sense that it tends to reject simplifying restrictions even when they are valid. There are two reasons for this. First, the use of SRS variances in testing restrictions tends to overstate the importance and significance of predictors involved. Second, our experience indicates that the associations of predictors with combined errors were far stronger than the corresponding associations with response errors. Thus sets of restrictions which failed to do significant damage to the model's goodness of fit for combined errors were even less likely to do so for response errors.

The predictors included in the household submodel were characteristics of the three possible actors involved (the subject individual, the interview respondent, and the reinterview respondent) plus all the memory interactions ($R_j \times X_j$) and intercept discussed in conjunction with equation 23.6 of Section 23.2.

Table 23.3 presents the log-likelihood values obtained for various specifications of the household and organizational submodels of combined measurement errors. The first column of numbers pertain to the household submodel. The first row presents the log-likelihood of the final parsimonious submodel which is −4,887.9 and involves 13 parameters. The χ^2 for the likelihood-ratio test of eight interview-reinterview respondent interactions is only seven and these parameters are eliminated. The third row indicates that when 11 additional memory interaction terms are added the log-likelihood value rises to −4,884.4. This implies a χ^2 of only seven with 11 degrees of freedom which means that we cannot reject the hypothesis that the apparent improvement in goodness of fit from the memory interactions is due solely to chance. These parameters are, therefore, also removed from the model.

Table 23.4. Polya-Aeppli Regression Analysis of Combined Errors, Parsimonious Combined Model[a]

Variable	Coefficients	Variable	Coefficients
λ_0	0.7786 (0.8558)	Log (Regional Office Error Rate)	0.7829** (0.2539)
Age of Subject	−0.0103** (0.0040)	Log (Month)	−0.4470 0.2939
Race	0.1103 (0.0883)	# Panels	0.4786** (0.1367)
Married Subject	0.2468 (0.1666)	Age of Interviewer	−0.0745* (0.0362)
Age of Respondent	0.0100** (0.0030)	Sex of Interviewer	−0·2004** (0.0785)
Married Respondent	−0.2216** (0.0828)	Race	0.1210 (0.0937)
Education of Respondent	−0.0188** (0.0052)	Job of Reinterviewer	0.6723 (0.7299)
Same Respondent	−0.0172 (0.0496)	Education of Reinterviewer	0.1198 (0.0671)
(Same Respondent) × (Wave)	−0.0585 (0.2771)	ρ^2	1.91%
(Age of Respondent) × (Same Respondent)	−0.0034 (0.0883)	log (L) (d.f.)	−4,846.0 (21)
Original Proxy	0.2167** (0.0623)	δ	0.5266** (0.0536)
Wave	0.0287 (0.0783)		

[a] Design corrected standard errors in parentheses.
*Significant at 5% level
**Significant at 1% level

Jumping ahead to Table 23.4 we see that the final household submodel contains three subject individual measures (age, marital status, and race), five respondent measures (age, marital status, education, wave, and whether proxy), and three memory interaction terms involving R, the indicator for the same respondent for the interview and the reinterview (R, $R \times$ age respondent, and $R \times$ wave). Older and unmarried subjects have lower measurement errors whereas older and unmarried respondents have higher measurements errors. Higher respondent education is associated with lower measurement error while proxy respondents have higher errors. The interaction terms suggest some weak memory effects, but primarily only for older respondents in the latter waves of the study.

The "organizational" submodel contains those predictors associated with either the administering regional office (RO) or the interviewing work force in the SIPP. The former consisted of an index constructed from the RO identifiers and measures of the organizational experience with SIPP and workload, while the latter consisted of a rather extensive list of demographic and job specific characteristics of the interviewers and reinterviewers.[5] The second column of figures in Table 23.3 presents the log-likelihood of the final organizational model along with that of one of the less restrictive alternative models investigated. This latter model included 11 interviewer-reinterviewer characteristics interactions and resulted in an increase in the log-likelihood function of only 4.4 units. The implied likelihood-ratio statistic of 8.8 (d.f. = 11) is well below the critical value and the simplifying restrictions were imposed.

The final organizational submodel contains the three "office" measures (the logarithms of the regional office error rate, and months in the field and number of panels being interviewed), three interviewer measures (sex, age, and race), and two reinterviewer measures (job classification and education).

23.4.2 Combined Model and Decompositions

With parsimonious specifications identified for the household and organizational submodels, we are now in a position to combine them into a single model and perform statistical decompositions of the explained error variance by source. In doing so, however, we must abandon the assumption of simple random sampling. The reason is that the sample is clustered by interviewers and reinterviewers within regional offices and design effects may well differ for the estimated coefficients of interviewer, office, respondent, and subject characteristics. If so, then any comparisons of estimated predictive power of the various sources must compensate for these differential design effects.

As noted in Section 23.2, we chose to compensate for possible differential design effects by basing our decomposition of effects on generalized Wald statistics constructed from a jackknifed estimate of the covariance matrix of the model's parameters. To do so, within each of the 12 regional offices interviewers are randomly assigned to one of three groups. Thirty-six pseudo-replicates are then created by sequentially

---

[5]An extensive set of interviewer characteristics was included in earlier versions of the model and found to contribute nothing to the explanation of either response or total errors. This set included: Experience with the SIPP, Experience with the Census Bureau, Job classification, Pay-grade, Education, Type A and Type D Nonresponse rates, and SIPP workload.

Table 23.5. Poisson Regression Analysis of Response Errors, Parsimonious Combined Model[a]

λ_0	0.1163 (0.9233)	Log (Regional Office Error Rate)	1.0242** (0.3246)
Age of Subject	−0.097 (0.0065)	Log (Month)	−0.5734 (0.3853)
Race	0.2243 (0.1344)	# Panels	0.3185 (0.1684)
Married Subject	0.4655* (0.2207)	Age of Interviewer	−0.0860 (0.0523)
Age of Respondent	0.0063 (0.0041)	Sex of Interviewer	−0·0667 (0.1152)
Married Respondent	−0.2652 (0.1095)	Race	−0.0092 (0.1323)
Education of Respondent	−0.0170** (0.0060)	Job of Reinterviewer	0.6297 (0.7140)
Same Respondent	−0.0435 (0.3271)	Education of Reinterviewer	0.0474 (0.0738)
(Same Respondent) × (Wave)	−0.0780 (0.0668)	ρ^2	1.45%
(Age of Respondent) × (Same Respondent)	−0.0003 (0.0026)	log (L) (d.f.)	−2,267.33 (20)
Original Proxy	0.2167** (0.0986)	δ	0.2919**
Wave	0.1328 (0.0985)		

[a] Design corrected standard errors in parentheses.
*Significant at 5% level
**Significant at 1% level

discarding all the interview-reinterview pairs for each of the 36 regional office subsamples of interviewers. The combined model is then estimated on the entire sample and on each of the 36 pseudo-replicates and the resulting parameter estimates are used to construct our design-corrected estimate of V as in equation 23.13 of Section 23.2.

Tables 23.4 and 23.5 present the results of these estimations for the Polya-Aeppli regression model of combined errors and the Poisson regression model of response errors, respectively. Design-corrected (jackknifed) standard errors are presented in the tables in parentheses below the coefficients. As noted above, the effects of respondent and subject age are of opposite signs. Each year increase in respondent age is associated with a one percent increase in combined errors whereas each year increase in subject age is associated with a one percent decrease. Since the SIPP is largely a labor force survey this suggests that the difficulty of the reporting task decreases with age as people's work patterns settle down. The ability of respondents to report these patterns consistently, however, decreases with age, as cognitive psychologists suggest.

Similarly, marital status also has differential impacts on combined errors. Married subjects have significantly higher response errors, whereas married respondents have lower response and combined error rates. In the SIPP questionnaire, items regarding joint income and assets are asked for married individuals, and therefore the difficulty of the reporting task is greater for married subjects than for nonmarried. Married individuals, however, seem to be better able to report their more complicated situations than are nonmarried people. The final respondent characteristic which has a significant impact on both response and combined error rates is education. Each year of education reduces combined error and response error rates by 1.7 and 1.88 percent, respectively. The corresponding effects of subject education were completely insignificant.

Interestingly, the race effects which were seen to be either significant or nearly significant in both the household and organizational submodels are not significant in the combined model. In part this is due to the design effects for both race measures included in the combined model, but it is also partly due to the size of the individual effects declining in the combined model. This is what one would expect if black interviewers were assigned to black respondents more frequently than were nonblack interviewers. Taken together interviewer and subject race are close to significant with a generalized Wald statistic of 3.5 with two degrees of freedom in the Polya-Aeppli model of combined errors. While not significant, this level of χ^2 is sufficiently high to provide some evidence of an effect.

The memory interactions (composed of the additive effect of having the same respondent in both interviews (R), and its interactions with age of respondent and wave) are, as a set, significant in the combined error model at the 95-percent level of confidence. Their Wald statistic is 8.41 with two degrees of freedom. As noted above, these coefficients suggest that memory of the interview response significantly increases observed consistency, but only in the latter waves of the study and only for older respondents.

With the exception of the reinterviewer measures and interviewer race, most of the office and interviewer measures found to be significant in the organizational submodel remain significant in the combined model. As a set they are highly significant with a generalized Wald statistic of 21.8 with two degrees of freedom. The pattern of effects on combined errors suggested by the coefficients is one in which combined errors decline at a decreasing rate for the first 12 months of the study when only one panel was in the field and then jumps to near its original level when the second, 1985, panel is introduced. One likely reason for this is that people in the regional offices responsible for the clerical edits of incoming interviews became better at their jobs with experience, but became heavily taxed when their workloads doubled as a result of the 1985 panel going into the field. Interviewer supervision also may have improved initially only to fall off when the workload increased.[6]

The only interviewer effects to remain significant in the Polya-Aeppli regression model of combined errors are interviewer age and sex. Female and older interviewers are seen to have significantly lower error rates than are younger or male interviewers.

Finally the shape parameter δ, which we have been treating primarily as a nuisance parameter, is similar in interpretation to the index of clumping in the Neyman A model. That it is positive and highly significant means that combined errors are highly clustered within interviews. The reason for this clustering is that several procedural errors in a sequence often result from a single initial error.

23.4.3 Decomposition of Explanatory Power by Source

Table 23.6 presents the decomposition of the explained response and combined measurement error variance by source as discussed in Section 23.2.3. These decompositions use the design-corrected generalized Wald statistics and corresponding partial likelihood-ratio indices obtained when groups of coefficients are removed from the model. The interpretation of these statistics is much the same as is the interpretation of the F-test of the partial-R^2 in ordinary regression analysis — it is a measure of the unique explanatory power of the set of variables in question over and above that effect which can be attributed to other variables in the model.

The first thing to note about the partial likelihood-ratio statistics of Table 23.6 is that they are all small — as are the SRS likelihood-ratio indices presented in Tables 23.4 and 23.5. While these statistics are typically lower than corresponding R^2's in linear models and should not be taken too seriously in evaluating the model, they are, nevertheless, so small as to indicate that most of the variance in measurement errors is

[6]Individual interviewer experience measures were not significant predictors of either response or procedural errors and their inclusion in earlier specifications did not change the results of overall experience with the survey and workload.

Table 23.6. Explanatory Power by Source of Error, Design-Corrected Generalized Wald Statistic

		Response Errors		Combined Errors	
		χ^2	ρ^2	χ^2	ρ^2
Subject	(d.f. = 3)	6.32	0.14%	10.08	0.10%
Respondent	(d.f. = 8)	50.45	1.10%	106.97	1.10%
Office	(d.f. = 3)	16.96	0.37%	24.92	0.26%
Interviewer	(d.f. = 5)	7.29	0.16%	47.97	0.50%

left unexplained by the predictors included in our model. This, in turn, suggests one of two possibilities — first that we have not included the proper predictors in the model or, second, that most of the measurement error in the SIPP is pure noise.

There are three additional general things to note about these design-corrected Wald statistics and partial likelihood-ratio indices. First, with the exception of the subject and interviewer effects for the Poisson model of response errors, each source of error contributes significantly to explaining measurement error rates. Second, the size and significance of the contribution from each source is greater for the Polya-Aeppli model of combined errors than it is for the Poisson model of response errors. Third, respondent characteristics contribute more than any other single source to the explanatory power of the models of both types of measurement error.

There are, however, some notable differences in the relative importance of the various sources for response and combined errors. Specifically, the role of the interviewer characteristics is relatively more important in the case of combined errors than in the case of pure response errors, while the role of the respondent is more important in response errors than in combined errors. This is exactly as we would expect since combined errors are composed of both response errors and procedural or skip-sequence errors. What is somewhat less intuitive is that the role of what we call 'office effects' (i.e., the regional office index, wave and number of panels) is relatively more important for response errors than for combined errors.[7]

[7] Had we used Wald statistics based on the assumption of simple random sampling we would have concluded incorrectly that office and interviewer characteristics, combined, were more important than respondent characteristics in determining measurement errors. The primary reason for this is that the SRS estimates fail to correct for the very large design effects associated with, in particular, the reinterviewer measures.

23.5 SUMMARY AND LIMITATIONS OF METHODS

In this chapter we have developed a model of measurement error which treats the entire interview-reinterview process as a single stochastic process. In doing so we analyze the number of interview-reinterview discrepancies at the individual level and have applied the analysis to data from the SIPP in order to estimate the relative importance of respondents, subject individuals, interviewers, and regional office characteristics as sources of measurement error. The use of count data yielded estimates which reflect the central tendency of the above effects across all questions in the reinterview. In the course of our analysis we found it useful to distinguish two types of measurement errors — response errors and procedural errors. Response errors, which appear in interview-reinterview data as discrepancies when the questions are asked in both interviews, were found to be relatively rare and, for the most part, independent within interviews.

Procedural errors, which appear when a response was not obtained for a question in either the interview or the reinterview (but not both), were found to be two and one-half times as prevalent as response errors and accounted for more than 70 percent of the combined errors in the SIPP reinterview data. Furthermore, unlike response errors, such procedural errors exhibit strong interdependencies within interviews. Because of the interdependencies introduced by procedural errors, the combined error process was found to follow the Polya-Aeppli distribution which is appropriate when errors occur in clusters where the number of clusters encountered is determined by a Poisson process and the number of errors within a cluster by a geometric distribution.

Once the appropriate stochastic process was determined for response and combined errors, regression models based on the corresponding distributions were developed and estimated by maximum-likelihood methods. The regressors employed corresponded to measured characteristics of the subject individual, the respondent, the interviewers and the administering office. The relative importance of these sources of error was estimated via a generalized Wald statistic which was corrected for the clustered nature of the sample. While respondent characteristics as a group were found to be most important in determining the level of both response and combined measurement error, the interviewer and office characteristics were also quite important — especially for combined measurement error.

The most serious limitation of our methods is that they rely on multivariate statistical control rather than randomization to purge the estimated effects of each source of measurement error of the confounding influences of other sources. The interviewer sex effect discussed above provides a perfect example of the uncertainties that lack of randomization impart to our conclusions. Specifically, male interviewers were found to have significantly higher combined errors than were female. Without randomized assignment it is impossible to say if this is truly an indication that male interviewers perform less well or whether, alternatively, it is a reflection of males being assigned to more difficult interviewing areas. Only to the extent that the subject and respondent characteristics included in our models account for this 'difficulty of the interview' would the interviewer characteristics effects be purged of the confounding assignment effect.

A related shortcoming of our analysis is that we were forced by both a lack of data and by the computational burden of our regression models to restrict our attention to relatively parsimonious sets of characteristics of each error source. The resulting source 'submodels' may therefore seem simplistic and incomplete. More complete specifications might well affect the estimates of the relative importance of the various sources of errors both because they more completely capture the individual source's contribution to measurement errors and because they do a better job of purging confounding effects. For these reasons we urge the reader to interpret our present results as suggestive of the possible relative importance of the various sources of measurement errors in survey.

Despite these uncertainties and qualifications there are a couple of implications of our findings for the way in which we allocate our resources to reducing measurement errors in surveys. First, response errors are infrequent relative to errors introduced as a result of skip sequencing and other causes of item nonresponse. This suggests that more thorough and innovative methods of pretesting skip sequences and/or the use of CATI methods would be worthwhile. Second, while the respondents appear from our decompositions to be the most important source of error, they are also the most expensive source to affect since there are so many of them. Significant portions of combined measurement error seems to be associated with interviewers and administering offices, and the behaviors of these are much more directly and cheaply altered than are respondents'. Our results suggest that, at least in the SIPP, substantial reductions in combined errors might be gained by modest increases in the resources allocated to the clerical edits conducted at the regional office level.

ACKNOWLEDGMENTS

The author would like to thank Dan Kasprzyk and Bob Clemons for their help in obtaining the data and Colm O'Muircheartaigh, Jim Lepkowski and Lynne Stokes for their helpful comments on earlier versions of the chapter.

SECTION E

MODELING MEASUREMENT ERRORS AND THEIR EFFECTS ON ESTIMATION AND DATA ANALYSIS

CHAPTER 24

APPROACHES TO THE MODELING OF MEASUREMENT ERROR

Paul Biemer
Research Triangle Institute

S. Lynne Stokes
The University of Texas at Austin

24.1 INTRODUCTION

The goal of this chapter is to provide the reader with a brief introduction to measurement error modeling. Emphasis will be given to the approaches taken in the subsequent chapters of this section. In reviewing each approach, the interpretation of the concepts of reliability, validity, bias, and variance will be examined under each model.

In the literature of measurement error modeling, there are two distinct developments of the theory. One stream of research, which shall be referred to as the "sampling perspective," views the measurement error in a survey response as a random variable arising from the sampling of a hypothetical error population. Thus the response to a particular survey item is the result of two "stages" of random sampling: the sampling of an individual unit from a finite population of individuals and the sampling of errors (which are typically modeled as additive) within individuals from infinite populations of errors. Here the goal is to evaluate the effect of measurement error on survey statistics.

A second stream of research, which shall be referred to as the "psychometric perspective," aims not at describing the measurement error at the individual unit level, but rather at describing the variance-covariance structure of the population taken as a whole. In this approach to modeling, the goal is to understand the relationship among multiple responses from a sample of units and to evaluate the correctness

Measurement Errors in Surveys.
Edited by Biemer, Groves, Lyberg, Mathiowetz, and Sudman.

ISBN: 0-471-53405-6

of individual responses. Measurement error models are specified as a means of explaining the observed intercorrelations among the responses and the relationship between responses and errors.

In our review of measurement error models, these two perspectives on modeling will be compared. We will show that despite the apparent differences in the models' motivation, terminology, and notation (see Chapter 1), they differ only in the assumptions made for their measurement error terms.

24.1.1 Samplers, Psychometricians, and True Values

The sampling approach to measurement error modeling began with the early work of Hansen, Hurwitz, and Madow (1953) and Sukhatme and Seth (1952). Their models focused on describing the effect of measurement error on estimators of a population parameter, such as a mean, calculated from a random sample from a finite population. In its simplest form, the model specifies that a single observation, y_j, from a randomly selected respondent j, is the sum of two terms: a true value μ_j and an error, ε_j. Mathematically, this is written as

$$y_j = \mu_j + \varepsilon_j. \tag{24.1}$$

For this model, μ_j is a random variable whose distribution depends on the sample design and it is assumed that:

1. Simple random sampling without replacement (SRSWOR) has been implemented to select the n units for the sample. This implies

a. $E(\mu_j) = \mu = \dfrac{1}{N}\sum_{i=1}^{N} \mu_i$

b. $\mathrm{Var}(\mu_j) = \sigma_\mu^2 = \dfrac{1}{N}\sum_{i=1}^{N} (\mu_i - \mu)^2$

c. $\mathrm{Cov}(\mu_j,\mu_{j'}) = -\dfrac{\sigma_\mu^2}{N-1}$, for $j \neq j'$ where N is the population size.

The distribution of the error variable, ε_j, is conceptual. It is assumed that, associated with each individual unit j, there is an infinite population of potential errors. The response, y_j, is the result of the individual's true value, μ_j, being distorted by an error, ε_j, which is added to μ_j when y_j is obtained. Thus ε_j represents a single random selection from the j-th unit's error distribution. Let $E(\cdot|j)$ and $\mathrm{Var}(\cdot|j)$ represent expectation and variance, respectively, over the j-th unit's hypothetical error distribution, and further assume for $j = 1, \ldots, N$,

2. $E(\varepsilon_j|j) = 0$

3. $\text{Var}(\varepsilon_j|j) = \sigma_j^2$

4. $\text{Cov}(\varepsilon_j, \varepsilon_{j'}) = 0$ for $j \neq j'$.

From assumption 2, one can show that $\text{Cov}(\mu_j, \varepsilon_{j'}) = 0$ for all j, j'.

Finally, it is common in the study of measurement errors to obtain two or more measures of μ_j from the j-th unit. For example, a reinterview may be conducted where the same questions are asked of the same individuals in order to measure the variance, σ_j^2, of the individual's error distribution. Alternatively, slight variations of the same question may be administered in the same interview. Hence, we denote by $y_{j\alpha}$ the α-th measure of μ_j for unit j and write

$$y_{j\alpha} = \mu_j + \varepsilon_{j\alpha} \tag{24.2}$$

where, in addition to assumptions (1) to (4), we add

5. $E(\varepsilon_{j\alpha}\varepsilon_{j\alpha'}|j) = 0, \ \alpha \neq \alpha'$.

Observations having the structure shown in (24.2), along with assumptions 1 through 5, shall be referred to in the sequel as conforming to Model (0). It will be shown that all the measurement error models in our review are variations of Model (0), where the variation is an elimination or relaxation of one or more of the five assumptions.

In addition to assumptions 1 through 5, two others are implied by the form of (24.2) and are thus true of Model (0). These are:

6. The measurement error is additive.

7. μ_j exists and is well-defined.

Multiplicative error structures have also been proposed (see, for example, Chapter 32); however, one may view this form of error as the case where $\varepsilon_{j\alpha} = \mu_j(e_j - 1)$ where e_j satisfies assumptions 2 through 5 and so is handled as a generalization of Model (0).

There are many cases where the existence of a "true" value is not tenable (see, for example, the Introduction to the volume). Psychometricians distinguish between two types of measurements or "scores": Platonic and non-Platonic (or classical). A Platonic true score is one for which the concept of a true value is plausible; physical measurements, personal demographic characteristics, and behavioral characteristics are examples. Classical true scores are those such as psychological states, attitudes, or knowledge, for which a true value cannot be well-defined.

To handle classical measurements, psychometricians assume a different origin for measurement error than for Platonic true scores. They assume that there exists a response distribution for the measurement $y_{j\alpha}$ which is associated with an individual j. Let μ_j denote the mean of the response distribution and let

$$\varepsilon_{j\alpha} = y_{j\alpha} - \mu_j; \tag{24.3}$$

that is, $\varepsilon_{j\alpha}$ is a "sampling deviation" for the α-th response (or measure) obtained from the j-th individual's response (or *propensity*) distribution. Since μ_j is the mean of the response distribution, we have $E(\varepsilon_{j\alpha}|j) = 0$. If the remaining assumptions of Model (0) hold, then the psychometrician's model for $y_{j\alpha}$ is equivalent to Model (0), with the only difference being the interpretation of μ_j. This model has been referred to in the literature as the "classical true score" model (see, for example, Lord and Novick, 1968).

An example will highlight this major difference between the two interpretations of μ_j in Model (0). Consider a survey in which a respondent's age (AGE) and satisfaction with life (SWL) are measured. AGE is a Platonic variable since it can be operationally defined as the number of whole years lived since birth, and μ_j is the true age of individual j. Under Model (0), we assume that the mean error in the reporting of AGE over hypothetical repetitions of the interview process under identical conditions is zero. By contrast, an individual's "true" satisfaction is not well-defined and is unverifiable by external, objective data. However, given a measure of life satisfaction, one may conceptualize an infinite number of applications of the question, each under identical conditions. In true score theory, an individual's true satisfaction is the mean response over the infinite number of trials.

Just as an individual may consistently underreport AGE, so too might he or she underreport SWL. Under the sampler's interpretation of μ_j, these *response biases* can be modeled by relaxing assumption 2 and allowing $E(\varepsilon_j|j) < 0$. For the psychometrician, the conditional expectation of ε_j is always 0. Instead, response bias is modeled by specifying that

$$y_{j\alpha} = \mu_j^* + \varepsilon_{j\alpha} \tag{24.4}$$

and

$$\mu_j^* = \mu_j + M_j \tag{24.5}$$

where M_j is interpreted as the effect of the "method" (for example, the question wording or structure) on the measurement of the true score μ_j (see Chapter 28). In either case,

$$E(y_{j\alpha}|j) = \mu_j + M_j$$

where $M_j = E(\varepsilon_{j\alpha}|j)$ in the sampling model. However, from the psychometric perspective, both μ_j^* and μ_j are means of propensity distributions.

Further, the variance of μ_j^* is made up of components of *valid variance*, $\mathrm{Var}(\mu_j)$, and *invalid variance*, $\mathrm{Var}(M_j)$ (see Chapter 28). Thus, in psychometric theory, response biases are modeled as components of variance, the magnitude of which can be assessed through covariance analysis. Equations (24.4) and (24.5) are *linear structural equation models* (see, for example, Bollen, 1989), which are discussed in greater detail in Section 24.5.

24.1.2 Measures of Measurement Error

We define the mean squared error of an estimator, say the sample mean $\bar{y}$, as $\mathrm{MSE}(\bar{y}) = E(\bar{y} - \mu)^2 = [\mathrm{Bias}(\bar{y})]^2 + \mathrm{Var}(\bar{y})$, where $\mathrm{Bias}(\bar{y}) = E(\bar{y}) - \mu$ and $\mathrm{Var}(\bar{y}) = E[\bar{y} - E(\bar{y})]^2$. Concern with mean squared error is consistent with the sampler's view that measurement errors are harmful if they affect the accuracy of survey estimates. Because SRSWOR is assumed under Model (0), we have

$$\mathrm{Bias}(\bar{y}) = E(\bar{\mu}) - \mu = 0$$

and

$$\begin{aligned}\mathrm{Var}(\bar{y}) &= \mathrm{Var}(\bar{\mu}) + \mathrm{Var}(\bar{\varepsilon}) \\ &= \left(1 - \frac{n-1}{N-1}\right)\frac{\sigma_\mu^2}{n} + \frac{\sigma_\varepsilon^2}{n} \qquad (24.6)\end{aligned}$$

where

$$\sigma_\varepsilon^2 = \frac{1}{N}\sum_{j=1}^{N} \sigma_j^2.$$

Since $\mathrm{Var}(\bar{\varepsilon}) \geq 0$, measurement error under Model (0) increases $\mathrm{MSE}(\bar{y})$ by the component σ_ε^2/n. Thus the sampler's aim is to reduce σ_ε^2.

The psychometrician's objective is to provide "good" measures of a true score or *construct*. The two major criteria for ascertaining the goodness of a measure are *validity* and *reliability*. The concept of validity is complex and numerous types of validity have been proposed. (See Groves, 1989, pp. 22–27 for a review of some of these.) Our concern here will be with *theoretical validity* (TV), defined as

$$\begin{aligned}TV &= \mathrm{Corr}(\mu_j, y_{j\alpha}) \\ &= \frac{\mathrm{Cov}(\mu_j, y_{j\alpha})}{[\mathrm{Var}(\mu_j)\mathrm{Var}(y_{j\alpha})]^{1/2}} \qquad (24.7)\end{aligned}$$

(Borhnstedt, 1983). TV reflects the degree to which the "indicator," $y_{j\alpha}$, measures the true score, μ_j. If $TV = 1$, then $y_{j\alpha}$ is exactly μ_j, apart from possible differences in measurement scale. For example, for AGE, $y_{j\alpha}$ may be expressed in months while μ_j is defined in years. Without transforming the scale of $y_{j\alpha}$, $y_{j\alpha} = 12\mu_j$; however, TV is still 1. A small TV is evidence of the failure of $y_{j\alpha}$ to measure μ_j. For Model (0),

$$TV = \frac{\sigma_\mu}{(\sigma_\mu^2 + \sigma_\varepsilon^2)^{1/2}}.$$

Reliability is defined as

$$R = 1 - \frac{\mathrm{Var}(\varepsilon_{j\alpha})}{\mathrm{Var}(y_{j\alpha})}, \tag{24.8}$$

that is, the proportion of the total variance, $\mathrm{Var}(y_{j\alpha})$, that is not due to measurement error. Under Model (0),

$$R = 1 - \frac{\sigma_\varepsilon^2}{\sigma_\mu^2 + \sigma_\varepsilon^2} = \frac{\sigma_\mu^2}{\sigma_\mu^2 + \sigma_\varepsilon^2}. \tag{24.9}$$

(In the sampling literature, $1 - R$ is called the *index of inconsistency*.) Note that $R = (TV)^2$ under Model (0); however, this relationship does not hold for more complex models. As an example, consider the case described by (24.4) and (24.5). Here

$$R = 1 - \frac{\sigma_\varepsilon^2}{\mathrm{Var}(y_j)} = \frac{\sigma_{\mu^*}^2}{\sigma_{\mu^*}^2 + \sigma_\varepsilon^2}$$

and

$$(TV)^2 = \frac{\sigma_\mu^2}{\sigma_{\mu^*}^2 + \sigma_\varepsilon^2}$$

where we have assumed that M_j is uncorrelated with μ_j and ε_j, for all j. In fact, $(TV)^2 \leq R$, which demonstrates that a measure may be reliable but still invalid. In Chapter 28 the relationship of validity and reliability under more complex measurement error structures is investigated.

Model (0) will be used as a starting point for the examination of more complex models which have been used for describing measurement error. They are needed because Model (0)'s assumptions are often not met. For example, assumption 4 implies that the correlation between the measured values of two different sample units is zero. This may not be true if the measurements on the two units were obtained by the same interviewer or if the two respondents somehow influenced each other's responses. Assumptions 1a through 1c hold for simple random sampling but not necessarily for more complex sample designs. Model (0) is inappropriate for categorical data under the interpretation of μ_j as a true value. In that case, assumption 2 does not hold, as we shall see in Section 24.4. Despite its limitations, though, Model (0) is a convenient starting point for examination of the effects of measurement error on estimation and data analysis.

24.2 INCORPORATING INTERVIEWER EFFECTS

Some of the earliest examinations of measurement errors in the survey research literature focused on evaluating the impact of interviewers on the data they collected. It was apparent that there was correlation among measured values collected by the same interviewer. In a SRSWOR design, this can be explained only by a violation of assumption 4 of Model (0). This effect is not unique to interviewers. Other common factors, such as coders and supervisors, may also induce correlation among the reported values for the units they are associated with. However in this section, we will refer to the common factor as an interviewer, while keeping in mind that there are other possible sources for correlated errors. We will discuss several models which were developed to describe this correlated component of measurement error. For each of them we must use notation which allows a sample unit to be associated with its interviewer. So we let $y_{ij\alpha}$ denote the measured value of the j-th unit of interviewer i's assignment on occasion α, and let $\varepsilon_{ij\alpha}$ denote the error in that measurement.

24.2.1 Hansen, Hurwitz, and Bershad (HHB) Model

Let us suppose that the responses from the n units in the sample are collected by I interviewers, the i-th of which has an assignment size m_i. Hansen, Hurwitz, and Bershad (1961) proposed a model which eased assumption 4 of Model (0) by replacing it with the following:

$$4'.\ \operatorname{Cov}(\varepsilon_{ij\alpha}, \varepsilon_{i'j'\alpha}) = 0 \qquad \text{for all } i \neq i',$$

$$= \rho_w \sigma_\varepsilon^2 \qquad \text{for all } i = i', j \neq j'$$

for $-1 \leq \rho_w \leq 1$.

Then

$$\bar{y}_\alpha = \frac{1}{n}\sum_{i=1}^{I} \sum_{j=1}^{m_i} y_{ij\alpha}$$

where

$$n = \sum_{i=1}^{I} m_i.$$

Under Model (0) with assumption 4′, Bias($\bar{y}_\alpha$) is still 0, but now

$$\begin{aligned}\mathrm{Var}(\bar{y}_\alpha) &= \frac{1}{n^2}\Bigg[\sum_i \sum_j \mathrm{Var}(y_{ij\alpha}) + 2\sum_i \sum_{j<j'} \mathrm{Cov}(y_{ij\alpha}, y_{ij'\alpha}) \\ &\quad + 2\sum_{i<i'} \sum_j \sum_{j'} \mathrm{Cov}(y_{ij\alpha}, y_{i'j'\alpha})\Bigg] \\ &= \left(1 - \frac{n}{N}\right)\frac{\sigma_\mu^2}{n} + \frac{\sigma_\varepsilon^2}{n} + \frac{1}{n^2}\left[\sum_{i=1}^{I} m_i(m_i - 1)\rho_w \sigma_\varepsilon^2\right]. \qquad (24.10)\end{aligned}$$

The latter term is called the *correlated component of response variance* of the sample mean. A parameter which is sometimes used to describe the magnitude of the correlated component is the intra-interviewer correlation coefficient,

$$\begin{aligned}\rho_y &= \frac{\mathrm{Cov}(y_{ij\alpha}, y_{ij'\alpha})}{\mathrm{Var}(y_{ij\alpha})} \\ &= \frac{\rho_w \sigma_\varepsilon^2}{\sigma_\mu^2 + \sigma_\varepsilon^2} = \rho_w(1 - R) \qquad (24.11)\end{aligned}$$

where R is the reliability defined in (24.9). Under conditions described in the next section, ρ_y can be estimated from survey data. These estimates are used to compare the severity of the interviewer effect across items or across surveys. When the interviewers' assignments are all of the same size, i.e., $m_i = m$ for all i, then, ignoring the finite population correction factor (fpc $= 1 - n/N$), (24.10) reduces to

$$\begin{aligned}\mathrm{Var}(\bar{y}_\alpha) &= \frac{1}{n}(\sigma_\mu^2 + \sigma_\varepsilon^2)[1 + (m - 1)\rho_y] \\ &= \frac{\mathrm{Var}(y_{ij\alpha})}{n}[1 + (m - 1)\rho_y]. \qquad (24.12)\end{aligned}$$

We can see that the presence of positively correlated errors inflates the variance of $\bar{y}_\alpha$ by the factor shown in the bracketed term of (24.12). When m is large, even weakly correlated measures can have a large impact on the precision of $\bar{y}_\alpha$. For example, suppose $R = 0.4$ and $\rho_w = 0.01$. Then by (24.11), $\rho_y = (0.01)(0.6) = 0.006$. For an interviewer assignment size of

$m = 100$, the increase in variance due to correlated interviewer errors is $99(0.006) = 0.59$ or 59%!

24.2.2 The Analysis of Variance (ANOVA) Model

In the ANOVA model, a specific form for the HHB measurement error term, $\varepsilon_{ij\alpha}$, is assumed. (From here on, we suppress the subscript α for notational convenience.) In particular, suppose that $\varepsilon_{ij} = b_i + e_{ij}$, where b_i is the systematic error due to the i-th interviewer and e_{ij} is a random error term which describes the difference between the observation y_{ij} and the term $\mu_{ij} + b_i$ for the j-th unit. Model (0) can now be written as

$$y_{ij} = \mu_{ij} + b_i + e_{ij}. \tag{24.13}$$

The assumptions for e_{ij} are the same as those made for ε_j in assumptions 2 through 5 of Model (0). That is, we assume that e_{ij} has conditional mean and variance 0 and σ_{ij}^2, respectively, where now the conditioning is on both the unit and the interviewer. Further, e_{ij} is uncorrelated with the other terms on the right side of (24.13), as well as with their counterparts for other units in the sample and in repeated trials. The b_i's may be assumed to be fixed or random, depending on the modeler's interest. If he or she is interested in a particular group of interviewers, for example the I interviewers who work the night shift in some telephone survey facility, then it is appropriate to consider b_i $(i = 1, \ldots, I)$ as fixed biases which are to be estimated and compared within the group. On the other hand, for one-time surveys and censuses one may not be interested in any particular set of interviewers but rather the effect of the process of interviewing. In such a case it is more appropriate to consider the set $\{b_1, \ldots, b_I\}$ as a random sample from an infinite population of interviewer effects. The random interviewer effects assumption is consistent with the HHB model discussed above and is often assumed in the literature. (See, for example, Chapter 23 and U.S. Bureau of the Census (1985).) Stokes and Yeh (1988) provide an analysis of fixed interviewer effects using empirical Bayes methods for identifying interviewers with aberrant b_i's.

For the random effects model, assume that $\{b_1, \ldots, b_I\}$ is a random sample from a population of interviewer effects having mean 0 and variance σ_b^2. Then $\text{Var}(y_{ij}) = \sigma_\mu^2 + \sigma_b^2 + \sigma_e^2$, where $\sigma_e^2 = \Sigma_i \Sigma_j \sigma_{ij}^2/n$ and

$$\begin{aligned}\text{Cov}(y_{ij}, y_{i'j'}) &= 0 \quad \text{if } i \neq i' \\ &= \sigma_b^2 \quad \text{if } i = i', j \neq j'.\end{aligned}$$

Assume for simplicity that $m_i = m$ for all i and $n = mI$. Since $E(\bar{y}) = \mu$ then $\text{Bias}(\bar{y}) = 0$. Further, if we can ignore the fpc, then

$$\mathrm{Var}(\bar{y}) = \frac{1}{n}(\sigma_\mu^2 + \sigma_b^2 + \sigma_e^2)[1 + (m - 1)\rho_y]$$

$$= \frac{\mathrm{Var}(y_{ij})}{n}[1 + (m - 1)\rho_y] \qquad (24.14)$$

which is identical to (24.12). But now ρ_y has the form

$$\rho_y = \frac{\sigma_b^2}{\sigma_\mu^2 + \sigma_b^2 + \sigma_e^2}.$$

Reliability is

$$R = \frac{\sigma_\mu^2}{\sigma_\mu^2 + \sigma_b^2 + \sigma_e^2}$$

which shows the effect of interviewer variability, σ_b^2, on reliability. Using (24.11), we can solve for the intra-interviewer correlation coefficient, provided that $R \neq 1$,

$$\rho_w = \frac{\rho_y}{1 - R}$$

$$= \frac{\sigma_b^2}{\sigma_b^2 + \sigma_e^2}$$

which is the proportion of measurement error variance that is due to interviewers. Finally we see that $(TV)^2 = R$ as before.

When interviewer effects are considered fixed, there is no change in variance as a result of interviewers. Rather the effect of interviewer error is seen in the bias as follows:

$$\mathrm{Bias}(\bar{y}) = \frac{1}{I}\sum_{i=1}^{I} b_i.$$

Likewise, R and TV are unchanged from their Model (0) forms.

If we write (24.13) as

$$y_{ij} = \mu + b_i + [(\mu_{ij} - \mu) + e_{ij}]$$

$$= \mu + b_i + e'_{ij} \qquad (24.15)$$

and add the assumptions $b_i \sim N(0, \sigma_b^2)$ (for b_i random) and $e_{ij} \sim N(0, \sigma_e^2)$, then the model would be recognizable as a one-way random effects ANOVA model. Estimation of the model parameters and their standard errors could then proceed as for the standard ANOVA model (see Kish, 1962). Note that the assumption that the combined errors, e'_{ij}, are i.i.d. random variables implies that the "sampling" deviations, $\mu_{ij} - \mu$, are also i.i.d. random variables. This assumption is satisfied if the n sample units are assigned completely at random to the I interviewers. This form of random assignment is referred to as the *interpenetration of assignments* (Mahalanobis, 1946). Interpenetration designs which restrict the randomization to subsets of interviewers or within subpopulations of units are available from the general ANOVA theory and are more feasible than full (unrestricted) interpenetration designs. Bailar (1983) provides a description of some interpenetration designs which are widely used. The model (24.15) can be readily extended to include effects for other types of personnel such as coders, editors, and supervisors. The literature on mixed ANOVA models provides a general estimation theory for such models (see Hartley and Rao, 1978).

24.2.3 Consequences of Interviewer Effects for Other Model (0) Assumptions

The introduction of within-item correlated error invalidates some other Model (0) assumptions. We have already seen that $E(\varepsilon_j|j)$ is not zero for either fixed or random interviewer effects since, in the notation of Section 24.2.2, $E(\varepsilon_{ij}|i) = b_i$. (Here, the expectation is taken over trials holding the interviewer fixed.) Indeed it may not be realistic to assume that $E(b_i) = 0$ (assumption 2); that is, that the biases of all interviewers cancel one another. For some survey questions, such as sensitive questions, the influence of interviewers on respondents is likely to be consistently in the same direction. In this case, $E(\varepsilon_{ij}) = B$, a constant, and (24.13) may be rewritten as

$$y_{ij} = \mu'_{ij} + b'_i + e_{ij}$$

where $\mu'_{ij} = \mu_{ij} + B$ and $b'_i = b_i - B$. The results of Section 24.2.2 now still apply if μ_{ij} and b_i are replaced everywhere by μ'_{ij} and b'_i, respectively. A further consequence of this modification is that $\text{Bias}(\bar{y}) = B$.

Another assumption which is questionable when interviewer effects enter the model is that $\text{Cov}(\mu_j, \varepsilon_{j'}) = 0$ for all j, j' which, as we noted in Section 24.1, is a consequence of assumption 2. In their model, Hansen, Hurwitz, and Bershad allowed for the possibility that μ_j and $\varepsilon_{j'}$ for two units j and j' may be correlated. They provide an example (p. 362 of their paper) where the units in the i-th interviewer's assignment may affect b_i so that b_i changes throughout the assignment. Consequently, interviewer i's "average" b_i may be correlated with the μ_{ij}, $j = 1, \ldots, m$. By incorporating this effect into the model, another variance component, called the *interaction variance* (Koch, 1973), is added to the variance in (24.10).

Another way of modeling this "interviewer–sample interaction" effect is to allow b_i to depend upon characteristics of the assignment, the data collection method, or the interviewer himself. Biemer and Stokes (1985) model such characteristics as fixed effects in the ANOVA model. They specify that $b_i \sim N(\beta_0 + \beta_1 x_{1i} + \ldots + \beta_p x_{pi}, \sigma_b^2)$ for $i = 1, \ldots, I$, where $\beta_0, \beta_1, \ldots, \beta_p$ are unknown regression coefficents and $x_1, \ldots, x_p$ are explanatory variables such as those mentioned above. The regression coefficients can be estimated by standard methods, if normality is assumed. This approach does not work when y_{ij} is categorical, since ANOVA assumptions are then not met. Methods that have been developed for this problem are discussed in Section 24.4.3.

24.3 EXTENSIONS TO COMPLEX SAMPLE DESIGNS

Several authors have addressed the problem of modeling measurement error when the sample design deviates from SRSWOR. See, for example, Koch (1973), Hartley and Rao (1978), Biemer and Stokes (1985), and Wolter (1985). Our approach here is similar to that of Koch (1973), and a variation of this method is used in Chapter 30, this volume.

Let U_j be an indicator variable for the j-th unit in the sample such that U_j is 1 if the unit is selected and 0 otherwise. Thus, $E(U_j) = \pi_j$, the probability that unit j is selected, and $E(U_j U_{j'}) = \pi_{jj'}$ is the joint probability of selecting both units j and j'. Consider estimators derived from the sample of the form

$$\hat{\theta} = \sum_{j=1}^{N} W_j U_j \mu_j \tag{24.16}$$

where the W_j's are fixed weights defined for every unit in the population. We further assume that $E(\hat{\theta}) = \theta$, the parameter to be estimated (i.e., $\hat{\theta}$ is unbiased). Let $E_1(\cdot)$ denote expectation over all possible samples selected by our sampling design and let $E_2(\cdot)$ denote expectation with respect to the distribution of ε_j. Define $\mathrm{Var}_1(\cdot)$ and $\mathrm{Var}_2(\cdot)$ analogously for the variance operator. We shall use the well-known decomposition of expectation and variance; viz., $E(\cdot) = E_1 E_2(\cdot)$ and $\mathrm{Var}(\cdot) = \mathrm{Var}_1 E_2(\cdot) + E_1 \mathrm{Var}_2(\cdot)$.

Consider the consequences of replacing μ_j by y_j in (24.16). Because the summation is over all units in the population, we must assume that a measurement error is attached to every characteristic, μ_i, in the population even though only n of these are observed. Then, let $\tilde{\theta}$ denote $\hat{\theta}$ with μ_j replaced by y_j; that is,

$$\tilde{\theta} = \sum_{j=1}^{N} W_j U_j y_j.$$

Under assumptions 2 through 5 of Model (0),

$$\begin{aligned} E(\tilde{\theta}) &= E_1 E_2(\tilde{\theta}) \\ &= E_1(\hat{\theta}) = \theta. \end{aligned}$$

Further,

$$\mathrm{Var}(\tilde{\theta}) = \mathrm{Var}_1 E_2(\tilde{\theta}) + E_1 \mathrm{Var}_2(\tilde{\theta}),$$

where the first term on the right is simply the variance due to the sampling design and the second term is the measurement error component of variance. In Wolter (1985, p. 383), the following results are shown:

$$\mathrm{Var}_1 E_2(\tilde{\theta}) = \sum_{j=1}^{N} W_j^2 \mu_j^2 \pi_j (1 - \pi_j) + 2 \sum\sum_{j<j'} W_j W_{j'} \mu_j \mu_{j'} (\pi_{jj'} - \pi_j \pi_{j'}) \tag{24.17}$$

and

$$E_1 \mathrm{Var}_2(\tilde{\theta}) = \sum_{j=1}^{N} W_j^2 \pi_j \sigma_j^2 + 2 \sum\sum_{j<j'} W_j W_{j'} \pi_{jj'} \sigma_{jj'} \tag{24.18}$$

where $\sigma_{jj'} = \mathrm{Cov}_2(\varepsilon_j, \varepsilon_{j'}) = E_2(\varepsilon_j \varepsilon_{j'}) - E_2(\varepsilon_j) E_2(\varepsilon_{j'})$. Hence, $\mathrm{Var}(\tilde{\theta})$ is the sum of (24.17) and (24.18), where, under Model (0), $\sigma_{jj'} = 0$. This form of $\mathrm{Var}(\tilde{\theta})$ is consistent with the results of Section 24.1 when $\tilde{\theta} = \bar{y}$ and SRSWOR is assumed. In that case,

$$\pi_j = \frac{n}{N} \text{ and } \pi_{jj'} = \frac{n(n-1)}{N(N-1)}$$

for all $j \neq j'$ and $W_j = 1/n$. Substitution into (24.17) and (24.18) with $\sigma_{jj'} = 0$ yields (24.6).

As a further illustration, consider single stage cluster sampling where the clusters are selected with probabilities π_j. Let y_j denote the total of the observations in the jth cluster and let μ_j denote the true total. Then, assuming $\sigma_{jj'} = 0$, consider the variance of

$$\hat{Y}_{HT} = \sum_{j=1}^{n} W_j y_j \text{ for } W_j = 1/\pi_j.$$

$\hat{Y}_{HT}$ is the Horvitz-Thompson estimator of the population total (Cochran, 1977). From (24.17) and (24.18) we have

$$\mathrm{Var}(\hat{Y}_{HT}) = \sum_{j=1}^{N} \frac{1 - \pi_j}{\pi_j} \mu_j^2 + 2 \sum\sum_{j<j'} \frac{\pi_{jj'} - \pi_j \pi_{j'}}{\pi_j \pi_{j'}} \mu_j \mu_{j'} + \sum_{j=1}^{N} \frac{\sigma_j^2}{\pi_j}. \tag{24.19}$$

Note that the contribution to the variance due to measurement error (the last term) is not equivalent to

$$\sigma_\varepsilon^2 = \frac{1}{N}\sum_{j=1}^{N} \sigma_j^2.$$

This shows that σ_ε^2, although essential for the psychometrical objective of reliability estimation, may be irrelevant for the sampling objective of estimator precision. For that purpose, it is important to note that those clusters having the smallest selection probabilities (usually the smallest units) may be the most damaging in terms of the precision of the survey estimates.

Now consider the estimation of the reliability R in cluster sampling. We shall illustrate that, in general, if the sample design is not taken into account appropriately in the estimation of σ_ε^2, biases may result and conclusions regarding the reliability and/or validity of a measurement may be incorrect. To illustrate, consider the usual estimator of σ_ε^2 for continuous survey data under SRSWOR, viz.

$$\hat{\sigma}^2 = \frac{1}{2n}\sum_{j=1}^{n} (y_{j1} - y_{j2})^2 \tag{24.20}$$

where (y_{j1}, y_{j2}) is a pair of independent and identically distributed measurements on unit j. Under Model (0), $E(\hat{\sigma}_\varepsilon^2) = \sigma_\varepsilon^2$. Under unequal probability sampling, however,

$$\begin{aligned} E(\hat{\sigma}_\varepsilon^2) &= E[\sum_{j=1}^{N} U_j(y_{j1} - y_{j2})^2/2n] \\ &= E_1[\sum_{j=1}^{N} U_j\, \sigma_j^2/n] \\ &= \sum_{j=1}^{N} \pi_j\, \sigma_j^2/n \end{aligned} \tag{24.21}$$

which is not σ_ε^2 unless $\pi_j = n/N$. It can be seen from this result that σ_ε^2 disproportionately weights the σ_j^2 toward those units having the largest probabilities of selection. In general, $\hat{\sigma}_\varepsilon^2$ is not relevant as an indicator of reliability unless either (a) the sample design is self-weighting or nearly so, or (b) the σ_j^2 are approximately equal.

24.4 MODELING CATEGORICAL DATA

24.4.1 General Results

Under Model (0), we assume $E(\varepsilon_j|j) = 0$ (assumption 2); that is, that the mean of each individual's hypothetical error distribution is zero. For categorical responses, this assumption leads to an inconsistency in the model assumptions. To see this, we shall consider dichotomous (or binary) responses since categorical data are usually reduced to this form. Suppose every unit in the population belongs to one of two possible classes corresponding to $\mu_j = 1$ or 0, where $\mu_j = 1$ indicates the category of interest. The error ε_j can then take on only three values $\{0, 1, -1\}$ as follows:

$$\text{For } \mu_j = 1,\ \varepsilon_j = 0 \text{ when } y_j = 1 \text{ and } \varepsilon_j = -1 \text{ when } y_j = 0;$$

and

$$\text{for } \mu_j = 0,\ \varepsilon_j = 0 \text{ when } y_j = 0 \text{ and } \varepsilon_j = 1 \text{ when } y_j = 1.$$

Define $\theta_j = P(\varepsilon_j = -1|\mu_j = 1)$ to be the probability that unit j belonging to class 1 is classified as in class 0, and define $\phi_j = P(\varepsilon_j = 1|\mu_j = 0)$ to be the probability that a unit belonging to class 0 is incorrectly classified as belonging to class 1. To replace assumption 2 in Model (0) we now assume:

$$2'.\ E(\varepsilon_j|j) = -\mu_j\,\theta_j + (1 - \mu_j)\,\phi_j \text{ for all } j.$$

Thus $E(\varepsilon_j|j)$ is not zero unless there is no misclassification error (i.e., $\theta = \phi = 0$). We shall denote Model (0) after replacing assumption 2 by 2′ by Model (0′) and consider alternative reparameterizations of this model.

If we reparameterize the model in terms of the mean of the propensity distribution (as in classical true score theory) we shall see that the Model (0) assumptions will hold for the new error term. Let $\mu_j^* = E(y_j|j)$, the mean of the propensity distribution for unit j. Then

$$\mu_j^* = \mu_j(1 - \theta_j) + (1 - \mu_j)\,\phi_j$$

for all j, and (24.1) may be rewritten as

$$y_j = \mu_j^* + \varepsilon_j^* \tag{24.22}$$

where $\varepsilon_j^* = \mu_j + \varepsilon_j - \mu_j^*$. Now, $E(\varepsilon_j^*|j) = \text{Cov}(\varepsilon_j^*, \mu_j) = \text{Cov}(\varepsilon_j^*, \varepsilon_{j'}^*) = 0$ and there are no further inconsistencies with the Model (0) assumptions. We shall now consider the structure of the error term in (24.22).

Note that ε_j^* can assume four values as shown in Table 24.1. For example, if $\mu_j = 1$, then $\mu_j^* = 1 - \theta_j$ and hence $\varepsilon_j^* = 1 + \varepsilon_j - (1 - \theta_j) = \varepsilon_j + \theta_j$. Thus ε_j^* is either θ_j or $-1 + \theta_j$ with probabilities $1 - \theta_j$ and θ_j, respectively. The normality assumption usually made for the error term in (24.1) (recall from Section 24.2.2) is obvously not valid for ε_j^*. This has led to the use of discrete distributions to model ε_j^*. Some of these will be discussed later.

We now consider the form of the quantities Bias($\bar{y}$), Var($\bar{y}$), R, and TV under Model (0′) with the reparameterization in (24.22). Define the following quantities: $\theta = E(\theta_j|\mu_j = 1)$, $\sigma_\theta^2 = \text{Var}(\theta_j|\mu_j = 1)$, $\phi = E(\theta_j|\mu_j = 0)$, $\sigma_\phi^2 = \text{Var}(\phi_j|\mu_j = 0)$, and $\mu^* = E(\mu_j^*) = E(\bar{y})$. Then it can be shown that

$$\text{Bias}(\bar{y}) = -\mu\theta + (1 - \mu)\phi \tag{24.23}$$

Table 24.1. Conditional Distribution of ε_j^*

ε_j^*	$P(\varepsilon_j^* \mid \mu_j)$
θ_j	$(1-\theta_j)\mu_j$
$-(1-\theta_j)$	$\theta_j\mu_j$
$-\phi_j$	$(1-\phi_j)(1-\mu_j)$
$(1-\phi_j)$	$\phi_j(1-\mu_j)$

$$\text{Var}(\bar{y}) = \left(\frac{N-n}{N-1}\right)\frac{\mu^*(1-\mu^*)}{n} \tag{24.24}$$

where $\mu^* = \mu + \text{Bias}(\bar{y}) = \mu(1-\theta) + (1-\mu)\phi$. An alternative expression for $\text{Var}(\bar{y})$ can be obtained by decomposing $\text{Var}(y_j)$ as $\text{Var}E(y_j|j) + E\text{Var}(y_j|j)$. The latter term in this sum is referred to as the *simple response variance* (*SRV*). (See, for example, Hansen, Hurwitz, and Pritzker, 1964.) It is the average of the variances of the individual propensity distributions in the population. The remaining term, $\text{Var}E(y_j|j)$, is referred to as the *sampling variance* (*SV*). We may now rewrite (24.24) as

$$\text{Var}(\bar{y}) = \left(\frac{N-n}{N-1}\right)\frac{SV}{n} + \frac{SRV}{n} \tag{24.25}$$

where

$$SV = \mu(1-\mu)(1-\theta-\phi)^2 + v_{\theta,\phi} \tag{24.26}$$

$$SRV = \mu\theta(1-\theta) + (1-\mu)\phi(1-\phi) - v_{\theta,\phi} \tag{24.27}$$

and $v_{\theta,\phi} = \mu\sigma_\theta^2 + (1-\mu)\sigma_\theta^2$. (See Biemer and Forsman, 1990 for the derivations.)

The reliability R and the theoretical validity TV now take the form

$$R = \frac{\text{Var}(\mu_j^*)}{\text{Var}(y_j)} = \frac{SV}{\mu^*(1-\mu^*)} \tag{24.28}$$

and

$$TV = \frac{\mu(-\mu)(1-\theta-\phi)}{[\mu(1-\mu)\mu^*(1-\mu^*)]^{1/2}}.$$

From these results, the following observations can be made:

i. $\bar{y}$ is unbiased for μ if and only if $\mu\theta = (1-\mu)\phi$; that is, if the expected number of false negative misclassifications in the sample is exactly equal to the expected number of false positive misclassifications.

ii. $\text{Var}(\bar{y})$ and $\text{MSE}(\bar{y})$ can actually be less than $\text{Var}(\bar{\mu})$ and $\text{MSE}(\bar{\mu})$, respectively. This can occur when $\mu^* < \mu < 0.5$ or $\mu^* > \mu > 0.5$.

iii. SRV, R, and TV all depend upon μ.
iv. $(TV)^2$ will not in general be R. However, if $v_{\theta,\phi} = 0$ then the numerator of $(TV)^2$ is $\mu(1 - \mu)SV$ and $(TV)^2 = R$.

Because of (i), it is unlikely that $\bar{y}$ will be unbiased for μ when there is misclassification error. However, if θ and ϕ are small, the bias may be negligible. A surprising result is that the MSE of the sample mean may actually be smaller in the presence of misclassification than in the absence of misclassification (ii above). It is easily shown that for $n = 1$, this occurs when $\text{Bias}(\bar{y})(1 - 2\mu) < 0$. We shall next illustrate the importance of these results in the estimation of measurement error components.

The simple response variance can be estimated in reinterview studies in which a sample of households is revisited and a second interview is conducted under conditions which are made as similar as possible to those of the first interview. The usual estimator of SRV is

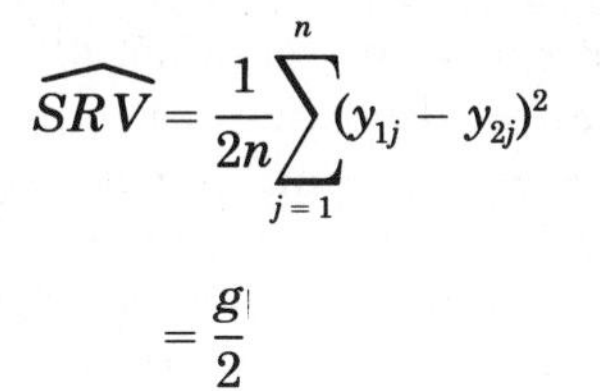

$$\widehat{SRV} = \frac{1}{2n}\sum_{j=1}^{n}(y_{1j} - y_{2j})^2$$

$$= \frac{g}{2}$$

where g is the gross difference rate, which was seen in (24.20). In a two-by-two table classifying each sample unit by its dichotomous interview and reinterview responses, g is simply the proportion of units in the off-diagonal cells; i.e., the proportion of disagreements in the sample. SRV, which is unbiased for SRV under Model (0′), is then used to estimate R or $1 - R$, the index of inconsistency. It is a common, but erroneous, practice to interpret SRV as an indicator of data quality, ignoring the influence of μ on the magnitude of SRV. For example, for a particular item, SRV may be estimated for different subpopulations and compared. Even if the error parameters θ, ϕ, and $v_{\theta,\phi}$ were identical across domains, but μ differed considerably, one would be led to erroneously conclude that the quality of the data from one subgroup exceeds that of the other.

To illustrate, in Table 24.2 we have computed SRV, $\text{Bias}(\bar{y})$, $\text{MSE}(\bar{y})/\text{MSE}(\mu)$ and R for the case in which $\theta = 0.05$, $\phi = 0.15$, $v_{\theta,\phi} = 0$, $n = 1$, and $0.1 \leq \mu \leq 0.9$. In this example, R ranges from 0.325 to 0.659 and, from (iv), $(TV)^2 = R$. $\text{Bias}(\bar{y})$ decreases to 0 as μ achieves the value $\mu = \phi/(\theta + \phi) = 0.75$ and $\text{MSE}(\bar{y})$ is less than $\text{MSE}(\bar{\mu})$ for $0.50 \leq \mu \leq 0.75$. Suppose that for a pretest of a survey item, a sample is drawn from a population with $\mu = 0.1$ while in the target population for the survey, $\mu = 0.4$. For the error rates of Table 24.2, reliability would be underestimated by roughly 50% in the pretest.

Table 24.2. Measurement Error Components as a Function of μ for $\theta = 0.05$ and $\phi = 0.15$

μ	SRV	Bias($\bar{y}$)	$\frac{\text{MSE}(\bar{y})}{\text{MSE}(\mu)}$	R or $(TV)^2$
0.1	0.120	0.13	2.16	0.33
0.2	0.112	0.11	1.41	0.48
0.3	0.104	0.09	1.17	0.57
0.4	0.096	0.07	1.06	0.62
0.5	0.088	0.05	1.00	0.65
0.6	0.080	0.03	0.975	0.66
0.7	0.072	0.01	0.981	0.65
0.8	0.064	−0.01	1.04	0.62
0.9	0.056	−0.03	1.27	0.51

Data analysis by ANOVA methods are not appropriate here because of the non-normality of the error ε_j^*. Some alternative approaches are discussed in the next section. Rao and Thomas (Chapter 31) consider the effects of measurement error on contingency table analysis and propose adjustments for the standard χ^2 test statistics. Their measurement error model is a generalization of Model (0′) to K categories.

24.4.3 Modeling Interviewer Effects with Dichotomous Data

Several authors have considered models designed specifically for describing interviewer effects for categorical items. For example, Stokes and Mulry (1987), Stokes and Hill (1985), and Pannekoek (1988) all consider models of the following form.

Rewrite ε_j^* in (24.22) as $\varepsilon_{ij}^* = b_i + e_{ij}$ where, as in Section 24.2.2, b_i is the i-th interviewer's systematic effect and $e_{ij} = y_{ij} - \mu_{ij}^* - b_i$ for the j-th respondent in the i-th interviewer's assignment. Then (24.22) becomes

$$y_{ij} = \mu_{ij}^* + b_i + e_{ij} \tag{24.29}$$

and, as in (24.13), $b_1, \ldots, b_I$ may be assumed to be either fixed constants or random variables, depending upon the study objectives. We can now rewrite the model to resemble the ANOVA model of Section 24.2.2 as follows:

$$y_{ij} = \mu^* + b_i + e'_{ij} \tag{24.30}$$

where $e'_{ij} = e_{ij} + \mu_{ij}^* - \mu^*$. For interpenetrated assignments, $E(e'_{ij}|i) = 0$, where $E(\cdot\,|i)$ denotes expectation over all possible random assignments of m_i respondents to interviewer i as well as the individual error distributions of the m_i respondents.

Let $p_i = P(y_{ij} = 1 | i)$, the probability the i-th interviewer records a one for any unit in his/her assignment. Note that $p_i = \mu^* + b_i$. Since y_{ij} is limited to either 0 or 1, e'_{ij} is restricted to $-p_i$ or $q_i = 1 - p_i$ with conditional probabilities q_i and p_i, respectively. Thus,

$$\begin{aligned}\text{Var}(e'_{ij}) &= E\text{Var}(e'_{ij}|i) + \text{Var}E(e'_{ij}|i) = E(p_i q_i) \\ &= \mu^*(1 - \mu^*) + \sigma_b^2 \qquad 24.31\end{aligned}$$

where $\sigma_b^2 = \text{Var}(b_i) = \text{Var}(p_i)$. Similarly it can be shown that $\text{Cov}(e'_{ij}, b_i) = \text{Cov}(e'_{ij}, e'_{i'j}) = E(b_i) = 0$ so that (24.30) is of the same form as the ANOVA model of Section 24.2.2 (that is, its components have the same first and second moments). However, it is not valid to assume normality for either the e'_{ij} or the b_i.

From Section 24.2.2, we have

$$\text{Var}(\bar{y}) = \left(\frac{N - n}{N - 1}\right)\frac{\mu^*(1 - \mu^*)}{n} + \frac{\mu^*(1 - \mu^*)\rho_y}{n^2}\sum_{i=1}^{I} m_i(m_i - 1) \qquad (24.32)$$

where

$$\rho_y = \frac{\sigma_b^2}{\mu^*(1 - \mu^*)}.$$

As in the previous section, $\mu^*(1 - \mu^*)$ may be further decomposed into SV and SRV components (see Biemer and Forsman, 1990). If $m_i = m$ for all i and the fpc can be ignored, then

$$\text{Var}(\bar{y}) = \frac{\mu^*(1 - \mu^*)}{n}[1 + (m - 1)\rho_y] \qquad (24.33)$$

which is a special case of (24.14). ρ_y depends upon μ, just as SRV does, which makes its interpretation difficult for dichotomous data. In U.S. Bureau of the Census (1985), the effect of μ on ρ_y is illustrated in a table similar to Table 24.2.

The analysis of variance estimator for σ_b^2 is unbiased for the model just considered (Stokes and Mulry, 1987). However, the non-normality of p_i, which we noted above, means that standard methods of interval estimation and testing are not valid. Stokes and Hill (1985) and Pannekoek (1988) proceed by assuming a beta distribution (having density $f(x) = B(\alpha,\beta)x^{\alpha-1}(1 - x)^{\beta-1}$ on $0 \leq x \leq 1$, where $B(\alpha,\beta)$ is the beta function) for p_i. Both α and β can be estimated, as can the standard errors of the estimates, using standard maximum likelihood methods. Since

$\sigma_b^2 = \text{Var}(p_i) = \mu^*(1 - \mu^*)(\alpha + \beta + 1)^{-1}$ (see, for example, Johnson and Kotz (1970), p. 40) and $\rho_y = \text{Var}(p_i)/\text{Var}(y_i) = (\alpha + \beta + 1)^{-1}$, these parameters can be estimated as well.

24.4.3 Models for Comparing Subpopulations

As we discussed in Section 24.2.3, comparisons of R and ρ_y among subpopulations is made difficult because of the dependence of these measures upon μ, the prevalence of the characteristic of interest. Anderson and Aitkin (1985) and Stokes (1988) considered this problem and proposed a new model for interviewer effects in these cases. Their model no longer assumes additive errors in the observed values, but instead assumes additivity in a latent, or unseen variable, which controls the observed value. We define y_{tij} to be either one or zero according to whether or not the j-th unit in the i-th interviewer's assignment in subpopulation t ($t = 1,\ldots,T$) is in a specified category or not. Then let z_{tij} denote a continuous unobservable variable such that

$$P(y_{tij} = 1|i) = P(z_{tij} > c_{tij}|i) \tag{24.34}$$

where c_{tij}, a constant for unit tij, is the threshold value for the category of interest. Then an ANOVA-like model is assumed for z_{tij}, viz.,

$$z_{tij}|i \sim G(\mu_t + a_i, \sigma^2) \tag{24.35}$$

where $\mu_t = \gamma_0 + \gamma_1 x_1 + \ldots + \gamma_p x_p$, G is a specified distribution (such as normal or logistic), σ is a measure of its variability, $\gamma_0,\ldots,\gamma_p$ are unknown regression coefficients and $x_1,\ldots,x_p$ are explanatory variables, and $a_i \sim N(0, \sigma_a^2)$ is the effect of the ith interviewer on the mean of the unobservable variable z_{tij}. Maximum likelihood estimators of $\gamma_0,\ldots,\gamma_p$ and σ_a^2 can be obtained. Anderson and Aitkin describe estimation for this model when the explanatory variables are covariates at the interviewer level, while Stokes describes its estimation for unit level covariates.

σ_a^2 and $\rho^* = \sigma_a^2/\text{Var}(z_{tij})$ are measures of the variability of interviewers, just as σ_b^2 and ρ_y are. However, ρ^* does not directly reflect the increase in $\text{Var}(\bar{y})$, which is due to interviewer variability, in the way that ρ_y did (in (24.13)) for the model considered in Section 24.2.2. Rather, σ_a^2 and ρ^* are direct measures of the increase in variance in the unobservable variable, z_{tij}. The reason ρ^* is of interest to us is that, under this model, it is not a function of μ_t, so it can be used for comparing the magnitude of interviewer effects across subpopulations.

Pannekoek (Chapter 25, this volume) considers a different approach to modeling the effect of interviewers on categorical variables. His goal was to make comparisons of the magnitude of interviewer effects across subpopulations which were defined at the interviewer level.

24.5 MODELS FOR MEMORY ERRORS

Memory (or recall) errors occur when the items in an inquiry relate to events that happened in the past and the respondents either fail to remember the events (*recall loss*) or place them in the wrong time periods (*telescoping*) or both. For example, in a health study, a respondent may forget a visit to the doctor or he or she may report the visit but remember that it occurred earlier or later than it really did. Memory errors may be a function of the amount of time between the inquiry and the time the event occurred. They can be a particular problem in panel surveys where the same events may be erroneously reported in more than one time period. This section describes some methods for modeling these types of errors.

In this book, memory effects are analyzed for two kinds of data: (a) *count data* which relates to the total number of times an event has occurred and (b) *duration data* which relates to the length of time between two events. Models which are appropriate for (a) may not be for (b) and vice-versa. We shall begin our study with a review of models for count data.

For count data, models have been proposed to describe the attenuation in an estimator, $\hat{Y}$, of the total number of events of some type that has occurred in a population. Suppose the sample of n units can be divided into T subsamples corresponding to T different *recall periods*, or the length of time between the period of reference and the time of the interview. For example, an interview may be conducted in July and reference the period of January through June. We may be interested in $T = 6$ recall periods corresponding to the six months of the reference period. Let $\hat{Y}_{kt}$ denote the estimator of the total for the k-th time period with a recall period of t. For example, in a panel survey, an estimator of the total for month k may be available from T panels, each panel having a different recall period so that the estimates $\hat{Y}_{k1}, \ldots, \hat{Y}_{kT}$ may be calculated for month k. The following model for the effect of memory errors on $\hat{Y}_{kt}$ has been proposed by Sudman and Bradburn (1973):

$$E(\hat{Y}_{kt}) = N_k \mu_k p(t) \qquad (24.36)$$

where

$$p(t) = \exp(-bt).$$

Here N_k is the population size, μ_k is the population mean at month k, and b (> 0) is a parameter reflecting the rate of decrease of the observed values within the reference period. $N_k\mu_k$ is the total number of events in the population in month k and $p(t)$ may be interpreted as the probability that an event is reported given that it has occurred. Tacitly assumed in the model are the following:

i. Events that are reported are reported for the correct time periods; that is, there is no telescoping or misplacement of events.

ii. The bias in $\hat{Y}_{kt}$ is zero at $t = 0$; that is, there is no other source of response bias.

In an attempt to satisfy (i), survey practitioners use *bounding interviews* in panel surveys, which are initial interviews conducted for the purpose of fixing a reference point for future interviews. While bounding interviews alleviate telescoping from outside to inside the reference period (or *external telescoping*), they cannot reduce *internal telescoping*, or the movement of events to incorrect time periods within a reference period. Telescoping of events is discussed further in Neter and Waksberg (1964). Assumption (ii) is an oversimplification, since other measurement errors can create a bias in $\hat{Y}_{k0}$. Nevertheless, if b were known, an improved estimator of $N_k\mu_k$ could be obtained by adjusting $\hat{Y}_{kt}$ for the bias, provided (24.36) holds.

Various generalizations of (24.36) have been proposed in the literature. Som (1973) specified $p(t) = \exp(-bt^2)$ and found that the model fit well to fertility data. Bailar and Biemer (1984) expressed $E(\hat{Y}_{kt})$ as $N_k\mu_{k1}p(t) + N_k\mu_{k2}$ where $\mu_k = \mu_{k1} + \mu_{k2}$. $N_k\mu_{k1}$ is the number of events which are subject to recall loss or the "forgettable events" and $N_k\mu_{k2}$ is the number of "unforgettable events," or the events reported regardless of the length of the recall period. They found that this model provided an improved fit over (24.36) for data from the U.S. National Crime Survey. They estimated the ratio of forgettable to total events by $r = N_k\hat{\mu}_{k1}/N_k\hat{\mu}_k$ and found that r ranged from 0.79 for assault to 0.58 for motor vehicle thefts, suggesting that a smaller proportion of motor vehicle thefts than assaults are subject to recall loss.

In the survey literature, models for memory error have focused almost exclusively on recall bias (see Groves, 1989, p. 428). However, using a variation of the model proposed by van Dosselaar (Chapter 26), we can model the effect of memory error on the variance — in particular, on reliability. Let $\mu_j(t)$ denote the true number of events experienced by the j-th sample unit t time periods before the interview. Let $y_j(t)$ denote the corresponding number of reported events for unit j. Then

$$y_j = \sum_{t=1}^{T} y_j(t)$$

is the number of events reported for the entire reference period and μ_j is the true number of events. Further assume that

$$y_j(t) = \mu_j(t)p_j(t) + \varepsilon_j(t) \tag{24.37}$$

where $p_j(t)$ is the probability that events occurring t time units from the time of the interview are reported and $\varepsilon_j(t)$ is a random error term uncorrelated with $\mu_j(t)p_j(t)$ and with zero mean. Thus we may rewrite (24.37) in terms of Model (0) as

$$y_j = \mu_j' + \varepsilon_j \tag{24.38}$$

where

$$\mu_j' = \sum_{t=1}^{T} \mu_j(t)p_j(t) \text{ and } \varepsilon_j = \sum_{t=1}^{T} \varepsilon_j(t).$$

The assumption $E(\varepsilon_j|j) = 0$ implies that, as in the previous model of recall error, there is no telescoping of events. Finally, we assume that $y_j(t)$ given $\mu_j(t)$ has a binomial distribution with parameters $\mu_j(t)$ and $p_j(t)$ and that $y_j(t)$ is uncorrelated with $y_j(t')$ for $t \neq t'$. Therefore

$$E[y_j(t)|\mu_j(t)] = \mu_j(t)p_j(t) \tag{24.39}$$

and

$$\text{Var}[y_j(t)|\mu_j(t)] = \mu_j(t)p_j(t)[1 - p_j(t)]. \tag{24.40}$$

In (24.28) we saw that reliability could be expressed as the ratio of sampling variance of the true scores to total variance. Thus the reliability of y_j is

$$R = \frac{\text{Var}E(y_j|j)}{\text{Var}(y_j)}$$

$$= 1 - \frac{E\text{Var}(y_j|j)}{\text{Var}(y_j)}. \tag{24.41}$$

From (24.40) and the independence of $y_j(t)$ and $y_j(t')$, we have

$$\text{Var}(y_j|j) = \sum_{t=1}^{T} \mu_j(t)p_j(t)[1 - p_j(t)]$$

and hence,

$$E\text{Var}(y_j|j) = \sum_{j=1}^{N} \sum_{t=1}^{T} \mu_j(t)\, p_j(t)[1 - p_j(t)]/N$$

$$= \sum_{t=1}^{T} \mu(t)\, p(t)[1 - p(t)]$$

if $p_j(t) = p(t)$ for all j. So R is dependent upon $p(t)$, $t = 1, \ldots, T$, the probability that events occurring t time units before the interview are reported. R is maximized when $p(t) = 1$ or 0 and is minimized when $p(t) = 0.5$. This clearly illustrates that R can be increased by shortening the recall period since $p(t)$ will increase as t is decreased.

Finally, the expected value of

$$\hat{Y}_t = N \sum_{j=1}^{n} y_j(t)/n,$$

the estimator of the total number of events occurring t time periods from the interview, is

$$E(\hat{Y}_t) = \sum_{j=1}^{N} \mu_j(t) p_j(t). \tag{24.42}$$

When $p_j(t) = p(t)$ for all j, we have

$$E(\hat{Y}_t) = p(t)\mu(t),$$

where

$$\mu(t) = \sum_{j=1}^{N} \mu_j(t).$$

Note that (24.42) is equivalent to (24.36) only if every respondent in the population has the same rate of recall loss.

Holt, McDonald, and Skinner (Chapter 32) model the measurement error in duration data in order to evaluate the impact of measurement error on event history analyses. They assume a multiplicative model for the reported duration of a state y_j, for example, an unemployment spell. Their model is

$$y_j = \mu_j(1 + \varepsilon_j) \tag{24.43}$$

where μ_j is the true duration and ε_j is a random error with zero mean and variance σ_j^2. It is assumed that ε_j is uncorrelated with μ_j as well as with other errors, $\varepsilon_{j'}$, $j \neq j'$. Their model is the analogy to Model (0) for multiplicative error structures. Note that (24.43) does not explicitly allow for time dependent errors as in (24.38). Holt, et al. model the time dependency by assuming that $\sigma_j^2 = t\sigma_{0j}^2$ where t is the time between the interview and the end of the duration and σ_{0j}^2 is a constant. Other forms of σ_j^2 are also considered in their chapter for modeling "covariate-related errors" and "duration-related errors."

24.6 STRUCTURAL EQUATION MODELS

In this section we attempt to provide notational and terminological links between the sampler's approach and the psychometrician's approach to measurement error modeling. It is not our aim to provide a summary of the field of structural equation modeling since such a summary can be found in Chapter 28. Bohrnstedt (1983) and Groves (1989, Chapter 1) also provide reviews of the psychometric approach to measurement error modeling. An application of the methodology is provided in Chapter 29. Rather, it is our hope that this section will aid the reader in recognizing the connections between the two modeling perspectives.

Structural equation methods have their roots in path analysis which was developed by Wright (1934). Path analysis uses path diagrams to represent the relationships among variables and errors. For example, suppose we are considering three alternative methods for assessing the value of μ_j, all of which are subject to measurement error. Suppose that each uses a different scale; say, a 10-point scale, a 20-point scale, and a 100-point scale. This situation may be represented by equations of the form

$$y_{j\alpha} = \lambda_\alpha \mu_j + \varepsilon_{j\alpha}. \tag{24.44}$$

If we arbitrarily choose a 10-point scale for our example, $\lambda_1 = 1$, $\lambda_2 = 2$, and $\lambda_3 = 10$. Figure 24.1 shows a path diagram for this example. In path diagrams, straight single-headed arrows represent a one-way causal influence from the variable at the base to the variable at the arrowhead. The variable coefficients are placed above the arrow.

In the field of psychometric research, measurements are rarely Platonic and variations and extensions of the classical true score model are usually assumed. Recall that this is Model (0) under the interpretation that μ_j is the mean of the propensity distribution for individual j. We can use equation (24.44) to define the three types of classical measures most often encountered in test theory: *parallel measures*, *tau-equivalent measures*, and *congeneric measures*. Parallel measures are those that have the same scale of measurement and the same reliability; that is, those for which $\lambda_\alpha = 1$ and $\text{Var}(\varepsilon_{j\alpha}) = \sigma_j^2$. Model (0) specifies that all measures are parallel. Measures on the same scale that have different reliabilities are called tau-equivalent; that is, $\lambda_\alpha = 1$ and $\sigma_{j\alpha}^2 \neq \sigma_{j\alpha'}^2$ for $\alpha \neq \alpha'$. Finally, measurements of μ_j on different scales with different reliabilities are congeneric measures. Interview-reinterview studies for which some questions in the original interview are reasked in the same way in the reinterview produce parallel measures if the survey conditions in the two interviews are identical and independent. In other studies, the same question is asked twice with two alternative measurement scales,

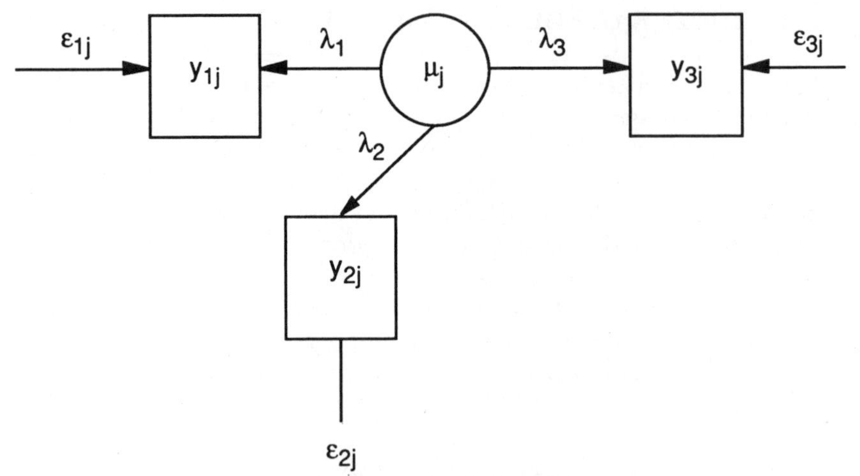

Figure 24.1. Example of a Path Diagram.

say a 10-point scale and a 100-point scale. For these studies, the congeneric measurements assumption may be more appropriate.

When two parallel or tau-equivalent measures are available, the reliability (R_α) of either measure may be estimated by their correlation. For congeneric measures, consistent estimates of R_α are obtainable if at least three measures are available. In this case, we say that R_α is identifiable under the congeneric model. Unfortunately, multiple measures rarely satisfy the assumptions of these simple models, including the less restrictive congeneric model. Reinterview studies have been used to assess the reliability of factual survey items (i.e., Platonic true scores). As discussed in Chapters 17 and 27, the effects of recall, recall loss, a change in respondent, and other realities invalidate a number of the assumptions made for the classical true score model. For subjective items, such as attitudes and opinions, retrospective reinterviews are inappropriate and researchers have resorted to repeating some version of the question within the same interview.

Alwin (1989) discusses some of the problems of obtaining multiple measures in a single cross-sectional survey and reviews some of the extensions of the classical true score model that have been proposed for coping with these problems. For example, if the question is altered each time it is repeated in order to disguise its redundancy, more error terms may be needed in the classical model to reflect the differences in the method of asking the question. The *multi-trait, multi-method model* of Werts and Linn (1970) and Jöreskog (1971) extends this idea to model the relationships between multiple measures of the same trait by different methods and of multiple traits by the same method in cross-sectional survey design. (See Chapter 28 for the specification of the model.)

Despite these important modeling innovations for the estimation of reliability in cross-sectional surveys, problems in the validity of the assumptions remain. Alwin (1989) discusses the advantages of using data from panel surveys. In a panel survey, a single variable can be measured at different points in time; thus the necessity for repeating a question in the same interview to measure reliability can be circumvented. Heise (1969) and Wiley and Wiley (1970) have introduced models, referred to as *simplex models*, which attempt to extract reliability information from the multiple measurements in panel surveys. Simplex models consist of two "submodels": one which describes the effect of time on the variable of interest and another which describes the effect of measurement error on the measurement of the variable at each timepoint. For a more detailed description of this approach, as well as additional references, the reader is referred to Chapter 28 and Alwin (1989).

All of the aforementioned models are special cases of a general class of models referred to as *structural equation models*. Bollen (1989, p. 319) describes structural equation models as a "synthesis" of regression and econometric modeling and factor analysis: "[The structural equation model] consists of a *measurement model* that specifies the relation of observed to latent variables and a *latent variable model* that shows the influence of latent variables on each other."

We can rewrite (24.44) in the structural equation modeling form as follows: Suppose that for each sample unit j ($= 1, \ldots, n$) we observe k possibly congeneric measures of the characteristic of interest. Since each measure may be translated and/or scaled differently, we write

$$\tau_{j\alpha} = \gamma_\alpha + \lambda_\alpha \mu_j \tag{24.45}$$

and

$$y_{j\alpha} = \tau_{j\alpha} + \varepsilon_{j\alpha} \tag{24.46}$$

for $\alpha = 1, \ldots, k$ and $j = 1, \ldots, n$. $\tau_{j\alpha} = E(y_{j\alpha} | j)$ or the *true score* for the α-th measure. Equation (24.46) may be further generalized, for instance, by adding a random component to $\tau_{j\alpha}$ referred to as the *specific component*. This component consists of the method effect (if any) and an interaction effect of the trait (or characteristic) and the method. Our formulation in (24.45), however, is adequate for the present purpose.

The model in (24.46) is typically expressed in standardized form in structural equation modeling literature. Let $y'_{j\alpha} = (y_{j\alpha} - \bar{y}_\alpha)/s_\alpha$ where

$$s_\alpha^2 = \sum_{j=1}^{n} (y_{j\alpha} - \bar{y}_\alpha)^2/(n-1).$$

Then (24.46) may be rewritten as

$$y'_{j\alpha} = \lambda_\alpha^s \tau'_{j\alpha} + \varepsilon'_{j\alpha} \tag{24.47}$$

where $\tau'_{j\alpha} = (\tau_{j\alpha} - \bar{\tau}_\alpha)/[\text{Var}(\tau'_{j\alpha})]^{1/2}$, $\varepsilon'_{j\alpha} = (\varepsilon_{j\alpha} - \bar{\tau}_\alpha)/s_\alpha$ and

$$\lambda^s_\alpha = \left[\frac{\text{Var}(\tau'_{j\alpha})}{s^2_\alpha}\right]^{1/2}. \quad (24.48)$$

Note that $\text{Var}(\tau'_{j\alpha}) = \text{Var}(\tau_{j\alpha})\left(\frac{n-1}{n}\right)$ (for negligible fpc) and

$$E(s^2_\alpha) = \text{Var}(y_{j\alpha}).$$

Thus the square of an estimate of the standardized *reliability coefficient*, λ^s_α, provides an estimate of R_α when

$$\frac{n-1}{n} \approx 1.$$

Saris and Andrews (Chapter 28) consider more complex forms for $\tau_{j\alpha}$ and $\varepsilon_{j\alpha}$ and provide a discussion of the various research designs which have been proposed for estimating their components.

Structural equation models are particularly useful for regression analysis when there are measurement errors in the dependent and/or the independent variables; that is, so-called *errors-in-variables* models. As an example, suppose we wish to study the relationship between two variables, say τ_1 and τ_2, which we express in mean deviation form as

$$\tau_{1j} = \beta\tau_{2j} + e_j \quad (24.49)$$

where β is a regression coefficient. (We have suppressed the prime used earlier to denote deviations from the mean.) We observe two measures of each of τ_1 and τ_2 as follows:

$$y_{1j\alpha} = \tau_{1\alpha} + \varepsilon_{1j\alpha}$$

and (24.50)

$$y_{2j\alpha} = \tau_{2\alpha} + \varepsilon_{2j\alpha}$$

for $\alpha = 1, 2$. Equations (24.49) and (24.50) are special cases of a general structural equation model given by

$$\boldsymbol{\tau} = \mathbf{B}t + \boldsymbol{\varepsilon}$$

and (24.51)

$$\mathbf{y}_l = \boldsymbol{\tau}_l + \boldsymbol{\varepsilon}_l, \; l = 1, \ldots, m$$

which is defined for each unit j in the sample. (Here we have suppressed the subscript j on the vectors for convenience.) In (24.51) $\boldsymbol{\tau} = [\tau_1, \ldots, \tau_m]$, the vector of true scores for the m variables; $\mathbf{B}$ is an $m \times m$ matrix of coefficients with zeros along the diagonal, which defines the relationships among the m variables; and $\boldsymbol{\varepsilon}$ is an $m \times 1$ vector of random errors. The model for $\boldsymbol{\tau}$ is called the *structural* or *latent variable model.* The measurement error model defines the relationship between the observations and the true scores. In (24.51), $\mathbf{y}_l$ is a $p_l \times 1$ vector of measurements on variable l, $\boldsymbol{\tau}_l$ is τ_l times a p_l-vector of 1's (for parallel or tau-equivalent measures) and $\boldsymbol{\varepsilon}_l$ is a $p_l \times 1$ vector of random measurement errors. We assume that $(\mathbf{I} - \mathbf{B})$ is nonsingular and that all errors are mutually uncorrelated within and between their vectors as well as with $\boldsymbol{\tau}$.

The equations in (24.51) may be generalized in a number of ways. For example, the measures need not be parallel as depicted in (24.50). Further, the true score vector $\boldsymbol{\tau}$ may depend upon other unmeasured (*exogenous*) latent variables as in factor analysis. Bollen (1989, Chapter 8) discusses these and other generalizations of (24.51). Fuller (Chapter 30, this volume) discusses the generalization of (24.51) for unequal probability sampling designs.

A number of software packages are available for estimating $\mathbf{B}$ and the variance components associated with (24.51). LISREL 7 (Jöreskog and Sörbom, 1989) and EQS (Bentler, 1985) are used by psychometricians and social scientists to analyze a very general class of structural equation models. (See Chapter 27 for an application of these models.) EVCARP (Schnell, Park, and Fuller, 1988) is known more widely among survey samplers for these analyses, especially when the data are from complex surveys. Fuller (Chapter 30) discusses the use of this package with an application.

24.7 CONCLUDING REMARKS

In this review of measurement error models, we have restricted the discussion to background for the other chapters in the section and have necessarily excluded a number of important approaches to modeling found in the contemporary measurement error literature. A more comprehensive review would include the latent trait (or Rasch) and latent class models (Johnson and Woltman (1987) and Johnson (1990)) and log-linear models (Tanur and Feinberg, 1989), which have recently been applied for the the analysis of survey measurement error. Besag (1977) and Biemer (1986) have used spatial models for the error term when the data have a spatial structure. Another model which shows potential for modeling measurement error is generalizability theory (see, for example, Shavelson and Webb, 1991).

The key differences among the various models that have been proposed for measurement error are found in their assumptions. We viewed these various models all as generalizations of a simple model we called Model (0). The usual indicators of error, variance, bias, reliability, and validity, have different interpretations under different sets of assumptions. There is a risk of misinterpretation of these measures when an inappropriate model is used. Determining which model to use in a given situation is a matter of deciding which assumptions are most appropriate for the measurement method used and the purpose of the modeling exercise. If there are dubious assumptions that have the potential for serious misinterpretations of the error study results, the approaches reviewed in the chapter could be used to relax the assumptions.

CHAPTER 25

A MIXED MODEL FOR ANALYZING MEASUREMENT ERRORS FOR DICHOTOMOUS VARIABLES

Jeroen Pannekoek
Netherlands Central Bureau of Statistics

25.1 INTRODUCTION

For measurements obtained by interviewing, it has been recognized for quite a long time that interviewers can have systematic effects on the results and a large number of articles concerning the analysis of interviewer effects have appeared in the literature. Two approaches have been used in analyzing interviewer effects. The first approach is to consider the interviewer effect to be fixed. Usually, in this approach, the effect of some variables measured at the interviewer level on the measurements is assessed (see e.g., Sudman and Bradburn, 1974). Examples of such explanatory variables are interviewer characteristics like age, sex, number of years of experience, etc., or, in an experimental setting where for instance the effects of different training methods or instructions are tested, dummy variables indicating to which experimental group an interviewer belongs. The second approach is to consider the interviewer effect as random measurement error (see e.g., the literature cited in Pannekoek, 1988). This random effect increases the variance of sample means and totals. Measurement of the interviewer variance component is necessary in order to obtain correct estimates of the variance of means and totals and can be useful for detecting "bad" items and for an optimal choice of the number of respondents per interviewer.

The views expressed in this chapter are those of the author and do not necessarily reflect the policies of the Netherlands Central Bureau of Statistics.

Measurement Errors in Surveys.
Edited by Biemer, Groves, Lyberg, Mathiowetz, and Sudman.

ISBN: 0-471-53405-6

Most investigators that use the second approach apply a one-way random effect analysis of variance model to estimate the interviewer variance component. The analysis of variance is appropriate for continuous response variables, but problems occur when the model is applied to categorical response variables (Chapter 24). For this reason Anderson and Aitkin (1985), Pannekoek (1988) and Stokes and Hill (1985) have used alternative methods that are suitable for categorical response variables. In this chapter a model will be described that combines the first and the second approach to modeling interviewer effects. This model extends the previous work on random effect models for categorical response variables by allowing not only for a random interviewer effect but also for fixed effects of variables measured at the interviewer level. Moreover, the random component is not restricted to being equal for all interviewers, but is allowed to vary across interviewers according to the values of explanatory variables. Two estimation methods will be used: maximum likelihood based on an assumption about the distribution of the random component and another method that does not require such an assumption. Using this model it will be shown that the presence of interviewer variance not only increases the variance of estimated proportions but also invalidates the usual χ^2 tests for hypotheses concerning the fixed effects. These effects of the influence of the interviewers are demonstrated in an example.

25.2 THE MODEL

Suppose we have a random sample of N individuals which have been assigned a score 1 for a specific category of a variable if they are classified as belonging to that category and a score 0 otherwise. Let y_{ij} be the score on that specific category of the jth individual, interviewed by the ith interviewer ($i=1, \ldots, I, j=1, \ldots, n_i$). For a given interviewer i, the y_{ij} are independent random variables with the same expectation $E(y_{ij}|i)=p_i$, say. To reflect the presence of random interviewer effects, it is usually assumed that the interviewers are randomly selected from a population of interviewers so that the p_i are realizations of a random variable P, say, with $E(P)=\pi$ and $\text{var}(P)=\sigma^2$. However, in order to incorporate fixed effects of interviewer characteristics in the model, a more general formulation is needed where the expected value π is not equal for all interviewers. Similarly, to allow for effects of interviewer characteristics on the interviewer variance, σ^2 should not be equal for all interviewers. Therefore, it is assumed here that each interviewer is randomly selected from a (possibly) different population so that the p_i are realizations of independent random variables P_i, say, with $E(P_i)=\pi_i$ and $\text{var}(P_i)=\sigma_i^2$. With these assumptions we have, unconditionally, $E(y_{ij})=EE(y_{ij}|i)=\pi_i$, $\text{cov}(y_{ij},y_{ij}')=\sigma_i^2$, $\text{var}(y_{ij})=\pi_i(1-\pi_i)$ and the intra-interviewer correlation for the ith interviewer is

$$\pi_i = \frac{\text{cov}(y_{ij},y_{ij'})}{\text{var}(y_{ij})} = \frac{\sigma_i^2}{\pi_i(1-\pi_i)}. \tag{25.1}$$

Hence, the assumption of a random interviewer effect introduces non-negative correlation between the scores of respondents interviewed by the same interviewer. The variance of the sum $y_i=\Sigma_j y_{ij}$ is

$$\text{var}(y_i)=n_i\pi_i(1-\pi_i)[1+(n_i-1)\rho_i] \tag{25.2}$$

which shows that, under this model, a positive correlation has the effect of increasing the variance of y_i relative to the uncorrelated case where $\text{var}(y_i)=n_i\pi_i(1-\pi_i)$.

For the special case where $\pi_i=\pi$ and $\rho_i=\rho$ for all i, this model is discussed more fully in Chapter 24. In this chapter however, the π_i and ρ_i are not assumed to be the same for all interviewers but are allowed to vary according to the values of some explanatory variables measured for each interviewer. To model the effects of explanatory variables on the expected values π_i, it is assumed that some function of π_i is a linear combination of the explanatory variables, which can be written as

$$f(\boldsymbol{\pi})=\mathbf{X}\boldsymbol{\beta} \tag{25.3}$$

where $\boldsymbol{\pi}=(\pi_1, \ldots, \pi_I)^{\mathrm{T}}$, $\boldsymbol{\beta}=(\beta_1, \ldots, \beta_R)^{\mathrm{T}}$ is a vector of regression coefficients and $\mathbf{X}$ is an $I\times R$ model matrix. The model matrix $\mathbf{X}$ is of full rank and contains for each interviewer the known scores of numerical variables and dummy variables representing the categories of categorical variables. Model (25.3) belongs to the class of generalized linear models (GLM) (McCullagh and Nelder, 1983). In the terminology of GLM's, $\mathbf{X}\boldsymbol{\beta}$ is called the linear predictor and f the link function. A suitable and frequently used link function for binary data is the logit function (Cox, 1970, Ch. 2), resulting in the logit model: $\ln[\pi_i/(1-\pi_i)]=\mathbf{x}_i^{\mathrm{T}}\boldsymbol{\beta}$, with $\mathbf{x}_i^{\mathrm{T}}$ being the ith row of $\mathbf{X}$. Or, equivalently, $\pi_i=\exp(\mathbf{x}_i^{\mathrm{T}}\boldsymbol{\beta})/[1+\exp(\mathbf{x}_i^{\mathrm{T}}\boldsymbol{\beta})]$. This link function maps the interval [0,1] onto the interval $[-\infty,+\infty]$ and hence ensures that no estimated value for π_i can lie outside the permissible range.

Similarly, the intra-interviewer correlation ρ_i may be allowed to vary across interviewers according to some generalized linear model. If we define $\boldsymbol{\rho}=(\rho_1, \ldots, \rho_I)^{\mathrm{T}}$, the model can be written as

$$g(\boldsymbol{\rho})=\mathbf{Z}\boldsymbol{\gamma} \tag{25.4}$$

with model matrix $\mathbf{Z}$, parameter vector γ and link function g. There is no tradition in choosing a link function for a correlation parameter. Since the permissible range of ρ_i is the same as that of π_i, the logit function has again the advantage that the range restriction on the estimated values is automatically satisfied. For this link function $\rho_i = \exp(\mathbf{z}_i^T\gamma)/[1+\exp(\mathbf{z}_i^T\gamma)]$. If the range restriction does not cause a problem, the identity link, resulting in $\rho_i = \mathbf{z}_i^T\gamma$, has the advantage of being more easily interpreted.

25.3 ESTIMATION PROCEDURES

In this section two estimation methods will be described for both the interviewer variance parameter $\boldsymbol{\rho}$ or γ and the parameter $\boldsymbol{\pi}$ or $\boldsymbol{\beta}$ describing the mean. The first method is maximum likelihood (ML) and is based on the assumption that the p_i are realizations of beta-distributed random variables. Although this distribution is very flexible in shape, this is a rather arbitrary assumption (but very convenient from a mathematical point of view). Alternative distributional assumptions can be used (Anderson and Aitkin, 1985) but are open to the same criticism. Therefore, as an alternative to ML based on a beta distribution we will also describe an estimation method due to Liang and Zeger (1986) that does not require a specification of the distribution of P_i.

25.3.1 Maximum Likelihood (ML)

The likelihood of the numbers of respondents that are classified as belonging to the category under study per interviewer, $y_i = \Sigma_j y_{ij}$, can be used for estimating both $\boldsymbol{\beta}$ and γ. This likelihood can be derived by noting that the joint distribution of y_i and P_i is the product of the distribution of P_i and the conditional distribution of y_i given $P_i = p_i$. The marginal distribution of y_i can then be obtained by integrating this joint distribution over p_i.

It is assumed that the distribution of P_i is given by

$$P_i \sim \text{beta}(a_i, b_i) \tag{25.5}$$

with $a_i > 0$, $b_i > 0$, $E(P_i) = a_i/(a_i + b_i) = \pi_i$ and $\text{var}(P_i) = \pi_i(1-\pi_i)/(a_i + b_i + 1) = \sigma_i^2$. Note that ρ_i can now be expressed in terms of the parameters a_i and b_i as $\rho_i = 1/(a_i + b_i + 1)$. The beta distribution of P_i combined with the binomial conditional distribution of y_i results in a beta-binomial (BB) marginal distribution of y_i (Johnson and Kotz, 1969, p. 79, 230). This distribution can be parameterized in a number of ways. Johnson and Kotz, for instance, use n_i, a_i, and b_i but also n_i, π_i, and α_i, with

$\alpha_i = 1/(a_i + b_i)$. For our purposes it is more convenient to use a parameterization in terms of the parameters n_i, π_i, and ρ_i so that we can write

$$y_i \sim BB(n_i, \pi_i, \rho_i) \tag{25.6}$$

with $E(y_i) = n_i\pi_i$ and $\text{var}(y_i)$ given by (25.2). If $a_i + b_i \rightarrow \infty$ while $a_i/(a_i + b_i) = \pi_i$ remains constant, the beta distribution degenerates to a "spike" at $a_i/(a_i + b_i)$, the beta-binomial distribution approaches the binomial (n_i, π_i) distribution and $\rho_i \rightarrow 0$.

Since the y_i are independent, the log-likelihood function for a given sample can be written as

$$LBB(\mathbf{n}, \boldsymbol{\pi}, \boldsymbol{\rho}) = \sum_{i=1}^{I} LBB_i\,(n_i, \pi_i, \rho_i) \tag{25.7}$$

where $\mathbf{n} = (n_1, \ldots, n_I)^{\mathrm{T}}$ and $LBB_i(n_i, \pi_i, \rho_i)$ denotes the contribution to the log-likelihood of the n_i observations for interviewer i. Note that the parameter vectors $\boldsymbol{\pi}$ and $\boldsymbol{\rho}$ together have $2I$ parameters whereas there are only I observed y_i values. Therefore restrictions on $\boldsymbol{\pi}$ and $\boldsymbol{\rho}$ are needed for identification. With such restrictions the log-likelihood can be maximized with respect to the parameters $\boldsymbol{\pi}$ and $\boldsymbol{\rho}$ by the Newton-Raphson algorithm. Pannekoek (1988) gives expressions for the required first and second order derivatives in the simple case where $\pi_i = \pi$ and $\rho_i = \rho$ for all i. If models of the form (25.3) and (25.4) are used for $\boldsymbol{\pi}$ and $\boldsymbol{\rho}$ (25.7) must be maximized with respect to the parameters $\boldsymbol{\beta}$ and $\boldsymbol{\gamma}$ which can be done by an adaptation of the procedure used by Pannekoek, which is described in the Appendix of this chapter.

25.3.2 Generalized Estimation Equations (GEE)

An estimator for $\boldsymbol{\beta}$ that does not require a specification of the distribution of P_i can be based on the estimating equations

$$\mathbf{D}^{\mathrm{T}}\mathbf{V}^{-1}(\mathbf{y} - \boldsymbol{\mu}) = \mathbf{0} \tag{25.8}$$

where $\mathbf{V} = \text{diag}\{\text{var}(y_i)\}$ is the expected covariance matrix of $\mathbf{y} = (y_1, \ldots, y_I)^{\mathrm{T}}\boldsymbol{\mu} = (\mu_1, \ldots, \mu_I)^{\mathrm{T}}$, $\mu_i = n_i\pi_i$, and $\mathbf{D} = \partial\boldsymbol{\mu}/\partial\boldsymbol{\beta}$. For the logit link function we have $\mathbf{D} = [\text{diag}\{n_i\pi_i(1 - \pi_i)\}]\mathbf{X}$. Equations (25.8) are a special case of the generalized estimating equations proposed by Liang and Zeger (1986). Since $\mathbf{V}$ depends on $\boldsymbol{\rho}$ (see (25.2)), a solution $\tilde{\boldsymbol{\beta}}$ to (25.8) will, in general, also depend on $\boldsymbol{\rho}$. If $\boldsymbol{\rho}$ is known $\tilde{\boldsymbol{\beta}}$ can be obtained by iteratively reweighted least squares according to

$$\tilde{\boldsymbol{\beta}}_{t+1} = \tilde{\boldsymbol{\beta}}_t + (\mathbf{D}^{\mathrm{T}}\mathbf{V}^{-1}\mathbf{D})^{-1}\mathbf{D}^{\mathrm{T}}\mathbf{V}^{-1}(\mathbf{y} - \boldsymbol{\mu}), \tag{25.9}$$

where $\tilde{\boldsymbol{\beta}}_t$ is the value of $\boldsymbol{\beta}$ in the tth iteration and $\mathbf{V}$, $\mathbf{D}$, and $\boldsymbol{\mu}$ are evaluated at $\boldsymbol{\beta}=\tilde{\boldsymbol{\beta}}_t$. Liang and Zeger show that $\tilde{\boldsymbol{\beta}}$ is consistent and asymptotically normal, even if $\boldsymbol{\rho}$ is replaced by a consistent estimator.

In the absence of interviewer effects, $\rho_i=0$ for all i and $\mathbf{V}$ becomes the covariance matrix of independent binomial distributions. In this case, it is well known (McCullagh and Nelder, 1983) that the likelihood equations for estimating $\boldsymbol{\beta}$ are given by (25.8) and that the iterative procedure (25.9) is equivalent to the Newton-Raphson procedure. If $\rho_i=\rho$ and $n_i=n$ for all i, then we can write $\mathbf{V}=c\mathbf{V}^*$, where $c=1+(n-1)\rho$ and $\boldsymbol{\beta}$ is the only unknown parameter in $\mathbf{V}^*$. For covariance matrices that have this multiplicative form, the estimating equations (25.8) are called quasi-likelihood equations and the corresponding estimator for $\boldsymbol{\beta}$ is the maximum quasi-likelihood (MQL) estimator (McCullagh and Nelder, 1983). From (25.8) it is apparent that in this special case, the estimator for $\boldsymbol{\beta}$ is independent of the value of ρ and is equal to the ML estimator based on binomial distributions. The variance of this MQL estimator for $\boldsymbol{\beta}$, however, is not independent of the value of ρ and exceeds the variance of the ML estimator for the case where $\rho=0$ by a factor c.

In order to apply (25.9), we need an estimator for $\boldsymbol{\rho}$. Liang and Zeger suggest a moment method for estimating the correlation vector $\boldsymbol{\rho}$, based on the standardized residuals $(y_{ij}-\pi_i)/[\pi_i(1-\pi_i)]^{\frac{1}{2}}$ of the binary observations y_{ij}. Specifically, if we define

$$d_{ijj'}=\frac{(y_{ij}-\pi_i)\,(y_{ij'}-\pi_i)}{\pi_i(1-\pi_i)} \tag{25.10}$$

we have $E(d_{ijj'})=\rho_i$ for $j\neq j'$ so that a moment estimator r_i for ρ_i is the average of the $d_{ijj'}$ for $j\neq j'$:

$$r_i=\frac{2}{n_i(n_i-1)}\sum_{j=1}^{n_i-1}\sum_{j'=j+1}^{n_i} d_{ijj'}. \tag{25.11}$$

Prentice (1988) extends the GEE approach to estimating γ. For our situation the GEE estimator $\tilde{\gamma}$ for γ is defined as the solution to

$$\mathbf{E}^{\mathrm{T}}\mathbf{W}^{-1}(\mathbf{r}-\boldsymbol{\rho})=\mathbf{0}, \tag{25.12}$$

with $\mathbf{E}=\partial\boldsymbol{\rho}/\partial\gamma$ and $\mathbf{W}$ the expected covariance matrix of $\tilde{\gamma}$ for $\boldsymbol{\beta}$ given. For the logit link $\mathbf{E}=[\mathrm{diag}\{\rho_i(1-\rho_i)\}]\mathbf{Z}$ and for the identity link $\mathbf{E}=\mathbf{Z}$. Assuming the $d_{ijj'}$ to be uncorrelated, $\mathbf{W}$ is a diagonal matrix with elements given by

$$W_i=\frac{2}{n_i(n_i-1)}\left\{1+\frac{\rho_i(1-2\pi_i)^2}{\pi_i(1-\pi_i)}-\rho_i\right\}. \tag{25.13}$$

With these specifications, an updating formula for γ can be derived analogous to (25.9) and an iterative procedure for estimating both $\boldsymbol{\beta}$ and γ

can be applied where in each iteration both $\boldsymbol{\beta}$ and $\boldsymbol{\gamma}$ are updated. Prentice (1988) gives an expression for the asymptotic (as $I \rightarrow \infty$) covariance matrix of $(I)^{\frac{1}{2}}[(\tilde{\boldsymbol{\beta}} - \boldsymbol{\beta})^T,(\tilde{\boldsymbol{\gamma}} - \boldsymbol{\gamma})^T]$.

If the logit link function for $\boldsymbol{\rho}$ is used, the algorithm may occasionally fail to converge. This occurs if some of the r_i are zero or negative, such that some elements of the vector $(\mathbf{E}^T\mathbf{W}^{-1}\mathbf{E})^{-1}\mathbf{E}^T\mathbf{W}^{-1}\mathbf{r}$ are ≤ 0. The elements of $\boldsymbol{\gamma}$ corresponding to these elements will then be tending to $-\infty$. This may be taken as evidence that a model that restricts the intra-interviewer correlations to be positive is inappropriate. If the identity link is used in this situation, the algorithm will converge but some of the elements of the estimated value of $\boldsymbol{\rho}$ will be zero or negative.

25.4 HYPOTHESES AND TEST STATISTICS

If there is no interviewer variance, the y_i are binomially distributed and hypotheses pertaining to the parameter $\boldsymbol{\beta}$ can conveniently be tested by the usual ML procedures based on the binomial distribution (see e.g., McCullagh and Nelder, 1983). Therefore, in this section, first a statistic will be described that can be used to test the assumption that the y_i are binomially distributed. This statistic is easy to calculate and will often be a useful first step in the analysis of interviewer variance. Then, tests of hypotheses concerning the parameters $\boldsymbol{\beta}$ and $\boldsymbol{\gamma}$ of the model discussed in the previous sections will be described and, finally, the consequences will be discussed of incorrectly applying the usual tests of hypotheses concerning $\boldsymbol{\beta}$ in the situation where interviewer variance is present.

The assumption that the y_i are binomially distributed is equivalent to the hypothesis

$$H_{0(1)}: \rho_i = 0 \text{ for all } i.$$

If this hypothesis is not rejected, it makes no sense to try to assess the effects of explanatory variables on the correlation parameter, and ML inference concerning $\boldsymbol{\beta}$ should be based on the binomial rather than on the beta-binomial distribution. The hypothesis $H_{0(1)}$ cannot be tested by a likelihood ratio test using the beta-binomial likelihood and corresponding ML estimators for the parameters since the distribution under $H_{0(1)}$ is not beta-binomial (ρ_i is not positive). Also, due to the restriction on the range of the ρ_i the standard result that the ML estimator is asymptotically normally distributed will not be very useful for constructing confidence intervals for estimated values for ρ_i near zero because convergence to a normal distribution can then be very slow. A suitable test for $H_{0(1)}$ is given by Tarone (1979), who derived a one-sided test of the binomial distribution against beta-binomial alternatives based on the asymptotically standard normally distributed statistic given by

$$z = \frac{\sum_{i=1}^{I} (y_i - n_i\hat{\pi}_i)^2[\hat{\pi}_i(1-\hat{\pi}_i)]^{-1} - \sum_{i=1}^{I} n_i}{\left\{2\sum_{i=1}^{I} n_i(n_i - 1)\right\}^{\frac{1}{2}}} \tag{25.14}$$

where $\hat{\pi}_i$ is the estimator of π_i under some model with $\boldsymbol{\beta}$ estimated using the binomial likelihood (see Prentice, 1986). If the GEE estimator is used for ρ_i, $H_{0(1)}$ can be tested using the asymptotic normal distribution of the estimator because this estimator is not restricted to be positive.

If $H_{0(1)}$ is rejected, then a model that includes parameters describing the intra-interviewer correlations is appropriate. Hypotheses concerning the effects of explanatory variables and their interactions on the interviewer correlations can then be formulated as

$$H_{0(2)}: \gamma = \gamma_0$$

where γ_0 is obtained by restricting some of the elements of γ to zero. Similarly, hypotheses concerning the fixed effects can be formulated by

$$H_{0(3)}: \boldsymbol{\beta} = \boldsymbol{\beta}_0$$

where $\boldsymbol{\beta}_0$ is obtained by restricting some of the elements of $\boldsymbol{\beta}$ to zero. If the beta-binomial model is used, the hypotheses $H_{0(2)}$ and $H_{0(3)}$ can be tested by likelihood-ratio tests. If the GEE estimator is used, the hypotheses $H_{0(2)}$ and $H_{0(3)}$ can be tested by a generalized chi-squared statistic, X^2, based on the asymptotic normality of the estimators $\tilde{\gamma}$ and $\tilde{\boldsymbol{\beta}}$. For $H_{0(2)}$ this statistic is given by

$$X_2^2 = (\tilde{\gamma} - \gamma_0)^T \mathbf{W}(\tilde{\gamma})^{-1}(\tilde{\gamma} - \tilde{\gamma}_0) \tag{25.15}$$

where $\mathbf{W}(\tilde{\gamma})$ is the covariance matrix of γ evaluated at $\gamma = \gamma_0$. The X^2-statistic for $H_{0(3)}$ is given by

$$X_3^2 = (\tilde{\boldsymbol{\beta}} - \tilde{\boldsymbol{\beta}}_0)^T \mathbf{V}(\tilde{\boldsymbol{\beta}})^{-1}(\tilde{\boldsymbol{\beta}} - \tilde{\boldsymbol{\beta}}_0) \tag{25.16}$$

where $\mathbf{V}(\tilde{\boldsymbol{\beta}})$ is the covariance matrix of $\tilde{\boldsymbol{\beta}}$ evaluated at $\boldsymbol{\beta} = \tilde{\boldsymbol{\beta}}_0$. Both X^2 statistics are distributed as χ^2 with degrees of freedom equal to the number of restricted parameters.

To show the effect of ignoring the interviewer variance on the likelihood ratio test of $H_{0(3)}$ we begin by assuming that there is no interviewer variance. Then, the y_i are binomially distributed and the likelihood ratio statistic based on this distribution is, under the null hypothesis, asymptotically equal to the quadratic form

$$[\boldsymbol{\pi}(\hat{\boldsymbol{\beta}}) - \boldsymbol{\pi}(\boldsymbol{\beta}_0)]^T \mathbf{V}[\boldsymbol{\pi}(\hat{\boldsymbol{\beta}})]^{-1}[\boldsymbol{\pi}(\hat{\boldsymbol{\beta}}) - \boldsymbol{\pi}(\boldsymbol{\beta}_0)] \tag{25.17}$$

where $\pi(\hat{\boldsymbol{\beta}})$ is the estimated value of $\boldsymbol{\pi}$ corresponding to $\hat{\boldsymbol{\beta}}$, $\boldsymbol{\pi}(\boldsymbol{\beta}_0)$ the value of $\boldsymbol{\pi}$ corresponding to $\boldsymbol{\beta}_0$, $\hat{\boldsymbol{\beta}}$ the maximum likelihood estimator for $\boldsymbol{\beta}$, $\boldsymbol{\beta}_0$ the true parameter under the null hypothesis and $\mathbf{V}[\boldsymbol{\pi}(\hat{\boldsymbol{\beta}})]$ the covariance matrix of $\boldsymbol{\pi}(\hat{\boldsymbol{\beta}})$ under the null hypothesis. This result, from which the asymptotic χ^2-distribution of the likelihood ratio test follows, can be established by a Taylor expansion of the likelihood ratio test statistic about the true $\boldsymbol{\beta}_0$ (see e.g., Cox and Hinkley, 1974). For the binomial distribution, $\mathbf{V}[\boldsymbol{\pi}(\hat{\boldsymbol{\beta}})]$ is a diagonal matrix with elements $\pi_i(\boldsymbol{\beta}_0)[1-\pi_i(\boldsymbol{\beta}_0)]/n_i$. Now, suppose that interviewer variance is present but ignored so that likelihood inference based on the binomial distribution is erroneously applied. Then, the maximum likelihood estimator for $\boldsymbol{\beta}$ is still a consistent (but not efficient) estimator for $\boldsymbol{\beta}$. The covariance matrix of $\boldsymbol{\pi}(\hat{\boldsymbol{\beta}})$, however, is in this case not $\text{diag}\{[\pi_i(\boldsymbol{\beta}_0)(1-\pi_i(\boldsymbol{\beta}_0))]\}/n_i$ since the presence of interviewer variance has the effect of increasing the variance by a factor $1+(n_i-1)\rho_i$. Therefore, the asymptotic χ^2 result does not hold in this case, and the value of the likelihood ratio test under $H_{0(3)}$ based on the binomial distribution can be expected to be too large when compared with a χ^2 variable. Consequently, the rejection rate of $H_{0(3)}$ will be too high. Rao and Scott (1984) show that in the presence of interviewer variance, the likelihood ratio test based on the binomial distribution is asymptotically distributed as a weighted sum of independent χ^2 random variables each with one degree of freedom.

25.5 EXAMPLE

In this example, data are used for about 5,000 respondents from a pilot survey on medical consumption. The questions for which interviewer effects will be analyzed are: the respondent's opinion on his or her own health in five categories ranging from 1 (very good) to 5 (bad); whether or not the respondent has contacted a general practitioner in the last three months; whether or not the respondent has contacted a specialist in the last three months; whether or not the respondent has used medicine in the last fourteen days and whether or not the respondent has any chronic diseases. Two methods of data collection were used (by different interviewers). The first method was face to face interviewing. With the second method the interviewer left a questionnaire with the respondent and collected the completed questionnaire at a later date. Each method was applied to a random subsample of about half the size of the total sample.

The respondents were not randomly assigned to interviewers. Therefore a random effect associated with interviewers cannot be interpreted as reflecting interviewer effects alone. Also regional differ-

Table 25.1. Estimates of Intra-interviewer Correlation (Standard Errors in Parentheses)

Category	Questionnaire				Face to face interview			
	ρ_Q^{ML}		ρ_Q^{GEE}		ρ_F^{ML}		ρ_F^{GEE}	
Health 1	0.041	(0.013)	0.030	(0.022)	0.132	(0.021)	0.132	(0.025)
2	0.038	(0.011)	0.034	(0.009)	0.104	(0.018)	0.087	(0.024)
3	0.008	(0.006)	0.001	(0.022)	0.015	(0.008)	0.014	(0.027)
4	0.009	(0.008)	0.007	(0.019)	0.028	(0.010)	0.043	(0.050)
5	0.006	(0.009)	0.004	(0.013)	0.055	(0.033)	0.023	(0.019)
General Practitioner	0.029	(0.012)	0.017	(0.016)	0.031	(0.011)	0.030	(0.016)
Specialist	0.018	(0.009)	0.014	(0.030)	0.034	(0.011)	0.030	(0.034)
Medicine	0.020	(0.010)	0.013	(0.023)	0.029	(0.010)	0.028	(0.021)
Chronic Disease	0.022	(0.009)	0.020	(0.028)	0.076	(0.016)	0.058	(0.036)

ences in the response will contribute to the estimate of the random effect. However, since the two subsamples were independent random samples from the complete sample, we may assume that the contribution of the regional differences is the same in both subsamples so that a larger interviewer variance for the face to face interview than for the questionnaire indicates that the interview is more susceptible to interviewer effects than the other method of data collection.

The two data collection methods will be considered as two categories of a discrete explanatory variable "Method," labeled F and Q corresponding to the face to face interview and the questionnaire, respectively. For each of the two categories the intra-interviewer correlation will be estimated. It will also be investigated whether the different data collection methods give rise to differences in the expected proportions and/or differences in the interviewer variance components. In other words, it will be investigated whether or not there is an effect of method on the proportions and/or interviewer variance components.

To answer these questions we begin with a model with $\boldsymbol{\beta}^T = (\beta_c, \beta_M)$, $\boldsymbol{\gamma}^T = (\gamma_c, \gamma_M)$ and $\mathbf{X}$ and $\mathbf{Z}$ each consisting of two columns, the first column containing 1's and the second containing 1's corresponding to interviewers in category F and 0's corresponding to interviewers in category Q. So $\rho_i = \rho_F$ and $\pi_i = \pi_F$, say, if interviewer i belongs to category F and $\rho_i = \rho_Q$ and $\pi_i = \pi_Q$, say, if otherwise. The ML and GEE estimates for ρ_Q and ρ_F are displayed in Table 25.1. These estimates were obtained using the identity link function for $\boldsymbol{\rho}$ and the logit link function for $\boldsymbol{\pi}$.

For all categories and both estimating procedures the estimated value of ρ is larger for the interview than for the questionnaire. This suggests that for these data the interview as a method of data collection is more susceptible to interviewer effects than the questionnaire.

Table 25.2. Likelihood Ratio and X_2^2 Tests for the Hypothesis $H_{0(2)}: \rho_Q = \rho_F$

Category	LR	p	X^2	p
Health 1	13.97	< 0.01	7.24	0.01
2	9.65	< 0.01	8.31	< 0.01
3	1.81	0.18	0.14	0.71
4	2.06	0.15	0.65	0.42
5	3.29	0.07	0.30	0.58
General Practitioner (GP)	0.03	0.87	0.29	0.59
Specialist	1.23	0.27	0.12	0.73
Medicine	0.41	0.52	0.22	0.64
Chronic Disease (CD)	8.96	< 0.01	0.77	0.38

The differences between the estimated values for the two estimation methods are not large for most of the 18 categories investigated. Large differences did occur, however, for some categories. For instance for the categories 4 and 5 of health in the interview and the category GP (general practitioner) in the questionnaire.

The estimated standard errors are quite different and in most cases much larger for the GEE estimates than for the ML estimates. If the normal approximation is used to test $H_{0(1)}$ for the GEE estimates, we find four of the estimated values to be significantly (at the 5 percent significance level) larger than zero; one (health category 2) for the questionnaire and three (health categories 1 and 2 and GP) for the interview. One reason for the large interviewer variance for the first two categories of health may be that the distinction between "very good" and "good" is not so clear and therefore the answers may more easily be influenced by the interviewers. The z test for the binomial distribution against beta- binomial alternatives leads to rejection (at the 5 percent level) of the binomial distribution for all categories except the categories 3, 4 and 5 of the health question in the questionnaire. Exactly the same conclusion is reached if the normal approximation for the ML estimators is used. That this is not always the case is apparent from the example in Pannekoek (1988) where the z test rejects $H_{0(1)}$ much more often than the confidence interval approach does.

The hypothesis that there is no method effect on the intra-interviewer correlations can be formulated as $H_{0(2)}: \gamma_M = 0$, which is equivalent to $H_{0(2)}: \rho_Q = \rho_F$. A likelihood ratio (LR) test for this hypothesis was obtained by comparing the maximized beta-binomial log-likelihoods of two models, one with parameters $\beta_c, \beta_M, \gamma_c$, and γ_M and one with parameters β_c, β_M, and γ_c. Also the GEE estimates of the parameters of these two models and the X_2^2 test for $H_{0(2)}$ were calculated. The results for both tests are given in Table 25.2.

Table 25.3. Binomial (B) and Beta-binomial (BB) Likelihood Ratio (LR) Tests and X_3^2 Tests for the Hypothesis $H_{0(3)}: \pi_Q = \pi_F$

Category	LR(B)	p	LR(BB)	p	X_3^2	p
Health 1	39.20	<0.01	7.62	0.01	6.00	0.01
2	54.52	<0.01	17.32	<0.01	15.79	<0.01
3	0.03	0.86	0.03	0.87	0.03	0.85
4	12.46	<0.01	8.08	0.01	6.74	0.01
5	1.84	0.18	2.49	0.12	2.11	0.15
General Practitioner	1.27	0.26	0.58	0.45	0.78	0.38
Specialist	8.77	<0.01	7.04	0.01	7.39	0.01
Medicine	4.24	0.04	2.97	0.09	3.50	0.06
Chronic Disease	1.75	0.19	1.06	0.30	1.14	0.29

Both tests indicate that the intra-interviewer correlation differs significantly between the two data collection methods for the first two categories of health. According to the likelihood ratio test, the difference is also significant for the category CD. That this is not so for the X_2^2 test is not surprising because Table 25.1 shows that, although the difference in the GEE estimates for this category is substantial, the estimated standard errors are quite large. For all categories, the p values are larger for the X_2^2 test than for the likelihood ratio test which also reflects the large standard errors of the GEE estimates.

The hypothesis that there is no method effect on the proportions is $H_{0(3)}: \beta_M = 0$ or, equivalently, $H_{0(3)}: \pi_Q = \pi_F$. For this hypothesis two likelihood ratio tests were used. The first test ignores the interviewer variance and is based on the two maximized binomial log-likelihoods with parameters β_c, β_M, and β_c, respectively. The second is based on the maximized beta-binomial log-likelihoods with parameters $\gamma_c, \gamma_M, \beta_c$, and β_M and γ_c, γ_M, and β_c, respectively. Also for this hypothesis a test based on the GEE estimates, X_3^2, was calculated. The results are provided in Table 25.3.

All three tests show that the estimated proportions differ significantly between the two data collection methods for three categories of health (1, 2 and 4) and the category 'specialist.' For the category 'medicine,' only the test ignoring the interviewer variance indicates a significant difference (at the 5 percent level). With one exception (health category 5), the values of LR(B) are larger than the values of the other two test statistics. The two largest differences between the LR(B) and the other tests occur for the two categories with the largest values of ρ, the first two categories of health. This illustrates the effect of interviewer variance on test statistics.

25.6 CONCLUSION

In this chapter a mixed model was described for the analysis of interviewer effects for proportions. This model can be used to assess the effects of explanatory variables measured at the interviewer level on both the intra-interviewer correlation and the expected value. Two estimation methods were used, maximum likelihood, based on the assumption that the random interviewer effect is beta distributed, and the generalized estimation equation (GEE) approach, which does not require such a distributional assumption. This method can be viewed as an extension of the quasi-likelihood method.

In an example, evidence was found for the existence of a substantive interviewer variance component for two out of five categories of an opinion question. Such evidence was much less prominent for four categories of factual questions. This difference between opinion and factual questions is frequently documented in the literature. The estimated standard errors for the maximum likelihood estimates of interviewer correlation were considerably smaller than those for the GEE estimates. This suggests that the GEE estimates can be considerably less efficient than the ML estimates. A drawback of the ML approach is that a specific distributional assumption is required and therefore this approach could benefit from methods that allow the checking of such an assumption.

APPENDIX: MAXIMUM LIKELIHOOD ESTIMATION OF $\boldsymbol{\beta}$ AND $\boldsymbol{\gamma}$

In order to apply the Newton-Raphson algorithm to obtain maximum likelihood estimates for the parameters $\boldsymbol{\beta}$ and $\boldsymbol{\gamma}$ of the model described in Section 25.2, the first and second order partial derivatives of the beta-binomial log-likelihood function with respect to these parameters are required. In this Appendix expressions for these derivatives will be given in terms of the derivatives with respect to the parameters π_i and ρ_i. The derivatives with respect to π_i and ρ_i can be obtained from the derivatives with respect to the parameters π_i and $\alpha_i = 1/(a_i + b_i)$ given by Pannekoek (1988) by noting that

$$\alpha_i = \rho_i/(1 - \rho_i), \frac{\partial L_i}{\partial \rho_i} = \frac{\partial L_i}{\partial \alpha_i}(1 - \rho_i)^{-2},$$

$$\frac{\partial^2 L_i}{\partial \rho_i^2} = \frac{\partial^2 L_i}{\partial \alpha_i^2}(1 - \rho_i)^{-4} - 2\frac{\partial L_i}{\partial \alpha_i}(1 - \rho_i)^{-3},$$

and

$$\frac{\partial^2 L_i}{\partial \pi_i \partial \rho_i^2} = \frac{\partial^2 L_i}{\partial \pi_i \partial \alpha_i^2}(1 - \rho_i)$$

where L_i is the contribution to the beta-binomial log-likelihood of the n_i observations for interviewer i.

The derivatives of $L = \Sigma L_i$ with respect to β_r can be expressed in terms of the derivatives of L with respect to π_i, the derivatives of π_i with respect to $\eta_i = \mathbf{x}_i^T \boldsymbol{\beta}$, and the derivatives of η_i with respect to β_r as follows

$$\frac{\partial L}{\partial \beta_r} = \sum_{i=1}^{I} \frac{\partial L_i}{\partial \pi_i} \frac{\partial \pi_i}{\partial \eta_i} \frac{\partial \eta_i}{\partial \beta_r} = \sum_{i=1}^{I} \frac{\partial L_i}{\partial \pi_i} \frac{\partial \pi_i}{\partial \eta_i} X_{ir} \tag{A.1}$$

and

$$\frac{\partial^2 L}{\partial \beta_r \partial \beta_{r'}} = \sum_{i=1}^{I} \left\{ \frac{\partial^2 L_i}{\partial \pi_i^2} \left[\frac{\partial \pi_i}{\partial \eta_i} \right]^2 + \frac{\partial L_i}{\partial \pi_i} \frac{\partial^2 \pi_i^2}{\partial \eta_i^2} \right\} X_{ir} X_{ir'}. \tag{A.2}$$

The derivatives $\partial L / \partial \gamma_s$ and $\partial^2 L / \partial \gamma_s \partial \gamma_{s'}$, can be obtained from (A.1) and (A.2) by replacing β_r by γ_s, $\beta_{r'}$ by $\gamma_{s'}$, π_i by ρ_i, η_i by $\xi_i = \mathbf{z}_i^T \gamma$, X_{ir} by Z_{is} and $X_{ir'}$ by $Z_{is'}$. The mixed partial derivatives can be expressed as

$$\frac{\partial^2 L_i}{\partial \beta_r \partial \gamma_s} = \sum_{i=1}^{I} \frac{\partial^2 L_i}{\partial \pi_i \partial \rho_i} \frac{\partial \pi_i}{\partial \eta_i} \frac{\partial \rho_i}{\partial \xi_i} X_{ir} X_{is}. \tag{A.3}$$

These partial derivatives depend on the choice of the link functions for π_i and ρ_i via the derivatives with respect to η_i and ξ_i. For the logit link function for π_i, we have

$$\frac{\partial \pi_i}{\partial \eta_i} = \frac{\exp(\eta_i)}{[1 + \exp(\eta_i)]^2} = \pi_i(1 - \pi_i) \tag{A.4}$$

and

$$\frac{\partial^2 \pi_i}{\partial \eta_i^2} = \frac{\exp(\eta_i)[1 - \exp(\eta_i)]}{[1 + \exp(\eta_i)]^3} = \pi_i(1 - \pi_i)(1 - 2\pi_i). \tag{A.5}$$

Similar expressions can be obtained for the logit link function for ρ_i. For the identity link function we have $\partial \pi_i / \partial \eta_i = \partial \rho_i / \partial \xi_i = 1$ and $\partial^2 \pi_i / \partial \eta_i^2 = \partial^2 \rho_i / \partial \xi_i^2 = 0$.

CHAPTER 26

MODELS FOR MEMORY EFFECTS IN COUNT DATA

Piet G. W. M. van Dosselaar
Netherlands Central Bureau of Statistics

26.1 INTRODUCTION

At the Netherlands Central Bureau of Statistics, the analysis of memory effects was initiated by Dirk Sikkel in the early eighties and since then several surveys have been analyzed with respect to memory errors. However no attempt has been made to formalize the procedures in a probabilistic setting. In this chapter we will present a general theory for the modeling of memory effects in count data from surveys without an accessible external gauging device, as well as some new models for reported counts of labor market transitions.

We consider a sample of respondents for which we would like to count the number of events of a certain type that occurred to them during a certain time interval. Information about these events is collected by using retrospective questions, and therefore our estimate of the number of experienced events may be affected by memory errors. Models for memory effects are developed to estimate the magnitude of the effects, to construct better survey estimates and, if possible, to develop correction procedures for the retrospective data.

The methods used to model memory effects are closely related to the type of gauge that is available. The word "gauge" refers to any measure of the "true" net or mean number of events that can serve as a standard of comparison for the net or mean number of retrospectively reported events. Section 26.2 of this chapter gives a classification of these methods

---

The views expressed in this chapter are those of the author and do not necessarily reflect the policies of the Netherlands Central Bureau of Statistics.

by type of gauging device. The remaining sections deal exclusively with the modeling of memory effects where the gauging device has to stem from the survey itself due to the absence of external sources. Whether a survey can furnish such a gauge depends on the combination of sample design (overlapping sample waves, for instance) and dependence on calendar time of the target variable. Although the ultimate model specification depends strongly on several survey characteristics, it appears that a general model can be derived, which can be extended to meet the goals and design of a particular survey. This general model consists of two components: the generation of events through a Poisson process and the conditional probability (as a function of elapsed time) that an event that actually took place is reported as well. For a particular survey this model can be extended by specifying these two components in more detail. This general model as well as some examples of its possible extensions will be treated in Section 26.3. Section 26.4 gives an example of how the general model is used in the process of building and estimating models for numbers of transitions between two labor market positions. The data on these transitions originate from the Dutch Continuing Labor Force Survey (CLFS). Subsection 26.4.1 describes the design of the CLFS. Subsection 26.4.2 formulates the memory effect models in terms of the general model. In subsection 26.4.3 these models are estimated using transformed duration data in a two-step procedure. The model that can be considered the optimum one on both practical and theoretical grounds is derived properly and estimated using untransformed duration data in a one-step procedure in subsection 26.4.4. Section 26.5 contains the main conclusions.

26.2 CLASSIFICATION OF METHODS BY TYPE OF GAUGING DEVICE

The methods that can be used to model memory effects can be classified into three different groups according to the type of gauge that is available. In most cases the index of interest is the relative rate of report with respect to the "true" population total or population mean of events as it is derived (estimated) from the gauge that is used.

Class 1: External Validating Records

Whenever we have access to existing records on the actual events experienced by our group of respondents, the effects of forgetting and telescoping can be studied by matching, as was done by for instance Sudman and Bradburn (1973) and Mathiowetz (1985). Since both survey and gauging device relate to the same respondents, analyses at an individual level are possible.

Class 2: Reference Survey of Better Quality
In this case we have at our disposal an external gauge, independent of our sample, in the form of a reference survey in which memory errors are (almost) absent. (See van Dosselaar, et al., 1989a.)

Class 3: No External Gauge
In most cases we do not have access to an external gauge and we must look for a gauging device within the survey itself. If seasonal effects are present in the data it is essential that the design provide us with measurements of the target variable with various time-lags between the time of occurrence of the events and the day of interview (e.g., overlapping samples). In the case that seasonal effects can be ruled out a priori, a one-time retrospective study would suffice to study memory effects. In both cases we need some information on the dates of the events and we have to make some assumptions with respect to the "true" total or mean number of events in our group of respondents. Two examples of this class of problems can be found in van Dosselaar, et al. (1989a). A general model for this "no external gauge" situation is developed in Section 26.3 of this chapter.

26.3 THE THEORY FOR CLASS 3 PROBLEMS

We have a sample of respondents for whom we would like to estimate the total number of events experienced in a certain period of time. The information we have on these events is retrospective and likely to be biased due to recall errors. We have no access to external validating records, also there is no reference survey of better quality to provide a reliable gauge. Therefore we have to look for a gauging device within the survey at hand. In this section a general model is developed that can be used as a starting point for any particular survey without external gauge, but with appropriate design and target variable.

26.3.1 The General Model

Suppose we have a sample of respondents interviewed at time τ about their experiences with events of a particular kind during time period $(t_0,\tau]$. The aim of the survey is to estimate the number of experienced events by this sample in the period $(t_0,\tau]$. It is assumed that

1. reported events and their occurrence dates have been reported correctly, and
2. events are distributed in time as a Poisson process.

Assumption 1 implies that overreporting is impossible and that telescoping effects are ignored. A necessary condition for assumption 2 to hold is that experienced events occur randomly and are rare in the population.

Let $\tilde{n}_T$ be the "true" number of events experienced by a randomly selected respondent in our sample in a relatively small interval $T=(t,t+s]\subset(t_0,\tau]$ and let $n_T(\tau)$ be the number of events in this interval reported at time τ. Let $\lambda(t)$ be the intensity of the underlying Poisson process of $\tilde{n}_T$ and let λ_T be the integral of $\lambda(t)$ over T. Informally, $\lambda(t)$ can be defined as the rate of occurrence of events (number of events per time unit) at time t and λ_T as the expected number of events in T. For a formal definition of the intensity of a Poisson process, see Snyder (1985). For the "true" number of events in T we have the following distribution:

$$P[\tilde{n}_T = k] = \frac{\lambda_T^k e^{-\lambda_T}}{k!}, \; k = 0, 1, 2, \ldots \tag{26.1}$$

For the distribution of the number of events in T reported at time τ we can write:

$$\begin{aligned} P[n_T(\tau) = k] &= \sum_{j=0}^{\infty} P[n_T(\tau) = k, \tilde{n}_T = k + j] \\ &= \sum_{j=0}^{\infty} P[n_T(\tau) = k | \tilde{n}_T = k + j] \, P[\tilde{n}_T = k + j] \\ &= \sum_{j=0}^{\infty} P[k \text{ out of } k + j \text{ events in } T \text{ reported at } \tau] \, P[\tilde{n}_T = k + j], \end{aligned} \tag{26.2}$$

for $k=0, 1, 2,\ldots$. Note that $j\geq 0$ as a consequence of assumption 1. Let us now concentrate on the probability that k out of the $k+j$ events that occurred in T are reported at τ. Due to the small length s of T we can treat all of the $k+j$ events in T as having equal probabilities of being reported at time τ and having occurrence times that are uniformly distributed over T. Let $h(u)$ be the probability that a randomly selected respondent reports an event that actually took place u time units before the interview. Note that $h(u)$ is defined to be only dependent on the time elapsed since the event and not on calendar time t or the intensity $\lambda(t)$. The probability that an event that occurred in T is reported at time τ is given by equation (26.3) and can be interpreted as the average value of the reporting probability $h(u)$ over the interval $[\tau-t-s,\tau-t)$:

$$H_T(\tau) = P[\text{event in } T \text{ is reported at } \tau] = \int_{\tau-t-s}^{\tau-t} \frac{h(u)}{s} \, du \tag{26.3}$$

where we take $h(u)$ to be a continuous function of u on the interval $[0,\infty)$. If we assume (assumption 3) that

$n_T(\tau)|\tilde{n}_T = m$ is distributed as a binomial random variable with parameters m and $H_T(\tau)$,

we obtain the following well-known result:

$$P[n_T(\tau) = k] = \sum_{j=0}^{\infty} \binom{k+j}{k} H_T(\tau)^k (1 - H_T(\tau))^j P[\tilde{n}_T = k + j]$$

$$= \sum_{j=0}^{\infty} \frac{(k+j)!\, H_T(\tau)^k (1 - H_T(\tau))^j}{k!\, j!} \frac{\lambda_T^{k+j} e^{-\lambda_T}}{(k+j)!}$$

$$= \frac{\lambda_T^k H_T(\tau)^k e^{-\lambda_T}}{k!} \sum_{j=0}^{\infty} \frac{(1 - H_T(\tau))^j \lambda_T^j}{j!}$$

$$= \frac{(\lambda_T H_T(\tau))^k e^{-(\lambda_T H_T(\tau))}}{k!}. \tag{26.4}$$

Assumption 3 will hold if, for example, the length s of interval T is small enough so that all events occurring in the interval have approximately the same probability of being reported and approximately the same probability of occurring at any time within the interval T. From equation (26.4) we learn that the number of events that happened in T and that are actually reported at τ is also Poisson distributed, but now with parameter $\lambda_T H_T(\tau)$. We can interpret $\lambda_T H_T(\tau)$ as the rate at which events are *reported* at time τ. This rate is seen to be the product of two rates: the rate of occurrence in the interval T and the probability of recall at τ of a given occurrence. This Poisson distribution for the reported number of events in a certain interval can serve as a starting point for the modeling of memory effects in a particular survey.

26.3.2 Possible Extensions of the General Model

For a particular survey the general model of subsection 3.1 can be specified in more detail. If the events are not too rare and sample sizes not too small, it can be used as a model at an individual level. Examples can be found in Sikkel (1985), where individual models are proposed for the reported numbers of contacts with the family doctor, the specialist and the dentist. In Sikkel (1990) some of these models are extended to study differences in medical consumption between several subpopulations. In those papers the intensity $\lambda(t)$ is treated as a Gamma distributed latent random variable, instead of assuming $\lambda(t)$ to be constant over the individuals. For these models the integrated intensity λ_T for any individual is simply the product of that individual's parameter value $\lambda(t)$ and the length s of the time period T under study. Several functional forms are proposed for the average forgetting probability $1 - H_T(\tau)$, which give some alternative descriptions of the possible dependence of the memory effects on time elapsed since the contact and on the number of previously reported contacts. Ultimately, the resulting Poisson-Gamma distributions are used to derive the distribution of so-called "profiles" of contacts. The models are estimated on retrospective data from a complete year of a continuing survey to neutralize seasonal effects and to provide for a gauge (i.e., no memory effects in data on a short period preceding the interview). These models illustrate the necessity of the assumption that the events under study occur randomly. Several models for the contacts with the family doctor and the specialist performed very well, whereas all models failed hopelessly for the more regular contacts with the dentist.

Whenever the number of reported events is rather small compared to the number of respondents or whenever the sample size is small, the general model can still be used, but now as a model at an aggregate level. Examples of this type of model can be found in van Dosselaar, et al. (1989a, 1989b), where several models for the reported numbers of transitions between the labor market states "employed" and "not employed" are proposed. Modified versions of these models and new estimation results will be presented in the next section of this chapter.

26.4 EXAMPLES FROM THE DUTCH CONTINUING LABOR FORCE SURVEY

26.4.1 Survey Design and Target Variables

In January 1987 the Netherlands Central Bureau of Statistics started the CLFS, which covers the noninstitutional population resident in the Netherlands. One of the survey goals is to publish figures on the labor market status (employed, unemployed and non-economically active) and on labor market dynamics. We will focus on transitions from one labor market status to another; for each transition the main aspects are the day of occurrence and the labor market status before and after the transition.

For the CLFS a stratified multistage sample (wave) of approximately 12,000 addresses is drawn every month. In July and August the sample size is half as large. Respondents are asked to give their employment histories for the year preceding the interview. For every respondent the past 12 months can be partitioned into periods of employment (up to a maximum of three jobs), unemployment, and inactivity. As a result of the

Wave	Month under review																									
	J	F	M	A	M	J	J	A	S	O	N	D	J	F	M	A	M	J	J	A	S	O	N	D	J	F	M
January	p	c	c	c	c	c	c	c	c	c	c	c	p														
February		p	c	c	c	c	c	c	c	c	c	c	c	p													
March			p	c	c	c	c	c	c	c	c	c	c	c	p												
April				p	c	c	c	c	c	c	c	c	c	c	c	p											
May					p	c	c	c	c	c	c	c	c	c	c	c	p										
June						p	c	c	c	c	c	c	c	c	c	c	c	p									
July							p	c	c	c	c	c	c	c	c	c	c	c	p								
August								p	c	c	c	c	c	c	c	c	c	c	c	p							
September									p	c	c	c	c	c	c	c	c	c	c	c	p						
October										p	c	c	c	c	c	c	c	c	c	c	c	p					
November											p	c	c	c	c	c	c	c	c	c	c	c	p				
December												p	c	c	c	c	c	c	c	c	c	c	c	p			
January													p	c	c	c	c	c	c	c	c	c	c	c	p		
February														p	c	c	c	c	c	c	c	c	c	c	c	p	
March															p	c	c	c	c	c	c	c	c	c	c	c	p

c=complete information available
p=only partial information available

Figure 26.1. Sample design of the continuing labor force survey

one year retrospection period and the periodicity of the survey, the waves are partly overlapping (see Figure 26.1). The interviews are distributed more or less homogeneously over the month for every wave. A respondent interviewed on, say, April 23, 1989 gives information on the period April 24, 1988–April 23, 1989. Hence, for any month under review information is available from respondents from thirteen different waves, where the first and last of these only give information on a part of that particular month. The survey design is described in more detail in van Bastelaer (1988).

Respondents are asked what their current main activities are, whether they are looking for a job, and whether they would be available if offered a job. Since retrospective questions on these topics cannot be formulated in the same wordings as the ones about current activities, validity problems other than recall errors may exist for retrospective data concerning availability and looking for a job. As availability and looking for a job are necessary conditions for the most frequently used definitions of unemployment, problems of validity may also exist for data concerning unemployment. Therefore, the models for memory effects have only been developed for the two state process, with states "employed" and "not employed." The latter category contains both unemployed and economically inactive respondents.

Even so, the problem of validity remains, because the survey questionnaire only allows for the three most recent jobs, so that information on the period before the third job is missing for respondents who report this maximum. Although the respondents who report three jobs constitute on the average as little as (roughly) 0.25 percent of the total response in a wave, they have great influence on the transition data. This influence can be quantified, as the respondents who report this maximum have to answer an additional question about the number of jobs they have had in the past year besides those already mentioned. Exclusion of these "3^+-jobs" appears to lead to a total number of reported transitions that is roughly 5 percent lower than it would have been, had they been included in the survey setup. These unreported "3^+-jobs," of course, are concentrated in the most distant part of the one-year retrospection period, where their omission leads to an extra decline in reported transitions of roughly 10 to 15 percent. Inclusion of this distant part of the retrospection period in the estimation process would result in a serious bias in the estimated memory effect parameters. To avoid the influence of this aspect of the survey setup on the estimation of the memory effect parameters, models have been developed and estimated for numbers of transitions reported with a maximum time lag of six months only.

A few words need to be said with regard to the possible presence of telescoping effects, which are ignored by the general model of Section 26.3 of this chapter. One should make a distinction between the telescoping of events into or out of the one-year retrospection period on the one hand, and the telescoping of events within the period on the other. The effects of the telescoping of events into or out of the one-year retrospection period are likely to be removed by only using data from the most recent half of the retrospection period (for the CLFS, this is the six months immediately preceding the interview), as this type of telescoping is located around the boundary of the retrospection period. The effect of the misplacement of dates within the retrospection period (internal telescoping) mainly depends on the average direction of these misplacements. In the case that the average direction of internal telescoping is either forward or backward, the estimates of the monthly intensities of the Poisson process might be biased. Whenever internal telescoping is present but without a specific direction (the misplacement of a random event has zero expectation), it might lead to an increase in the variances of the estimated intensities. There is some evidence (Mathiowetz, 1985) that both forward and backward telescoping are present in labor market data with magnitudes that are about equal, so that we can expect the estimated intensities to be nearly unbiased.

The time-period under study ranges from June 16, 1987 to June 15, 1988, divided into 12 one-month periods. Although the interviews are roughly equally distributed over the month for every sample wave, we

		Month of Reported Event (i)											
		1	2	3	4	5	6	7	8	9	10	11	12
Time Lag from Date of Report (j)	1	1	2	3	4	5	6	7	8	9	10	11	12
	2	2	3	4	5	6	7	8	9	10	11	12	13
	3	3	4	5	6	7	8	9	10	11	12	13	14
	4	4	5	6	7	8	9	10	11	12	13	14	15
	5	5	6	7	8	9	10	11	12	13	14	15	16
	6	6	7	8	9	10	11	12	13	14	15	16	17

Figure 26.2. Month of interview by month of reported event and time lag

will treat all respondents from a monthly sample as if they were interviewed on the 15th of that month.

In this section we will only use data from respondents who have reported at least one transition in the one-year retrospection period, assuming that respondents who do not report a transition have responded correctly. This assumption is based on the notion that, for most people, matters of employment and unemployment are of great importance. Moreover, a single labor market transition in the past year can be considered a very salient event, not likely to be forgotten.

26.4.2 Notation and Model Specifications

The two states of the process are indexed by x, where $x=0$ stands for labor market status "not employed" and $x=1$ stands for labor market status "employed." The one-month periods are denoted by T_i $(i=1, 2, \ldots, 12)$, where T_1 stands for the period "June 16, 1987–July 15, 1987", T_2 stands for the period "July 16, 1987–August 15, 1987" and so on. The different time lags with which the numbers in the two-way table are measured are indexed by j $(j=1,2,\ldots,6)$, where j stands for a time lag (in months) in the range $[j-1, j)$, the month of interview being $i+j-1$. Figure 26.2 displays the months of interview classified by month of reported event and time lag from date of report. Note that all pairs (i,j) for which $i+j-1$ takes on the same value concern the same wave, from now on denoted by S_{i+j-1}. In this situation we have $2 \times 12 \times 6$ "general models" where for each state x there are twelve intervals T_i of which the numbers of transitions are reported by six different sample waves S_{i+j-1}.

Let n_{xij} be the number of reported transitions from state x in T_i for wave S_{i+j-1} and let $\tilde{n}_{xij}$ be the corresponding "true" number of transitions. The n_{xij} coincide with $n_{xT_i}(i+j-1)$ as defined in the general model and the $\tilde{n}_{xij}$ coincide with $\tilde{n}_{xT_i}(i+j-1)$, where the argument $i+j-1$ is added to identify this particular group of respondents. Note that there was no need for this extra argument identifying the sample in the definition of the general model, since the general model is defined for one sample only. Now, however, we have $2 \times 12 \times 6$ "general models" with data from 17 ($=12+6-1$) different samples which have to be identified. These 144 general models are extended by specifying the integrated intensities λ_{T_i} and the reporting probabilities $H_{xT_i}(i+j-1)$ in more detail.

The intensity $\lambda_x(t)$ for a randomly selected individual is approximated by the constant λ_{xi} for $t \epsilon T_i$, so that the integrated intensity λ_{xT_i} for the sample that reports on T_i in interview month $i+j-1$ is the product of λ_{xi} and the total time spent in state x during T_i, aggregated over all the respondents from that sample. This total time spent in x during T_i by S_{i+j-1} is denoted by $\tilde{t}_{xij}$. In the sequel these $\tilde{t}_{xij}$ are called *durations* (note that these durations are not related to individual job length!).

The probability of reporting a transition from state x in period T_i that occurred between $j-1$ and j months ago corresponds to $H_{xT_i}(i+j-1)$ in the general model. Recall that $H_T(\tau)$ in the general model is defined to be only depending on the time lapse since the event. We see that $H_{xT_i}(i+j-1)$ only depends on $i+j-i=j$, and we will denote this conditional probability by H_{xj}.

For H_{xj} the following functional forms are considered:

1. $H_{xj} = H_{xj}$:unrestricted memory effects ($H_{x1} \equiv 1$);
2. $H_{xj} = 1 - \beta_x(j - \frac{1}{2})$:linear memory effects;
3. $H_{xj} = \exp[-\beta_x(j - \frac{1}{2})]$:exponential memory effects;
4. $H_{xj} = H_j$:state-independent memory effects ($H_1 \equiv 1$);
5. $H_{xj} = 1 - \beta(j - \frac{1}{2})$:state-independent linear memory effects;
6. $H_{xj} = \exp[-\beta(j - \frac{1}{2})]$:state-independent exponential memory effects;
7. $H_{xj} = 1$:no memory effects.

Models 1 and 4 are not based on a continuous function on the time-axis. They simply give the reporting rate in time lag period j relative to time lag period one. To provide for a gauge, the H_{x1} are fixed at one for these two models. The linear models are obtained by applying equation (26.3) with $h_x(u) = 1 - \beta_x u$. The exponential models are close approximations of the functions obtained when applying equation (26.3) with $h_x(u) = \exp(-\beta_x u)$ (the correct functions obtained by this procedure will be derived in subsection 26.4.4).

Under the model the n_{xij} are distributed according to a Poisson distribution which from (26.4) has parameter $\lambda_{xi} H_{xj} \tilde{t}_{xij}$ ($x=0,1$; $i=1,2,\ldots,12$ and $j=1,2,\ldots,6$). This is also equal to the expectation and the variance of n_{xij} under the model. For the estimation of the model parameters by the maximum likelihood method we need the joint likelihood for the entire $2 \times 12 \times 6$ table of reported transitions. By treating all n_{xij} as if they are mutually independent this joint likelihood is easily obtained as the product of the 144 marginal likelihoods. This assumed independence is guaranteed for data from different sample waves. Under the Poisson process the numbers on a wave diagonal are independent for each state x. The main problem that could arise with respect to the assumption of independence is the possible dependence of the pairs (n_{0ij}, n_{1ij}). One way to overcome this problem is to assume a special form of the bivariate Poisson distribution (e.g., Section 11.4 of Johnson and Kotz, 1969) for the pairs (n_{0ij}, n_{1ij}), and to write the joint likelihood for the entire table as the product over i and j of these bivariate distributions. However, for reasons of simplicity we choose to assume complete independence and write the joint likelihood as the product of the marginal distributions over all 144 cells. After estimating the model parameters, we will calculate Spearman's rank correlation coefficient for the paired normalized residuals of n_{0ij} and n_{1ij} to test whether these are correlated. The maximum likelihood procedure is carried out by maximizing the natural logarithm of the joint likelihood, using the Newton-Raphson algorithm.

To estimate the model parameters we need the "true" durations $\tilde{t}_{xij}$. Inspection of the reported durations t_{xij}, however, reveals a systematic effect of time lag: for $x = 0$ (not employed) the reported durations are increasing and for $x = 1$ (employed) they are decreasing as a function of j. Under assumption 1, the main explanation for this phenomenon is that respondents fail to report episodes of work. This explanation is consistent with the cognitive tasks the respondents have to perform during the interview. They first have to remember whether or not they had a job in a certain time-period, before the more detailed information with respect to the dates of job commencement and job termination is requested. The recall effects in reported durations are dealt with in two different ways. In subsection 26.4.3 the reported durations are transformed to eliminate the effects of time lag and the models are estimated using these corrected durations. In subsection 26.4.4 a form of time-correction is built into the model and estimation takes place using the original reported durations.

26.4.3 The Models Estimated with Transformed Durations

In this subsection models 1–6 are estimated using transformed durations. The t_{1ij} are transformed to $\tilde{t}_{1ij}$ by dividing them by a factor δ_j. The $\tilde{t}_{0ij}$ then are equal to $t_{0ij} + t_{1ij} - \tilde{t}_{1ij}$. The factors δ_j are calculated from a 12×6 table of reported durations $t_{1j}^{(w)}$ by wave and time lag according to equation (26.5), using reported durations from the first 12 waves:

$$\delta_j = \frac{\sum_{w=1}^{12} t_{1j}^{(w)}}{\sum_{w=1}^{12} t_{11}^{(w)}}, \qquad j = 1, 2, \ldots, 6 \tag{26.5}$$

Table 26.1. Model Fits for Different Models for H_{xj}

Model for Memory Effects	G^2	df	p-value
1: Unrestricted	137.83	110	0.037
2: Linear	147.89	118	0.033
3: Exponential	146.28	118	0.040
4: State-independent	138.53	115	0.067
5: State-independent linear	147.93	119	0.037
6: State-independent exponential	146.31	119	0.045
7: No memory effects	223.76	120	0.000

where $t_{1j}^{(w)}$ stands for the state 1 durations in the jth month before the interview as reported by wave w. Note that $t_{1j}^{(w)}$ coincides with $t_{1(w-j+1)j}$ for all pairs (w,j) for which $w-j+1$ is one of the integers 1,2, . . . ,12. Both numerator and denominator in equation (26.5) contain the reported durations for 12 calendar months, so that seasonal effects are neutralized. To understand the rationale behind the δ_j, divide both numerator and denominator in equation (26.5) by 12. Then, the numerator gives the average duration in state 1 in a one-month period, measured with a time lag in months in the range $[j-1,j)$. The denominator now gives this average duration measured with a time lag in months in the range [0,1). Hence, δ_j describes the effect of the time lag on the average reported durations, and by applying this correction factor we artificially inflate the durations from time-lag period j to bring them to the level of the durations reported in the first month before the interview, assuming that these are (almost) unaffected by recall error. The resulting vector δ of correction factors δ_j is given by (1, 0.971, 0.948, 0.918, 0.893, 0.863), clearly showing that reported durations decline with increasing time lag.

Models 1–6 are estimated for a $2 \times 12 \times 6$ table of reported transitions n_{xij} with a corresponding table of transformed durations $\tilde{t}_{xij}$. The model without memory effects (model 7) is estimated with untransformed durations. In Table 26.1 the likelihood ratio test statistic with corresponding degrees of freedom and p-values is given for all seven models.

Examining Table 26.1 we see that the p-value for all models but the one without memory effects (with a p-value of 0.000) is around 0.050. The p-value for the no memory effects model indicates that memory effects are present in the data. Comparing this model with the state-independent exponential model, we see that the addition of one parameter describing the memory effects decreases the value of the chi square statistic, denoted by G^2, by 70. The difference between the two models is highly significant. We also see that the state independent models do not differ significantly from their state dependent counterparts, which was not the case in van Dosselaar, et al. (1989a, 1989b), where untransformed durations were used to estimate the models.

The values for G^2 of the models that incorporate memory effects still are rather high, which could be due to the possible presence of a wave effect, not accounted for by the model. There are waves for which the numbers of reported transitions tend to be larger than the average and waves for which these numbers tend to be smaller than the average, probably due to the number of respondents in the wave who change jobs frequently. An analysis of normalized residuals by wave shows that there are indeed some waves for which the residuals differ significantly from the average.

Other possible explanations for high G^2-values include telescoping effects (i.e., misreporting of dates of employment), differences in distributions of interview days over the interview months possibly resulting in misclassifications of transitions with respect to the periods T_i, misclassification error in the classification of employment, interactions between calendar period and time-lag, or the possible invalidity of assumption 2. Neglecting the complex design (see Chapter 31) or possible dependencies in the data (see Gleser and Moore, 1985) could have led to inflated G^2-values.

Spearman's rank correlation coefficient ρ for the paired normalized residuals of n_{0ij} and n_{1ij} is roughly -0.05 for models 1 and 2, roughly 0.001 for models 3 through 6, and 0.26 for model 7. Performing Spearman's rank correlation test for independence, we see that the corresponding values of the test statistic $[=\rho(71)^{1/2}]$ lie outside the critical region (with $\alpha = 0.05$) for models 1–6. From these results it would seem that the assumed independence between n_{0ij} and n_{1ij} is safeguarded rather well for the models that fit the data best.

Table 26.2 shows the parameter estimates which describe the memory effects. The differences between the memory effect parameters for state 0 and state 1 in the state-dependent models are not significant. Table 26.3 contains the parameter estimates for λ_{0i} for all seven models and for a "model free" situation. This "model free" estimate is given by n_{0i1}/t_{0i1} and can be regarded as an estimate of the intensity when only the most recent data on period T_i are used. Under the assumption of no memory effects in the month preceding the interview these "model free" estimates are unbiased. The standard errors of these estimates given by $n_{0i1}^{\frac{1}{2}}/t_{0i1}$, however, are rather high. Looking at the estimates of λ_{0i} for models 1–6 we see that their values deviate neither much nor systematically from the "model free" ones, but they have considerably smaller standard errors. Standard errors of estimates are computed as the square root of the diagonal elements of the estimated covariance matrix of the parameter vector, which is equal to minus the inverse of the matrix of second order derivatives of the log-likelihood with respect to the model parameters. The estimated values for model 7 are clearly biased downwards.

Table 26.2. Parameter Estimates for Different Models for H_{xj}

Model for memory effects	Parameter state 0	Parameter state 1	Estimate state 0	Estimate state 1	Standard error state 0	Standard error state 1
1: Unrestricted	H_{02}	H_{12}	0.936	0.904	0.052	0.045
	H_{03}	H_{13}	0.814	0.833	0.046	0.042
	H_{04}	H_{14}	0.803	0.791	0.046	0.040
	H_{05}	H_{15}	0.806	0.789	0.046	0.039
	H_{06}	H_{16}	0.761	0.747	0.044	0.037
2: Linear	β_0	β_1	0.044	0.046	0.007	0.006
3: Exponential	β_0	β_1	0.053	0.055	0.010	0.009
4: State-independent	H_2		0.918		0.034	
	H_3		0.825		0.031	
	H_4		0.796		0.030	
	H_5		0.796		0.030	
	H_6		0.753		0.028	
5: State-independent linear	β		0.046		0.005	
6: State-independent exponential	β		0.054		0.006	

Table 26.3. Parameter Estimates for λ_{0i} in Different Models for Memory Effects (Standard Errors in Parentheses)

Model for H_{xj}	λ_{01}	λ_{02}	λ_{03}	λ_{04}	λ_{05}	λ_{06}
Model free	464 (85)	876 (117)	736 (76)	393 (54)	462 (58)	370 (56)
1: Unrestricted	477 (35)	884 (52)	805 (45)	424 (29)	431 (30)	411 (29)
2: Linear	468 (32)	865 (45)	792 (39)	418 (27)	425 (28)	405 (27)
3: Exponential	474 (33)	876 (47)	801 (41)	422 (28)	429 (29)	410 (28)
4: State-independent	480 (32)	889 (46)	810 (40)	426 (27)	434 (28)	414 (27)
5: State-indep. lin.	470 (31)	869 (42)	794 (37)	419 (26)	426 (27)	407 (26)
6: State-indep. exp.	476 (31)	879 (43)	804 (38)	424 (27)	431 (27)	411 (26)
7: No memory effects	363 (23)	676 (30)	636 (28)	333 (20)	337 (20)	321 (20)

Model for H_{xj}	λ_{07}	λ_{08}	λ_{09}	λ_{010}	λ_{011}	λ_{012}
Model free	597 (66)	330 (50)	296 (45)	343 (49)	385 (54)	334 (50)
1: Unrestricted	626 (37)	333 (25)	301 (24)	302 (23)	328 (24)	284 (22)
2: Linear	617 (33)	328 (23)	296 (22)	298 (22)	325 (23)	281 (21)
3: Exponential	624 (35)	332 (24)	299 (23)	301 (22)	329 (23)	284 (21)
4: State-independent	630 (34)	335 (23)	303 (22)	304 (22)	330 (22)	286 (21)
5: State-indep. lin.	619 (31)	329 (22)	297 (21)	299 (21)	326 (22)	282 (20)
6: State-indep. exp.	626 (32)	333 (23)	301 (22)	302 (22)	330 (22)	286 (20)
7: No memory effects	494 (24)	267 (18)	245 (17)	241 (17)	260 (17)	222 (15)

Note: Table entries have been multiplied by 10^5.

The estimation results presented in this subsection show that memory effects are present in these retrospective data on transitions and that models for memory effects can be used to obtain estimates for the intensities of the Poisson process generating the "true" number of transitions. Further, the estimates are efficient and nearly unbiased when compared to the "model free" estimates. The estimated memory effect parameters for the state-dependent models do not show a significant difference between the values for state 0 and state 1. These results are encouraging and support our theory that respondents mainly fail to report whole periods of employment.

We have estimated our models in a two-step procedure with the aid of transformed durations. To estimate and test these models properly and to examine whether the decline in reported employment durations can be linked to the decline in reported jobs, new models, incorporating a duration correction, are necessary. The mathematical derivation of such models is rather complicated. Therefore we will perform this task for the state independent exponential memory effects model only. We have chosen this particular model for both practical and theoretical reasons: its fit is as good or as bad as that of the others but it only uses one degree of freedom in modeling the memory effects, the state-independence is consistent with the respondents' cognitive task, the exponential forgetting curve is widely used in cognitive psychology, and, finally, the form of the function assures that the estimated probabilities H_{xj} will always be within the range [0,1].

26.4.4 The State-Independent Exponential Memory Effects Model with Incorporated Time Correction

In this subsection the state-independent exponential memory effects model with incorporated time correction is developed, estimated and interpreted. The specification of the reporting probability is derived again in complete agreement with the general model, whereas it remained an approximation in the corresponding model of the previous subsection. The decline in employment durations will be accounted for by the model, so that the model can be estimated with untransformed data in a one-step procedure.

For the derivation of this model, let us examine first what happens when an event occurs at time t. At time t the event is placed in memory and we want respondents to retrieve the information on the event under certain interview conditions at time $\tau > t$. The human memory is fallible and we assume that for every event experienced by our randomly selected respondent, there is a time span Δt after which the event is not retrieved from memory under the realized interview conditions of the survey. The exponential memory effects model states that Δt is exponentially distributed with parameter $\beta > 0$ (on the positive real numbers), that is $f_\beta(\Delta t) = \beta\exp(-\beta\Delta t)$. The probability that an event is reported s time units after its occurrence is given by $h(u) = \exp(-\beta u)$, which is obtained by integrating the density f_β of Δt over the interval (u,∞). This procedure can be applied with distributions $f(\Delta t)$ other than the exponential, and the general model thus can be made even more general by specifying $f(\Delta t)$ instead of $h(u)$, which can be derived from $f(\Delta t)$. In complete agreement with the general model the (integrated) probability H_j (subscript x for state has been omitted) of reporting an event that happened between $j-1$ and j months ago is given by the following equation:

$$H_j = \int_{u=j-1}^{j} e^{-\beta u} du = \frac{e^{-\beta(j-1)} - e^{-\beta j}}{\beta}. \tag{26.6}$$

For the estimation of the error in the measured durations, write these durations t_{xij} as $p_{xij}m_{ij}$, where m_{ij} is equal to $t_{0ij} + t_{1ij}$, and where p_{xij} stands for the proportion of m_{ij} spent in state x. Corresponding with this decomposition $\tilde{t}_{xij}$ should be written as $\tilde{p}_{xij}m_{ij}$, where $\tilde{p}_{xij}$ stands for the "true" proportion of m_{ij} spent in state x. To model the deviations of p_{xij} from $\tilde{p}_{xij}$, we will treat these proportions as measured and "true" probabilities, respectively, of occupying state x at time $i-\frac{1}{2}$ for a randomly selected respondent.

Let τ be the time of interview for such a randomly selected individual. Denote the "true" probability of occupying state x at time t by $\tilde{p}_x(t)$ and the probability of occupying state x at time t measured at time τ by $p_x(t;\tau)$. Under the assumption that reported jobs are remembered correctly we can write:

$$p_0(t;\tau) = \tilde{p}_0(t) + P[X(t;\tau) = 0, \tilde{X}(t) = 1] \tag{26.7}$$

where $X(t;\tau)$ and $\tilde{X}(t)$ denote the measured and true labor market position, respectively, of a randomly selected respondent at time t. In words, (26.7) states that the probability of a report at time τ of unemployment at time t is equal to the probability of truly being unemployed at time t plus the probability of truly being employed but falsely reporting unemployed at time t. The second probability on the right of equation (26.7) can be related to the reporting probability $h(u) = \exp(-\beta u)$ with u in the interval $(0,\tau-t)$. With $\tilde{X}(t) = 1$, our randomly selected respondent has a job at time t. Let W_{1t} be the waiting time between t and the last day of this work period, and let $\alpha_t(\cdot)$ be the density of W_{1t}. The second probability on the right of equation (26.7) can now be written as

$$P[X(t;\tau) = 0, \tilde{X}(t) = 1]$$

$$= \int_{s=0}^{\tau - t} P[X(t;\tau) = 0 | \tilde{X}(t) = 1, W_{1t} = s]\, \alpha_t(s)\, P[\tilde{X}(t) = 1]\, ds$$

$$= \int_{s=0}^{\tau - t} (1 - h(\tau - t - s))\, \alpha_t(s)\, \tilde{p}_1(t)\, ds. \tag{26.8}$$

The probabilities are only integrated over the interval $[0,\tau - t)$ because values of s greater than $\tau - t$ imply that the job held at time t is still held at the day of interview and is therefore reported with probability one. Substitution of equation (26.8) into equation (26.7) and rewriting the resulting equation in terms of $p_1(t;\tau)$ and $\tilde{p}_1(t)$ leads to the following equation linking the measured probability of being in state 1 to the "true" probability of being in that state:

$$p_1(t;\tau) = \tilde{p}_1(t) \left[1 - \int_{s=0}^{\tau - t} \alpha_t(s)\, (1 - h(\tau - t - s))\, ds \right]. \tag{26.9}$$

For our purpose of time correction it will be sufficient to have a waiting time distribution with only one parameter, independent of t. This implies that, for the moment, we will consider the process that governs job ending to be homogeneous (i.e., independent of calendar time). For such a process, it can be shown (see Cox and Lewis, 1966, Section 4.2) that the waiting time at any randomly selected time point t with $\tilde{X}(t) = 1$ is exponentially distributed with parameter $\alpha > 0$. Substitution of $\alpha_t(s) = \alpha \exp(-\alpha s)$ and $h(\tau - t - s) = \exp(-\beta(\tau - t - s))$ in equation (26.9) and integration of the resulting function over the interval $[0,\tau - t)$ yields equation (26.10), which gives us an expression for $p_1(t;\tau)$ in terms of $\tilde{p}_1(t)$, α, β, and $\tau - t$, viz.

$$p_1(t;\tau) = \tilde{p}_1(t) \left(\frac{\alpha}{\alpha - \beta} e^{-\beta(\tau - t)} - \frac{\beta}{\alpha - \beta} e^{-\alpha(\tau - t)} \right). \tag{26.10}$$

Treating p_{xij} as $p_x(i - \frac{1}{2}; i + j - 1)$ and $\tilde{p}_{xij}$ as $\tilde{p}_x(i - \frac{1}{2})$, we can write $\tilde{t}_{1ij}$ in the likelihood function as

$$\tilde{t}_{1ij} = \frac{t_{1ij}}{\dfrac{\alpha}{\alpha - \beta} e^{-\beta(j - \frac{1}{2})} - \dfrac{\beta}{\alpha - \beta} e^{-\alpha(j - \frac{1}{2})}} \tag{26.11}$$

whereas t_{0ij} is computed by subtracting $\tilde{t}_{1ij}$ from m_{ij} $(= t_{0ij} + t_{1ij})$. Using (26.4), we see that n_{xij} is Poisson distributed with parameter $\lambda_{xi} H_j \tilde{t}_{xij}$, with H_j as given by equation (26.6) and $\tilde{t}_{xij}$ as given by equation (26.11). Again, parameters are estimated by the maximum likelihood method using the Newton-Raphson algorithm to maximize the natural logarithm of the

Table 26.4. Parameter Estimates for the Exponential Memory Effects Model with Incorporated Time Correction

Parameter	Estimate	St. error	Parameter	Estimate	St. error
β	0.0539	0.0065	α	0.2372	0.0881
λ_{01}	464	32	λ_{07}	613	33
λ_{02}	857	44	λ_{08}	326	23
λ_{03}	786	39	λ_{09}	295	22
λ_{04}	414	27	λ_{010}	297	21
λ_{05}	421	28	λ_{011}	324	22
λ_{06}	401	27	λ_{012}	280	20

Note: Except for β and α, table entries have been multiplied by 10^5.

likelihood function, which is written as the product of the likelihoods for the $2 \times 12 \times 6$ separate cells in the table of reported numbers of transitions.

Estimation of this model leads to a value of G^2 of 145.75 with 118 degrees of freedom and a corresponding p-value of 0.042. The estimated model parameters are presented in Table 26.4.

The estimation results for this model are very encouraging and easily interpreted. As in Section 26.4.3, the estimated integrated intensities λ_{0i} have values close to the model free ones of that section and have very small standard errors. Under the model, we see that the time-span Δt, after which an event is no longer recalled, is exponentially distributed with parameter $\hat{\beta} = 0.0539$. Since an exponential distribution with parameter β has expectation $1/\beta$ and variance $1/\beta^2$ we find that under the realized interview conditions the average job of a respondent can be expected to be recalled up to 18.6 ($= 1/\hat{\beta}$) months after it has ended, with a variance of $(18.6)^2$. By the same argument we may conclude that for our group of respondents, who have reported at least one transition over the past year, the average job length is equal to 4.2 ($= 1/\hat{\alpha}$) months with a variance of $(4.2)^2$. This may seem short for an average job length, but the bulk of this group consists of people who can be labeled as "movers" on the labor market, who change jobs frequently. For a discussion of the concept of "movers" and the associated "mover-stayer" models, see for instance Langeheine and van de Pol (1990).

The incorporated time correction leads to a vector of correction factors δ_j with values (0.998, 0.988, 0.968, 0.943, 0.914, 0.882). These values, obtained by substitution of $\hat{\alpha}$ and $\hat{\beta}$ in the denominator of the right-hand side of equation (26.11), tend to be smaller than the ones computed a priori in subsection 26.4.3, but this could be due to the fact that in the a priori computation, some data were used that are not used here, and vice versa.

The normalized residuals on a wave diagonal tend to show the same sign (+ or −), which supports the hypothesis of wave effects being present in the data. This hypothesis is also supported by the table of estimated "true" proportions $\tilde{p}_{xij}$, which has both diagonals with proportions that are relatively high for almost all periods T_i and diagonals with proportions that are small for almost all T_i. Dividing the total number of expected transitions over a wave diagonal by the total number of reported transitions yields a quantity that can be interpreted as the effect of the wave that corresponds with the diagonal. Multiplying all expected numbers on each diagonal by the corresponding wave effect results in a considerable reduction of the value for Pearson's χ^2. The wave effects can be regarded as the scale factor in the scaled inhomogeneous Poisson process. (See Snyder, 1975, example 2.4.5.) Incorporating such wave effects by a random component will increase the fit, as measured by G^2, but will not affect the parameter estimates very much.

Spearman's rank correlation coefficient for the paired normalized residuals of n_{0ij} and n_{1ij} is equal to -0.014. The test for independence based on this coefficient demonstrates that the assumed independency between n_{0ij} and n_{1ij} is not likely to be violated.

26.5 CONCLUSIONS

The main conclusions of this chapter are that retrospective data on counts of events can be used to estimate the "true" number of counts of these events for the respondents in the survey, even in the case where no external gauging device is available. Provided we have a survey design and a target variable that satisfy certain conditions, models can be used to detect memory effects, to quantify the memory effects and to estimate the "true" number of counts efficiently. The memory effects themselves can be used to correct the table of reported events for recall loss. For a two-state process the estimated intensities can be used to estimate correct transition probabilities and calculate (period to period) transition tables. The exponential memory effects model with incorporated time correction might be used to set up an imputation scheme at record level to correct for recall loss. The development of such a procedure to select the most appropriate records for imputation requires further analysis.

CHAPTER 27

SIMPLE RESPONSE VARIANCE: ESTIMATION AND DETERMINANTS

Colm O'Muircheartaigh
London School of Economics and Political Science

27.1. INTRODUCTION

In this chapter we will consider the particular issue of response variance for a binary survey variable and the estimation of this variance where reinterview data are available. We wish to estimate the reliability of the responses and to understand what influences the magnitude of this reliability. An important feature of the work is to distinguish between the measurement error itself and problems which arise in its estimation.

As a starting point for the analysis we consider the standard survey model of response error and the estimation and interpretation issues which arise with this model. O'Muircheartaigh (1986) used data from the U.S. Current Population Survey (CPS) to estimate some of the characteristics of the measurement errors and from these derived some conclusions about the measurement process, but also identified some deficiencies in these conclusions.

We will then consider an alternative conceptualization of the measurement error using a log-linear model relating the errors to the characteristics of the measurement situation. We will reconsider the estimation problems in the light of the new model and reanalyze the CPS data in order to see whether the new approach helps to illuminate the underlying measurement process.

This chapter is based on data prepared when the author was an ASA/NSF Research Fellow at The U.S. Bureau of the Census (Grant #SES 8122051). The author would like to thank Karl Ashworth for his assistance in carrying out the analysis, and Daniel Hill for his helpful comments.

Measurement Errors in Surveys.
Edited by Biemer, Groves, Lyberg, Mathiowetz, and Sudman.

ISBN: 0-471-53405-6

The overall structure of models of response error is described in Chapter 24. It is useful, however, to describe briefly the standard model of response variance developed at the U.S. Bureau of the Census (Hansen, Hurwitz and Bershad, 1961). The basic model is

$$y_{jt} = y_j + b_j + e_{jt}. \tag{27.1}$$

The observation y_{jt} is the response obtained for individual j at trial t. The survey is assumed to be conceptually repeatable and t will take the values 1,2,.... etc for repetitions of the survey. The observation is considered to consist of a true value (y_j) for individual j and a response deviation which may be partitioned into a bias or systematic distortion (b_j) for individual j, and a variable response error (e_{jt}) that varies from trial to trial.

In practice we use the mean of sample s for a single trial

$$\bar{y}_{.t} = \sum_{j \in s} y_{jt}/n \tag{27.2}$$

to estimate the population mean of the true values

$$\bar{y} = \sum_{j=1}^{N} y_j/N. \tag{27.3}$$

Hansen, Hurwitz and Bershad are concerned only with the variance of the $\{e_{jt}\}$, which is the basic component of the variance of $\bar{e}_{.t}$. The variance of $\bar{e}_{.t}$ is the response variance of the estimate of the population mean. The simple response variance (SRV) is $E\,\mathrm{Var}(e_{jt}|j)$, i.e., the average variance of the e_{jt} for all elements in the population.

The data consist of the original interview and a reinterview for a set of individuals. Both interviews may be considered to be conducted under the same (or similar) essential survey conditions, and there is therefore no possibility of assessing the bias terms $\{b_j\}$ from the data.

Under the assumption of independent reenumeration, we have for each individual j two observations y_{j1} and y_{j2} where from (27.1)

$$\begin{aligned} y_{j1} - y_{j2} &= (y_j + b_j + e_{j1}) - (y_j + b_j + e_{j2}) \\ &= e_{j1} - e_{j2}. \end{aligned}$$

Thus if we calculate

$$s_e^2 = \frac{1}{2n}\sum_{j \in s}(y_{j1} - y_{j2})^2 \tag{27.4}$$

then

$$E(s_e^2) = \frac{1}{2}[\sigma_{e1}^2 + \sigma_{e2}^2]$$

since $\{y_{j1}\}$ and $\{y_{j2}\}$ are independent. Furthermore, if

$$\sigma_{e1}^2 = \sigma_{e2}^2 = \sigma_e^2$$

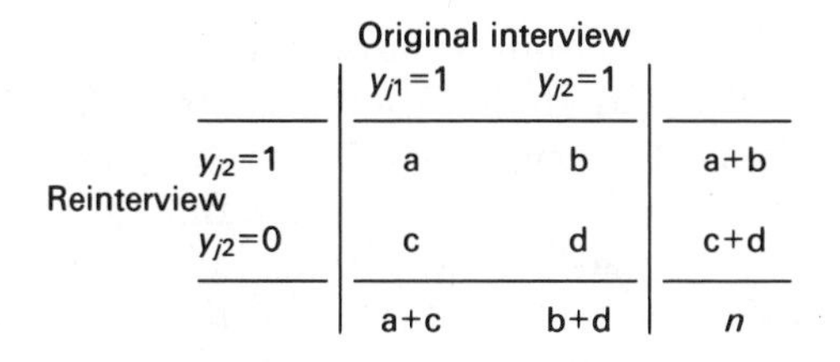

		Original interview $y_{j1}=1$	$y_{j2}=1$	
Reinterview	$y_{j2}=1$	a	b	a+b
	$y_{j2}=0$	c	d	c+d
		a+c	b+d	n

Figure 27.1. An interview–reinterview table.

then $E(s_e^2)=\sigma_e^2$, which is the simple response variance for the variable y.

Where y is a binary variable (such as "employed, not employed") the data from the two sets of interviews can be written as in Figure 27.1. Under the same assumptions as before, it can be shown that $(b+c)/n$, which is called the gross difference rate (*gdr*), is equal to $2\hat{\sigma}_e^2$.

The *gdr* is the proportion of individuals from whom different responses were obtained on the two occasions and has a strong intuitive appeal as a measure of reliability.

The discussion in Chapter 24 draws attention to the intuitive deficiencies in σ_e^2 in the case of binary variables. It is true that in most cases we would not consider this in the form presented there to be a plausible measure of response variance. It is nevertheless worth noting that the *SRV* even in this case is extremely plausible when expressed in the form of the *gdr*.

The data on which the analysis in this chapter is based come from the CPS reinterview program. The CPS is a monthly household sample survey conducted by the U.S. Bureau of the Census which provides estimates of labor force characteristics; in a given month each individual is classified according to labor force status. Any eligible respondent can report for all the individuals in the household. Each household is in the CPS sample for four months, then leaves the sample for eight months, and returns to the sample for a further four months. Thus at any given time the sample will contain households which will have been in the sample for 1,2,....,8 months. As part of the quality control procedures, a subsample of about 1 in 18 (about 5,000) households is reinterviewed. At the time of the original interview the interviewer does not know that the household will be reinterviewed. The person conducting the reinterview (the reinterviewer) is never the original interviewer, and is usually a senior interviewer or supervisor.

For 80 percent[1] of the reinterview sample a reconciliation between the original and reinterview responses is carried out as part of the

[1]Recently the CPS has changed to a 75%–25% split of reconciled to unreconciled reinterviews (see chapter 15 for details).

reinterview program. The reinterviewers for these cases have access to the original responses but are instructed not to consult them until after the reinterview has been completed. The reconciliation itself is carried out on a separate form. The term "contaminated" is used to describe these cases; the data used are the reinterview responses before the reconciliation operation is carried out. For the remaining 20 percent of the sample, no reconciliation is carried out and the reinterviewers have no access to the original responses; these are called the uncontaminated cases. The data consist of the interview-reinterview cases from January 1982 through January 1984 — a total of 126,122 cases. For some purposes the data for February 1984 through December 1984 are also used — a further 55,000 cases.

27.2 FACTORS AFFECTING THE ESTIMATED SIMPLE RESPONSE VARIANCE

In order to understand the process which produces the estimates of simple response variance, it is necessary to consider the factors which are likely to affect the values of the estimates. A diagrammatic representation is given in Figure 27.2. The most important distinction is that between the two sets of factors on the left, which affect the response variance itself, and the set of factors on the right, which affect not the response variance but its estimation. The estimates produced by the U.S. Bureau of the Census model (27.1) are affected by all three sets of factors.

If two independent observations were obtained for each question for each individual under identical survey conditions, then the estimated *SRV* would be an unbiased estimate of the actual *SRV* for the survey. Even in this case, however, it would be unwise to consider the *SRV* to be a single indivisible quantity.

Because of the way in which the data are collected for the CPS we may expect the quality of reporting to differ for different individuals. First, the data for some individuals will be based on self-reports and the data for other individuals will be based on reports by others (proxy reports). There is no prior justification for assuming that the reliability of reporting will be the same for these two situations. Second, the characteristics of the respondents (for instance, age, sex, position in the household) will vary and these variations may be related either to the quality of their self-reports or their proxy-reports; there may also be interactions in the case of proxy reports between the characteristics of the respondents and the subjects (those being reported on). Third, some of the interviews are conducted face to face (personal interviews) and others by telephone; this may also affect the response variance. Fourth,

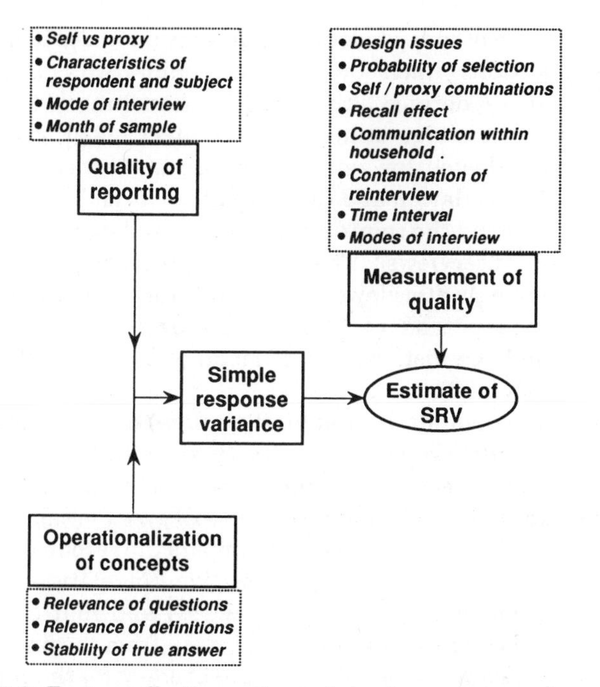

Figure 27.2. Factors affecting estimated simple response variance (*SRV*).

the reliability of the data may be related to the number of months for which the household has already been in the sample; there is evidence (Bailar (1975) on rotation group bias, for instance) that data quality is related to month in sample.

This set of factors must be considered in relation to the inherent measurement problems for the survey variables. Although a labor force characteristic such as whether a person is employed, unemployed or out of the labor force may seem to be a factual item and one that can easily be verified, there are people for whom labor force status is an attitude that cannot be verified from records; see Bailar and Rothwell (1984) for a discussion of this issue. There also may be interactions between the characteristics of respondents and the characteristics which influence the relevance of the questions and the stability of the true values for particular individuals.

The factors described above (those on the left of Fig. 27.2) relate to the magnitude of the actual response variance of the survey estimates. The set of factors on the right of Fig. 27.2, however, are factors which affect the estimate of response variance but do not affect the response variance itself. These factors reflect departures from the ideal (or assumed) situation of independent reenumeration under identical essential survey conditions. Each of them may invalidate to a greater or lesser degree the estimates derived from the estimation formula (27.4). Each of the factors involves design issues in the reinterview programme. First, the households selected for reinterview are not an equal probability sample from the population and the estimation formula should take this into account. Second, the reinterviews are not independent replications of the original interviews. There are some procedural differences between the original interviews and the reinterviews: the interviewers for the reinterviews tend to be more senior than the generality of interviewers, and the distribution of mode of interview (telephone vs. personal) may also differ.

These factors invalidate (or at least weaken) the assumption that the reliability of responses is the same in the two interviews. For instance, if the same individual responds to the questions in the first and second interviews there is a chance that he or she will give the same response on the two occasions because he or she remembers the response given on the first occasion. This would imply a positive correlation between the response deviation in the first interview (e_{j1}) and that on the second interview (e_{j2}). The availability of the response from the first occasion to the reinterviewer could have the same kind of effect for the contaminated 80 percent of the sample, as could communication between members of the household when different members respond on the two occasions. The effect might differ for different self/proxy combinations, and could be expected to differ for different time intervals between the first and second interviews.

These design and estimation issues are almost inextricably intertwined. The purpose of this chapter is to attempt to provide a framework within which they can be disentangled.

27.3 ALTERNATIVE MODELS

The basic characteristic in the examination of response errors is the response deviation d_{jt}. In the context of the response variance the dependent variable is e_{jt}, the variable response error. Our objective is to explain e_{jt}, in terms of the characteristics of the individual j, or of the characteristics of the respondent (if the respondent is different), or of other aspects of the measurement process.

It is difficult to model e_{jt} because under the model it is a random variable with an expected value of 0. Consequently we first considered using as the basis of the analysis the quantity σ_e^2, the SRV, or the absolute value of the response error e_{jt}. It is difficult, however, to model variation

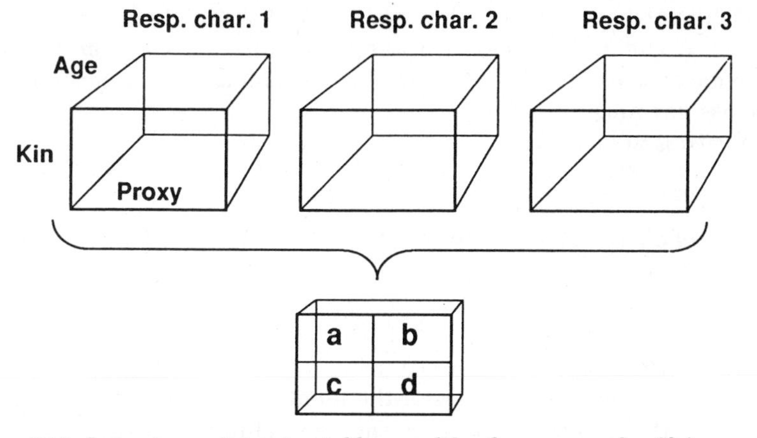

Figure 27.3. Interview–reinterview tables resulting from cross-classifying possible explanatory variables. Resp. char., response characteristic.

in either of these quantities satisfactorily. In the first place it is difficult to know whether an additive or a multiplicative model is the more appropriate; as the ideal value of $|e_{jt}|$ would be 0, if we want to model departures from zero an additive model would seem to be necessary, whereas a multiplicative model is intuitively more appealing.

The problems are exacerbated when we consider the case of binary variables. The response deviation d_{jt} can take only three possible values, -1, 0 and $+1$, and the response process can be defined by the probability of a false positive (ϕ) and the probability of a false negative (θ) (see Chapter 24). The data set can also be described completely as in Figure 27.1.

Consider a large table cross-classifying the possible explanatory variables and representing the sample in a multi-way contingency table. Such a table (see Figure 27.3) has in each cell the number of cases (frequency) with that combination of characteristics. Our interest, however, is not in the frequency in the cell but in the response variability demonstrated by the elements in the cell. In other words we have for every cell the 2×2 table representing the responses on the two trials. What we would like to do is to model the variation among the many 2×2 tables in terms of explanatory variables where these are the defining categories for the cell.

In the final section we discuss various possible ways of reparameterizing the 2×2 table. Here we consider a particular way of summarizing the information contained in the table. We already have a summary measure available to describe a 2×2 table. This is the *gdr* $[=(b+c)/n]$, which is of course equal to twice the estimate of the *SRV*. It is the proportion of disagreements between $t=1$ and $t=2$ and must take a value between 0 and 1. The measure itself therefore suffers from the same disadvantages as the SRV as the dependent quantity in modeling the response variance.

Writing the *gdr* as

$$g=(b+c)/n$$

the odds of disagreement between the observations for $t=1$ and $t=2$ can be written as

$$g/(1-g). \tag{27.5}$$

These odds vary from cell to cell. What we are trying to do is to explain this variation in terms of the identifying characteristics of the cells (i.e., the characteristics of the individuals in the cells or the measurement process represented by the combination of factors denoted by the cells). We postulate multiplicative effects on the odds. A simple example would be that the effect of being, say, younger multiplies the odds of disagreement between the first and second measurements by some factor.

It may be worth pointing out at this stage that there is no restriction on the factors which can be included in the model. They may include both the factors affecting the response variance itself (the left hand side of Figure 27.2) and the factors affecting its estimation (the right hand side of Figure 27.2). Since the model is multiplicative in the odds it is easier to express it as an additive model in the logarithms of the odds. Thus for a model with eight explanatory variables $A,B,C,$ ...... $,H$ the model can be written as

$$\log_e\left[\frac{g_{ijk\ldots p}}{1-g_{ijk\ldots p}}\right] = \beta^1 + \beta_i^A + \beta_j^B + \ldots + \beta_p^H$$

$$+ \beta_{ij}^{AB} + \ldots\text{etc} \tag{27.6}$$

where β_i^A represents the additive effect on the log odds of belonging to category i of variable A, etc.

27.4 THE DATA AND THE ESTIMATION PROCEDURE

In the form in which this model is set up, the data set being analyzed is the multiway table where the element within each cell is the ratio $g/(1-g)$. The model is an asymmetric model with a single dependent variable; it is in fact a logistic regression — a logit model with binomial error. There is a slight violation of the assumptions involved here as the data come from a process in which the odds of disagreement arise from two different sources — the probability of a false positive and the probability of a false negative.

As the data set consists of a table it is necessary to construct the table as the first stage in the analysis. The basic unit is the cell, and the quantity $g/(1-g)$ must be calculated for each cell. The analysis will then fit the specified model to the data and calculate the degree to which the model fits the data. This raises three issues. The first is the problem of empty cells in the table; the second is the evaluation of the model; the third is the interpretation of the results in terms of the original (raw) data.

The analysis was carried out using the GLIM program. The raw data set was very large (126,122 cases) and it was not convenient within GLIM to process the raw data. Thus each cross-tabulation was first produced outside GLIM and imported into the program. This meant that the set of variables for each analysis had to be determined in advance and the appropriate cross-tabulation produced.

For the examples presented in this chapter, the set of explanatory variables considered was confined to the set used in O'Muircheartaigh (1986). This set, described in Figure 27.4, provides an adequate range of variables and also permits a comparison of the results and their interpretation with the previous work.

Example 1

The first example is used simply to illustrate the application of the model and the interpretation of the parameters. The data set consisted of a cross-tabulation of the data by CONT, PROX, KIN, R1KIN, R2KIN and MODE and thus contained $2 \times 5 \times 5 \times 5 \times 5 \times 5 = 6{,}250$ cells.

Inspection of the list of explanatory variables makes it clear that some cells of the table contain what are called structural zeros, i.e., by definition it is impossible that there should be any observations in these cells. For example if PROX = 1 (self-reports on both occasions), then KIN must be equal to both R1KIN and R2KIN and the other cells in the subtable defined by PROX = 1 must be empty. In the analysis these cells must be specified as structural zeros in order to produce appropriate estimates for the parameters. It is also possible that there will be zero cells simply because of sampling fluctuations; these are called sampling zeros and are not specified in advance — the program makes appropriate allowance for them in the standard analysis. It is of course desirable that there should be few cells with low expected frequencies.

As a first stage in carrying out the analysis we fit the null model to the data. This essentially corresponds to finding out how much variabi-

Contamination (CONT)

1 Yes

2 No

Response status (PROX)

see figure 27.5

Relationship to head of household (KIN)

1 Head of household

2 Spouse of head

3 Other relative

The same categorization for the respondent at the first and second interview is denoted by R1KIN and R2KIN

Mode of Interview and reinterview (MODE)

1 Telephone/telephone

2 Face-to-face/telephone

3 Face-to-face/face-to-face

4 Telephone/face-to-face

5 Mixed in either interview

Age of subject (AGE)

1 24 years and under

2 25–34

3 35–44

4 44–54

5 55–64

6 65 years and over

Figure 27.4. Explanatory variables used in these analyses.

lity there is in the data and provides a benchmark against which to compare the fit of other models. The model could be specified as

$$\log\left[\frac{g}{1-g}\right] = \beta^1 . \tag{27.7}$$

Of the 6,250 cells in the cross-tabulation 1,075 were nonempty. Thus there are 1,074 degrees of freedom available in the table. This model simply fits a constant. The null model was fitted in the case of both the dependent variables, whether or not in the Civilian Labor Force (CLF) and whether or not unemployed (UNM).

The scaled deviance, which is a measure of the variation between the fitted and observed values in the cells of the table was 2,971 for CLF and 2,189 for UNM.

To illustrate the procedure we fit the main effects model

$$\log\left[\frac{g}{1-g}\right] = \beta^1 + \beta_2^{\mathrm{CONT}} + \beta_2^{\mathrm{PROX}} + \beta_3^{\mathrm{PROX}} + \beta_4^{\mathrm{PROX}} + \beta_5^{\mathrm{PROX}} . \tag{27.8}$$

In this model we allow the odds of disagreement between the original interviews and the reinterviews to be affected by the explanatory

Table 27.1. Estimates for the Model Incorporating the Main Effects of Contamination and Self/Proxy Response Status

	Dependent Variable			
	CLF		UNM	
Parameter	Estimates	s.e.	Estimates	s.e.
β^1	−3.94	0.034	−4.48	0.043
β_2^{CONT}	0.70	0.037	0.73	0.046
β_2^{PROX}	0.57	0.058	0.65	0.072
β_3^{PROX}	0.36	0.063	0.36	0.081
β_4^{PROX}	0.20	0.042	0.30	0.053
β_5^{PROX}	1.08	0.059	0.96	0.078

variables CONT and PROX. It is worth noting that the number of fitted terms for each of the explanatory variables is one less than the number of categories for that variable. Thus for CONT, which is binary, there is one term; for PROX, which has five categories, there are four terms. These two explanatory variables thus use up five degrees of freedom $[(2-1)+(5-1)]$.

The scaled deviance for the model was 2,307 for CLF and 1,788 for UNM with 1,069 degrees of freedom. The change in deviance is therefore 664 for CLF and 401 for UNM and the change in degrees of freedom is 5 in each case. Although the deviance itself cannot be tested directly, the difference between the deviances for the two models can be tested using chi-square. The two explanatory variables clearly contribute significantly to the explanation of the variation in the odds of disagreement.

Having established that the explanatory variables warrant inclusion in the model, Table 27.1 presents the parameter estimates and their estimated standard errors for the two dependent variables CLF and UNM. The sizes of the estimated standard errors relative to the parameter estimates reflect the conclusion above about the strongly significant effect of each of the explanatory variables.

The parameter estimates represent the deviation from the reference category for each variable. In these examples the reference category is the first category specified — for CONT it is the contaminated cases, for PROX it is the self/self category.

Taking as an example the parameter estimate β_2^{CONT}, the effect on the odds of disagreement of the reinterview not being contaminated is that the log odds are increased by 0.70 (CLF) or 0.73 (UNM). The odds are therefore multiplied by $e^{0.70}$ or $e^{0.73}$, i.e., 2.01 or 2.08. In other words the odds of disagreement between the first and second responses are approximately twice as high in the case of an uncontaminated reinterview as in the case of a contaminated one; put another way, the contamination depresses the estimate of response variability by about 50 percent.

27.5 THE EFFECTS OF MANY VARIABLES

The characteristics of the respondents in a survey have been shown to be related to the quality of the responses obtained. For fertility data, for instance, O'Muircheartaigh (1984) showed that respondent age and education were related to the size of the simple response variance. In addition to these extra-role characteristics of the respondent, the self/proxy response status may be important, as may the relationship between the respondent and the subject.

For CPS data the identity of the respondent is included on the record for each subject (the subject being the individual about whom the survey variables are being reported). Although the respondent's *characteristics* are not included on the records for the subjects for whom he or she responds, it is possible to obtain the information from the respondent's own subject record. This was done for the full data set.

Five factors in all are considered here: two subject-related factors — the relationship of the individual to the head of household (KIN) and the age of the individual (AGE); two interview-related factors — mode of interview and reinterview (MODE) and whether the reinterview was contaminated (CONT); and one reporting factor — the self/proxy status of the interview and the reinterview.

In the analysis carried out in O'Muircheartaigh (1986) each of these factors was considered; the tables published were confined to uncontaminated self-reports. The tables were presented in terms of *gdr* for each of the categories of respondent. Table 27.2 below extracts results for the two dependent variables for KIN, AGE, and MODE.

There is evidence from the table that consistency of reporting is higher for older respondents in the CPS. The results for UNM are those which show the clearest pattern. The estimated *gdr* decreases from a level of 5.02 for the youngest age group (age 24 and under) to a level of 0.54 for the oldest group (65 and over) and there is a monotonic trend between these extremes. For the labor force status variables generally, the highest variability occurs for the younger respondents and the lowest for the older respondents.

In the case of relationship of the respondent to the head of household (KIN) the most striking aspect is the very high *gdr* found for the 'other relatives' group. The values for this group are generally about twice as

Table 27.2. Gross Difference Rate for CLF and UNM by Relationship to Head of Household (KIN), Age, and Mode of Interview and Reinterview (MODE)

	Dependent	
Explanatory	CLF	UNM
KIN		
Head of household	0.036	0.029
Wife of head of household	0.039	0.017
Other relative	0.082	0.075
AGE		
24 and under	0.058	0.050
25–34	0.030	0.030
35–44	0.042	0.027
45–54	0.037	0.023
55–64	0.029	0.013
65 and over	0.031	0.005
MODE		
Telephone/telephone	0.027	0.016
Face to face/telephone	0.031	0.017
Face to face/face to face	0.050	0.044
Telephone/face to face	0.056	0.034
Mixed modes in either interview	0.033	0.018

large as for any other group and are worst for the two variables considered here — CLF and UNM.

In the case of MODE, the interest was in the comparison of face to face and telephone interviewing. Much of the interviewing in the CPS is done by telephone (from the interviewer's home) although the choice of mode is not randomized. In assessing the effect of mode, the cases must be classified by the combination of modes on the two occasions.

Since an even higher proportion of reinterviews is conducted by telephone, the two largest categories (modes 1 and 2) in the table both have a telephone reinterview. The difference in *gdr* is small but not negligible, the cases involving face to face interviews for the original interview have *gdr* about 10 percent greater than those conducted entirely by telephone. Modes 3 and 4 involve face to face reinterviews and have much higher *gdr* than modes 1 and 2. As it is probably true that households for which the reinterview is conducted in person differ in many ways from the other households, no firm conclusion can be reached about the relative consistency of reports obtained by face to face and telephone interviews. Nevertheless the results suggest that telephone interviewing is at least as reliable as face to face interviewing.

When separate logit models were fitted to the data, including each of the explanatory variables above one at a time, the results naturally confirmed the results above. This is so because the logit model simply reexpresses the relationship in the form of odds of disagreement rather than the proportion of disagreements. In terms of one-way or marginal analyses, therefore, the implications of the results will necessarily be the same.

The principal advantage of the log-linear formulation is that it permits the joint analysis of the effects of many variables. With the *gdr* as the measure of variability and using the cross-tabulation as the method of comparing effects, joint analysis by many variables quickly runs into difficulties in terms of cell sizes and the stability of the estimates. Once the model is specified as a logistic regression, however, there is no difficulty in estimating simultaneously the effects of many variables. For the dependent variable UNM we fitted a model which included six explanatory variables — AGE, KIN, CONT, PROX, R1KIN, and R2KIN. [The data were confined to three categories of KIN — (male) heads of household, wives of heads of household, and other relatives in the household, in order to simplify the interpretation of the results.]

$$\log\left[\frac{g}{1-g}\right] = \beta^{1} + \beta^{\mathrm{AGE}} + \beta^{\mathrm{KIN}} + \beta^{\mathrm{CONT}} + \beta^{\mathrm{PROX}} + \beta^{\mathrm{R1KIN}} + \beta^{\mathrm{R2KIN}} \qquad (27.9)$$

Table 27.3 shows the estimated parameters for the model (27.9) and also for each of the explanatory variables fitted independently. Column 1 of the table shows the marginal effect of each explanatory variable; column 2 shows the estimated effect when all six explanatory variables are included jointly in the model — i.e., the effects of each variable taking into account (or controlling) the effects of the other variables. The model given here includes only main effects and does not allow for the presence of interactions between the explanatory variables.

In the case of CONT, the effect of contamination on the estimate of response variability, the presence of the other variables in the model has no appreciable effect on the estimate of the parameter. Both when it is considered alone and jointly with the other variables the impact of contamination is to halve the estimated response variability.

The second and third explanatory variables considered in Table 27.3 are the age of the subject (AGE) and the relationship of the subject to the head of household (KIN). These variables are the most relevant variables of this kind available in the data set. Considering KIN in isolation we see that heads of household have the lowest response variability, wives of heads of household have almost the same, but other relatives in the household have a considerably higher response variability — their odds of discrepancy between the two interviews being more than three times higher than the other two groups. In the case of AGE there is a clear

Table 27.3. Estimated Parameters for the Logistic Regression of Odds of Disagreement on the Set of Explanatory Variables, Fitted Separately and Jointly

Marginal effect			Joint effects	
Estimate	e^{β}	Parameter	e^{β}	Estimate
0.00	1.00	β_1^{AGE}	1.00	0.00
−0.71	0.49	β_2^{AGE}	0.67	−0.40
−1.07	0.34	β_3^{AGE}	0.51	−0.68
−1.28	0.28	β_4^{AGE}	0.41	−0.88
−1.60	0.20	β_5^{AGE}	0.30	−1.21
−2.96	0.05	β_6^{AGE}	0.07	−2.64
0.00	1.00	β_1^{KIN}	1.00	0.00
0.07	1.07	β_2^{KIN}	0.94	−0.06
1.19	3.29	β_3^{KIN}	1.90	0.64
0.00	1.00	β_1^{CONT}	1.00	0.00
0.75	2.10	β_2^{CONT}	2.08	0.74

monotonic relationship between the age of the subject and the odds of discrepancy. The ratio of the odds for the youngest to the odds for the oldest is almost 20 to 1.

When the other predictors are taken into account, the pattern remains the same but the magnitude of the effects changes. Heads of household and their wives can still be seen to have comparable response variability and other relatives still have much higher response variability. The ratio of the odds of discrepancy for other relatives to the odds for heads of household and spouses is reduced to a little under two to one when age, etc. are taken into account. In the case of AGE the monotonic effect persists but the magnitude of the effects is reduced. The ratio of the odds for the youngest to the oldest is now a little under 14 to 1.

The analysis described above carries us a good deal farther in terms of understanding the determinants of response variability than previous analyses. The main advantage is that it is possible to consider the joint effects of many variables rather than confining ourselves to the analysis of the correlates of response variability one at a time. In the case of the joint analysis of AGE and KIN it is clear that, whereas there is an association between the two variables, each also has an effect separate from the other.

The other important variable fitted in the model above is PROX, the variable which measures the self/proxy response status for each of the responses in the data set. As this variable raises a number of particular issues, it is treated in the next section.

27.6 THE EFFECT OF SELF/PROXY REPORTING

Consider again the interview situation as described earlier in Figure 27.2. The data for some individuals (subjects) will be based on self-reports and for others will be based on reports by others (proxy reports). There is no prior basis for believing that the reliability of reporting will be the same for these two situations. Indeed, the conventional wisdom is that self-reports will be superior to proxy reports. A question of general interest to survey researchers, and of specific interest to CPS researchers, is whether survey data suffer in quality to the extent that some sampled individuals do not respond for themselves (i.e., to the extent that proxy responses are accepted in lieu of self-responses). Moore (1988) provides an interesting review of the literature in this area.

Whether a report is a self-report or a proxy report is a factor which may affect the quality of the report itself. In estimating the quality of the report, however, we need also take into account the extent to which the reinterview procedure departs substantially from the ideal of independent replication. Apart from procedural differences (including contamination, which affects the interviewer) there are two factors which may create a dependence between the response deviations in the first and second interviews. First, if the same individual responds to the questions on the first and second interviews, there is a chance that he or she will tend to give the same response on the two occasions because he or she remembers the response given on the first occasion. This would introduce a positive correlation between the response deviation on the original interview (e_{j1}) and that on the reinterview (e_{j2}). Second, even if the respondents are different on the two interviews, communication among members of the household could lead similarly to a positive between-interview correlation for the response errors.

In the CPS any eligible respondent can report for the whole household in either interview; thus some observations are self-reports and some are proxy reports. In terms of the interview-reinterview data being used in the estimation procedure, five categories of self/proxy combinations may usefully be identified — these are given in Figure 27.5.

In terms of the U.S. Bureau of the Census model of response variance, the impact of either memory or communication can be represented by a correlation coefficient ρ. A number of attempts have been made to estimate the value of ρ. Bailar (1968) in the context of the CPS and O'Muircheartaigh (1984) in the context of the World Fertility Survey considered the effect of increasing the time interval between interviews as a means of investigating ρ. Both studies found evidence for the presence of positive values of ρ but neither produced specific estimates. O'Muircheartaigh (1986), using the data considered in this chapter,

Original Interview	Reinterview	Code	Estimate of *SRV*
self	self	P_1	$s_{e_1}^2$
proxy	self	P_2	$s_{e_2}^2$
self	proxy	P_3	$s_{e_3}^2$
proxy	same proxy	P_4	$s_{e_4}^2$
proxy	different proxy	P_5	$s_{e_5}^2$

Figure 27.5. Self/proxy combinations.

presented a model for the estimation of ρ. The model provided two different quantitative estimates for ρ.

The model postulated that the simple response variance was different for self-reports and for proxy reports and that there was a between-interview correlation only for the reinterviews in which the respondent was the same as the respondent in the original interview. The correlation size was assumed to be the same for self- and proxy reports. Applying (27.4) to these assumptions yields:

$$
\begin{aligned}
E(s_{e_1}^2) &= \sigma_{e(s)}^2(1 - \rho) \\
E(s_{e_2}^2) &= [\sigma_{e(p)}^2 + \sigma_{e(s)}^2]/2 \\
E(s_{e_3}^2) &= [\sigma_{e(s)}^2 + \sigma_{e(p)}^2]/2 \\
E(s_{e_4}^2) &= \sigma_{e(p)}^2(1 - \rho) \\
E(s_{e_5}^2) &= \sigma_{e(p)}^2
\end{aligned}
\tag{27.10}
$$

These expected values provide two possible methods of estimating ρ. The first uses P_1, P_2, P_3, and P_4 and produces estimates of the order of 0.3 for ρ for all six of the labor force status variables considered. The second uses P_4 and P_5 and gives an alternative estimate of the order of 0.6. Two conclusions emerged from these results. First, the evidence is strong that there is a non-negligible between-trial correlation and that the usual estimate significantly underestimates the true *SRV*. The lower estimates assume that the cases in categories 2 and 3 are free from correlation. To the extent that there is communication within the household between the two respondents these estimates will be too low, thus it may be reasonable to assume that adjusting the estimates of *SRV* will underestimate *SRV*. Second, it is clear from the divergence between the two estimates of ρ that the simple model given by (27.1) and (27.10) is inadequate. No parsimonious model will ever represent the data perfectly, but it is clear that this model ignores some important terms.

Table 27.4. Values of *gdr* for the Five Self/Proxy Combinations, Together with the Logistic Parameter Estimates for the Marginal Effect of Self/Proxy Response Status, for the Variable UNM

Category	*gdr*	Estimate	e^{β}
self/self (P_1)	0.023	0.00	1.00
proxy/self (P_2)	0.041	0.52	1.68
self/proxy (P_3)	0.031	0.25	1.28
proxy/same proxy (P_4)	0.031	0.22	1.25
proxy/different proxy (P_5)	0.064	0.88	2.41

One possibility is to allow the between-trial correlation to differ for self-reports and proxy reports. This unfortunately does not help, as the data will not produce valid estimates of the two correlations jointly (i.e., estimates with $|\rho \leq 1|$). This suggests, as does an inspection of the data, that other characteristics of the individuals besides their self/proxy status must be introduced into the model.

The analysis in Section 27.5 shows that two important factors in determining the response variability of an observation are the age of the subject and the relationship of the subject to the head of household. Both these variables are related to the likely labor force status of the subject and also to the stability of that status.

In this section we reconsider the findings in O'Muircheartaigh (1986) using the model proposed in this chapter. The analysis is confined to the variable UNM.

Table 27.4 gives the parameter estimates derived from the model

$$\log\left[\frac{g}{1-g}\right] = \beta^{1} + \beta_{2}^{1\text{PROX}} + \beta_{3}^{1\text{PROX}} + \beta_{4}^{1\text{PROX}} + \beta_{5}^{1\text{PROX}} \tag{27.11}$$

and the corresponding values of *gdr* obtained from the 1986 analysis.

The results are fairly easy to interpret. The pattern of variation exhibited by the results is informative (and is similar for the five labor force status variables not presented here). In terms of *gdr* the lowest values of response variability are found in the P_1 (self/self) category, and these are uniformly lower than the values for the P_4 (proxy/same proxy) category. The two mixed categories P_2 (proxy/self) and P_3 (self/proxy) have similar levels of *gdr* for most variables, with a tendency for P_2 to be greater than P_3. The values of *gdr* for P_5 are very large relative to the other four categories.

In terms of the logistic model there is a parallel interpretation of the parameter estimates. The smallest value of beta occurs for P_1, and P_2 and

Table 27.5. Estimates of Effect of Self/Proxy Response Status for UNM

	Marginal effect		Effect in (27.9)	
Category	Estimate	e^{β}	Estimate	e^{β}
P_1 (self/self)	0.00	1.00	0.00	1.00
P_2 (proxy/self)	0.52	1.68	0.32	1.38
P_3 (self/proxy)	0.25	1.28	0.07	1.07
P_4 (proxy/same proxy)	0.22	1.25	−0.39	0.68
P_5 (proxy/different proxy)	0.88	2.41	−0.20	0.82

P_3 lie above P_1 and P_4. When expressed in this way, it is still clear that the response variability for P_5 is considerably larger than for the other categories.

This univariate analysis fails to take into account the other characteristics of the subjects and the respondents which may affect the response variability of the observations. From Section 27.5 we know that, in particular, the age and relationship to the head of the household of the subject are related to the degree of response variability. By fitting the model in equation (27.9) we can examine the way in which including these variables in the model (i.e., controlling for the effect of these variables) changes the interpretation of the effect of self/proxy response status. Table 27.5 gives the parameter estimates obtained when the six explanatory variables are fitted simultaneously. For comparison, the parameter estimates from fitting the one-variable model are given in the first column of the table.

The first point to note is that the relative response variability for the five self/proxy categories has been completely changed. The lowest variability is now to be found for P_4 (proxy/same proxy), and the second lowest for P_5 (proxy/different proxy). The three categories which include one or more self-reports have the highest response variability.

This reversal of the apparent effect of proxy reporting comes about because the analysis in Table 27.5 of the contrast between response status categories is controlled for some other characteristics of the data and these other characteristics are related to the level of response reliability. In particular the great majority of self-reports are from heads of household and wives of heads of household; these are types of respondents with relatively low response variability. Among proxy reports, on the other hand, about 40 percent of the cases belong to the "other relatives" category; this is a category with relatively high response variability. This imbalance in the distribution of self/proxy classes produces an artificially high estimate of response variability for the cases with proxy reports in one or other interview. The impact is particularly striking for the P_4 and P_5 groups, in which both interviews have proxy respondents.

This means that the marginal effect of proxy reporting (the effect estimated by a simple comparison of the response variabilities of the different classes of self/proxy status without taking the other characteristics of the cases into account) creates an entirely false picture of the true situation. This is an example of Simpson's paradox (for discussion see O'Muircheartaigh, 1986). The apparent superiority of self-reports suggested by Table 27.4 is merely an artifact of the relationship between the frequency distribution of self/proxy response status by type of subject and the different levels of reliability obtained for different types of subject.

There are a number of points worth noting about the results in columns 3 and 4 of Table 27.5. The category with the lowest variability is now P_4, the subjects for whom the same proxy respondent reported on the two occasions. The next lowest variability occurs for P_5, where there are also two proxy reports but the respondents are different on the two occasions. The highest variability is found for the classes P_2 and P_3, in which there is a mixture of self and proxy reporting. The self/self class (P_1) is now seen to have a higher variability than the proxy/proxy class.

In order to disentangle the influences at work here it is useful to reexpress what is happening in different terms. The first factor we would like to evaluate is the contrast between self and proxy reporting; thus we would like to estimate a self and a proxy effect. A factor which contaminates our estimate of this effect is memory (when the two occasions have the same respondent) and this memory effect may be different for self reports and for proxy reports. It is not unreasonable to postulate that memories of self reports will be stronger than memories of proxy reports.

The second factor is communication within the household. Communication is clearly possible in all situations. Of the three nonmemory situations, communication is least likely to be a factor in the case of P_5, where there are two different proxies; in any case the effect of communication is confounded in the estimation process. Of the two remaining situations we distinguish here between the case where the first report is a proxy report and the case where the first report is a self report. Where the first report is a self report, we argue that the impact of the communication may be to impress on the mind of the subsequent proxy respondent the answer(s) given by the self respondent so that there will be a positive between-interview correlation for the response deviations. Where the first report is given by a proxy respondent there is much less likely to be an effect of the communicated response on the subsequent self report.

To clarify the reasoning involved we express the five classes of self/proxy reports in different terms. Using notation analogous to that in the logistic model we will denote

effect of self-reporting by β_s
effect of memory (self) by $\beta_{m(s)}$
effect of proxy reporting by β_p
effect of memory (proxy) by $\beta_{m(p)}$
effect of communication by β_c.

We can now write the relationship between the five parameters β_1^{PROX} through β_5^{PROX} and the parameters above as

$$\beta_1^{\text{PROX}} = \beta_s + \beta_{m(s)}$$

$$\beta_2^{\text{PROX}} = (\beta_p + \beta_s)/2$$

$$\beta_3^{\text{PROX}} = (\beta_s + \beta_p)/2 + \beta_c$$

$$\beta_4^{\text{PROX}} = \beta_p + \beta_{m(p)}$$

$$\beta_5^{\text{PROX}} = \beta_p$$

where we expect the parameters $\beta_{m(s)}$, $\beta_{m(p)}$, and β_c to be negative.

Substituting the estimated values for β_1^{PROX} through β_5^{PROX} we can solve these equations for the five new parameters. We obtain the values

$$\beta_s = 0.84 \qquad e^{\beta_s} = 2.32$$

$$\beta_p = -0.20 \; e^{\beta_p} = 0.82$$

for self and proxy reporting.

The analysis confirms, as did the 1986 analysis, that self reporting is less reliable (more variable) than proxy reporting. This is disturbing, as both the conventional wisdom and common sense suggest that the best information comes from self reports. A fundamental problem with the present results arises from the nonexperimental nature of the selection of the self and proxy respondents. Respondents are self-selected and there may be biases in this process which invalidate the comparison. This would cause particular difficulty if the respondent characteristics that are the subject of the inquiry were related to the factors which determine the likelihood of self response. There is some evidence in the CPS data to support this. The probability of being found at home, and therefore of being chosen as a respondent, is clearly related to labor force status and perhaps also to ambiguity of labor force status. Furthermore the interviewers also have some control over the choice of respondent. It is possible, though not testable, that judicious selection among eligible respondents may exaggerate the quality of data provided by proxy respondents.

It is not suggested that response variability is a measure that describes all, or even the most important, aspects of data quality. It is possible that the response variability may be lower but that biases, for instance, may be greater. There is no evidence of this in the data here, but it is certainly a possibility. Mathiowetz and Groves (1985), however, in a study of telephone interviewing for the Health Interview Survey found "an overall tendency toward higher (better) proxy reports [which] runs directly counter to previous beliefs about self vs. proxy reports" (p. 96). A comparison of self and proxy reports in the Survey of Income and Program Participation (SIPP) based on an exhaustive record check provides evidence that self reports in that context were not superior to proxy reports.

There is an accumulation of evidence that suggests that in many situations self reports are not as clearly superior to proxy reports as we have previously imagined. It seems counter-intuitive that proxy reports should in general be superior other than in situations where there is a strong social desirability or threat element in the questions. Notwithstanding the qualifications expressed above, the evidence is now sufficiently striking that further controlled investigation is warranted. One hypothesis we would put forward is that the design of the survey instrument and its administration is not motivating the self reporter sufficiently to take advantage of the superior information available to him/her.

For the memory and communication effects we obtain the values

$$\beta_{m(s)} = -0.84 \quad e^{\beta_{m(s)}} = 0.43$$

$$\beta_{m(p)} = -0.19 \quad e^{\beta_{m(p)}} = 0.82$$

$$\beta_c = -0.25 \quad e^{\beta_c} = 0.78$$

If transformed to the parametrization of the earlier model these estimates would give

$$\rho_{m(s)} = 0.57$$

$$\rho_{m(p)} = 0.17$$

$$\rho_c = 0.22.$$

The estimates of $\beta_{m(s)}$, $\beta_{m(p)}$, and β_c clarify considerably the impact of the factors on the right hand side of Figure 27.2. First, they are intuitively acceptable (or plausible) values. The highest effect is for self memory; we would have expected this to be the case as a self report is more salient to the individual than would be a proxy report. The proxy memory effect is considerably lower but is still substantial. The effect of communication is very similar to the proxy memory effect.

It may be instructive to give an interpretation of the parameters under each of the two models.

The correlation coefficients $\rho_{m(s)}$ etc. express the memory and communication effects as a suppression of the simple response variance due to correlation between the response deviations on the two occasions. Thus the value of $\rho_{m(s)}$ of 0.57 implies that the correlation between the response deviations is such that when self reports are obtained on both occasions the estimated *SRV* underestimates the true *SRV* by 57 percent. When the two responses are from the same proxy respondent, the underestimation is of the order of 17 percent. Even when the two respondents are different, communication between the respondents on the two occasions can lead to a 22 percent underestimation of the *SRV*.

Under the log-linear model, the memory effect is seen as a reduction in the odds of finding a discrepancy between the responses on the two occasions. In a manner analogous to that in the *SRV* model, the fact that the questions are answered by the same individual on the two occasions means that the true odds of discrepancy will be underestimated by the proportion of discrepancies in the data. A value of $\beta_{m(s)}$ of -0.84, giving a value of 0.43 for e^{β}, implies that the expected value of the observed odds of discrepancy will be less than half the true odds of discrepancy. Similarly, the values of -0.19 for $\beta_{m(p)}$ (0.82 for e^{β}) and -0.25 for β_c (0.78 for e^{β}) quantify the suppression of discrepancies by proxy memory and communication, respectively.

Though it is possible to interpret the memory and communication effects under either of the two models, it is worth pointing out that the Census Bureau model would not permit the direct estimation of the parameters because of the need to take into account the joint effects of many variables in the estimation.

27.7 CONCLUSION

For more than thirty years the response variability in the Current Population Survey has been measured on the basis of the gross difference rate, *gdr*, presented by Hansen and his colleagues in 1961. Dealing as they were primarily with binary data the *gdr* provided a plausible and intuitive measure of the reliability of the responses. The *gdr* is the proportion of cases in which the reinterview response is different from the response in the original interview. It is also equal to twice the simple response variance (*SRV*) and therefore fits well into the usual survey model of response errors. Two particular issues are addressed in this chapter. The first is the problem of estimating the quality of the data given the shortcomings of the reinterviews as independent replicates of the original interviews. The second is the need to understand the determinants of data quality.

It happens that these issues are interrelated, and that in order to tackle them it is necessary to express the response process in a different way from that in which it is expressed in the Census Bureau model. The crucial difference is that instead of expressing response variability in terms of the proportion of discrepancies between the two interviews, it is expressed in terms of the odds of discrepancy. Though this may appear to be a slight difference it has very considerable implications for the analysis of data quality. In particular it makes it possible to construct a model in which both the factors affecting data quality and the factors affecting the measurement of data quality can be incorporated. We argue that this model is both more plausible and more useful than the standard model. It addresses the fundamental problems which arise in dealing with binary data in that the model is one specifically designed for such data and it makes it possible to incorporate into the analysis the structure of the data collection process.

The model provides an advance on earlier work in that it makes it possible to address all these issues simultaneously. It produces estimates of memory effects and communication effects, and distinguishes between memory for self reports and proxy reports. At the same time the analysis confirms the characteristics that are related to the magnitude of response variability, while controlling for the measurement process.

The analysis also provides some ideas for future work. It should be possible to reparameterize the model so that the memory and communication effects can be entered directly as parameters in the model rather than derived indirectly from the model parameters. The resulting estimates can then be compared directly (and, if desirable, included in hypothesis tests). It is possible to produce, and it may be possible to estimate, a model which conditions on true values. In such a model the probability of a false positive would be distinguished from the probability of a false negative; this would make the model a better representation of reality. Alternatively it might be possible to implement a model which would incorporate all the elements of the interview-reinterview table rather than the summary provided by $g/(1-g)$.

CHAPTER 28

EVALUATION OF MEASUREMENT INSTRUMENTS USING A STRUCTURAL MODELING APPROACH

Willem E. Saris
University of Amsterdam

Frank M. Andrews
University of Michigan

28.1 INTRODUCTION

In this chapter procedures for the evaluation of nonsampling errors in research based on survey or panel data will be discussed. In the social sciences many criteria have been discussed for evaluation of measurement instruments. We would like to mention some without the pretense of being complete: validity, invalidity, reliability, precision, measurement level, comparability of scores across respondents, and costs. Not all of these criteria will be discussed in this chapter. For more information on precision we refer to Cox III (1980) and Lodge (1982). With respect to measurement level we refer to Orth (1982) and Wegener (1982). Comparability of scores and costs has been evaluated by Saris (1988; see also Chapter 1 of this book).

We will further restrict the focus of this chapter to quality criteria for single questions and not composite scores. We also restrict the discussion to variables measured precisely enough to be treated as interval variables (Borgatta and Bohrnstedt, 1981). Labovitz (1970), Davison and Sharma (1988), and many others suggested that treating ordinal scales as if they were interval rarely produces misleading results.

We thank Ingrid Munck, Arne Kolstadt, Richard Bagozzi, and Brendan Bunting for their suggestions for improving the manuscript.

Measurement Errors in Surveys.
Edited by Biemer, Groves, Lyberg, Mathiowetz, and Sudman.

ISBN: 0-471-53405-6

On the other hand, Olssen (1979), van Doorn, et al. (1983) and many others have suggested that it can make quite a difference. There is also a vast literature on scaling of sets of categorical variables. We will not discuss this literature here. These restrictions mean that we will concentrate on measurement instruments consisting of single questions for which the response scale is precise enough to be treated as an interval variable. Many results discussed below could also be applied to composite scores, but we do not want to go into this discussion here.

In the older literature reliability and validity have been estimated using correlation coefficients (Cronbach, 1951; Campbell and Fiske, 1959). This approach has correctly been criticized for not being correct under all conditions. (See Werts and Linn (1970) and Althauser and Herberlein (1970).) Bollen (1989) provides a more general overview of these problems.

As an alternative, it has been suggested that structural models be used to estimate measurement parameters for the variables of interest. In these models certain parameters can be interpreted as indicators of validity, reliability, and invalidity. This approach has been suggested by many authors. One of the first papers using this approach explicitly was a paper by Heise and Bohrnstedt (1970). We can also refer to Werts and Linn (1970), Jöreskog (1971), Werts, Jöreskog and Linn (1972), Alwin (1974), Andrews and Withey (1974) and many others. More recently this approach has been used by Andrews (1984), Alwin (1989), and Saris (1990). Each of the last three papers takes different approaches in the evaluation of measurement instruments. Andrews followed the multitrait-multimethod approach originally introduced in the social sciences by Campbell and Fiske (1959). Alwin used a different model originally introduced in the social sciences by Heise (1969) and further developed by Wiley and Wiley (1970) which is known as the quasi simplex model (Jöreskog 1973). Saris (1990) suggested a third approach. His model is in a sense a combination of the two previous models. The basic form of this model was suggested by Jöreskog in 1971 in a paper on the congeneric test model, and it has been applied for the kinds of purposes discussed here by Saris (1982).

In the next sections we will start with an overview of the quality criteria we want to obtain for measurement instruments. Second, different designs will be discussed to estimate the parameters which are indicators for these criteria. Then an evaluation of the different designs will be made.

28.2 QUALITY CRITERIA FOR MEASUREMENT INSTRUMENTS

The literature on validity and measurement error is rich and varied, but also imprecise and inconsistent (Andrews, 1984). Therefore it is necessary that we say a few words about the different criteria we will use in this chapter. We think that some of the most important issues in measurement concern validity, invalidity, and reliability. About each of these concepts books have been written, and we do not pretend to give the final answer here. But at least we will try to provide some formal framework to indicate what we mean by each of these concepts.

In the social sciences we normally use questions to obtain answers, which are indicators of the latent variables of interest. In this process random measurement errors will always occur. These errors may be due to misunderstandings, typing and coding errors, wrong interpretations of the question, etc. This means that the scores on the latent variables of interest will never be known to us. We only obtain the responses to questions. Nevertheless, researchers are interested in the relationships between the latent variables of interest and not between observed responses. The common research practice in structural equation modeling (see Bollen, 1989 for a recent overview) is that researchers collect information from a sample of the population and derive from that sample estimates of the relationships between the different latent variables and between the latent variables and the observed variables. In this approach the observed as well as the latent variables are treated as random variables. We will do the same for our measurement models.

If we represent the unobserved variable of interest by the symbol F and the response given to the question using method i by y_i and the error by ε_i, we can formulate the simple model:

$$y_i = F + \varepsilon_i \tag{28.1}$$

where we assume $\text{cov}(F, \varepsilon_i) = 0$ and $E(\varepsilon_i) = 0$.

The first assumption means that we expect that in the population there is no correlation between the errors and the latent variables. This assumption is realistic if we can assume that no other variables related to the variable of interest have a direct influence on the responses. This is an essential assumption in all further models. If this assumption is not realistic one should explicitly include the omitted variables in the models in order to obtain consistent estimates of the parameters of interest. Such models have been suggested in Groves (1989) and Chapter 24.

The second assumption is somewhat arbitrary. It suggests that the expectation of the random error term in the population is zero. This assumption could have been ignored if we had introduced a constant term in (28.1) to take care of the difference in zero point on the scales of the observed and the latent variable. But the latent variable has no fixed scale so we can easily make the assumption that the scales have the same zero point. Note that only the observed variable and the error variable have an index i because F is assumed to be independent of the measurement procedure used. It would be very attractive if all our measurement instruments would satisfy this simple model. Unfortunately this is not the case. In the rest of this section some deviations from this simple model will be given. A first deviation from the specified model is that the scales of the observed and latent variables may be different. In (28.1) it is assumed that the scales of the latent and observed variables are the same and that the mean of the one is the mean of the other. This is of course a somewhat arbitrary decision because it is difficult to speak about the scale of an unobserved variable. However, later we will see that we have to take into account the fact that the arbitrarily chosen scale of the latent variable cannot be the same as the scales of all the observed variables. In that case we have to introduce scaling constants in equation (28.1) to correct for the differences between scales. Restricting ourselves to linear transformations, we get model (28.2):

$$y_i = a_i + b_i F + \varepsilon_i \tag{28.2}$$

with the same assumptions as before. The coefficients a and b are the scaling constants, which have an index because they can vary with the method used. Model (28.2) should be seen as a more general formulation of the simple model of equation (28.1).

Many studies have been done which suggest that the responses are not only affected by the opinion of the respondent and some random error but also by some other influences that are stable, in the sense that they appear again when the same measure is repeated. Such influences may include personal characteristics of the respondent or interviewer and/or aspects of the method used. Some of these factors produce effects on the distributions of the responses but, since we are mainly interested in relationships between variables, we will ignore these errors in this chapter. Other factors produce correlations between the responses on different questions independent of the method used. Although such factors invalidate measurement procedures we will not deal with them here because these factors are not specific for the method used but mainly for the traits which are measured.

We will concentrate in this study on variables which cause correlations between observed variables measured with the same method. This possibility was originally suggested by Campbell and Fiske (1959) but has been proposed by many others since then. Normally these factors are not directly measured. If we denote the sum of all these stable disturbance factors by D_i, we can adjust equation (28.2) by introducing this new variable as an additive factor:

$$y_i = a_i + b_i F + D_i + \varepsilon_i \tag{28.3}$$

where we assume $\text{cov}(F,D_i)=0$, $\text{cov}(F,\varepsilon_i)=0$, and $\text{cov}(D_i,\varepsilon_i)=0$.

If only one trait is studied, the D_i and ε_i cannot be distinguished, but we will show later that, by using certain designs, the variances of D_i and ε_i can be estimated. It is even possible to decompose the variance of D_i in a method-specific component M_i and a component representing the interaction between the trait and the method, also called the "unique component" of the method and the trait and denoted by U_i.

Using these different terms, the formulation becomes:

$$y_i = a_i + b_i F + g_i M_i + U_i + \varepsilon_i \tag{28.4}$$

for which we assume $\text{cov}(F,U_i)=0$, $\text{cov}(F,\varepsilon_i)=0$, $\text{cov}(U_i,\varepsilon_i)=0$, $\text{cov}(F,M_i)=0$, $\text{cov}(M_i,U_i)=0$, and $\text{cov}(M_i,\varepsilon_i)=0$.

The assumptions made are of the same type as we have mentioned in connection with equation (28.1). In the formulation of equation (28.4) we have added a parameter g_i to the method component M_i. This is an arbitrary choice but it makes the models with more than one observed variable more flexible as we will see shortly.

With only one observed variable for each method the two error terms, U_i and ε_i, cannot be estimated separately. Later, however, we will show that separation of the two terms is possible if repeated observations of the same response variable are available.

If we repeated the same question under exactly the same condition, the variation due to the method (M_i) and the unique disturbance (U_i) would remain the same. If also the variable of interest did not change, the only change in the response we could expect would be due to the random error component (ε_i), which can change through time. Combining the stable components in one term and designating them as T_i, we can write:

$$y_i = T_i + \varepsilon_i \tag{28.5}$$

and

$$T_i = a_i + b_i F + g_i M_i + U_i \tag{28.6}$$

where $\text{cov}(F,U_i)=0$, $\text{cov}(M_i,U_i)=0$, $\text{cov}(M_i,\varepsilon_i)=0$, $\text{cov}(F,\varepsilon_i)=0$, $\text{cov}(U_i,\varepsilon_i)=0$, $\text{cov}(F,M_i)=0$, and $\text{cov}(T_i,\varepsilon_i)=0$.

It should be clear that equation (28.4) is in complete agreement with equations (28.5) and (28.6), since substitution of (28.6) in (28.5) leads to (28.4).

Measurement quality parameters are conventionally presented as standardized coefficients (either product-moment correlation coefficients or effect coefficients from a standardized structural model). Shifting to standardized format, equations (28.5) and (28.6) can be written as follows:

$$y_i = h_i T_i + \varepsilon_i \tag{28.7}$$

and

$$T_i = b_i F + g_i M_i + U_i \tag{28.8}$$

with the same assumptions as above. In these equations the variables and the coefficients are standardized except the error term ε_i and the disturbance term U_i, which are normally not standardized but multiplied

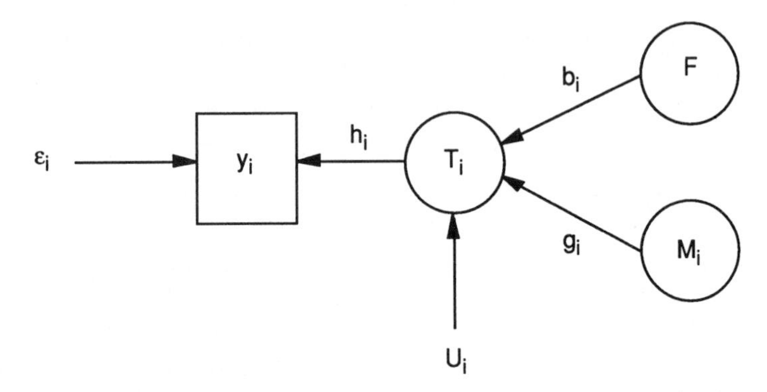

Figure 28.1. Composite path diagram of the measurement process for one trait and the method, M_i.

by a constant as can be seen very simply. From now on we assume that all parameters and variables are standardized except the disturbance terms in the equations. A path diagram that represents (28.7) and (28.8) is given in Figure 28.1.

Having specified a very general measurement model (equations 28.5 and 28.6 or 28.7 and 28.8), the different measurement quality criteria can be defined. We will begin with reliability. *Reliability* is often defined as the proportion of the variance in y_i that is stable across repeated implementations of the same measure. Using the unstandardized equation (28.5) we can write:

$$\text{reliability} = \text{var}(T_i)/\text{var}(y_i). \tag{28.9}$$

Alternatively, this is equal to the squared standardized regression coefficient of (28.7) since standardization of (28.5) shows that:

$$h_i = [\text{var}(T_i)/\text{var}(y_i)]^{\frac{1}{2}}. \tag{28.10}$$

The most basic conceptualization of reliability refers to the extent that two measures agree with each other. Thus it is no surprise that the reliability is numerically identical to the expected correlation between two measures (with equal reliabilities) of the same stable variable (T). In our notation, the correlation between two such measures will equal h_i^2.

We must, however, give a warning on this point. We cannot ignore the possibility that the trait (opinion or attitude of interest) has been changed between the two observations. In that case the correlation between the observed variables will underestimate the reliability. On the other hand the estimate of h_i^2 will still be a correct measure for test-retest reliability.

This conceptualization of reliability is in agreement with Lord and Novick (1968) if we set the stable part of the response variable equal to the "true score" as they call it. This term is sometimes taken too literally. We will, of course, never know the true value of a respondent on a latent variable; however, that is not what is meant by the term "true score." The "true score" only indicates the observed score minus the random measurement error or the observed score corrected for random measurement error.

The definition of reliability given above is not the only possible definition. We have given the definition of the test-retest or stability reliability. A different and, perhaps, more popular approach uses "internal homogeneity," "consistency," or "inter-judge" reliability. This form of reliability is defined as the correlation between two different measures which are supposed to measure the same variable. If these variables have F as common factor, this correlation, which we will denote as $r(y_i, y_{i'})$, is equal to

$$r(y_i, y_{i'}) = h_i b_i h_{i'} b_{i'}. \tag{28.11}$$

It will be clear that this definition of reliability is only in agreement with the test–retest reliability if $b_i = b_{i'} = 1$ and if both measures have equal test–retest reliability ($h_i = h_{i'} = h$). Only in that case is the correlation between two observed variables equal to h^2 and only in that case does this measure indicate the strength of the relationship between the true score of a variable and the observed variable.

A second alternative (Andrews, 1984 and Bollen, 1989) is the squared multiple correlation coefficient for the observed variable ($R^2_{yi \cdot F,M_i}$). For the squared multiple correlation we can write in our notation (Saris, 1982 or Hays, 1973):

$$R^2_{yi \cdot F,M_i} = (h_i b_i)^2 + (h_i g_i)^2. \tag{28.12}$$

This alternative homogeneity reliability measure is also a complex form which is only equal to the test-retest reliability if $b_i = 1$ and $g_i = 0$.

The formal specification of the last two definitions of reliability indicates that these reliability estimates are complex statistics, based on several different characteristics of the measurement model. This might be an argument to prefer the more simple definition of test–retest reliability.

Validity can also be defined in many different ways. Heise and Bohrnstedt (1970, p.107) suggest a definition of validity which is comparable with the reliability measure indicated above, namely, the squared standardized effect of the variable of interest on the true score for the response of interest; i.e., in the notation of equation (28.8) and Figure 28.1,

$$\text{validity} = b_i^2. \tag{28.13}$$

Given this definition of validity we will call the coefficient b_i the validity coefficient. If the variance of the true score explained by the variable of interest is the proportion valid variance, then 1 minus the explained variance is the *invalidity*. This invalidity can be decomposed in a proportion *method variance* (g_i^2) and a proportion *unique variance*: var(U_i). This follows immediately from equation (28.8).

Although this specification of measurement concepts seems logical, the approach is not frequently used and even then different terms have been used (Saris, 1982, Marsh and Hocevar, 1988 and Rindskopf and Rose, 1988). In order to avoid confusion (or to add to the confusion) let us call this validity "true score validity."

Validity is more commonly evaluated using the correlation between the latent variable (F) and the observed score (y_i) (see for example Groves (1989) and Andrews (1984)). Since the relationship between the indicator and the latent variable is used and not the relationship between the true score and the latent variable of interest, let us call this validity the indicator validity (IV).

Using simple path analysis it will be clear that

$$IV = (h_i b_i)^2. \tag{28.14}$$

The difference between the two measures of validity is clear from the definition. The true score validity (TV) is a pure measure, independent of the reliability of a measure. The IV is dependent on the reliability and the TV. The consequence of this is that the TV has a simpler interpretation.

If $TV = 1$, which is possible, then we know that the true score of a measure correlates perfectly with the variable of interest and can only differ from this variable by a scale transformation. This means that the observed variable and the variable of interest are really the same except for random measurement error and a possible linear transformation. If TV is less than 1 we know exactly how strong the relationship between the observed variable and the variable of interest is after correction for random measurement error. This shows that the TV gives a good quantitative measure of the validity of the observed variable (independent of random measurement error).

On the other hand, the IV will rarely be 1.0 as there is usually random measurement error. If IV is not 1.0 we never know if the deviation from 1.0 comes from unreliability or from invalidity. This means that no clear impression can be obtained of the validity of the observed variable independent of the random measurement error. Although the true score validity has the simpler interpretation and should, therefore, be preferred, we will see later that the IV also has an advantage.

Having defined the different quality criteria using this model and having made the connection between the parameters of the model and some of the criteria that have been used for evaluation of measurement instruments, we have to say that the parameters of this model cannot be estimated using only one observed variable. In this section the purpose was only to define the different concepts which will be used later.

28.3 DIFFERENT RESEARCH DESIGNS

In this section, we will discuss five different approaches used to estimate the quality of an instrument. If necessary, we will make brief references to other designs which are specific cases of the designs treated. After this discussion we will evaluate the different approaches.

28.3.1 The Quasi-Simplex Approach

The first design discussed here is the quasi-simplex approach (Heise, 1969). Heise suggested that this model is probably the most reasonable one to use for social science variables given that one has no batteries of questions which measure the same latent variable but that each question taps a different attitude or opinion. Given this situation, the only way one can get an identified model in order to estimate the quality of an instrument is the use of repeated observations. In this model a distinction between the variable of interest F and the true score T is not made because of identification problems, but it has been shown that, allowing for change of opinion or attitude, a design with three repeated observations is enough to get an identified model under mild restrictions (see Werts, et al., 1971). The basic model is presented in Figure 28.2 (with adaptations to match the notation used here).

Formally this model can be specified, assuming that only one method (question) is used over time, as follows:

$$y_1(t_i) = h_1(t_i)\ T_1(t_i) + \varepsilon_1(t_i) \qquad i = 1,2,3 \tag{28.15}$$

$$T_1(t_i) = d_{i,i\text{-}1} T_1(t_{i\text{-}1}) + U_1(t_i) \qquad i = 2,3 \tag{28.16}$$

where, for all k and l: $\text{cov}[T_1(t_k),\varepsilon_1(t_l)] = 0$; $\text{cov}[T_1(t_k),U_1(t_l)] = 0$ (for $k < l$); $\text{cov}[\varepsilon_1(t_k),\text{U}_1(t_l)] = 0$; $\text{cov}[\varepsilon_1(t_k),\varepsilon_1(t_l)] = 0$ (for $k \neq l$); $\text{cov}[U_1(t_k),U_1(t_l)] = 0$ (for $k \neq l$); and, for all t_i, $E[T_1(t_i)] = E[y_1(t_i)] = E[\varepsilon_1(t_i)] = 0$.

The first equation specifies the measurement model which is in fact the same as equation (28.5). The second equation of this model specifies a model for the possible changes of the variable T_1 over time.

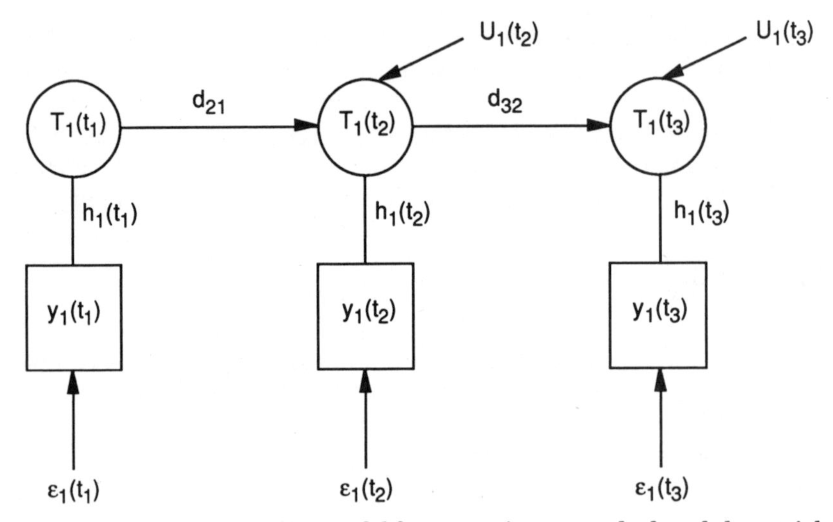

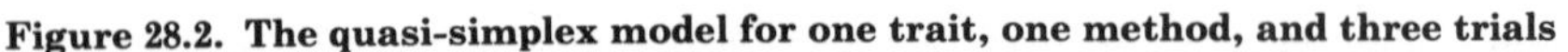

Figure 28.2. The quasi-simplex model for one trait, one method, and three trials

This specification is not sufficient to identify all the parameters. For this purpose one has two different possibilities. The solution suggested by Heise (1969) was the assumption that the standardized $h_1(t_i)$ coefficients are identical for each time point. This assumption means that one specifies that the reliability is constant over time. An alternative possibility is to work with unstandardized variables and to assume that $h_1(t_i) = 1$ at all time points and that the error variances remain the same through time.

The measurement equation is the same as equation (28.5). This means that only the reliability can be estimated with this design. The validity and invalidity cannot be estimated. Estimation of validity and invalidity is only possible if more than one method has been used. Therefore we have to conclude that this model does not provide all the information needed to evaluate a measurement instrument.

On the other hand the procedure is good if we want to determine the test-retest reliability of the variables. This approach has the advantage that it provides estimates of these reliability coefficients even under the condition of changing attitudes or opinions. The standardized h coefficient is the reliability coefficient of the measurement procedure. The only limitation of the quasi-simplex approach is that one has to assume that the reliabilities or measurement error variances at all points in time are the same if one has only three observations over time. If one has more than three observations, one can get an estimate of the reliability of the measurement procedure at any point in time except the first and the last point, which remain unidentified (Werts, et al., 1971).

A special case is the situation in which the latent variable does not change through time. In that case the model can be reduced to the parallel test model or tau equivalent model of the classical test theory (Lord and Novick, 1968). But this model has of course the same limitations as the quasi-simplex model: it provides only information about the reliability of instruments and not about the validity.

Given the limited possibilities of this design, we do not think that this approach is appropriate for evaluation of measurement instruments.

28.3.2 The Repeated Multi-Method Approach

As an alternative to the previous not completely satisfactory design, Saris (1990) has suggested the repeated multi-method (RMM) approach. A similar approach has been suggested by Jöreskog (1971) and by Saris (1982).

In this approach only one trait is measured, but at least two methods must be used, and observations must be made on at least two points in time with all measurement instruments. The two observations with the same instrument have to be within a short time interval so that it is reasonable to assume that the opinion or attitude has not changed during this interval. On the other hand, the time interval should be long enough so that the observations can be seen as independent observations of the same variable.

If one trait is measured by three methods at two points in time, the model described in Figure 28.3 can be used for description of the data obtained.

This model is completely in agreement with equations (28.5) and (28.6), which we have discussed before. $y_i(t_1)$ and $y_i(t_2)$ represent the two observations with method i of the variable of interest. T_i represents the true score obtained with the ith method. In order to obtain an identified model if only one latent variable is measured, at least three different methods have to be used. From a practical point of view (respondent burden and memory effects), this design is not the best. An alternative possibility is to measure two or more different traits. In that case one needs only two different methods which are used twice for each latent variable. The two methods do not have to be the same for each trait (Saris, 1990).

It can easily be checked that these models can be formulated by specifying for each observed variable equations (28.5) and (28.6). These models are identified and it is even possible to test them (Saris, 1990). In these models the reliability of the measurement instruments can be obtained from the squared standardized effect of T_i on $y_i(t_1)$ or $y_i(t_2)$ which is (h_i^2) or from the correlation between $y_i(t_1)$ and $y_i(t_2)$, which should be the same if the two observed variables satisfy the parallel test requirements

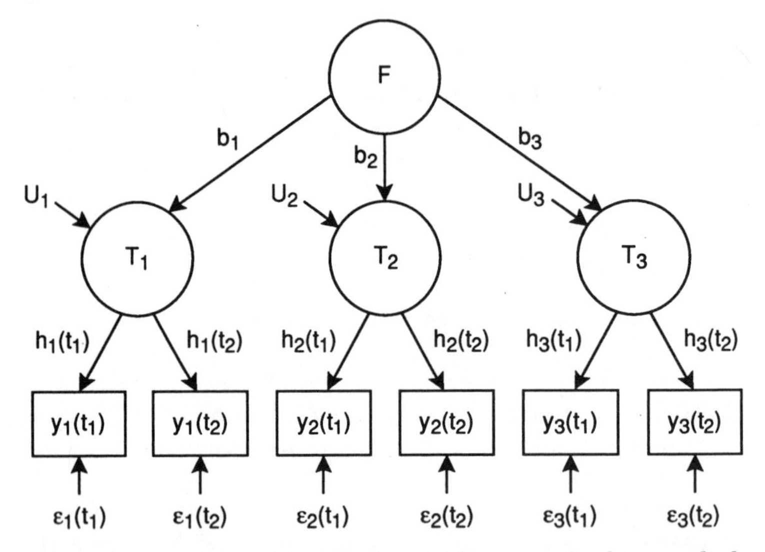

Figure 28.3. The repeated multi-method model for one trait, three methods, and two trials

(Jöreskog, 1971). In the terms of our earlier discussion of reliability, this model produces an estimate of test-retest reliability. As noted previously, a crucial assumption is that change has not occurred; to the extent it has, reliability will be underestimated.

The *true score validity coefficient* is, according to the previous definition, equal to the standardized b_i coefficient for each method. (In Figure 28.3 we see that the *indicator validity coefficient* will be the product of the paths from F to T_i and from T_i to y_i, and this product will equal $h_i b_i$.) The invalidity of each method is equal to the variance of the U_i terms. U_i terms.

The advantage of this design is that the design provides a good interpretable estimate of the reliability (test-retest) and validity (TV) of each measure. The disadvantages of this design are that the invalidity cannot be decomposed into method effects and unique components and one must assume that no change has occurred. Another complication is that many replications of the same question are necessary, which might lead to memory effects.

28.3.3 The Multitrait–Multimethod Approach

Campbell and Fiske (1959) originally suggested the use of different methods and different traits in order to get an impression of the quality of

Table 28.1 The Correlation Matrix for MTMM Design with Four Satisfaction Traits and Three Methods: Four Different Aspects of Life Are Evaluated in Three Different Ways Where y_{ji} = the Response for the jth Trait with ith Method

	y_{11}	y_{12}	y_{13}	y_{21}	y_{22}	y_{23}	y_{31}	y_{32}	y_{33}	y_{41}	y_{42}	y_{43}
y_{11}	1.00											
y_{12}	0.479	1.00										
y_{13}	0.462	0.467	1.00									
y_{21}	0.409	0.292	0.265	1.00								
y_{22}	0.330	0.382	0.298	0.763	1.00							
y_{23}	0.284	0.335	0.368	0.662	0.690	1.00						
y_{31}	0.467	0.395	0.324	0.317	0.371	0.255	1.00					
y_{32}	0.333	0.508	0.321	0.244	0.403	0.268	0.711	1.00				
y_{33}	0.272	0.419	0.406	0.216	0.312	0.335	0.675	0.663	1.00			
y_{41}	0.490	0.371	0.337	0.353	0.268	0.210	0.323	0.207	0.185	1.00		
y_{42}	0.337	0.468	0.342	0.210	0.281	0.195	0.227	0.281	0.231	0.654	1.00	
y_{43}	0.276	0.366	0.424	0.163	0.187	0.230	0.176	0.111	0.238	0.604	0.563	1.00

the measurement instruments being used. A typical correlation matrix which would be the result of such a study is presented in Table 28.1. In this case four traits are measured in three different ways. The traits are the satisfaction with three different aspects of life and a general satisfaction judgment. The three different methods are in sequence: a 10-point scale, a 100-point scale, and a five-points category. In this matrix the Pearson correlation coefficients are presented for the above-mentioned variables.

Originally, the correlation matrix itself was interpreted. Nowadays this matrix is not interpreted directly. Instead a causal model is specified to account for the correlations in the multitrait-multimethod matrix. We will do the same but we will not start with the commonly used model. Instead we start with a model which is more in line with our previous models. The model is presented in Figure 28.4.

The multitrait-multimethod (MTMM) design is different from the RMM approach due to the fact that each instrument is used only once. In Figure 28.4 we give a model for a design with only two traits and two methods to simplify the path diagram. Normally at least three traits and three methods are used. The presented model is different from the normally used model in that we make a distinction between the true score and the observed score as we did before. Normally this is not done because the model of Figure 28.4 is not identified. As only one observation is obtained with each instrument the random error term and the unique component cannot be distinguished. In order to solve this identification problem there are several options. Here we discuss two of them.

The most commonly chosen option to solve the identification

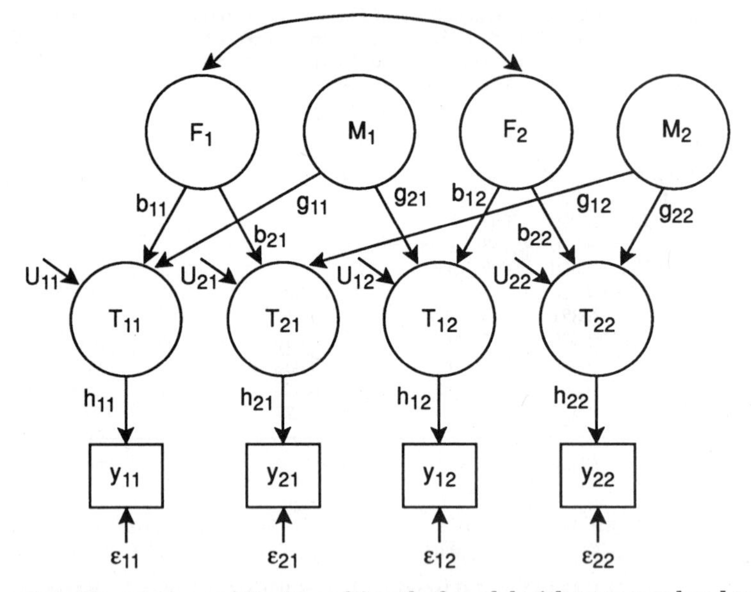

Figure 28.4. True score multitrait–multimethod model with two correlated traits and two uncorrelated methods

problem is that the true scores are ignored and the various factors and observed variables are assumed to affect each other directly (Alwin, 1974 and Andrews, 1984). This model is presented in Figure 28.5.

The model equations for this MTMM model can be derived from (28.7) and (28.8) by substitution of (28.8) in (28.7) and adding an extra subscript (j) in order to make a distinction between the different traits. This means that the effect of F_i on y_{ij}, denoted by ℓ_{ij} is an indirect effect which is equal to $h_{ij}b_{ij}$.

In the same way we can say that the method effect, denoted by m_{ij} is an indirect effect which is equal to $h_{ij}g_{ij}$. This classical MTMM model can be formulated as follows:

$$y_{ij} = \ell_{ij}F_j + m_{ij}M_i + W_{ij} \tag{28.17}$$

where for all i,j: $\text{cov}(F_j, W_{ij}) = 0$, $\text{cov}(M_i, W_{ij}) = 0$, $\text{cov}(F_j, M_i) = 0$, and $\text{var}(F_j) = \text{var}(M_i) = \text{var}(y_{ij}) = 1$.

Further, y_{ij} is the standardized response for the jth trait asked with the ith method, M_i represents the standardized variation which is due to the ith method, and W_{ij} is a disturbance term which consists of two components: U_{ij} and ε_{ij}.

We see that this approach leads to a model with parameters which are the *indicator validity* coefficient (ℓ_{ij}), an *attenuated method effect* (m_{ij}), and a residual (W_{ij}). An estimate of the reliability is obtained by adding the squared values of the validity and methods effects coefficients.

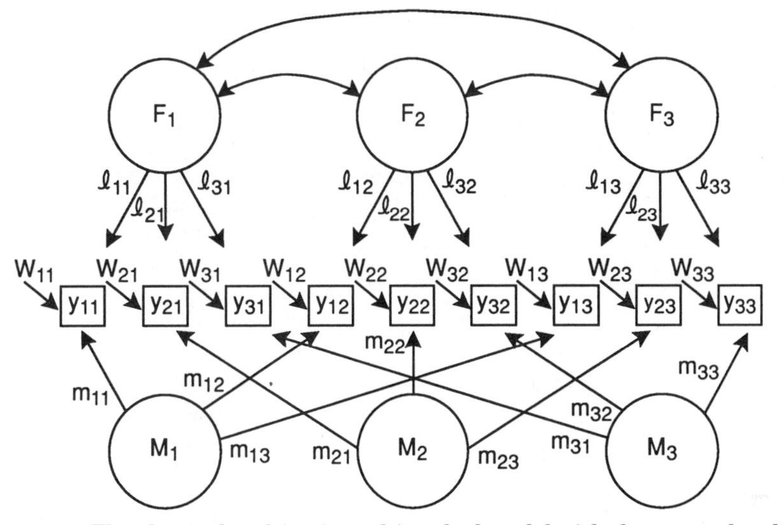

Figure 28.5. The classical multitrait–multimethod model with three correlated traits and three uncorrelated methods

Andrews et al. (1974, 1976, 1979, 1984) have used this approach in order to evaluate the instruments in several studies. The model used to estimate the parameters is the same as indicated above, with the additional restrictions that the method effects are identical for each method and that the correlations between the method factors are zero. The results of the estimation of the parameters of the model using LISREL 7 from the data of Table 28.1 are presented in Table 28.2.

First, we see that the measures of general satisfaction have a lower (indicator) validity. Secondly, we see that the categorical scale has lower validity than the other two types of scales. The method effects are somewhat comparable but they are all relatively small. The correlations produced by these method effects are all smaller than 0.15. The reliability estimates present a similar picture to that of the validity coefficients. This result is characteristic for the classical MTMM approach as we will see.

An alternative solution for the identification problem is obtained if it is possible to make the assumption that the unique components (U_i) in Figure 28.4 are zero. We will call this model the "true score model." In that case the model can be formulated in the following equations:

$$y_{ij} = h_{ij} T_{ij} + \varepsilon_i \qquad \text{for all } i, j \qquad (28.18)$$

and

$$T_{ij} = b_{ij} F_j + g_{ij} M_i \qquad \text{for all } i, j \qquad (28.19)$$

Table 28.2. The Estimated Values of the Parameters for the Data of Table 28.1 Using the Classical MTMM Model

	Satisfaction			
Method	General	House	Income	Contacts
		indicator validity coefficients (ℓ_j)		
10 pts	0.69	0.84	0.87	0.84
100 pts	0.73	0.88	0.82	0.78
category	0.66	0.79	0.81	0.72
		method effects (m_{ij})		
10 pts	0.32	0.32	0.32	0.32
100 pts	0.25	0.25	0.25	0.25
category	0.31	0.31	0.31	0.31
		reliability coefficients*		
10 pts	0.77	0.90	0.90	0.91
100 pts	0.77	0.92	0.88	0.82
category	0.73	0.84	0.86	0.78

* These coefficients are equal to $(\ell_{ij}^2 + m_{ij}^2)$.

where, for all i and j, $\text{cov}(F_j, U_{ij}) = 0$, $\text{cov}(F_j, \varepsilon_{ij}) = 0$, $\text{cov}(M_i, \varepsilon_{ij}) = 0$, and $\text{cov}(T_{ij}, \varepsilon_{ij}) = 0$.

All parameters of this "true score model" are identified and can be estimated. The results of the analysis of the data of Table 28.1 with this model are presented in Table 28.3.

The true score validity coefficient is equal to the standardized b coefficient. The method effect is the standardized g coefficient and the reliability is the standardized effect of the true score on the observed score h.

It turns out that all traits have approximately the same validity. This strongly contradicts the previous results where general life satisfaction had a much lower validity. The method effects of all measures are small but larger than in Table 28.2 due to correction for attenuation, and the method effects for the 100-point scales are systematically the smallest. Consequently the validity of this method is also the largest for all topics. The reliability of the measures of general satisfaction is much lower than the reliability of all other traits while also the category scale seems to have a bit lower reliability than the other two scales.

Comparing Table 28.2 and Table 28.3 illustrates some of the similarities and differences between these two parameterizations. First of all it can be shown mathematically that, ignoring computational differences, the reliability coefficients should be the same for the two parameterizations given the assumption that the unique variance is zero. Secondly, the validity coefficients and method effect will be higher

Table 28.3. The Estimated Values of the Parameters for the Data of Table 28.1 Using the "True Score MTMM Model"

	Satisfaction			
Method	General	House	Income	Contacts
		true score validity coefficients (b_{ij})		
10 pts	0.91	0.93	0.94	0.93
100 pts	0.94	0.96	0.96	0.95
category	0.90	0.93	0.93	0.92
		method effects (g_{ij})		
10 pts	0.42	0.36	0.35	0.36
100 pts	0.33	0.28	0.30	0.31
category	0.43	0.37	0.36	0.39
		reliability coefficients (h_{ij})*		
10 pts	0.76	0.90	0.93	0.90
100 pts	0.77	0.92	0.86	0.82
category	0.73	0.85	0.87	0.79

* These estimates of the reliability coefficients differ from the ones of Table 28.2 because of differences in the computations.

for the second parameterization than for the first due to correction for attenuation. The validity coefficients in Table 28.3 only reflect the relationship between F and T (in the notation of Figure 28.1), whereas the indicator validity coefficients of Table 28.2 reflect the linkage from F to y. A similar argument holds for the method effects.

Given the assumption of zero unique variance, the coefficients of the two tables have a simple relationship. From the results in one table the results for the other table can always be estimated because $\ell_{ij} = b_{ij}h_{ij}$. Multiplication of method effect and reliability for the same trait and method combination of Table 28.3 will lead to estimates of the method effects of Table 28.2 because $m_{ij} = g_{ij}h_{ij}$.

The calculations can also be done in the other direction because the estimates of the reliability coefficients should be the same, except for computational differences, for the different parameterizations. Thus, we also know that $b_{ij} = \ell_{ij}/h_{ij}$ and $g_{ij} = m_{ij}/h_{ij}$.

These results suggest that it is rather arbitrary whether we use the one or the other parameterization. The fit of the models will always be the same for both models (chi-square = 86.19 with df = 45 for both). It only makes a difference in the interpretation of the results. The roughly equal values across measures of the true score validity coefficients in Table 28.3 are in sharp contrast to the fact that general satisfaction showed much lower indicator validities than the other measures in Table 28.2. The reason for this difference is now easy to see: reliability estimates are lower for these measures than for the other measures and the indicator validity is a product of the true score validity and the reliability and is therefore lower than the true score validity itself. This example suggests rather clearly that the true score validity might be attractive as a concept due to the fact that it is not a mixed statistic and separates validity and reliability as different concepts. We will return to this point later.

28.3.4 The Multitrait-Multimethod-Multitime Approach

The multitrait-multimethod-multitime (MTMMMT) approach combines the good features of the RMM and MTMM designs and avoids their weaknesses. As the name implies, one obtains multitrait-multimethod data from the same people on at least two occasions. (In actual practice there seems to be little advantage in going beyond two occasions.)

The strength of the MTMMMT design is that it allows one to distinguish between random measurement error and stable unique disturbances (the U and ε terms of Figure 28.1). In the usual MTMM design, these two components occur together in the residual error term (the W of equation (28.17) and Figure 28.5). In the new approach the U term has been assumed to be zero, but with multiple implementations of the same measuring instrument this assumption is not necessary anymore. It becomes possible to unconfound these two components. Furthermore, it explicitly models real changes that may have occurred between the data collections, and hence no assumption on lack of change is needed, as was required for the RMM approach. Accordingly, the MTMMMT design has the capability to produce estimates of true score validity, test-retest reliability, the method effects and the unique variance without the need of possibly unrealistic assumptions which cannot be tested. It is the only design we know of that has this capability.

The diagram that represents the MTMMMT design is easy to describe, but its size and complexity suggest it not be drawn here. The diagram of the MTMM design shown in Figure 28.4 or 28.5 would be duplicated as many times as the data are collected, and additional factors would have to be added (one U factor for each measure). (An alternative format would be to omit the U factors and instead allow the residual error components pertaining to the same measure at different occasions to be correlated with one another; although equivalent, this alternative is not elegant because it does not explicitly distinguish between the U and ε factors.) Real change in the traits between data collections is handled in the same way as is shown in Figure 28.2 for the quasi-simplex approach.

For survey organizations doing panel studies, this design is not difficult to implement. However, it is as yet little used. Perhaps the first use of this design was by Andrews, who found that in the American national surveys he used, nearly all of the residual error appeared to be random error — i.e., stable unique disturbances constituted only very small portions of the variance of the 12 measures he was evaluating (Andrews, 1984, footnote 14). Rodgers (1989b) reports another implementation of this design and also found results that suggest that effects from stable unique factors were very weak.

The fact that early implementations of this design suggest that stable unique disturbances are not of major influence on survey measures is of great interest. If these results prove widely generalizable, the conceptual weakness of the MTMM design will be shown to be of only limited importance. However, such results can only be expected if the researcher ensures that the questions which are supposed to measure the same variable only vary with respect to the method used and not the content of the question itself. If there is any doubt about the assumption of zero stable unique variance which is especially the case when different measures are assumed to measure the same variable (Saris, 1982) then the MTMMMT design offers a useful means of assembling evidence for the importance of systematic unique components.

The MTMMMT approach is a particularly attractive design for assessing measurement quality. Its major drawback is practical (it requires panel data), not conceptual.

28.3.5 The Congeneric Test Approach

For completeness of this overview, still another approach should be mentioned: the congeneric test design and model (Jöreskog, 1971). In this approach only one opinion is studied with at least three methods. In this simple case the indicator validities are obtained but the method effects cannot be estimated and the residual errors confound three factors — random error, stable unique error, and methods effects. One can say that this design has the disadvantages of both the RMM model (no method effects) and the MTMM model (confounding of random and stable errors). Therefore we do not think that this design is a good candidate for evaluation studies even though it is frequently used.

28.4 CONCLUSION AND DISCUSSION

Although we think that the MTMMMT design deserves more attention than it gets, we have to admit that it can only be used in situations where panel data can be collected. This limits its applicability considerably even though we would like to recommend all researchers who do such research to use this design to evaluate the quality of their measurement instruments. This design provides all the required information with the most realistic assumptions.

On the other hand, when one cannot do panel research one cannot apply this approach and the only two reasonable candidates left are the RMM approach and the MTMM approach. It was shown that the RMM approach can provide estimates of all the different coefficients which have been specified as criteria for the quality of measurement instruments: test-retest reliability and true score validity and also the invalidity as a whole, but not the method effects. The estimation of this last coefficient would require that at least three theoretical variables be measured in three different ways with two repetitions. This would mean that we have to use an MTMMMT design within one survey, which is impractical because of the frequent repetition of the different questions (six times for each different topic).

The MTMM approach is less attractive with respect to the estimation of the true score validity and the test-retest reliability and the total invalidity. It can only provide this information if the unique variance is zero. Whether this is the case cannot be tested in this design. For this purpose one needs the MTMMMT design.

This suggests that the RMM approach is the most attractive design if one wants to evaluate instruments on the criteria of true score validity and test-retest reliability. There is, however, one more argument in favor of the MTMM approach which has not been mentioned yet. The MTMM approach provides us, under all conditions, with estimates of the effects of the latent variables on the observed variables, which we called indicator validity; the attenuated method effect; and the explained variance of the observed variables, which is called the homogeneity reliability (Andrews, 1984, Bollen, 1989). If relationships between different latent variables are studied, the indicator validity coefficient and method effect coefficient are necessary to obtain consistent estimates of the correlations between these theoretical variables. Using simple path analysis and the model of Figure 28.5 it follows that:

$$r(y_{11},y_{12}) = \ell_{11} r(F_1,F_2) \ell_{12} + m_{11} m_{12} \qquad (28.20)$$

and from this result follows:

$$r(F_1,F_2) = [r(y_{11},y_{12}) - m_{11} m_{12}]/[\ell_{11}\ell_{12}]. \qquad (28.21)$$

The first equation indicates that method effects inflate the observed correlations (compared with the correlation of interest) and that lack of validity deflates the same correlations. Equation (28.21) indicates that estimates of both effects are necessary if one would like to correct for measurement error in relationships and effects between latent variables.

Thus if we want to derive the correlation between the factors of interest we need estimates of the effects which can be obtained using the MTMM approach. The indicator validity coefficient can also be obtained with the RMM design but the method effect cannot be estimated except in very impractical and expensive studies. Thus, if the purpose is estimation of relationships between latent variables, the MTMM approach should be chosen.

But there is even more to say in favor of the MTMM approach. If the research design is properly chosen so that the unique variance is zero then the MTMM design can provide all the information which we prefer: the true score validity, the unattenuated method effects, and the test-retest reliability. Since the MTMM approach under this condition can serve both purposes and the RMM can approach only one, we recommend the use of the MTMM approach.

Although we think that the MTMM design is the most attractive one for our purposes, we have to admit that this approach is certainly not without problems. In Saris and van Meurs (1990) the authors have given a list of all the problems of the MTMM approach. Some of these are: (1) the presence of unique variance, (2) convergence problems of the programs, (3) instability of the method effects, (4) the problem of memory effects, (5) identification problems of different designs, (6) order of presentation effects, (7) the problem of a change of opinion during the interview, (8) testing MTMM models, and (9) robustness of estimates under nonnormal distribution. Although this list suggests that there are a lot of problems, the authors also present solutions to most of these. It would lead too far to discuss their solutions here.

With respect to the choice of a model to analyze MTMM data we can make the following comment. If the different measurement instruments do not measure the same variable (for example, when a different question is used) the unique variance will be relatively large. As this component cannot be detected in the MTMM approach, this unique variance will become a part of the error variance and consequently will bias the estimates of the validity and reliability in both models discussed. Under these circumstances both models are equally bad. Therefore we recommend the use of the MTMM approach especially for cases where the variables to be measured remain the same but only the measurement procedure is varied. In that case it is reasonable to assume that the different measures for the variables of interest really measure the same variable and it can be assumed that the unique variance is zero. But then it does not matter which model we use for estimation because we can get estimates of the same parameters using both models.

The choice of the model, however, has also consequences for the possible use of the parameters. On the one hand the traditional MTMM model has the advantage that it provides estimates of the direct links between observed variables and the factors (called indicator validity here), and between observed measures and a method factor, a major source of correlated error. With this information in hand the researcher has a better understanding about the nature of the observed relationships and can make appropriate estimates of the relationships between the latent variables as we have discussed above.

On the other hand we have made an argument here that it might be useful for evaluation of measurement instruments to use the parameters of the true score model, because the true score validity and disattenuated method effects are simpler statistical quantities, as a clear distinction between validity and reliability is made. In that case the interpretation of the parameters is simpler. Reporting these characteristics of measures provides other researchers with all the necessary information to evaluate measurement instruments and to correct for random and systematic measurement error in case they are interested in relationships between latent variables.

There is one more reason to choose this parameterization, which has not yet been mentioned but which connects with recent discussions in the literature. Several authors, for example Campbell and O'Connell (1967), Bentler and Lee (1978), Browne (1984), Cudeck (1988) and Bagozzi and Yi (1990), have suggested the possibility of multiplicative instead of additive effects of the methods on the correlations between the observed variables. The additive effect of the method factor on the correlations between the observed variables can clearly be seen in the result presented in equation (28.20). The correlation in that equation is equal to the sum of two components, one due to the traits and one due to the method factors. The above-mentioned authors suggest that this is not the only possible effect. Some suggest multiplicative effects. For example Bentler and Lee (1978) suggest that a rather restrictive model of this type could be specified as a special case of a second order factor analysis model. Wothke and Browne (1990) have recently indicated that a very general class of multiplicative models can be specified in this form. One of the attractions of the "true score model" for the MTMM data is that the additive and multiplicative models are both specific cases of the general model of Figure 28.4. One can see this model as one with an additive and a multiplicative effect of the method on the correlations. Depending on the restrictions introduced, a mixed model (equations 28.18 and 28.19), a pure multiplicative model, or a purely additive model can be specified.

This discussion suggests that the true score model, which is a second order factor analysis model for the MTMM data, has the advantage that it can provide estimates of the easily interpreted test-retest reliability as well as the unattenuated method effects and the true score validity, yet no repeated observations are necessary. It can also be used to test for additive and multiplicative method effects on the correlations. Thus, the new "true score" model for MTMM data developed in this chapter has many features which make it an attractive alternative model for evaluation of measurement instruments. The only requirement is that the MTMM studies be designed in such a way that the assumption of zero unique variance is a realistic one. In Saris and van Meurs (1990), one can find how this assumption can be satisfied.

CHAPTER 29

PATH ANALYSIS OF CROSS-NATIONAL DATA TAKING MEASUREMENT ERRORS INTO ACCOUNT

Ingrid M. E. Munck
Statistics Sweden

29.1 INTRODUCTION

With covariance structure models one can model and estimate measurement errors and take these errors into account in the statistical analysis of relationships. This is particularly useful when data quality varies among populations in a cross-national survey.

The development in this field of both statistical methodology and computer programming technology has now reached a point where it is feasible to do something about measurement errors in statistical analysis. The assumptions under which we work are much more realistic than earlier ones, and there is efficient software for both PC and for mainframe applications; see for example Bollen (1989).

29.1.1 A Cross-National Path Analysis Controlling for Measurement Errors

This chapter presents a cross-national application of a path model concerning the relationships between schooling and home status. The capacity of the LISREL program by Jöreskog and Sörbom (1989) to analyze data from several populations simultaneously has been used and found very appropriate for cross-national statistical analyses of data.

Measurement Errors in Surveys.
Edited by Biemer, Groves, Lyberg, Mathiowetz, and Sudman.

ISBN: 0-471-53405-6

We are concerned about the nature and size of measurement error effects. This is studied by means of

1. a comparison between the explanatory power of factors in a path model before and after measurement errors are taken into account, and

2. the estimation of an "international" path model in which country-specific measurement errors are controlled for and partialed out.

29.1.2 IEA Cross-National Survey

The application presented here is taken from the field of evaluation survey research in education. The history of the International Association for the Evaluation of Educational Achievement (IEA) dates back to the end of the 1950's when researchers from twelve countries decided to cooperate in a cross-national study of educational achievement (Husén, 1967). The success of this study in developing school achievement tests for use in various languages led to the establishment of a special non-profit, nongovernmental organization with the task of conducting and promoting educational survey research on an international scale.

During 1970 and 1971, data in the so-called Six-Subject Survey were collected in 21 countries (Peaker, 1975). Included were surveys of student achievement in six subject areas: science, reading comprehension, literature, civic education, and French and English as foreign languages. This chapter further analyzes some results from this massive study, which included about 250,000 students.

29.2 EMPIRICAL EFFECTS OF MEASUREMENT ERRORS — AN EXAMPLE

Numerous studies in education have demonstrated the impact of family environment on school achievement. The IEA international surveys have opened up the possibility of studying the relationships between home and school cross-nationally as each instrument is designed to measure the same thing in each of several different countries.

In the sociological tradition of status attainment models and path analysis, Bulcock, et al. (1977, 1978) used the IEA Six-Subject Survey data to analyze a resource model for science achievement. A student's scholastic achievement in the subject "Science" — encompassing physics, chemistry, and biology — was looked upon as a reflection of the dynamics of individual resource acquisition. The essential resources included in this model are home background resources, language

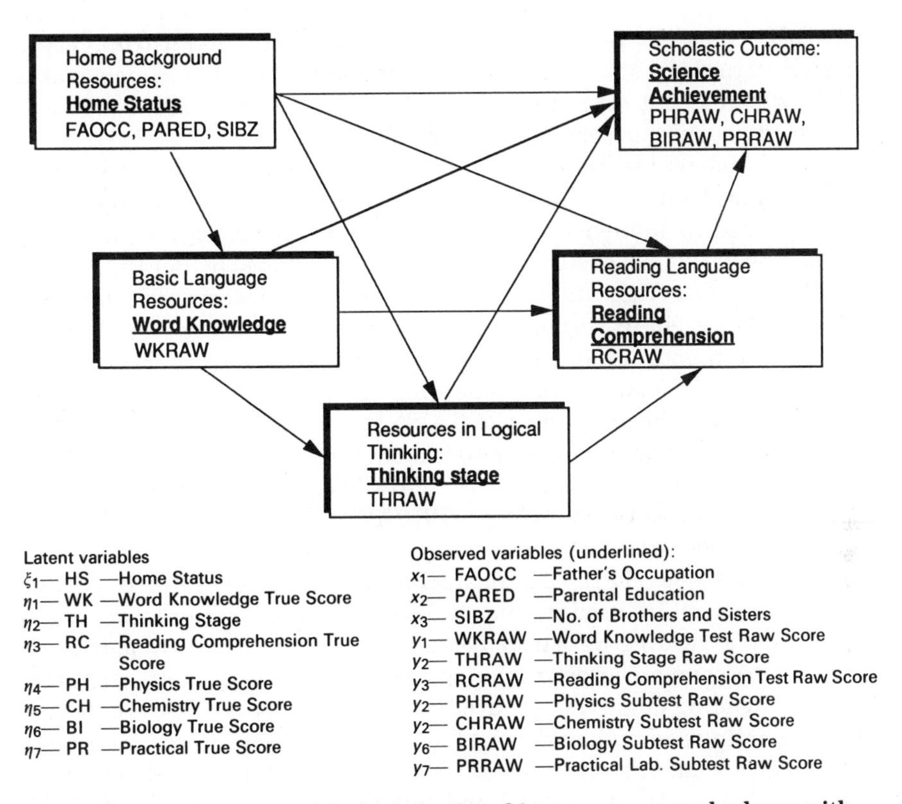

Figure 29.1. A resource model of scholastic achievement — causal scheme with concepts, latent-, and observed variables

resources, and resources in logical thinking, on which scientific understanding depends. The path diagram in Figure 29.1 depicts the recursive model defined by Bulcock, et al. and shows how these resources are interrelated and how they are assumed to explain directly and indirectly the variability in school achievement.

29.2.1 The Path Model

Corresponding to each concept of the theory there is a latent, unobserved variable defined with reference to the variables measured in the IEA study; see Figure 29.1.

In this study the different aspects of home background having an impact on students' school achievements are limited to proxy variables for home status. The indicators include: Father's Occupational Status

(FAOCC), Parental Education (PARED), and Family Size (SIBZ). These are all regarded as measuring the same latent variable, Home Status (HS), which is not directly measurable. In the model, this variable is regarded as an input, exogenous variable.

The student's basic language resources, or Word Knowledge (WK), are measured with a Word Knowledge test (WKRAW). Both resources are assumed to influence the student's capacity for Logical Thinking (TH). The variable measuring this aspect is developed by Bergling (1974) and labeled Thinking Stage (THRAW). The third intervening variable, through which family resources are converted into personal resources, is related to an additional aspect of the language resources, "reading as reasoning" or Reading Comprehension (RC) as measured by the IEA Reading Comprehension test (RCRAW).

Finally, the IEA Science Achievement test, composed by the subtests in Physics (PHRAW), Chemistry (CHRAW), Biology (BIRAW) and Practical Laboratory Knowledge (PRRAW), is the dependent variable measuring scholastic achievement in science. The corresponding latent variables are labeled PH, CH, BI, and PR. The ten observed variables and the latent variables used in this study are summarized in Figure 29.1.

From a reading research perspective, it has been of particular interest to compare cross-nationally the home background resources with the reading language resources in order to understand the extent to which reading comprehension mediates the effects of social background factors on a student's achievement. The Bulcock study used IEA data for 14-year-old boys from England and India. The ten variables described above were chosen to implement the model.

29.2.2 Reanalysis Including Measurement Errors

The reanalysis of the Bulcock model with structural equation models and latent variables makes it possible to include measurement errors explicit in the statistical model being analyzed. Generally, the measurement part of the model defines relationships between the observed and unobserved variables.

There are two alternative ways to specify the measurement model used here. First, when there are more than two indicators observed for the same underlying construct, the model parameters can be estimated. This is the case for the latent home background variable HS. In Figure 29.2 a path diagram for a three-indicator measure of HS is depicted.

Second, when external information about reliability of a measure is available, the measurement model parameters consist of inserted coefficients, so-called fixed parameters in LISREL, see Figure 29.3 (explained in full in Section 29.2.3).

All the dependent latent variables in the resource model will be

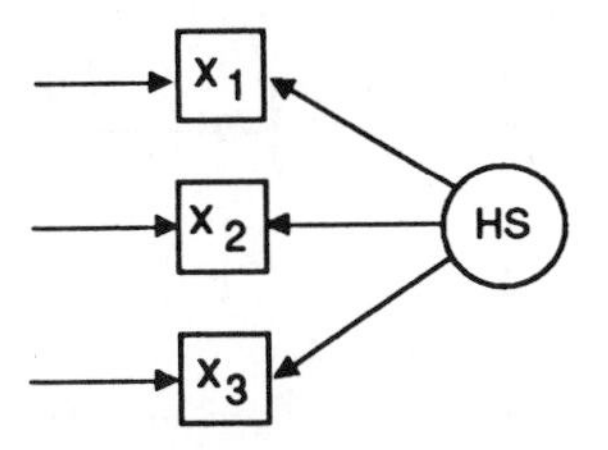

Figure 29.2. Path diagram for multiple indicator measurement model for home status HS. Observed variables: see Fig. 29.1.

estimated in this manner using item-analysis information about the composite scores, that is, the observed dependent y-variables.

In order to demonstrate the effects of including measurement errors, three alternative models all based on the same observed variables (reported in full in Munck, 1979) are compared:

Model 1 — path model for observed variables; no measurement errors taken into account.

Model 2 — path model with the measurement errors of all dependent and intervening variables inserted into the model. No measurement errors for the home background indicators, the exogeneous variables, are assumed.

Model 3 — path model with measurement errors of all independent as well as dependent variables taken into account. The home background indicators are assumed to be dependent of a latent Home Status variable, HS.

29.2.3 Inclusion of External Reliability Information

Before we turn to the specifications of Models 1, 2, and 3, we will look at the measurement parts of all the dependent variables and describe how these are expressed in terms of the available reliability information.

The simple measurement model for a latent variable η and only one observed variable y has two parameters, λ_y and θ_ε (in short λ and θ), which are both fixed parameters with assigned specified values; see Figure 29.3. The problem is to express these parameters as functions of the reliability coefficient of y, here denoted by r, and the variance of y, $\sigma^2(y)$. According to Lord and Novick (1968, p. 61), reliability can be defined as the squared correlation between the observed variable y and the latent variable η. This means that r can be expressed as

$$r = \rho_{y\eta}^2 = \frac{\lambda^2\sigma^2(\eta)}{\sigma^2(y)} \tag{29.1}$$

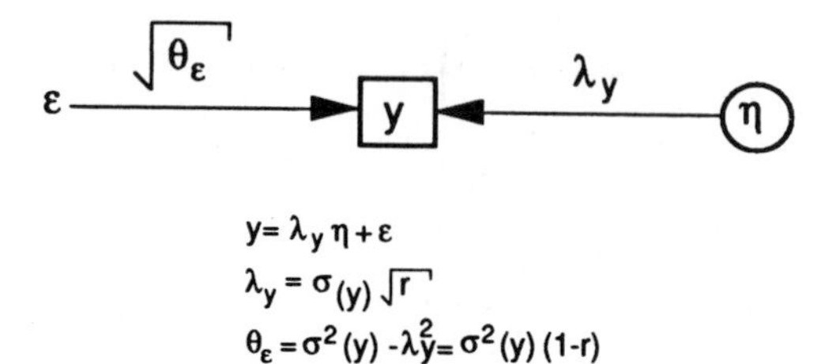

Figure 29.3. Path diagram for simple measurement model with one observed variable y

assuming that ε and η are uncorrelated, and that the means of y and η are zero (see the model specification in Figure 29.3). If the latent variable is assumed to be standardized (i.e., $\sigma^2(\eta) = 1$), we get

$$r = \frac{\lambda^2}{\sigma^2(y)} \tag{29.2}$$

and λ^2 can be interpreted as the part of the variance in y that can be explained by the latent variable η. The rest of the variance in y, not explained by the latent variable, is due to measurement error, which means that

$$\theta = \sigma^2(y)(1 - r). \tag{29.3}$$

If a correlation matrix is analyzed, all observed variables are considered standardized and the variance of the observed variable y is therefore equal to 1. The parameters λ and θ simply become $\sqrt{r}$ and $(1 - r)$, respectively. It is important to note that the cross-national analyses are performed on covariance matrices and that the variance of y is then estimated with the sample variance.

29.2.4 Specification of the Models

A covariance structure model is generally defined through the specification of a structural equation model for the latent variables, of measurement models for the relationships between observed and latent variables, and of accompanying assumptions about error variables and means. The three alternative models in this study all define a fully identified structural equation model but differ in the approach to the measurement models.

As we are not interested in studying differences between countries in mean achievement with this application, the structural equation does not include an intercept and the observed variables are corrected for their means, which implies that the means are zero.

For the first model, the structural equation model (see submodel 2 in the LISREL 7 manual, Jöreskog and Sörbom, 1989), is:

$$\mathbf{y} = \mathbf{B}\mathbf{y} + \mathbf{\Gamma}\mathbf{x} + \boldsymbol{\zeta} \tag{29.4}$$

where $\mathbf{B}(7 \times 7)$ and $\mathbf{\Gamma}(7 \times 3)$ are coefficient matrices and $\boldsymbol{\zeta}$ is a random vector of residuals. The covariance matrix for $\boldsymbol{\zeta}$, $\boldsymbol{\Psi}(7 \times 7)$, includes uncorrelated residuals among y_1, y_2, and y_3 and correlated residuals among the rest of the y-variables, the four science tests. This model is a fully recursive, fully saturated model with zero degrees of freedom and therefore fits the data perfectly. The independent variables x_1, x_2, and x_3 are regarded as a set of predetermined values (which can be handled with the Fixed-x option in the LISREL program). The further assumptions are: (a) $\boldsymbol{\zeta}$ is uncorrelated with $\mathbf{x}$; (b) $(\mathbf{I} - \mathbf{B})$ is nonsingular; and (c) the means for the residuals are zero.

The second model, including a measurement model for the y-variables, can be expressed in algebraic form as follows:

$$\boldsymbol{\eta} = \mathbf{B}\boldsymbol{\eta} + \mathbf{\Gamma}\mathbf{x} + \boldsymbol{\zeta} \tag{29.5}$$

$$\mathbf{y} = \mathbf{\Lambda}_y\boldsymbol{\eta} + \boldsymbol{\varepsilon} \tag{29.6}$$

where $\mathbf{B}(7 \times 7)$,$\mathbf{\Gamma}(7 \times 3)$ have the same dimensionality as in Model 1. $\mathbf{B}$ now contains the coefficients for the relationships between the mutually dependent latent variables, and $\mathbf{\Gamma}$ contains the coefficients for the relationships between the x-variables and the dependent latent variables. The diagonal of $\mathbf{\Lambda}_y$ is equal to $\sigma(y_i)\,(r_i)^{\frac{1}{2}}$, $i = 1, \ldots, 7$, with all fixed values. $\boldsymbol{\zeta}$ is a random vector of residuals and $\boldsymbol{\varepsilon}$ is a random vector of measurement errors. In this model the covariance matrix for $\boldsymbol{\zeta}$, $\Psi(7 \times 7)$, expresses the residuals for the latent dependent variables compared with the observed y-variables in Model 1. These residuals are assumed to be uncorrelated for η_1, η_2, and η_3 and correlated for the four science latent variables. The covariance matrix for $\boldsymbol{\varepsilon}$, $\mathbf{\Theta}_\varepsilon$, is assumed to be diagonal, which means that the measurement errors for the observed dependent variables are uncorrelated.

In the path diagram showing the results, the fixed LISREL parameters in Models 2 and 3 are marked with an asterisk, *, and to specify these parameters, external information is inserted according to formulas in Figure 29.3. The reliability coefficients are taken from Table 29.2 and further discussed in the next section.

In order to express the most elaborate model, Model 3, the general form of the covariance structure model is used,

$$\boldsymbol{\eta} = \mathbf{B}\boldsymbol{\eta} + \mathbf{\Gamma}\boldsymbol{\xi} + \boldsymbol{\zeta} \tag{29.7}$$

where the variables are defined in Figure 29.1, the coefficient matrix $\mathbf{B}(7 \times 7)$ is the same as for Model 2, and $\mathbf{\Gamma}(7 \times 1)$ contains the coefficients for the relationships between the independent latent variable Home Status and the dependent latent variables. The measurement models are:

$$\boldsymbol{y} = \boldsymbol{\Lambda}_y \boldsymbol{\eta} + \boldsymbol{\varepsilon} \tag{29.8}$$

$$\boldsymbol{x} = \boldsymbol{\Lambda}_x \boldsymbol{\xi} + \boldsymbol{\delta} \tag{29.9}$$

where $\boldsymbol{\Lambda}_y$ has all fixed parameters as in Model 2, and the parameters of $\boldsymbol{\Lambda}_x(3 \times 1)$ are all free except for λ_{11} which is chosen as scalar. One of the parameters in $\boldsymbol{\Lambda}_x$ has to be fixed in order to obtain an identified model and to fix the scale of the latent variable Home Status. $\boldsymbol{\Psi}(7 \times 7)$ and $\boldsymbol{\Theta}_\varepsilon$ are the same as in Model 2.

In this model, the exogenous latent Home Status variable, ξ, is defined as the common part of the x-variables and is measured without error. Through the use of several indicators measuring the same construct, we will be able to estimate the reliability of each of these indicators. In this application, the measures of Father's Occupation (x_1) and Parental Education (x_2) had something in common over and above the Home Status part. In other words, the measurement errors, δ_1 and δ_2, showed a significant correlation. In order to get a model with a satisfying goodness-of-fit, this correlation was included as an additional assumption, and the element $\theta_{2,1}$ in the covariance matrix for $\boldsymbol{\Theta}_\delta$ is assumed to be a free parameter whereas the rest of the cells under the diagonal are set at zero (see Section 29.2.5).

The further assumptions are: (a) the residual ζ is uncorrelated with the latent exogenous variable ξ; (b) the errors in measurement of the observed y-variables ε are uncorrelated with the latent dependent variables η; (c) the errors in measurement of the observed x-variables, δ, are uncorrelated with the latent exogenous variable ξ; (d) all error variables (ζ, ε, and δ) are mutually uncorrelated; (e) $(\mathbf{I}-\mathbf{B})$ is non-singular; and (f) the means for the latent variables and for the residuals are zero.

29.2.5 Design, Estimation, and Goodness-of-Fit

Data analysis with the LISREL program assumes an equal probability sampling design. Our data comes from IEA stratified two-stage samples, with the school as the primary and the student as the secondary sampling unit. To justify the use of LISREL, we have to assume that the theory formulated by Bulcock, is valid. This means that no extraneous factors should be omitted in the model (see Munck, 1979, for a discussion of such potential factors). The most important such factors were assumed to be nationality, age, and sex which are all controlled for in this study of 14-year-old boys from England.

The study also assumes compatible scales between countries, which implies that each variable, test-scores as well as home background indicators, is measured on an internationally valid interval scale. To work with genuine ordinal variables in simultaneous cross-national analysis is problematic and not discussed in this chapter.

LISREL can work with several kinds of parameter estimates and we have chosen to use maximum likelihood (ML) estimation. Several criteria for goodness-of-fit are available. Under the assumption of multivariate normality, ML estimation will provide a χ^2 measure of overall fit of the proposed model. We also include in the results below, for the analysis among countries, the goodness-of-fit index, GFI, which does not depend on the sample size.

Before we turn to the results we will report the goodness-of-fit result for Model 3 on 14-year-old boys in England, $\chi^2 = 19.8$, $df = 13$, and $p = 0.10$, which indicates an acceptable fit. Without the inclusion of the correlation between the measurement errors for x_1 and x_2, the model yields a goodness-of-fit test result which is significant at the 1 percent level. Model 1 and Model 2 have zero degrees of freedom and are therefore not testable.

29.2.6 Findings

The results from the three different models are presented in two ways:

1. The *predictive* value of the exogenous and intervening variables (i.e., the predictor variables), on Physics Achievement, Table 29.1, and

2. *Significant paths* for Model 2 and 3, Figure 29.4.

The predictive value is expressed in terms of the total causal effect TCE, generally defined in LISREL for ξ on η as $(\mathbf{I}-\mathbf{B})^{-1}\,\Gamma$. It is obtained from the reduced form equation:

$$\boldsymbol{\eta} = (\mathbf{I}-\mathbf{B})^{-1}\boldsymbol{\Gamma}\boldsymbol{\xi} + (\mathbf{I}-\mathbf{B})^{-1}\boldsymbol{\zeta}. \qquad (29.10)$$

For example, the TCE for HS on PH is the sum of the direct effect of HS on PH, γ_{41}, and the indirect effects are mediated by WK, TH, and RC; see Figure 29.4. TCE reflects the total effect on PH of a change in HS and is here taken as an estimate of the overall influence of home background on the physics achievement.

The TCE's are ranked in Table 29.1 in order to compare the simple model with the more elaborate models which take measurement errors into account. Although the results of Model 1 represent the same statistical solution as that in the Bulcock, et al. study, there are some discrepancies stemming from differences in variable definitions (Bul-

Table 29.1. Total Causal Effects (TCE) of Predictor Variables on Physics Achievement: the Bulcock Model and Models 1, 2, and 3. IEA Data for England, 14-year-old Boys, $N = 1{,}091$ (for Bulcock, et al. (1978) the number of cases ranged from 1,247 to 1,475 because of pair-wise deletion of missing cases)

Bulcock Model			Model 1		Model 2			Model 3		
Predictor	TCE	Rank Order	TCE	Rank Order	Predictor	TCE	Rank Order	Predictor	TCE	Rank Order
x_1 FAOCC	0.24	4	0.25	4	x_1 FAOCC	0.29	4			
x_2 PARED	0.05	6	0.04	6	x_2 PARED	0.04	6	ξ_1 HS	0.65	1
x_3 SIBZ	0.15	5	0.12	5	x_3 SIBZ	0.15	5			
y_1 WKRAW	0.46	1.5	0.42	2	η_1 WK	0.55	2	η_1 WK	0.34	3
y_2 THRAW	0.27	3	0.29	3	η_2 TH	0.36	3	η_2 TH	0.32	4
y_3 RCRAW	0.46	1.5	0.48	1	η_3 RC	0.62	1	η_3 RC	0.55	2

cock, et al. did not exclude the thinking stage items from the science test scores) and in handling missing observations.

When reliability coefficients are inserted into Model 1 to yield Model 2, the TCEs show an overall increase in value, although the rank ordering remains unchanged.

When Model 3 is applied to the same data set, the results change dramatically. The TCE value for HS (0.65) is now well above those for the other predictors; it is followed by RC (0.55), and WK and TH (both greater than 0.30).

The estimates of the direct effects in Models 2 and 3 are compared in Figure 29.4, which illustrates the significant paths.

For Model 2, it can be seen that among the x-variables, Father's Occupation (x_1) has the strongest direct effect 0.27 on WK. This so-called attenuated effect, not corrected for error in measurement, is decomposed in Model 3 in a reliability effect 0.49 of HS on x_1, and a disattenuated effect 0.57 of HS on WK (cf. Chapter 28 with the decomposition of the relationship between a factor and an observed response variable through the inclusion of a true score variable).

A large increase in the importance of the home factor appears as expected when the disattenuated effects in Model 3 are examined: the direct paths from HS to the primary and secondary language resources, WK and RC, are now 0.57 and 0.39 respectively, as compared to 0.27 and 0.12 respectively in Model 2 for the Father's Occupation indicator. The effect of the home factor on achievement is largely indirect, except for performance in practical work. Hence, it appears that the use of the multiple indicator measurement model for Home Status enables us to explain more of the variance in WK and RC than does the use of separate x-variables with zero measurement errors. The main consequence of the

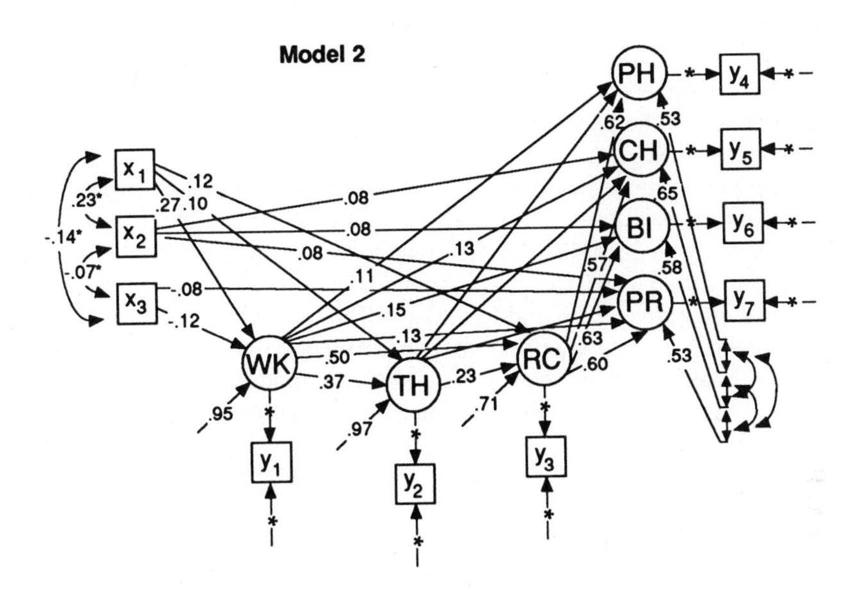

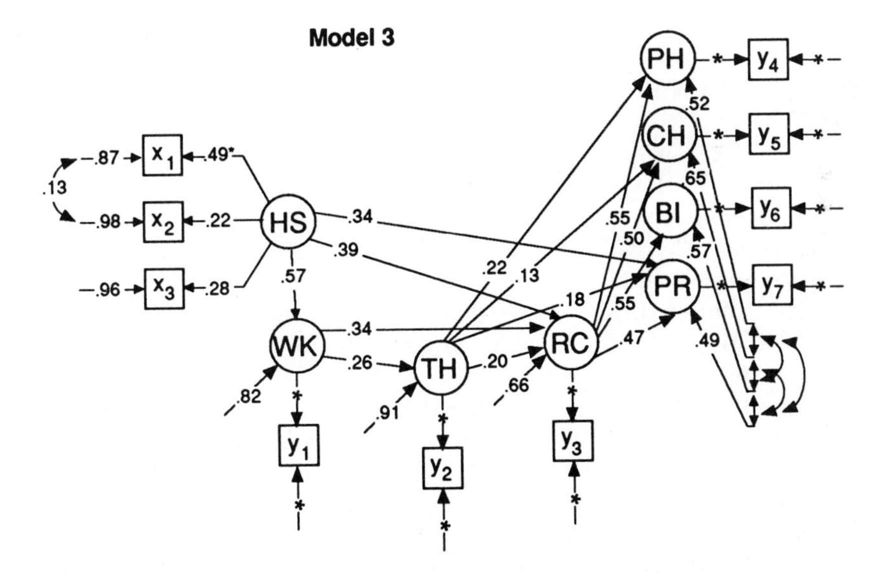

Figure 29.4 Path diagrams for Models 2 and 3 with significant paths for England, 14-year-old boys. Diagrams include paths with *t*-values greater than 2 only. Observed and latent variables: see Figure 29.1. *, fixed specified parameter.

increased importance of Home Status in the third model is that the interrelationships between WK, RC, and Science Achievement become weaker.

The interpretation of the results from modeling without measurement errors for the home variables, Bulcock's model and Models 1 and 2, can be summarized as follows:

The single most powerful explanation of achievement in science is the language resources. The argument that a child's progress through school is more a function of social class than ability is rejected. Bulcock, et al. concluded that the home and logical thinking resources were independently operative and that the different kinds of resources could be regarded as complementary rather than competing.

The results from the most elaborate model, Model 3, can be interpreted quite differently with regard to the role of the home in school achievement:

The influence of the home factor, which was the most powerful predictor overall, was mainly indirect. This finding is consistent with the resource conversion theory (cf. Fägerlind, 1975), according to which family resources are converted into personal resources such as language resources or logical thinking, which then in turn are utilized in the school learning process.

29.3 SEARCH FOR INVARIANT STRUCTURES OVER COUNTRIES

We now turn to the useful feature of the LISREL program which permits us simultaneously to analyze structural equation models in several populations. The validity of Model 3 will be studied using data for 14-year-old boys in England, Hungary, and Sweden.

The relationships between home background and school achievement will be tested for invariance between the countries, taking into account the varying quality of the data.

When a phenomenon is specified in its structural form, each relationship in the model then represents a causal link rather than a mere empirical association. The structural parameters in the equation represents the features of the mechanism that generates the observed variables (see Jöreskog and Sörbom, 1989, p. 1).

In all modeling there is a search for a *general* model. The consequence of that goal for structural equation models is that the parameters are "fundamental", or "invariant":

The structural form of the model is that parameterization — among the

Table 29.2. Reliability Coefficients (Kuder-Richardson 20) for the IEA Tests

Test	Number of items	England	Hungary	Sweden
Word Knowledge	40	0.833	0.860	0.672
Reading Comprehension	52	0.887	0.826	0.865
Physics	21	0.711	0.721	0.660
Chemistry	19	0.700	0.600	0.580
Biology	17	0.573	0.492	0.383
Practical	20	0.680	0.560	0.570

Sources: Thorndike, 1973: 54, 58. Comber and Keeves, 1973: 396.

> various possible ones — in which the coefficients are (relatively) unmixed, invariant, and autonomous. How do you know, if you have written a model in its structural form, rather than in some other form? Well, if the coefficients in the model are indeed relatively invariant across populations, somewhat autonomous, and not inseparable mixtures of the coefficients that "really" govern how the world works — then your model is actually in its "structural" form (Duncan, 1975: 151).

Therefore, if we can show a similarity cross-nationally in the relationships between the "true" variables in the path model discussed earlier, we will have arguments for a general, internationally valid model. Another way to express the problem is that we now turn to the criterion of replicability of the results to determine whether the correct model has been determined (Saris and Stronkhorst, 1984).

29.3.1 Data Quality

It is common practice in IEA studies to conduct pilot studies in every country and carefully design the instruments for measuring the same thing in the participating countries. This is done through an independent retranslation of the national instrument back into English and through carefully developed routines in the surveys. Country-specific reliabilities are calculated for scores and scales, and these are reported for the test score variables in Table 29.2.

It can be seen from Table 29.2 that Sweden had lower overall reliabilities than the two other countries and that the quality of data was generally higher for Word Knowledge and Reading Comprehension, than for the shorter subtests in Science Achievement.

Finally, the Thinking Stage Scale lacked reliability information and a constant $r = 0.80$ was inserted for all countries.

Table 29.3. Goodness-of-Fit Test Results for Model 3: Separate Analyses of Total Correlation Matrices and Within-School Correlation Matrices. 14-year-old boys, degrees of freedom = 13

	Total correlation matrix			Within-school correlation matrix		
Country	N	χ^2	p	N[a]	χ^2	p
England	1,091	19.8	0.100	984	15.8	0.260
Hungary	2,468	33.7	0.001	2,309	25.5	0.020
Sweden	793	23.5	0.036	699	19.1	0.120

[a] Degrees of freedom for within-school matrix = number of students in sample minus number of schools.

29.3.2 Validation of the Model

Model 3 was applied to the different samples, and the goodness-of-fit test results are given in Table 29.3.

The fit of the model is acceptable only for England; the p-values for analyses of the total correlation matrices are 0.10, 0.001, and 0.036 for England, Hungary, and Sweden, respectively. In order to adjust for differences in achievement level between schools, pooled within-school matrices were analyzed. The results of the subsequent goodness-of-fit tests show that all three countries gained an improved fit: both England and Sweden had achieved p-values greater than 0.10 and if the size of the Hungarian sample is taken into account, Model 3 fits the data of all three countries. Thus, it is the relative performance that is analyzed; a student receives a high score if he or she has performed well in relation to the other sampled students from the same school.

29.3.3 Invariance of the Measurement Structure

The theoretical measurement models in Model 3 were defined in equations (29.8) and (29.9). We saw from Table 29.2 that the reliability of the tests varied considerably from country to country and the problem now is to test whether this is the case for the latent home status variables as well.

The specific hypotheses to be tested are:

*H*0: No parameters are invariant over countries (except for the scalars λ_{11}) in Model 3.

*H*1: The relationships between the x-variables and the latent HS variable are invariant, i.e., Λ_x is invariant.

Table 29.4. Goodness-of-Fit Test Results for Model 3: Separate and Simultaneous Analyses According to Hypothesis *H*0 and *H*1. 14-year-old Boys

Country	Hypothesis	N	χ^2	df	p
England		984	15.8	13	0.260
Hungary		2,309	25.5	13	0.020
Sweden		699	19.1	13	0.120
All Countries	H0		60.4	39	0.015
All Countries	H1		96.1	43	0.000
Difference in χ^2 $\chi^2(H1) - \chi^2(H0)$			35.7	4	0.000

The analyses are based on within-school covariance matrices for England, Hungary, and Sweden, and the estimates of B and Γ in equation (29.7) contain unstandardized coefficients which depend upon the units in which the latent variables are scaled. The goodness-of-fit test results are reported in Table 29.4.

The χ^2-values for hypothesis *H*0 are merely the sum of the χ^2-values for the separate analyses. From this, we can conclude that Model 3 has a good fit. The difference between the χ^2-values for *H*1 and *H*0 is 35.7 with 4 degrees of freedom and $p = 0.001$, which is significant. We conclude that the three home status indicators, Father's Occupation, Parental Education, and Number of Siblings, vary in importance between the countries.

29.3.4 Towards an "International" Model

Now we turn to the inner structure (all paths from HS, WK, TH, and RC) of Model 3 and keep it invariant while the measurement parts remain country-specific. This means that corresponding parameters in **B** and **Γ**, equation (29.7), of the inner structure are specified as equal.

The goodness-of-fit indices, GFI, for the three countries are England 0.991, Hungary 0.996, and Sweden 0.986, which indicates good fit for each dataset to the restricted model. The overall goodness-of-fit test gave $\chi^2 = 136.66$, df $= 83$, supporting an "international" structure. The national deviations from this structure can to a large extent be explained by random disturbances.

The ML estimates for the international model of the direct effects between the latent variables are reported in Figure 29.5.

These estimates were calculated with equal weights for the three countries involved. The path between HS and TH was eliminated because it was quite insignificant in all countries and was also assumed

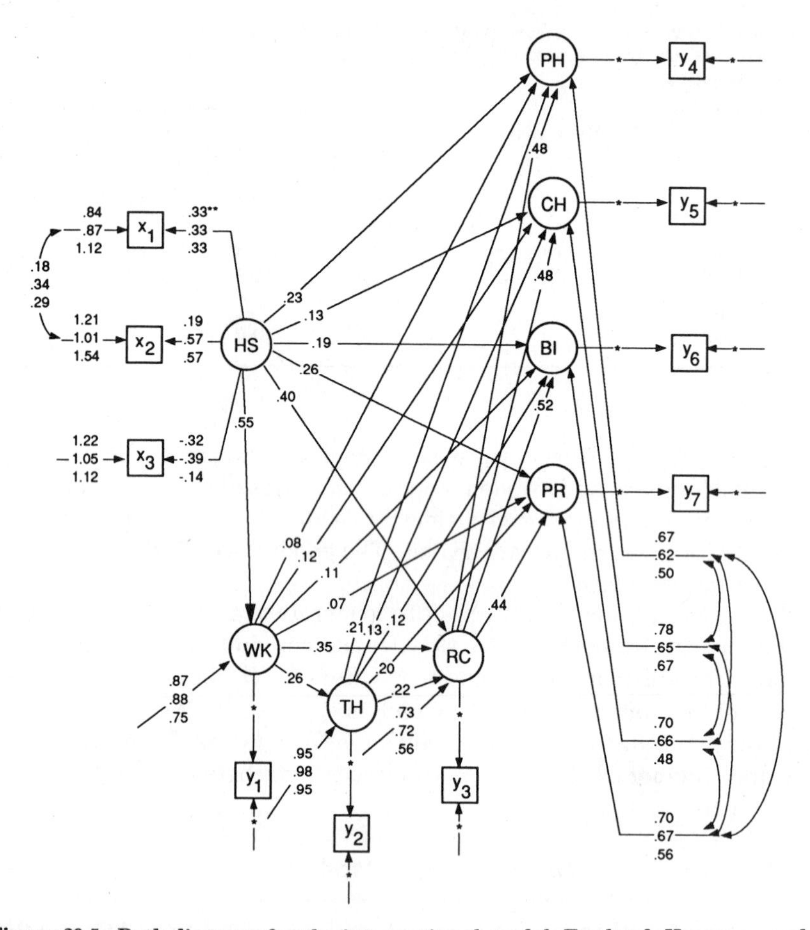

Figure 29.5. Path diagram for the international model. England, Hungary, and Sweden, 14-year-old boys. Observed and latent variables: see Figure 29.1. Simultaneous analysis of within-school covariance matrices, standardized solution. Estimates of country-specific parameters are given in the order of England, Hungary, and Sweden from top to bottom. *, fixed, country-specific parameter; **, fixed invariant parameter.

to be negligible according to the theory underlying the thinking stage variable. The solution reported is standardized to a common metric for the latent variables.

If one wants to trace the details in the fit of the model, the LISREL program offers a very useful possibility to estimate misspecifications in a model through so-called modification indices, MI. It turns out that the misspecifications in the "international" model are most apparent for paths from HS. These are reported in Table 29.5. The modification index gives a measure of predicted decrease in χ^2 if the specific parameter is set

Table 29.5. Some Modification Indices, MI, for the International Model. 14-year-old boys, $N_E = 984$, $N_H = 2{,}309$, $N_S = 699$. Simultaneous analysis on within-school covariance matrices

	England		Hungary		Sweden	
Path	Estimated Change	MI	Estimated Change	MI	Estimated Change	MI
HS→WK	−0.12	0.65	−0.07	2.04	+0.08	5.36
HS→TH	+0.11	4.54	−0.11	7.10	+0.01	1.22
HS→RC	−0.11	3.63	+0.13	5.78	−0.02	1.16
HS→PH	−0.04	0.45	−0.01	0.14	+0.00	0.89
HS→CH	−0.13	8.36	+0.14	14.44	−0.01	2.70
HS→BI	−0.06	0.16	−0.05	1.45	+0.03	3.14
HS→PR	+0.08	1.75	+0.01	0.00	−0.01	1.50

to be country-specific and the model is reestimated. These indices may be judged by means of a χ^2-distribution with 1 degree of freedom (Jöreskog and Sörbom, 1989).

It can be seen from Table 29.5 that the largest national deviations are found for the paths HS to CH. As the MIs are dependent on the sample size, and the *N*'s differ considerably between countries, they should be ranked within each country in order to point to the greatest national deviations from the international structure. The table also reports a prediction of the estimated change of a constrained path if the corresponding parameter is relaxed.

Finally, we demonstrate how the national deviations from the international structure can be expressed in the common metric from Figure 29.5. This is done in the values of estimated change in Table 29.5. From these results, we can trace paths that seem to be very stable among countries, for example, the path from HS and PH. The most unstable path from HS is found to be the path from HS to CH.

We have been able to estimate an overall stable international structure but the results also point to weaknesses in the data quality, especially in the indicators of home status, that should be improved in further research.

29.4 CONCLUSION AND DISCUSSION

We have been mainly concerned with the empirical evidence on the size and nature of measurement error effects arising from the varying quality in data for explanatory as well as outcome variables, on the one hand, and for different populations, on the other hand. In brief, the application of a path model formulated as a covariance structure model in a comparative analysis of three quite different societies has focused on two issues of a more general interest: (1) that the explanatory power of factors early in the causal scheme may rise considerably if measurement errors are taken into account, and (2) that quite complex general and stable structures can be identified if population-specific measurement errors are controlled for and partialed out.

These results have consequences for the design of measures in survey research. For example, if you choose the multiple-indicator measurement model, you need generally at least two reliable indicators to get an identified model. In practice, more than two indicators are often necessary to obtain a well-defined latent variable. This means that we have to allocate the data collection resources differently in many large scale surveys and often limit the study to fewer factors and use covariance structure modeling to measure them with an acceptable precision. In the case of IEA evaluation research, there has been an imbalance between the resources invested in the quality of the outcome achievement variables in the tests and those in the explanatory variables collected from questionnaires. The indication we get from this study is that weak explanatory power in many of these background variables can be attributed to bad precision in the instrument used.

Finally, we would like to point out some new developments in the field of covariance structure modeling which are of special relevance for the analysis of data from several populations.

In this study, we assumed that all random variables had zero means. This assumption can be relaxed in the Extended LISREL Model, which contains intercept terms in the relationships and mean values of the latent variables (Jöreskog and Sörbom, 1989). The achievement levels in the science tests for the three countries can then be estimated and country differences tested for significance after having taken measurement errors into account.

Another assumption was that data within countries were analyzed as if they were obtained from a single population, although the students in the sample belonged to different schools, possibly with different sets of parameter values. In Muthén (1989), the problem of latent variable modeling in heterogeneous populations is discussed, and models dealing with different structural slopes and structural residual variances on the within and between school levels are reported. This is an indication of the current developments in covariance modeling.

CHAPTER 30

REGRESSION ESTIMATION IN THE PRESENCE OF MEASUREMENT ERROR

Wayne A. Fuller
Iowa State University

30.1 INTRODUCTION

Regression is a technique often used in the analysis of survey data. The classical regression model is

$$y_t = \mathbf{x}_t' \boldsymbol{\beta} + q_t \tag{30.1}$$

where the q_t are independent $(0, \sigma^2)$ random variables independent of $\mathbf{x}_j$ for all t and j, $\mathbf{x}_t$ is a k-dimensional column vector of explanatory variables and β is the unknown parameter vector. Survey data typically violate the classical model in two important ways. If the survey is a clustered sample it is no longer reasonable to assume that the q_t within a cluster are independent. This fact has been recognized for some time. See, for example, Frankel (1971), Fuller (1975), and Fuller and Hidiroglou (1978). Furthermore, computer packages have been developed to compute regression coefficients and the standard errors of regression coefficients for complex survey data. Examples of such programs include SUPER CARP, PC CARP, and SUDAAN.

The second important way in which survey data violate the model is that one does not observe $(y_t, \mathbf{x}_t')$. Rather one observes

$$(Y_t, \mathbf{X}_t') = (y_t, \mathbf{x}_t') + (w_t, \mathbf{u}_t')$$

where $(w_t, \mathbf{u}_t')$ is the vector of measurement error. In survey sampling, the

This research was partly supported by Joint Statistical Agreement J.S.A 90-7 with the U.S. Bureau of the Census. I thank Joseph Croos for the computations of the design section and for the computations of the illustration.

Measurement Errors in Surveys.
Edited by Biemer, Groves, Lyberg, Mathiowetz, and Sudman.

ISBN: 0-471-53405-6

measurement error is often called response error because a primary source of error is the failure of respondents to give the same answer to a repeated question. That survey data are subject to measurement error is uniformly recognized by the community of survey samplers. An early description of the effect of measurement error on estimates of change is that of Bershad (1967). Examples of the incorporation of estimates of measurement error variances into analyses of survey data include Warren, et al. (1974), Rodgers (1989a), Chua and Fuller (1987), Poterba and Summers (1985), Abowd and Zellner (1985), and numerous articles in psychology and sociology journals. Despite this list of references, it seems fair to say that the fraction of researchers who explicitly recognize the presence of measurement error in their analyses is not large.

We begin by reviewing estimation of the vector $\boldsymbol{\beta}$ under the assumption that an estimator of the covariance matrix of the measurement error is available. The model for our investigation is

$$y_t = \mathbf{x}_t' \boldsymbol{\beta} + q_t \tag{30.2}$$

$$\mathbf{Z}_t' = (y_t, \mathbf{x}_t') + (w_t, \mathbf{u}_t') = \mathbf{z}_t' + \mathbf{a}_t' \tag{30.3}$$

where $\mathbf{x}_t$ is a k-dimensional column vector of unobserved explanatory variables, $\boldsymbol{\beta}$ is a $k \times 1$ vector of unknown parameters, q_t is the error in the equation, $\mathbf{Z}_t' = (Y_t, \mathbf{X}_t')$ is observed and $\mathbf{a}_t = (w_t, \mathbf{u}_t')'$ is the $(k+1)$-dimensional column vector of measurement error. If the model contains an intercept, the first entry in $\mathbf{x}_t$ is always a one. The remaining entries in $\mathbf{x}_t$ can be fixed or random. We let

$$\Sigma_{xx} = E\{\sum_{t=1}^{n} \mathbf{x}_t'\mathbf{x}_t\}$$

and assume Σ_{xx} is positive definite. Note that Σ_{xx} need not be a covariance matrix because some elements of $\mathbf{x}_t$ may be fixed. We begin by assuming

$$(w_t, \mathbf{u}_t')' \sim \text{ind.}(\mathbf{0}, \Sigma_{aa}) \tag{30.4}$$

independent of q_j for all t and j, where the notation $\sim$ind.$(\mathbf{0}, \Sigma)$ is read, "distributed independently with mean zero and covariance matrix Σ." We also assume $E\{\mathbf{a}_t | \mathbf{x}_t\} = \mathbf{0}$. There are some situations, such as binomial responses, where the measurement error is correlated with the true value. In most situations, the observations can be transformed to satisfy the assumption that the measurement error is uncorrelated with the true values. See Fuller (1987, p. 271) for the transformation.

The sum $q_t + w_t$ is sometimes denoted by e_t. If there is no measurement error in the explanatory variables, then $\Sigma_{uu} = \mathbf{0}$ and models (30.2)–(30.3) reduce to the classical regression model with error e_t.

The assumption that the measurement errors are independent may be violated in an interview study in which each interviewer contacts a number of respondents. The interview effect is analogous to a cluster effect in cluster sampling. We will discuss variance estimation for cluster samples. Our primary variance estimation scheme is also appropriate for heterogeneous error variance models.

It seems reasonable to assume that the measurement error is independent of the error in the equation. While it is possible for measurement error to be correlated with variables omitted from the model, the assumption that the measurement error is independent of equation error is a natural extension of the zero mean assumption.

As motivation for our estimation procedures, we recall the effect of measurement error on the ordinary least squares coefficient. Assume the univariate model

$$y_t = \beta_0 + \beta_1 x_t + q_t$$
$$X_t = x_t + u_t \tag{30.5}$$

where $(x_t, u_t, q_t) \sim \mathrm{NI}[(\mu_x, 0, 0), \mathrm{diag}(\sigma_{xx}, \sigma_{uu}, \sigma_{qq})]$. Let the ordinary least squares coefficient for a sample of n observations be

$$\hat{\gamma}_1 = \left[\sum_{t=1}^{n} (X_t - \bar{X})^2\right]^{-1} \sum_{t=1}^{n} (X_t - \bar{X})(Y_t - \bar{Y}). \tag{30.6}$$

Under model (30.5), the expected value of the ordinary least squares coefficient is

$$\gamma_1 = E\{\hat{\gamma}_1\} = \sigma_{XX}^{-1}\sigma_{XY} = [(\sigma_{xx} + \sigma_{uu})^{-1}\sigma_{xx}]\beta_1.$$

The normal distribution assumption makes this expectation exact but the expression for the expectation is a useful approximation for any distribution of x with second moments. The ratio $\sigma_{XX}^{-1}\sigma_{xx}$ is sometimes called the reliability ratio. Using data from the U.S. Department of Commerce (1975), Fuller (1987, p. 8) calculated ratios $\sigma_{XX}^{-1}\sigma_{xx}$ of 0.88 for education and 0.85 for income. With a ratio of 0.85 the ordinary least squares coefficient is biased towards zero by an amount equal to 15% of the true coefficient. The squared bias will be greater than the variance of the estimator in samples as small as 15. In samples of the size encountered in practice, the squared bias will be responsible for most of the squared error in the estimator.

The effect of measurement error on the vector of estimators for the k-dimensional model is not simple, but there is evidence that the biases may be sizeable. It has long been known that error in one variable can have a serious effect on the coefficient of another. Lord (1960) discusses this effect. A recent article in *Science* [Palca (1990)] explains how measurement error may be important in the interpretation of large scale health studies. Chua and Fuller (1987) and others have shown that measurement error can cause biases of 100% in estimates of gross change for labor status.

To see how the effect of measurement error can be magnified in a multivariate problem, assume that both education and income are included as explanatory variables in a regression. Let the variables be standardized so that the variance of observed income and the variance of observed education are both equal to one. Then the variance of true income is 0.85 and the variance of true education is 0.88. Assume that the measurement error in income is independent of the measurement error in education and that the correlation between observed income and observed education is 0.6. Assume that the vector of covariances of the dependent variable with education and income is (0.4235, 0.6000). Then the vector of true coefficients is

$$\begin{bmatrix} \beta_1 \\ \beta_2 \end{bmatrix} = \begin{bmatrix} 0.88 & 0.60 \\ 0.60 & 0.85 \end{bmatrix}^{-1} \begin{bmatrix} 0.4235 \\ 0.6000 \end{bmatrix} = \begin{bmatrix} 0 \\ 0.7058 \end{bmatrix} \tag{30.7}$$

where (β_1, β_2) are the coefficients for education and income, respectively, and the matrix with (0.88, 0.85) on the diagonal is the covariance matrix of true education and true income. This example is constructed so that the true marginal effect of education after income is zero. In the presence of measurement error, the expected value of the observed vector of ordinary least squares coefficients is

$$\begin{bmatrix} \gamma_1 \\ \gamma_2 \end{bmatrix} = \begin{bmatrix} 1.00 & 0.60 \\ 0.60 & 1.00 \end{bmatrix}^{-1} \begin{bmatrix} 0.4235 \\ 0.6000 \end{bmatrix} = \begin{bmatrix} 0.0992 \\ 0.5405 \end{bmatrix}.$$

The presence of measurement error has produced a positive value for the zero coefficient on education and a 23% bias in the coefficient for income.

This example illustrates a common empirical occurrence. A variable whose theoretical coefficient is zero often shows a significant coefficient in a regression containing independent variables measured with error. An excellent example of this phenomenon is described by Kelley (1973). See also Lord (1960) and Fuller and Hidiroglou (1978).

30.2 ESTIMATION

A sample of n observations on $\mathbf{Z}_t$ is not sufficient information to estimate the parameters of the model (30.2), (30.3) and (30.4). A second source of information is required. We first consider estimation of the error covariance matrix by repeated determinations on the same units. In survey sampling, this may take the form of a reinterview for a sample of individuals. The reinterview may be a part of the basic study or it may be a separate study. We assume that two determinations are made on d individuals.

In the physical sciences it is sometimes possible to make several independent identically distributed observations on the same material. Also, social scientists often devise sets of questions whose responses can be used as multiple estimates of a single underlying characteristic. Our procedure can be extended to either of these situations. Because obtaining more than two responses to an identical questionnaire is difficult with human respondents, the case of two responses is an important special case.

There is some feeling that it is impossible to obtain truly independent repeated observations from human respondents. However, it is well documented that repeated responses will differ considerably for almost all items. Also, if there is a bias in the response variance estimated from repeated determinations, we expect the response variance to be underestimated because of positive correlation between responses collected closely together in time. One obvious way for response variance to be overestimated in a replication study is to make mistakes in matching individuals.

For our purposes, we assume that two independent identically distributed observations $\mathbf{Z}_{tj}$, $j = 1, 2$, are available for $t = 1, 2, \ldots, d$ individuals. Let ζ_t^{-1} be the probability that the tth individual is included in the sample. Then

$$\mathbf{S}_{aa} = \left(2\sum_{t=1}^{d}\zeta_t\right)^{-1}\sum_{t=1}^{d}\zeta_t(\mathbf{Z}_{t1} - \mathbf{Z}_{t2})(\mathbf{Z}_{t1} - \mathbf{Z}_{t2})' \tag{30.8}$$

$$= \left(\sum_{t=1}^{d}\zeta_t\right)^{-1}\sum_{t=1}^{d}\zeta_t\mathbf{S}_{aatt}$$

where $2^{-1}(\mathbf{Z}_{t1} - \mathbf{Z}_{t2})(\mathbf{Z}_{t1} - \mathbf{Z}_{t2})' = \mathbf{S}_{aatt}$ is an estimator of the error covariance matrix Σ_{aa}. We note that if different individuals have different error covariance matrices, denoted by Σ_{aatt}, and if the d individuals are a probability sample from the population of individuals, then $\mathbf{S}_{aa}$ is a consistent estimator of the average covariance matrix, $\Sigma_{aa..} = E\{\Sigma_{aatt}\}$. In some situations one may be willing to assume a constant error covariance matrix and to set $\zeta = 1$ in the estimator.

We consider the estimation of β, given a sample of n observations on the $\mathbf{Z}$-vector and an estimator of Σ_{aa}, denoted by $\mathbf{S}_{aa}$, based upon d degrees of freedom. We begin by assuming that the sample of n observations is independent of the sample used to construct $\mathbf{S}_{aa}$. We do not require $\mathbf{S}_{aa}$ to be nonsingular. For example, if the model contains an intercept, the first entry in $\mathbf{x}_t$ will always be a one and the first row and column of $\mathbf{S}_{aa}$ will contain zeros. We assume that the sample is a stratified cluster sample, where $\mathbf{Z}_{ij\ell}$ is the ℓ-th element in cluster j of stratum i. If convenient, we will replace the triple subscript $ij\ell$ with the single subscript t. Thus, the sample mean of the observations is

$$\bar{\mathbf{Z}} = \left(\sum_{i=1}^{L} \sum_{j=1}^{n_i} \sum_{\ell=1}^{m_{ij}} \tau_{ij\ell} \right)^{-1} \sum_{i=1}^{L} \sum_{j=1}^{n_i} \sum_{\ell=1}^{m_{ij}} \tau_{ij\ell} \mathbf{Z}_{ij\ell}$$

$$= \left(\sum_{t=1}^{n} \tau_t \right)^{-1} \sum_{t=1}^{n} \tau_t \mathbf{Z}_t \tag{30.9}$$

where τ_t is the weight for the tth observation. The weight is generally equal to the reciprocal of the selection probability. In some situations, τ_t may be taken to be one. The sample moment matrix estimated with the weights is

$$\mathbf{M}_{ZZ} = \left(\sum_{t=1}^{n} \tau_t \right)^{-1} \sum_{t=1}^{n} \tau_t \mathbf{Z}_t \mathbf{Z}_t'. \tag{30.10}$$

Our estimator is constructed under the assumption that we desire an estimator of the population regression coefficient, where the population regression coefficient is the value one would compute if one observed the entire population of true values. Therefore, we desire an estimator of $\mathbf{M}_{zz}$, where $\mathbf{M}_{zz}$ is the second moment matrix of the true values, z_t, and $\mathbf{z}_t' = (y_t, \mathbf{x}_t')$ is defined in (30.3). We have

$$\mathbf{M}_{ZZ} = \left(\sum_{t=1}^{n} \tau_t \right)^{-1} \sum_{t=1}^{n} \tau_t (\mathbf{z}_t + \mathbf{a}_t)(\mathbf{z}_t + \mathbf{a}_t)'$$

and it follows that, under our assumptions,

$$\hat{\mathbf{M}}_{zz} = \mathbf{M}_{ZZ} - \mathbf{S}_{aa} \tag{30.11}$$

is a consistent estimator of $\mathbf{M}_{zz}$. Let $\hat{\mathbf{M}}_{yy}$, $\hat{\mathbf{M}}_{yx}$, and $\hat{\mathbf{M}}_{xx}$ be the submatrices of $\hat{\mathbf{M}}_{zz}$ obtained by partitioning $\hat{\mathbf{M}}_{zz}$ to conform with the partition $(y_t, \mathbf{x}_t')$ of $\mathbf{z}_t'$. Then

$$\hat{\beta} = \hat{\mathbf{M}}_{xx}^{-1} \hat{\mathbf{M}}_{xy} \tag{30.12}$$

is a consistent estimator of β.

The variance of the approximate distribution of $\hat{\beta}$ is obtained by using a Taylor expansion to derive

$$\hat{\beta} - \beta = (\mathbf{M}_{XX} - \mathbf{S}_{uu})^{-1}[(\mathbf{M}_{XY} - \mathbf{S}_{uw}) - (\mathbf{M}_{XX} - \mathbf{S}_{uu})\beta]$$

$$\approx \mathbf{M}_{xx}^{-1}(\mathbf{M}_{Xv} - \mathbf{S}_{uv}) \tag{30.13}$$

where $v_t = Y_t - \mathbf{X}_t'\beta$, $\mathbf{S}_{uv} = \mathbf{S}_{uw} - \mathbf{S}_{uu}\beta$,

$$\mathbf{M}_{Xv} = \left(\sum_{t=1}^{n} \tau_t\right)^{-1} \sum_{t=1}^{n} \tau_t \mathbf{X}_t v_t,$$

and $\mathbf{S}_{uu}$ and $\mathbf{S}_{uw}$ are submatrices of $\mathbf{S}_{aa}$. Under our assumptions,

$$E\{\mathbf{X}_t v_t\} = E\{\mathbf{u}_t v_t\} = E\{\mathbf{S}_{uv}\}. \tag{30.14}$$

It follows from (30.13) that the variance of the limiting distribution of $n^{1/2}(\hat{\beta} - \beta)$ is

$$\mathbf{M}_{xx}^{-1}\mathbf{V}\{\mathbf{M}_{Xv} - \mathbf{S}_{uv}\}\mathbf{M}_{xx}^{-1} = \mathbf{M}_{xx}^{-1}[\mathbf{V}\{\mathbf{M}_{Xv}\} + \mathbf{V}\{\mathbf{S}_{uv}\}]\mathbf{M}_{xx}^{-1}.$$

If the sampling scheme and the population are such that the elements of $n^{1/2}(\mathbf{M}_{ZZ} - \Sigma_{ZZ})$ and of $n^{1/2}(\mathbf{S}_{aa} - \Sigma_{aa})$ converge in distribution to a vector normal variable, then $n^{1/2}(\hat{\beta} - \beta)$ will converge to a normal vector. See Fuller (1987, Section 2.2 and Section 3.1) and Fuller (1975) for derivations of the approximate distribution.

The package EV CARP can be used to compute the estimator and to compute an estimate of the variance of the approximate distribution. EV CARP uses a distribution-free method to estimate the variance of $\mathbf{M}_{Xv}$ and estimates the variance of $\mathbf{S}_{uv}$ under the assumption that $\mathbf{S}_{aa}$ is a multiple of a Wishart matrix. To use EV CARP, the user must provide degrees of freedom for $\mathbf{S}_{aa}$ such that the approximation of the distribution of $\mathbf{S}_{aa}$ by a multiple of the Wishart distribution is reasonable. For example, if the variances of the diagonal elements of $\mathbf{S}_{aa}$ have been computed, a reasonable value for the degrees of freedom of the Wishart distribution is

$$d_{fs} = (k + 1)^{-1} 2 \sum_{i=1}^{k+1} [V\{s_{aaii}\}]^{-1} s_{aaii}^2$$

where $k+1$ is the dimension of $\mathbf{S}_{aa}$ and s_{aaii} is the i-th diagonal element of $\mathbf{S}_{aa}$. Recall that the variance of the estimator of variance with d degrees of freedom from a $N(\mu, \sigma^2)$ distribution is $2d^{-1}\sigma^4$.

The estimator of β and the approximate distribution of the estimator are different if the replicate observations used to compute $\mathbf{S}_{aa}$ are a subsample of the n observations used to estimate β. If there are replicate observations on the tth individual, we set

$$\mathbf{Z}_t = 0.5(\mathbf{Z}_{t1} + \mathbf{Z}_{t2}) \tag{30.15}$$

where $\mathbf{Z}_{t1}$ and $\mathbf{Z}_{t2}$ are the two observations on the tth individual. If the observation is not replicated, $\mathbf{Z}_t = \mathbf{Z}_{t1}$. Thus, the error variance for the mean of the two replicates is one half of the error variance of the unreplicated observations. If there are d replicated observations and $(n-d)$ nonreplicated observations, a consistent estimator of $\mathbf{M}_{zz}$ is

$$\hat{\mathbf{M}}_{zz} = \mathbf{M}_{ZZ} - (\sum_{t=1}^{n} \tau_t)^{-1}(\sum_{t=1}^{n} \tau_t - 0.5 \sum_{t=1}^{d} \tau_t)\mathbf{S}_{aa} \tag{30.16}$$

where $\mathbf{M}_{ZZ}$ is computed with the $\mathbf{Z}_t$ of (30.15) and we have assigned the first d subscripts to the replicated observations. The estimator of β is constructed by substituting the submatrices of (30.16) into (30.12).

The estimator of $\mathbf{M}_{zz}$ in (30.16) can also be written as

$$\hat{\mathbf{M}}_{zz} = (\sum_{t=1}^{n} \tau_t)^{-1} \sum_{t=1}^{n} \tau_t[\mathbf{Z}_t\mathbf{Z}_t' - \mathbf{S}_{aa} + v_t\{0.5\,\mathbf{S}_{aa} - \eta_t(\mathbf{S}_{aatt} - \mathbf{S}_{aa})\}] \tag{30.17}$$

where

$$\eta_t = (0.5 \sum_{j=1}^{d} \tau_j + \sum_{j=d+1}^{n} \tau_j)(\sum_{j=1}^{d} \tau_j)^{-1}\tau_t^{-1}\zeta_t(\sum_{j=1}^{d} \zeta_j)^{-1} \quad \text{for } t = 1, 2, \ldots, d$$

$$= 0 \text{ otherwise,}$$

$v_t = 1$ for a replicated observation and $v_t = 0$ for a nonreplicated observation. Expression (30.17) can be used to construct a distribution-free estimator of the variance of the approximate distribution of $\hat{\beta}$. An estimator of the variance of the approximate distribution of $\hat{\beta}$ is

$$\hat{\mathbf{V}}\{\hat{\beta}\} = (\sum_{t=1}^{n} \tau_t)^{-2}\, \hat{\mathbf{M}}_{xx}^{-1}\hat{\mathbf{G}}_{dd}\hat{\mathbf{M}}_{xx} \tag{30.18}$$

where

$$\hat{\mathbf{G}}_{dd} = \sum_{i=1}^{L} (n_i - 1)^{-1} n_i \sum_{j=1}^{n_i} (\mathbf{d}_{ij.} - \bar{\mathbf{d}}_{i..})(\mathbf{d}_{ij.} - \bar{\mathbf{d}}_{i..})',$$

$$\mathbf{d}_{ij\ell} = \tau_{ij\ell}(\mathbf{X}_{ij\ell}\hat{v}_{ij\ell} - \mathbf{S}_{uvij\ell}),\ \hat{v}_{ij\ell} = Y_{ij\ell} - \mathbf{X}_{ij\ell}\hat{\beta},$$

$$\mathbf{S}_{uvij\ell} = [\mathbf{S}_{aa} - v_t\{0.5\,\mathbf{S}_{aa} - (\mathbf{S}_{aatt} - \mathbf{S}_{aa})\}(\zeta_t \sum_{j=1}^{n} \tau_j)(\tau_t \sum_{j=1}^{d} \zeta_j)^{-1}](1,\ -\hat{\beta}')',$$

$$\mathbf{d}_{ij.} = \sum_{\ell=1}^{m_{ij}} \tau_{ij\ell}\mathbf{d}_{ij\ell}, \text{ and } \bar{\mathbf{d}}_{i..} = n_i^{-1} \sum_{j=1}^{n_i} \mathbf{d}_{ij.}.$$

See Fuller (1987, Ch. 3) and Fuller (1975) for derivations of the covariance matrix of the limiting distribution.

The variance expression (30.18) is a large sample expression derived under the mild assumptions traditional in survey sampling. The covariance matrix $\hat{\mathbf{G}}_{dd}$ is computed by complete analogy to the usual estimator of the covariance matrix of a vector of totals. Expression (30.18) and related expressions can be computed using the functionally related option of EV CARP. Hasabelnaby, et al. (1989) contains an application of some of these ideas.

We have assumed throughout that we use two independent observations to construct an unbiased estimator of the error variance. If the two determinations on the same individual are correlated, then the estimate of the measurement error variance is biased. The expectation of the estimate based on correlated determinations is $(1-\rho_{12})\,\Sigma_{uu}$, where ρ_{12} is the correlation between the two determinations. To illustrate the nature of this bias and its effect on estimated coefficients, we use the example associated with (30.7). Assume that two determinations with a correlation of $\rho_{12}=0.3$ are used to estimate the covariance matrix of measurement error. The population error covariance matrix is $\Sigma_{uu}=\text{diag}\,(0.12, 0.15)$. For $\rho_{12}=0.3$, the expected value of the estimated error covariance matrix is diag (0.084, 0.105). Therefore the vector of estimated coefficients based on this estimated covariance matrix will converge to

$$\begin{pmatrix} 0.936 & 0.600 \\ 0.600 & 0.895 \end{pmatrix}^{-1} \begin{pmatrix} 0.4235 \\ 0.6000 \end{pmatrix} = \begin{pmatrix} 0.0398 \\ 0.6437 \end{pmatrix}.$$

The estimator constructed with the biased covariance matrix is biased. The bias in the coefficients is about 0.4 of the bias in the ordinary least squares estimators. In this example, the bias reduction is less than $(1-\rho_{12})$ because the true x values are correlated.

30.3 ILLUSTRATIONS

We use a small constructed data set to illustrate the computations for a survey in which duplicate observations are made on a subsample of observations. The data are given in Table 30.1. The sample is a stratified sample with 16 observations in stratum one and 20 observations in stratum two. Duplicate observations, denoted by X_{t2}, are made on eight individuals in each of the two strata. The mean of the squared differences of the two determinations is 2 in stratum one and is 8 in stratum two. The estimated error variances are one half of the mean of the squared differences. The estimated error variance for a single determination on X for the population is

$$\hat{\sigma}_{uu}=(520)^{-1}[(320)1+(200)4]=2.153845.$$

The estimated variance of this estimator of σ_{uu} is

$$\hat{V}\{\hat{\sigma}_{uu}\}=(520)^{-2}\,[(320)^2(8)^{-1}(7)^{-1}8+(200)^2(8)^{-1}(7)^{-1}\,128]=0.3922.$$

The estimator $\hat{V}\{\hat{\sigma}_{uu}\}$ is the usual survey stratified estimator of variance where the eight observations in each stratum are the eight values of $0.5(X_2-X_1)^2$.

Table 30.1. Data for Example Calculations

Stratum	Cluster	Weight	Y	X_1	X_2	X_3	SQRT
1	1	20	2	2	2	2	0.71
1	2	20	4	4	2	3	0.71
1	3	20	4	2	4	3	0.71
1	4	20	2	4	4	4	0.71
1	5	20	2	2		2	1.00
1	6	20	2	4		4	1.00
1	7	20	4	2		2	1.00
1	8	20	4	4		4	1.00
1	9	20	6	4	4	4	0.71
1	10	20	4	6	4	5	0.71
1	11	20	4	4	6	5	0.71
1	12	20	6	6	6	6	0.71
1	13	20	4	4		4	1.00
1	14	20	4	6		6	1.00
1	15	20	6	4		4	1.00
1	16	20	6	6		6	1.00
2	1	10	9	4	4	4	1.41
2	2	10	3	8	4	6	1.41
2	3	10	3	4	8	6	1.41
2	4	10	9	8	8	8	1.41
2	5	10	3	4		4	2.00
2	6	10	3	8		8	2.00
2	7	10	9	4		4	2.00
2	8	10	9	8		8	2.00
2	9	10	6	7	7	7	1.41
2	10	10	12	11	7	9	1.41
2	11	10	12	7	11	9	1.41
2	12	10	6	11	11	11	1.41
2	13	10	6	7		7	2.00
2	14	10	6	11		11	2.00
2	15	10	12	7		7	2.00
2	16	10	12	11		11	2.00
2	17	10	5	6		6	2.00
2	18	10	5	10		10	2.00
2	19	10	11	6		6	2.00
2	20	10	11	10		10	2.00

The variance of an estimator of the variance for a normal distribution based upon d degrees of freedom is $2d^{-1}\sigma^4$, where σ^2 is the variance of the parent distribution. Therefore the estimated variance of our estimator is equivalent to the variance of a normal sample with 24 degrees of freedom because

$$0.3922 = 2(24)^{-1}(2.153845)^2.$$

The large degrees of freedom relative to the number of differences follows from the fact that our differences show less relative variability than normal differences and because we have a stratified sample.

We first estimate the regression of Y_t on X_{t1} without an adjustment for measurement error. The weighted regression estimator computed with EV CARP and the stratum weights is

$$\hat{Y}_t = \underset{(0.75)}{1.77} + \underset{(0.15)}{0.67}\, X_{t1}.$$

The standard errors are computed by formulas analogous to (30.16) without the correction for measurement error. The variance formula is given on page 67 of the EV CARP manual (see Schnell, et al., 1988).

We next compute an estimate of the regression coefficient under the assumption that X_{t1} is observed and that our estimate of the error variance comes from an independent sample. The input for EV CARP, under these assumptions, is composed of stratum identification, cluster identification, weight, (Y, X), the estimated error variance $(=2.15)$, and the normal equivalent degrees of freedom $(=24)$ for the estimated error variance. The estimated equation is

$$\hat{Y}_t = \underset{(1.32)}{0.01} + \underset{(0.26)}{0.99}\, x_t$$

where the standard errors are computed by EV CARP and the increase in variance associated with the estimated σ_{uu} is computed with the approximating degrees of freedom. If the error variance were known, the estimated standard errors would be 1.21 and 0.24 for the intercept and slope, respectively.

The correction of the estimated slope for the presence of measurement error produces an estimate that is about 50% larger than the ordinary least squares estimate. In this constructed example, the measurement error variance is about one-third of the variance of observed X_t. Both the ordinary least squares line and the measurement error line go through $(\bar{X}, \bar{Y}) = (5.38, 5.38)$. Because the slope of the measurement error line is greater than the slope of the ordinary least squares line, the intercept for the measurement error line is smaller than the intercept for the ordinary least squares line.

To compute the estimates using the replicated observations, we use the functionally related option of EV CARP. The column headed by X_3 in Table 30.1 contains observations of the type defined in (30.15). That is, X_3 is the mean of the available determinations on X. The variable in the column headed SQRT is the square root of the estimated error variance in X_3. This variable is $2^{-1/2}$ for replicated observations in stratum one, is one for unreplicated observations in stratum one, is $2^{1/2}$ for replicated observations in stratum two and is 2 for unreplicated observations in stratum one.

The weighted average of the error variances for X_3 is 1.6923 compared to the average of 2.1538 for X_1. Using the functionally related option of EV CARP, we enter SQRT as the square root of the error variance. The estimator of β is computed using $\hat{\mathbf{M}}_{zz}$ of (30.16). The estimated equation is

$$\hat{Y}_t = \underset{(1.27)}{0.01} + \underset{(0.25)}{0.99}\, x_t.$$

The standard errors are the square roots of variances that include a term reflecting the fact that the error variances were estimated.

30.4 INSTRUMENTAL VARIABLES

In Section 30.2, we discussed the use of information on the error covariance matrix in the estimation of the parameters of the regression model. In this section, we outline another estimation procedure for model (30.2), (30.3), (30.4). In addition to the assumptions of that model, we assume that a third vector, denoted by $\mathbf{H}_t$, is observed on each individual. It is assumed that

$$E\{(q_t, w_t, \mathbf{u}_t)|\mathbf{H}_t\} = \mathbf{0}. \tag{30.19}$$

For example, suppose that we are interested in estimating the coefficient for income in a simple regression. We might collect data on age, education, and occupation on the same individuals and be willing to assume that the response error in income is independent of the response errors in age, education, and occupation. We are also willing to assume that age, education, and occupation do not enter the explanatory equation. Formally, this is equivalent to assuming that age, education, and occupation are uncorrelated with the q_t of the model. Under these assumptions, it is possible to estimate β of (30.2).

Another example of an instrumental variable is the response to a second question constructed to be a measure, or indicator, of the same quantity. Multiple indicator models have been heavily used in psychology and sociology. See Goldberger and Duncan (1973). Groves (1989, Section 7.5.2 and Section 10.5) discusses the multiple indicator procedure.

The vector $\mathbf{H}_t$ may be observed subject to measurement error, but if $\mathbf{H}_t$ is subject to measurement error, the error in $\mathbf{H}_t$ must be uncorrelated with that in $\mathbf{Z}_t$. In order for $\mathbf{H}_t$ to be useful, it must be correlated with the true $\mathbf{x}_t$. Thus, we assume the population mean square due to the regression of $\mathbf{x}_t$ on $\mathbf{H}_t$,

$$\Sigma_{xH}\,\Sigma_{HH}^{-1}\,\Sigma_{Hx}, \tag{30.20}$$

is nonsingular. If an element of $\mathbf{X}_t$ is observed without error, that element is also an element of $\mathbf{H}_t$. For example, if there is an intercept in the model, the first element of $\mathbf{X}_t$ is always one and the first element of $\mathbf{H}_t$ is always one.

Under our model, the instrumental variable estimator is

$$\hat{\beta} = (\mathbf{M}_{XX} - \tilde{\gamma}\hat{\Sigma}_{rr22})^{-1}(\mathbf{M}_{XY} - \tilde{\gamma}\hat{\Sigma}_{rr21}) \tag{30.21}$$

where

$$\hat{\Sigma}_{rr} = \mathbf{M}_{ZZ} - \mathbf{M}_{ZH}\mathbf{M}_{HH}^{-1}\mathbf{M}_{HZ},$$

$\mathbf{M}_{ZZ}$, $\mathbf{M}_{HH}$, and $\mathbf{M}_{ZH}$ are defined as in (30.10), $\hat{\Sigma}_{rr}$ is partitioned to conform with the partition $(Y_t, \mathbf{X}_t')$ and $\tilde{\gamma}$ is the smallest root of

$$|\mathbf{M}_{ZZ} - \gamma\hat{\Sigma}_{rr}| = 0.$$

The estimator (30.21) is called the limited information maximum likelihood estimator in the econometrics literature. See Fuller (1987, p.50 and p.148) for derivations of the estimator in the measurement error context. Instrumental variable estimation has been used by economists and in some extended forms by biostatisticians. See the articles in *Statistics in Medicine* (Hasabelnaby, et al., 1989; Rosner, et al., 1989).

It is possible to use the program EV CARP to compute the instrumental variable estimator and to compute the variance of the approximate distribution for complex designs. The estimated variance for complex designs is

$$\hat{V}\{\hat{\beta}\} = (\mathbf{M}_{XX} - \tilde{\gamma}\hat{\Sigma}_{rr22})^{-1}\hat{\mathbf{G}}_{(IV)}(\mathbf{M}_{XX} - \tilde{\gamma}\hat{\Sigma}_{rr22})^{-1} \tag{30.22}$$

where $\hat{\mathbf{G}}_{(IV)}$ is the $\hat{\mathbf{G}}$ defined in (30.18) with

$$\mathbf{d}_{(IV)ij\ell} = \tau_{ij\ell}\hat{\mathbf{X}}'_{(IV)ij\ell}\hat{v}_{ij\ell}, \; \hat{\mathbf{X}}'_{(IV)ij\ell} = \mathbf{H}'_{ij\ell}\hat{\mathbf{M}}_{HH}^{-1}\hat{\mathbf{M}}_{HX},$$

and $\hat{v}_{ij\ell} = Y_{ij\ell} - \mathbf{X}'_{ij\ell}\hat{\beta}$. Details of the computations are given in the EV CARP manual (Schnell, et al., 1988).

The efficiency of instrumental variable estimation depends upon the correlation between the true values and the instruments. Recall our example in which age, education, and occupation are used as instruments for income. If all assumptions are satisfied, and if the squared multiple correlation between income and (age, education, and occupation) is 0.5, the variance of the instrumental variable estimator for a simple random sample is

$$V\{\hat{\beta}_{IV}\} = (0.5\sigma_{xx})^{-1}(\sigma_{qq} + \beta^2\sigma_{uu}). \tag{30.23}$$

Thus, the variance of the estimator of β is more than twice that of the ordinary least squares estimator constructed in the absence of measurement error.

To compare ordinary least squares and instrumental variable estimation in the presence of measurement error, assume that the error variance for the error in the explanatory variable is 0.15 of the variance of the observed explanatory variable. To compare alternative estimators of β, let $(\sigma_{xx}, \beta, \sigma_{qq}) = (1, 1, 1)$. Then the mean squared error of the ordinary least squares estimator for a random sample of 1,000 observations is

$$\mathrm{MSE}\{\hat{\gamma}_1\} \approx n^{-1}\sigma_{XX}^{-1}(\sigma_{YY} - \sigma_{XX}\sigma_{XY}^{2}) + (\sigma_{XX}^{-1}\sigma_{uu})^2 = 0.0180$$

where the first term is the variance and the second term is the squared bias. The variance of the instrumental variable estimator is

$$V\{\hat{\beta}_{IV}\} \approx (1{,}000)^{-1}(0.5)^{-1}\,(1.15) = 0.0023.$$

Thus, the instrumental variable estimator is about eight times as efficient as ordinary least squares for this sample size and parameter configuration.

Information on the error covariance matrix and instrumental variable information can be combined to obtain a more efficient estimator. Also if both types of information are available, it can be used to check some of the model assumptions. Hasabelnaby, et al. (1989) have discussed methods of combining information.

Under the assumption that the error distributions are symmetric, the estimated covariance between the estimator of β based on an estimated error covariance matrix and the instrumental variable estimator is

$$\mathbf{C}\{\hat{\beta}_{EV}, \tilde{\beta}_{IV}\} = \left(\sum_{t=1}^{n} \tau_t\right)^{-2} \hat{\mathbf{M}}_{xx}^{-1}\hat{\mathbf{G}}_{(IV)}(\mathbf{M}_{XX} - \tilde{\gamma}\boldsymbol{\Sigma}_{ss22})^{-1} \tag{30.24}$$

where $\hat{\beta}_{EV}$ denotes the estimator based on the estimated error covariance. If the error distributions are not symmetric, the **G**-term can be estimated with

$$\hat{\mathbf{G}}_{(EV,IV)} = \sum_{i=1}^{L} (n_i - 1)^{-1} n_i \sum_{j=1}^{n_i} (\mathbf{d}_{ij\cdot} - \bar{\mathbf{d}}_{i\cdot\cdot})(\mathbf{d}_{(IV)ij\cdot} - \bar{\mathbf{d}}_{(IV)i\cdot\cdot})' \tag{30.25}$$

where $\mathbf{d}_{ij\cdot}$ is defined in (30.18) and $\mathbf{d}_{(IV)ij}$ is defined in (30.22).

The combined estimator constructed by generalized least squares is

$$\ddot{\beta} = [(\mathbf{I}_k, \mathbf{I}_k)\tilde{\mathbf{V}}_{\beta\beta}^{-}(\mathbf{I}_k, \mathbf{I}_k)'][(\mathbf{I}_k, \mathbf{I}_k)\tilde{\mathbf{V}}_{\beta\beta}^{-}(\hat{\beta}'_{EV}, \tilde{\beta}'_{IV})'],$$

where $\tilde{\beta}_{IV}$ is the instrumental variable estimator, $\mathbf{I}_k$ is a $k \times k$ identity matrix, $\tilde{\mathbf{V}}_{\beta\beta}$ is the estimated covariance matrix of $(\hat{\beta}_{EV}, \tilde{\beta}_{IV})$, and $\tilde{\mathbf{V}}_{\beta\beta}^{-}$ is the generalized inverse of $\tilde{\mathbf{V}}_{\beta\beta}$. The generalized inverse is required if any of the explanatory variables, such as the intercept, are measured without error. An alternative method of computation is to only include the part of the instrumental variable estimator associated with the error prone observations in the computations.

30.5 DESIGN CONSIDERATIONS

In this section, we consider the problem of designing a survey whose objective is the estimation of a regression equation containing an explanatory variable subject to measurement error. We write the model as

$$\begin{aligned} y_t &= \beta_0 + \beta_1 x_t + q_t \\ (Y_{tj}, X_{tj}) &= (y_t, x_t) + (w_{tj}, u_{tj}) \end{aligned} \tag{30.26}$$

where $(q_t, w_{tj}, u_{tj})' \sim \text{NI}[\mathbf{0}, \text{diag}(\sigma_{qq}, \sigma_{ww}, \sigma_{uu})]$, and it is assumed that (q_t, w_{tj}, u_{tj}) is independent of x_i for all t, j, and i. The presence of the j subscript on (Y_{tj}, X_{tj}) means that it is possible to obtain multiple determinations on individuals.

In our calculations, we assume $\sigma_{xx} = 1$, $\beta_1 = 1$, and $\sigma_{ww} = 0$. Thus $\sigma_{ee} = \sigma_{qq} + \sigma_{ww} = \sigma_{qq}$. Only the assumption $\sigma_{ww} = 0$ results in a loss of generality. Any model with positive variance and $\beta_1 \neq 0$ can be reparameterized so that $\sigma_{xx} = 1$ and $\beta_1 = 1$. To consider nonzero variance for w and a possible covariance between w and u would add two additional parameters to the model. At the end of this section, we briefly discuss the effect of positive σ_{ww} on the conclusions.

We assume that funds sufficient for n interviews (determinations) are available. A portion of the funds will be used to conduct independent reinterviews on some of the respondents. We assume that it is not practical to interview a respondent more than twice. Thus, the total of n interviews is divided between $n - 2d$ individuals interviewed once and d individuals interviewed twice, where $0 < \delta \leq 0.5$ and $d = n\delta$. We require d to be positive in order to obtain an estimator of the error variance. We consider the estimator

$$\hat{\beta}_1 = [\alpha m_{\bar{X}\bar{X}} + (1 - \alpha) m_{XX} - (0.5\alpha + 1 - \alpha) S_{uu}]^{-1} \times [\alpha m_{\bar{X}Y} + (1 - \alpha) m_{XY}] \tag{30.27}$$

where

$$\bar{X}_{t.} = 0.5(X_{t1} + X_{t2}),$$

$$(m_{\bar{X}\bar{X}}, m_{\bar{X}Y}) = (d - 1)^{-1} \sum_{t=1}^{d} [(\bar{X}_{t.} - \bar{X}_{..})^2, (\bar{X}_{t.} - \bar{X}_{..})(Y_t - \bar{Y})],$$

$$(m_{XX}, m_{XY}) = (n - 2d - 1)^{-1} \sum_{t=d+1}^{n-d} [(X_{t1} - \bar{X}_{..})^2, (X_{t1} - \bar{X}_{..})(Y_t - \bar{Y})],$$

$$\bar{X}_{..} = (n - d)^{-1} \left[\sum_{t=1}^{d} \bar{X}_{t.} + \sum_{t=d+1}^{n-d} X_{t1} \right], \quad S_{uu} = (2d)^{-1} \sum_{t=1}^{d} (X_{t2} - X_{t1})^2,$$

Table 30.2. Optimum Values of δ for the Design with Partial Replication

	σ_{uu}/σ_{xx}						
σ_{ee}/σ_{xx}	0.02	0.05	0.10	0.15	0.25	0.5	1.0
0.05	0.107	0.236	0.438	0.500	0.500	0.500	0.500
0.10	0.079	0.176	0.320	0.461	0.500	0.500	0.500
0.25	0.053	0.120	0.218	0.308	0.484	0.500	0.500
0.50	0.038	0.089	0.163	0.229	0.352	0.500	0.500
1.00	0.027	0.065	0.121	0.172	0.263	0.482	0.500
2.00	0.020	0.047	0.089	0.152	0.197	0.352	0.500
4.00	0.014	0.034	0.065	0.093	0.146	0.263	0.484
10.00	0.009	0.022	0.042	0.061	0.097	0.177	0.320
20.00	0.006	0.015	0.030	0.044	0.070	0.130	0.236

and α is a constant that minimizes the variance of the estimator.

The duplicate observations serve two purposes. First, they provide the unbiased estimator of σ_{uu}, denoted by S_{uu}. Second, they reduce the variance of the error in the explanatory variable of the duplicated observations because the error variance of $\bar{X}_t$ is $0.5\sigma_{uu}$. Under the assumptions that the errors are normal and that $(\sigma_{xx}, \beta_1) = (1, 1)$, the approximate variance of $n^{1/2}(\hat{\beta}_1 - \beta_1)$ is

$$V\{n^{1/2}(\hat{\beta}_1 - \beta_1)\} = (1-\alpha)^2 A_1 + \alpha^2 A_2 + (1 - 0.5\alpha)^2 A_3 \tag{30.28}$$

where

$$\begin{aligned} A_3 &= 2\delta^{-1}\sigma_{uu}^2 \\ A_1 &= (1-2\delta)^{-1}[(\sigma_{ee}+\sigma_{uu})(1+\sigma_{uu}) + \sigma_{uu}^2] && \text{if } \delta < 0.5 \\ &= 0 && \text{if } \delta = 0.5 \\ A_2 &= \delta^{-1}[(\sigma_{ee} + 0.5\sigma_{uu})(1+0.5\sigma_{uu}) + 0.25\sigma_{uu}^2]. \end{aligned}$$

The value of α that minimizes the approximate variance is

$$\begin{aligned} \alpha &= (A_1 + A_2 + 0.25\,A_3)^{-1}(A_1 + 0.5\,A_3) && \text{if } \delta < 0.5 \\ &= 1 && \text{if } \delta = 0.5. \end{aligned}$$

Table 30.2 contains the optimum values of δ for a range of $(\sigma_{ee}, \sigma_{uu})$ combinations. The quantity $1-(\sigma_{ee}+1)^{-1}\sigma_{ee}$ is the squared correlation between Y and true x. Thus, the top line of the table corresponds to an R_{xY}^2 of about 0.95 and the last line corresponds to an R_{xY}^2 of about 0.05. The fraction of resources that should be allocated to replication increases as the percentage error variance in the explanatory variable increases and as the correlation between the true independent variable and the dependent variable increases.

For a large σ_{uu} and a high correlation, every individual in the sample should be observed twice ($\delta = 0.5$). If $\sigma_{uu}/\sigma_{xx} < 0.02$ and $R_{xY}^2 < 0.10$, it is

Table 30.3. Ratio of Variance for $\delta = 0.25$ to Variance for Optimum δ

	σ_{uu}/σ_{xx}						
σ_{xx}/σ_{ee}	0.02	0.05	0.10	0.15	0.25	0.50	1.00
0.05	1.08	1.00	1.07	1.21	1.42	1.68	1.87
0.10	1.13	1.02	1.01	1.10	1.32	1.62	1.84
0.25	1.19	1.08	1.00	1.01	1.13	1.48	1.77
0.50	1.23	1.13	1.03	1.00	1.03	1.32	1.67
1.00	1.26	1.17	1.08	1.03	1.00	1.14	1.52
2.00	1.28	1.21	1.13	1.06	1.01	1.03	1.32
4.00	1.29	1.24	1.17	1.12	1.05	1.00	1.13
10.00	1.31	1.27	1.21	1.17	1.10	1.02	1.01
20.00	1.31	1.28	1.24	1.20	1.15	1.06	1.00

optimal to replicate less than one percent of the individuals in the sample. For explanatory variables of the socio-economic type with $\sigma_{uu}/\sigma_{xx} \approx 0.15$ and an $R^2_{xY} \approx 0.50$, replicate observations on one fourth of the sample ($\delta = 0.20$) is nearly optimal. When one considers that multiple regressions generally have the effect of increasing σ_{uu}/σ_{xx}, a design in which one third of the observations are replicated ($\delta = 0.25$) becomes realistic.

Table 30.3 contains the ratio of the variance for a design with $\delta = 0.25$ to the variance for the optimal design. The design with $\delta = 0.25$ does very well down the diagonal of the table, from populations with high R^2_{xY} and high reliability ratios to populations with low R^2_{xY} and low reliability ratios. These are the types of combinations that we judge to be often observed in practice.

For a population with very low R^2_{xY} and a high reliability, the loss from using a design with $\delta = 0.25$ is about 30%. This is approximately the percentage reduction in the total number observed.

For a population with high R^2_{xY} and low reliability, the loss approaches 100% because the optimal design is to replicate all observations. We feel that Table 30.3 provides considerable support for a recommendation to make replicate observations on one third of the individuals in a general study.

It is worthwhile to compare the variance of $\hat{\beta}_1$ for the design with $\delta = 0.25$ to the mean squared error of the ordinary least squares estimator defined in (30.6). The efficiency of the two estimators is a function of n because the mean squared error of the ordinary least squares estimator has a squared bias component. The mean squared error of $\hat{\gamma}_1$, under our simplifying assumptions, is

$$MSE\{\hat{\gamma}_1\} = (\sigma_{XX}^{-1}\sigma_{uu})^2 + n^{-1}\sigma_{XX}^{-1}[1 + \sigma_{ee} - (1 + \sigma_{uu})^{-1}] \qquad (30.29)$$

Table 30.4. Ratio of Mean Squared Error of Ordinary Least Squares Estimator to Variance of Measurement Error Estimator for $n = 1{,}000$ and $\delta = 0.25$

	σ_{uu}/σ_{xx}						
σ_{ee}/σ_{xx}	0.02	0.05	0.10	0.15	0.25	0.50	1.00
0.05	4.90	16.92	34.99	46.90	58.19	58.77	42.59
0.10	3.13	11.45	26.64	38.26	50.95	54.84	41.17
0.25	1.78	6.07	15.91	25.25	37.99	46.34	37.62
0.50	1.27	3.63	9.75	16.43	27.27	37.61	33.27
1.00	1.00	2.23	5.69	9.87	17.74	28.00	27.58
2.00	0.86	1.48	3.31	5.68	10.63	18.94	21.16
4.00	0.79	1.09	2.01	3.26	6.07	11.69	14.84
10.00	0.75	0.85	1.19	1.69	2.87	5.61	8.06
20.00	0.74	0.77	0.91	1.14	1.71	3.13	4.66

where $\sigma_{XX} = (1 + \sigma_{uu})$. Table 30.4 contains the ratio of $MSE\{\hat{\gamma}_1\}$ to $V\{\hat{\beta}_1\}$ for $n = 1{,}000$. For the reasonable configuration of $\sigma_{uu}/\sigma_{xx} = 0.15$ and $R^2_{xY} = 0.50$, the design that uses one fourth of the total resources for replicate observations is ten times as efficient as the procedure of ordinary least squares applied to a larger sample. The efficiency is larger for larger samples and smaller for smaller samples.

The entries in Table 30.2 are constructed to minimize the variance of the measurement error estimator under the assumption of independent identically distributed determinations. If the two determinations used to estimate the error variance are correlated, the expectation of the estimator of the error variance is $(1 - \rho)\sigma_{uu}$, where ρ is the correlation between the two errors. The bias in the estimated error variance produces a biased estimator of the coefficient. The limiting value of $\hat{\beta}_1$ of (30.27) is

$$[\sigma_{xx} + \rho\sigma_{uu}]^{-1}\beta_1\sigma_{xx}.$$

To evaluate the effect of correlated errors on the variance of the estimated coefficient, we recall that the variance of the limiting distribution of the normalized estimated coefficient is (Fuller (1987, p. 108)),

$$n\,V\{\hat{\beta}_1\} = \sigma_{xx}^{-2}(\sigma_{xx}\sigma_{vv} + \sigma_{uu}\sigma_{vv} + \sigma_{uv}^2) + nd_f^{-1}\sigma_{xx}^{-2}(\beta_1^2\sigma_{uu}^2 + \sigma_{uv}^2)$$

where the second term comes from estimating σ_{uu} with an estimator based on d_f degrees of freedom, $\sigma_{uv} = -\beta_1\sigma_{uu}$ and $\sigma_{vv} = \sigma_{ee} + \beta_1^2\sigma_{uu}$.

Under the normal distribution assumption, we can find a set of parameters with correlated determinations that will yield the same set of sample moments as a set with uncorrelated determinations. Thus, if the set with uncorrelated determinations is $(\sigma_{uu}, \sigma_{xx}, \beta_1) = (\sigma_{uu}, 1, 1)$, then the set with correlation ρ is

$$\ddot{\sigma}_{uu}=(1-\rho)^{-1}\sigma_{uu},\ \ddot{\sigma}_{xx}=1-\rho\ddot{\sigma}_{uu},$$

and $\ddot{\beta}_1=(1-\rho\ddot{\sigma}_{uu})^{-1}$. For both sets of parameters, the estimated error variance is estimating σ_{uu} and the estimator of the variance of true x is estimating one. The set $(\ddot{\sigma}_{uu}, \ddot{\sigma}_{xx}, \ddot{\beta}_1)$ is appropriate for both the replicated and unreplicated portions of the sample because

$$1 = \ddot{\sigma}_{XX} - (1 - \rho)\ddot{\sigma}_{uu} \qquad \text{for the unreplicated sample;}$$

and

$$1 = \ddot{\sigma}_{\overline{XX}} - 0.5(1 - \rho)\ddot{\sigma}_{uu} \qquad \text{for the replicated sample,}$$

where $\ddot{\sigma}_{XX} = \ddot{\sigma}_{xx} + \ddot{\sigma}_{uu}$ and $\ddot{\sigma}_{\overline{XX}} = \ddot{\sigma}_{xx} + 0.5(1 + \rho)\ddot{\sigma}_{uu}$. It follows that the distribution of the estimator of the slope is the same for the two parameter sets. Hence, the δ of Table 30.2 minimizes the variance of the calculated estimator in the correlated error case.

To summarize, the presence of a correlation between the two determinations produces a bias in the estimated coefficient. If the correlation is unknown, it is still reasonable to use the measurement error procedure and the designs suggested by Table 30.2.

To simplify the derivation of the designs, we set the measurement error in the dependent variable equal to zero. If Y_t is observed subject to measurement error, w_t, and if duplicate observations are made on both Y and X, the optimal design is a function of the variance of w_t and of the correlation between u_t and w_t. The effect of obtaining duplicate observations on the error-prone dependent variable is to increase the optimal δ. For example, if $\sigma_{ww}=0.5\sigma_{ee}$ and $\sigma_{uw}=0$, the optimal δ is 0.35 for a population with $\sigma_{ww}+\sigma_{qq}=\sigma_{ee}=1.00$ and $\sigma_{uu}=0.15$. The optimal δ for a population with $\sigma_{ee}=\sigma_{qq}=1.00$ and $\sigma_{uu}=0.15$ is 0.17 (Table 30.2). The optimal δ for error-prone Y is somewhat smaller if the measurement errors are positively correlated. For example, if $\sigma_{ww}=0.5\sigma_{ee}$ and the correlation between u and w is 0.5, the optimal δ for a population with $\sigma_{ee}=\sigma_{ww}+\sigma_{qq}=1.00$ and $\sigma_{ww}=0.15$ is 0.25, still larger than the δ for a population with $\sigma_{ww}=0$. It seems fair to conclude that the optimal fraction to replicate is larger than that of Table 30.2 when Y is observed subject to measurement error and replicates are made on both Y and X. A δ of 0.25 provides good efficiency for a wide range of parameter configurations.

CHAPTER 31

CHI-SQUARED TESTS WITH COMPLEX SURVEY DATA SUBJECT TO MISCLASSIFICATION ERROR

J. N. K. Rao and D. Roland Thomas
Carleton University

31.1 INTRODUCTION

Standard methods for analyzing cross-classified categorical data are based on some form of chi-squared test procedure under loglinear models and simple random sampling with replacement. Unfortunately, many of the surveys used in practice have complex designs that involve stratification, unequal probability sampling, clustering and several stages of unit selection, and so yield data that do not satisfy the distributional assumptions underlying the standard analyses. Furthermore, it is often the case that the data items themselves are subject to classification errors that will bias the estimated cell probabilities and distort the properties of the hypothesis tests in various ways. There is now an extensive literature on methods for analyzing categorical data from complex surveys in the absence of misclassification, and on methods for dealing with classification errors under simple random sampling with replacement. The aim of this chapter is to unify the two approaches and provide methods that can deal with the practical reality of complex categorical survey data subject to misclassification. Both streams of research will be briefly reviewed, and then the strategy for a unified analysis will be described.

Three different analysis strategies have been proposed for incorporating the effect of the survey design in the analysis of categorical data in

The authors both wish to acknowledge research grant support from the Natural Sciences and Engineering Research Council of Canada.

Measurement Errors in Surveys.
Edited by Biemer, Groves, Lyberg, Mathiowetz, and Sudman.

ISBN: 0-471-53405-6

the absence of misclassification: (i) Wald tests based on weighted least squares (Koch, et al., 1975); (ii) the first and second order corrected chi-squared tests of Rao and Scott (1981, 1984); (iii) the jackknifed chi-squared procedure of Fay (1985). The Wald and the jackknifed chi-squared procedures require either cluster level data or, in the case of the Wald procedure, the full estimated covariance matrix. First order Rao-Scott corrections for loglinear models can be computed from published tables that additionally include estimated variances or design effects of cell estimates and their marginal totals, whereas the more accurate second order corrections require the full estimated covariance matrix. Simulation results have shown that Fay's jackknife tests and the Rao-Scott corrected tests perform well under quite general conditions. This chapter will focus primarily on the Rao-Scott procedures.

With reference to the literature on methods for dealing with misclassification under multinomial sampling or simple random sampling with replacement (SRS), several authors have considered the case of known misclassification probabilities. For multi-way tables, Korn (1981) explored the situations under which misclassification errors can be ignored without affecting the type I errors of standard loglinear model tests. In the case of models that are not preserved, appropriate chi-squared tests have been developed that take the misclassifications into account, e.g., the procedure proposed by Assakul and Proctor (1967) for testing independence in two-way tables. Mote and Anderson (1965) have shown that misclassification reduces the powers of standard chi-squared tests, irrespective of the effects of misclassification on type I error.

Except for the simple hypothesis of goodness-of-fit studied by Mote and Anderson (1965) under two different models of misclassification, double sampling methods must be used whenever the misclassification probabilities are not known. In double sampling (sometimes called reinterview surveys as in Chapter 15), the usual error-prone measurements are made on the first sample, while on the second, smaller sample, more expensive and relatively error-free measurements are made as well. Hochberg (1977) proposed a general approach to the analysis of such data using Wald tests based on weighted least squares. Chen (1979) and Espeland and Odoroff (1985), on the other hand, used loglinear models to describe both the misclassification process and the hypotheses of interest and developed chi-squared tests using the well-known EM algorithm to obtain the maximum likelihood estimates of cell probabilities.

In this chapter, chi-squared tests for complex survey data in the presence of misclassification will be considered under three scenarios: (i) when the misclassification errors can be ignored; (ii) when the misclassification probabilities are known, but misclassification errors cannot be ignored; (iii) when the misclassification probabilities are unknown.

Section 31.2 deals with the simplest of categorical data tests, namely, the goodness-of-fit test of a simple hypothesis. The problem of testing independence in two-way tables is considered in Section 31.3, assuming that the misclassification probabilities are known.

This chapter is mainly concerned with large sample (or asymptotic) design-based inferences on the finite population cell proportions. Such inferences nevertheless may also be considered as relating to asymptotic inferences on cell probabilities associated with an infinite superpopulation from which the finite population is drawn randomly.

31.2 SIMPLE GOODNESS-OF-FIT

Simple goodness-of-fit tests on cell proportions, p_i, in a one-way table arise in many applications. For example, in a household survey one might wish to test the hypothesis that the population distribution of a characteristic pertaining to household members is the same as that found in the previous census. The hypothesis of interest is $H_0: p_i = p_{0i}$, $i = 1, \ldots, K$, where K is the number of classes (or cells) and the p_{0i}'s are the hypothesized proportions.

31.2.1 Tests Incorporating the Effect of Survey Design in the Absence of Misclassification

Let $\hat{p}_i$ be a consistent estimator of p_i under a specified survey design. Typically, $\hat{p}_i$ is a ratio estimator, $\hat{p}_i = \hat{N}_i/\hat{N}$, where $\hat{N}_i$ is the sum of sample weights w_t of all units t belonging to category i, and $\hat{N}$ is the sum of all sample weights. Use of the sample (or design) weights ensures the consistency of $\hat{N}_i$ and $\hat{N}$ as estimators of the population counts N_i and $N = \Sigma N_i$, respectively. If our interest is confined to a specified domain D, then $\hat{N}$ refers to the sum of sample weights falling in D and $\hat{N}_i$ is the sum of sample weights of those units in D that are also in category i. For instance, in the Canadian Class Structure Survey (CCSS) example considered in Section 31.2.4, the domain D refers to the domain of salaried employees.

Let $\hat{\mathbf{V}}$ be a consistent estimator of the covariance matrix, $\mathbf{V}$, of the estimated proportions, $\hat{\mathbf{p}} = (\hat{p}_1, \ldots, \hat{p}_K)'$, obtained from the sample data by some appropriate method that accounts for the survey design. For example, for the CCSS data Rao and Thomas (1988) used the method of balanced repeated replication (BRR), due to McCarthy (1966), to estimate $\mathbf{V}$. For secondary data analyses from published tables, the full $\hat{\mathbf{V}}$ is not known but the cell design effects (deffs), $\hat{d}_i = \hat{v}_{ii}/[n^{-1}\hat{p}_i(1 - \hat{p}_i)]$, are often available, where $\hat{v}_{ii}$ represents the i-th diagonal element of $\hat{\mathbf{V}} = [\hat{v}_{ij}]$ and $n^{-1}\hat{p}_i(1 - \hat{p}_i)$ is the corresponding variance estimator under simple random sampling with replacement. Here n refers to either the overall sample size or the (random) sample size in the domain D.

Effect of Survey Design

The Pearson chi-squared statistic for testing $H_0{:}p_i = p_{0i}$, $i = 1, \ldots, K$, based on the estimated proportions, $\hat{p}_i$, is given by

$$X^2 = n\sum_{i=1}^{K} (\hat{p}_i - p_{0i})^2/p_{0i}. \tag{31.1}$$

Among sociologists, a popular method of constructing a Pearson statistic is first to scale the sample weights w_t so that they sum to n and then to use the scaled weighted up counts, $\tilde{n}_i$, in the customary formula:

$$\tilde{X}^2 = \sum_{i=1}^{K} (\tilde{n}_i - np_{0i})^2/(np_{0i}), \tag{31.2}$$

where $\tilde{n}_i$ is the sum of scaled weights, $\tilde{w}_t = nw_t/\hat{N}$, of all units t belonging to category i. This statistic is identical to X^2. Note that scaling is necessary, otherwise the use of weighted up counts $\hat{N}_i$ in place of $\tilde{n}_i$ would produce a statistic $\tilde{X}^2$ that is too large by a factor of $\hat{N}/n$.

We assume that $\hat{\mathbf{p}}_1 = (\hat{p}_1, \ldots, \hat{p}_{K-1})'$ is asymptotically $(K - 1)$-variate normal with mean vector $\mathbf{p}_{01} = (p_{01}, \ldots, p_{0,K-1})'$, and covariance matrix $\mathbf{V}_1$. The test statistic X^2 is then distributed asymptotically, under H_0, as a weighted sum, $\delta_1 W_1 + \ldots + \delta_{K-1}W_{K-1}$, of independent χ_1^2 variables W_i (Rao and Scott, 1981). The weights δ_i are the eigenvalues of a deff matrix given by $\mathbf{P}_{01}^{-1}\mathbf{V}_1$, where $\mathbf{P}_{01} = n^{-1}[\mathbf{D}_{\mathbf{p}01} - \mathbf{p}_{01}\mathbf{p}_{01}']$ is the covariance matrix under simple random sampling with replacement, and $\mathbf{D}_{\mathbf{p}01} = \text{diag}(p_{01}, \ldots, p_{0,K-1})$. The above result indicates that the survey design can have a substantial impact on the significance levels, i.e., treating X^2 as a χ_{K-1}^2 variable under H_0 can lead to an actual type I error rate, $Pr[X^2 > \chi_{K-1}^2(\alpha)]$, much bigger than the nominal level α, where $\chi_{K-1}^2(\alpha)$ is the upper α-point of χ_{K-1}^2, a χ^2 variable with $K - 1$ degrees of freedom. For instance, Brier (1980) studied constant deff clustering models for which $\mathbf{V}_1 = \lambda\mathbf{P}_1$ with $\lambda > 1$. In this case, $\delta_i = \lambda$ for all i and $Pr\,[X^2 > \chi_{K-1}^2(\alpha)] \approx Pr[\chi_{K-1}^2 > \lambda^{-1}\chi_{K-1}^2(\alpha)]$ which increases with λ. Although the constant deff model is somewhat unrealistic in the sample survey context, it clearly illustrates the effect of clustering on the significance level of the standard test based on X^2.

First-Order Corrections to X^2

A first-order correction to X^2 is given by

$$X_c^2 = X^2/\hat{\delta}.\tag{31.3}$$

where $\hat{\delta}.$, the mean of the estimated eigenvalues $\hat{\delta}_i$, reduces to

$$\hat{\delta}. = \frac{1}{K-1}\sum_{i=1}^{K}\frac{\hat{p}_i}{p_{0i}}(1-\hat{p}_i)\hat{d}_i. \tag{31.4}$$

Under H_0, the asymptotic expectation of X_c^2 equals $K - 1$, the expected value of χ^2_{K-1}, so that a first-order corrected test is obtained by treating X_c^2 as χ^2_{K-1}. An advantage of X_c^2 for secondary data analysis is that it depends only on the estimated cell proportions, $\hat{p}_i$, and the cell deffs, $\hat{d}_i$, which are either available with published tables or can be calculated from standard error tables (see, for example, U.S. Bureau of the Census (1978), Appendix L).

The first-order corrected test, obtained by referring X_c^2 to $\chi^2_{K-1}(\alpha)$, will tend to underestimate the upper percentage points of the true asymptotic distribution since the asymptotic variance of X_c^2 is larger than the variance of χ^2_{K-1}, unless all the δ_i's are equal. That is, the actual type I error rate will tend to be larger than the nominal level. However, this effect will be small if the variability among the δ_i's is not large.

Second-Order Correction to X^2

A better approximation to the asymptotic distribution of X^2 can be obtained if $\Sigma\hat{\delta}_i^2$ is also known. Rao and Scott (1981) proposed a second-order corrected test which can be implemented by treating

$$X_S^2 = \frac{X^2}{\hat{\delta}.(1+\hat{a}^2)} = \frac{X_c^2}{1+\hat{a}^2} \tag{31.5}$$

as χ^2_ν, a χ^2-variable with $\nu = (K-1)/(1+\hat{a}^2)$ degrees of freedom, under H_0, where $\hat{a}^2$ is the average squared deviation of the $\hat{\delta}_i$'s over $\hat{\delta}^2_.$

$$\hat{a}^2 = \Sigma(\hat{\delta}_i - \hat{\delta}.)^2/[(K-1)\hat{\delta}^2_.]. \tag{31.6}$$

Noting that

$$\Sigma\hat{\delta}_i^2 = n^2\sum_{i=1}^{K}\sum_{j=1}^{K}\hat{v}_{ij}^2/(p_{0i}p_{0j}),$$

we can readily calculate $\hat{a}^2$ and ν from $\hat{\mathbf{V}}$. Under H_0, the asymptotic mean and variance of X_S^2 are the same as the mean and variance of χ^2_ν.

Alternative Tests

An asymptotically correct test of H_0 is obtained by treating the Wald statistic

$$X_W^2 = (\hat{\mathbf{p}}_1 - \mathbf{p}_{01})'\hat{\mathbf{V}}_1^{-1}(\hat{\mathbf{p}}_1 - \hat{\mathbf{p}}_{01}) \tag{31.7}$$

as a χ^2_{K-1} variable, where $\hat{\mathbf{V}}_1$ is the estimated covariance matrix of $\hat{\mathbf{p}}_1$. The Wald test requires the knowledge of $\hat{\mathbf{V}}$, as in the case of X_S^2.

A serious concern with the Wald test is its sensitivity to the stability of $\hat{\mathbf{V}}_1^{-1}$, the inverse of the estimated covariance matrix (Fay, 1985). A simulation study by Thomas and Rao (1987) showed that the Wald test provides poor control of type I error rate unless the degrees of freedom associated with $\hat{\mathbf{V}}_1$ are much greater than the degrees of freedom for the hypothesis. The applicability of Wald tests is therefore somewhat limited, especially for multi-way tables.

An alternative to the Wald test and the second-order corrected test is the jackknife test (Fay, 1985) which requires survey estimates at the level of primary sampling units or of replicates as in the case of BRR. Simulation results (Thomas and Rao, 1987) showed that both the jackknife test and the second-order corrected test perform well in terms of both control of test level and power.

31.2.2 Methods for Dealing with Misclassification under SRS

Suppose now that errors in classification are present. Let $a_{ij}(i \neq j)$ be the probability that an observation which truly belongs to the ith category is erroneously assigned to the jth category, and let a_{ii} be the corresponding probability of correct classification. Let π_i be the probability of classification of an observation to the ith category when errors in classification are present, and let p_i be the true probability that an observation belongs to the ith category, so that

$$\boldsymbol{\pi} = \mathbf{A}'\mathbf{p} \tag{31.8}$$

where $\boldsymbol{\pi} = (\pi_1, \ldots, \pi_K)'$, $\mathbf{p} = (p_1, \ldots, p_K)'$and $\mathbf{A} = [a_{ij}]$ is the $K \times K$ matrix with elements a_{ij}. It follows from (31.8) that $H_0{:}\mathbf{p} = \mathbf{p}_0$ is equivalent to testing $H_0'{:}\boldsymbol{\pi} = \boldsymbol{\pi}_0 = \mathbf{A}'\mathbf{p}_0$, provided $\mathbf{A}$ is nonsingular, where $\mathbf{p}_0 = (p_{01}, \ldots, p_{0,K})'$.

Mote and Anderson (1965) illustrated the effect of ignoring errors in classification using the usual Pearson statistic

$$X^2 = \sum_{i=1}^{K} (n_i - np_{0i})^2/(np_{0i}) \tag{31.9}$$

where n_i is the observed counts in the ith category ($\Sigma n_i = n$). They showed that X^2 in general leads to inflated type I error rates. They also proposed an alternative test of H_0, for the case of known $\mathbf{A}$, which treats

$$X'^2 = \sum_{i=1}^{K} (n_i - n\pi_{0i})^2/(n\pi_{0i}) \tag{31.10}$$

as a χ^2_{K-1} variable. This test is asymptotically correct, but its asymptotic power with misclassification is less than the asymptotic power of X^2 with no misclassification, i.e., the effect of misclassification is to reduce the limiting power of the test.

For the case of unknown **A**, Mote and Anderson (1965) proposed two different models: (I) the misclassification rate is the same for all classes; (II) misclassification occurs only in adjacent categories and at a constant rate. Model II applies only in the case of ordered categories. For case I,

$$\pi_{0i} = p_{0i} + \theta(1 - Kp_{0i}) \tag{31.11}$$

where θ is an unknown parameter $(0 < \theta < 1/K)$ and $K \geq 3$. Note that $\pi_{0i} \neq p_{0i}$ for at least one i unless all $p_{0i} = 1/K$. We assume that at least two of the p_{0i} are not equal to $1/K$. For case II,

$$\begin{aligned} \pi_{01} &= p_{01}(1-\theta) + p_{02}\theta \\ \pi_{0i} &= p_{0,i-1}\theta + p_{0i}(1-2\theta) + p_{0,i+1}\theta, \quad i = 2, \ldots, K-1 \\ \pi_{0K} &= p_{0,K-1}\theta + p_{0K}(1-\theta) \end{aligned} \tag{31.12}$$

where θ is an unknown parameter $(0 < \theta < [2 + 2\cos(\pi/K)]^{-1} \leq \frac{1}{2})$.

It is easy to show that the maximum likelihood estimation (m.l.e.) of θ for case I is given by the solution of

$$\Sigma'[n_i/(\hat{\theta} - q_{0i})] = 0, \tag{31.13}$$

where $q_{0i} = p_{0i}/(Kp_{0i} - 1)$, $p_{0i} \neq 1/K$ and the summation in (31.13) is over those i for which $p_{0i} \neq 1/K$. A procedure for locating and evaluating the consistent root $\hat{\theta}$ is described by Mote and Anderson (1965).

Similarly, for case II the m.l.e. of θ is given by the solution of

$$\sum_{i=1}^{K} [n_i/(\hat{\theta} - q_{0i})] = 0 \tag{31.14}$$

where

$$\begin{aligned} q_{01} &= p_{01}/(p_{01} - p_{02}) \\ q_{0i} &= p_{0i}/(2p_{0i} - p_{0,i-1} - p_{0,i+1}), \quad i = 2, \ldots, K-1 \\ q_{0K} &= p_{0K}/(p_{0K} - p_{0,K-1}). \end{aligned}$$

If the denominator of any q_{0i} vanishes, the corresponding $n_i/(\hat{\theta} - q_{0i})$ in (31.14) is omitted.

An asymptotically correct test of H_0 treats

$$X'^2 = \sum_{i=1}^{K} (n_i - n\hat{\pi}_{0i})^2/(n\hat{\pi}_{0i}) \tag{31.15}$$

as a χ^2_{K-2} variable, where $\hat{\pi}_{0i}$ is obtained from π_{0i} by substituting $\hat{\theta}$ for θ.

31.2.3 Tests Incorporating the Effect of Survey Design in the Presence of Misclassification

Case of known **A**

For the case of known **A**, we consider

$$X'^2 = n\sum_{i=1}^{K} (\hat{p}_i - \pi_{0i})^2/\pi_{0i} \tag{31.16}$$

which is obtained from (31.10) by substituting the survey estimate $\hat{p}_i$ for n_i/n. In this case, the survey estimate $\hat{\mathbf{p}}$ is a consistent estimate of the category probabilities $\boldsymbol{\pi}$ under misclassification. Noting that $\hat{\mathbf{p}}$ is approximately multivariate normal with mean vector $\boldsymbol{\pi}_0$ and covariance matrix $\mathbf{V}$, under H_0, it readily follows from Rao and Scott (1981) that a first-order corrected test treats

$$X_c'^2 = X'^2/\hat{\delta}'_. \tag{31.17}$$

as a χ^2_{K-1} variable, where $\hat{\delta}'_.$ is given by (31.4) with p_{0i} changed to π_{0i}. Similarly, the second-order corrected test treats

$$X_S'^2 = X'^2/\hat{\delta}'_.(1 + \hat{a}'^2)] = X_c'^2/(1 + \hat{a}'^2) \tag{31.18}$$

as $X^2_{\nu'}$, where $\hat{a}'^2$ is given by (31.6) with p_{0i} changed to π_{0i}, and $\nu' = (K-1)/(1 + \hat{a}'^2)$.

Case of unknown **A**

For general survey designs, we do not have an m.l.e. of θ under model I or model II, because of difficulties in getting appropriate likelihood functions. As a result, we employ a "pseudo" m.l.e., $\hat{\theta}$, obtained from (31.13) or (31.14) by replacing n_i/n with the survey estimator $\hat{p}_i$ (see Rao and Scott, 1984). The asymptotic consistency of $\hat{\mathbf{p}}$ ensures that $\hat{\theta}$ and the resulting smoothed estimates $\hat{\pi}_{0i} = \pi_{0i}(\hat{\theta})$ are also.

The uncorrected test is based on

$$X'^2 = n\sum_{i=1}^{K} (\hat{p}_i - \hat{\pi}_{0i})^2/\hat{\pi}_{0i}\,. \tag{31.19}$$

It is shown in Appendix 31.A that a first-order correction to X'^2, under either model I or model II, is given by

$$X_c'^2 = X'^2/\hat{\delta}'_. \tag{31.20}$$

where the correction factor $\hat{\delta}'_.$ is the mean of the estimated nonzero eigenvalues, $\hat{\delta}'_1, \ldots, \hat{\delta}'_{K-2}$, of the matrix $n(\hat{\mathbf{F}}'\mathbf{D}_{\hat{\pi}0}^{-1}\hat{\mathbf{F}})\hat{V}$, where

$$\hat{\mathbf{F}} = \mathbf{I} - \frac{1}{\hat{\Delta}}(\hat{\boldsymbol{\pi}}_0 - \mathbf{p}_0)(\hat{\boldsymbol{\pi}}_0 - \mathbf{p}_0)'\mathbf{D}_{\hat{\pi}0}^{-1}$$

with

$$\hat{\Delta} = \sum_{i=1}^{K} (\hat{\pi}_{oi} - p_{0i})^2/\hat{\pi}_{0i}$$

and $\mathbf{D}_{\hat{\pi}0} = \text{diag}\,(\hat{\pi}_{01}, \ldots, \hat{\pi}_{0K})$. Further, $\hat{\delta}'_{\cdot}$ may be reduced to

$$\hat{\delta}'_{\cdot} = \frac{1}{K-2}\Big[\sum_{i=1}^{K} \frac{\hat{p}_i}{\hat{\pi}_{0i}}(1-\hat{p}_i)\hat{d}_i - \text{deff}\left(\sum \frac{p_{0i}}{\pi_{0i}}\hat{p}_i\right)\Big] \tag{31.21}$$

with

$$\text{deff}\left(\sum \frac{p_{0i}}{\pi_{oi}}\hat{p}\right) = \hat{\Delta}^{-1}\sum_{i=1}^{K}\sum_{j=1}^{K} (p_{0i}/\hat{\pi}_{0i})(p_{0j}/\hat{\pi}_{0j})\hat{v}_{ij}.$$

The first-order corrected test of H_0 treats X'^2_c as χ^2_{K-2}, and it requires the knowledge of the full $\hat{\mathbf{V}}$. A conservative test that depends only on the cell deffs, $\hat{d}_i$, however, may be obtained by omitting the deff term in (31.21), i.e., by using

$$\hat{\delta}_{\cdot}(a) = \frac{1}{K-2}\sum \frac{\hat{p}_i}{\hat{\pi}_{0i}}(1-\hat{p}_i)\hat{d}_i. \tag{31.22}$$

Suppose $\hat{a}'^2 = \Sigma\,(\hat{\delta}'_i - \hat{\delta}'_{\cdot})^2/[(K-2)\hat{\delta}'^2_{\cdot}]$ is also known. Then a second-order corrected test can be implemented by treating

$$X'^2_S = \frac{X'^2}{\hat{\delta}'_{\cdot}(1+\hat{a}'^2)} = \frac{X'^2_c}{1+\hat{a}'^2} \tag{31.23}$$

as χ^2 with $\nu' = (K-2)/(1+\hat{a}'^2)$ degrees of freedom, under H_0.

31.2.4 Example

Rao and Thomas (1988) applied the tests in Section 31.2.1 to data from the Canadian Class Structure Survey (CCSS), a national survey of the economically active adult population of Canada. A variable called employment class, in which each respondent's occupation is categorized by its degree of management control and autonomy, was considered. A test of goodness-of-fit was then formulated as $H_0{:}p_i = p_{0i}$, $i = 1, \ldots, K = 5$, where p_i's are the unknown population proportions and p_{0i}'s are the corresponding known proportions from a U.S. survey. Note that the U.S. estimates were treated as fixed, i.e., as if they were population

Table 31.1. Estimated and Hypothesized Proportions, and Deffs for Categories of Employment Class: Canadian Class Structure Survey (n = 1,463)

	Estimated $\hat{p}_i$ (Canada)	Hypothesized p_{0i} (U.S.)	Cell Deff $\hat{d}_i$
Decision-making managers	0.141	0.148	1.10
Advisor-managers	0.034	0.052	1.96
Supervisors	0.118	0.149	1.31
Semi-autonomous workers	0.191	0.110	1.44
Workers	0.516	0.541	1.87

proportions, in order to implement a goodness-of-fit test; strictly speaking, H_0 constitutes a test of homogeneity of proportions.

Table 31.1, taken from Rao and Thomas (1988), gives the estimated and hypothesized proportions, $\hat{p}_i$ and p_{0i}, and the estimated cell deffs, $\hat{d}_i$. Rao and Thomas (1988) also present the full estimated covariance matrix, $\hat{\mathbf{V}}$, obtained by the method of BRR. Assuming no misclassification, the following values of the test statistics were obtained:

$$X^2 = 106.8,\ X_c^2 = 72.3, \text{ and } X_S^2 = 63.0.$$

Further, $\hat{\delta}. = 1.48$, $\hat{a} = 0.43$ and $\nu = 3.38$. Assuming that the second-order corrected test X_S^2 achieves the nominal level of 5%, the approximate type I error rates of X^2 and X_c^2 were obtained as 18% and 6%, respectively. The corrections to X^2 are therefore necessary, and the first-order corrected test based on X_c^2 is satisfactory in terms of controlling the type I error rate. Comparing $X_S^2 = 63.0$ to $\chi^2_{3.38}(0.05) = 8.45$, the upper 5% point of χ^2 with $\nu = 3.38$ degrees of freedom, or $X_c^2 = 72.3$ to $\chi_4^2(0.05) = 9.45$ if only the cell deffs, $\hat{d}_i$, were available, we may conclude that H_0 is not tenable.

Turning to tests in the presence of misclassification, we obtained the following values, assuming the same p_{0i}'s:

Model I.

$$\hat{\theta} = 0.0057,\ \hat{\delta}'_. = 1.27,\ \hat{\delta}'_.(a) = 1.92,\ \hat{a}' = 0.47,\ \nu' = 2.46$$
$$X'^2 = 104.4,\ X_c'^2 = 82.0,\ X'^2/\hat{\delta}'_.(a) = 54.4,\ X_S'^2 = 67.6.$$

Model II.

$$\hat{\theta} = 0.1003,\ \hat{\delta}'_. = 1.17,\ \hat{\delta}'_.(a) = 1.73,\ \hat{a}' = 0.36,\ \nu' = 2.66$$
$$X'^2 = 43.6,\ X_c'^2 = 37.1,\ X'^2/\hat{\delta}'_.(a) = 25.2,\ X_S'^2 = 32.8.$$

Assuming that the second-order corrected test, $X_S'^2$, achieves the nominal level of 5%, the approximate type I error rates of X'^2, $X_c'^2$ and $X'^2/\hat{\delta}'_{\cdot}(a)$ are 11.6%, 6.1%, and 1.3% respectively under Model I, and 9.2%, 5.7%, and 1.2%, respectively under Model II. It is clear that the first-order corrected test, depending only on the cell deffs, $\hat{d}_i$, is quite conservative here, thus leading to loss in power. Moreover, the effect of survey design on X'^2 is less severe than the effect on X^2 in the absence of misclassification (compare the type I error rate of 11.6% under Model I or 9.2% under Model II to 18% in the absence of misclassification).

Under the more realistic Model II, the second-order corrected test statistic $X_S'^2 = 32.8$ ($\nu' = 2.66$) compared to $X_S^2 = 63.0$ ($\nu = 3.38$), the value in the absence of misclassification. Thus, the effect of misclassification is to reduce the power of the test, as noted by Mote and Anderson (1965). On the other hand, under Model I for which $\hat{\theta} \approx 0$, $X_S'^2 = 67.6$ ($\nu' = 2.46$) is close to $X_S^2 = 63.0$ ($\nu = 3.38$), the value in the absence of misclassification.

31.2.5 A Model-Free Approach: Double Sampling

We now turn to the use of double sampling to test the hypothesis of simple goodness-of-fit without assuming any model of misclassification. In double sampling, the error-prone measurements are made on a large first-phase sample of size m, selected according to a specified design, and the expensive, error-free measurements are then made on a smaller subsample of size n selected according to another specified design (usually SRS or stratified SRS). Rao (1973) developed the relevant theory for the case of SRS in the first phase and stratified SRS in the second phase. Shrout and Newman (1989) applied this theory to estimate the prevalence of rare disorders. Kott (1990) obtained appropriate variance estimators when the first-phase sample is a sample of clusters selected via stratified SRS, and the secondary units within sampled clusters are restratified based on their characteristics and subsampled in the second phase via SRS from each new stratum. Särndal and Swensson (1987) developed a general theory under arbitrary designs in both phases.

Tests Under SRS in Both Phases

Tenenbein (1972) considered SRS in both phases, and obtained m.l.e.'s of the true cell probabilities, p_i, and their standard errors. Let m_j be the number of units classified into category j in the first phase sample, using the error-prone measuring instrument. Also, let n_{ij} be the number of units in the second-phase sample originally classified into category j whose true (second phase) category is i. Then, the m.l.e. of the p_i's and the misclassification probabilities, a_{ij}, are given by

$$\hat{p}_i = \sum_{j=1}^{K} \hat{\pi}_j(n_{ij}/n_{+j}) \tag{31.24}$$

and

$$\hat{a}_{ij} = (\hat{\pi}_j/\hat{p}_i)(n_{ij}/n_{+j}) \tag{31.25}$$

where $n_{+j} = \Sigma_i n_{ij}$ and $\hat{\pi}_j = m_j/m$. The estimated variances of the m.l.e.'s $\hat{p}_i$, $i = 1, \ldots, k - 1$ have been derived by Tenenbein (1972), using Taylor series arguments. A straightforward extension yields the covariances between estimates $\hat{p}_i$ and $\hat{p}_\ell$, $i \neq \ell$. Thus the estimated covariance matrix of $\hat{\mathbf{p}}_1 = (\hat{p}_1, \ldots, \hat{p}_{K-1})'$ is given by $\mathbf{V}_1^*$ with elements

$$v_{ii}^* = \text{estvar}(\hat{p}_i) = \frac{1}{n}\left[\hat{p}_i - \hat{p}_i^2 \sum_{j=1}^{K} \frac{\hat{a}_{ij}^2}{\hat{\pi}_j}\right] + \frac{\hat{p}_{ij}^2}{m}\left[\sum_{j=1}^{K} \frac{\hat{a}_{ij}^2}{\hat{\pi}_j} - 1\right] \tag{31.26}$$

and

$$v_{i\ell}^* = \text{estcov}(\hat{p}_i, \hat{p}_\ell) = -\frac{1}{n}\hat{p}_i\hat{p}_\ell \sum_{j=1}^{K} \frac{\hat{a}_{ij}\hat{a}_{\ell j}}{\hat{\pi}_j} + \frac{\hat{p}_i\hat{p}_\ell}{m}\left[\sum_{j=1}^{K} \frac{\hat{a}_{ij}\hat{a}_{\ell j}}{\hat{\pi}_j} - 1\right], i \neq \ell \tag{31.27}$$

where estcov denotes the estimated covariance.

Consider the simple chi-squared statistic

$$X^2 = n \sum_{i=1}^{K} (\hat{p}_i - p_{0i})^2/p_{0i}$$

$$= (\hat{\mathbf{p}}_1 - \mathbf{p}_{01})'\mathbf{P}_{01}^{-1}(\hat{\mathbf{p}}_1 - \mathbf{p}_{01}) \tag{31.28}$$

for testing $H_0: p_i = p_{0i}$, $i = 1, \ldots, K$. The asymptotic distribution of X^2, under H_0, is no longer χ^2_{K-1} since the covariance matrix of $\hat{\mathbf{p}}_1$ is different from $\mathbf{P}_{01}$. However, a first-order correction to X^2 is readily obtained, assuming that $\hat{\mathbf{p}}_1$ is approximately $(K - 1)$-variate normal with mean vector $\mathbf{p}_{01}$ and estimated covariance matrix, $\mathbf{V}_1^*$, for sufficiently large subsamples. Following Rao and Scott (1981), a first-order corrected test treats

$$X_c^{*2} = X^2/\delta_\cdot^*, \tag{31.29}$$

as a χ^2_{K-1} variable, where $\delta_\cdot^*$ is the mean of the eigenvalues, $\delta_1^*, \ldots, \delta_{K-1}^*$, of $\mathbf{P}_{01}^{-1}\mathbf{V}_1^*$. Note that

$$(K - 1)\delta_\cdot^* = n \sum_{i=1}^{K} v_{ii}^*/p_{0i}. \tag{31.30}$$

A second-order corrected test can also be implemented by treating

$$X_S^{*2} = X_c^{*2}/(1 + a^{*2}) \qquad (31.31)$$

as $\chi^2_{\nu^*}$ under H_0, where

$$a^{*2} = \Sigma(\delta_i^* - \delta_\cdot^*)^2/[(K-1)\delta_\cdot^{*2}] \qquad (31.32)$$

and $\nu^* = (K-1)/(1 + a^{*2})$.

31.2.6 Tests Incorporating the Effect of Survey Design

Suppose now that the survey design is different from SRS in at least one of the two phases. Let $\hat{\mathbf{p}}_1 = (\hat{p}_1, \ldots, \hat{p}_{K-1})'$ now represent a consistent estimator of $\mathbf{p}_1$ under the specified two-phase design. Also, let $\hat{\mathbf{V}}_1$ be a consistent estimator of the covariance matrix of $\hat{\mathbf{p}}_1$. Then, a first-order corrected test is obtained by treating

$$X_c^2 = X^2/\hat{\delta}_\cdot \qquad (31.33)$$

as a χ^2_{K-1} variable, where $\hat{\delta}_\cdot$ is the mean of the eigenvalues, $\delta_1, \ldots, \delta_{K-1}$, of $\mathbf{P}_{01}^{-1}\hat{\mathbf{V}}_1$. It may be noted that

$$(K-1)\hat{\delta}_\cdot = n\sum_{i=1}^{K} \hat{v}_{ii}/p_{0i} = \sum_{i=1}^{K} \frac{\hat{p}_i}{p_{0i}}(1 - \hat{p}_i)\hat{d}_i \qquad (31.34)$$

where $\hat{v}_{ii} = \text{estvar}(\hat{p}_i)$ and $\hat{d}_i = \hat{v}_{ii}/[n^{-1}\hat{p}_i(1 - \hat{p}_i)]$ is the deff associated with the i-th cell. Hence, the first-order corrected test depends only on the estimated proportions, $\hat{p}_i$, and their standard errors under the specified two-phase design or their deffs.

A second-order corrected test can also be implemented if the estimated covariances, $\hat{v}_{i\ell} = \text{estcov}(\hat{p}_i, \hat{p}_\ell)$, are also available. It is obtained by treating

$$X_S^2 = X_c^2/(1 + \hat{a}^2) \qquad (31.35)$$

as χ^2_ν under H_0, where

$$\hat{a}^2 = \Sigma(\hat{\delta}_i - \hat{\delta}_\cdot)^2/[(K-1)\hat{\delta}_\cdot^2] \qquad (31.36)$$

and $\nu = (K-1)/(1 + \hat{a}^2)$.

Example

Tenenbein (1972) gives an example of sampling inspection, involving SRS in both phases. A first-phase sample of size $m = 207$ is selected and the units are classified into three categories, defective, satisfactory, and superior, based on a subjective evaluation of the quality of the produced units which cost \$1 per unit. A second-phase subsample of size $n = 58$ is then selected and classified into the same three categories based on objective criteria for determining the true category of a unit costing \$9 per unit. The estimated proportions and the associated covariance matrix are given by

$$\hat{\mathbf{p}} = (0.209, 0.474, 0.317)'$$

and

$$\mathbf{V}^* = \begin{bmatrix} 0.1666 & -0.1344 & -0.0320 \\ -0.1344 & 0.2260 & -0.0916 \\ -0.0320 & -0.0916 & 0.1235 \end{bmatrix} \times 10^{-2}.$$

Also,

$$\hat{\mathbf{A}} = \begin{bmatrix} 1.000 & 0 & 0 \\ 0.221 & 0.745 & 0.034 \\ 0 & 0 & 1.00 \end{bmatrix}.$$

We obtained the following values of the test statistics, assuming $\mathbf{p}_0 = (0.1, 0.6, 0.3)'$:

$$X^2 = 8.5, \text{ and } X_c^{*2} = 11.9.$$

Further, $\delta_\cdot^* = 0.71$, $a^{*2} = 0.28$, and $v^* = 1.56$. Assuming that the second-order corrected test, X_S^{*2}, achieves the nominal level of 5%, the approximate type I error rates of X^2 and X_c^{*2} were obtained as 2.6% and 6.7% respectively. The Pearson test statistic, X^2, is conservative in this example since the estimated proportions, $\hat{p}_i$, under double sampling are more efficient than the proportions, n_{i+}/n, based only on the second-phase sample, where $n_{i+} = \Sigma_j n_{ij}$. The first-order corrected test, X_c^{*2}, on the other hand, leads to somewhat inflated type I error rate.

The above data will now be used to illustrate the effect of survey design. We will treat $\hat{\mathbf{p}}$ as the survey estimate of $\mathbf{p}$ with covariance matrix $\hat{\mathbf{V}}$. Since we do not have an empirical estimate $\hat{\mathbf{V}}$ for a design that is different from SRS in one or both phases, we created a covariance matrix $\hat{\mathbf{V}}$ that exhibits the variance inflation typical of complex designs involving clustering. We write the spectral decomposition of $\mathbf{V}^*$, the covariance matrix for designs having SRS in both phases, as $\mathbf{V}^* = \mathbf{Q}^*\mathbf{\Lambda}^*\mathbf{Q}^{*\prime}$. Here $\mathbf{Q}^*$ is an orthonormal matrix of eigenvectors and $\mathbf{\Lambda}^*$ is a diagonal matrix of eigenvalues λ_i^*, $i = 1, 2, 3$, in decreasing order with $\lambda_3^* = 0$. A covariance matrix that mimics the effects of cluster sampling can be constructed by setting

$$\hat{\mathbf{V}} = \mathbf{Q}^*\mathbf{\Lambda}^*\mathbf{B}\,\mathbf{Q}^{*\prime}$$

where $\mathbf{B}$ is a diagonal matrix of inflation factors having diagonal elements b_i, with $b_i \geq 1$, $i = 1,2$ and $b_3 = 0$. As noted by Thomas (1990), covariance matrices of this form satisfy the constraint that the elements of $\hat{\mathbf{p}}$ sum to one. For the example, the inflation factors have been set at $b_1 = 2.0$, $b_2 = 1.5$. Now treating $\hat{\mathbf{p}}$ as the survey estimate of $\hat{\mathbf{p}}$ with estimated covariance matrix $\hat{\mathbf{V}}$, we obtained the following values of the test statistics, assuming $\mathbf{p}_0 = (0.1, 0.6, 0.3)'$:

$$X^2 = 8.5, \quad X_c^{*2} = 11.9, \text{ and } X_c^2 = 7.2.$$

Further, $\hat{\delta}. = 1.17$, $\hat{a}^2 = 0.31$, and $\nu = 1.53$. Assuming that the second-order corrected test, X_S^2, achieves the nominal level of 5%, the approximate type I error rates of X^2, X_c^{*2} and X_c^2 were obtained as 9.8%, 17.6%, and 6.9%, respectively. Ignoring the design and using X_c^{*2} leads to substantial inflation of type I error rate, and to a lesser extent with X^2. The first-order corrected test, X_c^2, taking account of the survey design, performs much better, although it too leads to a somewhat inflated type I error rate.

31.3 TWO-WAY CONTINGENCY TABLES

Consider a two-way cross classification of variables A and B, having I and J categories respectively. Let $p_{ij} = N_{ij}/N$ be the population proportion in the (i,j)th cell, estimated by $\hat{p}_{ij} = \hat{N}_{ij}/\hat{N}$ where $\hat{N}_{ij}$ is the sum of sample weights of all units belonging to the (i,j)th cell and $\hat{N}$ is the sum of sample weights, as before. Also, let $\hat{p}_{i+} = \Sigma_j \hat{p}_{ij}$ and $\hat{p}_{+j} = \Sigma_i \hat{p}_{ij}$ be estimates of the row and column margin proportions, p_{i+} and p_{+j}, respectively. The hypothesis of independence can be stated as

$$H_0 : p_{ij} = p_{i+}p_{+j}, \quad i = 1, \ldots, I; \quad j = 1, \ldots J.$$

It can also be expressed as $H_0 : u_{12(ij)} = 0$ for all (i,j) in the saturated log linear model

$$\ln p_{ij} = \mu_{ij} = \tilde{u} + u_{1(i)} + u_{2(j)} + u_{12(ij)} \tag{31.37}$$

where the parameters $u_{1(i)}$, $u_{2(j)}$, and $u_{12(ij)}$ are constrained by $\Sigma_i u_{1(i)} = 0$, $\Sigma_j u_{2(j)} = 0$, $\Sigma_i u_{12(ij)} = 0$ for all j, $\Sigma_j u_{12(ij)} = 0$ for all i, and $\tilde{u}$ is a normalizing factor to ensure that $\Sigma_i \Sigma_j p_{ij} = 1$ (see e.g., Bishop, et al., 1975b, p. 24). In matrix notation, (31.37) may be written as

$$\boldsymbol{\mu} = \tilde{u}\mathbf{1} + \mathbf{X}_1\boldsymbol{\theta}_1 + \mathbf{X}_2\boldsymbol{\theta}_2 \tag{31.38}$$

with $\mathbf{X}_2'\mathbf{X}_1 = 0$, $\mathbf{X}_1'\mathbf{1} = \mathbf{0}$, and $\mathbf{X}_2'\mathbf{1} = \mathbf{0}$, where $\boldsymbol{\mu} = (\mu_{11}, \ldots, \mu_{1J}; \ldots; \mu_{I1}, \ldots, \mu_{IJ})'$, $\boldsymbol{\theta}_1$ is the $(I + J - 2)$-vector of parameters $u_{1(1)}, \ldots, u_{1(I-1)}; u_{2(1)}, \ldots, u_{2(J-1)}$ with associated model matrix $\mathbf{X}_1$ consisting of $+1$'s, 0's, and -1's, and $\boldsymbol{\theta}_2$ is the $(I-1)(J-1)$-vector of parameters $u_{12(11)}, \ldots, u_{12(1,J-1)}; \ldots; u_{12(I-1,1)}, \ldots, u_{12(I-1,J-1)}$ with associated model matrix $\mathbf{X}_2$ similar to $\mathbf{X}_1$. The hypothesis H_0 may be expressed as $H_0: \boldsymbol{\theta}_2 = \mathbf{0}$.

31.3.1 Tests Incorporating the Effect of Survey Design in the Absence of Misclassification

The Pearson chi-squared statistic for testing H_0, based on the estimated cell proportions $\hat{p}_{ij}$, is given by

$$X^2 = n\sum_{i=1}^{I}\sum_{j=1}^{J} (\hat{p}_{ij} - \hat{p}_{i+}\hat{p}_{+j})^2/(\hat{p}_{i+}\hat{p}_{+j}). \tag{31.39}$$

Rao and Scott (1984) showed that X^2 is distributed asymptotically, under H_0, as a weighted sum, $\delta_1 W_1 + \ldots + \delta_{(I-1)(J-1)} W_{(I-1)(J-1)}$, of independent χ_1^2 variables W_t, where the δ_t's are the eigenvalues of the deff matrix

$$\boldsymbol{\Gamma} = (\tilde{\mathbf{X}}_2'\mathbf{P}\tilde{\mathbf{X}}_2)^{-1}(\tilde{\mathbf{X}}_2'\mathbf{V}\tilde{\mathbf{X}}_2) \tag{31.40}$$

$$\tilde{\mathbf{X}}_2 = [\mathbf{I} - \mathbf{X}_1(\mathbf{X}_1'\mathbf{P}\mathbf{X}_1)^{-1}\mathbf{X}_1'\mathbf{P}]\mathbf{X}_2.$$

Further, $\mathbf{V}$ is the covariance matrix of $\hat{\mathbf{p}} = (\hat{p}_{11}, \ldots, \hat{p}_{1J}; \ldots; \hat{p}_{I1}, \ldots, \hat{p}_{IJ})'$ and $\mathbf{P} = (\mathbf{D}_\mathbf{p} - \mathbf{pp}')/n$ with $p_{ij} = p_{i+}p_{+j}$ is the corresponding covariance matrix under simple random sampling with replacement, where $\mathbf{D}_\mathbf{p} = \text{diag}\,(p_{11}, \ldots, p_{IJ})$ and $\mathbf{p} = (p_{11}, \ldots, p_{1J}; \ldots; p_{I1}, \ldots, p_{IJ})'$.

A first-order correction to X^2 is given by

$$X_c^2 = X^2/\hat{\delta}. \tag{31.41}$$

where $\hat{\delta}.$, the mean of the estimated eigenvalues, $\hat{\delta}_t$, reduces to

$$\hat{\delta}. = (I-1)^{-1}(J-1)^{-1}\left[\sum_{i=1}^{I}\sum_{j=1}^{J}\frac{\hat{p}_{ij}(1-\hat{p}_{ij})}{\hat{p}_{i+}\hat{p}_{+j}}\hat{d}_{ij} - \sum_{i=1}^{I}(1-\hat{p}_{i+})\hat{d}_{A(i)} - \sum_{j=1}^{J}(1-\hat{p}_{+j})\hat{d}_{B(j)}\right]. \tag{31.42}$$

The correction factor $\hat{\delta}.$ depends only on the estimated cell deffs

$$\hat{d}_{ij} = \text{estvar}(\hat{p}_{ij})/[n^{-1}\hat{p}_{ij}(1-\hat{p}_{ij})]$$

and the marginal row and column deffs

$$\hat{d}_{A(i)} = \text{estvar}(\hat{p}_{i+})/[n^{-1}\hat{p}_{i+}(1 - \hat{p}_{i+})]$$

and

$$\hat{d}_{B(j)} = \text{estvar}(\hat{p}_{+j})/[n^{-1}\hat{p}_{+j}(1 - \hat{p}_{+j})]$$

respectively, where estvar denotes the estimate of variance. An advantage of X_c^2 is that it can be simply implemented from published tables providing the estimated proportions $\hat{p}_{ij}$, $\hat{p}_{i+}$, $\hat{p}_{+j}$ and the associated deffs or standard errors.

The first-order corrected test is obtained by referring X_c^2 to $\chi^2_{(I-1)(J-1)}(\alpha)$, the upper α-point of a χ^2 variable with $(I - 1)(J - 1)$ degrees of freedom. As before, the actual type I error rate of the test will tend to be larger than the nominal level.

If the average squared deviation of the estimated eigenvalues is also known, then a second-order corrected test of H_0 can be implemented by treating

$$X_S^2 = \frac{X^2}{\hat{\delta}.(1 + \hat{a}^2)} = \frac{X_c^2}{1 + \hat{a}^2} \tag{31.43}$$

as χ^2_ν, a χ^2 variable with $\nu = (I - 1)\ (J - 1)/(1 + \hat{a}^2)$ degrees of freedom, where

$$\hat{a}^2 = \Sigma(\hat{\delta}_t - \hat{\delta}.)^2/[(I - 1)(J - 1)\hat{\delta}.^2].$$

31.3.2 Method for Dealing with Misclassification under SRS

Suppose now that errors of classification are present. Let a_{sitj} be the probability that an observation which truly belongs to the (s,t)-th cell is erroneously assigned to the (i,j)th cell. Let π_{ij} be the probability of classification of an observation to the (i,j)th cell when errors of classification are present, and let p_{ij} be the true probability that an observation belongs to the (i,j)th cell, so that

$$\pi_{ij} = \sum_{s=1}^{I} \sum_{t=1}^{J} p_{st} a_{sitj}. \tag{31.44}$$

In matrix notation, (31.44) may be written as

$$\boldsymbol{\pi} = \mathbf{A}'\mathbf{p},$$

where $\boldsymbol{\pi} = (\pi_{11}, \ldots, \pi_{1J}; \ldots; \pi_{I1}, \ldots, \pi_{IJ})'$, and $\mathbf{A}$ is the $(IJ) \times (IJ)$ matrix with elements a_{sitj}. The rows of **A** are indexed by the subscript pairs (s,t) listed in lexicographic order, and the columns of **A** are similarly indexed by the subscripts (i,j).

The usual Pearson statistic for testing the hypothesis of independence, $H_0: p_{ij} = p_{i+}p_{+j}$, ignoring errors of classification, is given by

$$X^2 = \sum_{i=1}^{I} \sum_{j=1}^{J} (n_{ij} - n_{i+}n_{+j}/n)^2/(n_{i+}n_{+j}/n), \tag{31.45}$$

where n_{ij} is the observed count in the (i,j)th cell, and $n_{i+} = \Sigma_j n_{ij}$, $n_{+j} = \Sigma_i n_{ij}$ are the row and column marginal totals of n_{ij}. If the errors of classification in the row direction are independent of errors in the column direction, i.e., if a_{sitj} can be decomposed in the form $a_{sitj} = b_{si}c_{tj}$, then Assakul and Proctor (1967) show that H_0 is equivalent to $H_0': \pi_{ij} = \pi_{i+}\pi_{+j}$. As a result, the usual test which treats X^2 as χ^2 with $(I-1)(J-1)$ degrees of freedom will be asymptotically correct, i.e., it is unaffected by errors of classification. However, the limiting power is reduced due to errors of classification. Mote and Anderson (1965) showed that X^2 is unaffected by misclassification if errors are only in the row direction or only in the column direction.

We now turn to the case of nonindependent errors, but assume that the **A**-matrix is known; Assakul and Proctor (1967) used reinterview data to construct the **A**-matrix. In this case, the Pearson statistic X^2, in general, leads to inflated type I error rates.

Noting that H_0 may be expressed as $H_0: \boldsymbol{\theta}_2 = \mathbf{0}$ in the loglinear model (31.38), we may write the model in terms of $\boldsymbol{\pi}$ as

$$\mathbf{g}(\boldsymbol{\pi}) = \ln(\mathbf{A}'^{-1}\boldsymbol{\pi}) = \tilde{u}\mathbf{1} + X_1\boldsymbol{\theta}_1 \tag{31.46}$$

where $\ln(\mathbf{A}'^{-1}\boldsymbol{\pi})$ is an IJ-vector with elements $\ln(\mathbf{b}_t'\boldsymbol{\pi})$, and $\mathbf{b}_t'$ is the tth row of $\mathbf{A}'^{-1}$. Scott, et al. (1990) considered models of the form (31.46) for general $\mathbf{g}(\boldsymbol{\pi})$, and proposed an iterative scheme involving weighted least squares to calculate the maximum likelihood estimates of $\tilde{u}$, $\boldsymbol{\theta}_1$, and hence of $\boldsymbol{\pi}$. Further details are available from the authors. Assakul and Proctor (1967) proposed a modified minimum chi-square method to estimate $\boldsymbol{\pi}$.

Denoting the m.l.e of $\boldsymbol{\pi}$ by $\hat{\boldsymbol{\pi}}$, an asymptotically correct test of H_0 is obtained by treating

$$X'^2 = \sum_{i=1}^{I} \sum_{j=1}^{J} (n_{ij} - n\hat{\pi}_{ij})^2/(n\hat{\pi}_{ij}) \tag{31.47}$$

as a χ^2 variable with $(I-1)(J-1)$ degrees of freedom.

If the **A**-matrix is unknown and the errors of classification are nonindependent, then double sampling methods must be used to obtain an appropriate test of independence (see Section 31.3.5).

31.3.3 Tests Incorporating the Effect of Survey Design in the Presence of Misclassification: Known A

For the case of independent misclassification errors, we consider the test statistic (31.39). Since H_0 is equivalent to H_0' in this case, it readily follows that the first-order corrected test (31.41) and the second-order corrected test (31.43) can be used without any modification.

Turning to the case of nonindependent errors, we consider the test statistic

$$X'^2 = n\sum_{i=1}^{I}\sum_{j=1}^{J}(\hat{p}_{ij} - \hat{\pi}_{ij})^2/\hat{\pi}_{ij} \tag{31.48}$$

which is obtained from (31.47) by substituting the survey estimate $\hat{p}_{ij}$ for n_{ij}/n and using the "pseudo" m.l.e., $\hat{\boldsymbol{\pi}}$. The latter estimator is obtained by replacing n_{ij}/n with the survey estimate $\hat{p}_{ij}$ in Scott, et al.'s (1990) iterative scheme. The true probabilities, $\mathbf{p}$, are estimated by $\mathbf{A}'^{-1}\hat{\boldsymbol{\pi}}$.

Under H_0, the test statistic X'^2 is asymptotically distributed as a weighted sum, $\delta_1' W_1 + \dots + \delta_{(I-1)(J-1)}' W_{(I-1)(J-1)}$, of independent χ_1^2 variables W_t, where the δ_t''s are the eigenvalues of the deff matrix

$$\boldsymbol{\Gamma} = n(\tilde{\mathbf{X}}_2'\mathbf{PAD}_\pi^{-1}\mathbf{A}'\mathbf{P}\tilde{\mathbf{X}}_2)^{-1}(\tilde{\mathbf{X}}_2'\mathbf{PAD}_\pi^{-1}\mathbf{VD}_\pi^{-1}\mathbf{A}'\mathbf{P}\tilde{\mathbf{X}}_2) \tag{31.49}$$

with

$$\tilde{\mathbf{X}}_2 = [\mathbf{I} - \mathbf{X}_1(\mathbf{X}_1'\mathbf{PAD}_\pi^{-1}\mathbf{A}'\mathbf{PX}_1)^{-1}\mathbf{X}_1'\mathbf{PAD}_\pi^{-1}\mathbf{A}'\mathbf{P}]\mathbf{X}_2,$$

$\mathbf{D}_\pi = \text{diag}(\pi_{11}, \dots, \pi_{IJ})$ and $p_{ij} = p_{i+}p_{+j}$. The proof of this result is given in Appendix 31.B. If errors of classification are absent, then $\mathbf{A} = \mathbf{I}$, $\boldsymbol{\pi} = \mathbf{p}$ and (31.49) reduces to (31.39) obtained by Rao and Scott (1984), noting that $\mathbf{D}_\mathbf{p}^{-1}\mathbf{p} = \mathbf{1}$, $\mathbf{V1} = \mathbf{0}$ and $\mathbf{PD}_\mathbf{p}^{-1}\mathbf{P} = \mathbf{P}/n^2$. The estimated eigenvalues, $\hat{\delta}_t'$, are obtained from (31.49) by substituting $\hat{\boldsymbol{\pi}}$ for $\boldsymbol{\pi}$, $\hat{\mathbf{V}}$ for $\mathbf{V}$ and $\mathbf{A}'^{-1}\hat{\boldsymbol{\pi}}$ for $\mathbf{p}$.

A first-order corrected test is implemented by treating

$$X_c'^2 = X'^2/\hat{\delta}_\cdot' \tag{31.50}$$

as a χ^2 variable with $(I-1)(J-1)$ degrees of freedom, where $\hat{\delta}_\cdot'$ is the mean of the estimated eigenvalues, $\hat{\delta}_t'$. Similarly, a second-order corrected test is implemented by treating

$$X_S'^2 = \frac{X'^2}{\hat{\delta}_\cdot'(1 + \hat{a}'^2)} = \frac{X_c'^2}{1 + \hat{a}'^2} \tag{31.51}$$

as a χ^2 variable with $(I-1)(J-1)$ degrees of freedom, where

$$\hat{a}'^2 = \Sigma(\hat{\delta}_t' - \hat{\delta}_\cdot')^2/[(I-1)(J-1)\hat{\delta}_\cdot'^2].$$

It may be noted that $\hat{\delta}_\cdot'$ cannot be expressed simply in terms of estimated

Table 31.2. Survey Estimates of Proportions for the Housing Data Example of Assakul and Proctor (1967); n = 10,895

House Condition	Plumbing: Has all facilities	Plumbing: Lacks some
Sound	0.7173	0.1266
Deteriorating	0.0685	0.0876

cell deffs and the marginal row and column deffs, unlike $\hat{\delta}.$ in the case of no misclassification errors.

31.3.4 Example

Assakul and Proctor (1967) analyzed a 2×2 table on housing conditions, namely general house condition (sound/deteriorating) by condition of plumbing (has all facilities/lacks some). Using housing census data and the results of a reinterview study published by the U.S. Bureau of the Census, they estimated the misclassification matrix $\mathbf{A}$ as

$$\hat{\mathbf{A}} = \begin{bmatrix} 0.936 & 0.010 & 0.051 & 0.003 \\ 0.186 & 0.518 & 0.053 & 0.243 \\ 0.638 & 0.031 & 0.301 & 0.030 \\ 0.110 & 0.434 & 0.043 & 0.413 \end{bmatrix}.$$

With this misclassification matrix, they then tested the independence of house condition and plumbing condition using data from a different substratum. Survey estimates of the four cell proportions, $\hat{\mathbf{p}} = (\hat{p}_{11}, \hat{p}_{12}, \hat{p}_{21}, \hat{p}_{22})'$, are shown in Table 31.2. Assakul and Proctor (1967) estimated the true cell proportions $\mathbf{p}$ using the estimate $\hat{\mathbf{A}}$ of the misclassification matrix, and the modified minimum chi-squared procedure. Treating $\hat{\mathbf{A}}$ as fixed and using (31.47), they obtained a test of the independence hypothesis which is asymptotically valid under the multinomial assumption.

We now use these data to illustrate the theory of Section 31.3.3. Since the true covariance matrix of the data of Table 31.2 is not known, we used the procedure of Section 31.2.6 to create a covariance matrix $\hat{\mathbf{V}}$ that exhibits the variance inflation (relative to multinomial data) typical of complex designs involving clustering. Using the spectral decomposition of $\hat{\mathbf{P}} = (\mathbf{D}_{\hat{\mathbf{p}}} - \hat{\mathbf{p}}\hat{\mathbf{p}}')/n$, the multinomial form of the covariance matrix corresponding to the survey estimates $\hat{\mathbf{p}}$, written as $\hat{\mathbf{P}} = \mathbf{Q}\mathbf{\Lambda}\mathbf{Q}'$, we constructed the covariance matrix

Table 31.3. Analysis of Misclassified Housing Data with Inflated Covariance Matrix $\hat{\mathbf{V}}$

Strategy	Statistics[1]	Pseudo M.L.E. of True Probabilities π
Ignoring design and ignoring misclassification ($\mathbf{A}=\mathbf{I}$)	$X^2=1{,}443.6$	$(0.6631, 0.1808, 0.1227, 0.0334)'$
Accounting for design, ignoring misclassification. ($\mathbf{A}=\mathbf{I}$)	$X^2=1{,}443.6$ $\hat{\delta}_. = \hat{\delta}_1 = 3.12$ $X_c^2 = 461.9$	$(0.6631, 0.1808, 0.1227, 0.0334)'$
Ignoring design, but accounting for misclassification. ($\mathbf{A}=\hat{\mathbf{A}}$)	$X'^2=52.5$	$(0.6523, 0.2280, 0.0888, 0.0309)'$
Accounting for the design, and the misclassification. ($\mathbf{A}=\hat{\mathbf{A}}$)	$X'^2=52.5$ $\hat{\delta}'_. = \hat{\delta}'_1 = 1.78$ $X_c'^2 = 29.4$	$(0.6523, 0.2280, 0.0888, 0.0309)'$

[1] All test statistics should be referred to a chi-squared distribution on one degree of freedom.

$$\hat{\mathbf{V}} = \mathbf{Q}\Lambda\mathbf{B}\mathbf{Q}'$$

with $\mathbf{B} = \mathrm{diag}(2.0, 1.5, 1.0, 0)$. This mimics the effect of cluster sampling with realistic inflation factors $b_1 = 2.0$, $b_2 = 1.5$, and $b_3 = 1.0$. Complete results are shown in Table 31.3. The pseudo m.l.e. estimates of the true probabilities, p_{ij}, under misclassification shown in the table were obtained by means of the iterative scheme of Scott, et al. (1990), while those ignoring misclassification are the standard estimates, $\hat{p}_{i+}\hat{p}_{+j}$.

The first row of Table 31.3 gives the value of X^2 corresponding to equation (31.39) when both design and misclassification are ignored, and agrees with the result of $X^2 = 1{,}444$ quoted by Assakul and Proctor (1967). The second row shows that ignoring misclassification but accounting for the effect of the design via the first-order correction of equation (31.41) reduces the test statistic to $X_c^2 = 461.9$. It should be noted that for the current 2×2 example, the second-order correction (31.43) is the same as the first-order correction (31.41) since there is only one eigenvalue, $\hat{\delta}_1$, and hence $\hat{a} = 0$. The third row of Table 31.3 shows that ignoring the effect of the design but accounting for misclassification via equation (31.47) reduces the test statistic from $X^2 = 1{,}444$ to $X'^2 = 52.5$, a large drop that shows the dramatic effect of misclassification in this example. This result, based on the m.l.e. of $\boldsymbol{\pi}$, is in close agreement with the corresponding test value of $X'^2 = 51.5$ given by Assakul and Proctor (1967), based on their modified minimum chi-squared estimate of $\boldsymbol{\pi}$.

The final row of Table 31.3 gives the results of simultaneously accounting for the design and the misclassification. In this case, $\hat{\delta}'_{\cdot} = \hat{\delta}'_1 = 1.78$, corresponding to the deff matrix (31.49). The resulting first-order corrected statistic (equation (31.50)) is $X'^2_c = 29.4$. Though misclassification clearly has the strongest biasing effect on X^2 in this example, it is clear that both misclassification and design must be considered together. Ignoring the realistic effects of survey design and considering only misclassification can lead to a serious distortion of test results.

31.3.5 The Case of Unknown A: Double Sampling

We again consider a double sampling design in which relatively cheap but error-prone measurements are made on a large first-phase sample of size m, selected according to a possibly complex design, and the expensive, relatively error-free measurements are then made on a smaller subsample of size n selected according to another specified design.

Let $\hat{\mathbf{p}}$ now represent a consistent estimator of $\mathbf{p}$ under the specified two-phase design, and let $\hat{\mathbf{V}}$ be a consistent estimator of the covariance matrix of $\hat{\mathbf{p}}$, again under the double sampling design. As noted in Section 31.2.5, some results on variance estimation are available in the literature.

A Pearson chi-squared statistic for testing the null hypothesis of independence is given by (31.39), where n now refers to the size of the second sample, and the $\hat{p}_{ij}$ are elements of $\hat{\mathbf{p}}$, the estimate of the true proportion $\mathbf{p}$ under the double sampling design. It immediately follows that the standard Rao-Scott corrections can be applied to this statistic. The deff matrix is estimated by inserting $\hat{\mathbf{p}}$ for $\mathbf{p}$ and the double sampling variance estimator $\hat{\mathbf{V}}$ for $\mathbf{V}$ in (31.40). The first- and second-order corrected tests are then computed using (31.41) through (31.43). Thus, the first-order corrected test can be implemented from published tables providing double sampling estimates of p_{ij}, p_{i+}, p_{+j} and the associated deffs or standard errors.

31.4 SUMMARY AND CONCLUSIONS

This chapter has reported on an investigation of chi-squared test procedures for categorical data when the data are subject to two possible violations of standard assumptions. First, the data may arise from complex surveys in which case the assumption of multinomially distributed counts, valid under simple random sampling, will no longer apply. Second, the data themselves may be subject to classification errors which will bias the estimated category probabilities and distort the properties of hypothesis tests in various ways. The discussion has focused on two testing problems, namely goodness-of-fit of a simple hypothesis, and independence in a two-way table. For each of these problems, a review has been presented of the effects of a complex survey design on standard chi-squared tests in the absence of misclassification, as well as a review of the effects of misclassification on standard chi-squared tests under simple random sampling. The different approaches required for the case of both known and unknown misclassification probabilities have been described. Finally, a unified approach to testing has been developed that accounts for the effect of complex survey designs in the presence of misclassification. This synthesis of the two previous lines of research, which constitutes the primary contribution of this chapter, is summarized in some detail below.

31.4.1 Simple Goodness-of-Fit Tests

Results pertaining to simple goodness-of-fit tests will be summarized first. When the misclassification probabilities are known, first and second-order corrected chi-squared tests of the Rao and Scott (1981, 1984) type can be readily constructed using consistent survey estimates of the cell probabilities under misclassification. Appropriate formulae are given by equations (31.18) and (31.19). For the case of unknown misclassification probabilities, two distinct testing strategies have been considered, the first a model-based approach that can be directly applied to the complex survey data, and the second a model-free approach in which the misclassification probabilities are estimated from data generated via a double sampling scheme.

Model-Based Approach

This approach is based on two models of the misclassification process originally investigated by Mote and Anderson (1965) for the case of simple random sampling. It uses the test statistic (31.19) which involves the consistent survey estimate $\hat{\mathbf{p}}$ together with a consistent pseudo m.l.e. of the category probabilities under misclassification. Though the statistic (31.19) does not have a large sample chi-squared distribution under complex sampling designs, first and second order corrected chi-squared tests have been constructed. Appropriate formulae are given by (31.20) and (31.23). Unlike the case of known misclassification probabilities, the first order corrections require knowledge of the full covariance matrix of survey estimates. Though a conservative version of the first-order correction can be constructed that requires knowledge only of cell variances or design effects, this version appears from the example to be overly conservative. For the example data considered, the model that allows for misclassification only into adjacent categories appears to be the most realistic. It also appears from the example that the effect of misclassification is to reduce the power of the corrected test procedure, a result that is consistent with the findings of Mote and Anderson (1965) for data from simple random samples.

Model-Free Approach: Double Sampling

This approach assumes that a subsample of the original sample is taken using more expensive and relatively error-free measurements. Under this design it is possible to obtain consistent estimates of the misclassification probabilities and the true category proportions. It is shown in this chapter that even when simple random sampling is used in both phases (Tenenbein, 1972), standard tests of the chi-squared type no longer have the chi-squared distribution asymptotically. It is demonstrated in the example that a standard test based on the assumption of multinomially distributed counts is quite conservative. Use of a first- or second-order corrected tests specified by (31.29) and (31.31) provides a test having type I error close to the nominal value. When one or both of the sampling phases uses a complex design, an estimate is required of the covariance matrix of the double sampling estimator of the vector of true category probabilities. Though some results on variance estimation are available in the literature, this aspect of the problem has not been pursued in this study. An example of the likely effect of survey design has been provided using an artificially constructed, though realistic, covariance matrix. In the example, the effect of the survey design counteracts the above noted tendency to conservativeness of the naive test, resulting in a type I error rate of 9.8% when the naive test is used in the complex data case. Again, first-order and second-order corrected tests specified by (31.33) and (31.35) provide appropriate type I error protection.

31.4.2 Independence in Two-Way Tables

The corrections for chi-squared tests of independence under complex survey sampling have been combined with Assakul and Proctor's (1967) chi-squared test procedures for misclassified data under simple random sampling for the case of known misclassification probabilities. For tests on two-way tables, it is difficult to find plausible models of misclassification to parallel those used for the goodness-of-fit case. When the misclassification probabilities are unknown, therefore, the only practical way to proceed with the analysis of two-way (and higher-way) tables is via double sampling.

Independent Classification Errors

Accounting for both survey design and misclassification is straightforward when the misclassification probabilities are independent. In this case, because the type I error rate of the standard chi-squared test is unaffected by misclassification, the usual Rao and Scott (1984) corrected chi-squared tests can be applied without any modification.

Nonindependent Classification Errors: Misclassification Probabilities Known

When the misclassification probabilities are known, the chi-squared procedure developed by Assakul and Proctor (1967) must be modified to: (i) use the survey-based estimate of the category proportions under misclassification; (ii) use a pseudo maximum likelihood estimate of the misclassified category proportions, under the hypothesis of independence of the true category proportions. The latter estimates have been obtained using an iterative scheme developed by Scott, Rao and Thomas (1990). The resulting modified statistic (31.48) does not have a chi-squared distribution asymptotically, but first- and second-order corrections have been developed that simultaneously account for the effect of the survey design and the misclassification. The appropriate formulae are given by (31.50) and (31.51). It should be noted that these expressions are formulated in general terms and can be applied to more complex hypotheses on two-way tables, as well as to hypotheses on multi-way tables, provided that the corresponding misclassification probabilities are known or independently estimable. For the test of independence, the example originally given by Assakul and Proctor (1967) clearly shows that under realistic levels of design-induced extra-multinomial variation, both survey design effects and misclassification must be considered jointly. Ignoring either can lead to serious distortions of test results.

Misclassification Probabilities Unknown: Double Sampling

Provided that the double sampling estimates of the true probabilities are available, together with corresponding variance and covariance estimates or only variance estimates (or deffs), again under the double sampling design, the standard Rao-Scott theory for complex designs can be applied directly. In this case, the analytic difficulties of the known **A** case are exchanged for the estimation difficulties of double sampling.

APPENDIX

31.A. First-Order Correction to X'^2: Simple Goodness-of-Fit

The uncorrected test statistic X'^2, given by (31.19), is asymptotically equivalent to

$$X'^2 = n(\hat{\mathbf{p}} - \boldsymbol{\pi}_0)'\mathbf{F}'\mathbf{D}_{\boldsymbol{\pi}0}^{-1}\mathbf{F}(\hat{\mathbf{p}} - \boldsymbol{\pi}_0) \tag{31.A1}$$

where

$$\mathbf{F} = \mathbf{I} - (\Sigma b_i^2/\pi_{0i})^{-1}\mathbf{b}\mathbf{b}'\mathbf{D}_{\boldsymbol{\pi}0}^{-1}$$

and

$$\mathbf{b} = (\partial\boldsymbol{\pi}_0/\partial\theta) = (\partial\pi_{01}/\partial\theta, \ldots, \partial\pi_{0K}/\partial\theta)'$$

(see, for example, Bishop, et al., 1975b, p. 517). Now noting that $\hat{\mathbf{p}} - \boldsymbol{\pi}_0$ is asymptotically normal with mean vector $\mathbf{0}$ and covariance matrix $\mathbf{V}$, under H_0, it readily follows from (31.A1) that X'^2 is distributed asymptotically as a weighted sum, $\delta_1'W_1 + \ldots + \delta_{K-2}'W_{K-2}$, of independent χ_1^2 variables W_i, where $\delta_1', \ldots, \delta_{K-2}'$ are the nonzero eigenvalues of the matrix $n(\mathbf{F}'\mathbf{D}_{\boldsymbol{\pi}0}^{-1}\mathbf{F})\mathbf{V}$.

A first order correction to X'^2 is given by $X_c'^2 = X'^2/\hat{\delta}_\cdot'$ where $\hat{\delta}'$is the estimate of the mean of the nonzero eigenvalues, namely $\delta_\cdot' = (K-2)^{-1}\text{tr}[n(\mathbf{F}'\mathbf{D}_{\boldsymbol{\pi}0}^{-1}\mathbf{F})\mathbf{V}]$, where tr denotes the trace operator. Now using (31.11) and (31.12), we may write for both models

$$\mathbf{F}'\mathbf{D}_{\boldsymbol{\pi}0}^{-1}\mathbf{F} = \mathbf{D}_{\boldsymbol{\pi}0}^{-1} - \Delta^{-1}\mathbf{D}_{\boldsymbol{\pi}0}^{-1}(\boldsymbol{\pi}_0 - \mathbf{p}_0)(\boldsymbol{\pi}_0 - \mathbf{p}_0)'\mathbf{D}_{\boldsymbol{\pi}0}^{-1}$$

where $\Delta = \Sigma(\pi_{0i} - p_{0i})^2/\pi_{0i}$. It now follows that

$$\delta_\cdot' = (K-2)^{-1}[n\Sigma v_{ii}/\pi_{0i} - n\Delta^{-1}\text{Var}\{\Sigma(p_{0i}/\pi_{0i})\hat{p}_i\}] \tag{31.A2}$$

where Δ/n is the variance of $\Sigma(p_{0i}/\pi_{0i})\hat{p}_i$ under simple random sampling. Estimating the terms in (31.A2) we get $\hat{\delta}_\cdot'$ given by (31.21).

31.B. Derivation of Equation (31.49)

With reference to equation (31.46), let $\boldsymbol{\pi} = \mathbf{f}(\boldsymbol{\theta}_1)$ and $\mathbf{F} = \partial\mathbf{f}/\partial\boldsymbol{\theta}_1'$. With this notation, and $\hat{\mathbf{F}} = \mathbf{F}(\hat{\boldsymbol{\theta}})$, the "pseudo" likelihood equations reduce to $\hat{\mathbf{F}}'\mathbf{D}_{\boldsymbol{\pi}}^{-1}\hat{\mathbf{p}} = \mathbf{0}$, with solution $\boldsymbol{\theta}_1 = \hat{\boldsymbol{\theta}}_1, \boldsymbol{\pi} = \hat{\boldsymbol{\pi}} = \mathbf{f}(\hat{\boldsymbol{\theta}}_1)$. Following the arguments of Rao and Scott (1984), it can be shown that

$$\hat{\mathbf{p}} - \mathbf{f}(\hat{\boldsymbol{\theta}}_1) = \mathbf{B}[\hat{\mathbf{p}} - \mathbf{f}(\boldsymbol{\theta}_1)] + o_p(n^{-1/2}) \tag{31.B1}$$

where $\mathbf{B} = [\mathbf{I} - \mathbf{F}(\mathbf{F}'\mathbf{D}_{\boldsymbol{\pi}}^{-1}\mathbf{F})^{-1}\mathbf{F}'\mathbf{D}_{\boldsymbol{\pi}}^{-1}]$. On substituting (31.B1) into equation (31.47) it follows that the statistic X'^2 is asymptotically equivalent to

$$\bar{X}'^2 = n[\hat{\mathbf{p}} - \mathbf{f}(\boldsymbol{\theta}_1)]'\mathbf{B}'\mathbf{D}_\pi^{-1}\mathbf{B}[\hat{\mathbf{p}} - \mathbf{f}(\boldsymbol{\theta}_1)]. \tag{31.B2}$$

From (31.46), and noting that $\mathbf{1}'\boldsymbol{\pi} = 1$, we get $\mathbf{F} = n\mathbf{A}'\mathbf{P}\mathbf{X}_1$ from which it can be shown that

$$\mathbf{B}'\mathbf{D}_\pi^{-1}\mathbf{B} = \mathbf{D}_\pi^{-1}\mathbf{B} = \mathbf{D}_\pi^{-1}[\mathbf{I} - \mathbf{A}'\mathbf{P}\mathbf{X}_1(\mathbf{X}'_1\mathbf{P}\mathbf{A}\mathbf{D}_\pi^{-1}\mathbf{A}'\mathbf{P}\mathbf{X}_1)^{-1}\mathbf{X}_1'\mathbf{P}\mathbf{A}\mathbf{D}_\pi^{-1}].$$

Let $\mathbf{y} = n^{1/2}\mathbf{D}_\pi^{-1/2}(\hat{\mathbf{p}} - \boldsymbol{\pi})$ and $\mathbf{X}_1^* = n\mathbf{D}_\pi^{-1/2}\mathbf{A}'\mathbf{P}\mathbf{X}_1$. Then (31.B2) becomes

$$\bar{X}'^2 = \mathbf{y}'[\mathbf{I} - \mathbf{X}_1^*(\mathbf{X}_1^{*\prime}\mathbf{X}_1^*)^{-1}\mathbf{X}_1^{*\prime}]\mathbf{y}. \tag{31.B3}$$

Since $\mathbf{y}$ and $\mathbf{X}_1^*$ are orthogonal to $\mathbf{D}_\pi^{1/2}\mathbf{1}$, it follows that the right hand side of (31.B3) can be written as $\mathbf{y}'[\mathbf{X}_2^*\mathbf{X}_2^{*\prime}\mathbf{X}_2^*)^{-1}\mathbf{X}_2^{*\prime}]\mathbf{y}$ where $\mathbf{X}_2^*$ is the orthogonal completion of $(\mathbf{D}_\pi^{1/2}\mathbf{1}, \mathbf{X}_1^*)$, i.e., it has maximum column rank and satisfies

$$\mathbf{X}_2^{*\prime}\mathbf{X}_1^* = \mathbf{0}, \quad \mathbf{X}_2^{*\prime}\mathbf{D}_\pi^{1/2}\mathbf{1} = \mathbf{0}. \tag{31.B4}$$

If, corresponding to the definition of $\mathbf{X}_1^*$ given above, we write $\mathbf{X}_2^*$ in the form $\mathbf{X}_2^* = n\mathbf{D}_\pi^{-1/2}\mathbf{A}\mathbf{P}\tilde{\mathbf{X}}_2$, then $\mathbf{X}_2^*$ satisfies the second condition in (31.B4). $\tilde{\mathbf{X}}_2$ must therefore be chosen to satisfy

$$\tilde{\mathbf{X}}_2'(\mathbf{P}\mathbf{A}\mathbf{D}_\pi^{-1}\mathbf{A}'\mathbf{P})\mathbf{X}_1 = \mathbf{0}. \tag{31.B5}$$

The choice

$$\tilde{\mathbf{X}}_2 = [\mathbf{I} - \mathbf{X}_1(\mathbf{X}_1'\mathbf{P}\mathbf{A}\mathbf{D}_\pi^{-1}\mathbf{A}'\mathbf{P}\mathbf{X}_1)^{-1}\mathbf{X}_1'\mathbf{P}\mathbf{A}\mathbf{D}_\pi^{-1}\mathbf{A}'\mathbf{P}]\mathbf{X}_2 \tag{31.B6}$$

clearly satisfies (31.B5), where $\mathbf{X}_2$ is chosen so that $(\mathbf{1},\mathbf{X}_1,\mathbf{X}_2)$ is nonsingular. Thus, $\bar{X}'^2$ can be written as

$$\bar{X}'^2 = \mathbf{Z}'(\tilde{\mathbf{X}}_2'\mathbf{P}\mathbf{A}\mathbf{D}_\pi^{-1}\mathbf{A}'\mathbf{P}\tilde{\mathbf{X}}_2)^{-1}\mathbf{Z} \tag{31.B7}$$

where $\mathbf{Z} = n^{1/2}\tilde{\mathbf{X}}_2'\mathbf{P}\mathbf{A}\mathbf{D}_\pi^{-1}(\hat{\mathbf{p}} - \boldsymbol{\pi})$. If we assume that $\hat{\mathbf{p}} - \boldsymbol{\pi}$ is approximately normal with mean vector $\mathbf{0}$ and covariance matrix $\mathbf{V}$, for sufficiently large samples, then using standard results on the distribution of quadratic forms (see Rao and Scott, 1984), $\bar{X}'^2$ and hence X'^2 will be distributed as the weighted sum of χ_1^2 variates specified in the text preceding equation (31.49). The weights δ_i', $i = 1, \ldots, (I-1)(J-1)$, will be eigenvalues of the deff matrix

$$\mathbf{D} = \mathbf{Q}\mathbf{C}(\mathbf{Z}) \tag{31.B8}$$

where $\mathbf{Q}$ is the kernel of the quadratic form (31.B7) and $\mathbf{C}(\mathbf{Z})$, the covariance matrix of $\mathbf{Z}$, is given by

$$\mathbf{C}(\mathbf{Z}) = n\tilde{\mathbf{X}}_2'\mathbf{P}\mathbf{A}\mathbf{D}_\pi^{-1}\mathbf{V}\mathbf{D}_\pi^{-1}\mathbf{A}'\mathbf{P}\tilde{\mathbf{X}}_2. \tag{31.B9}$$

Hence equation (31.49). This result in fact holds for any multiway table that can be described in the form (31.46), provided the misclassification matrix $\mathbf{A}$ is known.

CHAPTER 32

THE EFFECT OF MEASUREMENT ERROR ON EVENT HISTORY ANALYSIS

D. Holt, J. W. McDonald and C. J. Skinner
University of Southampton

32.1 INTRODUCTION

The various biological, social, and economic processes which evolve over an individual's life cycle have long been of interest in the social and life sciences. Examples of such processes include those of physical and psychological development, education, family formation and childbearing, social mobility, and job mobility. For a given individual, the outcome of such processes can often be represented by an event history: a record of the occurrence and timing of specific events, at each of which the individual experiences a change of state, for example, a change of marital status or of employment status. In recent years increasing numbers of social surveys have collected such event history data on, for example, education, fertility, employment, and migration. The availability of methods of analysis and their applications to such data have also increased greatly (e.g., Tuma and Hannan, 1984; Allison, 1984; Heckman and Singer, 1985; Blossfeld, et al., 1989).

Much of the event history data collected in surveys is, however, retrospective and, as described in Chapter 8, such data are particularly prone to measurement error. The aim of this chapter is to investigate how such measurement error can affect methods of event history analysis. We hope this investigation will be of use to analysts of event history data who are concerned about how measurement error might distort their analyses. The nature of errors of retrospective measurement can vary greatly between applications and therefore we have chosen to focus on just two examples: age at menarche (first menstruation) and duration of unemployment. These two examples typically involve

Measurement Errors in Surveys.
Edited by Biemer, Groves, Lyberg, Mathiowetz, and Sudman.

ISBN: 0-471-53405-6

somewhat different types of data collection as well as different types of measurement error. While we concentrate on these two examples, we do attempt to couch our discussion in a framework which has general application.

Event history analysis is concerned very generally with transitions between different states, with durations of time spent in states, and with the relation between these transitions and durations and individual characteristics. In our study we shall for simplicity restrict attention to just one spell in a given state and the dependence of its duration on covariates which are fixed in time. It seems clear from studies, as described in Chapter 8, of errors in recall of timing of events, such as telescoping, that such errors may lead to very misleading estimates of duration distributions. What seems less clear and what will be the main object of study here is the effect of such errors on estimates of the effects of covariates on duration distributions. While theoretically some bias is likely to be introduced it does not seem obvious in advance whether this bias will be negligible in practice.

Our method of assessing the effect of measurement error is as follows. First we review in Section 32.2 standard methods of analyzing the effects of covariates on duration distributions using survey data. We make a number of simplifying assumptions which are commonly made in practice. Then in Section 32.3 we describe some applications of these methods to two examples, age at menarche and duration of unemployment. In Section 32.4 we describe the possible kinds of measurement error in event history data collected in surveys, and review the literature on studies of measurement error for our two examples. Next we attempt, in the main part of this chapter, to use the results of these studies to conduct a simulation study of the performance of the methods described in Sections 32.2 and 32.3 in the presence of realistic types and amounts of measurement error. Some introductory remarks are made in Section 32.5 and the numerical results are given in Section 32.6.

We only allow for measurement error in the timing and reporting of events but not in the covariates. It is well known that covariate measurement error can lead to very misleading results in event history analysis (Prentice, 1982; Pepe, et al., 1989; Gong, et al., 1990), as it does in regression analysis (see Chapter 30). Some studies which consider the effect of errors in measured event times are Carroll, et al. (1978), Peters (1988) and Trivellato and Torelli (1989). Hoem (1985) also gives some related discussion. One other topic which we do not address is that of unobserved heterogeneity. This has received extensive attention in the social science literature (e.g., Heckman and Singer, 1984) but raises estimation problems which would have led, in our view, to more complication than illumination in our simulation study. There is a close

INITIAL EVENT	TERMINAL EVENT	COVARIATES
birth	menarche	socio-economic factors, number of siblings
start of first spell of unemployment	end of first spell of unemployment	age, sex, occupation
birth	first marriage	sex, education
first marriage	first birth	age at first marriage, education
end of full-time education	first full-time employment	social class, sex, education

Figure 32.1. Examples of event histories and covariates.

connection between unobserved heterogeneity and covariate measurement error, as noted by Gong, et al. (1990), which would seem to merit further investigation.

32.2 EVENT HISTORY ANALYSIS

32.2.1 Event Histories

We assume that for each individual i in the population:

i. an initial event occurs at time I_i;
ii. a terminal event occurs at time $I_i + T_i$;
iii. there is an associated vector of covariates $\mathbf{x}_i$.

We refer to the period between I_i and $I_i + T_i$ as the *spell* and the length T_i of the spell as the *duration time* for the ith individual. Our objective is to study how T_i depends on the value of the vector of covariates $\mathbf{x}_i$. In this simple situation the *event history* for individual i is defined just by the values of I_i, T_i, and $\mathbf{x}_i$. Some examples of such event histories and associated covariates are shown in Figure 32.1. The first two of these examples, menarche and unemployment, are described in greater detail in Section 32.3.

Note that for some individuals it is possible that the spell never occurs; for example, some individuals never have a first marriage. Also, for those for whom the initial event occurs, it is possible that the terminal event never occurs, for example, some married women never have a first birth. For simplicity we avoid these possibilities by excluding such individuals from the population of interest. Furthermore it is possible that for some individuals repeated spells may occur, for example,

individuals may experience more than one spell of unemployment. We shall also avoid this possibility by restricting attention to just one of the spells, the first.

32.2.2 Event History Models

There is only space here for the briefest exposition of event history models; some introductory texts were referred to earlier. In the simple setting considered here an event history model may be viewed as a regression model in which the duration time T_i is the dependent variable and $\mathbf{x}_i$ is the vector of explanatory variables.

For simplicity we assume that the duration times follow a Weibull distribution with a constant *power parameter* and a *scale parameter* which depends on the covariates via the function $\exp(-\mathbf{x}_i'\boldsymbol{\beta})$, where $\boldsymbol{\beta}$ is a vector of regression coefficients. The model may be expressed in a form analogous to a conventional regression model by the equation

$$\log T_i = (\mathbf{x}_i'\boldsymbol{\beta} + \sigma u_i)/\alpha, \tag{32.1}$$

where u_i is a random term following a standard extreme value distribution (Blossfeld, et al., 1989, page 52).

One reason for choosing the Weibull distribution is that it permits a simple alternative representation of the model in terms of the hazard function of T_i. The *hazard function* $h(t; \mathbf{x}_i)$ is defined as

$$h(t; \mathbf{x}_i) = \lim_{dt \to 0} \text{pr}(t < T_i \leq t + dt | T_i > t)/dt,$$

that is, $h(t; \mathbf{x}_i)dt$ is approximately the conditional probability that the terminal event occurs in the time interval $(t, t+dt]$ given that it has not occurred up to time t, where dt is taken as "infinitesimally small."

The hazard function for the Weibull model above is

$$h(t, \mathbf{x}_i) = \alpha\, t^{\alpha - 1}\, e^{\mathbf{x}_i'\boldsymbol{\beta}}, \tag{32.2}$$

(Blossfeld et al., 1989, page 52). This model has the property that for two individuals with $\mathbf{x}$ values $\mathbf{x}_1$ and $\mathbf{x}_2$, the ratio of hazard functions

$$h(t; \mathbf{x}_1)/h(t; \mathbf{x}_2) = e^{(\mathbf{x}_1 - \mathbf{x}_2)'\boldsymbol{\beta}}$$

is the same at all times t. For this reason this model is referred to as a *proportional hazards model.* If one of the $\mathbf{x}$ variables x_j is a dummy variable taking only the values 0 or 1 and if β_j is the corresponding regression coefficient then it follows from (32.2) that $\exp(\beta_j)$ is the ratio of hazards for two individuals who differ with respect to x_j but have common

values of all the remaining x variables. This ratio $\exp(\beta_j)$ is called the *relative risk* in epidemiology.

32.2.3 Measurement Schemes

So far we have described the nature of event histories in the population and some models for these event histories. Data on all event histories for every individual in the population will rarely be available to the researcher, however. Instead individuals are sampled from the population and the event histories of individuals in the sample are often 'censored' by the temporal nature of the observation scheme. While the models described in Section 32.2.2 are commonly employed in many areas of application of event history analysis, the kinds of observation schemes arising in sample surveys tend to be distinct from those arising in standard areas of application such as clinical trials or reliability studies in engineering. In the latter applications, observation typically begins at time I even though, for some individuals, observation may end at some time $C \leq I+T$ so that the duration time is right censored, that is, it is only known that $T > C-I$. In sample surveys matters tend to be more complicated because observation seldom starts at time I for all individuals and so left censoring may also occur.

In general subjects are included in a sequence of one or more surveys. On the first survey, data is collected retrospectively for a fixed initial reference period or as far back as a specific event such as birth or marriage. For subsequent surveys, data is collected retrospectively for the newly completed inter-survey period. The state (for example, employed/unemployed) for each subject at the time of each survey is also collected. From the series of observations the entire spell (that is, initial event to terminal event) can be obtained, subject to censoring. We refer to the period from the start of the initial reference period to the final survey in the series as the *complete reference period.* We distinguish between observing a general sample from the population and a state-based sampling scheme.

Observation Scheme A: General Population Sample

A representative sample of individuals is selected from the whole population. All questions relate to a complete reference period, which starts at C_1 and ends at C_2 (the final survey). For each sampled individual the covariate values are recorded; the times I and $I+T$ are also recorded if they occur between times C_1 and C_2 but not otherwise. We permit the possibility that all initial events in the population occur after time C_1 so that a complete retrospective history is collected. A common way in which this scheme arises is when a single survey takes place at time C_2 and retrospective information is obtained about whether the initial and/or terminal event occurred since time C_1 and, if so, at what exact time the events occurred. For ease of exposition, later sections will assume that this simple form has been employed.

Observation Scheme B: State-Based Sample with Follow-up Survey(s)
A representative sample of individuals is selected for a survey at time C_1 from the subpopulation of individuals who are in between initial and terminal events, that is, $I \leq C_1 \leq I+T$. For example, a sample of those currently unemployed is selected. Values of covariates and I are recorded retrospectively at this survey. The sample is then followed up until a final survey at time C_2 when it is noted whether $I+T \leq C_2$ and, if so, the value of $I+T$ is recorded.

This scheme often arises when the spell occurs relatively rarely in the population, such as in the unemployment example below, and where it would be inefficient in cost terms to collect data on large numbers of individuals who do not experience the spell. For ease of exposition, we shall assume in later sections that only two surveys are used at times C_1 and C_2.

For each scheme, different kinds of information about the dates or times of events may be given. Ideally, the exact dates of events are collected (continuous time data). However, with a sequence of surveys we often only know the date of the survey just prior to the occurrence of an event and the date of the survey just after the event. This results in *interval-censored data*, where we only know a lower bound and an upper bound for the time the event occurred. This form of data also results from the grouping or discretization of continuous data, for example, durations only given in *completed* years. We refer to this type of data as *grouped time data*. One extreme case of interval-censored data is termed *current status data* (discussed in detail later), where all the data are either right or left censored.

32.2.4 Model Fitting

The form of the data (that is, continuous time, grouped time, current status) has implications for the model fitting procedures since for each of these forms of data the likelihood functions are different (e.g., see Peters, 1988; Nelson, 1978, 1982). Methods of fitting the Weibull proportional hazards model assuming continuous time data and no measurement error are now considered briefly.

Observation Scheme A: General Population Sample
Following Heckman and Singer (1984, page 103) we suppose left-censored spells are omitted. Given our assumption that durations are identically distributed given **x**, this omission of data does not lead to any bias, just some loss of efficiency (note that we assume the measurement of **x** is not curtailed by the observation scheme).

Maximum likelihood estimates of the parameters and using the data from both uncensored and right-censored spells may be obtained by conventional methods. For the theory see Cox and Oakes (1984, page 81). For implementation via the computer package GLIM see Aitkin and Clayton (1980) and Roger and Peacock (1982).

Observation Scheme B: State-Based Sample with Follow-Up Survey
The standard estimation procedure for Scheme A cannot be used in this case because the dependence between the duration times and the censoring times would lead to bias (James, 1989). Instead conditional maximum likelihood methods as described by Lancaster (1979) may be used.

Current Status Data
The current status method of data collection yields only right- and left-censored data on the age at some event, for example, menarche. Maximum likelihood methods for this kind of data are described by Diamond and McDonald (1991).

32.3 TWO APPLICATIONS OF EVENT HISTORY ANALYSIS TO SURVEY DATA

32.3.1 Age at Menarche

There has been a vast amount of information collected about physiological and psychological changes associated with puberty. First menstruation or menarche defines the start of the potential reproductive years. Differences in the mean age at menarche between and within populations has long been of interest to human biologists and demographers. Marshall and Tanner (1986) discuss methods of data collection on age at menarche and conclude that "the method of data collection by recollection is clearly unsatisfactory, but it was used in nearly all the older surveys." The most common way of collecting data on age at menarche is asking each respondent her precise age at the time of questioning and whether she has yet begun to menstruate. An affirmative answer (age at menarche <age at survey) yields left-censored data, while a negative answer (age at menarche > age at survey) results in right-censored data. Such data are termed *current status data* in demography and *status quo data* in the medical/human biology literature (e.g., see Aw and Tye, 1970; Atwood and Taube, 1976) Rather than base an analysis on unreliable retrospectively reported age at menarche data, most analysts prefer to analyze the more reliable current status data.

Diamond and McDonald (1991) analyzed current status age at menarche data and found that age at menarche increased with number of siblings in the family. The estimated median age at menarche was 13.15 for zero siblings, 13.23 for one sibling, 13.51 for two, 13.71 for three, 13.66 for four and 13.84 for five or more siblings.

32.3.2 Unemployment Spells

The most common observation scheme seems to be Scheme B in which a sample of unemployed are selected at time C_1 and then followed up until time C_2. The prototype study of this kind is that of Lancaster (1979) who analyzes data on 479 unskilled unemployed persons and follows them over five weeks. He fits a Weibull proportional hazards model where the covariates are age, local unemployment rate and replacement ratio, the ratio of the income individuals receive in unemployment to that which they received in their previous jobs. Lynch (1985) conducts a similar analysis of data on young unemployed who were reinterviewed after six months.

Examples of studies employing Scheme A are Burdett, et al. (1984, 1985). Assuming exponentially distributed duration times in unemployment they avoid having to omit left censored spells. In contrast, Short and Woodrow (1985) in an analysis of the Survey of Income and Program Participation do not assume exponentially distributed durations but do omit left-censored spells.

There are also analyses for other observation schemes. Narendranathan and Nickell (1985) analyze a cohort study of 2,160 unemployed men, where observation begins unusually at time I. Nickell (1979) analyzes retrospective data from a cross-section of unemployed who are not then followed up; this analysis requires certain assumptions about the rate of entry of individuals into unemployment.

All the above papers deal with the dependence of duration times on covariates. There is also a large literature on the simpler problem of estimating duration distributions without covariates (e.g., Kiefer, et al.,

1985), as well as various extensions to more complex issues such as joint dependence of spell duration and wage at exit from unemployment on covariates (Kiefer and Neumann, 1979).

32.4 ERRORS IN MEASUREMENT OF EVENT HISTORIES

32.4.1 Types of Errors

Observation scheme A generally involves asking respondents the following retrospective questions:

i. Has the initial event ever occurred?
ii. When did it occur?
iii. Has the terminal event occurred?
iv. When did it occur?

If the answers to (i) or (iii) are 'no' then the remaining questions are skipped. For current status data, question (iv) on the timing of the terminal event is omitted. Sometimes the occurrence of the initial event is assumed, for example, when it is birth. Time in (ii) and (iv) can be measured in several ways, for example, age at occurrence, date of occurrence or time between occurrence and survey. Any of the measurements may be recorded continuously or grouped into intervals.

For observation scheme B the sample is selected from the population for whom the answers to (i) and (iii) are 'yes' and 'no' respectively at a given point in time. These answers may be available from the sampling frame or may require a screening survey. Question (ii) is then asked at the first survey which takes place at time C_1 and questions (iii) and (iv) at the follow-up survey at time C_2.

Errors in the measured responses can occur for any of questions (i)-(iv). Errors in questions (i) and (iii) involve either failure to report an event which did occur or the reporting of an event which did not occur. Errors in questions (ii) and (iv) involve misreporting the timing of an event.

32.4.2 Age at Menarche

The initial event in this example is birth and its occurrence is assumed. The timing of this initial event may be based on the respondent's reported age at survey or date of birth. While the existence of age reporting errors is well known, it will be assumed here that age at survey or date of birth is reported accurately.

We know of no published evidence of errors in reporting occurrence of the terminal event, that is, women failing to report menarche when it has occurred or vice versa, although such errors would not be implausible, especially around the usual age of menarche if, for example, reporting menarche embarrasses the respondent.

A number of studies of errors in reported age at menarche do, however, exist. The following studies compared age at menarche from medical records with the value reported retrospectively (the sample size n and average number R of years between the survey and menarche are given in parentheses): Livson and McNeill (1962) ($n=43$, $R=17$), Damon, et al. (1969) ($n=60$, $R=19$), Damon and Bajema (1974) ($n=143$, $R=39$), Bergsten-Brucefors (1976) ($n=339$, $R=$about 4) and Bean, et al. (1979) ($n=160$, $17<R<53$). Broadly similar results were found. Treating the value in the medical records as the true value and hence defining the measurement error as the difference between the reported value and this value, the errors were generally found to be symmetrically distributed around zero with little evidence of bias. Neither did the errors seem to be correlated with the true value (Livson and McNeill, 1962). There was evidence of heaping of reported values at whole years (Livson and McNeill, 1962) as well as some evidence of heaping of errors at 12 months (Bergsten-Brucefors, 1976). As expected, the magnitude of the errors tend to be positively related to length of recall. For recall of about four years, 63 percent of errors were less than three months and the correlation between true and reported age was 0.81 (Bergsten-Brucefors, 1976). For recall of 39 years, only 50 percent of errors were less than six months and the correlation had declined to 0.61 (Damon and Bajema, 1974).

32.4.3 Unemployment Duration

One survey employing observation scheme A is the U.S. Work Experience Survey (WES) with a 12-month retrospective reference period. Morgenstern and Barrett (1974) compared estimated person-years unemployed from the WES with estimates from the U.S. Current Population Survey (CPS), which uses a one week reference period, and found that the WES tended to understate unemployment by about 3 percent for men and 23 percent for women. Horvath (1982) repeated this analysis with later data and found an average understatement of 19 percent. He also broke down annual estimates into two six-month estimates and concluded that there was much greater understatement in the WES for a 6–12 months recall period than for a 0–6 months period. These results do not distinguish, however, between different spells of unemployment. Mathiowetz and Duncan (1988) found in a validation study of the Panel Study of Income Dynamics that measured mean person-years of unemployment again understated the "true" amount determined from records, but this understatement was minor compared to the error in reported spells. On average only 34 percent of all spells of unemployment were reported in the survey, decreasing from 63 percent of spells of 29 or more weeks to just 25 percent of spells of one week.

Mathiowetz (1985) offers some evidence of telescoping, both backward and forward, of the reported timing of both initial and terminal events. More evidence of such effects comes from studies which employ observation scheme B and ask the unemployed how long they have been unemployed. Bowers and Horvath (1984) and Poterba and Summers (1984) investigated errors in responses to this question indirectly by comparing responses in successive months of the CPS. They found substantial evidence of inconsistency and of a negative relationship between reporting error and duration; short durations seemed to be overreported, perhaps because unemployment is confused with job search, whereas longer durations may have been underreported. Related findings for Italian and Canadian Labor Force Survey data were reported by Trivellato and Torelli (1989) and Lemaitre (1988), respectively. Trivellato and Torelli (1989) demonstrated substantial heaping around reported durations of 6 and 12 months and multiples thereof. Horrigan (1987) describes similar heaping in CPS weekly data.

32.5 MEASUREMENT ERROR AND EVENT HISTORY ANALYSIS

In Section 32.2 we described procedures for estimating α and $\boldsymbol{\beta}$ given data on the duration times T_i and the covariates $\mathbf{x}_i$. We now consider properties of these procedures when the true durations T_i are replaced by the observed durations T_i^{obs}. We consider various plausible models for generating measurement error *in the duration times*, namely, covariate-related errors, recall-related errors and duration-related errors. Each is described in detail later.

In order to fix ideas, consider the loglinear form of (32.1) with a multiplicative error structure

$$T_i^{obs} = T_i(1 + \varepsilon_i),$$

where ε_i denotes the measurement error. For this multiplicative-error

Model Parameter	Estimate
β_0 constant (group 1)	−34.35
β_1 (group 2 versus group 1)	−0.55
β_2 (group 3 versus group 1)	−0.68
α power parameter	13.19

Group	Size	Mean age at menarche (years)	Relative risk
Group 1: 1 sibling	400	13.1	1.000
Group 2: 2–3 siblings	400	13.5	0.577
Group 3: 4+ siblings	200	13.8	0.507

Figure 32.2. Parameter values for age at menarche simulations.

structure, the relationship between T_i^{obs} and $\mathbf{x}_i$ retains the log-linear form of (32.1) but with an additive error term $\delta_i = \log(1+\varepsilon_i)$:

$$\log T_i^{obs} = (\mathbf{x}_i{}'\beta + \sigma u_i)/\alpha + \delta_i. \tag{32.3}$$

In some very special circumstances, measurement error will not induce bias in standard point estimators of $\boldsymbol{\beta}$. Thus, if ε_i is independent of both T_i and $\mathbf{x}_i$, and the regression parameters are estimated by least-squares regression of log T_i^{obs} on $\mathbf{x}_i$ then measurement error will not induce bias in the least-squares estimators (apart from the constant term) but will only inflate their variances. However, least-squares estimation is seldom used in this context since it is inefficient and requires modification to handle censored data (Lawless, 1982, pages 328–335).

In general, measurement error will lead to inconsistencies in the standard estimators of $\boldsymbol{\beta}$. Given the form of (32.3) it seems likely that the greater bias will be induced when the measurement error ε_i is related to the covariates $\mathbf{x}_i$. The magnitude of this bias will be investigated empirically in the next section.

32.6 NUMERICAL ILLUSTRATION

32.6.1 Age at Menarche: Framework

Data were generated according to a known event history model to induce the observations that would be obtained according to the sampling scheme, and to superimpose on these observations the impact of various types of measurement errors. Throughout all numerical investigations we use a data set comprising 400, 400 and 200 subjects from three groups. The basic parameters of the Weibull model and coefficients for the groups are given in Figure 32.2 together with the implied average age at menarche for each group. These values were chosen to reflect the size of effects found in Cole and McDonald (1989) for groups defined by number of siblings.

Births were simulated using a uniform distribution over a five year interval ($0 \leq I \leq 5$) independently of sibship size. Age at menarche for each subject was simulated from the Weibull distribution with parameter values given in Figure 32.2. Thus all three groups have the same power parameter (13.19) but different scale parameters.

We consider two basic survey situations and related sources of measurement error.

Survey at $C_2 = 25$ Years (No Censoring)
If the five year cohort is surveyed at $C_2 = 25$, the youngest women will be at least 20 years old and all respondents will have experienced menarche so that no event histories are censored. Observation scheme A in Section 32.2 describes this case with the three special characteristics: that the population is defined to be a particular cohort, the observation point C_2 is so late that no censoring occurs, and the question of left censoring does not arise and so no consequent loss of efficiency is incurred. For this situation the event histories may be analyzed using the Weibull model described in Section 32.2.4.

The literature on reporting errors for age at menarche suggests that these are symmetric about the true value and, to a first approximation, normally distributed, although there is some evidence of additional symmetric errors of 12 months. We have assumed a normal random error as an additive term for duration so that the observed (reported) age at menarche, T_i^{obs} is given by

$$T_i^{obs} = T_i + \varepsilon_i \text{ where } \varepsilon_i \sim N(0, \sigma_i^2).$$

We have chosen σ_i in two ways: (i) group-related measurement error and (ii) age-at-menarche related measurement error. In the first case $\sigma_i = \sigma_g$ for the three groups, $g = 1, 2, 3$. The values used are:

Standard Deviations of Group-Related Measurement Error

Case	σ_1	σ_2	σ_3
G1	0.6	0.6	0.6
G2	0.3	0.3	0.3
G3	0.8	0.8	0.4
G4	0.4	0.6	0.8

Cases G1 and G2 assume a constant measurement error variance for all three groups whereas G3 and G4 do not. A standard deviation of 0.6 years will result in about two thirds of all reported ages being within seven months of the true value, whereas a standard deviation of 0.3 years will result in a corresponding value of less than four months.

For age-at-menarche related measurement error we assume that reporting error will be at a maximum at age 13.5 years and smaller for women who experienced menarche at a particularly young or old age. The expression for σ_i used is

$$\sigma_i = \sigma \exp\{-(T_i - 13.5)^2/8\}.$$

We have taken two values for σ as follows:

Standard Deviations of Age-Related Measurement Error

Case	σ
A1	0.6
A2	0.4

Survey at $C_2 = 16$ Years (Censored Durations)

In this case all respondents will be aged between 11 and 16 years when surveyed and about half will not have experienced menarche and so their event histories will be censored. Both of the estimation methods described for the survey at $C_2 = 25$ are applicable and in addition a third approach is available. This is to use the current status of each subject at the time of the survey. In this case the dependent variable is whether or not each respondent has experienced menarche at the time of the survey and no information is used about the age at which menarche occurred for those who have already experienced it (other than that it must be less than the respondent's age at interview). The method of analysis is referred to in Section 32.2.4 and described more fully in Diamond, et al. (1986) and Diamond and McDonald (1991).

We have assumed that the current status of women is correctly reported and that a reporting error can occur only for those women who have reached menarche at the date of the survey. We assume that this recall error has larger variance the longer the length of recall. Thus we introduce measurement error related to R_i, the length of time between menarche and the survey:

$$R_i = C_2 - T_i - I_i,$$

for which the observed (reported) value is

$$R_i^{obs} = R_i \times \varepsilon_i \text{ where } \varepsilon_i \sim N(1, \sigma_i^2).$$

The observed (reported) duration is given by

$$T_i^{obs} = C_2 - I_i - R_i^{obs}.$$

We choose values $\sigma_i = \sigma_g$, $g = 1, 2, 3$ for the three sibship size groups which are consistent with those reported in the literature as follows:

Table 32.1. Average Bias of Estimators of Weibull Regression Parameters (β) with Measurement Error Compared to Without Measurement Error

		Type of Data					
		Continuous Time		Grouped Time		Current Status	
Type of Error		β_1	β_2	β_1	β_2	β_1	β_2
	G1	−0.087	−0.113	−0.106	−0.134	—	—
Group	G2	−0.027	−0.034	−0.033	−0.042	—	—
Related	G3	−0.180	−0.255	−0.174	−0.242	—	—
	G4	−0.014	0.028	−0.042	−0.028	—	—
Age	A1	−0.071	−0.096	−0.071	−0.094	—	—
Related	A2	−0.041	−0.052	−0.046	−0.058	—	—
Recall	R1	−0.008	−0.008	−0.001	−0.002	0	0
Related	R2	−0.009	−0.013	−0.000	−0.002	0	0
	R3	−0.018	−0.022	−0.003	−0.004	0	0

Standard Deviations of Recall-Related Measurement Error

Case	σ_1	σ_2	σ_3
R1	0.175	0.175	0.175
R2	0.125	0.125	0.125
R3	0.175	0.150	0.125

For each survey date and each type of measurement error separately, 100 simulations of a data set of size 1,000 were carried out. For each simulation the dates of birth and age at menarche were simulated according to the uniform and Weibull distributions, respectively. Each of the appropriate estimation methods was used on the data set and then the measurement error was introduced according to the various sources described above. The estimation procedures were then applied again to the resulting data. Properties of procedures such as the means and variances of estimators were estimated from the 100 replications. In addition, since the estimation methods were all cases of generalized linear models and could be fitted using GLIM, the model-based variance estimators could also be compared with the variances obtained using the simulation replications.

32.6.2 Age at Menarche: Results Table

Table 32.1 contains estimates of the bias for the various estimators when measurement error is present. These were obtained by subtracting the parameter estimates when measurement error occurs from the corresponding estimates for the same data but obtained before the measurement error was introduced.

As one would expect, the bias of the estimators is most severe when the measurement error is largest. The bias of the grouped-time estimator is just as severe as for the Weibull estimator and the worst case is when the measurement error variance is related to the group membership (G3). In this case, the first group has the largest measurement error variance and so the estimate of the scale parameter for this group is attenuated most. The other groups have smaller measurement error variances and are less attenuated towards zero. The net effect is that the contrast effect is diminished and the estimates of covariate effects are shrunken towards zero. In contrast, the case of G4, where the measurement error variances are in the reverse order to G3, results in the covariate parameter being relatively unbiased or biased in the opposite direction.

We note that the size of the bias is appreciable. The true parameter values are -0.55 and -0.68 with relative risk values of $\exp(-0.55)=0.577$ and $\exp(-0.68)=0.507$. For case G3 the corresponding Weibull estimates are -0.3787 and -0.4275 with relative risk estimates of 0.685 and 0.652. Thus, the relative risk parameter estimates are attenuated by 19 percent and 29 percent respectively towards a relative risk of one. Attenuation towards one would also occur if the relative risks were greater than one.

For the recall-related measurement error, when $C_2=16$ years only half of the respondents have achieved menarche. Hence, only half of the intervals are subject to measurement error and, in consequence, the bias of parameter estimates is very small. The current-status estimator has no bias since the misreported duration is not used in the estimation procedure.

Table 32.2 contains the variances of the estimators obtained from the between replication variation. In parentheses are the average variance estimates from the generalized linear model fitting procedures. As expected, the grouped-time estimators are less efficient than the continuous-time estimators. Variances of estimators are generally smaller when measurement error is present. An additional point is that for the continuous-time estimator the model-based variance (in parentheses) is generally smaller than the between replicate variance, suggesting that the model-based estimator of variance is slightly biased. The grouped-time procedure does not seem to be affected by this and the model-based estimator of variance is more satisfactory.

For the recall-related measurement error, when $C_2=16$ years the variances of the parameter estimates are larger than for $C_2=25$ years, even when there is no measurement error. This is because of the loss of efficiency due to half of the observations being censored.

Table 32.2 Variances of Estimators for Weibull Regression Parameters Calculated from 100 Replications for Both Observation Schemes and Various Measurement Error Models*

		Type of Data					
		Continuous Time		Grouped Time		Current Status	
$C_2 = 25$ Years		β_1	β_2	β_1	β_2	β_1	β_2
No Error		681 (496)	891 (740)	808 (594)	11 (897)	— —	— —
Group Related	G1	610 (496)	945 (741)	597 (565)	838 (844)	— —	— —
	G2	575 (496)	938 (741)	533 (584)	863 (877)	— —	— —
	G3	533 (495)	789 (739)	547 (560)	788 (821)	— —	— —
	G4	443 (498)	860 (750)	444 (567)	798 (871)	— —	— —
Age Related	A1	626 (495)	888 (740)	691 (574)	10 (862)	— —	— —
	A2	588 (495)	783 (740)	673 (583)	906 (877)	— —	— —
$C_2 = 16$ Years							
No Error		825 (922)	1,416 (1,471)	887 (988)	1,421 (1,586)	1,478 (1,587)	2,153 (2,374)
Recall Related	R1	934 (924)	1,501 (1,468)	1,054 (988)	1,549 (1,580)	1,780 (1,586)	2,559 (2,365)
	R2	705 (920)	1,271 (1,471)	897 (982)	1,358 (1,588)	1,348 (1,586)	1,909 (2,384)
	R3	726 (924)	1,341 (1,476)	743 (987)	1,377 (1,591)	1,306 (1,589)	1,908 (2,374)

*All entries are multiplied by 10,000. Entries in parentheses are average estimated variances from the model fitting procedure for the 100 replications. Entries for the no measurement error case are based upon 300 replications.

In general, the continuous-time estimator is most efficient, followed by the grouped-time estimator and then the current-status estimator. This reflects the loss of information in using grouped time or current status data. The tradeoff between biased estimators due to measurement error and the loss of efficiency due to using less efficient methods will depend on the size of the measurement error and the size of the data set. For the simulation examples presented, the squared bias for $C_2 = 25$ years is of the same order of magnitude as the variance or larger. For $C_2 = 16$ years, the bias terms are much smaller and variance is the dominating factor.

32.6.3 Unemployment: Framework

Data were generated for 100 cases for the date of the initial event (becoming unemployed), the covariate x (age of the person) and the date of the terminating event (becoming employed) according to a known Weibull event history model. Hence, the length of time between the initial event and the first survey (U_1) and the length of the subsequent period to the second survey or the terminal event (U_2) were derived. The second survey was taken to be one year after the first. The date of the initial event was uniformly distributed over a three year period before the first survey and only if the person was still unemployed at the survey date was the case included in the survey data set. This rejective approach mirrors the selection process of state-dependent observation at the time of the first survey in accordance with observation scheme B (state-dependent selection). The true parameter values of the event history model were:

$$\begin{aligned}
&\beta_0 \text{ constant (scale parameter)} &&= -0.25\\
&\beta_1 \text{ coefficient for age} &&= 0.64\\
&\alpha \text{ power parameter} &&= 0.80
\end{aligned}$$

Measurement error was superimposed onto U_1 and U_2 according to a multiplicative mechanism. Measurement error was assumed to be related to the length of the recall period from the relevant survey. For example, for U_{1i}, the observed duration was $U_{1i}^{obs} = U_{1i}\ (1 + \varepsilon_{1i})$ where $\varepsilon_{1i} \sim N(0, \sigma_{1i}^2)$. The impact of measurement error was considered for four situations, $k = 1, 2$:

Case

M1 $\sigma_{ki} = 0$

M2 $\sigma_{ki} = 0.08$

M3 σ_{ki} proportional to age so that, for example, σ_{ki} at age 30 equals 0.08, at age 20 equals 0.053, and at age 40 equals 0.107.

M4 σ_{ki} inversely proportional to age so that, for example, σ_{ki} at age 30 equals 0.08, at age 20 equals 0.107, and at age 40 equals 0.053.

For both U_{1i}^{obs} and U_{2i}^{obs} a further condition was imposed: the size of

Table 32.3. Unemployment — Average Bias of Estimators with Measurement Error Compared to Without Measurement Error. Standard error in ( )*

Measurement Error on U_2		σ_{2i}											
Measurement Error on U_1		M1			M2			M3			M4		
		β_0	β_1	α	β_0	β_1	α	β_0	β_1	α	β_0	β_1	α
	M1	0	0	0	435	−120	12	42	20	−70	992	−296	26
					(269)	(48)	(57)	(224)	(59)	(55)	(335)	(51)	(62)
	M2	20	8	−10	453	−111	2						
		(30)	(1)	(1)	(267)	(49)	(58)						
σ_{1i}	M3	23	15	−15				62	35	−21			
		(35)	(1)	(1)				(226)	(61)	(56)			
	M4	34	4	−9							1,024	−292	17
		(35)	(1)	(1)							(330)	(52)	(63)

* All entries are multiplied by 1,000.
β_0, constant parameter. β_1, coefficient for age. α, power parameter.

the measurement error could not move the event across either of the survey dates C_1 or C_2. For these cases the measurement error was truncated to keep the event in the correct temporal order with respect to the surveys.

The whole process was repeated for 90 simulations using the same starting seeds for random number generation. Thus comparisons between different measurement error assumptions are conditional on the same age x_i and true duration values U_{1i} and U_{2i} and the impact of the measurement error is more accurately identified.

32.6.4 Unemployment: Results

Table 32.3 contains the bias of the parameter estimates due to measurement error, made from the same simulated data before and after measurement error is introduced. Each estimate of bias is based upon 90 simulations of 100 subjects each, and standard errors for the estimates of bias are also given.

The bias is very small in almost all cases. This is particularly so for the coefficient of the covariate (age) and the power parameter of the Weibull distribution. The bias of the constant (scale) parameter of the Weibull event history model is slightly larger. For analyses based on 100 subjects, the bias effects are much smaller than the standard error of the parameter estimates and this would continue to be so unless the sample size was very much larger.

The measurement error on U_1 (the period of unemployment before the first survey) has virtually no effect compared to measurement error on U_2 (the duration of unemployment after the first survey). This is so despite the fact that U_1 is potentially much larger than U_2 since the latter is bounded between the two surveys which were one year apart for the simulation studies. Since the measurement error is proportional to the length of recall this too could be larger for U_1. The method of analysis treats the observed value of U_1 as a covariate and effectively conditions on this.

The effect of age-related measurement error is to increase the bias of parameter estimators when the measurement error standard deviation is inversely proportional to age. There are two reasons for this: (i) the age distribution is skewed towards younger people so that more cases have an increased measurement error and (ii) the event history model implies shorter durations for unemployment for younger people. This implies that the length of recall from the second survey for the date of regaining employment is longer and, hence, the size of the measurement error will be larger since the measurement error is proportional to this recall period.

32.7 DISCUSSION AND CONCLUSIONS

The essential purpose of event history analysis is to model the duration spent in a given state and to take account of covariates that affect the duration distribution. Very often, data used to model the event history is obtained from two or more surveys carried out to include the same subjects. Between the surveys and prior to the first survey, information is collected retrospectively. Survey practitioners recognize that such data collection methods result in observations which are subject to a variety of sources of measurement error.

We have considered the effect of measurement errors on the time of the initial and terminal events and hence on the duration in a state. We have not considered the effect of measurement error on the covariates. For ordinary regression, measurement error on the dependent variable does not lead to inconsistency of the regression estimators but for event history analysis this is not the case.

The numerical results demonstrate situations where the bias effect of measurement error is important and other cases where it is not. Specifically, the age at menarche example demonstrates that retrospective measurement error when the survey takes place a considerable time after the event of interest can have an important effect. When the survey takes place closer to the event of interest so that about half of the cases are censored, the bias effect is much less important. To some extent this is because the length of recall is shorter so measurement errors are smaller. However, we have not considered misreporting of the state (that is, women who have not experienced menarche reporting that they have and vice versa). Censored observations are not subject to the measurement error that we considered in this example. Further work is needed to investigate the effect of misreporting of the state and to explore the combined effect of misreporting duration and state.

The unemployment example illustrates the case of a state-based sample with follow-up survey. The effects of the measurement error are small although further work is needed to establish the conditions when this is generally the case. One of the main points of interest is the small effect caused by measurement error for the initial retrospective period of unemployment before the initial survey. As for age at menarche, we have not considered misreporting of state. Under observation scheme A, a general population survey, whole spells of unemployment could be omitted by errors of recall. The literature suggests that this is more likely for short periods of unemployment. Further research is needed to explore the impact of this on event history analysis.

ACKNOWLEDGMENTS

Research by D. Holt and C. J. Skinner was supported by ESRC grant R000 23 2522. Research by C. J. Skinner was also partly supported by Joint Statistical Agreement 90-7, U.S. Bureau of the Census and Cooperative Agreement 68-3A75-0-9, Soil Conservation Service, USDA, while at Iowa State University.

REFERENCES

Abowd, J.M., and Zellner, A., "Estimating Gross Labor Force Flows," *Journal of Business and Economic Statistics*, Vol. 3, 1985, pp. 254–283.

Adams, E.W., "Elements of a Theory of Inexact Measurement," *Philosophy of Science*, Vol. 32, 1965, pp. 205–228.

Adams, E.W, "On the Nature and Purpose of Measurement," *Synthese*, Vol. 16, 1966, pp. 125–169.

Agar, M., and Hobbs, J., "Interpreting Discourse: Coherence and the Analysis of Ethnographic Interviews," *Discourse Processes*, Vol. 5, No. 1, 1982, pp. 1–32.

Agar, M.V., and Hobbs, J., "Natural Plans: Using AI Planning in the Analysis of Ethnographic Interviews," *Ethos*, Vol. 1, No. 1/2, 1983, pp. 33–48.

Aitkin, M.A., and Clayton, D.G., "The Fitting of Exponential, Weibull and Extreme Value Distributions to Complex Censored Survival Data Using GLIM," *Applied Statistics*, Vol. 29, 1980, pp. 156–163.

Aitkin, M.A., and Longford, N.T., "Statistical Modelling Issues in School Effectiveness Studies," *Journal of the Royal Statistical Society, A,* Vol. 149, 1986, pp. 1–43.

Akkerman, A.E., "De 5PFT (Five Personality Factors Test)," University of Amsterdam, unpublished report, 1977.

Alba, J.W., and Hasher, L., "Is Memory Schematic?" *Psychological Bulletin*, Vol. 93, No. 2, 1983, pp. 203–231.

Alba, J.W., and Hutchinson, J.W., "Dimensions of Consumer Expertise," *Journal of Consumer Research*, Vol. 13, 1987, pp. 411–454.

Allen, J.D., "Evaluation of Aerial Photography as a Technique for Estimating Citrus Fruit Yield," U.S. Department of Agriculture Staff Report, Statistical Reporting Service, 1977.

Allen, J.D., "A Look at the Remote Sensing Applications Program of the National Agricultural Statistics Service," *Journal of Official Statistics*, Vol.6, No. 4, 1990, pp. 393–410.

Allison, P.D., *Event History Analysis: Regression for Longitudinal Event Data*, Beverley Hills: Sage, 1984.

Althauser, R.P., and Herberlein, T.A., "Validity and the Multitrait-Multimethod Matrix," in E.F. Borgatta, and G.W. Bohrnstedt (eds.), *Sociological Methodology 1970*, 1970, pp. 151–169, San Francisco: Jossey-Bass.

Alwin, D.F., "Approaches to the Interpretation of Relationships in the Multitrait-Multimethod Matrix," *Sociological Methodology 1973–74*, 1974, pp. 79–105, San Francisco: Jossey-Bass.

Alwin, D.F., "Problems in the Estimation and Interpretation of the Reliability of Survey Data," *Quality and Quantity*, Vol. 23, 1989, pp. 277–331.

Andersen, R., Kaspar, J., and Frankel, M., *Total Survey Error*, San Francisco: Jossey-Bass, 1979.

Anderson, A.B., Basilevsky, A., and Hum, D.P.J., "Measurement: Theory and Techniques," in P.H. Rossi, J.D. Wright, and A.B. Anderson (eds.), *Handbook of Survey Research*, 1983, pp. 231–287, Academic Press.

Anderson, B.A., Silver, B.D., and Abramson, P., "The Effects of Race of Interviewer on Measures of Electoral Participation by Blacks," *Public Opinion Quarterly*, Vol. 52, 1988a, pp. 53–83.

Anderson, B.A., Silver, B.D., and Abramson, P., "The Effects of the Race of Interviewer on Race-related Attitudes of Black Respondents," *Public Opinion Quarterly*, Vol. 52, 1988b, pp. 289–324.

Anderson, D.A., and Aitkin, M., "Variance Component Models with Binary Response: Interviewer Variability," *Journal of the Royal Statistical Society, B*, Vol. 47, 1985, pp. 203–210.

Andrews, F.M., "Estimating the Construct Validity and Correlated Error Components of the Rated Effectiveness Measures," in F.M. Andrews (ed.), *Scientific Productivity*, Chapter 14, 1979, London: Cambridge University Press.

Andrews, F.M., "Construct Validity and Error Components of Survey Measures: A Structural Modelling Approach," *Public Opinion Quarterly*, Summer, Vol. 48, No. 2, 1984, pp. 409–422.

Andrews, F.M., and Crandall, R., "The Validity of Measures of Self-reported Well-being," *Social Indicators Research*, Vol. 3, 1976, pp. 1–19.

Andrews, F.M., and Withey, S.B., "Assessing the Quality of Life as People Experience It," paper presented at the Annual Meeting of the American Sociological Association, Montreal, 1974.

Aquilino, W.S., and Losciuto, L.A., "Interview Mode Effects in Drug Use Surveys," *Public Opinion Quarterly*, Vol. 54, 1990, pp. 362–395.

Argyle, M., *Social Interaction*, London: Methuen, 1969.

Assakul, K., and Proctor, C.H., "Testing Independence in Two-Way Contingency Tables with Data Subject to Misclassification," *Psychometrika*, Vol. 32, No. 1, 1967, pp. 67–76.

Astin, A.W., Green, K.C., Korn, W.S., Schalit, M., and Berz, E.R., *The American Freshman: National Norms for Fall 1988*, Los Angeles: The Higher Education Research Institute, 1988.

Atkinson, J., "Understanding Formality: The Categorization and Production of 'Formal' Interaction," *The British Journal of Sociology*, Vol. 33, No. 1, 1982, pp. 86–117.

Atwood, C.L., and Taube, A., "Estimating Mean Time to Reach a Milestone, Using Retrospective Data," *Biometrics*, Vol. 32, 1976, pp. 159–172.

Aw, E., and Tye, C.Y., "Age of Menarche of a Group of Singapore Girls," *Human Biology*, Vol. 42, 1970, pp. 329–336.

Ayidiya, S.A., and McClendon, M.J., "Response Effects in Mail Surveys," *Public Opinion Quarterly*, Vol. 54, 1990, pp. 229–247.

Back, K.W., and Gergen, K.J., "Idea Orientation and Ingratiation in the Interview: A Dynamic Model of Response Bias," *Proceedings of the Social Statistics Section*, American Statistical Association, 1963, pp. 284–288.

Bagozzi, R.P., and Yi, Y., "Assessing Method Variance in Multitrait-Multimethod Matrices: The Case of Self Reported Affect and Perceptions at Work," *Journal of Applied Psychology*, Vol. 75, 1990, pp. 547–560.

Bailar, B.A., "Recent Research in Reinterview Procedures," *Journal of the American Statistical Association*, Vol. 63, No. 1, 1968, pp. 41–63.

Bailar, B.A., "Bias in the CPS Reinterview Sample," U.S. Bureau of the Census, Washington, DC, unpublished memorandum, 1974.

Bailar, B.A., "The Effects of Rotation Group Bias on Estimates from Panel Surveys,' *Journal of the American Statistical Association*, Vol. 70, 1975, pp. 23–30.

Bailar, B.A., "Interpenetrating Subsamples," in N.L. Johnson and S. Kotz (eds.), *Encyclopedia of Statistical Sciences*, Vol. 4, 1983, pp. 197–201.

Bailar, B.A., and Biemer, P.P., "Some Methods for Evaluating Nonsampling Error in Household Censuses and Surveys," in P.S.R.S. Rao and J. Sedransk (eds.), *W.G. Cochran's Impact on Statistics*, 1984, pp. 253–274, New York: John Wiley and Sons.

Bailar, B.A., and Dalenius, T., "Estimating the Response Variance Components of the U.S. Bureau of the Census' Survey Model, "*Sankhyā B*, 1969, pp. 341–360.

Bailar, B.A., and Rothwell, N.D., "Measuring Employment and Unemployment," in C.F. Turner and E.A. Martin (eds.), *Surveying Subjective Phenomena*, Vol. 2, 1984, pp. 129–142, New York: Russell Sage Foundation.

Bailar, B.A., Bailey, L., and Stevens, J., "Measures of Interviewer Bias and Variance," *Journal of Marketing Research*, Vol. 14, 1977, pp. 337–343.

Bailey, L., Moore, T.F., and Bailar, B.A., "An Interviewer Variance Study for the Eight Impact Cities of the National Crime Survey Cities Sample," *Journal of the American Statistical Association*, Vol. 73, No. 1, 1978, pp. 16–23.

Barath, A., and Cannell, C., "Effect of Interviewer's Voice Intonation," *Public Opinion Quarterly*, Vol. 40, No. 3, 1976, pp. 370–373.

Bartlett, F.C., *Remembering: A Study in Experimental and Social Psychology*, Cambridge, England: Cambridge University Press, 1932.

Battaglia, R., "1983 Soybean Objective Yield Nonsampling Error Research Study," U.S. Department of Agriculture, SRS Staff Report No. YRB-85–01, 1985.

Bauman, L., and Adair, E., "The Use of Ethnographic Interviewing to Inform Questionnaire Construction," paper presented at the Annual Meeting of the American Association for Public Opinion Research, St. Petersburg, FL, May, 1989.

Bean, J.A., Leeper, J.D., Wallace, R.B., Sherman, B.M., and Jagger, H., "Variations in the Reporting of Menstrual Histories," *American Journal of Epidemiology*, Vol. 109, 1979, pp. 181–185.

Belson, W.A., *The Design and Understanding of Survey Questions*, Aldershot, England: Gower, 1981.

Belson, W.A., *Validity in Survey Research*, Aldershot, England: Gower, 1986.

Belson, W.A., and Duncan, J.A., "A Comparison of the Checklist and the Open Response Questioning Systems," *Applied Statistics*, Vol. 11, 1962, pp. 120–132.

Bentler, P.V., "Theory and Implementation of EQS: A Structural Equations Program," Los Angeles: BMDP Statistical Software, 1985.

Bentler, P., and Lee, S., "Statistical Aspects of a Three Mode Factor Analysis Model," *Psychometrika*, Vol. 43, 1978, pp. 343–352.

Bergling, K., *The Development of Hypothetico-Deductive Thinking in Children*, Stockholm: Almqvist & Wiksell International, and New York: John Wiley & Sons, 1974.

Bergsten-Brucefors, A., "A Note on the Accuracy of Recalled Age at Menarche," *Annals of Human Biology*, Vol. 3, 1976, pp. 71–73.

Berk, M.L., and Bernstein, A.M., "Interviewer Characteristics and Performance on a Complex Health Survey," *Social Science Research*, Vol. 17, 1988, pp. 239–251.

Bershad, M.A., "Gross Changes in the Presence of Response Errors," Washington, DC, U.S. Bureau of the Census, unpublished memorandum, 1967.

Besag, J., "Errors-in-Variables Estimation for Gaussian Lattice Schemes," *Journal of the Royal Statistical Society, B*, Vol. 39, 1977, pp. 73–78.

Bickart, B.A., Blair, J., Menon, G., and Sudman, S., "Cognitive Aspects of Proxy Reporting of Behavior," in M. Goldberg, G. Gorn, and R. Pollay (eds.), *Advances in Consumer Research, Vol. XVII*, 1990, Provo, UT: Association for Consumer Research.

Biemer, P.P., "The Use of Spatial Models for the Estimation of Nonsampling Errors in Censuses," *Proceedings of the U.S. Bureau of the Census Annual Research Conference*, U.S. Bureau of the Census, Washington, DC, 1986, pp. 7–20.

Biemer, P.P., "Measuring Data Quality," in R.M. Groves, P.P. Biemer, L. E. Lyberg, J.T. Massey, W.L. Nicholls II, and J. Waksberg (eds.), *Telephone Survey Methodology*, 1988, pp. 273–282, New York: Wiley.

Biemer, P.P., and Forsman, G., "On the Quality of Reinterview Data with Application to the Current Population Survey," *Proceedings of the U.S. Bureau of the Census, Annual Research Conference*, U.S. Bureau of the Census, Washington, DC, 1990.

Biemer, P.P., and Stokes, S.L., "Optimal Design of Interviewer Variance Experiments in Complex Surveys," *Journal of the American Statistical Association*, Vol. 80, 1985, pp. 158–166.

Biemer, P.P., and Stokes, S.L., "The Optimal Design of Quality Control Samples to Detect Interviewer Cheating," *Journal of Official Statistics*, Vol. 5, No. 1, 1989, pp. 23–40.

Bigsby, F.G., "Forecasting Corn Ear Weight Using Surface Area and Volume Measurements: A Preliminary Report," National Agricultural Statistics Service, Research Report No. SRB-89–05, 1989.

Billiet, J., and Loosveldt, G., "Interviewer Training and Quality of Responses," *Public Opinion Quarterly*, Vol. 52, No. 2, 1988, pp. 190–211.

Bingham, W., and Moore, B., *How to Interview*, New York: Harper and Row, 1924.

Bishop, D., Jeanrenaud, C., and Lawson, K., "Comparison of a Time Diary and Recall Questionnaire for Surveying Leisure Activities," *Journal of Leisure Research*, Vol. 7, No.1, 1975a, pp. 73–80.

Bishop, G.F., "Think-aloud Responses to Survey Questions: Some Evidence on Context Effects in Surveys," paper presented at the National Opinion Research Center Conference on Context Effects, Chicago, IL, 1986.

Bishop, G.F., "Experiments with Middle Response Alternatives in Survey Questions," *Public Opinion Quarterly*, Vol. 51, 1987, pp. 220–232.

Bishop, G.F., "Think-aloud Responses to Survey Questions: Illustrations of a New Qualitative Technique," paper presented to the American Association for Public Opinion Research, St. Petersburg, FL, May 1989.

Bishop, G.F., Oldendick, R.W, and Tuchfarber, A.J., "What Must My Interest in Politics Be if I Just Told You 'I Don't Know'," paper presented to the American Association for Public Opinion Research, Hunt Valley, MD, 1982.

Bishop, G.F., Oldendick, R.W., and Tuchfarber, A.J., "Effects of Filter Questions in Public Opinion Surveys," *Public Opinion Quarterly*, Vol. 47, 1983, pp. 528–546.

Bishop, G.F., Hippler, H.J., Schwartz, N., and Strack, F., "A Comparison of Response Effects in Self-administered and Telephone Surveys," in R.M. Groves, P.P. Biemer, L.E. Lyberg, J.T. Massey, W.L. Nicholls II, and J. Waksberg (eds.), *Telephone Survey Methodology,* 1988, pp. 321–340, New York: Wiley.

Bishop, Y.M.M., Fienberg, S.E., and Holland, P.W., *Discrete Multivariate Analysis*, Cambridge, MA: MIT Press, 1975b.

Bissell, A.F., "How Reliable is Your Capability Index?", *Applied Statistics*, 1990, Vol. 39, pp. 331–340.

Blair, E., "More on the Effects of Interviewer's Voice Intonation," *Public Opinion Quarterly,* Vol. 41, No. 4, 1977–78, pp. 544–548.

Blair, E., and Burton S., "Cognitive Processes Used by Survey Respondents in Answering Behavioral Frequency Questions," *Journal of Consumer Research*, Vol. 14, 1987, pp. 280–288.

Blair, E., Sudman, S., Bradburn, N., and Stocking, C.B., "How to Ask Questions about Drinking and Sex: Response Effects in Measuring Consumer Behavior," *Journal of Marketing Research*, Vol. 14, 1977, pp. 316–321.

Blankenship, A.B., "The Influence of the Question Form upon the Response in a Public Opinion Poll," *Psychological Record*, Vol. 3, 1940, pp. 345–422.

Blossfeld, H.P., Hamerle, A., and Mayer, K.U., *Event History Analysis*, Hillsdale, New Jersey: Lawrence Erlbaum, 1989.

Bodenhausen, G.V., and Wyer, R.S., "Social Cognition and Social Reality: Information Acquisition and Use in the Laboratory and the Real World," in H.J. Hippler, N. Schwarz, and S. Sudman (eds.), *Social Information Processing and Survey Methodology*, 1987, pp. 6–41, New York: Springer Verlag.

Bohrnstedt, G.W., "Measurement," in P.H. Rossi, R.A. Wright, and A.B. Anderson (eds.), *Handbook of Survey Research*, 1983, pp. 70–122, New York: Academic Press.

Bollen, K.A., *Structural Equations with Latent Variables*, New York: John Wiley and Sons, 1989.

Bond, D.C., "Nonsampling Errors from Lab Procedures of the Wheat Objective Yield Survey," U.S. Department of Agriculture, Statistical Reporting Service, Staff Report No. AGES841218, 1985.

Borgatta, E.F., and Bohrnstedt, G.W., "Level of Measurement: Once Over Again," in G.W. Bohrnstedt, and E.F.Borgatta (eds.), 1981, *Social Measurement: Current Issues*, Beverly Hills: Sage.

Boruch, R., and Cecil, J., *Methods for Assuring Privacy and Confidentiality of Social Research Data*, Philadelphia: University of Pennsylvania Press, 1979.

Bower, G.H., and Gilligan, S.G., "Remembering Information Related to One's Self," *Journal of Research in Personality*, Vol. 13, 1977, pp. 420–432.

Bowers, N., and Horvath, F.W., "Keeping Time: An Analysis of Errors in the Measurement of Unemployment Duration," *Journal of Business and Economic Statistics*, Vol. 2, 1984, pp. 140–149.

Boyd, H.W., Jr., and Westfall, R., "Interviewers as a Source of Error in Surveys," *Journal of Marketing*, Vol. 19, 1955, pp. 311–324.

Bradburn, N.M., *The Structure of Psychological Well-Being*, Chicago: Aldine, 1969.

Bradburn, N.M., "Response Effects," in P.H. Rossi, J.D. Wright, and A.B. Anderson (eds.), *Handbook of Survey Research*, 1983, New York: Academic Press.

Bradburn, N.M., and Mason, W.M., "The Effect of Question Order on Responses," *Journal of Market Research*, Vol. 1, 1964, pp. 57–61.

Bradburn, N.M., and Miles, C., "Vague Quantifiers," *Public Opinion Quarterly*, Vol. 43, No. 1, 1979, pp. 92–101.

Bradburn, N.M, Rips, L.J., and Shevell, S.K., "Answering Autobiographical Questions: The Impact of Memory and Inference on Surveys," *Science*, Vol. 236, 1987, pp. 157–161.

Bradburn, N.M., Sudman, S., and Associates, *Improving Interviewing Methods and Questionnaire Design: Response Effects to Threatening Questions in Survey Research*, San Francisco: Jossey-Bass, 1979.

Brehm, J., and Brehm, S., *Psychological Reactance: A Theory of Freedom and Control*, New York: Academic Press, 1981.

Brenner, M., "Aspects of Conversational Structure in the Research Interview," in P. Werth, (ed.), *Conversation and Discourse: Structure and Interpretation*, 1981a, pp. 19–40, London: Croom Helm.

Brenner, M., "Patterns of Social Structure in the Research Interview," in M. Brenner, (ed.), *Social Method and Social Life*, 1981b, pp. 115–158, New York: Academic Press.

Brenner, M., "Skills in the Research Interview," in M. Argyle, (ed.), *Social Skills and Work*, 1981c, pp. 28–58, London and New York: Methuen.

Brenner, M., "Response Effects of Role-restricted Characteristics of the Interviewer," in W. Dijkstra, and J. van der Zouwen (eds.), *Response Behaviour in the Survey-interview*, 1982, pp. 131–165, London: Academic Press.

Brier, S.S., "Analysis of Contingency Tables Under Cluster Sampling," *Biometrika*, Vol. 67, No.3, 1980, pp. 591–596.

Briggs, C., *Learning How to Ask*, Cambridge: Cambridge University Press, 1986.

Brooks, C.A., and Bailar, B.A., "An Error Profile: Employment as Measured by the Current Population Survey," Statistical Policy Working Paper 3, Office of Federal Statistical Policy and Standards, U.S. Department of Commerce, 1978.

Brown, N.R., Rips, L.J., and Shevell, S.K., "The Subjective Dates of Natural Events in Very-Long Term Memory," *Cognitive Psychology*, Vol. 17, 1985, pp. 139–177.

Browne, M.W., "The Decomposition of Multitrait-Multimethod Matrices," *British Journal of Mathematical and Statistical Psychology*, Vol. 30, 1984, pp. 62–83.

Buckley, W., *Sociology and Modern Systems Theory*, Englewood Cliffs, NJ: Prentice-Hall, 1967.

Bulcock, J.W., Clifton, R., and Beebe, M., "A Language Resource Model of Science Achievement: Comparisons Between 14-year-olds in England and India," Stockholm: Institute of International Education, University of Stockholm, Research Report No. 25, 1977.

Bulcock, J.W., Clifton, R., and Beebe, M., "Reading Competency as a Predictor of Scholastic Performance: Comparison Between Industrialized and Third World Nations," paper presented at the Sixth IRA World Congress on Reading, International Reading Association, Newark, NJ, 1978.

Burdett, K., Kiefer, N.M., and Mortensen, D.T., "Earnings, Unemployment and the Allocation of Time over Time," *Review of Economic Studies*, Vol. 51, 1984, pp. 559–578.

Burdett, K., Kiefer, N.M., and Sharma, S., "Layoffs and Duration Dependence in a Model of Turnover," *Journal of Econometrics*, Vol. 28, 1985, pp. 51–69.

Burt, R.S., Fischer, M.G., and Christman, K.P., "Structures of Well-Being; Sufficient Conditions for Identification as Restricted Covariance Models," *Sociological Methods and Research*, Vol. 8, 1979, pp. 111–120.

Burt, R.S., Wiley, J.A., Minor, M.J., and Murray, J.R., "Structure of Well-Being, Form, Content, and Stability over Time," *Sociological Methods and Research*, Vol. 6, 1978, pp. 365–407.

Button, G., "Answers as Interactional Products: Two Sequential Practices Used in Interviews," *Social Psychology Quarterly*, Vol. 50, No. 2, 1987, pp. 160–171.

Byford, R.G., "Advances in Voice Data Collection," *Transactions of 44th Annual Quality Congress*, American Society for Quality Control, 1990, pp. 592–600.

Campanelli, P., Martin, E.A., and Salo, M.T., "Survey vs. Ethnographic Interviewing of Hard-to-reach Populations," paper presented at the International Survey Measurement Error Conference, Tucson, AZ, 1990.

Campbell, D.T., and Fiske, D.W., "Convergent and Discriminant Validation by the Multitrait-Multimethod Matrix," *Psychological Bulletin*, Vol. 56, No. 2, 1959, pp. 81–105.

Campbell, D.T., and O'Connell, E.J., "Method Factors in Multitrait-Multimethod Matrices: Multiplicative Rather than Additive," *Multivariate Behavioral Research*, Vol. 2, 1967, pp. 409–426.

Cannell, C.F., "Reporting of Hospitalization in the Health Interview Survey, *Vital and Health Statistics*, Series 2, No. 6, Public Health Service, Washington, DC: U.S. Government Printing Office, 1965.

Cannell, C.F., "Some Comments on Methodological Issues," paper presented at the Questionnaire Design Advisory Conference, U.S. Bureau of Labor Statistics, January 1987.

Cannell, C.F., and Fowler, F.J., "Comparison of a Self-Enumerative Procedure and a Personal Interview: A Validity Study," *Public Opinion Quarterly*, Vol. 27, 1963, pp. 250–264.

Cannell, C.F., and Fowler, F.J., "A Note on Interviewer Effect in Self-enumerative Procedures," *American Sociological Review*, Vol. 29, 1964, pp. 276.

Cannell, C.F., and Henson, R., "Incentives, Motives, and Response Bias," *Annals of Economic and Social Measurement*, Vol. 3, No. 2, 1974, pp. 307–317.

Cannell, C.F., and Kahn, R., "Interviewing," in G. Lindzey, and E. Aronson (eds.), *The Handbook of Social Psychology*, Vol 2., 1968, pp. 526–595, Reading, Mass.: Addison-Wesley.

Cannell, C.F., and Oksenberg, L., "Observation of Behavior in Telephone Interviews," in R. Groves, P. Biemer, L. Lyberg, J. Massey, W. Nicholls II, and J. Waksberg (eds.), *Telephone Survey Methodology*, 1988, pp. 475–495, New York: Wiley.

Cannell, C.F., and Robison, S., "Analysis of Individual Questions," in L. Lansing, S. Withey, and A. Wolfe, (eds.), *Working Papers on Survey Research in Poverty Areas*, Chapter 11, Ann Arbor, MI.: Institute of Social Research, The University of Michigan, 1971.

Cannell, C.F., Fowler, F.J., and Marquis, K.H., "The Influence of Interviewer and Respondent Psychological and Behavioral Variables on the Reporting in Household Interviews," *Vital and Health Statistics*, Series 2, No. 26, Washington, DC: U.S. Government Printing Office, 1968.

Cannell, C.F., Lawson, S.A., and Hauser, D.L., *A Technique for Evaluating Interviewer Performance*, Ann Arbor, MI: The University of Michigan, 1975.

Cannell, C.F., Marquis, K.H., and Laurent, A., "A Summary of Studies of Interviewing Methodology," *Vital and Health Statistics*, Series 2, No. 69, Rockville MD: National Center for Health Studies, 1977a.

Cannell, C.F., Miller, P.V., and Oksenberg, L.F., "Research on Interviewing Techniques," in S. Leinhardt (ed.), *Sociological Methodology*, 1981, pp. 389–437, San Francisco: Jossey-Bass.

Cannell, C.F., Oksenberg, L., and Converse, J.M., *Experiments in Interviewing Techniques: Field Experiments in Health Reporting: 1971–1977*, Hyattsville MD: NCHSR, 1977b.

Cannell, C.F., Groves, R., Magilavy, L., Mathiowetz, N., and Miller, P., "An Experimental Comparison of Telephone and Personal Health Interview Surveys," *Vital and Health Statistics*, Series 2, No. 106, Washington, DC: U.S. Government Printing Office, 1987.

Cannell, C.F., Kalton, G., Oksenberg, L., Bischoping, K., and Fowler, F., "New Techniques for Pretesting Survey Questions," Final Report for grant HS05616, National Center for Health Services Research, Ann Arbor, MI, University of Michigan, unpublished report, 1989.

Cantril, H., *Gauging Public Opinion*, Princeton, NJ: Princeton University Press, 1944.

Cantril, H., *The Human Dimension: Experiences in Policy Research*, New Brunswick, NJ: Rutgers University Press, 1967.

Carroll, G.R., Hannan, M.T., Tuma, N.B., and Warsavage, B., "The Impact of Measurement Error in the Analysis of Log-linear Rate Models: Monte Carlo Findings," Technical Report 69, Laboratory for Social Research, Stanford University, 1978.

Cartwright, N., *How the Laws of Physics Lie*, Oxford, New York, 1983. Two reviews by M.J. Moravcsik (*Scientometrics*, 8, 1985, pp. 141–144), and Malcolm R. Forster (*Philosophy of Science*, 52, 1985, pp. 478–480.)

Casady, R.J., Snowden, C.B., and Sirken, M.G., "A Study of Dual Frame Estimators for the National Health Interview Survey," *Proceedings of the Section on Survey Research Methods*, American Statistical Association, 1981, pp. 444–447.

Cash, W.S., and Moss, A.J., "Methodology Study for Determining the Optimum Recall Period for the Reporting of Motor Vehicle Accidental Injuries," *Proceedings of the Section on Social Statistics*, American Statistical Association, 1969, pp. 364–378.

Chase, D.R., and Harada, M., "Response Error in Self-Reported Recreation Participation," *Journal of Leisure Research*, Vol. 16, No. 4, 1984, pp. 322–329.

Chen, T.T., "Log-linear Models for Categorical Data with Misclassification and Double Sampling," *Journal of the American Statistical Association*, Vol. 74, No. 366, 1979, pp. 481–488.

Chu, A., Eisenhower, D., Hay, M., Morganstein, D., and Neter, J., "Measuring Recall Error in Self-Reported Recreational Activities," paper presented at the International Conference on Measurement Errors in Surveys, Tucson, AZ, 1990.

Chua, T.C., and Fuller, W.A., "A Model for Multinomial Response Error Applied to Labor Flows," *Journal of the American Statistical Association*, Vol. 82, 1987, pp. 46–51.

Churchill, L., *Questioning Strategies in Sociolinguistics*, Rowley, Mass.: Newbury House, 1978.

Cicourel, A., *Method and Measurement in Sociology*, New York: Free Press, 1964.

Cicourel, A., *Theory and Method in a Study of Argentine Fertility*, New York: Wiley, 1974.

Cicourel, A., "Interviews, Surveys, and the Problem of Ecological Validity," *American Sociologist*, Vol. 17, 1982, pp. 11–20.

Cicourel, A., "The Interpenetration of Communicative Contexts: Examples from Medical Encounters," *Social Psychology Quarterly*, Vol. 50, No. 2, 1987, pp. 217–226.

Cisin, I., "Community Studies on Drinking Behavior," *Annals of the New York Academy of Sciences*, Vol. 107, 1963, pp. 607–612.

Cisin, I., "The Variance of Prevalence Estimates," in J. Rittenhouse (ed.), *Developmental Papers: Attempts to Improve the Measurement of Heroin Use*, National Institute on Drug Abuse, Rockville, MD, unpublished report, 1980.

Cisin, I., and Parry, H., "Sensitivity of Survey Techniques in Measuring Illicit Drug Use," in J. Rittenhouse (ed.), *Developmental Papers: Attempts to Improve the Measurement of Heroin Use*, National Institute on Drug Abuse, Rockville, MD, unpublished report, 1980.

Clark, H., "Responding to Indirect Speech Acts," *Cognitive Psychology*, Vol. 11, 1979, pp. 430–477.

Clark, H., "Language and Language Users," in G. Lindzey, and E. Aronson, (eds.), *Handbook of Social Psychology*, Vol. 2, 1985, pp. 179–232, New York: Random House.

Clark, H., and Marshall, C., "Definite Reference and Mutual Knowledge," in A. Joshi, B. Webber, and I. Webber (eds.), *Elements of Discourse Understanding*, 1981, pp. 10–63, Cambridge: Cambridge University Press.

Clark, H.H., and Schober, M.F., "Asking Questions and Influencing Answers", in J. Tanur (ed.), *Questions about Questions: Inquiries into the Cognitive Basis of Surveys*, New York: Russell Sage, 1990.

Clark, J., and Tifft, L., "Polygraph and Interview Validation of Self-reported Deviant Behavior," *American Sociological Review*, Vol. 31, 1966, pp. 516–523.

Clayton, R.L., and Winter, D.L.S., "Voice Recognition and Voice Response Applications for Data Collection in a Federal/State Establishment Survey," *Proceedings of Military and Government Speech Tech '89 Media Dimensions*, 1989.

Cochran, W.G., *Sampling Techniques*, 3rd edition, New York: John Wiley & Sons, 1977.

Cohen, J., "A Coefficient of Agreement for Nominal Scales," *Educational and Psychological Measurement*, Vol. 210, 1960, pp. 37–46.

Cohen, J., "Weighted Kappa: Nominal Scale Agreement with Provision for Scaled Disagreement or Partial Credit," *Psychological Bulletin*, Vol. 70, 1968, pp. 213–222.

Cohen, J., and Cohen, P., *Applied Multiple Regression for the Behavioral Sciences*, Hillsdale, NJ: Lawrence Erlbaum, 1983.

Cole, M.J., and McDonald, J.W., "Bootstrap Goodness-of-Link Testing in Generalized Linear Models," in A. Decarli, B.J. Francis, R. Gilchrist, and G.U.H. Seeber (eds.), *Statistical Modelling: Proceedings of GLIM89 and 4th International Workshop on Statistical Modelling*, 1989, pp. 89–94, New York: Springer Verlag.

Collins, M., "Interviewer Variability: A Review of the Problem," *Journal of the Market Research Society*, 1980, pp. 77–95.

Comber, L.C., and Keeves, J.P., *International Studies in Evaluation I: Science Education in Nineteen Countries: An Empirical Study*, Stockholm: Almqvist & Wiksell and New York: John Wiley & Sons, 1973.

Converse, J.M., "Strong Arguments and Weak Evidence: The Open/Closed Questioning Controversy of the 1940s," *Public Opinion Quarterly*, Vol. 48, 1984, pp. 267–282.

Converse, J.M., *Survey Research in the United States*, Berkeley: University of California Press, 1987.

Converse, J.M., and Presser, S., *Survey Questions*, 1986, pp. 30–31, Beverly Hills: Sage Publications, Inc.

Converse, P., "Attitudes and Non-attitudes: Continuation of a Dialogue," in E. Tufte (ed.), *The Quantitative Analysis of Social Problems*, 1970, pp. 168–189, Reading, MA: Addison-Wesley.

Cowan, C.D., Murphy, L.R., and Weiner, J., "Effects of Supplemental Questions on Victimization Estimates from the National Crime Survey," *Proceedings of the Section on Survey Research Methods*, American Statistical Association, 1978, pp. 277–282.

Cox, B., Elliehausen, G., and Wolken, J., "Surveying Small Businesses about Their Finances," *Proceedings of the Section on Survey Research Methods*, American Statistical Association, 1989, pp. 553–557.

Cox, B., Elliehausen, G., and Wolken, J., "Measurement and Response Problems in Business Surveys," paper presented at Joint Statistical Meetings, American Statistical Association, Section on Survey Research Methods, Anaheim, CA, 1990.

Cox, D.R., *Analysis of Binary Data*, London: Chapman and Hall, 1970.

Cox, D.R., and Hinkley D.V., *Theoretical Statistics,* London: Chapman and Hall, 1974.

Cox, D.R., and Lewis, P.A.W., *The Statistical Analysis of Series of Events,* New York: John Wiley & Sons, 1966.

Cox, D.R., and Oakes, D., *Analysis of Survival Data*, London: Chapman and Hall, 1984.

Cox III, E.P., "The Optimal Number of Response Alternatives for a Scale: A Review," *Journal of Marketing Research*, Vol. XVII, 1980, pp. 407–422.

Cronbach, L.J., "Coefficient Alpha and the Internal Structure of Tests," *Psychometrika*, Vol. 16, 1951, pp. 93–96.

Crowder, R.G., *Principles of Memory and Learning*, Hillsdale, NJ: Lawrence Erlbaum Associates, 1976.

Cudeck, R., "Multiplicative Models and MTMM Matrices," *Journal of Educational Statistics*, Vol. 13, 1988, pp. 131–147.

Damon, A., and Bajema, C.J., "Age at Menarche: Accuracy of Recall After 39 Years," *Human Biology*, Vol. 46, 1974, pp. 381–384.

Damon, A., Damon, S.T., Reed, R.B., and Valadian, I., "Age at Menarche of Mothers and Daughters with a Note on Accuracy of Recall," *Human Biology*, Vol. 41, 1969, pp. 161–175.

Davis, H.L., Hoch, S.J., and Ragsdale, E.K.E., "An Anchoring and Adjustment Model of Spousal Predictions," *Journal of Consumer Research*, Vol. 13, 1986, pp. 25–37.

Davis, J.A., and Smith, T.W., *General Social Surveys, 1972–1990: Cumulative Codebook*, Chicago: NORC, 1990.

Davison, M.L., and Sharma A.R., "Parametric Statistics and Levels of Measurement," *Psychological Bulletin*, Vol. 104, 1988, pp. 137–144.

Dawes, R.M., and Smith, T., "Attitude and Opinion Measurement," in G. Lindzey, and E. Aronson (eds.), *Handbook of Social Psychology,* Vol. 2, 1985, New York: Random House.

Dean, J., and Whyte, W., "How Do You Know if the Informant is Telling the Truth?," *Human Organization*, Vol. 17, No. 2, 1958, pp. 34–38.

De Jong-Gierveld, J., "Developing and Testing a Model of Loneliness," *Journal of Personality and Social Psychology*, Vol. 53, 1987, pp. 119–128.

De Leeuw, E.D., "Data Quality in Mail, Telephone, and Face to Face Surveys: A Quantitative Review," Technical Report No. 1A, Response Effects in Surveys, Vrije Universiteit, Amsterdam, 1990a.

De Leeuw, E.D., "Data Quality in Mail, Telephone, and Face-to-face Surveys: A Mode Comparison in The Netherlands," Vrije Universiteit, Amsterdam, unpublished manuscript, 1990b.

De Leeuw, E.D., and van der Zouwen, J., "Data Quality in Telephone and Face to Face Surveys: A Comparative Meta Analysis," in R.M. Groves, P.P. Biemer, L.E. Lyberg, J.T. Massey, W.L. Nicholls II, and J. Waksberg (eds.), *Telephone Survey Methodology*, 1988, pp. 283–299, New York: Wiley.

De Leeuw, J., and Kreft, G.G., "Random Coefficient Models for Multilevel Analysis," *Journal of Educational Statistics*, Vol. 11, 1986, pp. 57–85.

DeMaio, T.J., "Approaches to Developing Questions," *Statistical Policy Working Paper 10*, Washington, DC: Statistical Policy Office, U.S. Office of Management and Budget, 1983.

DeMaio, T.J., "Social Desirability and Survey Measurement: A Review," in C.F. Turner, and E. Martin (eds.), *Surveying Subjective Phenomena*, 1984, pp. 257–282, New York: Russell Sage.

Deming, W.E., "On Errors in Surveys," *American Sociological Review*, Vol. 9, No. 4, 1944, pp. 359–369.

Deming, W.E., *Some Theory of Sampling*, New York: Wiley, 1950.

Diamond, I.D., and McDonald, J.W., "Analysis of Current-Status Data," in J. Trussell, R. Hankinson, and J. Tilton (eds.), *Demographic Applications of Event History Analysis*, 1991, Clarendon Press: Oxford.

Diamond, I.D., McDonald, J.W., and Shah, I.H., "Proportional Hazards Models for Current Status Data: Application to the Study of Differentials in Age at Weaning in Pakistan," *Demography*, Vol. 23, No. 4, 1986, pp. 607–620.

Dijkstra, W., "How Interviewer Variance Can Bias the Results of Research on Interviewer Effects," *Quality and Quantity*, Vol. 17, 1983, pp. 179–187.

Dijkstra, W., "Interviewing Style and Respondent Behaviour: An Experimental Study of the Survey-interview," *Sociological Methods and Research*, Vol. 16, 1987, pp. 309–334.

Dijkstra, W., Ormel, J., and Willige, G., "Oorzaken van interviewer variantie in survey onderzoek (Causes of Interviewer Variance in Survey Interviews)," *Mens en Maatschappij*, Vol. 54, 1979, pp. 270–291.

Dijkstra, W., van der Veen, L., and van der Zouwen, J., "A Field Experiment on Interviewer-Respondent Interaction," in M. Brenner, J. Brown, and D. Canter, (eds.), *The Research Interview: Uses and Approaches*, 1985, pp. 37–63, London: Academic Press.

Dijkstra, W., and van der Zouwen, J., "Styles of Interviewing and the Social Context of the Survey-Interview," in H. Hippler, N. Schwarz, and S. Sudman (eds.), *Social Information Processing and Survey Methodology*, 1987, pp. 200–211, New York: Springer Verlag.

Dijkstra, W., and van der Zouwen, J., "Types of Inadequate Interviewer Behavior in Survey-Interviews," in W. Saris, and I. Gallhofer, (eds.), *Sociometric Research, Volume 1, Data Collection and Scaling*, 1988, pp. 24–35, New York: St. Martin's Press.

Dillman, D.A., *Mail and Telephone Surveys: The Total Design Method*, New York: Wiley, 1978.

Dillman, D.A., "Mail and Other Self-administered Questionnaires," in P. Rossi, R.A. Wright, and B.A. Anderson (eds.), *Handbook of Survey Research*, 1983, pp. 359–377, Academic Press.

Dillman, D.A., and Mason, R.G., "The Influence of Survey Method on Question Response," paper presented at the Annual Meeting of the American Association for Public Opinion Research, Delavan, WI, 1984.

Dippo, C., Coleman, J., and Jacobs, C., "Evaluation of the 1972–73 Consumer Expenditure Survey," *Proceedings of the Section on Social Statistics*, American Statistical Association, 1977, pp. 486–491.

Duncan, O.D., *Introduction to Structural Equation Models*, New York: Academic Press, 1975.

Dutka, S., and Frankel, L.R., "Observation Techniques in Store Auditing," *Agricultural Economics Research*, Vol. XII, No. 3, 1960, U.S. Department of Agriculture.

Dutka, S., and Frankel, L.R., "Let's Not Forget About Response Error," *Modern Marketing Series*, Vol. 12, 1975, New York: Audits and Surveys.

Eisenberg, W., and McDonald, H., "Evaluating Workplace Injury and Illness Records: Testing a Procedure," *Monthly Labor Review*, April 1988, pp. 58–60.

Eisenhart, C., "Realistic Evaluation of the Precision and Accuracy of Instrument Calibration Systems," *Journal of Research of the National Bureau of Standards*, 1963, pp. 161–187.

Eisenhower, D., Frankenberry, M., Carra, J., and Stroup, C., "Assessment of Household Exposure to Common Consumer Products, Quality of the Data," paper presented at the annual meetings of American Public Health Association, Las Vegas, NV, 1986.

Ericsson, K.A., and Simon, H.A., "Verbal Reports as Data," *Psychological Review*, Vol. 87, 1980, pp. 215–251.

Espeland, M.A., and Odoroff, C.L., "Log-Linear Models for Doubly Sampled Categorical Data Fitted by the EM Algorithm," *Journal of the American Statistical Association*, Vol. 80, No. 391, 1985, pp. 663–670.

Faulkenberry, G.D., and Tortora, R.D., "A Study of Measurement Error Suitable for a Classroom Example," *The American Statistician*, Vol. 33, 1979, pp. 19–22.

Fay, R.E., "A Jack-Knifed Chi-Squared Test for Complex Samples," *Journal of the American Statistical Association*, Vol. 80, No. 394, 1985, pp. 148–157.

Fecso, R., "Methods of Collecting Agricultural Statistics in the People's Republic of China," *Proceedings of the Section on Survey Research Methods*, American Statistical Association, 1984, pp. 335–340.

Fecso, R., "Sample Survey Quality: Issues and Examples from an Agricultural Survey," *Proceedings of the Section on Survey Research Methods*, American Statistical Association, 1986, pp. 586–591.

Federal Committee on Statistical Methodology, "Quality in Establishment Surveys," *Statistical Policy Working Paper 15*, Washington, DC: Statistical Policy Office, U.S. Office of Management and Budget, 1988.

Fee, J.L.F., "Symbols in Survey Questions: Solving Problems of Multiple Wording Meanings," *Political Methodology*, Vol. 7, 1981, pp. 71–95.

Feldman, J.M., "Constructive Processes in Survey Research: Explorations in Self-generated Validity," in N. Schwarz and S. Sudman (eds.), *Context Effects in Social and Psychological Research*, 1991, New York: Springer Verlag.

Feldman, J.M., and Lynch, J.G., Jr., "Self-Generated Validity: Effects of Measurement on Belief, Attitude, Intention and Behavior," *Journal of Applied Psychology*, Vol. 73, 1988, pp. 421–435.

Feldman, J.J., Hyman, H., and Hart, C.W., "A Field Study of Interviewer Effects on the Quality of Survey Data," *Public Opinion Quarterly*, Vol. 15, 1951, pp. 734–761.

Fellegi, I.P., "Response Variance and Its Estimation," *Journal of the American Statistical Association*, Vol. 59, 1964, pp. 1016–1041.

Fellegi, I.P., "An Improved Method of Estimating the Correlated Response Variance," *Journal of the American Statistical Association*, 1974, pp. 496–501.

Fischhoff, B., "Hindsight ≠ Foresight: The Effect of Outcome Knowledge on Judgment under Uncertainty," *Journal of Experimental Psychology: Human Performance and Perception*, 1975, pp. 288–299.

Fishburne, P., "Survey Techniques for Studying Threatening Topics: A Case Study on the Use of Heroin," Ph.D. Dissertation, Department of Sociology, New York University, 1980.

Fisher, R.P., and Quigley, K.L., "Applying Cognitive Theory in Public Health Investigations: Enhancing Food Recall with the Cognitive Interview," in J. Tanur (ed.), *Questions about Questions: Inquiries into the Cognitive Basis of Surveys*, 1990, New York: Russell Sage.

Forsman, G., "Analys av återintervjudata," (Analysis of Reinterview Data), Urval, No. 19, Statistics Sweden, 1987.

Forsman, G., "Early Survey Models and Their Use in Survey Quality Work," *Journal of Official Statistics*, Vol. 5, 1989, pp. 41–55.

Forsyth, B.H., Hubbard, M.L., and Lessler, J.T., "A Method for Identifying Cognitive Properties of Survey Items," paper presented to the American Association for Public Opinion Research, Lancaster, PA, May 1990.

Fowler, F.J., "Education, Interaction, and Interviewer Performance," Ph.D. Dissertation, University of Michigan, 1966.

Fowler, F.J., "The Effect of Unclear Terms on Survey-based Estimates," in F.J. Fowler, (ed.) *Conference Proceedings, Health Survey Research Methods*, 1989, pp. 9–12, Washington, DC: National Center for Health Services Research.

Fowler, F.J., *Survey Research Methods*, Applied Social Research Methods Series, Volume 1, Sage, 1990.

Fowler, F.J., and Mangione, T.W., "The Value of Interviewer Training and Supervision," Final Report to the National Center for Health Services Research, Grant #3-R18-HS04189, 1985.

Fowler, F.J., and Mangione, T.W., "Reducing Interviewer Effects on Health Survey Data," Washington, DC: National Center for Health Services Research, 1986.

Fowler, F.J., and Mangione, T.W., *Standardized Survey Interviewing*, Newbury Park, CA: Sage Publications, 1990.

Francik, E., and Clark, H., "How to Make Requests That Overcome Obstacles to Compliance," *Journal of Memory and Language*, Vol. 24, 1985, pp. 560–568.

Francisco, C.A., Fuller, W.A., and Fecso, R., "Statistical Properties of Crop Production Estimators," *Survey Methodology*, Vol. 13, No. 1, 1987, pp. 45–62.

Frankel, L.R., "The Role of Accuracy and Precision in Sampling Surveys," in N.L. Johnson, and H. Smith, Jr. (eds.), *New Developments in Survey Sampling*, New York, Wiley, 1969. Also in *Modern Marketing Series*, Vol. 7, Audits and Surveys, New York, 1970.

Frankel, M., *Inference from Survey Samples*, Ann Arbor, Michigan: University of Michigan, 1971.

Freeman, J., and Butler, E.W, "Some Sources of Interviewer Variance in Surveys," *Public Opinion Quarterly*, Vol. 40, 1976, pp. 69–71.

Fuller, W.A., "Simple Estimators for the Mean of Skewed Populations," U.S. Bureau of the Census, unpublished report, 1970.

Fuller, W.A., "Regression Analysis for Sample Surveys," *Sankhya, C*, Vol. 37, 1975, pp. 117–132.

Fuller, W.A., *Measurement Error Models*, New York: Wiley, 1987.

Fuller, W.A., and Hidiroglou, M.A., "Regression Estimation after Correcting for Attenuation," *Journal of the American Statistical Association*, Vol. 73, 1978, pp. 99–104.

Fägerlind, I., *Formal Education and Adult Earnings: A Longitudinal Study of the Economic Benefits of Education*, Stockholm: Almqvist & Wiksell International, 1975.

Galtung, J., *Theory and Methods of Social Research*, London: George Allen and Unwin, 1967.

Garfinkel, H., and Sacks, H., "On Formal Structures of Practical Actions," in J. McKinney and E. Tiryakian (eds.), *Theoretical Sociology: Perspectives and Development*, 1970, pp. 337–366, New York: Appleton-Century-Crofts.

Gems, B., Ghosh, D., and Hitlin, R., "A Recall Experiment: Impact of Time on Recall of Recreational Fishing Trips," *Proceedings of the Section on Survey Research Methods*, American Statistical Association, 1982, pp. 372–375.

Ghosh, D.N., "Optional Recall for Discrete Events," *Proceedings of the Section on Survey Research Methods*, American Statistical Association, 1978, pp. 615–616.

Gibson, C.O., Shapiro, G.M., Murphy, L.R., and Stanko, G.J., "Interaction of Survey Questions as it Relates to Interview-Respondents Bias," *Proceedings of the Section on Survey Research Methods*, American Statistical Association, 1978.

Gieseman, R.W., "The Consumer Expenditure Survey: Quality Control by Comparative Analysis," *Monthly Labor Review*, March 1987, pp. 8–14.

Gleser, L.J., and Moore, D.S., "The Effect of Positive Dependence on Chi-squared Tests for Categorical Data," *Journal of the Royal Statistical Society*, B, Vol. 47, 1985, pp. 459–465.

Goffman, E., "Replies and Responses," in *Editors Forms of Talk*, 1981, pp. 5–77, Philadelphia: University of Pennsylvania Press.

Gold, M., "Undetected Delinquent Behavior," *Journal of Research in Crime and Delinquency*, Vol. 3, 1966, pp. 27–46.

Goldberger, A.S., and Duncan, O.D., (eds.), *Structural Equation Models in the Social Sciences*, New York: Seminar Press, 1973.

Goldstein, H., *Multilevel Models in Educational and Social Research*, London: Griffin, 1987.

Goldstein, H., "Multilevel Modelling of Survey Data," paper presented at the Institute of Statisticians conference on Survey Design, Methodology and Analysis, University of Essex, 1990.

Goldstein, H., "Nonlinear Multilevel Models, with an Application to Discrete Response Data," *Biometrika*, Vol. 78, 1991, pp. 45–51.

Gong, G., Whittemore, A.S., and Grosser, S., "Censored Survival Data with Misclassified Covariates: A Case Study of Breast Cancer Mortality," *Journal of the American Statistical Association*, Vol. 85, 1990, pp. 20–27.

Goudy, W.J., and Potter, H.R., "Interview Rapport: Demise of a Concept," *Public Opinion Quarterly*, Vol. 39, 1975–1976, pp. 530–543.

Gove, W., and Geerken, M., "Response Bias in Surveys of Mental Health: An Empirical Investigation," *American Journal of Sociology*, Vol. 82, 1977, pp. 1289–1317.

Gower, A., and Zylstra, P., "The Use of Qualitative Methods in the Design of a Business Survey," paper presented at the International Conference on Measurement Errors in Surveys, Tucson, AZ, 1990.

Gray, P.G., "Examples of Interviewer Variability Taken from Two Sample Surveys," *Applied Statistics*, Vol. 5, 1956, pp. 73–85.

Greenberg, B., and Petkunas, T., "An Evaluation of Edit and Imputation Procedures Used in the 1982 Economic Censuses in Business Division," 1982 Economic Censuses and Census of Government Evaluation Studies, U.S. Department of Commerce, 1987, pp. 85–98.

Greenberg, B., and Surdi, R., "A Flexible and Interactive Edit and Imputation System for Ratio Edits," *Proceedings of the Section on Survey Research Methods*, American Statistical Association, 1984, pp. 421–426.

Greenberg, B., Abul-Ela, A., Simmons, W., and Horvitz, D., "The Unrelated Questions Randomized Response Model Theoretical Framework," *Journal of the American Statistical Association*, Vol. 64, 1969, pp. 520–539.

Grice, H.P., "Logic and Conversation," in P. Cole and J.L. Morgan (eds.), *Syntax and Semantics: Volume 3, Speech Acts,* 1975, New York: Academic Press.

Grootaert, C., "The Use of Multiple Diaries in a Household Expenditure Survey in Hong Kong," *Journal of the American Statistical Association*, Vol. 81, No. 396, 1986, pp. 938–944.

Groves, R.M., "Actors and Questions in Telephone and Personal Interview Surveys," *Public Opinion Quarterly*, Vol. 43, 1979, pp. 190–205.

Groves, R.M., *Survey Errors and Survey Costs*, New York: John Wiley, 1989.

Groves, R.M., and Fultz, N., "Gender Effects Among Telephone Interviewers in a Survey of Economic Attributes," *Sociological Methods and Research*, Vol. 14, No. 1, 1985, pp. 31–52.

Groves, R.M., and Kahn, R.L., *Surveys by Telephone: A National Comparison with Personal Interviews*, Academic Press, 1979.

Groves, R.M., and Lepkowski, J.M., "Dual Frame, Mixed Mode Survey Designs," *Journal of Official Statistics*, 1985, pp. 263–286.

Groves, R.M., and Magilavy, L.J., "Estimates of Interviewer Variance in Telephone Surveys," *Proceedings of the Section on Survey Research Methods,* American Statistical Association, 1980, pp. 622–627.

Groves, R.M., and Magilavy, L.J., "Measuring and Explaining Interviewer Effects," *Public Opinion Quarterly*, Vol. 50, 1986, pp. 251–256, Reprinted in E. Singer and S. Presser, *Survey Research Methods*, Chicago: University of Chicago Press, 1989.

Groves, R.M., and Nicholls II, W.L., "The Status of Computer-Assisted Telephone Interviewing: Part II-Data Quality Issues," *Journal of Official Statistics*, 1986, pp. 117–134.

Groves, R.M., Biemer, P.P., Lyberg, L.E., Massey, J.T., Nicholls II, W.L., and Waksberg, J. (eds.), *Telephone Survey Methodology*, New York: Wiley, 1988.

Groves, R.M., Cannell, C., Maynard, D., Polanyi, L., Salo, M., and Schaeffer, N., "Improving Data Quality from the Survey Interview Through Studies of Interaction," Ann Arbor, MI, University of Michigan, unpublished report, 1990.

Groves, R.M., Magilavy, L.J., and Mathiowetz, N.A., "The Process of Interviewer Variability Evidence from Telephone Surveys," *Proceedings of the Section on Survey Research Methods*, American Statistical Association, 1981, pp. 438–443.

Hagenaars, J.A., and Heinen, T.G., "Effects of Role-Independent Interviewer Characteristics on Responses," in W. Dijkstra, and J. van der Zouwen, (eds.), *Response Behavior in the Survey Interview*, 1982, pp. 91–130, London: Academic Press.

Handlin, O., *Truth in History*, Cambridge, MA: Belknap, 1979.

Hansen, M.H., Hurwitz, W.N., and Bershad, M.A., "Measurement Errors in Censuses and Surveys," *Bulletin of the International Statistical Institute*, Vol. 38, No. 2, 1961, pp. 359–374.

Hansen, M.H., Hurwitz, W.N., and Madow, W.G., *Survey Methods and Theory, Vol. I: Methods and Applications, Vol. II: Theory*, New York: John Wiley and Sons, 1953.

Hansen, M.H., Hurwitz, W.N., and Pritzker, L., "The Estimation and Interpenetration of Gross Differences and the Simple Response Variance," in C.R. Rao (ed.), *Contributions to Statistics*, 1964, pp. 111–136, Calcutta: Pergamon Press Ltd.

Hansen, M.H., Hurwitz, W.N., Marks, E.S., and Mauldin, W.P., "Response Errors in Surveys," *Journal of the American Statistical Association*, Vol. 46, 1951, pp. 147–190.

Hanson, R.H., and Marks, E.S., "Influence of the Interviewer on the Accuracy of Survey Results," *Journal of the American Statistical Association*, Vol. 53, 1958, pp. 635–655.

Hanuschak, G., "Initial Report on the Quality of Agricultural Survey Programs," in *Seminar on the Quality of Federal Data, Statistical Policy Working Paper 20*, National Technical Information Service, Springfield, VA., 1991.

Hartley, H.O., and Rao, J.N.K., "Estimation of Nonsampling Variance Components in Sample Surveys", in N.K. Namboodiri (ed.), *Survey Sampling and Measurement*, 1978, pp. 35–43, New York: Academic Press.

Hasabelnaby, N.A., Ware, J.H., and Fuller, W.A., "Indoor Air Pollution and Pulmonary Performance: Investigating Errors in Exposure Assessment," *Statistics in Medicine*, Vol. 8, 1989, pp. 1109–1126.

Hatchet, S., and Schuman, H., "Race of Interviewer Effects Upon White Respondents," *Public Opinion Quarterly*, Vol. 39, 1975, pp. 523–528.

Hauck, W.W., and Donner, A., "Wald's Test as Applied to Hypotheses in Logit Analysis," *Journal of the American Statistical Association*, Vol. 72, No. 360, 1977, pp. 851–853.

Hauser, J., "Testing the Accuracy, Usefulness, and Significance of Probabilistic Choice Models: An Information-Theoretic Approach," *Operations Research*, Vol. 26, 1978, pp. 406–421.

Hays, W.L., *Statistics for the Social Sciences*, 2nd edition, London: Holt, Rinehart and Winston, 1973.

Heckman, J.J., and Singer, B., "Economic Duration Analysis," *Journal of Econometrics*, Vol. 24, 1984, pp. 63–132.

Heckman, J.J., and Singer, B., *Longitudinal Analysis of Labor Market Data*, Cambridge: Cambridge University Press, 1985.

Heise, D., "Separating Reliability and Stability in Test-retest Correlation," *American Sociological Review*, 34, 1969, pp. 93–101.

Heise, D., and Bohrnstedt, G.W., "Validity, Invalidity and Reliability," in E.F. Borgatta, and G.W. Bohrnstedt (eds.), *Sociological Methodology*, 1970, San Francisco: Jossey Bass.

Hendricks, W.A., "Theoretical Aspects of the Use of the Crop Meter," U.S. Department of Agriculture, Agricultural Marketing Service, 1942.

Henson, R.M., *Effects of Instructions and Verbal Modelling on Health Information Reporting*, Ann Arbor, MI: Survey Research Center, University of Michigan, 1973.

Henson, R.M., Cannell, C.F., and Lawson, S., "Effects of Interviewer Style on Quality of Reporting in a Survey Interview," *Journal of Psychology*, Vol. 93, 1976, pp. 221–227.

Heritage, J., "A Change-of-State Token and Aspects of Its Sequential Placement," in J. Atkinson, and J. Heritage (eds.), *Structure of Social Action*, 1984, pp. 299–345, Cambridge: Cambridge University Press.

Heritage, J., "Recent Developments in Conversation Analysis," *Sociolinguistics*, Vol. 15, No. 1, 1985, pp. 1–19.

Herniter, J., "An Entropy Model of Brand Purchase Behavior," *Journal of Marketing Research*, 1973, pp. 361–375.

Hippler, H.J., and Schwarz N., "Response Effects in Surveys," in H.J. Hippler, N. Schwarz, and S. Sudman (eds.), *Social Information Processing and Survey Methodology*, 1987, New York: Springer Verlag.

Hippler, H.J., and Schwarz, N., "No Opinion Filters: A Cognitive Perspective," *International Journal of Public Opinion Research*, Vol. 1, 1989, pp. 77–87.

Hippler, H.J., Schwarz, N., and Noelle-Neumann, E., "Response Order Effects and Dichotomous Questions: The Impact of Administration Mode," paper presented to the American Association for Public Opinion Research, St. Petersburg, FL, 1989.

Hippler, H.J., Schwarz, N., and Noelle-Neumann, E., "Response Order Effects in Survey Measurement: Cognitively Elaboration and the Likelihood of Endorsement," paper presented to the American Association for Public Opinion Research, Lancaster, PA, May 1990.

Hippler, H.J., Schwarz, N., and Sudman, S. (eds.), *Social Information Processing and Survey Methodology*, 1987, New York: Springer Verlag.

Hoch, S.J., "Perceived Consensus and Predictive Accuracy: The Pros and Cons of Projection," *Journal of Personality and Social Psychology*, Vol. 53, No. 2, 1987, pp. 221–234.

Hochberg, Y., "On the Use of Double Sampling Schemes in Analyzing Categorical Data with Misclassification Errors," *Journal of the American Statistical Association*, Vol. 72, No. 360, 1977, pp. 914–921.

Hochstim, J.R., "A Critical Comparison of Three Strategies of Collecting Data from Households," *Journal of the American Statistical Association*, Vol. 62, 1967, pp. 976–989.

Hoem, J.M., "Weighting, Misclassification, and Other Issues in the Analysis of Survey Samples of Life Histories," in J.J. Heckman, and B. Singer (eds.), *Longitudinal Analysis of Labor Market Data*, 1985, pp. 249–293, Cambridge: Cambridge University Press.

Hormuth, S.E., and Brückner, E., "Telefoninterviews in Sozialforschung und Sozialpsychologie, Ausgewhälte Probleme der Stichprobengewinnung, Kontaktierung und Versuchsplanung (Telephone Interviews in Social Research and Social Psychology: Selected Problems in Sampling, Contacting the Respondent, and Study Design)," *Kölner Zeitschrift für Soziologie und Sozialpsychologie*, Vol. 37, 1985, pp. 526–545.

Horrigan, M.W., "Time Spent Unemployed: A New Look at Data from the CPS," *Monthly Labor Review*, Vol. 110, No. 7, 1987, pp. 3–15.

Horvath, F.W., "Forgotten Unemployment: Recall Bias in Retrospective Data," *Monthly Labor Review*, Vol. 105, No. 3, 1982, pp. 40–43.

Horvitz, D.G., and Folsom, R., "Methodological Issues in Medical Care Expenditure Surveys," *Proceedings of the Section on Survey Research Methods*, American Statistical Association, 1980, pp. 21–29.

Horvitz, D.G., Shah, B., and Simmons, W., "The Unrelated Question Randomized Response Model," *Proceedings of the Section on Social Statistics*, American Statistical Association, 1967, pp. 65–72.

Houseman, E.E., and Lipstein, B., "Observations and Audit Techniques for Measuring Retail Sales," *Agricultural Economics Research*, Vol. XII, No. 3, 1960, U.S. Department of Agriculture.

Hox, J.J., "Zelf-evaluatie, assertiviteit, en stemming (Self- evaluation, Assertiveness and Mood)," Amsterdam, University of Amsterdam, unpublished report, 1978.

Hox, J.J., "Het gebruik van hulptheorieen bij operationalisering (Using Auxiliary Theories for Operationalization), Ph.D. dissertation, 1986, University of Amsterdam, Amsterdam.

Huber, P.W., "Pathological Science in Court," *Daedalus*, Vol. 119, No. 4, 1990, pp. 97–118.

Hubbard, M.L., Caspar, R.A., and Lessler, J.T., "Respondent Reactions to Item Count Lists and Randomized Response," *Proceedings of the Section on Survey Research Methods*, American Statistical Association, 1989, pp. 544–549.

Huddleston, H.F., "Sampling Techniques for Measuring and Forecasting Crop Yields,' Washington, D.C., Economics, Statistics, and Cooperative Service, U.S. Department of Agriculture, 1978.

Husén, T. (ed.), *International Study of Achievement in Mathematics: A Comparison of Twelve Countries, Volumes I and II*, Stockholm: Almqvist & Wiksell and New York: John Wiley & Sons, 1967.

Hyman, H., "Do They Tell the Truth?" *Public Opinion Quarterly*, Vol. 8, 1944, pp. 557–559.

Hyman, H., Cobb, W.J., Feldman, J., Hart, C.W., and Stember, C., *Interviewing in Social Research*, Chicago: University of Chicago Press, 1954. Reprinted in 1975.

Hymes, D.H., "Models of the Interaction of Language and Social Life," in J. Gumperz, and D. Hymes (eds.), *Directions in Sociolinguistics: The Ethnography of Communication*, 1972, pp. 35–71, New York: Holt, Rinehart, and Winston.

Hymes, D.H., *Foundations in Sociolinguistics: An Ethnographic Approach*, Philadelphia: University of Pennsylvania Press, 1974.

Irvine, J., "Formality and Informality in Communicative Events," *American Anthropologist*, Vol. 81, No. 4, 1979, pp. 773–790.

Irwin, J.O., Cochran, W.G., and Wishart, J., "Crop Estimation and Its Relation to Agricultural Meteorology," *Journal of Royal Statistical Society, Supplement*, Vol. V, No. I, 1938, pp. 1–45.

Israel, G.D., and Taylor, C.L., "Can Response Order Bias Evaluation?" *Evaluation and Program Planning*, Vol. 13, No. 4, 1990, pp. 365–371.

Jabine, T.B., King, K.E., and Petroni, R.J., *Survey of Income and Program Participation: Quality Profile*, U.S. Bureau of the Census, Washington, DC, 1990.

Jabine, T.B., Straf, M.L., Tanur, J.M., and Tourangeau, R. (eds.), *Cognitive Aspects of Survey Methodology: Building a Bridge Between Disciplines*, Washington, DC: National Academic Press, 1984.

Jacobs, E., Jacobs, C., and Dippo, C., "The U.S. Consumer Expenditure Survey," *Bulletin of the International Statistical Institute*, Paris, 1989, pp.123–142.

James, I.R., "Unemployment Duration: Modelling and Estimation with Particular Reference to the Australian Longitudinal Survey," *Australian Journal of Statistics*, Vol. 31A, 1989, pp. 197–212.

Jefferson, G., "Side Sequences," in D. Sudnow (ed.), *Studies in Social Interaction*, 1972, pp. 294–338, New York: Free Press.

Jefferson, G., "A Case of Precision Timing in Ordinary Conversation: Over-Lapped Tag-Positioned Address Terms in Closing Sequences," *Semiotica*, Vol. 9, 1973, pp. 47–96.

Jefferson, G., "Error Correction as an Interactional Resource," *Language in Society*, Vol. 3, 1974, pp. 181–199.

Jefferson, G., "A Technique for Inviting Laughter and Its Subsequent Acceptance/Declination," in G. Psathas (ed.), *Everyday Language: Studies in Ethnomethodology*, 1979, pp. 79–96, New York: Irvington.

Jefferson, G., "On 'Troubles-Premonitory' Response to Inquiry," *Sociological Inquiry*, Vol. 50, Nos. 3–4, 1980, pp. 153–185.

Jefferson, G., "On the Organization of Laughter in Talk about Troubles," in J. Atkinson, and J. Heritage (eds.), *Structure of Social Action*, 1984a, pp. 346–369, Cambridge: Cambridge University Press.

Jefferson, G., "Notes on a Systematic Deployment of the Acknowledgement Tokens 'Yeah' and 'Mm hmm'," *Papers in Linguistics*, Vol. 17, 1984b, pp. 197–216.

Jefferson, G., "On the Interactional Unpackaging of a 'Gloss'," *Language in Society*, Vol. 14, 1985, pp. 435–466.

Jefferson, G., "Notes on 'Latency' in Overlap Onset," *Human Studies*, Vol. 9, 1986, pp. 153–183.

Jefferson, G., Sacks, H., and Schegloff, E., "Notes on Laughter in the Pursuit of Intimacy,"in G. Button, and J. Lee (eds.), *Talk and Social Structure*, 1987, pp. 152–205, Clevedon, Avon, England: Multilingual Matters.

Jobe, J.B, and Mingay, D.J., "Cognitive Research Improves Questionnaires," *American Journal of Public Health*, Vol. 79, No. 8, 1989, pp. 1053–1055.

Jobe, J.B., and Mingay, D.J., "Cognitive Laboratory Approach to Designing Questionnaires for Surveys of the Elderly," *Public Health Reports*, Vol. 105, 1990, pp. 518–524.

Johnson, N.L., and Kotz, S., *Discrete Distributions*, New York: Wiley, 1969.

Johnson, N.L., and Kotz, S., *Distributions in Statistics: Continuous Univariate Distributions-2*, New York: Houghton Mifflin, 1970.

Johnson, R.A., "Measurement of Hispanic Ethnicity in the U.S. Census: An Evaluation Based on Latent Class Analysis," *Journal of the American Statistical Association*, Vol. 85, 1990, pp. 58–65.

Johnson, R.A., and Wichern, D.W., *Applied Multivariate Statistical Analysis*, Second Edition, Englewood Cliff: Prentice Hall, 1982.

Johnson, R.A., and Woltman, H.F., "Evaluating Census Data Quality Using Intensive Reinterviews: A Comparison of U.S. Census Bureau Methods and Rasch Methods," in C.C. Clogg (ed.), *Sociological Methodology*, 1987, pp. 185–204, Washington, D.C.: American Sociological Association.

Johnston, L.D., and Bachman, J.G., *Monitoring the Future: Questionnaire Responses for the Nation's High School Seniors-1975*, Ann Arbor, MI: Institute for Social Research, 1980.

Jones, E., Farina, A., Hastorf, A., Markus, H., Miller, D., and Scott, R., *Social Stigma: The Psychology of Marked Relationships*, New York: W. H. Freeman & Co., 1984.

Joseph, J.G., Emons, C.A., Kessler, R.C., Wortman, C.B., O'Brien, K., Hocker, W.T., and Schaefer, C., "Coping with the Threat of AIDS: An Approach to Psychosocial Assessment," *American Psychologist*, Vol. 34, 1984, pp. 1297–1302.

Juster, T.F., "Response Errors in the Measurement of Time Use," *Journal of the American Statistical Association*, Vol. 81, No. 394, 1986, pp. 390–402.

Jöreskog, K.G., "Statistical Analysis of Congeneric Tests," *Psychometrika*, Vol. 36; 1971, pp. 109–133.

Jöreskog, K.G., "Analyzing Psychological Data by Structural Analysis of Covariance Matrices," in C. Atkinson, H.Krantz, R.D.Luce, and P.Suppes (eds.), *Comtemporary Developents in Mathematical Psychology*, 1973, pp. 1–56, San Francisco:Freeman.

Jöreskog, K.G., and Sörbom, D., *LISREL 7: A Guide to the Program and Applications*, SPSS International B.V., The Netherlands, 1989.

Kahneman, D., and Tversky, A., "Subjective Probability: A Judgment of Representativeness," *Cognitive Psychology*, Vol. 3, 1971, pp. 450–454.

Kahneman, D., and Tversky, A., "On the Psychology of Prediction," *Psychological Review*, Vol. 80, 1973, pp. 237–251.

Kalton, G., *Compensating for Missing Survey Data*, Ann Arbor, MI, Institute for Social Research, The University of Michigan, 1983.

Kalton, G., Kasprzyk, D., and McMillen, D.B., "Nonsampling Errors in Panel Surveys," in D. Kasprzyk, G.J. Duncan, G. Kalton, and M.P. Singh (eds.), *Panel Surveys*, 1989, pp. 249-270, New York: Wiley.

Kane, E., and Schuman, H., "Open Survey Questions as Measures of Personal Concern with Issues: A Reanalysis of Stouffer's Communism, Conformity, and Civil Liberties," in P. Marsden (ed.), *Sociological Methodology*, 1991, Oxford: Basil Blackwell.

Kaplan, S., "Appropriate Responses to Inappropriate Questions," in A. Joshi, B. Webber, and I. Sag (eds.), *Elements of Discourse Understanding*, 1981, pp. 127–144, Cambridge: Cambridge University Press.

Katti, S.K., and Gurland, J., "The Poisson Pascal Distribution," *Biometrics*, Vol. 17, 1961, pp. 527–538.

Katz, D., and Kahn, R., *The Social Psychology of Organizations*, New York: John Wiley and Sons, Inc., 1966.

Kelley, J., "Causal Chain Models for the Socioeconomic Career," *American Sociological Review*, Vol. 38, 1973, pp. 481–493.

Kemsley, W.F.F., "The Household Expenditure Enquiry of the Ministry of Labour: Variability in the 1953–54 Enquiry," *Applied Statistics*, Vol. 10, No. 3, 1961, pp. 117–135.

Kemsley, W.F.F., and Nicholson, J.L., "Some Experiments in Methods of Conducting Family Expenditure Surveys," *Journal of the Royal Statistical Society*, Vol. 123, No. 307, 1960, pp. 307–328.

Kemsley, W.F.F., Redpath, R.U., and Holmes, M., *Family Expenditure Survey Handbook*, Office of Population and Surveys, London, Her Majesty's Stationary Office, London, 1980.

Kiefer, N.M., and Neumann, G.R., "An Empirical Job-Search Model, with a Test of the Constant Reservation Wage Hypothesis," *Journal of Political Economy*, Vol. 87, 1979, pp. 89–107.

Kiefer, N.M., Lundberg, S.J., and Neumann, G.R., "How Long is a Spell of Unemployment? Illusions and Biases in the Use of CPS Data," *Journal of Business and Economic Statistics*, Vol. 3, 1985, pp. 118–128.

Kilss, B., and Alvey, W. (eds.), *Statistical Uses of Administrative Records: Recent Research and Present Prospects*, Statistics of Income Division, Internal Revenue Service, Washington, DC, 1984.

Kinsey, A., Pomeroy, W., and Martin, C., *Sexual Behavior in the Human Male*, Philadelphia: W.B. Saunders, 1948.

Kiregyera, B., "Types and Some Causes of Nonsampling Errors in Household Surveys in Africa," *Journal of Official Statistics*, Vol. 3, No. 4, 1987, pp. 349–358.

Kish, L., "Studies of Interviewer Variance for Attitudinal Variables," *Journal of the American Statistical Association*, Vol. 57, 1962, pp. 92–115.

Kish, L., *Survey Sampling*, New York: John Wiley and Sons, 1965.

Knight, K., and McDaniel, R., *Organizations: An Information Systems Approach*, Belmont, CA: Wadsworth Publishing Company, Inc., 1979.

Koch, G.G., "An Alternative Approach to Multivariate Response Error Models for Sample Survey Data with Applications to Estimators Involving Subclass Means," *Journal of the American Statistical Association*, Vol. 68, 1973, pp. 906–913.

Koch, G.G., and Lemeshow, S., "An Application of Multivariate Analysis in Complex Surveys," *Journal of the American Statistical Association*, Vol. 67, 1972, pp. 780–782.

Koch, G.G., Freeman, D.H., and Freeman, J.L., "Strategies in the Multivariate Analysis of Data from Complex Surveys," *International Statistical Review*, Vol. 93, No. 1, 1975, pp. 59–78.

Korn, E.L., "Hierarchical Log-Linear Models Not Preserved by Classification Errors," *Journal of the American Statistical Association*, Vol. 76, No. 373, 1981, pp. 110–113.

Kott, P.S., "Variance Estimation When a First Phase Area Sample Is Restratified," *Survey Methodology*, Vol. 16, No. 1, 1990, pp. 99–104.

Krech, D., and Crutchfield, R.S., *Theory and Problems of Social Psychology,* New York: McGraw Hill, 1948.

Kreft, G.G., DeLeeuw, J., and Kim, K.-S., "Comparing Four Different Statistical Packages for Hiearchical Linear Regression: Genmod, HLM, ML2, and VARCL," Los Angeles: Center for Research on Evaluation, Standards, and Student Testing, UCLA, CSE Report, 1990.

Krosnick, J.A., "The Impact of Cognitive Sophistication and Attitude Importance on Response and Question Order Effects," in N. Schwarz, and S. Sudman (eds.), *Context Effects in Social and Psychological Research,* 1991a, New York: Springer Verlag.

Krosnick, J.A., "Cognitive Sophistication as a Requisite of Question and Response Order Effects in Attitude Measurement," in N. Schwarz, and S. Sudman (eds.), *Context Effects in Social and Psychological Research*, 1991b, New York: Springer Verlag.

Krosnick, J.A., "Response Strategies for Coping with the Cognitive Demands of Attitude Measures in Surveys," *Applied Cognitive Psychology*, Vol. 5, 1991c, pp. 213–236.

Krosnick, J.A., and Alwin, D.F., "An Evaluation of a Cognitive Theory of Response-Order Effects in Survey Measurement," *Public Opinion Quarterly*, Vol. 51, 1987, pp. 201–219.

Krosnick, J.A., and Berent, M.K., "The Impact of Verbal Labeling of Response Alternatives and Branching on Attitude Measurement Reliability in Surveys," paper presented at the Annual Meeting of the American Association for Public Opinion Research, Lancaster, PA, 1990.

Kuiper, N.A., and Rogers, T.B., "Encoding of Personal Information: Self-Other Differences," *Journal of Personality and Social Psychology*, Vol. 37, 1979, pp. 499–514.

Labovitz, S., "The Assignment of Numbers to Rank Order Categories," *American Sociological Review*, Vol. 35, 1970, pp. 515–525.

Lachman, R., Lachman, J.L., and Butterfield, E.C., *Cognitive Psychology and Information Processing: An Introduction*, Hillsdale, NJ: Erlbaum, 1979.

Lamale, H.H., *Study of Consumer Expenditures, Incomes, and Savings, Methodology of the Survey of Consumer Expenditures in 1950*, Wharton School of Finance and Commerce, University of Pennsylvania, 1959.

Lancaster, A., "Econometric Methods for the Duration of Unemployment," *Econometrica*, Vol. 47, 1979, pp. 939–956.

Landis, J.R., and Koch, G., "A Review of Statistical Methods in the Analysis of Data Arising from Observer Reliability Studies," *Statistica Neerlandica*, Vol. 29, 1976, pp. 101–123 and 151–161.

Langeheine, R., and van de Pol, F., "A Unifying Framework for Markov Modeling in Discrete Space and Discrete Time," *Sociological Methods and Research*, Vol. 18, 1990, pp. 416–441.

Larsen, S.F., and Plunkett K., "Remembering Experienced and Reported Events," *Applied Cognitive Psychology*, Vol. 1, 1987, pp. 15–26.

LaVange, L.M., and Folsom, R.E., "Development of NCS Error Adjustment Models," Research Triangle Institute, unpublished report, 1985.

Lawler, E., and Rhode, J., *Information and Control in Organizations*, Pacific Palisades, CA: Goodyear Publishing Company, 1976.

Lawless, J.F., *Statistical Models and Methods for Lifetime Data*, New York: Wiley, 1982.

Lazarsfeld, P.F., "The Controversy Over Detailed Interviews — An Offer for Negotiation," *Public Opinion Quarterly*, Vol. 8, 1944, pp. 38–60.

Lee, E.S., Forthofer, R.N., and Lorimor, R.J., *Analyzing Complex Survey Data*, Newbury Park, CA: Sage, 1989.

Lemaitre, G., "A Look at Response Errors in the Labour Force Survey," *Canadian Journal of Statistics*, Vol. 16, 1988, pp. 127–141.

Leonard, M., "CPS Reinterview Quality Control Results for 1988," U.S. Bureau of the Census, Washington, D.C., unpublished report, 1990.

Lessler, J.T., "Measurement Errors in Surveys," in C.F. Turner, and E. Martin (eds.), *Surveying Subjective Phenomena*, Vol. 2, 1984, pp. 405–440, New York: Russel Sage Foundation.

Lessler, J.T., "Use of Laboratory Methods and Cognitive Science for the Design and Testing of Questionnaires," Statistics Sweden, unpublished research report, 1987.

Lessler, J.T., and Sirken, M.G., "Laboratory-Based Research on the Cognitive Aspects of Survey Methodology," *Milbank Memorial Fund Quarterly/Health and Society*, Vol. 63, 1985, pp. 565–581.

Lessler, J.T., Kalsbeek, W.D., and Folsom, R.E., "A Taxonomy of Survey Errors, Final Report," Research Triangle Institute project 255U-1791–03F, Research Triangle Institute, 1981.

Lessler, J.T., Tourangeau, R., and Sarter, W., "Questionnaire Design in the Cognitive Research Laboratory," *Vital and Health Statistics*, Series 6, No. 1, 1989, Washington DC: National Center for Health Statistics.

Leyes, J., "An Administrative Record Paradigm: A Canadian Experience," *Seminar on the Quality of Federal Data, Statistical Policy Working Paper 20*, National Technical Information Service, Springfield, VA., 1991.

Liang, K., and Zeger S.L., "Longitudinal Data Analyses Using Generalized Linear Models," *Biometrika*, Vol. 73, 1986, pp. 13–22.

Linton, M., "Transformation of Memory in Everyday Life," in U. Neisser (ed.), *Memory Observed*, 1982, pp. 77–92, San Francisco: W. H. Freeman and Company.

Little, R., and Rubin, D.B., *Statistical Analysis with Missing Data*, Wiley and Sons, 1987.

Little, R., and Smith, P., "Editing and Imputation for Quantitative Survey Data," *Journal of the American Statistical Association*, Vol. 82, No. 397, 1987, pp 58–68.

Livson, N., and McNeill, D., "The Accuracy of Recalled Age of Menarche," *Human Biology*, Vol. 34, 1962, pp. 218–221.

Locander, W., Sudman, S., and Bradburn, N., "An Investigation of Interview Method, Threat, and Response Distortion," *Journal of the American Statistical Association*, Vol. 71, No. 354, 1976, pp. 269–275.

Lodge, M., *Magnitude Estimation*, New York: Sage, 1982.

Loebl, A.D., "Accuracy and Relevance and the Quality of Data," in G.E. Liepins and V.R.R. Uppuluri (eds.), *Data Quality Control: Theory and Pragmatics*, 1990, pp. 131–141, New York: Marcel Dekker.

Lofland, J., *Deviance and Identity*, Englewood Cliffs, NJ: Prentice-Hall, 1969.

Loftus, E.F., "Leading Questions and the Eyewitness Report," *Cognitive Psychology*, Vol. 7, 1975, pp. 560–572.

Loftus, E.F., "Shifting Human Color Memory," *Memory and Cognition*, Vol. 5, 1977, pp. 696–699.

Loftus, E.F., *Memory: Surprising New Insights into How We Remember and Why We Forget*, Reading, MA: Addison-Wesley, 1980.

Loftus, E.F., and Fathi, D.C., "Retrieving Multiple Autobiographical Memories," *Social Cognition*, Vol. 3, 1985, pp. 280–295.

Loftus, E.F., and Marburger, W., "Since the Eruption of Mt. St. Helens, Has Anyone Beaten You Up? Improving the Accuracy of Retrospective Reports with Landmark Events," *Memory and Cognition*, Vol. 11, 1983, pp. 114–120.

Loftus, E., Smith, K.D., Linger, M.R., and Fiedler, J., "Memory and Mismemory for Health Events," in J. Tanur (ed.), *Questions about Questions: Inquiries into the Cognitive Basis of Surveys*, 1990, New York: Russell Sage.

Longford, N.T., "A Quasi-likelihood Adaptation for Variance Component Analysis," *Proceedings of the Section on Statistical Computing*, American Statistical Association, 1988, 137–142.

Longford, N.T., *VARCL-manual*, Princeton, NJ: ETS, 1990.

Lord, F.M., "Large-sample Covariance Analysis When the Control Variable is Fallible," *Journal of the American Statistical Association*, Vol. 55, 1960, pp. 307–321.

Lord, F., and Novick, M.R., *Statistical Theories of Mental Test Scores*, Reading, MA, Addison-Wesley, 1968.

Lyberg, I., "Sampling, Nonresponse, and Measurement Isses in the 1984–85 Swedish Time Budget Survey," *Proceedings of the Fifth Annual Research Conference*, U.S. Bureau of the Census, 1989, pp. 210–231.

Lyberg, L., "Nonresponse Problems in the 1978 Swedish Household Expenditure Survey," *Series on Methodological Problems in Statistics on Individuals and Households*, No. 12, 1980, Statistics Sweden (in Swedish).

Lynch, L.M., "State Dependency in Youth Unemployment: A Lost Generation," *Journal of Econometrics*, Vol. 28, 1985, pp. 71–84.

Mahalanobis, P.C., "Recent Experiments in Statistical Sampling in the Indian Statistical Institute," *Journal of the Royal Statistical Society*, Vol. 109, 1946, pp. 325–378.

Mahalanobis, P.C., and Sen, S.B., "On Some Aspects of the Indian National Sample Survey," Bulletin of the International Statistical Institute, Vol. 34, 1953, pp. 5–14.

Majulan Singapura, "Report on the Household Expenditure Survey 1982/83," Singapore Department of Statistics, March 1985, p. 10.

Marlaire, C., and Maynard, D., "Standardized Testing as an Interactional Phenomenon," *Sociology of Education*, Vol. 63, 1990, pp. 83–101.

Marquis, K.H., "Effects of Race, Residence, and Selection of Respondent on the Conduct of the Interview," in J. Lansing, S. Withey, and A. Wolfe (eds.), *Working Papers on Survey Research in Poverty Areas*, 1971a, Ann Arbor, MI: Institute for Social Research, The University of Michigan.

Marquis, K.H., "Purpose and Procedure of the Tape Recording Analysis," in J. Lansing, S. Withey, and A. Wolfe (eds.), *Working Papers on Survey Research in Poverty Areas*, 1971b, Ann Arbor, MI: Institute for Social Research, The University of Michigan.

Marquis, K.H., and Cannell, C.F., "A Study of Interviewer-Respondent Interaction in the Urban Employment Survey," Ann Arbor: Survey Research Center, University of Michigan, research report, 1969.

Marquis, K.H., and Cannell, C.F., "Effect of Some Experimental Techniques on Reporting in the Health Interview," *Vital and Health Statistics*, Series 2, No. 41, 1971, Washington, D.C.: U.S. Government Printing Office.

Marquis, K.H., Cannell, C.F., and Laurent, A., "Reporting Health Events in Household Interviews: Effects of Reinforcement, Question Length, and Reinterviews," *Vital and Health Statistics*, Series 2, No. 45, 1972, Washington, D.C.: U.S. Government Printing Office.

Marsh, H.W., and Hocevar, D., "A New Powerful Approach to Multitrait-Multimethod Analyses: Applications of Second Order Confirmatory Factor Analysis," *Journal of Applied Psychology*, Vol. 73, 1988, pp. 107–117.

Marshall, W.A., and Tanner, J.M., "Puberty," in F. Falkner, and J.M. Tanner (eds.), *Human Growth*, Vol. 2, 1986, pp. 171–209, New York: Plenum.

Mason, W.M., Wong, G.Y., and Entwhisle, B., "Contextual Analysis Through the Multilevel Linear Model," in S. Leinhardt (ed.), *Sociological Methodology 1983*, 1984, San Francisco: Jossey-Bass.

Mathiowetz, N.A., "The Problem of Omissions and Telescoping Error: New Evidence from a Study of Unemployment," *Proceedings of the Section on Survey Research Methods*, American Statistical Association, 1985, pp. 482–487.

Mathiowetz, N.A., and Cannell, C.F., "Coding Interviewer Behavior as a Method of Evaluating Performance," *Proceedings of the Section on Survey Research Methods*, American Statistical Association, 1980, pp. 525–528.

Mathiowetz, N.A., and Duncan G.J., "Out of Work, Out of Mind: Response Errors in Retrospective Reports of Unemployment," *Journal of Business and Economic Statistics*, Vol. 6, 1988, pp. 221–229.

Mathiowetz, N.A., and Groves, R.M., "The Effects of Respondent Rules on Health Survey Reports," *American Journal of Public Health*, Vol. 75, No. 6, 1985, pp. 639–644.

Matthews, R.V., "An Overview of the 1985 Corn, Cotton, Soybean, and Wheat Objective Yield Survey," U.S. Department of Agriculture, National Agricultural Statistics Service, Staff Report, 1985.

Matthews, R.V., "Head Count Differences Between Sample Plots in the 1983–1985 Wheat Objective Yield Surveys," U.S. Department of Agriculture, National Agricultural Statistics Service, Staff Report No. YRB-86–09, 1986.

May, J., "Questions as Suggestions: The Pragmatics of Interrogative Speech," *Language and Communication*, Vol. 9, No. 4, 1989, pp. 227–243.

Maynard, D., and Clayman, S., "The Diversity of Ethnomethodology," in W.R. Scott and J. Blake (eds.), *Annual Review of Sociology*, Vol. 17, 1991, pp. 385–418, Palo Alto, CA: Annual Reviews, Inc.

Maynard, D., and Marlaire, C., "The Interactional Substrate of Educational Testing," paper presented at the Conference on Video Analysis, Guilford, England: University of Surrey, 1987.

McCarthy, P.J., "Replication: An Approach to the Analysis of Data from Complex Surveys," *Vital and Health Statistics*, Series 2, No. 14, 1966, Washington, D.C.: National Center for Health Statistics.

McClendon, M.J., and O'Brien, D.J., "Question Order Effects and the Sense of Well-Being," paper presented to the American Association for Public Opinion Research, Buck Hill Falls, PA, 1983.

McCullagh, P., and Nelder, J.A., *Generalized Linear Models*, London: Chapman and Hall, 1983.

McGeoch, J.A., "Forgetting and the Law of Disuse," *Psychology Review*, Vol. 15, 1932, pp. 352–370.

McHoul, A., "Why There Are No Guarantees for Interrogators," *Journal of Pragmatics*, Vol. 11, 1987, pp. 455–471.

McPhee, J., "A Reporter at Large/Looking for a Ship—II," *The New Yorker*, 2, April, 1990, pp. 46–86.

Means, B., and Loftus, E.F., "When Personal History Repeats Itself: Decomposing Memory for Recurring Events," *Applied Cognitive Psychology*, Vol. 5, 1991, pp. 297–318.

Means, B., Mingay, D. J., Nigam, A., and Zarrow, M., "A Cognitive Approach to Enhancing Health Survey Reports of Medical Visits," in M. Grunberg, P. Morris, and R. Sykes (eds.) *Practical Aspects of Memory: Current Research*, 1988, Chichester: Wiley.

Means, B., Nigam, A., Zarrow, M., Loftus, E.F., and Donaldson, M.S., "Autobiographical Memory for Health-Related Events," *Vital and Health Statistics*, Series 6, No. 2, 1989, Washington DC: National Center for Health Statistics.

Menon, G., Sudman, S., Bickart, B., Blair, J., and Schwarz, N., "The Use of Anchoring Strategies by Proxy Respondents in Answering Attitude Questions," paper presented at the Annual Conference of the American Association for Public Opinion Research, Lancaster, PA, 1990.

Mersch, M.L., and Dyke, T.C., "Construction and Use of Quality Control Tables," *Technical Conference Transactions*, American Society for Quality Control, 1978, pp. 360–364.

Merton, R., Fiske, M., and Kendall, P., *The Focused Interview: A Manual of Problems and Procedures*, Second edition, New York: The Free Press, 1990.

Miller, J., "Complexities of the Randomized Response Solution," *American Sociological Review*, Vol. 46, 1981, pp. 928–930.

Miller, J., "A New Survey Technique for Studying Deviant Behavior," Ph.D. Dissertation, Sociology Department, The George Washington University, 1984.

Miller, J., "The Nominative Technique: A New Method of Estimating Heroin Prevalence," in B. Rouse, N. Kozel, and L. Richards (eds.), *Self Report Methods of Estimating Drug Use: Meeting Current Challenges to Validity*, NIDA Research Monograph 57, 1985, Washington, DC: U.S. Government Printing Office.

Miller, J., and Cisin, I., "Measuring the Hidden Prevalence of Heroin Use," in J. Rittenhouse (ed.), *Developmental Papers: Attempts to Improve the Measurement of Heroin Use*, National Institute on Drug Abuse, Rockville, MD, unpublished report, 1980.

Miller, J., Cisin, I., and Harrell, A., "A New Technique for Surveying Deviant Behavior: Item-Count Estimates of Marijuana, Cocaine, and Heroin," paper presented at the Annual Meeting of the American Association for Public Opinion Research, St. Petersburg, FL, 1986a.

Miller, J., Cisin, I., Courtless, T., and Harrell, A., "Valid Self-Reports of Criminal Behavior: Applying the Item Count Approach," National Institute of Justice, Washington, DC, unpublished report, 1986b.

Mingay, D.J., and Greenwell, M.T., "Memory Bias and Response Order Effects," *Journal of Official Statistics*, Vol. 5, 1989, pp. 253–263.

Mishler, E., "Meaning in Context: Is There Any Other Kind?" *Harvard Educational Review*, Vol. 49, No. 1, 1979, pp. 1–19.

Mishler, E., *Research Interviewing*, Cambridge, Mass.: Harvard, 1986a.

Mishler, E., "The Analysis of Interview Narratives," in T. Sarbin, (ed.), *Narrative Psychology: The Storied Nature of Human Conduct*, 1986b, pp. 233–255, New York: Praeger.

Mockovak, W.P., "New Directions in Census Training," *Journal of Official Statistics*, Vol. 5, No. 2, 1989, pp. 281–292.

Moerman, M., "The Use of Precedent in Natural Conversation," *Semiotica*, Vol. 9, No. 3, 1973, pp. 193–218.

Molenaar, N.J., "Response Effects of Formal Characteristics of Questions," in W. Dijkstra, and J. van der Zouwen (eds.), *Response Behavior in the Survey Interview*, 1982, New York: Academic Press.

Molenaar, N.J., "Formuleringseffecten in survey-interviews: een non-experimenteel onderzoek (Wording Effects in Survey Interviews: A Non-experimental Study)," Dissertation, Amsterdam: VU-Uitgeverij, 1986.

Moore, J.C., "Self/Proxy Response Status and Survey Response Quality," *Journal of Official Statistics*, Vol. 4, No. 2, 1988, pp. 155–172.

Morgenstern, R.D., and Barrett, N.S., "The Retrospective Bias in Unemployment Reporting by Sex, Race and Age," *Journal of the American Statistical Association*, Vol. 69, 1974, pp. 355–357.

Morton-Williams, J., "The Use of 'Verbal Interaction Coding' for Evaluating a Questionnaire," *Quality and Quantity*, Vol. 13, 1979, pp. 59–75.

Morton-Williams, J., and Sykes, W., "The Use of Interaction Coding and Follow-up Interviews to Investigate Comprehension of Survey Questions," *Journal of the Market Research Society*, 1978, Vol. 26, No. 2, 1984, pp. 109–127.

Mosteller, F., "Nonsampling Errors," in H. Kruskal, and J. M. Tanur (eds.), *National Encyclopedia of Statistics*, 1978, New York: Free Press/Macmillan.

Mote, V.L., and Anderson, R.L., "An Investigation of the Effect of Misclassification on the Properties of Chi-square Tests in the Analysis of Categorical Data," *Biometrika*, Vol. 52, Nos. 1 and 2, 1965, pp. 95–109.

Moynihan, D.P., *Letter to New York*, April 20, 1990, pp. 2–3.

Mueller, J.E., "Choosing Among 133 Candidates," *Public Opinion Quarterly*, Vol. 34, 1970, pp. 395–402.

Munck, I.M.E., "Model Building in Comparative Education: Applications of the LISREL Method to Cross-national Survey Data," *IEA Monograph Studies*, No. 10, 1979, Stockholm: Almqvist & Wiksell International.

Murdock, B.B., and Walker, K.D., "Modality Effects in Free Recall," *Journal of Verbal Learning and Verbal Behavior*, Vol. 8, 1969, pp. 665–676.

Murphy, L.R., and Cowan, C.D., "Effects of Bounding on Telescoping in the National Crime Survey," *Proceedings of the Social Statistics Section*, American Statistical Association, 1976, pp. 633–638.

Muthén, B.O., "Latent Variable Modeling in Heterogeneous Populations," *Psychometrika*, Vol. 54, 1989, pp. 557–585.

Narendranathan, W., and Nickell, S., "Modelling the Process of Job Search," *Journal of Econometrics*, Vol. 28, 1985, pp. 85–101.

Nathan, G., Sirken, M., Willis, G., and Esposito, J., "Laboratory Experiments on the Cognitive Aspects of Sensitive Questions," paper presented at the International Conference on Measurement Errors in Surveys, Tucson, AZ, November 1990.

National Research Council, "Counting Illnesses and Injuries in the Workplace: Proposals for a Better System," in E. Pollack, and D. Keimig (eds.), *Panel on Occupational Safety and Health Statistics*, Committee on National Statistics, Washington DC: National Academy Press, 1987.

Nelson, D.C., "Soybean Objective Yield Destructive Counting Study," Washington, DC, U.S. Department of Agriculture, Statistical Reporting Service, Staff Report No. AGESS801218, 1980.

Nelson, W., "Life Data Analysis for Units Inspected Once for Failure (Quantal Response Data)," *IEEE Transactions on Reliability*, Vol. R-27, 1978, pp. 274–279.

Nelson, W., *Applied Life Data Analysis*, New York: Wiley, 1982.

Neter, J., "Measurement Errors in Reports of Consumer Expenditures," *Journal of Marketing Research*, Vol. VII, 1970, pp. 11–25.

Neter, J., and Waksberg, J., "A Study of Response Errors in Expenditure Data from Household Interviews," *Journal of the American Statistical Association*, Vol. 59, 1964, pp. 18–55.

Neter, J., and Waksberg, J., "Response Error in Collection of Experimental Data vs Household Interviews: An Experimental Study," *U.S. Bureau of the Census Technical Papers*, No. 11, U.S. Government Printing Office, Washington, DC, 1965.

Nicholls II, W.L., "Computer-Assisted Interviewing--Research, Development, and Implementation," U.S. Bureau of the Census Planning Committee for Computer Assisted Interviewing, unpublished manuscript, 1987.

Nicholls II, W.L., and Groves, R.M., "The Status of Computer-Assisted Telephone Interviewing: Part I — Introduction and Impact on Cost and Timeliness of Survey Data," *Journal of Official Statistics*, 1986, pp. 93–115.

Nickell, S., "Estimating the Probability of Leaving Unemployment," *Econometrica*, Vol. 47, 1979, pp. 1249–1266.

Nieto de Pascual, J., "1958 Corn Production Study," Ames, Iowa, Statistical Laboratory, Iowa State College, Agricultural Experiment Station Project No. 1207, 1959.

Nisbett, R.E., and Ross, L., *Human Inference: Strategies and Shortcomings of Social Judgement*, Englewood Cliffs, N.J.: Prentice-Hall, 1980.

Noelle-Neumann, E., "Wanted: Rules for Wording Questions," *Public Opinion Quarterly*, Vol. 34, 1970, pp. 191–201.

Nunnally, J.C., *Psychometric Theory*, New York: McGraw Hill, 1978.

Näsholm, H., Lindström, H.L., and Lindkvist, H., "Response Burden and Data Quality in the Swedish Family Expenditure Surveys," *Proceedings of the Fifth Annual Research Conference*, U.S. Bureau of the Census, 1989, pp. 501–514.

O'Muircheartaigh, C.A., "Response Errors in an Attitudinal Sample Survey," *Quality and Quantity*, Vol. 10, 1976, pp. 97–115.

O'Muircheartaigh, C.A., "Methodology of the Response Errors Project," World Fertility Survey, Scientific Report 28, 1982.

O'Muircheartaigh, C.A., "The Pattern of Response Variance in the Peru Fertility Survey," World Fertility Survey, Scientific Report 45, The Hague: International Statistical Institute, 1984.

O'Muircheartaigh, C.A., "Correlates of Reinterview Response Inconsistency in the Current Population Survey," *Proceedings of the Second Annual Research Conference*, U.S. Bureau of the Census, 1986, pp. 208–234.

O'Muircheartaigh, C.A., and Marckward, A.M., "An Assessment of the Reliability of World Fertility Study Data," *Proceedings of the World Fertility Survey Conference*, Vol. 3., 1980, pp. 305–379, The Hague: International Statistical Institute.

Oksenberg, L., "Analysis of Monitored Telephone Interviews," Report to the U.S. Bureau of the Census for JSA 80–23, Ann Arbor, MI, Survey Research Center, The University of Michigan, 1981.

Oksenberg, L., and Cannell, C., "Some Factors Underlying the Validity of Self-Report," *Bulletin of the International Statistical Institute*, 1977, pp. 325–346.

Oksenberg, L., and Cannell, C., "Effects of Interviewer Vocal Characteristics on Nonresponse," in R.M. Groves, P.P. Biemer, L.E. Lyberg, J.T. Massey, W.L. Nicholls II, and J. Waksberg (eds.), *Telephone Survey Methodology*, 1988, pp. 257–272, New York: Wiley.

Oksenberg, L., Cannell, C., and Kalton, G., "New Methods for Pretesting Survey Questions," in C. Cannell, L. Oksenberg, G. Kalton, K. Bischoping, and F. Fowler (eds.), *New Techniques for Pretesting Survey Questions*, 1989, pp. 30–62, Ann Arbor, MI, Survey Research Center, University of Michigan, unpublished report.

Oksenberg, L., Coleman, L., and Cannell, C., "Interviewers' Voices and Refusal Rates in Telephone Surveys," *Public Opinion Quarterly*, Vol. 50, No. 1, 1986, pp. 97–111.

Olssen, U., "On the Robustness of Factor Analysis Against Crude Classification of the Observation," *Multivariate Behavioral Research*, Vol. 14, 1979, pp. 485–500.

Orth, B., "The Theoretical and Empirical Study of Scale Properties of Magnitude Estimation and Category Rating Scales," in B. Wegener (ed.), *Social Attitudes and Psychological Measurement*, 1982, pp. 351–379, Hillsdale, Erlbaum.

Orvell, M., *The Real Thing: Imitation and Authenticity in American Culture, 1880–1940*, Chapel Hill: University of North Carolina Press, 1989.

Ostrom, T.M., and Upshaw, H.S., "Psychological Perspective and Attitude Change," in A.C. Greenwald, T.C. Brock, and T.M. Ostrom (eds.), *Psychological Foundations of Attitudes*, 1968, pp. 217–242, New York: Academic Press.

Paget, M., "Experience and Knowledge," *Human Studies*, Vol. 6, 1983, pp. 67–90.

Palca, J., "Getting to the Heart of the Cholesterol Debate," *Science*, Vol. 247, 1990, pp. 1170–1171.

Palmisano, M., "The Application of Cognitive Survey Methodology to an Establishment Survey Field Test," *Proceedings of the Section on Survey Research Methods*, American Statistical Association, 1988, pp. 179–184.

Pannekoek, J., "Interviewer Variance in a Telephone Survey," *Journal of Official Statistics*, Vol. 4, 1988, pp. 375–384.

Parry, H., and Crossley, H., "Validity of Responses to Survey Questions," *Public Opinion Quarterly*, Vol. 14, 1950–51, pp. 61–80.

Parton, M., *Surveys, Polls and Samples*, New York: Harper, 1950.

Payne, S.L., *The Art of Asking Questions*, Princeton, NJ: Princeton University Press, 1951.

Peaker, G.F., *International Studies in Evaluation VIII: An Empirical Study of Education in Twenty-One Countries: A Technical Report*, Stockholm: Almqvist & Wiksell International and New York: John Wiley & Sons, 1975.

Pearl, R.B., "Methodology of Consumer Expenditure Surveys," *U.S. Bureau of the Census Working Papers*, No. 27, U.S. Government Printing Office, Washington, DC, 1968.

Pearl, R.B., "Reevaluation of the 1972–73 U.S. Consumer Expenditure Survey," *U.S. Bureau of the Census Technical Papers*, No. 46, U.S. Government Printing Office, Washington, DC, July 1979.

Peckham, J., Sr., *The Wheel of Marketing*, A.C. Nielsen Publication, 1978.

Pennie, K., "Consumer Expenditure Quality Control Reinterview Results for October 1986 -December 1987," Washington, DC, U.S. Bureau of the Census, unpublished report, 1989.

Pense, R.B., "1980 Soybean Destructive Counting Study," Washington, D.C., U.S. Department of Agriculture, Statistical Reporting Service, Staff Report No. AGESS810915, 1981.

Pepe, M.S., Self, S.G., and Prentice, R.L., "Further Results on Covariate Measurement Errors in Cohort Studies with Time to Response Data," *Statistics in Medicine*, Vol. 8, 1989, pp. 1167–1178.

Peters, H.E., "Retrospective Versus Panel Data in Analyzing Lifecycle Events," *Journal of Human Resources*, Vol. 23, 1988, pp. 488–513.

Pettigrew, T.F., *A Profile of the Negro American*, New York: Van Nostrand, 1964.

Petty, R.E., and Cacioppo, J.T., *Attitudes and Persuasion: Classic and Contemporary Approaches*, Dubuque, Iowa: W.C. Brown Company Publishers, 1982.

Petty, R.E., and Cacioppo, J.T., *Communication and Persuasion: Central and Peripheral Routes to Attitude Change*, New York: Springer Verlag, 1986.

Phillips, D., and Clancy, K., "Response Bias in Field Studies of Mental Illness," *American Sociological Review*, Vol. 35, 1970, pp. 503–515.

Phipps, P.A., and Tupek, A.R., "Assessing Measurement Errors in a Touchtone Recognition Survey," paper presented at the International Conference on Measurement Errors in Surveys, Tucson, AZ, November 1990.

Plewis, I., Creeser, R., and Mooney, A., "Reliability and Validity of Time Budget Data: Children's Activities Outside School," *Journal of Official Statistics*, Vol. 6, 1990, pp. 411–420.

Poikolainen, K., and Karkkainen, P., "Diary Gives More Accurate Information About Alcohol Consumption than Questionnaire," *Ireland, Drug and Alcohol Dependence*, Vol. 11, 1983, pp. 209–216.

Pollack, K., and Bek, Y., "A Comparison of Three Randomized Response Models for Quantitative Data," *Journal of the American Statistical Association*, Vol. 71, 1976, pp. 884–886.

Pomerantz, A., "Agreeing and Disagreeing with Assessments: Some Features of Preferred/Dispreferred Turn Shapes," in J. Atkinson, and J. Heritage (eds.), *Structure of Social Action*, 1984a, pp. 57–101, Cambridge: Cambridge University Press.

Pomerantz, A., "Pursuing a Response," in J. Atkinson, and J. Heritage (eds.), *Structure of Social Action*, 1984b, pp. 152–163, Cambridge: Cambridge University Press.

Poterba, J.M., and Summers, L.H., "Response Variation in the CPS: Caveats for the Unemployment Analyst," *Monthly Labor Review*, Vol. 107, No. 3, 1984, pp. 37–43.

Poterba, J.M., and Summers, L.H., "Adjusting the Gross Change Data: Implications for Labor Market Dynamics," *Proceedings of the Conference on Gross Flows in Labor Force Statistics*, Washington, D.C.,U. S. Bureau of the Census and U. S. Bureau of Labor Statistics, 1985, pp. 81–95.

Prentice, R.L., "Covariate Measurement Errors and Parameter Estimation in a Failure Time Regression Model," *Biometrika*, Vol. 69, 1982, pp. 331–342.

Prentice, R.L., "Binary Regression Using an Extended Beta-binomial Likelihood, with Discussion of Correlation Induced by Covariate Measurement Errors," *Journal of the American Statistical Association*, Vol. 81, 1986, pp. 321–327.

Prentice, R.L., "Correlated Binary Regression with Covariates Specific to Each Binary Observation," *Biometrics*, Vol. 44, 1988, pp. 1033–1048.

Presser, S., "Can Changes in Context Reduce Vote Over-Reporting in Surveys?" *Public Opinion Quarterly*, Vol. 54, 1990, pp. 586–593.

Psathas, G., "Some Sequential Structures in Direction-Giving," *Human Studies*, Vol. 9, 1986, pp. 231–246.

Rao, J.N.K., "On Double Sampling for Stratification and Analytical Surveys," *Biometrika*, Vol. 60, No. 1, 1973, pp. 125–133.

Rao, J.N.K., and Scott, A.J., "The Analysis of Categorical Data from Complex Sample Surveys: Chi-Squares Tests for Goodness of Fit and Independence in Two-Way Tables," *Journal of the American Statistical Association*, Vol. 76, No. 374, 1981, pp. 221–230.

Rao, J.N.K., and Scott, A.J., "On Chi-Squared Tests for Multi-Way Tables with Cell Proportions Estimated from Survey Data," *Annals of Statistics*, Vol. 12, No. 1, 1984, pp. 46–60.

Rao, J.N.K., and Thomas, D.R., "The Analysis of Cross-Classified Categorical Data from Complex Sample Surveys," *Sociological Methodology*, Vol. 18, 1988, pp. 213–269.

Rao, M.V.S., and Vidwans, S.M., "Household Budget Surveys in Developing Countries," *Bulletin of the International Statistical Institute*, 1989, pp. 143–167.

Rasinski, K.A., "The Effect of Question Wording on Public Support for Government Spending," *Public Opinion Quarterly*, Vol. 53, 1989, pp. 388–394.

Raudenbusch, S., and Bryk, A.S., "A Hierarchical Model for Studying School Effects," *Sociology of Education*, Vol. 59, 1986, pp. 1–17.

Redfern, P., "Which Countries Will Follow the Scandinavian Lead in Taking a Register-Based Census of Population?" *Journal of Official Statistics*, Vol. 2, No. 4, 1986, pp. 415–424.

Redpath, R., "The Family Expenditure Survey: Some Problems of Collecting Data," *Survey Methodology Bulletin*, Office of Population Censuses and Surveys, Her Majesty's Stationary Office, London, 1987, pp. 12–24.

Reese, S.D., Danielson, W.A., Shoemaker, P.J., Chang, T.K., and Hsu, H.L., "Ethnicity-of-Interviewer Effects Among Mexican-Americans and Anglos," *Public Opinion Quarterly*, Vol. 50, 1986, pp. 563–572.

Reiser, B.J., Black, J.B., and Abelson, R.P., "Knowledge Structures in the Organization and Retrieval of Autobiographical Memories," *Cognitive Psychology*, Vol. 17, 1985, pp. 89–137.

Reiser, M., Fecso, R., and Chua, M., "Some Panel Aspects of the Objective Yield Survey," *Proceedings of the Section on Survey Research Methods*, American Statistical Association, 1989, pp. 469–474.

Reiser, M., Fecso, R., and Taylor, K., "A Nested Error Model for the Objective Yield Survey," *Proceedings of the Section on Survey Research Methods*, American Statistical Association, 1987.

Rholes, W.S., Riskind, J.H., and Lane, J.W., "Emotional States and Memory Biases: Effects of Cognitive Priming and Mood," *Journal of Personality and Social Psychology*, Vol. 52, 1987, pp. 91–99.

Rice, S.A., "Contagious Bias in the Interview," *American Journal of Sociology*, Vol. 35, 1929, pp. 420–423.

Rice, S.C., Wright, R.A., and Rowe, B., "Development of Computer Assisted Personal Interview for the National Health Interview Survey," *Proceedings of the Section on Survey Research Methods*, American Statistical Association, 1988, pp. 397–400.

Rindskopf, D., and Rose, T., "Some Theory and Applications of Confirmatory Second Order Factor Analysis," *Multivariate Behavioral Research*, Vol. 23, 1988, pp. 51–67.

Ring, E., "Asymmetrical Rotation," *European Research*, Vol. 3, 1975, pp. 111–119.

Rittenhouse, J., and Sirken, M., "A Note on Networks, Nominations and Multiplicity as Contributory to Heroin Estimation," in J. Rittenhouse (ed.), *Developmental Papers: Attempts to Improve the Measurement of Heroin Use in the National Survey*, National Institute on Drug Abuse, Rockville, MD, unpublished report, 1980.

Robinson, D., and Rhode, S., "Two Experiments with an Anti- Semitism Poll," *Journal of Abnormal Social Psychology*, Vol. 41, 1946, pp. 136–144.

Rodgers, W.L., "Comparisons of Alternative Approaches to the Estimation of Simple Causal Models from Panel Data," in D. Kasprzyk, G. Duncan, G. Kalton, and M. P. Singh (eds.), *Panel Surveys*, 1989a, New York: Wiley.

Rodgers, W.L., "Reliability and Validity in Measures of Subjective Well-being," paper presented at the International Conference on Social Reporting, Wissenschaftszentrum, Beslin, September 18–20, 1989b.

Rodgers, W.L., Herzog, A.R., and Andrews, F.M., "Interviewing Older Adults: Validity of Self-report of Satisfaction," *Psychology and Aging*, Vol. 3, 1988, pp. 264–272.

Roger, J.H., and Peacock, S.B., "Fitting the Scale as a GLIM Parameter for Weibull, Extreme Value, Logistic and Log-logistic Regression Models with Censored Data," *GLIM Newsletter*, Vol. 6, 1982, pp. 30–37.

Rogers, T.B., Kuiper, N.A., and Kirker, W.S., "Self-Reference and the Encoding of Personal Information," *Journal of Personality and Social Psychology*, Vol. 35, 1977, pp. 677–688.

Rohrmann, B., "Empirische Studien zur Entwicklung von Antwortskalen für die Sozialwissenschaftliche Forschung (Empirical Studies of the Development of Response Scales for Social Science Research)", *Zeitschrift für Sozialpsychologie*, Vol. 9, 1978, pp. 222–245.

Roper, B.W., "The Subtle Effects of Context," *The Public Perspective*, Vol. 1, 1990, p. 25.

Roshco, B., "The Polls: Polling on Panama," *Public Opinion Quarterly*, Vol. 42, No. 4, 1978, pp. 551–562.

Rosner, B., Willett, W.C., and Spiegleman, D., "Correction of Logistic Regression Relative Risk Estimates and Confidence Intervals for Systematic Within-Person Measurement Error," *Statistics in Medicine*, Vol. 8, 1989, pp. 1051–1069.

Ross, S., "Rethinking Thinking," *Modern Maturity*, February-March, 1990, pp. 52–58.

Rothschild, B.B., and Wilson, L.B., "Nationwide Food Consumption Survey 1987: A Landmark Personal Interview Survey Using Laptop Computers," *Proceedings of the Fourth Annual Research Conference*, U.S. Bureau of the Census, 1988, pp. 347–356.

Royston, P.N., "Using Intensive Interviews to Evaluate Questions," in F.J. Fowler (ed.), *Conference Proceedings, Health Survey Research Methods*, 1989, pp. 3–8, Washington, DC: National Center for Health Services Research.

Royston, P.N., Bercini, D., Sirken, M., and Mingay, D., "Questionnaire Design Laboratory," *Proceedings of the Section on Survey Research Methods*, American Statistical Association, 1986, pp. 703–707.

Rugg, D., "Experiments in Wording Questions II," *Public Opinion Quarterly*, Vol. 5, 1941, pp. 91–92.

Rugg, D., and Cantril, H., "The Wording of Questions in Public Opinion Polls," *Journal of Abnormal and Social Psychology*, Vol. 37, 1942, pp. 469–495.

Rugg, D., and Cantril, H., "The Wording of Questions," in H. Cantril, and research associates in the Office of Public Opinion Research at Princeton University (eds.), *Gauging Public Opinion*, 1947, pp. 23–50, Princeton: Princeton University Press.

Rustemeyer, A., "Measuring Interviewer Performance in Mock Interviews," *Proceedings of the Social Statistics Section*, American Statistical Association, 1977, pp. 341–346.

Sacks, H., "On the Preferences for Agreement and Contiguity in Sequences in Conversation," in G. Button, and J. Lee (eds.), *Talk and Social Organization*, 1987, pp. 54–69, Clevedon, Avon, England: Multilingual Matters, Ltd.

Sacks, H., Schegloff, E., and Jefferson, G., "A Simplest Systematics for the Organization of Turn Taking in Conversation," in J. Schenkein (ed.), *Studies in the Organization of Conversational Interaction*, 1984, pp. 7–56, New York: Academic Press.

Sanderson, F.H., *Methods of Crop Forecasting*, Cambridge, Harvard University Press, 1954.

Saris, W.E., "Different Questions, Different Variables," in C.Fornell (ed.), *Second Generation of Multivariate Analysis*, Vol. 2, 1982, pp. 78–96, New York: Praeger.

Saris, W.E., *Variation in Response Functions: A Source of Measurement Error*, Amsterdam: Sociometric Research Foundation, 1988.

Saris, W.E., "Technological Revolution in Data Collection," *Quality and Quantity*, Vol. 23, 1989, pp. 333–349.

Saris, W.E., "Models for Evaluation of Measurement Instruments," in W.E. Saris, and A. van Meurs (eds.), *Evaluation of Measurement Instruments by Meta-analysis of Multitrait Multimethod Studies*, Chapter 2, 1990, Amsterdam: North Holland.

Saris, W.E., and Stronkhorst, H., *Causal Modelling in Nonexperimental Research*, Amsterdam: Sociometric Research Foundation, 1984.

Saris, W.E., and van Meurs, A., *Evaluation of Measurement Instruments by Meta-analysis of Multitrait Multimethod Studies*, Amsterdam: North Holland, 1990.

Sarle, C.F., "Adequacy and Reliability of Crop Yield Estimates," Washington, DC, U.S. Department of Agriculture, Technical Bulletin 311, 1932.

Schaeffer, N.C., "Evaluating Race-of-Interviewer Effects in a National Survey," *Sociological Methods and Research*, Vol. 8, 1980, pp. 400–419.

Schaeffer, N., and Thomson, E., "The Discovery of Grounded Uncertainty: Developing Questions about Strength of Fertility Motivation," in P. Marsden (ed.), *Sociological Methodology*, Vol. 21, Oxford: Basil Blackwell, in press.

Schegloff, E., "Notes on a Conversational Practice: Formulating Place," in D. Sudnow (ed.), *Studies in Social Interaction*, 1972, pp. 75–119, New York: Free Press.

Schegloff, E., "Discourse as an Interactional Achievement: Some Uses of 'Uh-Huh' and Other Things that Come Between Sentences," in D. Tannen (ed.), *Analyzing Discourse: Text and Talk*, 1982, pp. 71–93, Washington, DC: Georgetown University Press.

Schegloff, E., Comment, *Journal of the American Statistical Association*, Vol. 85, No. 409, 1990, pp. 248–250.

Schegloff, E., Jefferson, G., and Sacks, H., "The Preference for Self-Correction in the Organization of Repair in Conversation," *Language*, Vol. 53, No. 2, 1977, pp. 361–382.

Schnell, D., Park, H.J., and Fuller, W.A., "EV CARP," Statistical Laboratory, Iowa State University, Ames, Iowa, 1988.

Schreiner, I., "Reinterview Results from the CPS Independent Reconciliation Experiment (Second Quarter 1978 through Third Quarter 1979)," Washington, DC, U.S Bureau of the Census, unpublished report, 1980.

Schreiner, I., "Final Results of the CPS/CATI Phase II Reinterview Study (January - December 1988)," CATI Research Report No. CPS-5, U.S. Bureau of the Census, Washington, D.C., unpublished report, 1989.

Schreiner, I., Pennie, K., and Newbrough, J., "Interviewer Falsification in Census Bureau Surveys," *Proceedings of the Section on Survey Research Methods*, American Statistical Association, 1988, pp. 491–496.

Schuman, H., "Survey Research," in E. Aronson, and G. Lindzey (eds.), *Handbook of Social Psychology (Vol. 1)*, 1985, Reading, MA: Addison-Wesley.

Schuman, H., and Converse, J. M., "The Effects of Black and White Interviewers on Black Responses in 1968," *Public Opinion Quarterly*, Vol. 35, 1971, pp. 44–68.

Schuman, H., and Presser, S., "Question Wording as an Independent Variable in Survey Analysis," *Sociological Methods and Research*, Vol. 6, 1977, pp. 151–176.

Schuman, H., and Presser, S., *Questions and Answers in Attitude Surveys: Experiments on Question Form, Wording, and Context*, New York: Academic Press, 1981.

Schuman, H., and Scott, J., "Problems in the Use of Survey Questions to Measure Public Opinion," *Science*, Vol. 236, 1987, pp. 957–959.

Schuman, H., Kalton, G., and Ludwig, J., "Context and Continuity in Survey Questionnaires," *Public Opinion Quarterly*, Vol. 47, 1983, pp. 112–115.

Schwarz, N., "Assessing Frequency Reports of Mundane Behaviors: Contributions of Cognitive Psychology to Questionnaire Design," in C. Hendrick, and M. S. Clark (eds.), *Research Methods in Personality and Social Psychology*, 1990a, Beverly Hills, CA: Sage.

Schwarz, N., "What Respondents Learn from Scales: Informative Functions of Response Alternatives," *International Journal of Public Opinion Research*, Vol. 2, 1990b, pp. 274–285.

Schwarz, N., and Bienias, J., "What Mediates the Impact of Response Alternatives on Frequency Reports of Mundane Behaviors?" *Applied Cognitive Psychology*, Vol. 4, 1990, pp. 61–72.

Schwarz, N., and Hippler, H.J., "What Response Scales May Tell Your Respondents," in H.J. Hippler, N. Schwarz, and S. Sudman (eds.), *Social Information Processing and Survey Methodology*, 1987, New York: Springer-Verlag.

Schwarz, N., and Scheuring, B., "Judgments of Relationship Satisfaction: Inter- and Intraindividual Comparisons as a Function of Questionnaire Structure," *European Journal of Social Psychology*, Vol. 18, 1988, pp. 485–496.

Schwarz, N., and Strack, F., "The Survey Interview and the Logic of Conversation," paper presented at the annual meetings of the American Association for Public Opinion Research, Toronto, Canada, 1988.

Schwarz, N., and Strack, F., "Context Effects in Attitude Surveys: Applying Cognitive Theory to Social Research," in M. Hewstone and W. Stroebe, (eds.), *European Review of Social Psychology*, Vol. 2, 1991, pp. 31–50, Chichester: Wiley.

Schwarz, N., Hippler, H.J., and Noelle-Neumann, E., "A Cognitive Model of Response Order Effect in Survey Measurement," in N. Schwarz, and S. Sudman (eds.), *Context Effects in Social and Psychological Research*, 1991, New York: Springer Verlag.

Schwarz, N., Münkel, T., and Hippler, H.J., "What Determines a Perspective? Contrast Effects as a Function of the Dimension Tapped by Preceding Questions," *European Journal of Social Psychology*, Vol. 20, 1990b, pp. 357–361.

Schwarz, N., Strack, F., and Mai, H.P., "Assimilation and Contrast Effects in Part-whole Question Sequences: A Conversational Logic Analysis," *Public Opinion Quarterly*, Vol. 55, 1991, pp. 3–23.

Schwarz, N., Hippler, H.J., Deutsch, B., and Strack, F., "Response Scales: Effects of Category Range on Reported Behavior and Comparative Judgments," *Public Opinion Quarterly*, Vol. 49, 1985, pp. 388–395.

Schwarz, N., Hippler, H.J., Noelle-Neumann, E., and Münkel, T., "Response Order Effects and Long Lists: Primacy Recency in Asymmetric Contrast Effects," paper presented to the American Association for Public Opinion Research, St. Petersburg, FL, 1989.

Schwarz, N., Strack, F., Hippler, H.J., and Bishop, G., "Psychological Sources of Response Effects in Surveys: The Impact of Administrative Mode," in J. Jobe, and E.F. Loftus (eds.), *Cognitive Aspects of Survey Methodology*, Special issue of *Applied Cognitive Psychology*, in press.

Schwarz, N., Strack, F., Müller, G., and Chassein, B., "The Range of Response Alternatives May Determine the Meaning of the Question: Further Evidence on Informative Functions of Response Alternatives," *Social Cognition*, Vol. 6, 1988, pp. 107–117.

Schwarz, N., Bless, H., Bohner, G., Harlacher, U., and Kellenbenz, M., "Response Scales as Frames of Reference: The Impact of Frequency Range on Diagnostic Judgment, *Applied Cognitive Psychology*, Vol. 5, 1991, pp. 37–50.

Schwarz, N., Knäuper, B., Hippler, H. J., Noelle-Neumann, E., and Clark, L., "Rating Scales: Numeric Values May Change the Meaning of Scale Labels," *Public Opinion Quarterly*, in press.

Schönpflug, W., and Büch, B., "Psychische Prozesse beim Psychologischen Skalieren. V. Subjektive Beurteilungen von Skalenkategorien," *Psychologische Beiträge*, Vol. 12, 1970, pp. 384–392.

Scott, A.J., Rao, J.N.K., and Thomas, D.R., "Weighted Least-Squares and Quasi-likelihood Estimation for Categorical Data under Singular Models," *Linear Algebra and its Applications*, Vol. 127, 1990, pp. 427–447.

Shainin, P., "The Tools of Quality; Part III: Control Charts," *Quality Progress*, Vol. XXIII, No. 8, 1990, pp. 79–82.

Shavelson, R.J., and Webb, N.M., *A Primer on Generalizability Theory*, Newbury, Park, CA: Sage Publications, Inc., 1991.

Shewhart, W.A., *Statistical Method from the Viewpoint of Quality Control*, Graduate School, Department of Agriculture, 1939.

Shimizu, I., and Bonham, G., "Randomized Response in a National Survey," *Journal of the American Statistical Association*, Vol. 73, 1978, pp. 35–39.

Short, K.S., and Woodrow, K.A., "An Exploration of the Applicability of Hazard Models in Analyzing the Survey of Income and Program Participation: Labor Force Transitions," *Proceedings of the Section on Social Statistics*, American Statistical Association, 1985, pp. 345–350.

Shrout, P.E., and Newman, S.C., "Design of Two-Phase Prevalance Surveys of Rare Disorders," *Biometrics*, Vol. 45, No. 3, 1989, pp. 549–555.

Siemiatycki, J., "A Comparison of Mail, Telephone, and Home Interview Strategies for Household Health Surveys," *American Journal of Public Health*, Vol. 69, 1979, pp. 238–245.

Sikkel, D., "Models for Memory Effects," *Journal of the American Statistical Association*, Vol. 80, 1985, pp. 835–841.

Sikkel, D., "Retrospective Questions and Group Differences," *Journal of Official Statistics*, Vol. 6, 1990, pp. 165–177.

Silberstein, A.R., "Recall Effects in the U.S. Consumer Expenditure Interview Survey," *Journal of Official Statistics*, Vol. 5, No. 2, 1989, pp. 125–142.

Silberstein, A.R., "First Wave Effects in the U.S. Consumer Expenditure Interview Survey," *Survey Methodology*, Vol. 16, No. 2, 1990, pp. 293–304.

Silberstein, A.R., and Jacobs, C.A., "Symptoms of Repeated Interview Effects in the Consumer Expenditure Interview Survey," in D. Kasprzyk, G. Duncan, G. Kalton, and M.P. Singh (eds.), *Panel Surveys*, 1989, pp. 289–303, New York: Wiley.

Simon, H., *Models of Man*, New York: Wiley, 1957.

Simon, H.A., and Kaplan, C.A., "Foundations of Cognitive Science" in M.J. Posner (ed.), *Foundations of Cognitive Science*, 1989, Boston: MIT Press.

Singer, E., Frankel, M.R., and Glassman, M.B., "The Effect of Interviewer Characteristics and Expectations on Response," *Public Opinion Quarterly*, Vol. 47, 1983, reprinted in E. Singer, and S. Presser (eds.), *Survey Research Methods*, 1989, Chicago: University of Chicago Press.

Singer, E., and Kohnke-Aguirre, L., "Interviewer Expectation Effects: A Replication and Extension," *Public Opinion Quarterly*, Vol. 43, 1979, pp. 245–260.

Singer, E., and Presser, S. (eds.), *Survey Research Methods*, Chicago: University of Chicago Press, 1989a.

Singer, E., and Presser, S., "The Interviewer," in E. Singer, and S. Presser (eds.), *Survey Research Methods,* 1989b, pp. 245–246, Chicago: University of Chicago Press.

Singer, E., Strack, F., and Mai, P., "The Effect of Interviewer Characteristics and Expectations on Response," *Public Opinion Quarterly*, Vol. 47, 1983, pp. 68–83.

Sirken, M., "Network Surveys of Rare and Sensitive Conditions," *Advances in Health Survey Research Methods*, Hyattsville, MD, National Center Health Statistics, 1975.

Smith, R., "1984 AHS-MS Reinterview Results," U.S. Bureau of the Census, Washington, DC, unpublished report, 1987.

Smith, T.W., "In Search of House Effects: A Comparison of Responses to Various Questions by Different Survey Organizations," *Public Opinion Quarterly*, Vol. 42, 1978, pp. 443–463.

Smith, T.W., "Can We Have Confidence in Confidence? Revisited," in D.F. Johnston (ed.), *Measurement of Subjective Phenomena*, 1981, Washington, D.C., Government Printing Office.

Smith, T.W., "House Effects: A Comparison of the 1980 General Social Survey and the 1980 American National Election Study," *Public Opinion Quarterly*, Vol. 46, 1982a, pp. 54–68.

Smith, T.W., "Conditional Order Effects," *GSS Methodological Report*, No. 20, Chicago: NORC, 1982b.

Smith, T.W., "An Experimental Comparison of Clustered and Scattered Scale Items," *Social Psychology Quarterly*, Vol. 46, 1983a, pp. 163–168.

Smith, T.W., "Children and Abortions: An Experiment in Question Order," *GSS Methodological Report*, No. 27, Chicago: NORC, 1983b.

Smith, T.W., "A Preliminary Analysis of Methodological Experiments on the 1984 GSS," *GSS Methodological Report,* No. 30, Chicago: NORC, 1984.

Smith, T.W., "Unhappiness on the 1985 GSS: Confounding Change and Context," *GSS Methodological Report* No. 34, Chicago: NORC, 1986.

Smith, T.W., "Ballot Position: An Analysis of Context Effects Related to Rotation Design," *GSS Methodological Report,* No. 55, Chicago, NORC, 1988a.

Smith, T.W., "Counting Flocks and Lost Sheep: Trends in Religious Preference Since World War II," *GSS Social Change Report*, No. 26, Chicago: NORC, 1988b.

Smith, T.W., "Rotation Designs on the GSS," *GSS Methodological Report*, No. 52, Chicago: NORC, 1988c.

Smith, T.W., "Timely Artifacts: A Review of Measurement Variation in the 1972–1988 GSS," *GSS Methodological Report*, No. 56, Chicago: NORC, 1988d.

Smith, T.W., "Some Thoughts on the Nature of Context Effects," paper presented at the First Nagshead Conference on Cognition and Survey Methodology, Nags Head, NC, 1989.

Smyth, M.M., Morris, P.E., Levy, P., and Ellis, A.W., *Cognition in Action*, Hillsdale, NJ: Erlbaum, 1987.

Snyder, D.L., *Random Point Processes*, New York: John Wiley & Sons, 1985.

Som, R.K., *Recall Lapse in Demographic Inquiries*, New York: Asia Publishing House, 1973.

Spencer, B., "Errors in True Values," paper presented at the International Conference on Measurement Errors in Surveys, Tucson, AZ, 1990.

Stanton, J.L., and Tucci, L.A., "The Measurement of Consumption: A Comparison of Surveys and Diaries," *Journal of Marketing Research*, Vol. XIX, 1982, pp. 274–277.

Steele, R.J., "Corn Objective Yield: Operational vs. Non-invasive Maturity Category Determinations," Washington, DC, U.S. Department of Agriculture, National Agricultural Statistics Service, Staff Report SRB-87–04, 1987.

Stein, N. L., and Nezworski, T., "The Effects of Organization and Instructional Set on Story Memory," *Disclosure Processes*, Vol. 1, 1978, pp. 177–193.

Steinkamp, S.W., "Some Characteristics of Effective Interviewers," *Journal of Applied Psychology*, Vol. 50, 1966, pp. 487- 492.

Stokes, S.L., "Estimation of Interviewer Effects for Categorical Items in a Random Digit Dial Telephone Survey," *Journal of the American Statistical Association*, Vol. 83, 1988, pp. 623–630.

Stokes, S.L., and Hill, J.R., "Modeling Interviewer Variability for Dichotomous Variables," *Proceedings of the Section on Survey Research Methods*, American Statistical Association, 1985, pp. 344–348.

Stokes, S.L., and Mulry M.H., "On the Design of Interpenetration Experiments for Categorical Data Items," *Journal of Official Statistics*, Vol. 3, 1987, pp. 389–401.

Stokes, S.L., and Yeh, M., "Searching for Causes of Interviewer Effects in Telephone Surveys," in R.M. Groves, P.P. Biemer, L.E. Lyberg, J.T. Massey, W.L. Nicholls II, and J. Waksberg (eds.), *Telephone Survey Methodology*, 1988, New York: John Wiley.

Strack, F., and Martin L.L., "Thinking, Judging and Communicating: A Process of Context Effects in Attitude Surveys," in H.J. Hippler, N. Schwarz, and S. Sudman (eds.) *Social Information Processing and Survey Methodology*, 1987, New York: Springer Verlag.

Strack, F., and Schwarz, N., "Communicative Influences in Standardized Question Situations: The Case of Implicit Collaboration," in K. Kiedler, and G. Semin (eds.) *Language and Social Cognition*, Beverly Hills: Sage, in press.

Strobel, D., "Determining Outliers in Multivariate Surveys by Decomposition of a Measure of Information," *Proceedings of the Section on Business and Economic Statistics*, American Statistical Association, 1982, pp. 31–35.

Strobel, D., "The Measure of Dispersion between Observed and Estimated Values and Its Application as an Alternative to Least Squares Regression," unpublished report, 1984.

Strongman, K.T., and Russell, P.N., "Salience of Emotion in Recall," *Bulletin of the Psychonomic Society*, Vol. 24, 1986, pp. 25–27.

Strube, G., "Answering Survey Questions: The Role of Memory," in H.J. Hippler, N. Schwarz, and S. Sudman (eds.), *Social Information Processing and Survey Methodology*, 1987, pp. 86–101, New York: Springer Verlag.

Suchman, L., and Jordan, B., "Interactional Troubles in Face-to- Face Survey Interviews," *Journal of the American Statistical Association*, Vol. 85, 1990, pp. 232–241.

Sudman, S., "On the Accuracy of Recording of Consumer Panels," *Journal of Marketing Research*, Vol. 1, 1964, pp. 14–20 and Vol. 1, 1964, pp. 69–83.

Sudman, S., "Reducing Response Error in Surveys," *The Statistician*, Vol. 29, No. 4, 1980, pp. 237–273.

Sudman, S., and Bradburn, N.M., "Effects of Time and Memory Factors on Response in Surveys," *Journal of the American Statistical Association*, Vol. 68, 1973, pp. 805–815.

Sudman, S., and Bradburn, N.M., *Response Effects in Surveys*, Chicago: Aldine, 1974.

Sudman, S., and Bradburn, N.M., *Asking Questions: A Practical Guide to Questionnaire Design*, 1982, San Francisco: Jossey-Bass.

Sudman, S., and Ferber, R., "Experiments in Obtaining Consumer Expenditures by Diary Methods," *Journal of the American Statistical Association*, Vol. 66, No. 336, 1971, pp. 725–735.

Sudman, S., and Ferber, R., "A Comparison of Alternative Procedures for Collecting Consumer Expenditure Data for Frequently Purchased Items," *Journal of Marketing Research*, Vol. XI, 1974, pp. 128–135.

Sudman, S., and Schwarz, N., "Contributions of Cognitive Psychology to Advertising Research," *Journal of Advertising Research*, Vol. 29, No. 3, 1989, pp. 43–53.

Sudman, S., Finn, A., and Lannom, L., "The Use of Bounded Recall Procedures in Single Interviews," *Public Opinion Quarterly*, Vol. 48, 1984, pp. 520–524.

Sudman, S., Blair, E., Bradburn, N., and Stocking, C., "Estimates of Threatening Behavior Based on Reports of Friends," *Public Opinion Quarterly*, Vol. 41, 1977a, pp. 261–264.

Sudman, S., Bradburn, N., Blair, E., and Stocking, C., "Modest Expectations: The Effect of Interviewers' Prior Expectations on Responses," *Sociological Methods and Research*, Vol. 6, No. 2, 1977b, pp. 171–182.

Sudman, S., Menon, G., Blair, J., and Bickart, B., "The Effect of Level of Participation in Proxy Reporting," *Proceedings of the Section on Survey Research Methods*, American Statistical Association, 1990.

Suen, H.K., and Ary, D., *Analyzing Quantitative Behavioral Observation Data*, Hillsdale, NJ: Lawrence Erlbaum Associates, 1988.

Sukhatme, P.V., and Seth, G.R., "Nonsampling Errors in Surveys," *Journal of Indian Society of Agricultural Statistics*, Vol. 5, 1952, pp. 5–41.

Suppes, P., Krantz, D.M., Luce, R., and Tversky, A., *Foundations of Measurement*, Vol. I, New York: Academic Press, 1971.

Suppes, P., Krantz, D.M., Luce, R., and Tversky, A., *Foundations of Measurement*, Vol. II, New York: Academic Press, 1989.

Suppes, P., Krantz, D.M., Luce, R., and Tversky, A., *Foundations of Measurement,* Vol. III, New York: Academic Press, 1990.

Survey Research Center, *Interviewer's Manual,* Revised edition, Ann Arbor, MI, Survey Research Center, Institute for Social Research, The University of Michigan, 1976.

Swan, G.E., and Denk, C.E., "Dynamic Models for the Maintenance of Smoking Cessation: Event History Analysis of Late Relapse," *Journal of Behavioral Medicine*, Vol. 10, 1987, pp. 527–554.

Swan, G.E., Denk, C.E., Parker, S.D., Carmelli, D., Furze, C.T., and Rosenman, R.H., "Risk Factors for Late Relapse in Male and Female Exsmokers," *Addictive Behaviors*, A13, 1988, pp. 253–266.

Sykes, W., and Collins, M., "Effects of Mode of Interview: Experiments in the U.K.," in R.M. Groves, P.P. Biemer, L.E. Lyberg, J.T. Massey, W.L. Nicholls II, and J. Waksberg (eds.), *Telephone Survey Methodology*, 1988, pp. 301–320, New York: Wiley.

Sykes, W., and Morton-Williams, J., "Evaluating Survey Questions," *Journal of Official Statistics*, Vol. 3, No. 2, 1987, pp.191–207.

Särndal, C.E., and Swensson, B., "A General View of Estimation for Two Phases of Selection with Applications to Two-Phase Sampling and Nonresponse," *International Statistical Review*, Vol. 55, No. 3, 1987, pp. 279–294.

Tanur, J., and Feinberg, S., "Combining Cognitive and Statistical Approaches to Survey Design," *Science*, Vol. 243, 1989, pp. 1017–1022.

Tarnai, J., and Dillman D., "Public Opinion of the Washington State Safety Belt Use Law," A Statewide Survey Conducted by the Social and Economic Sciences Research Center, Washington State University, Pullman, WA, 1988.

Tarnai, J., and Dillman, D., "Questionnaire Context as a Source of Response Differences in Mail vs. Telephone Surveys," in N. Schwartz and S. Sudman (eds.), *Order Effects in Social and Psychological Research*, New York: Springer Verlag, in press.

Tarone, R.E., "Testing the Goodness of Fit of the Binomial Distribution," *Biometrika*, Vol. 66, 1979, pp. 585–590.

Tenenbein, A., "A Double Sampling Scheme for Estimating from Misclassified Multinomial Data with Application to Sampling Inspection," *Technometrics*, Vol. 14, No. 1, 1972, pp. 187–202.

Thelen, W., "CPS Type A Noninterview Rates in Subsequent Months for Households in Reinterview – April 1978 through January 1979," Washington, DC, U.S. Bureau of the Census, unpublished report, 1979.

Thissen, C., "Scan Data: Collecting and Maintaining Quality Information," Food Marketing Institute, Washington, DC, 1990.

Thomas, D.R., "Analysis of Categorical Data Under an Overdispersion Model of the Covariance Matrix," Ottawa, Carleton University, School of Business, Working Paper Series 90–14, 1990.

Thomas, D.R., and Rao, J.N.K., "Small-Sample Comparisons of Level and Power for Simple Goodness-of-Fit Statistics Under Cluster Sampling," *Journal of the American Statistical Association*, Vol. 82, No. 398, 1987, pp. 630–636.

Thorndike, R.L., *International Studies in Evaluation III: Reading Comprehension in Fifteen Countries*, Stockholm: Almqvist & Wiksell International and New York: John Wiley and Sons, 1973.

Tourangeau, R., "Cognitive Sciences and Survey Methods," in T. Jabine, E. Loftus, M. Straf, J. Tanur, and R. Tourangeau (eds.), *Cognitive Aspects of Survey Methodology: Building a Bridge Between Disciplines*, 1984, Washington, DC: National Academy of Science.

Tourangeau, R., "Attitude Measurement: A Cognitive Perspective," in H.J. Hippler, N. Schwartz, and S. Sudman (eds.), *Social Information Processing and Survey Methodology*, 1987, New York: Springer Verlag.

Tourangeau, R., and Rasinski, K.A., "Context Effects in Attitude Surveys," unpublished NORC Report, 1986.

Tourangeau, R., and Rasinski, K.A., "Cognitive Processes Underlying Context Effects in Attitude Measurement," *Psychological Bulletin*, Vol. 103, 1988, pp. 299–314.

Tourangeau, R., Rasinski K.A., and D'Andrade, R., "Attitude Structure and Belief Accessibility," *Journal of Experimental Social Psychology*, Vol. 26, 1990, pp. 1–48.

Tourangeau, R., Rasinski, K.A., Bradburn, N.M., and D'Andrade, R., "It's Not Just What You Ask but When You Ask It," Unpublished NORC Report, 1988.

Tourangeau, R., Rasinski, K.A., Bradburn, N.M., and D'Andrade, R., "Carryover Effects in Attitude Surveys," *Public Opinion Quarterly*, Vol. 53, No. 4, 1989, pp. 495–525.

Tracy, P., and Fox, J., "The Validity of Randomized Response for Sensitive Measurements," *American Sociological Review*, Vol. 46, 1981, pp. 187–200.

Tremblay, A., and Fecso, R., "A Test for Bias in the Location of Sunflower OY Units," U.S. Department of Agriculture, National Agricultural Statistics Service, Research Report No. SRB-88-04, 1988.

Trewin, D., and Lee, G., "International Comparisons of Telephone Coverage," in R.M. Groves, P.P. Biemer, L.E. Lyberg, J.T. Massey, W.L. Nicholls II, and J. Waksberg (eds.), *Telephone Survey Methodology*, 1988, pp. 9–24, New York: Wiley.

Trilling, L., "Reality in America," in L. Trilling (ed.), *The Liberal Imagination*, 1951, New York: Viking.

Trivellato, U., and Torelli, N., "Longitudinal Analysis of Unemployment Duration from a Household Survey with Rotating Sample: A Case Study with Italian Labor Force Data," *Proceedings of the Fifth Annual Research Conference*, U.S. Bureau of the Census, 1989, pp. 408–427.

Tucker, C., "Interviewer Effects in Telephone Surveys," *Public Opinion Quarterly*, Vol. 47, 1983, pp. 84–95.

Tucker, C., "An Analysis of the Dynamics in the CE Diary Survey," *Proceedings of the Section on Social Statistics*, American Statistical Association, 1986, pp. 18–27.

Tucker, C., "The Estimation of Instrument Effects on Data Quality in the Consumer Expenditure Diary Survey," paper presented at the International Conference on Measurement Errors in Surveys, Tucson, AZ, November 1990.

Tucker, C., and Bennett, C., "Procedural Effects in the Collection of Consumer Expenditure Information," *Proceedings of the Section on Survey Research*, American Statistical Association, 1988, pp. 256–261.

Tucker, C., Vitrano, F., Miller, L., and Doddy, J., "Cognitive Issues and Research on the Consumer Expenditure Diary Survey," paper presented at the Annual Conference of the American Association for Public Opinion Research, St. Petersburg, FL, 1989.

Tulving, E., "Episodic and Semantic Memory" in E. Tulving, and W. Donaldson (eds.), *Organization of Memory*, 1972, pp. 381–403, New York: Academic Press.

Tulving, E., "Ecphoric Processes in Recall and Recognition" in J. Brown (ed.), *Recall and Recognition*, 1975, London: Wiley.

Tulving, E., *Elements of Episodic Memory*, New York: Oxford University Press, 1983.

Tulving, E., and Thomson, E.M., "Encoding Specificity and Retrieval Processes in Episodic Memory," *Psychological Review*, Vol. 80, 1973, pp. 352–373.

Tuma, N.B., and Hannan, M.T., *Social Dynamics: Models and Methods*, New York: Academic Press, 1984.

Turner, C.F., and Martin, E. (eds.), *Surveying Subjective Phenomena*, Vols. 1 and 2, New York: Russell Sage, 1984.

Turner, R., "Inter-week Variations in Expenditure Recorded During a Two-week Survey of Family Expenditure," *Applied Statistics*, Vol. 10, No. 3, 1961, pp. 136–146.

Tversky, A., "Elimination by Aspects: A Theory of Choice," *Psychological Review*, Vol. 79, 1972, pp. 281–299.

Tversky, A., and Kahneman, D., "Availability: A Heuristic for Judging Frequency and Probability," *Cognitive Psychology*, Vol. 5, 1973, pp. 207–232.

U.S. Bureau of the Census, "The Current Population Survey Reinterview Program, Some Notes and Discussion," Technical Paper No. 6, 1963, U.S. Government Printing Office, Washington, DC.

U.S. Bureau of the Census, "Evaluation and Research Program of the U.S. Censuses of Population and Housing, 1960: Accuracy of Data on Population Characteristics as Measured by Reinterviews," Series ER 60, No. 4, 1964, Washington DC: Government Printing Office.

U.S. Bureau of the Census, "Accuracy of Data for Selected Housing Characteristics as Measured by Reinterviews," Evaluation and Research Program, Series PHC(E)-10, Washington, DC, 1975.

U.S. Bureau of the Census, "The Current Population Survey: Design and Methodology," Technical Paper No. 40, Washington, DC, 1978.

U.S. Bureau of the Census, "Enumerator Variance in the 1970 Census," Evaluation and Research Program, Series PHC(E)-13, Washington, DC, 1979.

U.S. Bureau of the Census, "Evaluation of Censuses of Population and Housing," STD-ISP-TR-5, Washington, DC: U.S. Government Printing Office, 1985.

U.S. Bureau of the Census, "Revised Monthly Retail Sales and Inventory, January 1980 Through December 1989," *Current Business Reports*, BR89-R, U.S. Government Printing Office, Washington, DC, 1990.

U.S. Bureau of Labor Statistics, "Consumer Expenditure Survey, 1987," *BLS Bulletin*, No. 2354, U.S. Government Printing Office, Washington, DC, June 1990.

U.S. Department of Commerce, "1970 Census of Population and Housing: Accuracy of Data for Selected Population Characteristics as Measured by the 1970 CPS-Census Match," Series PHE(E)-11, Washington, DC: U.S. Government Printing Office, 1975.

U.S. Department of Justice, Federal Bureau of Investigation, "Crime in the United States, 1989," Washington, DC: U.S. Government Printing Office, 1989.

U.S. National Center for Health Statistics, "Health Interview Responses Compared with Medical Records," *Vital and Health Statistics*, Series 2, No. 7, 1965, Washington, DC: Government Printing Office.

U.S. National Center for Health Statistics, "Interview Data on Chronic Conditions Compared with Information Derived from Medical Records," *Vital and Health Statistics*, Series 2, No. 23, 1967, Washington, DC: Public Health Service.

U.S. National Center for Health Statistics, "Comparability of Age on the Death Certificate and Matching Census Record: United States-May-August 1960," *Vital and Health Statistics*, Series 2, No. 29, 1968, Washington DC: Government Printing Office.

U.S. National Center for Health Statistics, "Comparability of Marital Status, Race, Nativity and Country of Origin in the Death Certificate and Matching Census Record: United States-May-August 1960," *Vital and Health Statistics*, Series 2, No. 34, 1969, Washington DC: Government Printing Office.

U.S. National Center for Health Statistics, "An Experimental Comparison of Telephone and Personal Health Interview Surveys," *Vital and Health Statistics*, Series 2, No. 106, 1987, Washington, DC: Government Printing Office.

U.S. Office of Management and Budget, *Standard Industrial Classification Manual*, U.S. Government Printing Office, Washington, DC, 1972.

U.S. Office of Management and Budget, "Report on Statistical Uses of Administrative Records," *Statistical Policy Working Paper 6*, National Technical Information Service, Springfield, Virginia, 1980.

U.S. Office of Management and Budget, "Computer-Assisted Survey Information Collection," *Statistical Policy Working Paper 19*, National Technical Information Service, Springfield, Virginia, 1990.

U.S. Public Health Service, "The Health Consequences of Smoking: Cancer and Chronic Lung Disease in the Workplace. A Report of the Surgeon General," DHHS Pub. No. 85–50207, 1985, Washington, DC: U.S. Government Printing Office.

Uniform Code Council, *UPC Symbol Specification Manual*, Dayton, Ohio, 1986.

United Nations, "Estimation of Crop Areas and Yields in Agricultural Statistics," Food and Agricultural Organization, FAO Economic and Social Development Paper No. 22, 1982, Rome.

Van Bastelaer, A., "The Continuous Labour Force Survey in the Netherlands," paper presented at the Conference on Methodische Fragen von Stichproben am Beispiel des Mikrozensus, Wiesbaden, F.R.G., 1988.

Van Bastelaer, A., Kerssemakers, F., and Sikkel, D., "Data Collection with Hand-held Computers: Contributions to Questionnaire Design," *Journal of Official Statistics*, Vol. 4, 1988, pp. 141–154.

Van Dijk, T., *Macrostructures: An Interdisciplinary Study of Global Structures in Discourse, Interaction, and Cognition*, Hillsdale, NJ: Lawrence Erlbaum Associates, 1980.

Van Doorn, L., Saris, W.E., and Lodge, M., "Discrete or Continuous Measurement: What Does It Matter?" *Kwantitatieve Methoden* Vol. 10, 1983, pp.104–121.

Van Dosselaar, P.G.W.M., van den Hurk, G.J.H.H., and Israëls, A.Z., "Analysis of Memory Effects at the Netherlands Central Bureau of Statistics," *Bulletin of the International Statistical Institute*, Vol. 53, Book 2, 1989a, pp. 383–402.

Van Dosselaar, P.G.W.M., van den Hurk, G.J.H.H., and Israëls, A.Z., "Memory Effects in Retrospective Surveys," *CBS Select*, Vol. 5, 1989b, pp. 167–180, The Hague: Staatsuitgeverij.

Vitrano, F.A., Hubble, D.L., and Vacca, E.A., "Cognitive Issues and Reporting Level Patterns from the CE Diary Operational Test," *Proceedings of the Section on Survey Research Methods*, American Statistical Association, 1988, pp. 262–267.

Vogel, F., "A Research Report on Michigan Tart Cherries," Washington, DC, Staff Report, U.S. Department of Agriculture, Statistical Reporting Service, 1970.

Wagenaar, W.A., "My Memory: A Study of Autobiographical Memory Over Six Years," *Cognitive Psychology*, Vol. 18, 1986, pp. 225–252.

Waksberg, J., and Valliant, R., "Final Report on the Evaluation and Calibration of NEISS," Westat, Inc. for Consumer Products Safety Commission, 1978.

Warner, S., "Randomized Response: A Survey Technique for Eliminating Evasive Answer Bias," *Journal of the American Statistical Association*, Vol. 60, 1965, pp. 63–69.

Warner, S., "The Linear Randomized Response Model," *Journal of the American Statistical Association*, Vol. 66, 1971, pp. 884–888.

Warren, F.B., "Corn Yield Validation Studies, 1953–83," Washington, DC, U.S. Department of Agriculture, Statistical Reporting Service, Staff Report No. YRB-85–07, 1985.

Warren, F.B., "Comparisons Between Objective Yield Survey and Farmers Reported Yields of Corn," Washington, DC, U.S. Department of Agriculture, National Agricultural Statistics Service, Staff Report YRB-86-04, 1986.

Warren, R.D., White, J.K., and Fuller, W.A., "An Errors-in-Variables Analysis of Managerial Role Performance," *Journal of the American Statistical Association*, Vol. 69, 1974, pp. 886–893.

Warwick, D., and Lininger, C., *The Sample Survey: Theory and Practice*, New York: McGraw-Hill, 1975.

Webb, E., Campbell, D., Schwartz, R., and Sechrest, L., *Unobtrusive Measures: Nonreactive Research in the Social Sciences*, Chicago: Rand McNally, 1966.

Wegener, B., *Social Attitudes and Psychophysical Measurement*, Hillsdale: Earlbaum, 1982.

Wegener, B., Faulbaum, F., and Maag, G., "Die Wirkung von Antwortvorgaben bei Kategorialskalen (The Impact of Verbal Response Alternatives on Measurement with Rating Scales)," *ZUMA-Nachrichten*, Vol. 10, 1982, pp. 3–20.

Weinberg, E., "Data Collection: Planning and Management," in P. Rossi, D.A. Wright, and A.B. Anderson (eds.), *Handbook of Survey Research*, 1983, pp. 329–358, Academic Press.

Weiss, C., "Validity of Welfare Mothers' Interview Response," *Public Opinion Quarterly*, Vol. 32, 1968, pp. 622–633.

Weiss, C.H., "Interaction in the Research Interview: The Effects of Rapport on Response," *Proceedings of the Social Statistics Section*, American Statistical Association, 1970, pp. 17–20.

Werking, G., Tupek, A., and Clayton, R., "CATI and Touchtone Self-Response Applications for Establishment Surveys," *Journal of Official Statistics*, Vol. 4, 1988, pp. 349–362.

Werner, O., and Schoepfle, M., *Systematic Fieldwork Volume 1, Foundations of Ethnography and Interviewing*, Newbury Park, CA: Sage, 1987.

Werts, C.E., and Linn, R.L., "Path Analysis: Psychological Examples," *Psychological Bulletin*, Vol. 74, 1970, pp. 193–212.

Werts, C.E., Jöreskog, K.G., and Linn, R.L., Comment on "The Estimation of Measurement Error in Panel Data," *American Sociological Review*, Vol. 36, 1971, pp. 110–113.

Werts, C.E., Jöreskog, K.G., and Linn, R.L., "A Multitrait-Multimethod Model for Studying Growth," *Educational Psychological Measurement*, 1972, pp. 655–678.

Wiese, W.L., Smith, M.W., and Glennon, B.M., *Atomic Transition Probabilities, Vol. I: Hydrogen Through Neon*, National Bureau of Standards, Washington, DC, 1966.

Wiggins, R.D., Longford, N., O'Muircheartaigh, C.A., "A Variance Components Approach to Interviewer Effects," London School of Economics, unpublished manuscript, 1989.

Wigton, W.E., and Kibler, W.E., "New Methods for Filbert Objective Yield Estimation," *Agricultural Economics Research*, Vol. 24, No. 2, 1972.

Wikman, A., "Svarsprecisionen i surveyundersökningar om levnadsförhållanden (Response Precision in Surveys of Living Conditions)" Stockholm, Statistics Sweden, unpublished report, 1980.

Wildt, A.R., and Mazis, M.B., "Determinants of Scale Responses: Label Versus Position," *Journal of Marketing Research*, Vol. 15, 1978, pp. 261–267.

Wiley, D.E., "The Identification Problem for Structural Equations Models with Unmeasured Variables," in A.S. Goldberger, and O.D. Duncan (eds.), *Structural Equation Models in the Social Sciences*, 1973, pp. 69–83, New York; Seminar Press.

Wiley, D.E, and Wiley, J.A., "The Estimation of Measurement Error in Panel Data," *American Sociological Review*, Vol. 35, Number 1, 1970, pp. 112–117.

Williams, J.A., Jr., "Interviewer-Respondent Interaction: A Study of Bias in the Information Interview," *Sociometry*, Vol. 27, 1964, pp. 338–352.

Willis, G.B., Royston, P., Bercini, D., "Problems with Survey Questions Revealed by Cognitively-based Interviews," *Proceedings of the 5th Annual Research Conference*, U.S. Bureau of the Census, 1989, pp. 345–360.

Wilson, T., Wiemann, J., and Zimmerman, D., "Models of Turn Taking in Conversational Interaction," *Journal of Language and Social Psychology*, Vol. 3, No. 3, 1984, pp. 159–183.

Winer, B.J., *Statistical Principles in Experimental Design*, 2nd Edition, New York: McGraw Hill, 1971.

Wiseman, F., Moriarty, M., and Schafer, M., "Estimating Public Opinion with the Randomized Response Model," *Public Opinion Quarterly*, Vol. 39, 1975–76, pp. 507–513.

Wolfson, N., "Speech Events and Natural Speech: Some Implications for Sociolinguistic Methodology," *Language in Society*, Vol. 5, No. 1, 1976, pp. 189–209.

Wolter, K.W., *Introduction to Variance Estimation*, New York: Springer Verlag, 1985.

Wood, R.A., "Optimum Plot Size for Winter Wheat," Washington, DC, U.S. Department of Agriculture, Statistical Reporting Service, Staff Report, 1972.

Woolsey, T. D., "Results of the Sick-Leave Memory Test of October, 1952," Department of Health, Education, and Welfare unpublished memorandum, 1953.

Wothke, W., and Browne, M.W., "The Direct Product Model for the MTMM Matrix Parameterized as a Second Order Factor Analysis Model," *Psychometrika*, Vol. 55, 1990, pp. 255–262.

Wright, S., "The Method of Path Coefficients," *Annals of Mathematical Statistics*, Vol. 5, 1934, pp. 161–215.

Wyer, R.S., *Cognitive Organization and Change: An Information Processing Approach*, Hillsdale, NJ, Erlbaum, 1974.

Zarkovich, S. S., *Quality of Statistical Data*, Food and Agricultural Organization of the United Nations, Rome, 1966.

Zdep, S., and Rhodes, I., "Making the Randomized Response Technique Work," *Public Opinion Quarterly*, Vol. 40, 1976–77, pp. 531–537.

INDEX

Page numbers in *italics* refer to tables, figures, and exhibits.

A

B

C

D

F

G

H

I

N

O

P

Q

R

S

T

U

V

W

Y